Molecular Reaction Dynamics in Gases, Liquids and Interfaces

Assisi, Italy
25–27 June 2012

FARADAY DISCUSSIONS

Volume 157, 2012

RSC Publishing

The Faraday Division of the Royal Society of Chemistry, previously the Faraday Society, founded in 1903 to promote the study of sciences lying between Chemistry, Physics and Biology.

EDITORIAL STAFF

Editor
Philip Earis

Deputy editor
Jane Hordern

Development editor
Heather Montgomery

Senior publishing editor
Anna Pendlebury

Publishing editors
Sarah Dixon, Carrie Mowatt, Catherine Pridmore

Publishing assistants
Aliya Anwar, Ella Mitchell, Claire Sissen

Publisher
Niamh O'Connor

Faraday Discussions (Print ISSN 1359-6640, Electronic ISSN 1364-5498) is published 6 times a year by the Royal Society of Chemistry, Thomas Graham House, Science Park, Milton Road, Cambridge, UK CB4 0WF. Volume 158 ISBN-13: 978-1-84973-449-3

2012 annual subscription price: print+electronic £709, US $1,322; electronic only £673, US $1,256. Customers in Canada will be subject to a surcharge to cover GST. Customers in the EU subscribing to the electronic version only will be charged VAT. All orders, with cheques made payable to the Royal Society of Chemistry, should be sent to RSC Distribution Services, c/o Portland Customer Services, Commerce Way, Colchester, Essex, UK CO2 8HP.
Tel +44 (0) 1206 226050;
E-mail sales@rscdistribution.org

If you take an institutional subscription to any RSC journal you are entitled to free, site-wide web access to that journal. You can arrange access *via* Internet Protocol (IP) address at www.rsc.org/ip. Customers should make payments by cheque in sterling payable on a UK clearing bank or in US dollars payable on a US clearing bank.

US Postmaster: send address changes to *Faraday Discussions*, c/o Mercury Airfreight International Ltd., 365 Blair Road, Avenel, NJ 07001. All despatches outside the UK by Consolidated Airfreight.

PRINTED IN THE UK

Faraday Discussions documents a long-established series of *Faraday Discussion* meetings which provide a unique international forum for the exchange of views and newly acquired results in developing areas of physical chemistry, biophysical chemistry and chemical physics.

Molecular Reaction Dynamics in Gases, Liquids and Interfaces

Faraday Discussions

www.rsc.org/faraday_d

A General Discussion on Molecular Reaction Dynamics in Gases, Liquids and Interfaces was held in Assisi, Italy on 25th, 26th and 27th June 2012.

RSC Publishing is a not-for-profit publisher and a division of the Royal Society of Chemistry. Any surplus made is used to support charitable activities aimed at advancing the chemical sciences. Full details are available from www.rsc.org

CONTENTS

ISSN 1359-6640; ISBN 978-1-84973-448-6

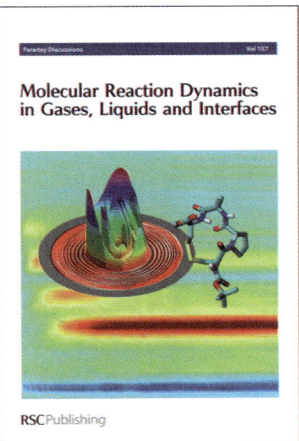

Cover
Artist's view of recent developments in the field of reaction dynamics: from a flux map for crossed molecular beam reactive scattering of F atoms with HD molecules (courtesy of Xueming Yang) to the structure of a small peptide studied by 2D infra-red spectroscopy (courtesy of Peter Hamm), both superimposed on a time-dependent infra-red absorption spectrum for DCN products of reaction of CN radicals with d_{12}-cyclohexane in solution in chloroform (courtesy of Andrew Orr-Ewing).

Image reproduced by permission of Prof. Andrew Orr-Ewing and Prof. Piergiorgio Casavecchia.

Società Chimica Italiana

IOP | Institute of Physics
Molecular Physics Group

CONCLUDING REMARKS

ADDITIONAL INFORMATION

Molecular reaction dynamics across the phases: similarities and differences

F. Fleming Crim*

Received 8th August 2012, Accepted 9th August 2012
DOI: 10.1039/c2fd20123b

This introduction to the 157th Faraday Discussion describes features of bimolecular reaction and photodissociation in gases and liquids and at interfaces. Two unifying ideas are the concepts of a transition state on a single potential energy surface and of a conical intersection between two surfaces, the former being important in bimolecular reactions and the latter often being important in photodissociation. State-resolved studies of the reactions of methane and its isotopologues with F, Cl, and Br illustrate many aspects of bimolecular reactions including the ability of excitation in vibrational modes to enhance or inhibit a reaction and to control the cleavage of selected bonds. There are clear parallels between those gas-phase reactions and the dissociative chemisorption of methane and its isotopologues on metal surfaces. Similarly, features such as relative reaction rates and energy disposal patterns observed in gas-phase reactions largely carry over into reactions in solution. New experiments comparing photolysis in the gas phase and in solution show that there are again many similarities in the processes in the two environments. Although the influence of the surroundings in those cases is subtle, there are situations in which the surroundings can produce a much larger effect on the photolysis by hindering the dissociation and initiating isomerization.

1. Introduction

A Faraday Discussion with the expansive topic of *Molecular Reaction Dynamics in Gases, Liquids, and Interfaces* involves diverse aspects of chemical transformations, and, fortunately, a few organizing principles unite those strands. One is the notion that many chemical transformations occur through a transition state, a high energy configuration along the pathway between reactants and products. Another principle, which has come to the fore more recently, is the idea that excited state reactions, such as photodissociation and photoisomerization, often involve geometries in which two different electronic states form a conical intersection. Viewed in the proper coordinate system, this conical intersection is a special point at which the two states are degenerate but away from which they have different energies.[1] As with a transition state, the motion of the molecule in the vicinity of a conical intersection often controls the dynamics of the reaction.

Understanding the motion that carries a system through those special configurations and the consequences of that motion for energy consumption and disposal lies at the heart of molecular reaction dynamics. This introduction to the Discussion presents examples of reactions in gases and liquids and at interfaces with an eye toward identifying the similarities and differences among them. We examine

Department of Chemistry, University of Wisconsin - Madison, Madison, Wisconsin 53706, USA.
E-mail: fcrim@chem.wisc.edu

separately the dynamics of bimolecular reactions and of photodissociation, both of which appear prominently in the articles and commentary in this Discussion.

2. Bimolecular reaction

The influence of the surroundings is the essential difference between reactions in gases and those in liquids and at interfaces. Interactions with the surroundings potentially alter the relative energies of the reactants, the products, and, perhaps most important, the transition state.

Fig. 1 illustrates three cases. In the first case, shown in Fig. 1(a), the solid curve is the energy along the reaction coordinate for a gas-phase reaction of two molecules. The dashed blue curve is an energy profile for a reaction in solution. It shows a slight stabilization of the reactants and products along with a greater stabilization of the transition state by the surrounding solvent. The result is a lower energy barrier to the reaction because of the greater stabilization of the transition state. The essential

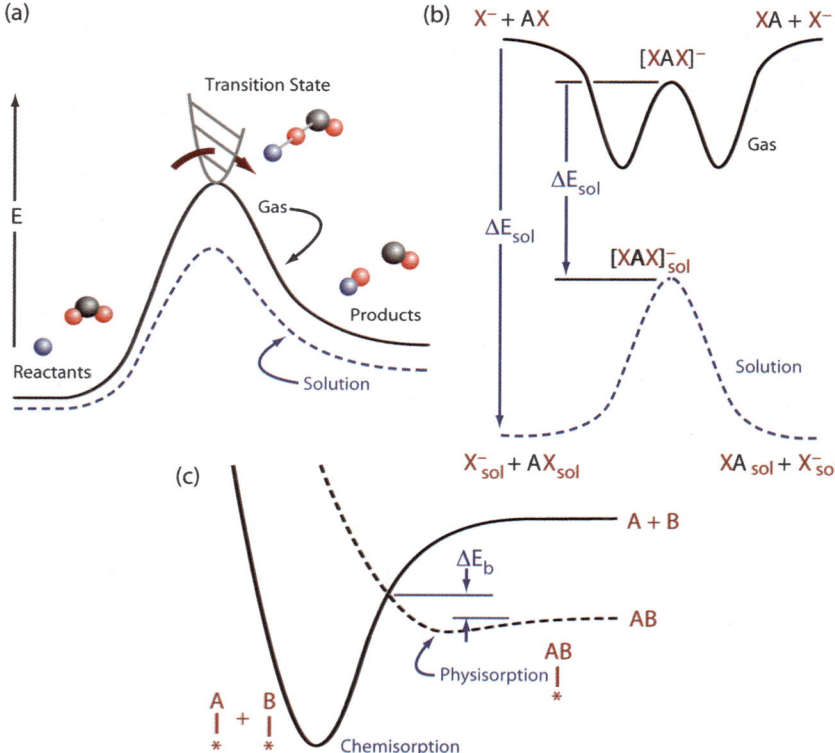

Fig. 1 (a) Energy profiles for a bimolecular reaction in the gas phase (solid black curve) and for a reaction in a liquid (dashed blue curve). In this example, the solvent lowers the energy of the reactants and products slightly but stabilizes the transition state by a larger amount. (b) Energy profiles for a gas-phase bimolecular ion-molecule reaction (solid black curve) and for the same ion-molecule reaction in a liquid (dashed blue curve). The transition state lies below the asymptotes in the gas phase, and there are bound minima along the reaction coordinate. The solvent stabilizes the reactants and products much more than the transition state, producing a very different energy profile in solution. (c) Potential energy curves for physisorption and dissociative chemisorption of the reactant AB. Reaction can occur by physisorption of the precursor AB followed by dissociation from that shallow well to form strongly bound A and B species on the surface. The minimum energy required is ΔE_b.

shape of the curve remains the same with small, but potentially significant, changes in the overall energetics. Another factor that is not obvious from examining the reaction profile alone is the role that the solvent might play in limiting motion through the transition state or redistributing energy in the reactants and products, ideas that enter into more sophisticated models of reaction in solution.[2,3]

The situation is very different for the symmetric ion-molecule reaction having the energy profile in Fig. 1(b). The solid curve has a deep attractive well as the ion approaches the molecule. In this example, the entrance-channel complex is so strongly bound that it "submerges" the transition state below the energy of the reactants and products. However, solvation changes the situation dramatically by stabilizing the reacting anion, X^-, much more that the transition state, as the dashed blue curve shows. The distribution of the charge over more atoms makes the solvent stabilization of the transition state smaller than that of the reactants. Interaction with the surroundings removes the entrance-channel well and produces a more conventional profile with the transition state lying well above the energy of the reactants and products.

The one-dimensional profiles shown in Fig. 1 are minimum energy paths across a multidimensional potential energy hypersurface, and the topology of that surface is critically important to energy consumption and disposal. The concepts of late and earlier barriers along with the kinematic effects of different mass combinations are another central organizing principle in molecular reaction dynamics.[4] These ideas are touchstones for this Discussion, and testing their limits in larger systems is a current theme in molecular reaction dynamics.

Descriptions of reactions at interfaces often use these same ideas, and surface reactions present other potential environmental effects as well. The two potential energy curves shown in Fig. 1(c) illustrate the simplest role that the surface can play. The dashed curve shows the energy profile for physisorption of the AB molecule in a shallow well, and the solid curve shows the energetics for dissociative chemisorption into two surface bound species A and B, which correlate to high energy dissociated fragments in the gas phase. Thus, the physisorbed molecule is a potential precursor in the dissociative chemisorption reaction, and the energy required to move from one region to another is the effective barrier, E_b, in this simple example. However, there may be larger barriers associated with the dissociation, and the reaction could involve displacement of surface atoms as well as changing the geometry of the dissociating molecule.

2.1 Gas-phase reactions: the CH₄ poster child

Abstraction of a hydrogen atom from the simplest alkane, CH_4, and its partially deuterated isotopologues illustrates varied aspects of molecular reaction dynamics,[5] and there are several examples in this Discussion. Fig. 2 shows the energetics for the abstraction of a hydrogen atom from methane to form a hydrogen halide and a methyl radical

$$X + CH_4 \rightarrow HX + CH_3$$

which vary dramatically with the identity of the halogen atom, X. The strong HF bond makes the reaction to form HF exothermic by 11 150cm^{-1} (133 kJ mol^{-1}), but the reaction to form HCl is modestly endothermic by 500 cm^{-1} (6 kJ mol^{-1}). By contrast, the still weaker HBr bond makes that reaction endothermic by 4300 cm^{-1} (50 kJ mol^{-1}). Thus, this series of reactions offers the possibility of examining both the consumption of energy in bond- and mode-selective reactions and the disposal of energy into product vibrations. Correlations between barrier location and thermochemistry show that exothermic reactions typically have early barriers (reactant-like transition states) and endothermic reactions typically have late barriers (product-like transition states). For triatomic systems, translational energy

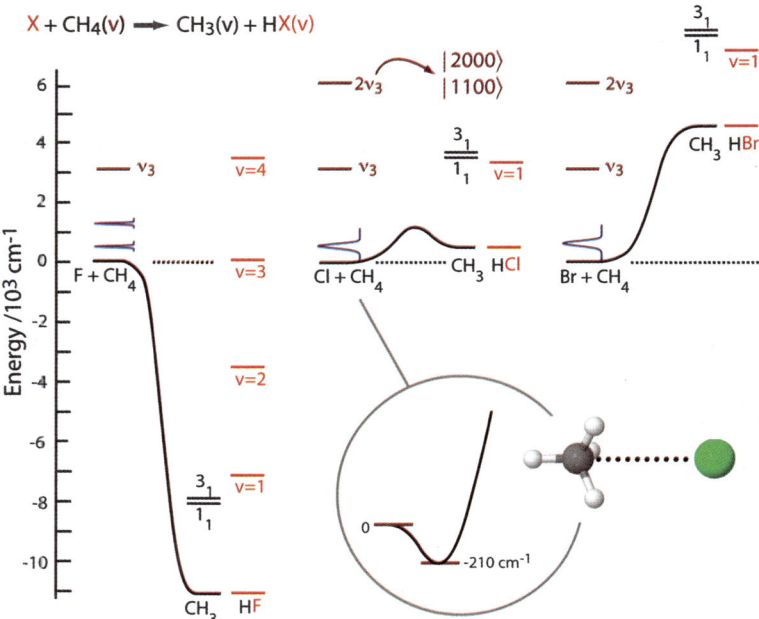

Fig. 2 Energetics for the reaction of a halogen atom with CH_4. The horizontal lines indicate the energies of the symmetric C–H stretching states (v_3) of CH_4 (dark red), the energy levels of the symmetric and antisymmetric C–H stretching states (1_1 and 3_1) of CH_3 (black), and the vibrational levels of the hydrogen halide HX (red). The inset shows the calculated van der Waals minimum for the case of Cl + CH_4. There are similar minima for all three reactions.

promotes reaction most effectively in the former case and vibrational energy in the latter case.[4] The inset in Fig. 2 illustrates another potentially important feature of these reactions, the presence of a shallow van der Waals well in the entrance channel. Its calculated depth[6] is about 210 cm^{-1} for the case of chlorine shown in the figure, and the well depth is more than 100 cm^{-1} for each of these reactions.[7]

The energy available for populating different product states differs markedly for the three reactions. There is enough energy available in the reaction with F to populate vibrational states of HF up to $v = 3$ or to excite the CH_3 product to high vibrational states. Fig. 2 shows the energies of only the two lowest C–H stretching states of methyl, the symmetric C–H stretch v_1 and the anti-symmetric C–H stretch v_3, indicated by the notation 1_1 and 3_1. The endothermic reactions of CH_4 with Cl and Br cannot produce vibrationally excited products without additional translational or vibrational energy in the reactants.

There are now examples of all three of these atoms reacting with vibrationally excited methane,[8-11] and the figure shows the energies of CH_4 reactants containing one (v_3) and two quanta ($2v_3$) of antisymmetric C–H stretching excitation. Initial excitation of two quanta of C–H stretch, as shown for the Cl and Br reactions, introduces an additional aspect of initial state preparation. It is possible to excite eigenstates corresponding to either one quantum of excitation in each of two bonds, the $|1100\rangle$ state in local-mode notation,[12,13] or two quanta in one bond, the $|2000\rangle$ state. As described below, these initial states react differently despite being very close in energy, an example of vibrational-mode-selective chemistry.

2.1.1 Translational energy.

Translational energy allows reactions to surmount the energy barrier, as in the case of Cl and Br reactions with CH_4, or provides energy to populate higher lying product states, as in the case of the exothermic reaction of F with CH_4. The barrier for the reaction of Cl is low enough that a relatively small amount of translational energy from photolysis of a precursor can overcome it. The reaction of translationally energized Cl atoms with a methane isotopologue, CH_3D, illustrates this behaviour. The reaction has two channels, abstraction of either H or D,

$$Cl + CH_3D \begin{cases} DCl + CH_3 \\ HCl + CH_2D \end{cases}$$

that have different energy barriers arising from the difference in the zero-point energy of the C–H and C–D bonds. At the transition state, the bond that breaks is unbound and, thus, has no zero-point energy. Consequently, the transition state for breaking a C–H bond, with its larger zero-point energy, lies lower than that for breaking a C–D bond because the zero-point energy "disappears" in the transition state. This difference in barrier heights is the origin of the primary kinetic isotope effect, and Fig. 3(a) shows the relative size of the barriers for the two channels in the reaction of Cl with CH_3D.

The figure illustrates the approach we use for studying the role of both translational and vibrational energy in the reaction.[14–16] These experiments photolyze Cl_2

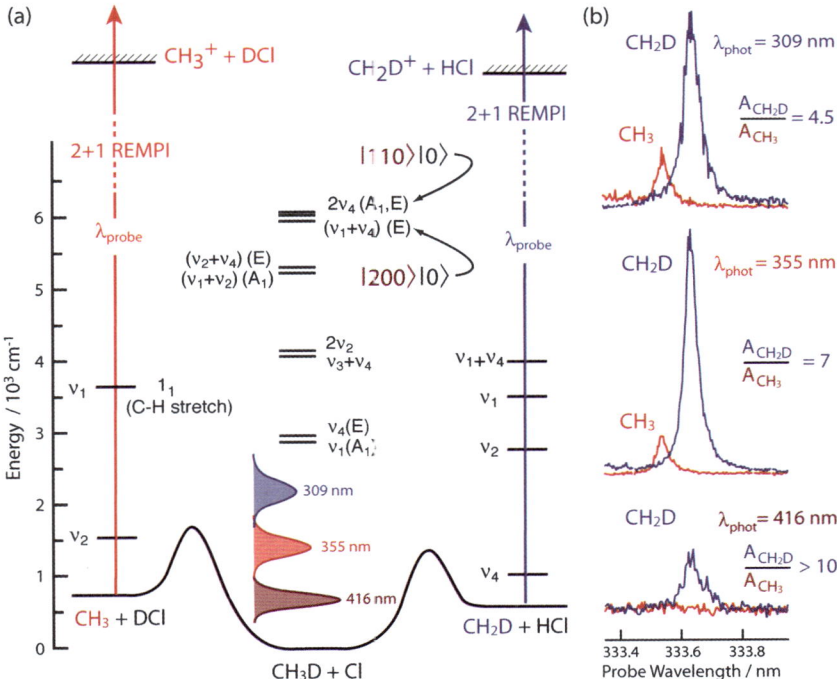

Fig. 3 (a) Energy profile along the reaction coordinates for Cl reacting with CH_3D to produce either CH_3 + DCl or CH_2D + HCl showing vibrational levels of CH_3D as well as vibrational levels of the CH_3 or CH_2D products detected by (2+1) REMPI. The distributions show the relative translation energy in the system following photolysis of Cl_2 with 416 nm (dark red), 355 nm (red), or 309 nm (blue) light. (b) Relative REMPI signals for the CH_2D product (blue) and the CH_3 (red) product at each photolysis wavelength. The relative amount of the CH_2D product grows as the translational energy decreases. At the lowest energy, the CH_3 is undetectable.

to generate Cl atoms in a free-jet expansion that also contains CH_3D, and they monitor individual vibrational states of both the CH_2D and CH_3 reaction products by resonance enhanced multiphoton ionization (REMPI). In these (2+1) REMPI measurements, two photons excite a resonant intermediate state, and another photon ionizes the molecule for detection in a time-of-flight mass spectrometer. As described below, it is also possible to excite the vibrational eigenstates of CH_3D shown in the diagram in order to observe the influence of vibrational excitation on the course of the reaction. First we consider the influence of translational energy alone.

The three distributions in Fig. 3(a) show the relative translational energy, E, between Cl and CH_3D for photolysis of Cl_2 with three different wavelengths of light.[17] Photolysis with 416-nm light gives a most probable relative translational energy of $E_{416} = 610$ cm^{-1}, photolysis with 355-nm light gives 1280 cm^{-1}, and photolysis with 309-nm light gives 1970 cm^{-1}.[18] This threefold change in energy alters the branching between the two channels substantially, as Fig. 3(b) shows. The curves are the REMPI signals as a function of the probe wavelength for the CH_3 product (red) and the CH_2D product (blue). At the highest energy ($\lambda_{phot} = 309$ nm), the area of the CH_2D signal is about 4.5 times larger than that of CH_3, but it increases to 7 times larger for the next lower energy photolysis ($\lambda_{phot} = 355$ nm). At the lowest photolysis energy ($\lambda_{phot} = 416$ nm), we observe only the CH_2D product. These variations suggest that the lowest energy photolysis places the system well below the barrier to breaking the C–D bond (to produce CH_3) and that the highest energy photolysis places the system above both barriers.

This conclusion is consistent with a calculation by Cazko, et al.[6] that finds the barrier to cleaving the C–H bond in the Cl + CH_4 reaction to be about 1200 cm^{-1}. Correcting that value by the 360 cm^{-1} difference in zero-point energies of a C–H and a C–D bond gives an estimate of 1560 cm^{-1} for the height of the barrier to cleaving the C–D bond, as shown in the figure. Thus, we expect that few, if any, of the collisions carry the system over the barrier at the lowest translational energy (610 cm^{-1}). It is likely that H-atom tunnelling is the dominant pathway in that case and that reaction to cleave the C–D bond is negligible. At the next higher translational energy (1280 cm^{-1}), a sizeable fraction of the collisions have enough energy to pass over the barrier to breaking the C–H bond, but only a few have enough to surmount the barrier for breaking the C–D bond. The CH_3 products likely come from tunnelling while the CH_2D products come primarily from reaction over the barrier. Finally, at the highest translational energy both channels are open, and the observed ratio of products is consistent with statistical partitioning between the two channels.[17,19]

2.1.2 Vibrational energy.
Vibration is another degree-of-freedom available for initial excitation in the methane reactant. The reaction of F with CH_4 provides a striking example of the influence that vibrational energy can have even on a very exothermic reaction. In this Discussion, Kawamata, et al. detect the CH_3 fragment from the reaction using ion imaging and infer the energy content of the unobserved HF product.[8] They find HF vibrational states populated up to the energetic limit shown in Fig. 2. Most significantly, however, they observe that excitation of the antisymmetric C–H stretch (ν_3) in CH_4 inhibits the C–H bond cleavage. When reaction does occur, it preserves the vibration and forms CH_3 with a quantum of C–H stretch excitation. The authors suggest that the transition state in this early-barrier reaction resembles the reactants and that a geometry with a stretched C–H bond does not lie along the minimum energy path.

Reactions of Cl with methane and its isotopologues have provided clear examples of both mode- and bond-selected chemistry in which vibrational excitation directs the reaction along a particular course.[5] Although there are some subtleties, two general trends emerge from studies of all the isotopologues. One is that exciting an eigenstate with substantial C–H or C–D stretching character in a mixed isotopologue leads to preferential cleavage of that bond.[15] The other is that the bonds that do not break largely act as spectators, preserving their initial excitation in the products.[20,21] As the energy level diagram for CH_3D in Fig. 3(a) shows, one quantum of C–H stretching excitation (ν_1 or ν_4) places the system above the barrier, and in most cases, there is significant translational energy from the photolysis as well. Even though there is background reaction from translationally energized molecules without vibrational excitation, adding vibrational excitation increases the reaction cross section markedly. However, reactive scattering experiments in CHD_3 show that vibration is no more effective than an equivalent amount of energy in translation.[22] Both forms of energy accelerate the reaction, but only vibrational excitation controls the outcome.

The reaction of CH_3D with Cl also illustrates mode-selective reaction. Exciting the symmetric C–H stretch (ν_1) or the antisymmetric C–H stretch (ν_4) leads to preferential cleavage of the C–H bond to produce CH_2D, as illustrated in Fig. 3(a). Vibrational action spectra obtained by monitoring the CH_2D product show that molecules with the symmetric stretching mode excited are about six times more reactive than those with the antisymmetric mode excited, even though the two states differ primarily in the phase of their vibrations and lie within 100 cm^{-1} of each other. The key to this mode selectivity is the evolution of the initially excited eigenstate during the collision.[16]

The bond selectivity and the spectator behaviour of the surviving bonds shows that the perturbation of the incoming Cl atom does not mix the vibrational states extensively although it changes the relative reactivity of the two states. Models of the interaction of vibrationally excited water with Cl[23] and of vibrationally excited methane with a Ni surface[24] as well as more detailed electronic structure calculations of the interaction of Cl with CH_3D[16] provide a picture of the evolution of the initially excited state. As a Cl atom approaches a CH_3D molecule along a C–H bond, the symmetric stretching vibration changes from the movement of all three atoms together to the vibration of the bond directed toward the incoming Cl. By contrast, the antisymmetric stretching vibration evolves into motion of the two C–H bonds directed away from the incoming atom, making that vibration less reactive. In fact, there is a relatively simple symmetry argument that leads to this outcome for approach along a C–H bond.[16] This result points to the importance not only of the nature of the initially prepared vibrations but also of their evolution during the reaction. Even in the absence of extensive state mixing (collision induced intramolecular energy redistribution), excited states can evolve in subtle ways that alter their reactivity.

The reaction of CH_4 with Br is another example of mode selectivity that is strongly reminiscent of the first examples of such behaviour in the reaction of H or Cl with H_2O.[25–27] As Fig. 2(c) shows, one quantum of C–H stretching excitation provides too little energy to surmount the barrier, and, even with an additional 1300 cm^{-1} of translational energy, the total energy is barely enough for the reaction to occur. However, two quanta of excitation in the antisymmetric stretch ($2\nu_3$) provide enough energy for the reaction. As described above, there are two spectroscopically distinct transitions in the $2\nu_3$ region that excite either the local-mode state containing one quantum in each of two different C–H bonds, $|1100\rangle$, or the one with two quanta in one C–H bond, $|2000\rangle$. In the former case, there is not enough energy in either bond to overcome the barrier, and we observe no reaction.[11,19] In the later case, however, we observe reaction to produce ground vibrational state methyl radicals. This mode-selective reaction is another manifestation of the independence of the potentially reacting bonds. There is not enough energy in any single bond to overcome the reaction barrier even though there is enough energy in the molecule. Because the interaction with the Br atom does not transfer energy between the bonds, only the state that initially has its vibrational energy in a single bond reacts. This behaviour is the other side of the spectator picture in which the reacting bond does not communicate with the non-reacting bond.

2.2 Reactions in solution: H-atom abstraction

Transferring bimolecular, gas-phase reactions into solution presents challenges and opportunities. A molecule in a gas at a pressure of 1 Torr collides roughly every 100 ns, but in a gas having the density of a liquid, a molecule collides every 100 fs. Thus, observing events on the timescale of the interval between interactions requires time resolution of about 100 fs. Consequently, one must trade the high spectral resolution possible in gas-phase measurements for high time resolution in liquid-phase experiments. However, it is still possible to observe populations in different vibrational states in favourable cases and to answer some of the same questions that gas-phase molecular reaction dynamics explores.

Hydrogen-atom abstraction reactions from alkanes and substituted alkanes

$$X + RH \rightarrow HX + R \cdot$$

are attractive targets for which there are some gas-phase data for comparison. In the gas-phase measurements on CH_4 described above, quantum-state resolved detection of the methyl radical is a key. In solution, it is possible to follow the reaction using transient absorption of the HX product, and pioneering experiments by Hochstrasser first used that approach.[28,29] These experiments produce a reactive radical photolytically and monitor the reaction product with transient infrared absorption. Changing the identity of the attacking species makes the range of thermochemistry shown in Fig. 2 potentially available for such reactions, but hydrogen atom abstraction by Cl and by CN are the two most extensively studied examples. In the first experiments, Hochstrasser studied the reactions of Cl with cyclohexane[28] and CN with chloroform,[29] and the CN reactions showed the first evidence of vibrationally excited products in the case of deuterochloroform. More recent experiments have built on those studies.

It is also possible to follow a reaction by observing the loss of one of the reactants. Fig. 4(a) shows a scheme for an experiment probing either the loss of the Cl reactant or the formation of the HCl product in the reaction of Cl with an alkane.[30] After photolyzing Cl_2 with a pulse of 350 nm light, we probe the Cl atom, which forms a weakly bound complex with the solvent, with a 330 nm pulse that excites the strongly bound charge-transfer state. Monitoring this transient absorption as a function of the delay between the photolysis and probe pulse for the reaction of pentane with Cl in a CH_2Cl_2 solution yields the decay curves shown in the top part of Fig. 4(b). The decay is faster for larger pentane concentrations, and the concentration dependence of the decay yields the bimolecular rate constant. Similarly, it is possible to monitor the ground-vibrational-state HCl product by transient absorption of a 3.6 μm infrared pulse. The growth of that signal, shown in the bottom of Fig. 4(b), exactly mirrors the decay of the Cl atom.

A set of experiments probing the loss of Cl in reactions with alkanes, alcohols, and chloroalkanes provides a set of rate constants for comparison to gas-phase results.[31] The alkanes react at the diffusion limited rate, but the alcohols have a slightly smaller rate, reflecting a small activation barrier for the reaction. However, the chloroalkanes provide the most informative comparison with the rates of gas-phase reactions. The reaction rates for the chloroalkanes vary from nearly diffusion-limited to almost an order-of-magnitude slower, reflecting the changes in the barrier to reaction with increasing substitution. There are structure–activity relationships for reactions of the chloroalkanes with Cl developed for modelling the rates of atmospheric reactions.[32,33] These gas-phase rules reproduce the relative reaction rates in solution very well,[31] showing that the variation of the activation energy among the reactants in solution follows that in the gas phase. Thus, it seems that the gas-phase relative reaction rates are a sound predictor of the relative rates in solution, at least in the weakly interacting solvent we used.

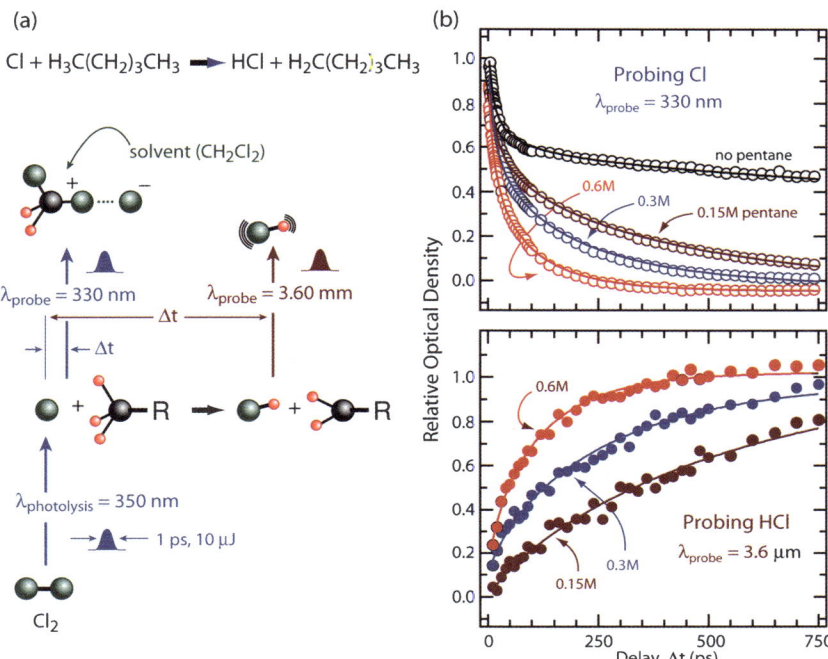

Fig. 4 (a) Scheme for time-resolved study of the bimolecular reaction of Cl with an alkane in solution. Photolysis of Cl_2 generates Cl atoms that react with pentane in CH_2Cl_2 solution, and transient absorption of the Cl-solvent complex to a charge-transfer state follows the evolution of the Cl atom. A short infrared pulse follows the appearance of the HCl product of the H-atom abstraction reaction. (b) Time-evolution of the Cl reactant (top) and HCl product (bottom).

The reactions of CN with alkanes are perhaps an even more interesting possibility for comparison because the stronger H–CN bond makes the reaction exothermic enough to excite vibrations of the product, as studies of CN reactions in the gas phase demonstrate.[34–36] Recent experiments by Orr-Ewing and coworkers use broadband infrared detection of the HCN formed in the reaction of photolytically produced CN radicals in solution.[37,38] Fig. 5 shows the results of their measurements on the reaction of CN with cyclohexane

$$CN + c\text{-}C_6H_{12} \rightarrow HCN(v_1\ v_2\ v_3) + c\text{-}C_6H_{11}.$$

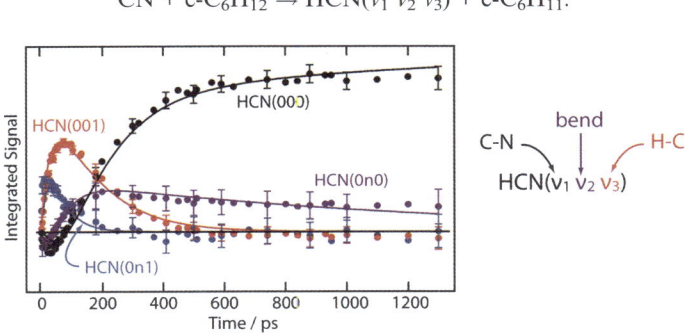

Fig. 5 Time evolution of vibrational-state populations of HCN formed in the reaction of CN with cyclohexane in dichloromethane solution. The vibrationally excited products, HCN(001) and HCN(0n1), appear initially. Subsequent vibrational relaxation populates the ground vibrational states. (Adapted from ref. 37 with permission of AAAS.)

The vibrational quantum numbers of HCN denoting the three normal modes correspond to the C–N stretch (v_1), the HCN bend (v_2), and the C–H stretch (v_3). The time-resolved, broadband infrared spectra show that the reaction initially forms HCN with an excited C–H stretch along with some bending excitation. The time-evolution of each of those states shown in Fig. 5 clearly demonstrates the evolution of the vibrationally excited HCN. Both stretch-excited products, HCN(001), and stretch-bend-excited products, HCN(0n1), appear in the first 60 ps. Vibrational relaxation eventually moves molecules into the ground state, HCN(000), over about 200 ps. Previous measurements assigned this time constant to reaction,[29,39] but it is clear that those experiments observed vibrational relaxation after a much faster reaction, a cautionary point for studies of nascent products in solution.

These experiments provide a comparison of reaction dynamics in solution, not just rates, with those in gases. The initial production of vibrationally excited molecules is consistent with the observations in gases although the gas-phase reaction produces higher levels of excitation. Clearly the presence of the solvent does not alter the energy release dramatically but rather makes more subtle differences in the course of the reaction.[37,38] A combination of experimental advances and new insights is opening the door to studies of molecular reaction dynamics in solution.

2.3 Reactions on metal surfaces: the CH$_4$ poster child redux

The dissociative chemisorption of methane on a metal surface

$$CH_4 \rightarrow CH_3(ads) + H(ads)$$

offers the possibility of observing mode- and bond-selective chemistry in yet another environment, and there are striking parallels with the reaction dynamics in the gas phase. Elegant surface scattering experiments using molecular beams of vibrationally excited methane and its isotopologues[40-53] have established the relative efficacy of translational and vibrational energy in promoting the dissociation and have demonstrated both mode- and bond-selective reactions of vibrationally excited molecules. For example, in the case of methane on Ni(100),[48] the symmetric stretching vibration (v_1) promotes dissociative adsorption about ten times more effectively than the antisymmetric stretching vibration (v_3). In addition, these experiments show that translational energy normal to the surface also enhances the reaction. A translational energy increment of about 50 kJ mol^{-1} increases the probability of dissociative chemisorption as effectively as the 36 kJ mol^{-1} of vibrational energy provided by excitation of the symmetric stretch. Both the mode-selectivity and the effect of increased translational energy parallel the behaviour described above for single collision reactions of methane with Cl.

There are parallels in the bond-selectivity of the gas-phase and surface reactions as well. The first demonstration of bond-selective dissociative chemisorption used the reaction of subsurface D atoms with the chemisorbed species to show the preferential cleavage of the initially excited bond in CHD$_3$ on a Ni(111) surface.[45] In these experiments, CHD$_3$ molecules with an initially excited C–H stretching vibration always dissociate to leave CD$_3$ radicals on the surface, as shown by the exclusive production of CD$_4$ in reaction with the subsurface D atoms. An extensive study presented in this Discussion[54] expands the scope of experiments on bond-selective dissociative chemisorption by using Reflection Absorption Infrared Spectroscopy (RAIRS) to interrogate the species on the surface after dissociative chemisorption of methane and its isotopologues. The generality of this surface spectroscopy allows Beck and coworkers to examine the entire series of methane molecules with increasing deuteration: CH$_4$, CH$_3$D, CH$_2$D$_2$, CHD$_3$, and CD$_4$. Fig. 6 shows a portion of one of their RAIRS spectra comparing the absorbed species produced from CH$_3$D without vibrational excitation (lower trace) to those from CH$_3$D with C–H stretching excitation (upper trace). The data show that chemisorption cleaves

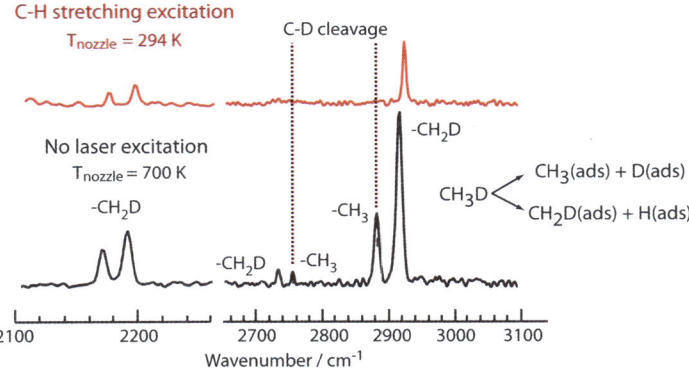

Fig. 6 Reflection Absorption Infrared Spectra (RAIRS) of adsorbed species on Pt(111) from dissociative chemisorption of CH_3D with energy in translation (bottom trace, black) or in C–H stretching vibration (top trace, red). Translational energy causes both the C–H and C–D bonds to break, but vibrational excitation of the C–H stretch leads solely to cleavage of that bond. (Adapted from ref. 54 with permission of the Royal Society of Chemistry.)

only the C–H bond when it is initially vibrationally excited but that both the C–H and C–D bonds break when the energy is in relative translation.

These experiments "ring the changes" on all of the isotopologues and paint a consistent picture of the effect of vibrational excitation, showing that dissociative chemisorption cleaves the vibrationally excited bond, in analogy to bimolecular gas-phase reactions. The interaction with the surface is complicated, and these detailed results present a challenge to theory in understanding the evolution of the excited molecule as it approaches the surface and the nature of the transition state through which the reaction occurs. It seems likely that the evolution of the excited state during the interaction with the surface has much in common with the evolution as the vibrationally excited molecule approaches an atom in the gas phase.

3.   Photodissociation

Electronic photodissociation involves promoting a molecule to an excited electronic state from which it either dissociates directly or on which it evolves to reach another electronic state from which it dissociates. As described in the introduction, some of the most important regions on the excited-state surface are those where two electronic states having the same energy form a conical intersection. Conical intersections have emerged as a unifying idea in excited state chemistry.[55–58] Fig. 7 shows one-dimensional cuts through the potential energy surfaces of several molecules in which conical intersections are important. As the figure shows, multiple surfaces can participate, and, in all of these cases, there is one state in which excitation into an antibonding orbital initiates the excited state dynamics.

3.1   Conical intersections and product states

Moving an electron into a σ* orbital in ammonia makes the system dissociative along the N–H coordinate, as shown in Fig. 7(a). Interactions with higher-lying states create an excited-state barrier, and beyond that barrier the excited state falls in energy to correlate with ground-state NH_2 and H fragments. Along the way, it crosses the ground state of NH_3, which correlates with excited-state NH_2, and this crossing forms a conical intersection. The surfaces in Fig. 8 show the structure along two coordinates, one N–H bond length, R, and the umbrella bending coordinate, θ. The barrier between the two pyramidal minima in the ground electronic state is at

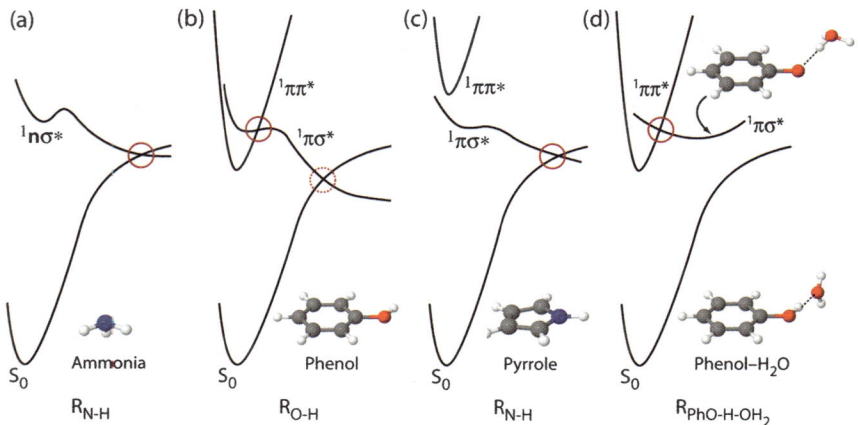

Fig. 7 Potential energy curves for (a) ammonia, (b) pyrrole, (c) phenol, and (d) the phenol-water complex. The red circles mark the conical intersections in each system.

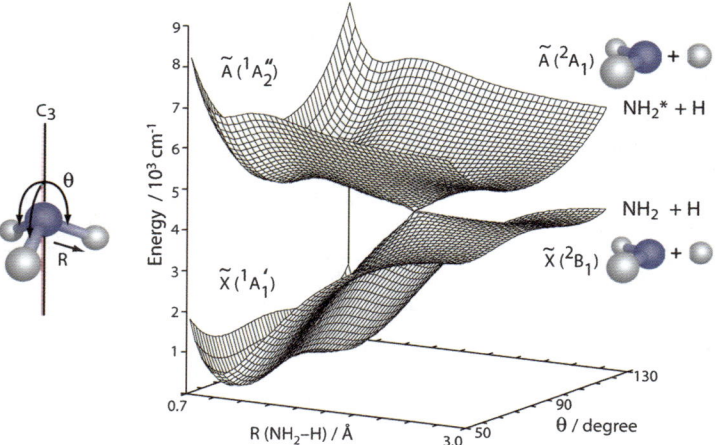

Fig. 8 Calculated ground- and first excited-state potentials of ammonia. The coordinates are R, the distance along one of the N–H bonds, and the angle, θ, the out-of-plane bending angle.

the planar geometry, $\theta = 90°$, and motion along that coordinate is the familiar inversion tunnelling. In planar ammonia, the excited state and the ground state, which correlate with different electronic states of the products, do not interact because they have different symmetries. However, in non-planar geometries, they do interact because their symmetries are the same. The two adiabatic states separate in energy, forming the conical intersection shown in the figure, which is particularly easy to see along these coordinates. A more general description uses two different coordinates that are not always so easy to identify in terms of bond coordinates.[1,56,58] However, locating the conical intersection constrains only those two coordinates and leaves the others free.

The curves for phenol in Fig. 7(b) illustrate a recurring theme in the excited state dynamics of heteroaromatic molecules.[59] Both excited states involve the promotion of an electron from the bonding π orbital of the aromatic ring, but they differ in the destination orbital. The $\pi\pi^*$ state has that electron in the antibonding pi orbital, and the $\pi\sigma^*$ state has that electron in an antibonding sigma orbital. The transition to the $\pi\pi^*$ state is generally stronger (the transition to the $\pi\sigma^*$ state is symmetry forbidden in benzene), and in many cases vibronic coupling of the two states provides the oscillator strength for the transition to the $\pi\sigma^*$ state. The curves for phenol have two conical intersections, one between the two excited states, $\pi\pi^*$ and $\pi\sigma^*$, and one between the $\pi\sigma^*$ state and the ground state S_0. The branching between the states at each of these intersections is critical in the dissociation dynamics.[58,60] It determines the populations of the electronic states of the dissociation products, as revealed, for example, by high resolution, H-atom Rydberg state time-of-flight measurements.[59]

The curves for pyrrole in Fig. 7(c) illustrate a different situation involving the same three states. In this case, the location of the $\pi\pi^*$ state at an energy well above that of the $\pi\sigma^*$ state removes one of the conical intersections, and excitation of the $\pi\pi^*$ state does not lead directly to the intersection and dissociation on the $\pi\sigma^*$ state. Thus, the relative energies of these interacting states is a controlling feature of the photodissociation that explains trends among various molecules.[59]

It is also possible to alter the relative energies of the states by interaction with an adduct, as the calculated potentials for the complex of phenol with water[61] shown in Fig. 7(d) illustrate. A single hydrogen-bonded water molecule alters the energy of the $\pi\sigma^*$ state to remove the intersection between that state and the ground state. In fact, there is stable minimum on the $\pi\sigma^*$ state in which the hydrogen atom resides on the water molecule. This change in the excited-state potential raises the possibility that the surroundings could alter dissociation pathways substantially in solution by changing the energies of interacting excited states, a possibility that experiments are beginning to explore.[62]

3.2 Dissociation of isolated molecules and the influence of vibrations

There is a rich history of probing asymptotic product states, using techniques such as Rydberg tagging and ion imaging, to infer details of the dissociation dynamics. One of the keys is detecting one product in a specific state with sufficient speed resolution to infer the distribution of the unobserved products among their quantum states. For example, in this Discussion, Kable and coworkers describe experiments using ion imaging to study the three-body dissociation of acetaldehyde.[63]

A means of influencing excited-state dynamics is to prepare a molecule on the excited state with selected initial nuclear motions. Exciting vibrations in ground-state molecules brings new Franck–Condon factors into play and permits excitation to new regions of the excited-state surface. Early experiments show that these schemes can control the cleavage of a selected bond in HOD, for example.[64] The photodissociation of vibrationally excited ammonia molecules even makes it possible to influence the relative yield of ground-state and excited-state NH_2 products, reflecting the branching between the two states at conical intersections.[65,66]

The experiments on the photodissociation of vibrationally excited ammonia prepare NH_3 molecules with either a quantum of symmetric N–H stretching excitation or a quantum of antisymmetric N–H stretching excitation. As Fig. 8 shows, formation of ground-state ($^2B_1$) NH_2 products makes about 10 000 cm^{-1} more energy available than formation of excited-state ($^2A_1$) NH_2. In the latter case, the extra energy resides in electronic excitation of the product. Experiments detecting the translational energy of the products show that photodissociation of molecules from the antisymmetric N–H stretching state forms slowly recoiling H atoms and suggest that the partner is the electronically excited product NH_2^*.[65] Ion imaging experiments confirmed this supposition by detecting the recoiling H atoms with sufficient speed resolution to show the undetected NH_2 product has the rotational state structure of NH_2^*.[66]

The situation for photodissociation from the symmetric N–H stretching state is very different. The H-atom recoil distribution shows that photolysis forms NH_2 in its ground electronic state with a kinetic-energy distribution that is very similar to that for photolysis of ground-state NH_3. Thus, only initial excitation of the antisymmetric stretching vibration promotes the adiabatic decomposition in which the dissociation evolves on the upper surface. Theoretical calculations show that in the excited state the antisymmetric N–H stretching vibration carries the system away from the conical intersection shown in Fig. 8 towards another one where non-adiabatic transitions to the lower surface are less efficient. This behaviour in part comes from the development of angular momentum in the vibrationally excited molecule.[67] These experiments show that exploring different parts of the excited-state surface can substantially alter the decomposition dynamics in systems with conical intersections. Measurements show a similar, but less dramatic, effect in the decomposition of vibrationally excited phenol.[68]

3.3 Photodissociation in solution

3.3.1 Dissociation of prototypical aromatic molecules in solution.
Several aromatic molecules are prototypes because of their conical intersections, illustrated in Fig. 7, and because of the detailed gas-phase studies by Ashfold and coworkers.[59] Thus, they are good targets for photodissociation in solution, and in this Discussion, Bradforth, Ashfold, and coworkers present such a ground breaking, time-resolved experiment.[62] As described above, observing the early time dynamics of the decomposition in solution relies on good time resolution rather than the high energy resolution used in gas-phase studies. They study the photodissociation of isolated phenol and p-methylthiophenol, using Rydberg tagging schemes, and compare their results to those for the photodissociation of these same two molecules in cyclohexane solution, probed using transient absorption spectroscopy.

The calculated potential energy curves[62] for the two molecules in Fig. 9 are qualitatively similar. However, the relative energies of the conical intersections differ because the $\pi\sigma^*$ state crosses the bound $\pi\pi^*$ state at a lower relative energy in p-methylthiophenol than in phenol. Comparing the solution and gas-phase results shows that most aspects of the initial dissociation dynamics are the same in cyclohexane solution and in an isolated molecule. As in isolated molecules, the energy of the initial excitation relative to the crossing of the $\pi\pi^*$ and $\pi\sigma^*$ states strongly

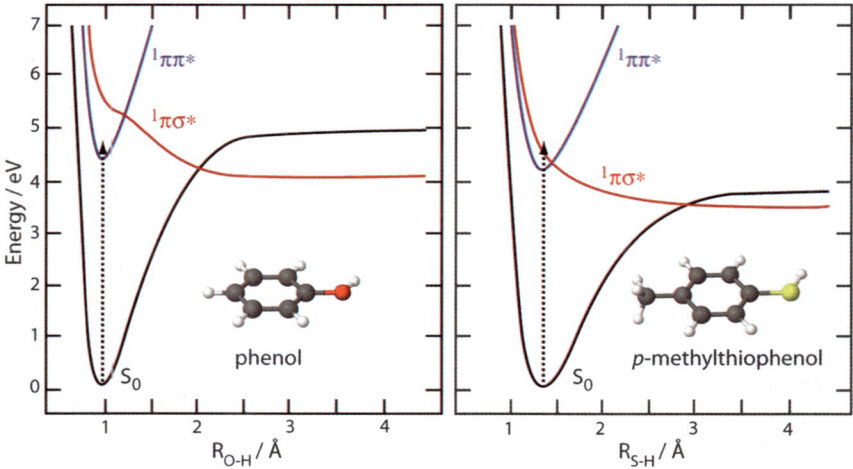

Fig. 9 Calculated potential energy curves for the first three singlet states of phenol (left) and p-methylthiophenol (right). The vertical dotted arrow indicates the energy of a 267-nm photon. (Adapted from ref. 62 with permission of the Royal Society of Chemistry.)

influences the dissociation dynamics. Because the dissociative $\pi\sigma^*$ state lies at a relatively low energy in p-methylthiophenol, excitation with a 267-nm photon leads to ultrafast dissociation for molecules both in solution and in the gas phase. However, the $\pi\sigma^*$ state lies above the energy provided by a 267-nm photon in phenol, and the ultrafast solution-phase measurements directly observe the slow dissociation in about 1 ns that state-resolved measurements suggest also occurs in the gas-phase. The dynamics are not completely indifferent to the phase in which the dissociation occurs, and hydrogen-bonded dimers and other oligomers may play a role in solution.[62]

3.3.2 Solvent-induced isomerization.

The surrounding solvent can also exert more direct influence on photodissociation in solution. In the gas phase, placing a molecule in a repulsive excited state generally causes the fragments to separate unimpeded by further interactions. There are examples of roaming in isolated molecules, in which relatively direct dissociation competes with a rearrangement followed by decomposition to the same products. The common feature in these systems is the presence of a channel leading to different products that is near enough in energy to allow the system to explore that channel but not dissociate. For example, there is a radical decomposition channel for formaldehyde, $H_2CO \rightarrow H + HCO$, that facilitates roaming in the molecular dissociation channel, $H_2CO \rightarrow H_2 + CO$. In this case, the hydrogen atom that explores the radical channel returns to form H_2 in the molecular channel rather than depart completely.[69,70]

The surroundings can hinder separation of the products in solution, frustrating the dissociation and turning direct photolysis into isomerization. The most likely consequence of such caging is recombination to restore the original molecule, but if there are other stable arrangements of the atoms, it can induce isomerization into that geometry. The photolysis of halomethanes is an example.[71,72] Gas-phase photodissociation of bromoform, $CHBr_3$, cleaves the C–Br bond to produce free Br and a halomethyl radical

$$CHBr_3 \rightarrow CHBr_2 + Br$$

which separate completely. In solution, however, the departing Br atom can return to the radical and either reform bromoform or, less frequently, bind to one of the Br atoms on the radical to form iso-bromoform, $CHBr_2$-Br. Fig. 10(a) shows the structure of the isomer that contains a Br–Br bond. There are many examples of formation of iso-halomethanes, and some of them are intermediates in reactions with olefins.[73]

The ground-state potential energy surface and corresponding contour map in Fig. 10(a) show the two minima and the minimum-energy pathway between them, which passes over a barrier that lies below the dissociation asymptote. Calculations show that isomerization along this path occurs by stretching one bond to a halogen atom and swinging it around to another halogen atom.[74,75] The location of this barrier below the nearby dissociation asymptote is reminiscent of the situation that produces roaming in isolated molecules. In this case, however, interactions with the surroundings are a key component of the isomerization, which does not follow the minimum-energy path.

The upper curve in Fig. 10(b) shows the excited state of bromoform on which direct dissociation occurs in the gas phase. In solution, by contrast, the separating fragments encounter the surroundings, which provide the "cage repulsion" marked in the figure. The departing atom returns to explore new configurations including that of the isomer. We have followed this isomerization in solution using ultrafast, broadband transient infrared spectroscopy. The 30 cm^{-1} difference in the C–H stretching transitions in bromform and iso-bromoform allows us to follow the evolution of both species.[74] The isomer appears in less than 10 ps, and its spectrum evolves over about 100 ps as the vibrationally excited isomer relaxes.

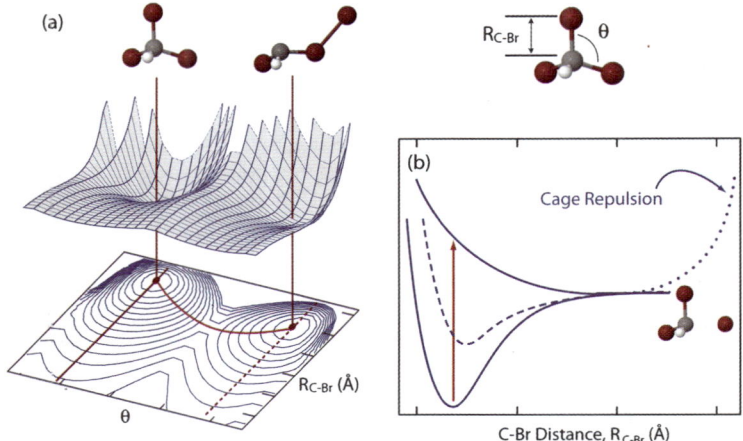

Fig. 10 (a) Sketch of the potential energy surface for ground state bromoform, CHBr$_3$. The structure in the upper right shows the coordinate system. There is a deep minimum for the bromoform (left) and a shallower minimum for the isomer, *iso*-bromoform (right). The barrier along the minimum energy path connecting the two lies below the dissociation asymptote. (b) Cuts through the ground-state surface for bromoform and *iso*-bromoform and through the excited-state surface for bromoform. The dotted curve qualitatively represents the repulsion by the surrounding molecules in solution.

4. Conclusions

Conceptual and technical advances are enabling studies of reaction dynamics in gases and liquids and at interfaces with sufficient detail to allow comparison of closely related reactions in different environment. One example described in this Discussion is the fate of methane and its isotopologues prepared in selected vibrational states. There are studies of their reactions with halogen atoms in the gas phase and of their dissociative chemisorption on a metal surface. Initial vibrational excitation influences the course of the reaction similarly, producing mode- and bond-selected chemistry in both cases. Another example is bimolecular reaction in solution, where the relative reaction rates and the energy disposal mirror analogous gas-phase reactions. The surroundings also influence photodissociation to different degrees depending on the system. In one example, the photolysis of a pair of isolated, aromatic molecules is similar to that of the same two molecules in solution, albeit with some subtle differences arising from solvation. For those two molecules, the energy at which there is a conical intersection between two excited states is the dominant factor in both environments. By contrast, there are examples where the surroundings exert a much stronger influence by inhibiting the separation of fragments and producing isomers. The key to all of these studies is examining systems in sufficient detail in multiple environments to permit detailed comparisons. As the contributions to this Discussion illustrate, increasingly sophisticated experimental approaches and theoretical descriptions open the door to ever more incisive studies in increasing complex systems.

Acknowledgements

The U.S. National Science Foundation and the U.S. Department of Energy sponsor different aspects of the work from the University of Wisconsin - Madison described here. The National Science Foundation (CHE-0910917) supports the research on bimolecular reactions of translationally and vibrationally energized molecules in the gas-phase and the research on bimolecular reactions and cage recombination

This journal is © The Royal Society of Chemistry 2012

in solution. The Office of Basic Energy Sciences of the Department of Energy (DE-FG02-86ER13500) supports the research on photodissociation of vibrationally excited molecules.

References

1 D. R. Yarkony, *Acc. Chem. Res.*, 1998, **31**, 511–518.
2 H. A. Kramers, *Physica*, 1940, **7**, 284–304.
3 R. F. Grote and J. T. Hynes, *J. Chem. Phys.*, 1980, **73**, 2715–2732.
4 J. C. Polanyi, *Acc. Chem. Res.*, 1972, **5**, 161–168.
5 F. F. Crim, *Proc. Natl. Acad. Sci. U. S. A.*, 2008, **105**, 12654–12661.
6 G. Czako and J. M. Bowman, *J. Chem. Phys.*, 2012, **136**, 044307.
7 M. Cheng, Y. Feng, Y. K. Du, Q. H. Zhu, W. J. Zheng, G. Czako and J. M. Bowman, *J. Chem. Phys.*, 2011, **134**, 191102.
8 H. Kawamata, W. Zhang and K. Liu, *Faraday Discussions*, 2012, **157**, DOI: 10.1039/C2FD20004J.
9 W. R. Simpson, A. J. Orr-Ewing and R. N. Zare, *Chem. Phys. Lett.*, 1993, **212**, 163–171.
10 H. Kawamata and K. P. Liu, *J. Chem. Phys.*, 2010, **133**, 124304.
11 A. E. Berke, E. H. Volpa and F. F. Crim, *J. Phys. Chem. A*, 2012, (in preparation).
12 M. S. Child and L. Halonen, *Adv. Chem. Phys.*, 1984, **57**, 1–58.
13 M. S. Child, *Acc. Chem. Res.*, 1985, **18**, 45–50.
14 S. Yoon, S. Henton, A. N. Zivkovic and F. F. Crim, *J. Chem. Phys.*, 2002, **116**, 10744–10752.
15 S. Yoon, R. J. Holiday and F. F. Crim, *J. Chem. Phys.*, 2003, **119**, 4755–4761.
16 S. Yoon, R. J. Holiday, E. L. Sibert and F. F. Crim, *J. Chem. Phys.*, 2003, **119**, 9568–9575.
17 A. E. Berke, C. J. Annesley, E. H. Volpa and F. F. Crim, *J. Phys. Chem. A*, 2012, (in preparation).
18 W. J. van der Zande, R. Zhang, R. N. Zare, K. G. McKendrick and J. J. Valentini, *J. Phys. Chem.*, 1991, **95**, 8205.
19 A. E. Berke, Ph.D., University of Wisconsin - Madison, 2012.
20 R. J. Holiday, C. H. Kwon, C. J. Annesley and F. F. Crim, *J. Chem. Phys.*, 2006, **125**, 133101.
21 C. J. Annesley, A. E. Berke and F. F. Crim, *J. Phys. Chem. A*, 2008, **112**, 9448–9453.
22 S. Yan, Y. T. Wu, B. L. Zhang, X. F. Yue and K. P. Liu, *Science*, 2007, **316**, 1723–1726.
23 J. R. Fair, D. Schaefer, R. Kosloff and D. J. Nesbitt, *J. Chem. Phys.*, 2002, **116**, 1406–1416.
24 L. Halonen, S. L. Bernasek and D. J. Nesbitt, *J. Chem. Phys.*, 2001, **115**, 5611–5619.
25 A. Sinha, M. C. Hsiao and F. F. Crim, *J. Chem. Phys.*, 1991, **94**, 4928–4935.
26 A. Sinha, J. D. Thoemke and F. F. Crim, *J. Chem. Phys.*, 1992, **96**, 372–376.
27 J. D. Thoemke, J. M. Pfeiffer, R. B. Metz and F. F. Crim, *J. Phys. Chem.*, 1995, **99**, 13748–13754.
28 D. Raftery, M. Iannone, C. M. Phillips and R. M. Hochstrasser, *Chem. Phys. Lett.*, 1993, **201**, 513–520.
29 D. Raftery, E. Gooding, A. Romanovsky and R. M. Hochstrasser, *J. Chem. Phys.*, 1994, **101**, 8572–8579.
30 L. Sheps, A. C. Crowther, S. L. Carrier and F. F. Crim, *J. Phys. Chem. A*, 2006, **110**, 3087–3092.
31 L. Sheps, A. C. Crowther, C. G. Elles and F. F. Crim, *J. Phys. Chem. A*, 2005, **109**, 4296–4302.
32 S. M. Aschmann and R. Atkinson, *Int. J. Chem. Kinet.*, 1995, **27**, 613–622.
33 S. M. Senkan and D. Quam, *J. Phys. Chem.*, 1992, **96**, 10837–10842.
34 G. A. Bethardy, F. J. Northrup and R. G. Macdonald, *J. Chem. Phys.*, 1996, **105**, 4533–4549.
35 L. R. Copeland, F. Mohammad, M. Zahedi, D. H. Volman and W. M. Jackson, *J. Chem. Phys.*, 1992, **96**, 5817.
36 V. R. Morris, F. Mohammad, L. Valdry and W. M. Jackson, *Chem. Phys. Lett.*, 1994, **220**, 448–454.
37 S. J. Greaves, R. A. Rose, T. A. A. Oliver, D. R. Glowacki, M. N. R. Ashfold, J. N. Harvey, I. P. Clark, G. M. Greetham, A. W. Parker, M. Towrie and A. J. Orr-Ewing, *Science*, 2011, **331**, 1423–1426.
38 R. A. Rose, S. J. Greaves, T. A. A. Oliver, I. P. Clark, G. M. Greetham, A. W. Parker, M. Towrie and A. J. Orr-Ewing, *J. Chem. Phys.*, 2011, **134**, 244503.
39 A. C. Crowther, S. L. Carrier, T. J. Preston and F. F. Crim, *J. Phys. Chem. A*, 2008, **112**, 12081–12089.

40 L. B. F. Juurlink, P. R. McCabe, R. R. Smith, C. L. DiCologero and A. L. Utz, *Phys. Rev. Lett.*, 1999, **83**, 868–871.
41 L. B. F. Juurlink, R. R. Smith and A. L. Utz, *Faraday Discuss.*, 2000, **117**, 147–160.
42 L. B. F. Juurlink, R. R. Smith and A. L. Utz, *J. Phys. Chem. B*, 2000, **104**, 3327–3336.
43 R. R. Smith, D. R. Killelea, D. F. DelSesto and A. L. Utz, *Science*, 2004, **304**, 992–995.
44 L. B. F. Juurlink, R. R. Smith, D. R. Killelea and A. L. Utz, *Phys. Rev. Lett.*, 2005, **94**, 208303.
45 D. R. Killelea, V. L. Campbell, N. S. Shuman and A. L. Utz, *Science*, 2008, **319**, 790–793.
46 D. R. Killelea, V. L. Campbell, N. S. Shuman and A. L. Utz, *J. Phys. Chem. C*, 2008, **112**, 9822–9827.
47 R. D. Beck, P. Maroni, D. C. Papageorgopoulos, T. T. Dang, M. P. Schmid and T. R. Rizzo, *Science*, 2003, **302**, 98–100.
48 P. Maroni, D. C. Papageorgopoulos, M. Sacchi, T. T. Dang, R. D. Beck and T. R. Rizzo, *Phys. Rev. Lett.*, 2005, **94**, 246104.
49 R. Bisson, M. Sacchi, T. T. Dang, B. Yoder, P. Maroni and R. D. Beck, *J. Phys. Chem. A*, 2007, **111**, 12679–12683.
50 R. Bisson, M. Sacchi and R. D. Beck, *J. Chem. Phys.*, 2010, **132**, 094702.
51 R. Bisson, M. Sacchi and R. D. Beck, *Phys. Rev. B: Condens. Matter Mater. Phys.*, 2010, **82**, 121404.
52 B. L. Yoder, R. Bisson and R. D. Beck, *Science*, 2010, **329**, 553–556.
53 J. Higgins, A. Conjusteau, G. Scoles and S. L. Bernasek, *J. Chem. Phys.*, 2001, **114**, 5277–5283.
54 L. Chen, H. Ueta, R. Bisson and R. D. Beck, *Faraday Discussions*, 2012, **157**, DOI: 10.1039/C2FD20007D.
55 F. Bernardi, M. Olivucci and M. A. Robb, *Chem. Soc. Rev.*, 1996, **25**, 321–328.
56 D. R. Yarkony, *J. Phys. Chem. A*, 2001, **105**, 6277–6293.
57 W. Domcke, L. Seidner and G. Stock, *Springer Ser. Chem. Phys.*, 1998, **63**, 491–495.
58 B. G. Levine and T. J. Martinez, *Annu. Rev. Phys. Chem.*, 2007, **58**, 613–634.
59 M. N. R. Ashfold, B. Cronin, A. L. Devine, R. N. Dixon and M. G. D. Nix, *Science*, 2006, **312**, 1637–1640.
60 Z. G. Lan, W. Domcke, V. Vallet, A. L. Sobolewski and S. Mahapatra, *J. Chem. Phys.*, 2005, **122**, 224315.
61 A. L. Sobolewski and W. Domcke, *J. Phys. Chem. A*, 2001, **105**, 9275–9283.
62 Y. Zhang, T. A. A. Oliver, M. N. R. Ashfold and S. E. Bradforth, *Faraday Discussions*, 2012, **157**, DOI: 10.1039/C2FD20043K.
63 G. de Wit, B. R. Heazlewood, M. S. Quinn, A. T. Maccarone, K. Nauta, S. A. Reid, M. J. T. Jordan and S. H. Kable, *Faraday Discussions*, 2012, **157**, DOI: 10.1039/C2FD20015E.
64 R. L. Vander Wal, J. L. Scott and F. F. Crim, *J. Chem. Phys.*, 1990, **92**, 803.
65 A. Bach, J. M. Hutchison, R. J. Holiday and F. F. Crim, *J. Phys. Chem. A*, 2003, **107**, 10490–10496.
66 M. L. Hause, Y. H. Yoon and F. F. Crim, *J. Chem. Phys.*, 2006, **125**, 174309.
67 D. R. Yarkony, *J. Chem. Phys.*, 2004, **121**, 628–631.
68 M. L. Hause, Y. H. Yoon, A. S. Case and F. F. Crim, *J. Chem. Phys.*, 2008, **128**, 104307.
69 J. M. Bowman and B. C. Shepler, *Annu. Rev. Phys. Chem.*, 2011, **62**, 531–553.
70 N. Herath and A. G. Suits, *J. Phys. Chem. Lett.*, 2011, **2**, 642–647.
71 X. Zheng, W.-H. Fang and D. L. Phillips, *J. Chem. Phys.*, 2000, **113**, 10934–10946.
72 X. M. Zheng and D. L. Phillips, *Chem. Phys. Lett.*, 2000, **324**, 175–182.
73 D. L. Phillips, W. H. Fang, X. Zheng, Y. L. Li, D. Wang and W. M. Kwok, *Curr. Org. Chem.*, 2004, **8**, 739–755.
74 T. J. Preston, M. A. Shaloski and F. F. Crim, 2012, (in preparation).
75 T. J. Preston, Ph.D., University of Wisconsin - Madison, 2012.

Three-state surface hopping calculations of acetaldehyde photodissociation to CH_3 + HCO on *ab initio* potential surfaces

Bina Fu, Yongchang Han and Joel M. Bowman*

Received 24th January 2012, Accepted 29th February 2012
DOI: 10.1039/c2fd20010d

We report Trajectory Surface Hopping (TSH) calculations of CH_3CHO photodissociation involving three electronic states, S_1, T_1, and S_0, with a focus on the radical products CH_3 + HCO, which can be formed from both T_1 and S_0. We use previously reported potential energy surfaces and spin–orbit couplings for T_1 and S_0 and report a new potential and spin–orbit coupling for S_1 here. Roughly 32 000 trajectories are performed at energies corresponding to seven photolysis wavelengths between 372 and 230 nm. Motivated by recent experiments, we examine the branching ratio of the T_1 to S_0 pathways as a function of photolysis energy. We also present the relative translational energy and CH_3 vibrational energy distributions from these pathways at a photolysis energy of 100 kcal mol^{-1}, formed from both the T_1 and S_0 potentials. As with standard quasiclassical trajectory calculations, violation of zero-point energy for products also occurs in TSH calculations. This is shown to be a serious issue for this branching ratio and one of several methods considered to deal with this issue is shown to give satisfactory results.

1. Introduction

The photodissociation of acetaldehyde is a complex process with many possible reaction channels.[1] There have been several recent state-resolved experiments focusing on the radical products CH_3 + HCO, the dominant products of this dissociation.[3–8] Following excitation to S_1, CH_3CHO^* intersystem crosses to T_1, where these products can be directly formed, and/or further intersystem crosses to S_0 where they can also be formed. A realistic schematic of these potentials and dynamics is given in Fig. 1, based on calculations that we report and discuss in detail in the next section.

The branching between these electronic pathways products and the different final state dynamics from these two states have been the focus of experiment and represent a major challenge to theory, which up to now has not explicitly considered coupling between these surfaces. Chen and Fang reported a DFT/B3LYP and CASSCF study of some relevant stationary points on these three potential energy surfaces (PESs).[9] They did locate a high-energy conical intersection between S_1 and S_0 which is above the energies of relevance of to the work reported here.

The experimental determination of the electronic branching ratio to CH_3 + HCO is obviously challenging and must make use of any significant differences in observables, such as the translational energy distribution and/or the internal energy distribution of the products from each electronic pathway. Here experiment has relied on the existence of an exit channel barrier on T_1, found in *ab initio* calculations,[10–13] but

Department of Chemistry and Cherry L. Emerson Center for scientific computation Emory University, Atlanta, Georgia 30322. E-mail: jmbowma@emory.edu

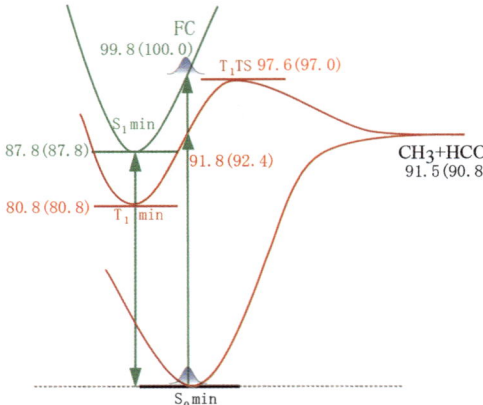

Fig. 1 Schematic of the 3-state electronic energetics of the photodissociation of CH_3CHO to $CH_3 + HCO$, based on the present full-dimensional potential energy surfaces and direct MRCI + Q/aVTZ calculations, in parentheses. The S_1 and T_1 PESs were shifted so that the energies of the respective minima agree with the *ab initio* ones. The vertical excitation energies from the PESs indicated from the S_0 global minimum to T_1 and S_1 are also indicated and as seen they agree well MRCI + Q/aVTZ energies. As indicated the dissociation on the T_1 PES has a barrier, whereas the dissociation on S_0 is barrierless.

a barrierless dissociation on S_0.[1,2,4,14] This important difference leads to expected and observed differences in the translational energy distributions of the radical products, for these states. The relative translational energy distribution has been calculated using (low-level) direct-dynamics initiated from the T_1 barrier to $CH_3 + HCO$[11,12] and also in much more extensive quasiclassical trajectory calculations (QCT) done on a limited, but full-dimensional high-level *ab initio* T_1 PES.[13]

It is also worth noting that these radical products, or more precisely the incipient formation of these radicals, on S_0 is the source of the "roaming" dynamics[1,13–17] to form the molecular produces, $CH_4 + CO$ on S_0.[1] Our work on CH_3CHO photodissociation began with a focus on roaming, stimulated by experiments by Houston and Kable, who deduced correctly that roaming is quite significant in CH_3CHO.[18] Further experiments confirmed and quantified the extent of roaming in CH_3CO.[19,20] We developed a global potential energy surface for S_0,[2,14] and performed extensive quasiclassical trajectory calculations (QCT) using it.[2,14,20,21] This PES, which is used here, also describes numerous other channels, including $CH_3 + HCO$.

In addition to photodissociation studies, there is also considerable interest in the bimolecular reaction $O(^3P)+C_2H_4$, which yields many products and also clear evidence of intersystem crossing from T_1 to S_0.[22] These experiments, which access different regions of the PESs compared to the photodissociation ones, have also attracted extensive theoretical work.[23–25] Very recently, we reported a global PES for T_1 and spin–orbit (SO) coupling to S_0, that describes these reaction dynamics. These PESs and the SO coupling have been used in trajectory surface hopping (TSH) calculations of this reaction in a recent joint experimental (with the Casavechia group)-theoretical paper.[26]

Here we augment the S_0 and T_1 PESs and SO coupling with a new full-dimensional S_1 PES limited in to the Franck–Condon region (from S_0) and also the SO coupling between S_1 and T_1. These PESs and SO couplings are used in 3-state TSH calculations focused on the $CH_3 + HCO$ electronic branching ratio as a function of photolysis energy, the relative translational energy and CH_3 vibrational energy distributions at one energy where this branching ratio is approximately 0.5. A brief description of the details of these calculations is given in the next section, followed by a description of the S_1 PES. Results and discussion follow and we conclude with a brief summary and remarks for future work.

2. Potential energy surfaces and spin–orbit coupling

Details about the potential energy surfaces for S_0 and T_1 have been published previously,[1,2,21,26] so only a short summary of those is given here. For S_0 roughly 200 000 *ab initio* energies were calculated using a combination of CCSD(T)/aug- cc-pVTZ (AVTZ) and MRCI/cc-pVTZ(VTZ) method and basis. For T_1 roughly 60 000 CCSD(T)/AVTZ electronic energies were used. For both PESs linear least-squares fitting was done using a basis of permutationally invariant polynomials, as described in detail elsewhere.[27,28] These PESs are highly complex describing numerous reaction

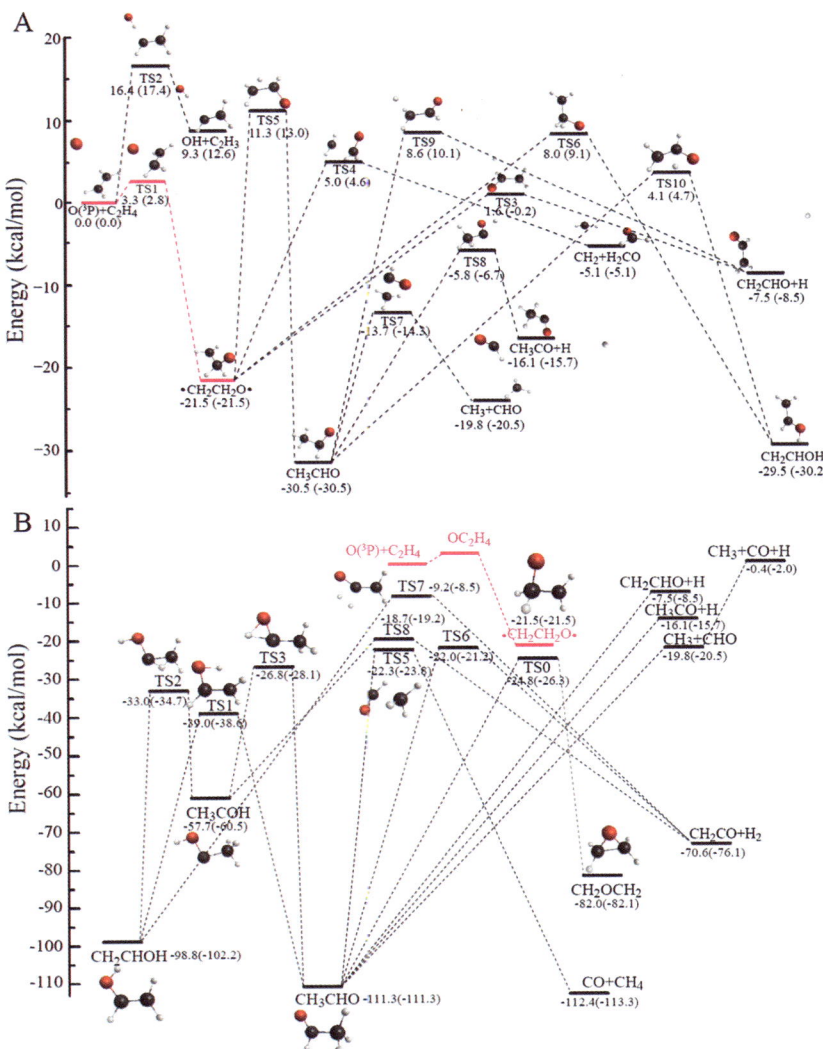

Fig. 2 Schematic of the S_0 and T_1 potential energy surfaces used in the dynamics calculations. The numbers in parentheses are from RCCSD(T)/aVTZ calculations. Note the zero energy in these figures is for $O(^3P)+C_2H_4$.

channels. A schematic of the stationary points and energies of these PESs shown in Fig. 2 gives some indication of the complexity of each PES.

The PES for S_1 has not been reported previously and we do that here. The energies we consider in the dynamics calculations are insufficient for reaction from S_1 (giving electronically excited state products) and so our focus is on the region of S_1 around the global minimum; this is also the important Franck–Condon region for excitation from S_0. We used the efficient complete-active-space-second-order multireference perturbation theory (CASPT2) method for the description of the open-shell singlet character of the S_1 PES. The orbitals for the CASPT2 calculations were taken from state averaged complete active space self consistent field (SA-CASSCF) calculations. The active space used in both the CASSCF and CASPT2 calculations was 6 electrons in 6 orbitals, and was chosen because it gave a consistent description of the wavefunction across the regions of the PES of interest. The S_0, T_1, and S_1 states were all included with equal weights in the SA-CASSCF calculations while the S_0 and S_1 states were included in the CASPT2 calculations. The AVDZ, VTZ and AVTZ basis sets were tested in benchmark calculations on the stationary points. The VTZ set was found to be the best compromise between accuracy and efficiency, and was used for the computation of the PES. All the *ab initio* calculations were done by the quantum chemistry program MOLPRO.[29]

In total 20 021 CASPT2/AVTZ data points were used to fit the S_1 PES. The configuration-space sampling was mainly done by a mixture of grid sampling and random displacement of Cartesian coordinates about stationary points. Some configurations were also selected from the trajectories initiated on the S_1 minimum. The PES was fitted, as were the S_0 and T_1 PESs, using a basis of permutationally invariant polynomials in Morse-like variables in all internuclear distances, $y_{ij} = \exp(-r_{ij}/\lambda)$, where $\lambda = 2.0$ bohr. The fit included all basis functions up to total polynomial order 5 for a total of 2655 coefficients. The weighted RMS of the fitted surface is 3.5 kcal mol^{-1} for energies up to 62 kcal mol^{-1} relative to global minimum of S_1.

A schematic representation of the S_1 PES is given in Fig. 3. Energies, relative to the S_1 minimum, and structures are shown for the present PES. The energies are also compared to those calculated by CASPT2/VTZ and MRCI(Q)/AVTZ theories and basis. We can see the PES represents the saddle points of the dissociation of CC and CH bonds quite well, which are about 22.1 and 31.4 kcal mol^{-1} above the minimum. These numbers are fortuitously closer to the MRCI(Q)/aVTZ eneriges than to the CASPT2/VTZ ones. Since the electronically excited products on S_1, $CH_3 + HCO$ or $CH_3CO + H$, are not accessible in the current work, no sampling of the configuration-space in those regions was done. Thus, the present PES is limited to the Franck–Condon region from S_0, as noted already. Structures are

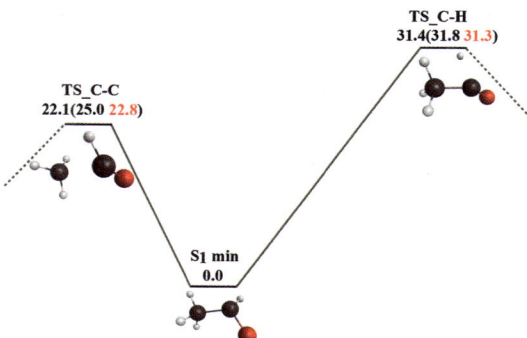

Fig. 3 Schematic of the S_1 PES used in the present study. The energies are shown relative to the S1 minimum, and those shown by black and red fonts in the parentheses are obtained from CASPT2/VTZ and MRCI(Q)/aVTZ calculations, respectively.

also indicated at the minimum and two saddle points. As seen, the relevant structure at the minimum is a distorted version of the S_0 CH$_3$CHO equilibrium structure.

Finally, an adjustment to the absolute energies of S_1 was done to yield accurate excitation energies from S_0 and T_1. Specifically, the S_1 PES was shifted so that the energy of S_1 minimum is in agreement with calculated MRCI(Q)/AVTZ value (87.8 kcal mol^{-1}) above the S_0 minimum. The resulting energies (without ZPE energies), relative to the S_0 global minimum, of these three PESs were summarized graphically in Fig. 1. As indicated, the vertical excitation energy from S_0 to T_1 is 91.8 kcal mol^{-1}, and that from S_0 to S_1 is 99.8 kcal mol^{-1}, which are in good agreement with MRCI(Q)/AVTZ values of 92.4 kcal mol^{-1} and 100.0 kcal mol^{-1}, respectively. The properties of S_0 and T_1 are not changed as shown before. The potential energies of the CH$_3$ + HCO channel on S_0 and T_1 are accurately given as 91.5 kcal mol^{-1}, relative to S_0 minimum, compared to the result of 90.8 kcal mol^{-1} calculated by RCCSD(T)/AVTZ theory/basis; this method/basis was used to generate the data which were used in the fits of the S_0 and T_1 PESs. As noted already, there exists a saddle-point barrier to form the CH$_3$ + HCO products on T_1, and no barrier on S_0 for that channel. As seen, the barrier height on the T_1 PES is 97.6 kcal mol^{-1} relative to the S_0 minimum, in excellent agreement with the RCCSD(T)/AVTZ value of 97.0 kcal mol^{-1}.

The spin–orbit calculations were done with efficient 3-state CASSCF/VTZ calculations. Roughly 3200 calculations were done mostly in the Franck–Condon region *i.e.*, distorted CH$_3$CHO configurations on S_1 and also in regions relevant to the CH$_3$ + HCO channel. As an examination of Fig. 2 shows the CH$_3$CHO minimum on T_1 is surrounded by large barriers, except for the channel of interest here, CH$_3$ + HCO where, as already noted, a small but significant exit channel barrier exists. So it was not necessary to consider the full range of configuration on T_1 for the SO coupling calculations. This restricted region on T_1 also governs where hops to S_0 occur and so SO couplings between T_1 and S_0 are also sensibly restricted to this region. In this region the S_1/T_1 SO coupling is 5–10 cm^{-1} and roughly 65–70 cm^{-1} between T_1 and S_0.

3. Trajectory surface-hopping calculations

The trajectory surface hopping (TSH) method employed is a modified version of Tully's Fewest-Switches Trajectory Surface-Hopping algorithm[30] with the extension of the time uncertainty algorithm,[31] as implemented in the *ANT09* program,[32] which we use. Similar calculations were reported by us recently for 3-electronic state H$_2$CO photodissociation.[33] The present calculations are virtually identical in method to those and so we refer the reader to ref. 33 for details that are omitted here.

The diabatic representation is used for the SO-coupled three states of CH$_3$CHO, and the potential energy matrix is expressed as

$$\begin{bmatrix} V_{S_0} & V_{S_0 T_1}^{SO} & 0 \\ V_{S_0 T_1}^{SO} & V_{T_1} & V_{S_1 T_1}^{SO} \\ 0 & V_{S_1 T_1}^{SO} & V_{S_1} \end{bmatrix}, \quad (1)$$

where V_{S_0}, V_{T_1} and V_{S_1} are the PESs of S_0, T_1, and S_1 respectively, $V_{S_0 T_1}^{SO}$ is the spin–orbit coupling between S_0 and T_1 and $V_{S_1 T_1}^{SO}$ denotes the spin–orbit coupling between S_1 and T_1.

Initial conditions for the trajectories on S_1 were generated by sampling from a ground vibrational-state Wigner distribution on S_0, as described in detail elsewhere for H$_2$CO.[33] Adjustments were made to the angular velocities to enforce zero total angular momentum for the trajectories. The total energy is given by the sum of the photolysis energy plus the (harmonic) zero-point energy of CH$_3$CHO(S_0), 35 kcal mol^{-1}; the former correspond to wavelengths of 372, 349, 329, 308, 301,

287, 273, 260 and 230 nm. These initial conditions and a simplified schematic of the 3-state energetics were given in Fig. 1. Note that the energy gap between the S_1 and T_1 minima is much smaller than the one between T_1 and S_0. These differences turn out to be important for the surface hopping dynamics

Final conditions are the usual ones employed in standard QCT calculations, *i.e.*, the internal energies of the products and the relative translational energy. The well-known issue of violation of zero-point energy does arise in a significant way here, because both products are polyatomic molecules, compared to products atom plus molecule products, where obviously ZPE violation can only occur in the polyatomic product. The issue has attracted much attention and the reader is referred to the literature for a review of it and suggestions for dealing with it.[34–43] Here, we will consider three simple approaches. One ignores the issue, the second and third approaches discard trajectories that violate the ZPE constraint; however, using two different criteria.[21] In one, which we have termed the "hard constraint", trajectories are discarded if either product has less than ZPE. In the second, termed the "soft constraint", trajectories are discarded if the sum of the vibrational energies of the polyatomic products is less than the sum of the product ZPEs.

At each energy roughly 3500 trajectories were performed, for a total of roughly 32 000 trajectories. The TSH calculations are very computationally intensive and the coupled trajectories were run to a maximum propagation time of 50 ps, roughly 10^6 time steps. In order to achieve this feasible length of propagation very time consuming multiple hopping between S_1 and T_1 was avoided by not permitting transitions from T_1 to S_1 subsequent to an S_1 to T_1 hop. This was tested and found to be an efficient and accurate strategy to increase efficiency. Also, some independent, single-surface, trajectory calculations on S_0 were run for up to 1 ns to accurately determine final state properties of the CH_3 + HCO products from S_0.

Initially, we ran the TSH calculations with constant values of the spin–orbit coupling of 10 and 65 cm^{-1} for $V_{S_1 T_1}^{SO}$ and $V_{S_0 T_1}^{SO}$, respectively, in the strong interaction region and zero coupling when the C–C distance or one of the C–H distances is larger than 5.0 bohr. However, the computational effort for hopping between S_1 and T_1 was prohibitive because of the long propagation time required for trajectories to hop to T_1 and so we artificially increased $V_{S_1 T_1}^{SO}$ to 65 cm^{-1}. Also, as noted already, we did not allow hops back to S_1 from T_1. These expedient measures appear justified to us because no photochemistry takes place on S_1, at the energies considered here and also because we are not attempting to calculate accurate lifetimes of the photochemical processes. (At some time in the future, and certainly in response to an experimental measurement of the lifetime, we would be very interested in doing this.)

4. Results and discussion

The first set of results we present is the distribution of energy gaps at the hops from S_1 to T_1 and T_1 to S_0; these are given in Fig. 4. As seen, they are much larger for T_1 to S_0 transitions than for S_1 to T_1 transitions. This is essentially a consequence of the much smaller energy difference between the S_1 and T_1 PESs in the energetically restricted Franck–Condon region. As seen, there are hops with very small gaps, which obviously occur near the PES crossing region; however, these are a minor set.

As noted in the Introduction, the focus here is on the radical products CH_3 + HCO, which can result from dissociation on T_1 or S_0. Also as noted already there is a barrier to these products on T_1 whereas the dissociation on S_0 is barrierless. Thus, we consider a range of photolysis energies that spans the opening of the T_1 channel with the goal of examining the branching ratio *versus* photolysis energy. We also examine and contrast the translational and internal state distributions of these products. As already noted in the literature, the dynamics on S_0 and T_1 are expected to be quite different, owing to the existence of an exit channel barrier to these products on T_1.

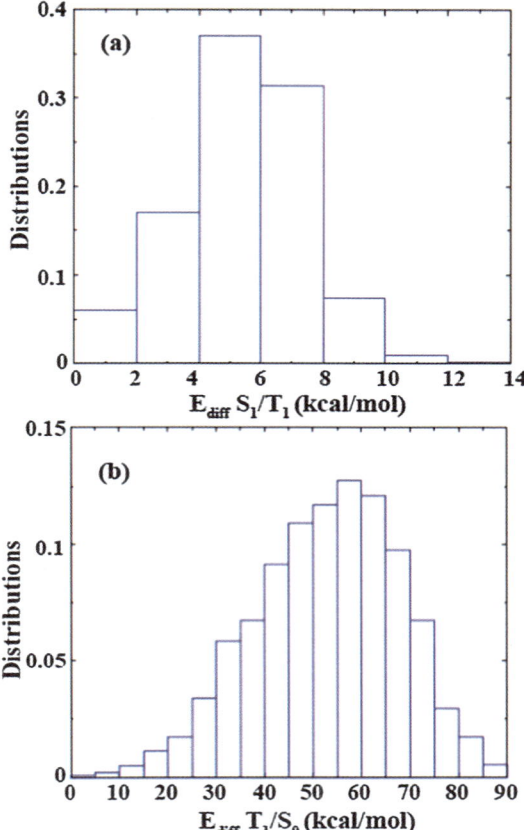

Fig. 4 Distribution of energy gaps at hops between S_1 and T_1 and T_1 and S_0 for trajectories run at a total energy corresponding to 308 nm.

Before presenting results we need to discuss the fact of zero-point energy (ZPE) violation in trajectory calculations. This is a well-known issue in QCT calculations and although less noted, also in TSH calculations. In both calculations the initial conditions are generated from a quasi or semi-classical quantum state, typically the ground vibrational state, with ZPE. After the completion of any trajectory the products are formed with any internal energy, subject to the usual conservation laws. This can be problematic for processes that are in the vicinity of the correct quantum threshold energy at which products must be in at least the ground vibrational state. A number of "fixes" for trajectories that produce products with less than zero-point energy are currently used. Generally they all result in discarding trajectories depending on what we have termed a "soft" or "hard" ZPE constraint. The former is applied if the sum of the vibrational energies of the products is less than the sum of their ZPEs and the latter is applied if any product has less than ZPE. For polyatomic products the "hard" constraint can often result in discarding many trajectories. The soft constraint, *prima facie*, appears to have no theoretical justification does produce results that are surprisingly accurate, see for example quasiclassical trajectory calculations of the $Cl + CH_4 \rightarrow CH_3 + HCl$ reaction.[44,45]

In the present TSH calculations we find this to be a serious issue for the radical products from T_1 in the interesting threshold region. The issue is present but less severe on S_0 and we will return to this difference below. Returning now to Fig. 1, the conventional Transition State Theory (without tunneling) estimate for the photolysis threshold energy for $CH_3 + HCO$ from T_1 is given by the difference between the vibrationally adiabatic ground state barrier, *i.e.*, the barrier indicated plus the local ZPE, minus the S_0 CH_3CHO ZPE. From the PESs and using the harmonic approximation to estimate the ZPEs this threshold energy is 93.1 kcal mol^{-1}. (This is 1.9 kcal mol^{-1} above the benchmark value of a 91.2 kcal mol^{-1}.[13,23]) For dissociation on S_0 the threshold energy is just the difference in energy of $CH_3(0) + HCO(0) - CH_3CHO(0)$, where we use "(0)" to indicate the ZPE. This threshold energy is 87.1 kcal mol^{-1}. So, since the trajectories are all initiated with $CH_3CHO(0)$, formation of the radical products at photolysis energies less than 87.1 kcal mol^{-1} clearly violates the correct quantum threshold energy. This is also the rigorous energetic threshold for T_1; however, as noted above the dynamical threshold energy (without tunneling) is 93.1 kcal mol^{-1}.

With these threshold energies in mind, consider the results shown in Fig. 5 for the calculated T_1:S_0 branching ratio to form the $CH_3 + HCO$ products, as a function of photolysis energy. Clearly, the ratio with no ZPE constraint produces a very unphysical result, with a threshold well below even the energetic one of 87.1 kcal mol^{-1}. This indicates that products are formed from both T_1 and S_0 with less than the total correct ZPE. The hard ZPE constraint yields a threshold energy around 96 kcal/mole and a very gradual increase in the branching ratio. The soft constraint ratio has a threshold energy of roughly 92 kcal mol^{-1}, which is in (surprisingly?) good accord with the ZPE-included estimate of 93 kcal mol^{-1}. This is discussed below. Further, the rise in the branching ratio is fairly sharp with increasing photolysis energy. This result appears to be in good accord with recent experiments of Kable and co-workers[8] and Banares and co-workers.[5] The former experiments are somewhat indirect in that they measure the HCO product from CH_3CDO, which they argue can only originate from S_0 and so the quenching of that product is correlated with the opening of the T_1 channel to form CH3 + DCO. The latter experiments reported the $CH_3 + HCO$ translational energy distributions as a function of photolysis energies. Both report a threshold for T_1 at roughly 90 kcal mol^{-1}. This energy is slightly below the soft constraint threshold energy, and is consistent with the PES T_1 barrier being higher than the benchmark value by roughly 2 kcal mol^{-1}, as noted above.

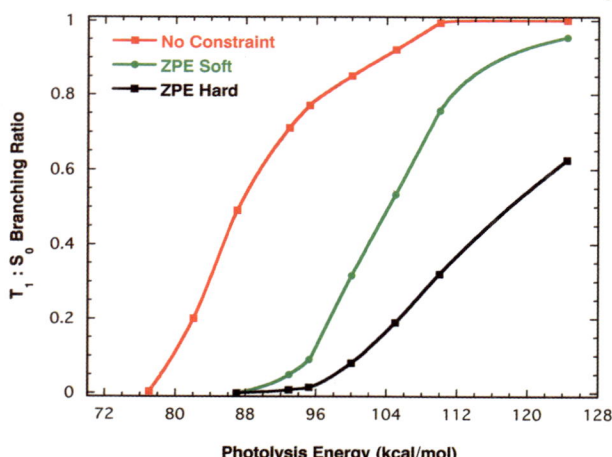

Fig. 5 T_1:S_0 branching ratio for the CH3 + HCO products as a function of photolysis energy for trajectories with no zero-point energy constraint (no ZPE) and the soft constraint (soft ZPE), as described in detail in the text.

Next consider the translational energy distributions from T_1 and S_0 at a photolysis energy of 100 kcal mol^{-1} (corresponding to 286 nm), where the soft constraint branching ratio is roughly 0.5. These are shown in Fig. 6 for the no and soft constraint cases. First, note that these two sets of results are nearly identical for S_0, which indicates that very few trajectories dissociating from S_0 violate the soft ZPE constraint. This is not the case for T_1, where there are significant differences for E_t greater than 10 kcal mol^{-1}. In general, trajectories that produce products with less than their ZPE generally transport the excess energy to relative translations

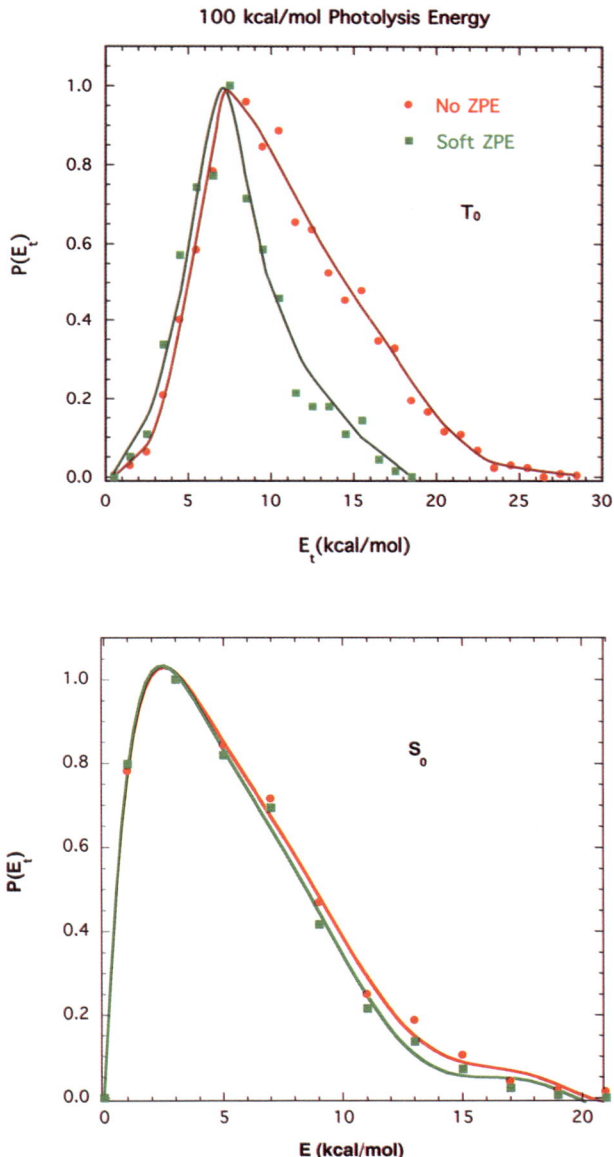

Fig. 6 CH$_3$–HCO relative translational energy distributions from T_1 and S_0 for the indicated energy.

energy, which results in hotter translational energy distributions, as seen for the T_1 trajectories with no constraint. Nevertheless, both sets of trajectories indicate hotter translational energy distribution from T_1 than is found for S_0, as expected in accord with experiments.[6] At a photolysis energy of 91 kcal mol^{-1} Banares and co-workers report a CH_3 translational energy distribution with a peak that corresponds to 8 kcal mol^{-1} for the relative E_t, in excellent agreement with the result shown in Fig. 6, albeit for slightly higher photolysis energy.[6]

The corresponding CH_3 vibrational energy distributions from T_1 and S_0 are shown in Fig. 7. As seen, the distribution from T_1 is colder than the one from S_0, by roughly 6 kcal mol^{-1}. This is in accord with expectations. Note that the average vibrational energy from T_1 is somewhat below the (harmonic) ZPE of 18 kcal mol^{-1}, indicating a very cold vibrational distribution. This result is in good accord with the recent QCT calculations of Thompson et al.,[13] who ran thousands of trajectories from the T_1 barrier, with initial ZPE in the real-frequency normal modes. Note the trajectories run here correspond to very different initial conditions and so the results of the two sets of trajectories are not quantitatively comparable. The present results for the T_1 distribution are also in accord with recent experiments of Banares and co-workers,[6] who report an average CH_3 vibrational energy of the ZPE at a photolysis energy where T_1:S_0 branching ratio is roughly 0.55. At lower photolysis energies, where S_0 dominates, that group reported an average CH_3 vibrational energy of roughly 1.5 kcal mol^{-1} above the ZPE, in good agreement with the present calculations from which find an average roughly 2 kcal mol^{-1} above the ZPE for the S_0 channel.

Overall, the present surface hopping calculations account well for the detailed observations of the CH_3 + HCO products from the photodissociation of CH_3CHO over the range of photolysis energies where the T_1:S_0 branching ratio changes from zero to one. The issue of forming products with less than ZPE is a significant one, especially for the T_1 channel, where there is a moderate amount of kinetic energy released owing to the exit channel barrier. The soft ZPE constraint, applied here, appears to give the most reasonable results for both the T_1:S_0 branching ratio and translational energy distributions. This is an extremely simple constraint to apply and it is one that we have applied previously, also with good results.[44,45] It is quite interesting, as noted above, that the threshold energy for the T_1 channel using this

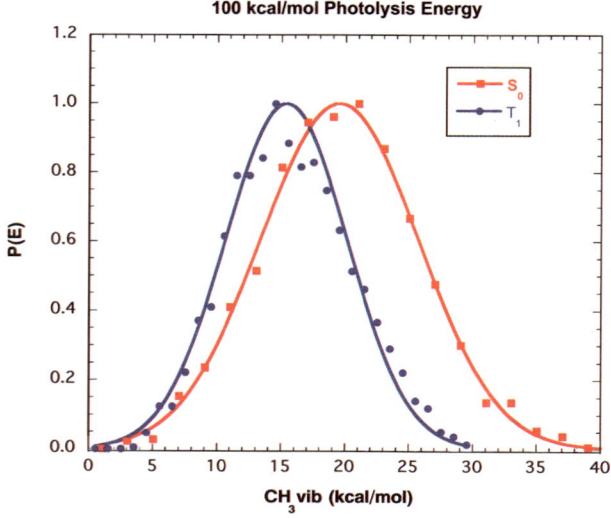

Fig. 7 CH_3 vibrational energy distributions from T_1 and S_0.

constraint is in good accord with the barrier including the total ZPE at the barrier. Does this suggest that this constraint is enforcing a similar constraint at the T_1 barrier? We believe so because the barrier is located in the exit channel and thus in a region of small curvature. We know from both theory and quantum calculations of atom + diatom reactions that the dynamics is vibrationally adiabatic at barrier that occur in regions of small reaction path curvature.[45–49] So looked at in reverse, the reaction of $CH_3(0) + HCO(0)$ to form CH_3CHO on T_1 should be vibrationally adiabatic at the now-entrance channel barrier. Since each ZPE should be conserved at the barrier, clearly the softer condition that the sum of these ZPE energies should evolve adiabatically to the T_1 barrier should be valid. Based on this, we believe that the soft constrain does enforce, at least approximately, the total ZPE constraint at the T_1 barrier and thus, obtaining the correct threshold energy does not appear to be fortuitous.

The soft constraint can result in discarding trajectories, for example if both poly-atomic products are produced with less than the ZPE, then clearly the sum will also be less than the total ZPE. This can occur if the products are produced with a large relative kinetic energy at the "expense" of vibrational energy, even though there is sufficient total energy to form both products with ZPE. This is especially an issue for the T_1 channel because of the exit channel barrier. We briefly mention a proposal to deal with this case. It is to transfer the amount of energy needed to bring both products to their respective ZPEs from the relative kinetic energy and then to accept that trajectory provided the resulting, lowered relative translational energy is sufficient to overcome the exit channel barrier. This proposal, however, raises the question about whether should add the local ZPE to the barrier. Based on the discussion above this does appear to be justified and so that would be the recommendation. We plan to investigate this suggestion in the future.

Finally we note that other products are formed from the photodissociation of CH_3CHO, especially at the higher energies, e.g., $CH_4 + CO$ and roaming to these products, etc., and these will be reported elsewhere.

5.   Summary and conclusions

We reported Trajectory Surface Hopping calculations of the three-state photodissociation of CH_3CHO as a function of photolysis energy, using full dimensional ab initio potential energy surfaces and spin–orbit couplings. The new PES for S_1 was reported here. Most surface hops were found to occur between S_1 and T_1 and T_1 and S_0 with large energy gaps, owing to the constraints of the dynamics in the Franck–Condon region of the initial excitation and Wigner distribution of the initial state on S_0. The branching ratio to form the radical products $CH_3 + HCO$ from T_1 and S_0 was the focus of these calculations. Signature relative translational energy and CH_3 vibrational energy distributions were presented. Results using several treatments for so-called zero-point energy violation of the products were presented. Those using the soft constraint were found to be in the best accord with experiment.

The present calculations highlight the issue of zero-point leak in trajectory surface hopping where it was found to significantly affect the fundamental $T_1:S_0$ branching ratio to products. A new suggestion was made to deal with this and it remains to be tested. Finally, we hope that the availability of PESs and spin–orbit couplings will stimulate semi-classical and approximate quantum calculations of the dynamics of this photodissociation.

Acknowledgements

We thank Scott Kable (SK) and Paul Houston for discussions. We also thank SK for sending a preprint of ref. 8. Finally, we thank the Department of Energy (DE-FG02-97ER14782) for financial support.

References

1 For a review see, J. M. Bowman and B. C. Shepler, *Annu. Rev. Phys. Chem.*, 2011, **62**, 531.
2 Y.-C. Han, B. C. Shepler and J. M. Bowman, *J. Phys. Chem. Lett.*, 2011, **2**, 1715.
3 H. A. Cruse and T. P. Softley, *J. Chem. Phys.*, 2005, **122**, 124303.
4 B. R. Heazlewood, S. J. Rowling, A. T. Maccarone, M. J. T. Jordan and S. H. Kable, *J. Chem. Phys.*, 2009, **130**, 054310.
5 L. Rubio-Lago, A. Garcia-Vela, A. Arregui, G. A. Amaral and L. Bañares, *J. Chem. Phys.*, 2009, **131**, 174309.
6 G. A. Amaral, A. Arregui, L. Rubio-Lago, J. D. Rodriguez and L. Bañares, *J. Chem. Phys.*, 2010, **133**, 064303.
7 B. R. Heazlewood, A. T. Maccarone, D. U. Andrews, D. L. Osborn, L. B. Harding, S. J. Klippenstein, M. J. T. Jordan and S. H. Kable, *Nat. Chem.*, 2011, **3**, 443.
8 D. U. Andrews, B. R. Heazlewood, A. T. Maccarone, T. Conroy, R. J. Payne, M. J. T. Jordan and S. H. Kable, Photoisomerisation of acetaldehyde between 310 and 330 nm: using H/D exchange in CH_3CDO to probe photochemical and photophysical processes, preprint.
9 S. Chen and W.-H. Fang, *J. Chem. Phys.*, 2009, **131**, 054306.
10 J. S. Yadav and J. D. Goddard, *J. Chem. Phys.*, 1986, **84**, 2682.
11 M. N. D. S. Cordeiro, E. Martinez-Nunez, A. Fernandez-Ramos and S. A. Vazquez, *Chem. Phys. Lett.*, 2003, **375**, 591.
12 Y. Kurosaki and K. Yokoyama, *Chem. Phys. Lett.*, 2003, **371**, 568.
13 K. C. Thompson, D. L. Crittenden, S. H. Kable and M. J. T. Jordan, *J. Chem. Phys.*, 2006, **124**, 044302.
14 B. C. Shepler, B. J. Braams and J. M. Bowman, *J. Phys. Chem. A*, 2007, **111**, 8282.
15 D. Townsend, S. A. Lahankar, S. K. Lee, S. D. Chambreau, A. G. Suits, X. Zhang, J. Rheinecker, L. B. Harding and J. M. Bowman, *Science*, 2004, **306**, 1158.
16 A. G. Suits, *Acc. Chem. Res.*, 2008, **41**, 873.
17 J. M. Bowman and A. Suits, *Phys. Today*, 2011, **64**, 33.
18 P. L. Houston and S. H. Kable, *Proc. Natl. Acad. Sci. U. S. A.*, 2006, **103**, 16079.
19 L. Rubio-Lago, G. A. Amaral, A. Arregui, J. G. Izquierdo, F. Wang, D. Zaouris, T. N. Kitsopoulos and L. Bañares, *Phys. Chem. Chem. Phys.*, 2007, **9**, 6123.
20 B. R. Heazlewood, M. J. T. Jordan, S. H. Kable, T. M. Selby, D. L. Osborn, B. C. Shepler, B. J. Braams and J. M. Bowman, *Proc. Natl. Acad. Sci. U. S. A.*, 2008, **105**, 12719.
21 B. C. Shepler, B. J. Braams and J. M. Bowman, *J. Phys. Chem. A*, 2007, **111**, 8282.
22 P. Casavecchia, G. Capozza, E. Segoloni, F. Leonori, N. Balucani and G. G. Volpi, *J. Phys. Chem. A*, 2005, **109**, 3527 and the references therein.
23 T. L. Nguyen, L. Vereecken, X. J. Hou, M. T. Nguyen and J. Peeters, *J. Phys. Chem. A*, 2005, **109**, 7489 and references therein.
24 W. Hu, G. Lendvay, B. Maiti and G. C. Schatz, *J. Phys. Chem. A*, 2008, **112**, 2093.
25 A. C. West, J. S. Kretchmer, B. Sellner, K. Park, W. L. Hase, H. Lischka and T. L. Windus, *J. Phys. Chem. A*, 2009, **113**, 12663.
26 B. Fu, Y.-C. Han, J. M. Bowman, L. Angelucci, N. Balucani, F. Leonori and P. Casavecchia, *Proc. Natl. Acad. Sci. U. S. A.*, 2012, **109**, 9733.
27 B. J. Braams and J. M. Bowman, *Int. Rev. Phys. Chem.*, 2009, **28**, 577.
28 J. M. Bowman, G. Czakó and B. Fu, *Phys. Chem. Chem. Phys.*, 2011, **13**, 8094.
29 H.-J. Werner , *et al.*, MOLPRO, version 2008.1, a package of ab initio programs.
30 J. C. Tully, *J. Chem. Phys.*, 1990, **93**, 1061.
31 A. Jasper and D. G. Truhlar, *Chem. Phys. Lett.*, 2003, **369**, 60.
32 Z. H. Li, A. W. Jasper, D. A. Bonhommeau, R. Valero, D. G. Truhlar, *ANT2009*, University of Minnesota, Minneapolis, MN, 2009.
33 B. Fu, B. C. Shepler and J. M. Bowman, *J. Am. Chem. Soc.*, 2011, **133**, 7957.
34 J. M. Bowman, B. Gazdy and Q. Sun, *J. Chem. Phys.*, 1989, **91**, 2859.
35 W. H. Miller, W. L. Hase and C. L. Darling, *J. Chem. Phys.*, 1989, **91**, 2863.
36 G. H. Peslherbe and W. L. Hase, *J. Chem. Phys.*, 1994, **100**, 1179.
37 K. F. Lim and D. A. McCormack, *J. Chem. Phys.*, 1995, **102**, 1705.
38 D. Bonhommeau and D. G. Truhlar, *J. Chem. Phys.*, 2008, **129**, 014302.
39 A. J. C. Varandas, *Chem. Phys. Lett.*, 1994, **225**, 18.
40 G. Czakó, A. L. Kaledin and J. M. Bowman, *J. Chem. Phys.*, 2010, **132**, 164103.
41 M. L. González-Martínez, W. Arbelo-González, J. Rubayo-Soneira, L. Bonnet and J. C. Rayez, *Chem. Phys. Lett.*, 2008, **463**, 65.
42 G. Czakó and J. M. Bowman, *J. Chem. Phys.*, 2009, **131**, 244302.
43 L. Bonnet and J. Espinosa-García, *J. Chem. Phys.*, 2010, **133**, 164108.
44 G. Czakó and J. M. Bowman, *Science*, 2011, **334**, 343.
45 G. Czakó and J. M. Bowman, *J. Chem. Phys.*, 2012, **136**, 044307.

46 W. H. Miller, N. C. Handy and J. E. Adams, *J. Chem. Phys.*, 1980, **72**, 99–112.
47 E. Pollak, *J. Chem. Phys.*, 1981, **74**, 5586.
48 B. C. Garrett, D. G. Truhlar, J. M. Bowman and A. F. Wagner, *J. Phys. Chem.*, 1986, **90**, 4305.
49 A. F. Wagner and J. M. Bowman, *J. Chem. Phys.*, 1987, **86**, 1976.

Reaction dynamics of temperature-variable anion water clusters studied with crossed beams and by direct dynamics

R. Otto,[†a] J. Xie,[b] J. Brox,[c] S. Trippel,[d] M. Stei,[a] T. Best,[a]
M. R. Siebert,[b] W. L. Hase[b] and R. Wester[*a]

Received 1st February 2012, Accepted 27th February 2012
DOI: 10.1039/c2fd20013a

We present a study of the different product channels in the reactions of OH^- and $OH^-(H_2O)$ with methyl iodide over a range of collision energies. Direct dynamics classical trajectory simulations are employed to obtain an atomistic comparison with the experimental results. For the experiments we have combined a crossed beam ion imaging setup with a multipole rf ion trap. The trap allows us to prepare the molecular and cluster ions with a controlled internal temperature and thus provides well-defined initial conditions for reaction experiments at low collision energy. Changing the internal temperature of the cluster ions was found to have a profound effect on their reactivity.

1 Introduction

To explore the reaction dynamics of atoms and molecules, crossed beam experiments have been established as a standard method for many years.[1-5] If the collisions are studied at low internal and external energies they can provide insight in the quantum dynamics of gas phase chemistry. The technique of velocity map imaging, pioneered by Eppink and Parker,[6] allows for the kinematically complete detection of reaction products. As the scattered products are recorded for all angular directions at once, the data acquisition occurs at a substantially increased rate compared to setups with movable detector systems. Recently it became possible to extend the crossed beam imaging technique to slow ion-neutral collisions with the development of a pulsed velocity map imaging spectrometer.[7-9] In combination with a full three dimensional particle detection scheme these experiments allow for the study of vibrationally resolved reaction dynamics of small ion-molecule systems.[10]

While our previous work focused on experiments using atomic ions, it is of great interest to study systems of increased complexity that involve larger ionic species such as small organic molecules or clusters. Many techniques are available today to produce these species in the gas phase, among them plasma discharges, laser desorption and electrospray ionisation. However, the internal energy of the molecules created in these sources is often very high and usually not very well defined,[11,12]

[a] Institut für Ionenphysik und Angewandte Physik, Universität Innsbruck, Technikerstraße 25, A-6020 Innsbruck, Austria. E-mail: roland.wester@uibk.ac.at
[b] Department of Chemistry and Biochemistry, Texas Tech University, Lubbock, Texas 79409-1061, USA
[c] Physikalisches Institut, Universität Freiburg, Hermann-Herder-Straße 3, 79104 Freiburg, Germany
[d] Center for Free Electron Lasers, DESY, Notkestrasse 85, 22706 Hamburg, Germany

† Present address: Department of Chemistry and Biochemistry, University of California, San Diego, 9500 Gilman Drive, La Jolla, 92093-0340, CA, USA

which makes them unsuitable as reactants for ion imaging experiments with well-defined energetics.

Here we present a new experimental approach to study reactive collisions of complex molecular ions and clusters with neutral molecules. It combines crossed beam ion imaging with a cold multipole radiofrequency ion trap that is used to cool the molecular species to a well defined internal temperature prior to reactive scattering. We apply this scheme to carry out crossed molecular beam studies on the reactions

$$OH^- + CH_3I \rightarrow \text{products} \tag{1}$$

$$OH^-(H_2O) + CH_3I \rightarrow \text{products} \tag{2}$$

using full three-dimensional velocity map imaging. This allows us to analyse the branching ratio of different reaction channels at different temperatures. Furthermore, differential scattering cross section images are presented for several product channels.

In concert with experiments classical trajectory simulations have been used to investigate the dynamics of gas-phase S_N2 reactions like those studied here.[9,13-17] The simulations have made important predictions concerning the reaction dynamics and have provided atomic-level interpretations of experimental measurements. The use of direct dynamics has greatly expanded the applicability of the simulations.[18]

The paper is organised in the following way: in section 2 we present the new combination of a velocity map imaging spectrometer with a multipole ion trap. In section 3, the different reaction channels that arise from reaction (1) and (2) are presented, which have been studied at two different internal temperatures of the reactant cluster ion. Classical trajectory direct dynamics simulations that provide a comparison with the experiments and an atomistic understanding of the observed reaction dynamics are presented in section IV.

2   Experimental setup

The major components of the experimental setup are shown in Fig. 1. The setup consists of a pulsed ion source, an octupole radio-frequency ion trap, a pulsed neutral beam source and a pulsed ion imaging spectrometer. The trap is operated in a tandem-time-of-flight configuration,[9] which uses time-of-flight mass spectrometry for loading selected ion species in the trap and extracting them after a thermalisation with an inert buffer gas towards the ion imaging region. Here, the ions are decelerated to well defined kinetic energies before they are crossed with a neutral gas jet from a supersonic expansion. The ionic products arising from reactive collisions between the ion package and the neutral beam are than mapped onto a time- and position-sensitive detector using a pulsed field velocity map imaging spectrometer.

2.1   Ion production and trapping

The ion production takes place in a pulsed plasma discharge ion source from a suitable precursor gas. For the production of OH^- and $OH^-(H_2O)$ clusters a mixture of 10% NH_3 in 90% Ar bubbled through a 30% ammonia solution has been expanded in a $100\mu s$ pulse at a stagnation pressure of 1 bar. Behind the nozzle of the valve a ring electrode is placed which is switched to -400 V for a short time interval of 10–20 μs. The time of ignition of the plasma is stabilised by a short 1 keV electron pulse from a self-built electron gun.

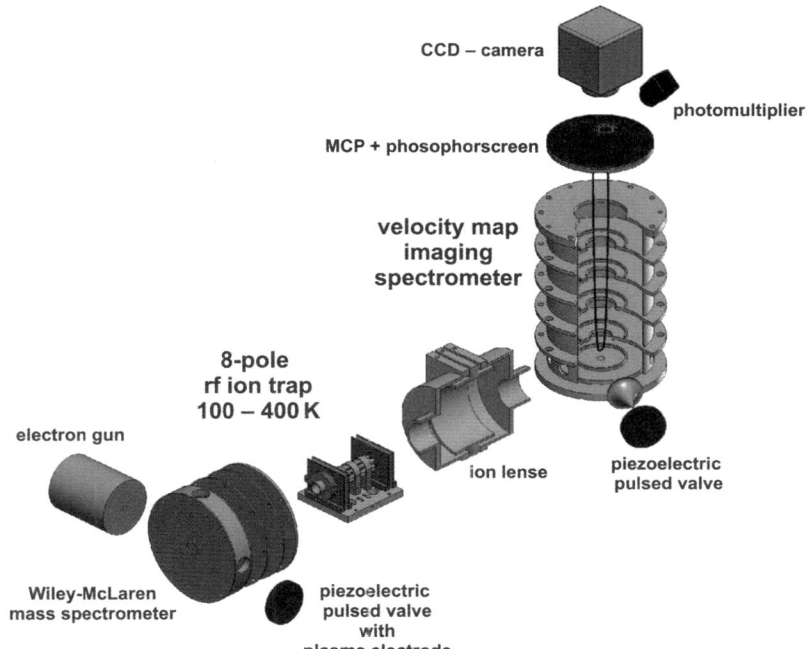

Fig. 1 The experimental setup is a combination of a multipole rf ion trap and a velocity map imaging spectrometer for crossed molecular beam studies with internally cold molecular and cluster ions.

The ions created in the gas jet travel between the field plates of a Wiley–McLaren time-of-flight mass spectrometer. Its four plates are switched to 250 V, 220 V, 100 V and 0 V respectively to accelerate the ions perpendicularly out of the jet. A set of cylindrical ion lenses is used to collimate the ion beam towards the ion trap. A pair of deflectors is used to steer the beam in horizontal and vertical direction. The ions are mass selected according to their time of flight and are decelerated before they enter the trap.

2.2 Octupole trap as source for cold ions

In order to accumulate and cool molecular ions before the scattering process we store them in a multipole radiofrequency ion trap. These traps have shown to be excellent devices for cooling external and internal degrees of freedom of complex ionic systems *via* collisions with an inert buffer gas.[19,20] The effective trapping potential for a multipole of order n at a radial position r can be expressed as

$$V^*(r) = \frac{1}{4}\frac{n^2(qV_0)^2}{m\Omega^2 r_0^2}\left(\frac{r}{r_0}\right)^{2n-2} + q\Phi_0 \tag{3}$$

where V_0 denotes the amplitude of an oscillating rf-field, m is the mass of a test particle with charge q in an electric field oscillating at frequency Ω, and Φ_0 is a non-oscillating dc potential. A large order n has been shown to be favourable to reduce the effect of rf-heating, which is crucial to achieve extremely low temperatures.[21] At the same time, a large n causes a flat effective trapping potential with steep walls, leading to a large extension of the trapped ion cloud. Ultimately this limits the translational energy resolution that can be achieved for the extracted ion package due to phase space conservation arguments. Since the main purpose for the ion trap in the presented experiments is to act as a source for a crossed beam experiment, here an octupole ion trap was realised as a compromise between good buffer gas cooling and a narrow confinement of the ion cloud.

The ion trap consists of eight cylindrical rods with a diameter of $d = 2.5$ mm that are implanted alternately into two copper sideplates. They have an inscribed diameter of $D = 7.5$ mm at a total trap length of 36 mm. The sideplates are connected to the two opposite phases $\pm V_0 \cos(\omega t)$ of a self-built rf generator. The rf rods are surrounded by three electrostatic ring electrodes (shaping electrodes) which allow one to apply a dc offset potential inside the ion trap to push the ion cloud towards one end of the trap. Axial confinement is achieved using cylindrical endcaps that are switched from an attractive potential (ions enter the trap) to a repulsive potential (ions are inside the trap).

The entire trap resides on a static platform potential (about 230 V) that approximately matches the kinetic energy the ions have when they leave the source region. Hence, approaching the trap decelerates the ions to 1–2 eV. A short and intense buffer gas pulse is applied into the trap volume to finally stop and thermalise the arriving ions.

The ion trap is mounted on a dewar tube which is filled with liquid nitrogen to cool the entire trap assembly. In equilibrium between the heat transport through the dewar's stainless steal base plate and the heat input (mainly due to blackbody radiation, but also caused by thermal contact with wires) a trap temperature of 100 K is achieved. Alternatively, the entire assembly can be heated to temperatures above room temperature.

After a short storage period of 20 ms, which is estimated based on the Langevin collision rate to be long enough to thermalise all internal degrees of freedom of the trapped species, the ions are extracted towards the imaging spectrometer.

2.3 Crossed beam imaging spectrometer and neutral beam source

The ion package leaves the trap region with a kinetic energy of the trap's platform potential. In order to tune the collision energy with the neutral beam the ions are decelerated again before entering the imaging spectrometer. This is achieved by putting the entire spectrometer onto another variable platform potential. A set of lenses is used to keep the ion package collimated upon the deceleration and while entering the imaging region. In order to measure the kinetic energy distribution of the parent ion beam, the decelerated ion package is imaged using the velocity map imaging spectrometer. An energy spread of the ion package of between 100 and 200 meV (FWHM) can be routinely achieved for the decelerated ion package.

Once in the spectrometer the ion package collides with a supersonic beam from a pulsed neutral beam source.[8] In the experiments presented here a mixture of 10% methyl iodide and 90% helium was expanded at a stagnation pressure of 0.8 bar and skimmed using a 300 μm skimmer. Although it is well known, that with helium as a carrier gas it is problematic to achieve low jet temperatures it was chosen here to avoid clustering of CH_3I in the jet during the expansion. The same holds for the quite low stagnation pressure of the expansion. In addition the valve was heated to a temperature of 330 K to avoid clogging of the nozzle by the methyl iodide.

The velocity distribution of the neutral gas jet is determined by ionisation of the molecules in the scattering region using a short 1–2 keV electron pulse from a self-built electron gun and subsequent imaging of the ions. From the velocity spread of the neutral jet we find a translational temperature of 130 K under the employed beam conditions. This shows that the CH_3I is not strongly cooled in the expansion, as expected for the low stagnation pressure of the molecular beam. We estimate the rotational degrees of freedom to be in equilibrium with the translational temperature. For the vibrational temperature the nozzle temperature poses an upper limit, so that we expect it to be found between 130 and 330 K.

After individual diagnostics of the two beams they are crossed between the field plates of the spectrometer. Two microseconds after the bunch crossing the field plates are pulsed to their extraction voltages to map the ion products onto a position- and time-sensitive detector system.[22]

3 Direct dynamics simulations

Direct dynamics simulations[24] were performed to study the atomic-level mechanisms for reactions (1) and (2) and to compare the results with the experimental measurements. Different levels of electronic structure theory were considered for the simulations and following the previous work for the $F^- + CH_3I$ reaction[16,25] DFT/B97-1/ECP/d was chosen (the ECP/d basis set has been described previously[25]). Heats of reaction calculated for reactions (6) and (7) (see below) with different electronic structure methods are listed in Table 1 and compared with experiment.

The simulations were performed at a collision energy of 2.0 eV with the following vibrational and rotational temperatures: CH_3I, $T_{vib} = 330$ K and $T_{rot} = 130$ K; and OH^- and $OH^-(H_2O)$, $T_{vib} = T_{rot} = 100$ K. Standard sampling methods[26] were used to select trajectories with these initial conditions. The direct dynamics trajectories were calculated using a 6th-order symplectic algorithm.[27] A total of 937 and 900 trajectories were calculated for reactions (1) and (2), respectively. The trajectories were integrated for a total time of 0.5 ps for reaction (1) and 1.0 ps for reaction (2). Reactive trajectories were identified by animating the trajectories and determining their atomic-level motions. The simulations were performed with the VENUS/NWChem software package.[28–31]

4 Results

In the following we present the results from experiment and simulation for the different product channels that arise from reactions (1) and (2). In most of the presented experiments the reactant ions are thermalised to a buffer gas temperature of 100 K. In that way all internal degrees of freedom of the reactant ions are cooled prior to the reactive collisions. This temperature has also been assumed in the simulations. The collision energy for the two reactions has been varied between 0.15 and 2.5 eV.

The three-dimensional velocity map imaging detector allows us to collect all the different ionic reaction products in the same measurement. The branching ratios into the different reaction channels are analysed using the time-of-flight information of the detected product species. Furthermore, the influence of the internal temperature of the reactant ions can be studied by increasing the temperature of the ion trap and this is demonstrated by comparing results for 100 to 300 K internal cluster temperature. A lot of additional insight into the reaction dynamics that lead to

Table 1 Comparisons of heats of reactions for different electronic structure methods. The ECP/d basis set value is given in normal text, the aug-cc-pVTZ-pp value is in parentheses. Calculated energies include harmonic zero-point energy (ZPE) corrections. The experimental value is the 0 K heat of reaction.[23]

	Reaction 6	reaction 7
MP2	−58.36 (−56.52)	34.93 (40.84)
B97-1	−64.87 (−63.56)	28.14 (30.15)
B3LYP	−64.48 (−62.92)	26.35 (28.53)
B2PLYP	−62.10 (−59.46)	29.06 (31.88)
Experimental	−66.44	26.32

the individual product channels are gained from the two-dimensional velocity distributions that represent the differential scattering cross section.

4.1 The unsolvated reaction

4.1.1 Product branchings. Shown in Fig. 2 is a typical time-of-flight mass spectrum obtained for reaction (1) at a collision energy of 2.0 eV. At this relative energy three different reaction channels are observed that form I^-, CH_2I^- and IOH^- products respectively. All data shown are collected in the same measurement cycle and therefore all product species are formed under identical beam overlap conditions. For one given relative collision energy the fractional contributions of all product channels can therefore be compared in terms of a relative cross section. However, as a change in the collision energy influences the beam overlap the data for one product channel obtained at different relative energies can not be compared in that way. In the following we therefore present branching ratios rather than relative cross sections.

In order to extract the fractional contribution of each channel the area under each peak in the spectra is fitted. The sum of all contributions is subsequently normalised to unity. In this way the branching ratio of the different product channels can be obtained. Shown in Fig. 3 is the branching ratio for the products of reaction (1) as a function of the relative collision energy. One observes the I^- and CH_2I^- products for all collision energies. Both of these reaction channels are exothermic by 2.8 eV and 0.2 eV respectively. While the I^- products are formed in a nucleophilic substitution reaction, the CH_2I^- anions represent the products of a proton transfer from CH_3I to OH^-.

The third reaction channel that is observed for reaction (1) leads to the formation of IOH^- products. This channel can be identified with a stripping of the iodine from the CH_3 group. It features a threshold behaviour which is characteristic for an endothermic process. To describe this threshold we fit the data with the relation

$$\sigma \propto \left(1 - \frac{E_T}{E}\right), E > E_T \tag{4}$$

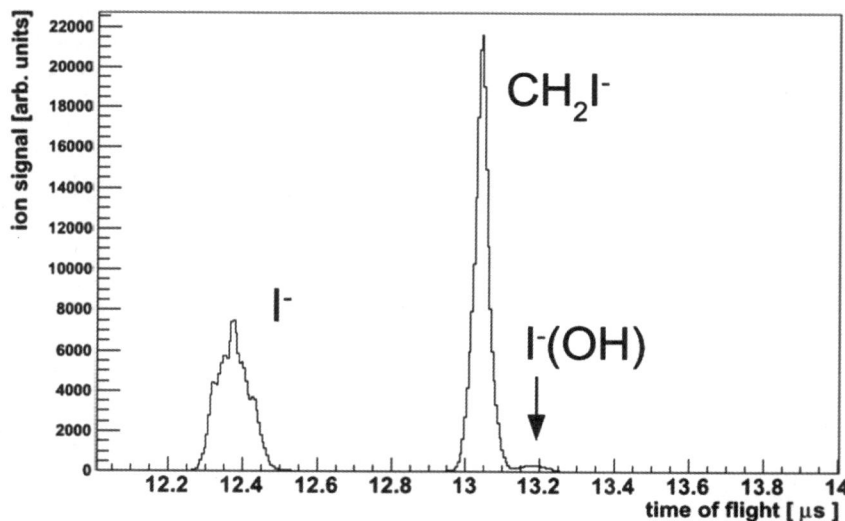

Fig. 2 Typical time-of-flight trace showing the products for the OH^- + CH_3I reaction at 2 eV relative collision energy.

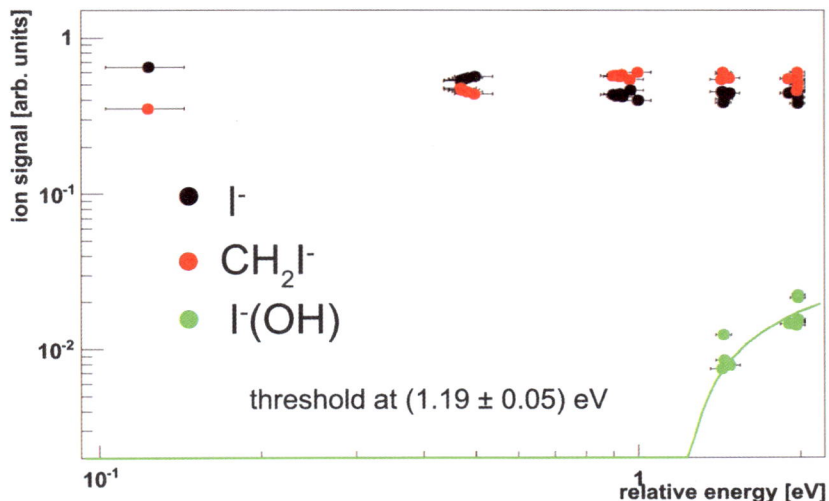

Fig. 3 Branching ratios for the different product channels observed in the OH⁻ + CH₃I reaction as a function of relative collision energy between 0.15 and 2 eV. The threshold behaviour of the endothermic IOH⁻ channel has been fitted using eqn (4).

which arises from the condition that the available kinetic energy has to overcome a centrifugal barrier and the potential energy of the reaction.[32] The fit yields a threshold energy of $E_T = 1.2$ eV which is in contrast to the calculated endoergicity of the reaction of 0.3 eV (see next section). To explain this observation we have to conclude that the system has to overcome an activation barrier that has not been identified, yet.

4.1.2 Trajectory simulations. The three product ions CH₂I⁻, I⁻, and IOH⁻ were observed in the simulations at a collision energy of 2 eV, in agreement with the experiments. These ions can be formed by the following four pathways:

$$OH^- + CH_3I \rightarrow CH_2I^- + H_2O - 4.23 \tag{5}$$

$$OH^- + CH_3I \rightarrow I^- + CH_3OH - 64.87 \; (-66.44) \tag{6}$$

$$OH^- + CH_3I \rightarrow I^- + CH_2 + H_2O + 28.14 \; (+26.32) \tag{7}$$

$$OH^- + CH_3I \rightarrow IOH^- + CH_3 + 6.80 \tag{8}$$

The first energy (in kcal mol⁻¹) given for these reactions is the DFT energy with harmonic zero-point energy (ZPE) corrections and the second, in parentheses, is the experimental 0 K heat of reaction.[23]

Reaction (8) forms two doublet radicals and formation of these products is not accurately described by the restricted DFT/B97-1 method used for the trajectories. To study these events the trajectories with atomistic motions leading to the products of reaction (8) were halted when the restricted DFT potential energy surface began to differ from that of unrestricted DFT. These trajectories were then rerun with the unrestricted DFT/B97-1 method.

Surprisingly the reaction channel

$$OH^- + CH_3I \rightarrow I^-(H_2O) + CH_2 + 18.18 \ (+16.36) \qquad (9)$$

which is energetically possible as well has been observed neither in the experiments nor in the trajectory calculations. The experimental energy for this reaction was determined from the value for reaction (7) by using the 9.96 kcal mol^{-1} DFT/B97-1/ECP/d energy for I$^-$(H$_2$O) complexation.

The largest impact parameter b at which reaction was observed is 5 Å. The reaction probabilities P_r for reactions (5)-(8) were calculated at b of 0, 1, 2, 3, 4, 4.5, and 5 Å, and the equation $\sigma_r = \int P_r(b) 2\pi b db$ was numerically integrated to obtain a total reaction cross section of 13 ± 3 Å^2. The respective individual cross sections for reactions (5), (6), (7), and (8) are 8.0 ± 1.2, 4.2 ± 1.2, 0.74 ± 0.32, and 0.08 ± 0.06 Å^2. From the simulated cross sections the relative ratio of the CH$_2$I$^-$: I$^-$: IOH$^-$ ions is about $0.61 : 0.37 : 0.02$ in very good agreement with the experimental findings of $0.55(10) : 0.43(10) : 0.02(1)$ (see Fig. 3). The values in brackets denote the estimated experimental accuracies.

Absolute cross section measurements at this collision energy on the related reactions of OH$^-$ with CH$_3$Cl and CH$_3$Br found cross sections of 1.9 Å^2 (CH$_3$Cl) and 1.7 Å^2 (CH$_3$Br) for the nucleophilic substitution channel, compared to 4.2 Å^2 in the present simulation, and 3.0 Å^2 (CH$_3$Cl) and 2.7 Å^2 (CH$_3$Br) for the proton transfer channel, compared to 8.0 Å^2 in the present simulation.[33]

The total reaction probability *versus* impact parameter is plotted in Fig. 4. The probability is $\sim 20\%$ for $b = 0$–4 Å, but then drops dramatically with increase in b. The probability of reaction (5) *versus* b follows the pattern in Fig. 4. In contrast, reactions (7) and (8) only occur at small b, and the probability of reaction (6) peaks at $b = 4$ Å. The atomistic motions for two trajectories, following reactions (6) and (7), are shown in Fig. 5.

4.2 The monosolvated reaction

4.2.1 Branching ratios. If a single water molecule is attached to the reactant anion many more product species are observed. A typical mass spectrum at a collision energy of 0.35 eV is shown in Fig. 6.

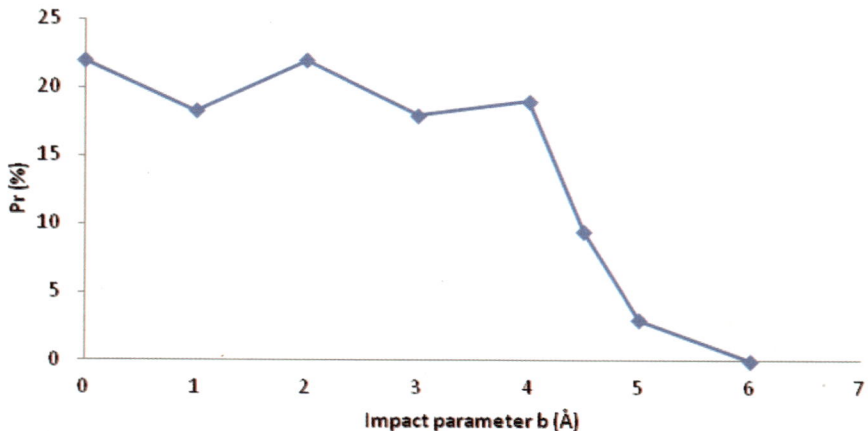

Fig. 4 Reaction probability as a function of the impact parameter for the reaction of OH$^-$ with CH$_3$I at 2.0 eV relative collision energy, calculated with direct dynamics trajectory simulations. The connecting line is plotted for guidance.

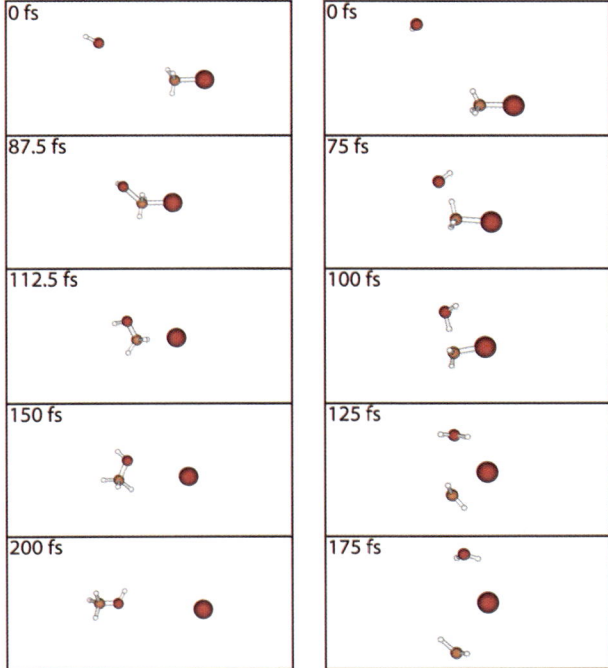

Fig. 5 Images of selected trajectories for reaction (6), left panel, and reaction (7), right panel as a function of time. The trajectories are calculated in full-dimensional direct dynamics trajectory simulations.

In the $OH^-(H_2O)$ reaction the nucleophilic substitution can lead to two different ionic products, I^- and $I^-(H_2O)$. Two more reaction channels are observed in the mass spectra. The first can be clearly identified as a proton transfer reaction leading to unsolvated CH_2I^- products. The second channel can be interpreted either as a proton transfer reaction forming solvated $CH_2I^-(H_2O)$ products or as a ligand switching mechanism that forms $OH^-(CH_3I)$ products. We exclude this channel to be $I^-(CH_3OH)$ for reasons given below.

At low collision energies another peak has been observed in the mass spectra, shown in Fig. 6. We assign this peak to be $I^-(CH_3OH)(H_2O)$. It is no longer observable at collision energies above 0.5 eV. The formation of this product species must involve a third particle to carry away the reactions excess energy, since the lifetime of a collision complex with few hundred meV of internal energy is expected to be only in the order of picoseconds[34] and not microseconds, which would make it undetectable in our setup. We assume that this complex is either created in a secondary, stabilising collision in the supersonic jet or in a collision with a CH_3I reactant molecule that is clustered to another molecule in the jet, which then carries away excess energy. Both events are unlikely but may still lead to the small observed signal of $I^-(CH_3OH)(H_2O)$.

Shown in Fig. 7 are the product branching ratios for reaction (2) as a function of the relative collision energy. The contribution from the $I^-(CH_3OH)(H_2O)$ which accounts only for a fraction of 10^{-3} to the entire ion signal is not shown. As in the OH^- case, the nucleophilic substitution reaction forming I^- products is the dominant process over the whole energy range studied in this experiment. In the monosolvated system this reaction is less exoergic, which reflects the binding energy of the water molecule to the reactant OH^- ion. The formation of the solvated product

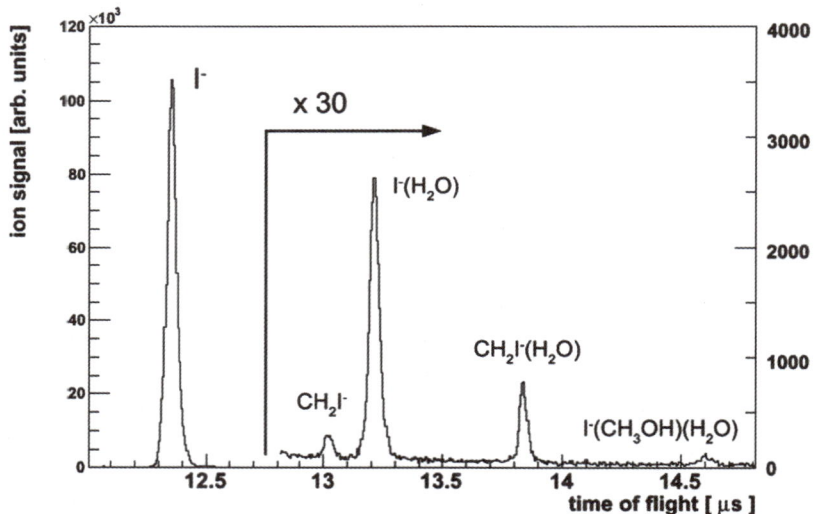

Fig. 6 Typical time-of-flight trace for the OH⁻(H₂0) + CH₃I reaction at 0.35 eV relative collision energy. At a flight time of 14.6 μs a small contribution from an I⁻(CH₃OH)(H₂O) cluster can be observed.

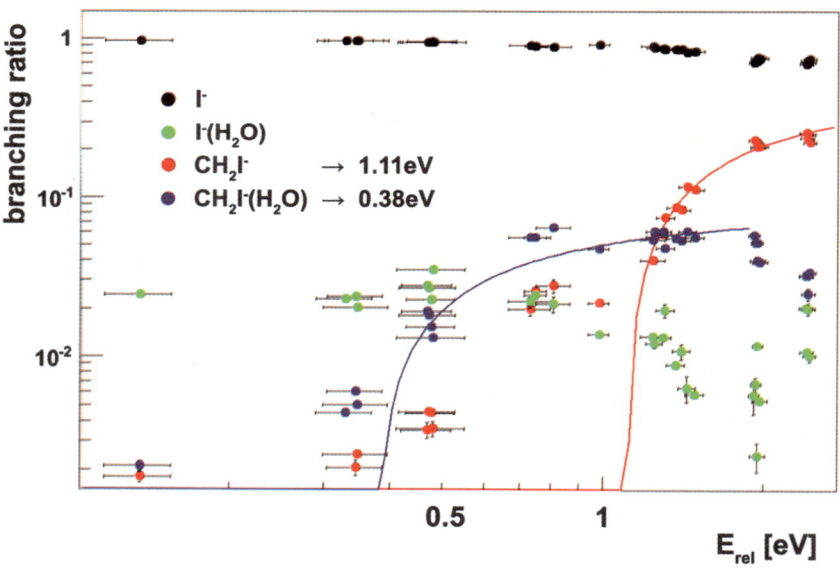

Fig. 7 Branching ratios for the different product channels observed in the OH⁻(H₂O) + CH₃I reaction as a function of relative collision energy. The threshold behaviour of the endothermic proton transfer channels has been fitted using eqn (4).

is stronger exoergic by 0.5 eV due to the stabilisation energy of the water molecule. However, although energetically favourable we find that the formation of I⁻(H₂O) products is 20-fold suppressed compared to unsolvated I⁻ products, as can be seen from the branching ratios in Fig. 7. This implies that despite a lower exoergicity the nucleophilic substitution shows a clear preference for unsolvated product formation. As can be seen from Fig. 7 this strong suppression of solvated products is not found for the proton transfer channel. We will use the two dimensional velocity distributions in section 4.2.4 to further elucidate this finding.

In the case of the CH_2I^- formation, going from reaction (1) to reaction (2), *i.e.* solvating the reactant OH^- ion, completely changes the energetics of the proton transfer from a slightly exoergic to a clearly endoergic process. Both proton transfer channels we observe here show a clear threshold behaviour which can again be described using eqn (4). A fit to the data yields appearance energies of 1.1 eV and 0.4 eV for the CH_2I^- and the $CH_2I^-(H_2O)/OH^-(CH_3I)$ channel respectively. We estimate an accuracy for these values of 0.1 eV due to the uncertainties in the relative collisions energies. The clear signature of an endoergic process provides the basis for excluding that the channel with mass 159 amu be the formation of $I^-(CH_3OH)$ as this would be a strongly exoergic process.

For both endothermic proton-transfer channels we observe a minor contribution of reaction products at collision energies that are significantly lower than the derived appearance energies. These contributions account for less than 1% of the entire signal and are most likely caused by hot reactants from either the neutral jet or the ion beam. For the CH_2I^- channel we also assume to have small admixtures of OH^- in the parent $OH^-(H_2O)$ ion beam. These are explained as the products of a break-up of the $OH^-(H_2O)$ cluster in collisions with the residual background gas (typical pressures in the trap region are 10^{-6} mbar). With an estimated mean free path length of 100 m we assume a collision probability for the clusters of 10^{-3} in the region between the ion trap and the imaging spectrometer.

While the contribution from the unsolvated CH_2I^- channel keeps increasing with rising collision energy, we observe a maximum in the yield of $CH_2I^-(H_2O)$ products at around 1.5–2 eV. This might be caused by new fragmentation channels, that open up at these high collision energies.

For technical reasons we restricted this study to products that arrive at later flight times than the $OH^-(H_2O)$ reactant ions and hence did not detect possible OH^- products together with the other products discussed here. Studies on the related CH_3Cl and CH_3Br systems show that this collision induced dissociation channel

$$OH^-(H_2O) + CH_3X \rightarrow OH^- + H_2O + CH_3X$$

becomes dominant at energies above 1 eV.[33]

4.2.2 Changing the ion's internal temperature. One of the charming features of the new combination of an octupole ion trap with a crossed beam apparatus is the possibility to change the internal temperature of the ion ensemble prior to reaction. To demonstrate this we performed the scattering experiment of reaction (2) described in the previous sections with a trap temperature of 300 K instead of 100 K. The same procedure as described above is used to map out the branching ratios of the different product channels.

Shown in Fig. 8 are the product branchings for the 300 K measurement (closed symbols) together with the 100 K data (open symbols) for comparison. The $I^-(H_2O)$ and IOH^- channels are omitted for clarity. The appearance energies of the two endoergic proton transfer channels are again fitted using eqn (4). The fits of the 300 K data (solid line) are shown together with those of the 100 K data (dashed lines). We determine the appearance energies of the two channels to be 0.7 eV for CH_2I^- and 0.2 eV for the complex with mass 159 amu. Both energies differ significantly from the 100 K appearance energies of 1.1 eV and 0.4 eV.

In order to understand the large shift in appearance energies with cluster temperature of 0.4 and 0.2 eV, respectively, we consider the difference in internal energy of the $OH^-(H_2O)$ cluster at 100 K and 300 K. To determine the amount of internal energy of the cluster we sum up the contribution from the rotational and vibrational degrees of freedom

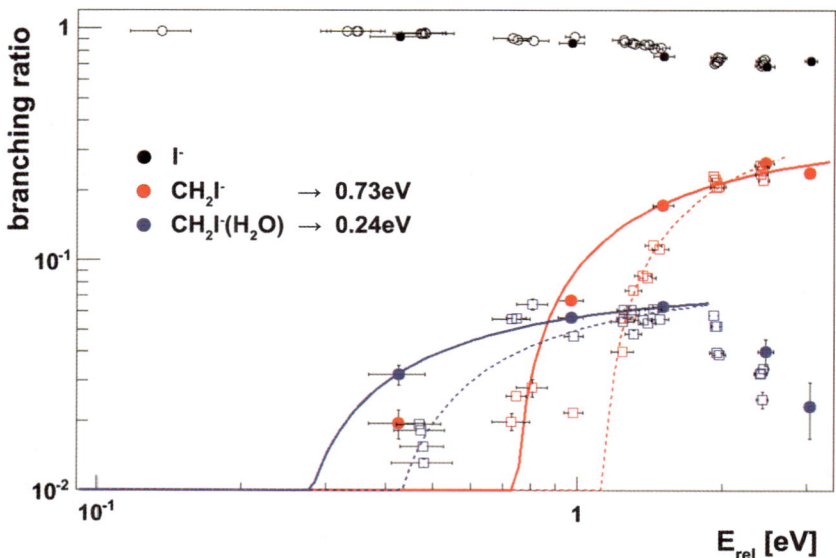

Fig. 8 Comparison of the branching ratios of the $OH^-(H_2O) + CH_3I$ reaction channels for two different temperatures of the ion trap. The data for the 100 K trap (open symbols and dashed line) are shown together with the data for 300 K (closed symbols and solid line). The threshold behaviour of the endothermic proton transfer channels has been fitted using eqn (4).

$$E_{int} = E_{rot} + E_{vib} \qquad (10)$$

From a structure calculation of the cluster geometry (see section IV) we find rotational constants for the cluster of 10.3, 0.303 and 0.301 cm^{-1}, so that we approximately describe the system as an oblate symmetric top molecule. We have verified that a quantum mechanical treatment of the rotational energy of this system converges against the classical case where the energy is given by $E_{rot} = 3/2kT$, so that we find a rotational energy for the 100 K and 300 K case of 100 cm^{-1} and 300 cm^{-1} respectively.

In contrast to the rotational levels the vibrational spacing in the cluster is large and only a few vibrational levels are populated at low temperatures. Calculated energies[35] of the first four vibrational levels are listed in Table 2. Due to a tunnelling between two identical $OH^-(H_2O)$ conformers each level is split into an upper and a lower tunnelling doublet states, indicated by the + and − column in Table 2. Analysing the Boltzmann statistics for the vibrational degrees of freedom at 100 K and 300 K yields a contribution to the internal energy of 17 cm^{-1} and 155 cm^{-1} respectively.

Table 2 Energy of the first four vibrational levels of $OH^-(H_2O)$ obtained in *ab initio* calculations.[35] The + and − states represent levels that are split due to a tunnelling between two possible conformers

Mode	+	−
torsion	141	215
OO stretch	515	540
wag	576	606
rock	465	528

Summing up the two contributions we find a difference in internal energy between 100 K and 300 K clusters of about 340 cm^{-1} or 40 meV. This difference is a factor of 10 and 5, respectively, lower than the observed shifts in appearance energies for the two studied temperatures. A possible explanation for this striking effect would be that the population of the first vibrationally excited mode significantly enhances the reactivity for the proton transfer mechanism. A clarification of the involved dynamics will require further experimental and theoretical investigations beyond the scope of this work.

4.2.3 Trajectory simulations. The formation of three product ions were observed in the trajectory calculations:

$$OH^-(H_2O) + CH_3I \rightarrow I^- + CH_3OH + H_2O - 40.48 \qquad (11)$$

$$OH^-(H_2O) + CH_3I \rightarrow CH_2I^- + 2\ H_2O + 25.43 \qquad (12)$$

$$OH^-(H_2O) + CH_3I \rightarrow [CH_3\cdots I\cdots OH]^- + H_2O + 1.48 \qquad (13)$$

The energies given for these reactions (in kcal mol^{-1}) are those calculated with the DFT/B97-1 method used for the simulations and include ZPE corrections. Interestingly, good agreement has been found between the calculated endothermicity of the proton transfer reaction (12) and the value obtained from the fit to the appearance energy for the 100 K cluster temperature. Also based on this finding, the lower appearance energy for 300 K has to be attributed to internally excited reactants that react with much higher probability than ground state reactants.

For reaction (13), the system is trapped in the $[CH_3\cdots I\cdots OH]^-$ complex well at the end of the simulation time (1 ps). The lifetime of the $[CH_3\cdots I\cdots OH]^-$ complex may be attributed in part to energy lost when the water molecule dissociates. If the internal energy remaining in the $[CH_3\cdots I\cdots OH]^-$ complex is higher than one or more of its barriers for dissociation, after the water molecule dissociates, the complex is expected to dissociate to $CH_3I + OH^-$ or $CH_3OH + I^-$ [Reaction (11)], if given enough time. It is important to note that the $[CH_3\cdots I\cdots OH]^-$ complex has the same mass as $CH_2I^-(H_2O)$, a mass observed experimentally.

Overall, the dynamics simulations produce ions that are not solvated, i.e., reactions (11), (12), and (13) all form products in which water has dissociated. This tendency is consistent with experiments at higher collision energies (as mentioned earlier, experimentally, solvated ions are 20-fold suppressed relative to their unsolvated counterparts). Preliminary dynamics results at lower collision energies indicate that a higher proportion of solvated ions are formed, which is also consistent with the experimental results.

For reactive scattering at a relative collision energy of 2.0 eV the largest impact parameter at which reaction was observed is 4.5 Å. Reaction probabilities were determined for impact parameter b of 0, 1, 2, 3, 4, and 4.5 Å. The total reaction cross section is 5.5±1.4 Å^2, with respective individual cross sections of 2.3±0.8, 1.8±0.6, and 1.4±0.5 Å^2 for reactions (11), (12) and (13). For each reaction channel the probability decreases with increasing impact parameter.

Atomistic details of two trajectories, for reactions (11) and (12) are shown in Fig. 9. For reaction (11) the water molecule remains attached to OH^- as reaction occurs and then is dissociated from the system after CH_3OH is formed. For the reaction (12) trajectory, the water molecule is first detached from OH^-, which then abstracts a proton from CH_3I.

4.2.4 Two dimensional velocity distributions. Much more information about the reaction mechanisms that underlie the different channels can be gained from evaluating the full three dimensional information $v(x,r,\theta)$ of each scattering event. Here we illustrate this with results for three different product ions for reaction (2), CH_2I^-, $CH_2I^-(H_2O)/OH^-(CH_3I)$, and $I^-(H_2O)$. A comparison of the velocity distributions for these three channels is shown in Fig. 10 for a relative collision energy of 1.5 eV. The dynamics of the I^- channel are subject of a separate study.[37]

Each velocity image in Fig. 10 is obtained from the three-dimensional record of up to 5×10^5 scattering events. The events are transformed in the centre of mass frame such that the relative velocity vector is aligned along the x-axis and integrated over the cylindrical θ-coordinate weighting each event with $1/v_r$. The circles in the image represent spheres of the same translational energy, spaced at 0.5 eV intervals. The outermost circle marks the kinematical cutoff for the velocity distribution, *i.e.* the maximum amount of translational energy available given by the relative translational energy plus the exoergicity ΔH of the reaction.

For the products of the unsolvated proton transfer channel we observe scattering in the forward direction (left panel in Fig. 10). The non-isotropic angular distribution implies that the reaction proceeds as a direct process on a timescale that is much shorter than the rotational period of transient reaction complexes. A fast and direct mechanism as we find it here for the transfer of the proton from the

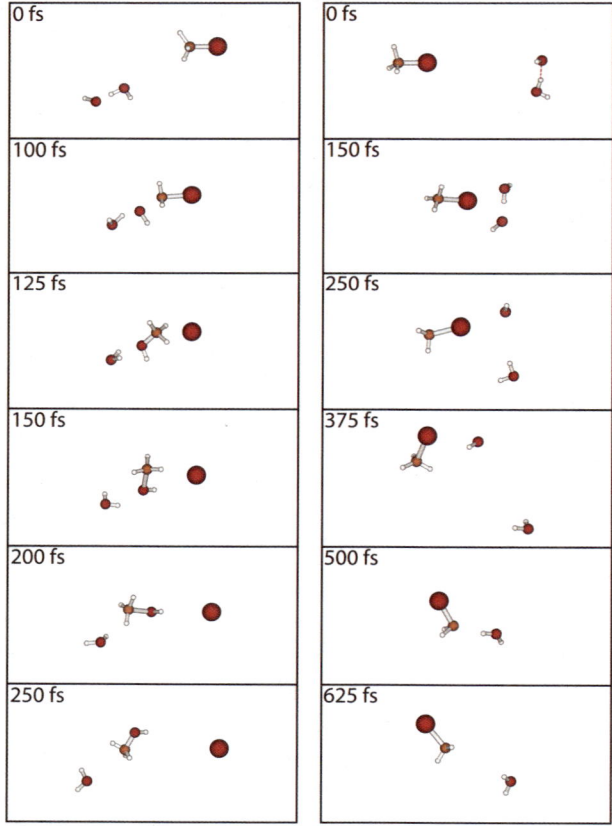

Fig. 9 Images of selected trajectories for reaction (11), left panel, and reaction (12), right panel as a function of time. The trajectories are calculated in full-dimensional direct dynamics trajectory simulations.

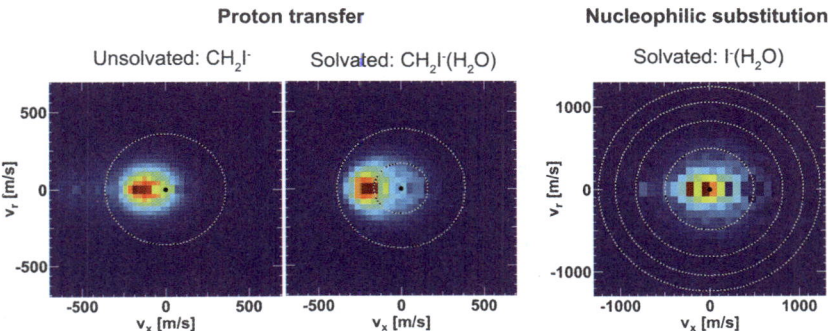

Fig. 10 Two-dimensional velocity distribution showing the differential scattering cross section for the two proton transfer channels and the solvated nucleophilic substitution channel of reaction $OH^-(H_2O) + CH_3I$ at a relative collision energy of 1.5 eV.

CH_3I to the $OH^-(H_2O)$ is rather typical for exothermic barrier-less proton transfer reactions.[36] Also for the solvated proton transfer forming $CH_2I^-(H_2O)$ (centre panel in Fig. 10) we observe products that are scattered into the forward direction. This again points to a fast and direct stripping mechanism of the proton.

It has to be noted that about 10% of the CH_2I^- products are found outside the kinematic cutoff at high velocities in forward direction. These are likely to stem from the OH^- contribution in the parent $OH^-(H_2O)$ beam due to in-beam fragmentation, as explained above. Although we assume the admixture of OH^- only to be in the order of 10^{-3} the different cross sections for the two reactions have to be considered. Based on former cross section measurements on the $OH^- + CH_3Cl$ and $OH^- + CH_3Br$ reactions[33] we estimate the cross section for the $OH^- + CH_3I \rightarrow CH_2I^- + H_2O$ reaction to be about 100 times larger than that for the $OH^-(H_2O) + CH_3I \rightarrow CH_2I^- + 2 H_2O$ reaction (Note that the admixtures of OH^- travel at the same velocity as their parent ions and therefore feature a much smaller beam energy).

For the solvated nucleophilic substitution channel completely different dynamics are observed. Here, the $I^-(H_2O)$ (right panel in Fig. 10) features a roughly isotropic distribution of slow product ions only. This points to a complex mediated reaction mechanism with a complex lifetime that is at least as large as the rotational period of the complex.

The different reaction dynamics in the solvated proton transfer and the solvated nucleophilic substitution can be understood by looking at the steric configuration of the product molecules that are formed. In the reaction forming $I^-(H_2O)$ the water molecule has to undergo a complicated re-arrangement to attach to the leaving I^-, *e.g.* by migrating around the central methyl group. In the course of the proton transfer reaction, however, it already finds itself in its final configuration when it attaches to the backside of the CH_3I.

5 Conclusion

The present experiment describes and employs a combination of velocity map ion imaging with a multipole radio frequency ion trap to study slow ion-neutral collisions. This combination advances crossed-beam reaction dynamics experiments, as it allows one to control the internal degrees of freedom of charged molecular and cluster reactants for reactive scattering experiments.

Using this setup we have studied the different product channels in the reactions of OH^- and $OH^-(H_2O)$ with methyl iodide between 0.15 and 2 eV of collision energy. The branching ratio for the formation of the different products is extracted for each energy. For the $OH^- + CH_3I$ reaction at 2 eV collision energy the measured branching rations are compared with direct dynamics simulations and a very good quantitative agreement is found. In the $OH^-(H_2O) + CH_3I$ system a long-lived meta-stable reaction complex is found in the simulations, which will eventually decay into the observed products.

Changing the internal temperature of the $OH^-(H_2O)$ complex, we have found a profound change in appearance energy for the observed proton transfer channels. This can not be explained by the increased internal excitation, but points to a vibrational state-specific enhancement of the reaction rate for the first excited vibrational mode of the cluster. Insight into the different dynamics of proton transfer and nucleophilic substitution is obtained by comparing the measured differential scattering cross sections for these two product channels.

The combination of the temperature-variable ion trap with crossed beam ion imaging opens up several new routes for reactive scattering experiments. In the future we expect to obtain more information on reaction dynamics under microsolvation with size-selected clusters and on reactivity of complex molecular ions as a function of their internal temperature.

Acknowledgements

We thank the University of Freiburg, where the measurements presented here have been carried out, for supporting this research. This work has been supported by the Deutsche Forschungsgemeinschaft under contract No. WE 2592/3-2. R.O. acknowledges support by the Landesgraduiertenförderung Baden-Württemberg and M.St. acknowledges support by the Marie Curie Initial Training Network ICONIC. The research at Texas Tech University is based upon work supported by the National Science Foundation under Grant No. CHE-0957416 and the Robert A. Welch Foundation under Grant No. D-0005. Support was also provided by the High Performance Computing Center (HPCC) at Texas Tech University (TTU), under the direction of Philip W. Smith, the Texas Advanced Computing Center (TACC) of the University of Texas at Austin, and the TTU Department of Chemistry & Biochemistry cluster Robinson whose purchase was funded by the National Science Foundation under the CRIF-MU Grant No CHE-0840493.

References

1 Y. T. Lee, *Science*, 1987, **236**, 793.
2 P. Casavecchia, *Rep. Prog. Phys.*, 2000, **63**, 355–414.
3 K. Liu, *Annu. Rev. Phys. Chem.*, 2001, **52**, 139.
4 M. Qiu, Z. Ren, L. Che, D. Dai, S. Harich, X. Wang, X. Yang, C. Xu, D. Xie and M. Gustafsson, *et al.*, *Science*, 2006, **311**, 1440.
5 X. Wang, W. Dong, C. Xiao, L. Che, Z. Ren, D. Dai, X. Wang, P. Casavecchia, X. Yang and B. Jiang, *et al.*, *Science*, 2008, **322**, 573.
6 A. T. J. B. Eppink and D. H. Parker, *Rev. Sci. Instrum.*, 1997, **68**, 3477–3484.
7 J. Mikosch, U. Frühling, S. Trippel, D. Schwalm, M. Weidemüller and R. Wester, *Phys. Chem. Chem. Phys.*, 2006, **8**, 2990–2999.
8 J. Mikosch, S. Trippel, R. Otto, C. Eichhorn, P. Hlavenka, M. Weidemuller and R. Wester, *J. Phys.: Conf. Ser.*, 2007, **88**, 12025–12025.
9 J. Mikosch, S. Trippel, C. Eichhorn, R. Otto, U. Lourderaj, J. X. Zhang, W. L. Hase, M. Weidemüller and R. Wester, *Science*, 2008, **319**, 183–186.
10 S. Trippel , *et al.*, in preparation.
11 V. Gabelica, E. Schulz and M. Karas, *J. Mass Spectrom.*, 2004, **39**, 579–593.
12 V. Gabelica and E. Pauw, *Mass Spectrom. Rev.*, 2005, **24**, 566–587.
13 S. R. Vande and W. L. Hase, *J. Am. Chem. Soc.*, 1989, **111**, 2349–2351.
14 S. R. Vande and W. L. Hase, *J. Chem. Phys.*, 1990, **93**, 7962–7980.
15 W. L. Hase, *Science*, 1994, **266**, 998–1002.
16 J. X. Zhang, J. Mikosch, S. Trippel, R. Otto, M. Weidemüller, R. Wester and W. L. Hase, *J. Phys. Chem. Lett.*, 2010, **1**, 2747–2752.
17 P. Manikandan, J. Zhang and W. L. Hase, *J. Chem. Phys. A*, 2012, **116**, 3061–3080.
18 L. Sun and W. L. Hase, *Rev. Comput. Chem.*, 2003, **19**, 79–146.

19 D. Gerlich, *Adv. Chem. Phys.*, 1992, **82**, 1–176.
20 R. Wester, *J. Phys. B: At., Mol. Opt. Phys.*, 2009, **42**, 154001.
21 O. Asvany and S. Schlemmer, *Int. J. Mass Spectrom.*, 2009, **279**, 147–155.
22 S. Trippel, M. Stei, R. Otto, P. Hlavenka, J. Mikosch, C. Eichhorn, A. Lourderaj, J. Zhang, W. L. Hase, M. Weidemüller and R. Wester, *J. Phys.: Conf. Ser.*, 2009, **194**, 012046.
23 D. R. Lide, *CRC Handbook of Chemistry and Physics*, 84th Edition, CRC Press, 2003.
24 L. Sun and W. L. Hase, in *Born–Oppenheimer Direct Dynamics Classical Trajectory Simulations*, John Wiley & Sons, Inc.. 2003, pp. 79–146.
25 J. X. Zhang and W. L. Hase, *J. Phys Chem. A*, 2010, **114**, 9635–9643.
26 G. Peslherbe, H. Wang and W. Hase. *Adv. Chem. Phys.*, 1999, **105**, 171–201.
27 C. Schlier and A. Seiter, *Comput. Phys. Commun.*, 2000, **130**, 176–189.
28 W. Hase, R. Duchovic, X. Hu, A. Komornicki, K. Lim, D. Lu, G. Peslherbe, K. Swamy, S. Linde and A. Varandas, *et al.*, *QCPE Bull*, 1996, **16**, 43.
29 X. Hu, W. Hase and T. Pirraglia, *J. Comput. Chem.*, 1991, **12**, 1014–1024.
30 E. Bylaska, W. De Jong, N. Govind, K. Kowalski, T. Straatsma, M. Valiev, D. Wang, E. Apra, T. Windus and J. Hammond, *et al.*, *Pacific Northwest National Laboratory, Richland, Washington*, 2007, **99352**, 0999.
31 M. Valiev, E. Bylaska, N. Govind, K. Kowalski, T. Straatsma, H. Van Dam, D. Wang, J. Nieplocha, E. Apra, T. Windus and W. A. de Jong, *Comput. Phys. Commun.*, 2010, **181**, 1477–1489.
32 R. Levine, *Molecular reaction dynamics*, Cambridge Univ Pr, 2005.
33 P. M. Hierl, J. F. Paulson and M. J. Henchman. *J. Phys. Chem.*, 1995, **99**, 15655–15661.
34 J. Mikosch, R. Otto, S. Trippel, C. Eichhorn, M. Weidemüller and R. Wester, *J. Phys. Chem. A*, 2008, **112**, 10448–10452.
35 A. B. McCoy, X. C. Huang, S. Carter and J. M. Bowman, *J. Chem. Phys.*, 2005, **123**, 064317.
36 L. Li, L. Yue and J. Farrar, *J. Chem Phys.*, 2006, **124**, 124317.
37 R. Otto, J. Brox, S. Trippel, T. Best and R. Wester, *Nat. Chem.*, 2012, **4**, 534.

Multi-path variational transition state theory for chemical reaction rates of complex polyatomic species: ethanol + OH reactions†

Jingjing Zheng and Donald G. Truhlar*

Received 29th January 2012, Accepted 29th February 2012
DOI: 10.1039/c2fd20012k

Complex molecules often have many structures (conformations) of the reactants and the transition states, and these structures may be connected by coupled-mode torsions and pseudorotations; some but not all structures may have hydrogen bonds in the transition state or reagents. A quantitative theory of the reaction rates of complex molecules must take account of these structures, their coupled-mode nature, their qualitatively different character, and the possibility of merging reaction paths at high temperature. We have recently developed a coupled-mode theory called multi-structural variational transition state theory (MS-VTST) and an extension, called multi-path variational transition state theory (MP-VTST), that includes a treatment of the differences in the multi-dimensional tunneling paths and their contributions to the reaction rate. The MP-VTST method was presented for unimolecular reactions in the original paper and has now been extended to bimolecular reactions. The MS-VTST and MP-VTST formulations of variational transition state theory include multi-faceted configuration-space dividing surfaces to define the variational transition state. They occupy an intermediate position between single-conformation variational transition state theory (VTST), which has been used successfully for small molecules, and ensemble-averaged variational transition state theory (EA-VTST), which has been used successfully for enzyme kinetics. The theories are illustrated and compared here by application to three thermal rate constants for reactions of ethanol with hydroxyl radical—reactions with 4, 6, and 14 saddle points.

I. Introduction

The progress in density functionals that allows a quantitative treatment of the potential energy surfaces of complex reactions[1-3] places new demands on reaction rate theory for maintaining a high level of accuracy in the treatment of the dynamics. Complex molecules often have many structures (conformations) of the reactants and the transition states, and these structures may be connected by coupled-mode torsions and pseudorotations; some but not all structures may have hydrogen bonds in the transition state or reagents. A quantitative theory of the reaction rates of complex molecules must take account of these structures, their coupled-mode nature, their qualitatively different character, and the possibility of merging reaction paths at high temperature.

Department of Chemistry and Supercomputing Institute, University of Minnesota, Minneapolis, Minnesota 55455-0431. E-mail: truhlar@umn.edu

† Electronic supplementary information (ESI) available: Cartesian coordinates of optimized geometries. See DOI: 10.1039/c2fd20012k

Transition state theory[4,5] (TST) provides an efficient way to obtain thermal reaction rate constants. Transition state theory has a long history, and it is widely understood to be a very general theory of reaction rates.[4–10] Early doubts about the reasonableness of its assumptions have been largely erased by incorporating variational transition states and multidimensional tunneling approximations and by comparing the results to accurate quantum mechanical calculations on simple reactions and to experiment, especially kinetic isotope effects, for a variety of processes, both simple and complex.[9–13] Advances in electronic structure theory have allowed one to obtain accurate enough potential energy surfaces that the hope of calculating absolute reaction rates is sometimes achieved for simple systems and is becoming within reach for more and more complicated systems. Practical methods have been proposed for defining sequences of generalized transition states and optimizing variational transition states within the sequences.[14–25] At this stage of theoretical development, the chief sources of concern, aside from the accuracy of the potential energy surface, when applying transition state theory to practical problems are issues such as the choice of reaction coordinate, the treatment of anharmonicity, and ensemble averaging over all contributing structures of the reactants and the generalized transition states.

Reactions of small molecules without torsions often have only a single conformation of the reactants and a single conformation of the transition state. Textbook presentations of transition state theory are often limited to this case, and the relevant partition functions are often treated as products of separable translational, rotational, vibrational, and electronic contributions, with the vibrational modes treated independently of each other and of rotation.[26] At the other extreme are complex reactions in solution or in enzymes. Here one must deal directly with free energies and free energies of activation that correspond to essentially uncountable numbers of conformations of the reagents, solvent molecules, and possibly catalysts. Intermediate between these extremes are gas-phase reactions of molecules with torsions. Here, as compared to small-molecule kinetics, there can be an appreciable conformational contribution to the free energies, and some vibrational modes may be strongly coupled to one another and to overall rotation. However, as compared to reactions in liquids or in enzymes, the number of conformations, although large, is countable. We have recently developed two formulations[27,28] of variational transition state theory, including multidimensional tunneling contributions, that are applicable to the case of gas-phase molecules with nonseparable torsions. We presented them in their most primitive form, and we applied the first[27] to reactions with up to 262 distinguishable conformations of the transition state[29] and the second to a reaction with four distinguishable conformations of the transition state.[28]

The first new formulation of TST is a coupled-mode theory called multi-structural variational transition state theory[27] (MS-VTST), and the second, multi-path VTST[28] (MP-VTST), extends it to multiple reaction paths to include a more complete treatment of the differences in the multi-dimensional tunneling paths and their contributions to the reaction rate. The new theories use a multi-faceted configuration-space dividing surface to define the variational transition state. Note that MP-VTST and MS-VTST both include the contribution of all the reaction paths to the total reactive flux, but MS-VTST does this in a more approximate way. MS-VTST is a special case of MP-VTST. MS-VTST has been presented elsewhere[27] and applied to unimolecular[27,28] and bimolecular reactions.[30,31] The MP-VTST was presented for and applied to unimolecular reactions.[28] Here we present MP-VTST for bimolecular reactions, explain the additional approximations that reduce it to MS-VTST, and then apply both MP-VTST and MS-VTST to three reactions.

We will also discuss the challenges in applying TST to calculate thermal rate constants accurately, especially for complex systems. A key challenge is interfacing methods for calculating potential energy surfaces to dynamical steps requiring a semi-global potential energy surface, such as the treatment of variational effects, quantum effects, and anharmonicity. Some calculations in the literature have achieved good agreement with experimental data without considering all or some of these important effects or by using inaccurate potential energy surfaces because the agreement between experiments and calculations is sometimes achieved by error cancellation in the calculations, but to make theory reliably predictive one cannot rely on error cancellation, and we need to understand the possible sources of error and the best available strategies to mitigate these errors.

The reactions studied in the present work are three hydrogen abstraction reactions of the various sites of ethanol with hydroxyl radicals,

$$CH_3CH_2OH + OH \rightarrow CH_3CHOH + H_2O \qquad (R1a)$$

$$CH_3CH_2OH + OH \rightarrow CH_2CH_2OH + H_2O \qquad (R1b)$$

$$CH_3CH_2OH + OH \rightarrow CH_3CH_2O + H_2O \qquad (R1c)$$

These reactions are important elementary steps in biofuel combustion and have been extensively studied experimentally[32–41] and theoretically.[40,42,43] The overall reaction rate constant, k_1, is the sum of k_{1a}, k_{1b}, and k_{1c}, and the IUPAC[44] recommendation for this quantity is $6.70 \times 10^{-18} \, T^2 \exp(511/T)$ cm^{3}molecule^{-1}s^{-1} over the temperature range 216–599 K. The branching ratios are less well established.[39] Although these reactions have been studied by several groups using TST,[40,42,43] there is still much to learn, and in the present work we will present a more detailed analysis and discussion of these reactions based on our newly developed multi-path and multi-structural versions of VTST, and we will use these reactions as examples to show the challenges in obtaining accurate thermal rate constants of complex polyatomic systems using state-of-art electronic structure and dynamics methods.

II. Computational methods

II. A. Electronic structure methods

The conformational structures of ethanol, the hydroxyl radical, and the saddle points are optimized by various density functionals including M08-HX,[45] M08-SO,[45] M05-2X,[46] and M06-2X.[47] The 6-31+G(d,p) basis set is used for all geometry optimizations and for the straight direct dynamics calculations except that the ma-TZVP[48] basis set is used for M08-HX and M08-SO calculations for R1a to test the sensitivity to basis set.

The first grid used for density functional integrations is a pruned grid based on 99 radial shells around each atom and 974 angular points in each shell. This grid is used for the optimization and frequency calculation for all the stationary points and for the reaction paths of R1a and R1c. The reaction paths for R1b require a finer grid, and we used a grid that has 96 radial shells around each atom and a spherical product angular grid having 32 θ points and 64 φ points in each shell.

The best estimate (BE) classical barrier height is obtained by a two-step procedure. The first step is to estimate the complete basis set (CBS) limit of the coupled-cluster singles and doubles method with quasiperturbative inclusion of connected triples (CCSD(T)).[49] The second step is to add a finite basis set (FBS) correction for a higher-level (HL) treatment of electron correlation energy. Thus

$$V_{BE}^{\ddagger} = V_{CBS}^{\ddagger}(CCSD(T)) + \Delta V_{FBS}^{\ddagger} \qquad (1)$$

$$\Delta V_{FBS}^{\ddagger} = V^{\ddagger}(CCSDT(2)_Q/FBS) - V^{\ddagger}(CCSD(T)/FBS) \qquad (2)$$

where $V^{\ddagger}$ denotes the classical barrier height. Usually the CBS limit is obtained by extrapolating a few finite basis set results, but here we used a more efficient method, namely CCSD(T)-F12b[50,51] with a may-cc-pVQZ[52] basis set, to approximate the CCSD(T)/CBS limit. We also calculated CCSD(T)-F12a/aug-cc-pVTZ energies to check the convergence with respect to the basis set.

The FBS correction accounts for the effect of higher excitations by using the CCSDT(2)$_Q$[53] method with a finite basis set. The basis set for use in eqn (2) can be smaller than that for eqn (1) because the higher-level corrections show faster basis set convergence. The finite basis set used in eqn (2) is the maug-cc-pVDZ[54] basis set. To check this assumption, we also calculate $\Delta V^{\ddagger}_{HL}$ by using a smaller basis set, cc-pVDZ.

Restricted open-shell Hartree–Fock orbitals are used for all the coupled cluster calculations. In all coupled cluster calculations (CCSD(T), CCSD(T)-F12a, CCSD(T)-F12b, and CCSDT(2)$_Q$), correlation of core electrons is not included.

To gauge the importance of multi-reference character in the stationary-point electronic wave functions, we computed the T_1 diagnostic.[55] It has been suggested that significant multi-reference character, and hence a concomitant higher than usual inaccuracy of CCSD(T) calculations, is indicated by a T_1 diagnostic of 0.02 or greater for closed-shell systems[55] and by a T_1 diagnostic of 0.045 or greater for open-shell systems.[56–58]

For computing vibrational partition functions, all Hessians and force constants (in either internal or Cartesian coordinates) are multiplied by λ^2_{ZPE}, where λ_{ZPE} is a previously determined scale factor[59] that, when the frequencies computed from scaled Hessians are used in the harmonic oscillator (HO) formula, should make the zero-point energy more accurate. The scale factor accounts for systematic errors in the electronic structure method but also, significantly, for the difference (on average) between zero-point energies computed by the HO formula and those including anharmonicity. For this reason, when results given below are labeled as HO or local harmonic (LH), the label and the language refer to the use of the HO formulas; the results themselves are quasiharmonic because of the use of scaling to account (approximately) for anharmonicity. But this mainly includes the anharmonicity in the high-frequency modes because it is the high-frequency modes that dominate the zero-point energy. Anharmonicity due to low-frequency torsions is accounted for by the method discussed in the next subsection.

All the density functional calculations were performed by the *Gaussian 09*[60] package with the *MNGFM* version 5.1[61] module. The CCSD(T)-F12a/b calculations were carried out by the *Molpro*[62] program, and the *NWChem*[63] program was used for CCSDT(2)$_Q$ calculations.

II. B. Partition functions

Ethanol and the transition states of the three reactions each have multiple conformational structures (minima or saddle points, respectively, on the potential energy surface) caused by internal rotations (torsions). We will treat the anharmonicity associated with these torsions by the multi-structural method including torsional anharmonicity[64] (MS-T). In the MS-T calculations, the translational partition function and electronic partition function are separable from the conformational–rotational–vibrational partition function.

The MS-T method can be applied to both stable species and transition states. We will label a general species as α, which can be a reactant (R_i with $i = 1$ or 2) or the saddle point $\ddagger$. A reactant has F vibrational modes, where F is the number of internal coordinates, and a transition state, being a hypersurface dividing reactants from products, has only $F - 1$ vibrations. In general, let j label the distinguishable conformational structures of a species, and let J be the total number of such conformational structures. We will use k and K to denote the structure number and number of structures when we are specifically referring only to the transition state. A transition state with K structures is a multifaceted dividing surface with K facets. Because we use curvilinear coordinates,[17] facet k is a curved hypersurface locally orthogonal to the reaction path through saddle point k, as shown in Fig. 1. We will let U_j^α denote the equilibrium potential energy of structure j; for transition states it is the lowest potential energy in facet k. The structure with the lowest value of U_j^α or U_k^α for a given species α is labeled $j = 1$ or $k = 1$, and U_1^α is set equal to zero. Then all other U_j^α are measured with respect to U_1^α. Throughout the entire article all partition functions for species α are calculated with the zero of energy at U_1^α. We will denote the (zero-point-inclusive) ground-state energy of structure j as $\tilde{U}_j^\alpha$.

The conformational–rotational–vibrational partition function is calculated as

$$Q_{\mathrm{con-rovib}}^{\mathrm{MS-T},\,\alpha} = \sum_{j=1}^{J} Q_j^{\mathrm{S},\,\alpha} \tag{3}$$

where $Q_j^{\mathrm{S},\alpha}$ is the contribution of structure j of species α and is given by[64]

$$Q_j^{\mathrm{S},\,\alpha} = Q_{\mathrm{rovib},j}^{\mathrm{SS-HO},\,\alpha} Z_j^\alpha \prod_{\tau=1}^{t_\alpha} f_{j,\tau}^\alpha \tag{4}$$

where t_α is the number of torsions in species α, Z_j^α is a factor for guiding the MS-T scheme to the correct high-temperature limit (within the parameters of the model), $f_{j,\tau}^\alpha$ is a torsional anharmonicity factor based on the internal coordinates, which in conjunction with Z_j^α adjusts the harmonic partition function of structure j for the presence of the torsional motion τ, and the single-structure (SS) rotational-vibrational partition function of structure j using the harmonic oscillator approximation is

$$Q_{\mathrm{rovib},j}^{\mathrm{SS-HO},\alpha} = Q_j^{\mathrm{rot},\alpha} \exp(-U_j^\alpha/k_\mathrm{B}T) Q_j^{\mathrm{HO},\alpha} \tag{5}$$

where $Q_j^{\mathrm{rot},\alpha}$ is classical rotational partition function of structure j, k_B is Boltzmann's constant, T is temperature, and $Q_j^{\mathrm{HO},\alpha}$ is the local-harmonic-oscillator vibrational partition function calculated at structure j.

We define the MS-T torsional anharmonicity factor for each species as

$$F^{\mathrm{MS-T},\,\alpha} = \frac{Q_{\mathrm{con-rovib}}^{\mathrm{MS-T},\,\alpha}}{Q_{\mathrm{rovib},\,1}^{\mathrm{SS-HO},\,\alpha}} \tag{6}$$

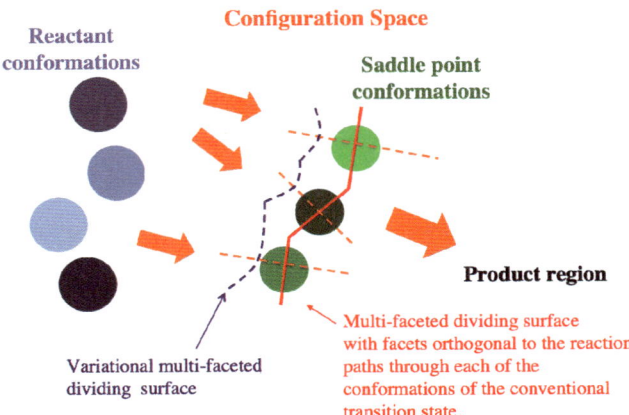

Fig. 1 Multi-faceted dividing surfaces: conventional transition state in red; variational transition state in blue.

where $Q_{\text{rovib},1}^{\text{SS-HO},\alpha}$ is the single-structure rotational-vibrational partition function calculated at the global minimum (structure 1 by definition). Torsional anharmonicity has two aspects. First, the torsions usually generate multiple structures. Second, low barriers along the torsional potential energy profiles mean that the contributions of the structures cannot simply be added together as independent quasiharmonic oscillators, but rather one must account for the merging of the wells as the available energy exceeds the torsional barriers. To separate the effects of multiple-structure anharmonicity, $F^{\text{MS},\alpha}$, from torsional-potential anharmonicity, $F^{\text{T},\alpha}$, we write eqn (6) as

$$F^{\text{MS-T},\alpha} = F^{\text{MS},\alpha} F^{\text{T},\alpha} \tag{7}$$

where

$$F^{\text{MS},\alpha} = \frac{Q_{\text{con-rovib}}^{\text{MS-LH},\alpha}}{Q_{\text{rovib},1}^{\text{SS-HO},\alpha}} \tag{8}$$

$$F^{\text{T},\alpha} = \frac{Q_{\text{con-rovib}}^{\text{MS-T},\alpha}}{Q_{\text{rovib},1}^{\text{SS-LH},\alpha}} \tag{9}$$

We define a reaction torsional anharmonicity factor for a bimolecular reaction as

$$F^{\text{MS-T}} = \frac{F^{\text{MS-T},\ddagger}}{F^{\text{MS-T,React1}} F^{\text{MS-T,React2}}} \tag{10}$$

where React1 and React2 are the two reactants of a bimolecular reaction. For the reactions studied in the present work, the OH radical has no torsion so that $F^{\text{MS-T,OH}} = 1$, and the torsional anharmonicity activation factor can be written as

$$F^{\text{MS-T}} = \frac{F^{\text{MS-T},\ddagger}}{F^{\text{MS-T, ethanol}}} \tag{11}$$

The MS-T partition functions were calculated by the *MSTor*[65,66] program.

The overall partition function of species α is obtained by multiplying $Q_{\text{con-rovib}}^{\text{MS-T},\alpha}$ by the electronic partition function Q_{elec}^{α} and the translational partition function.

II. C. Dynamics: MP-VTST

First we establish notation by considering single-structure VTST, which can be written as

$$k^{\text{VTST}} = \frac{k_{\text{B}}T}{h} \frac{Q_{\text{elec}}^{\ddagger}}{\Phi_{\text{trans}}^{\text{R}} \prod_{i=1}^{2} Q_{\text{elec}}^{\text{R}_i} Q_{\text{con-rovib},1}^{\text{SS-HO},\text{R}_i}} \exp\left(-V^{\ddagger}/k_{\text{B}}T\right) \kappa \Gamma Q_{\text{con-rovib},1}^{\text{SS-HO},\ddagger} \tag{12}$$

where h is Planck's constant; $\Phi_{\text{trans}}^{\text{R}}$ is a relative translational partition function per volume of the two reactants; $V^{\ddagger}$ is the classical barrier height, that is, the potential energy of the lowest-energy saddle point relative to the lowest reactant; κ is a tunneling transmission coefficient that accounts multidimensional tunneling and nonclassical reflection; and Γ is a recrossing transmission coefficient given by the ratio of the flux through the dynamical bottleneck along the reaction path and the flux through the conventional transition state dividing surface at saddle point $k = 1$. Note that all quantities in eqn (12) except k_{B}, h, and $V^{\ddagger}$ depend on temperature, but dependence on temperature is not shown as an argument in this article. When eqn (12) is applied with κ set equal to unity, the result is called canonical variational theory (CVT), and when it is applied with a multidimensional tunneling (MT) approximation, it is called CVT/MT where MT can be ZCT, SCT, LCT, μOMT, or LAT (denoting respectively the zero-curvature tunneling approximation,[67] the small-curvature tunneling approximation,[68] the large-curvature tunneling approximation,[69–72] the microcanonically optimized multidimensional tunneling approximation,[72] and the least-action tunneling approximation[73,74]). The CVT result is sometimes called quasiclassical to denote that quantum effects are included on the bound motions of the transition state but not on the reaction coordinate.

The meaning of the recrossing coefficient Γ is illustrated by writing eqn (12) as

$$k^{\text{VTST}} = \frac{k_B T}{h} \frac{Q_{\text{elec}}^{\ddagger}}{\Phi_{\text{trans}}^R \prod_{i=1}^{2} Q_{\text{elec}}^{R_i} Q_{\text{con-rovib,1}}^{\text{SS-HO},R_i}} \exp(-V^{\ddagger}/k_B T) \kappa Q_{\text{con-rovib,1}}^{\text{SS-HO, VTS}} \tag{13}$$

where VTS denotes that the partition function is evaluated not at the saddle point but rather at the canonical variational transition state; thus we say that Γ accounts for variational effects. In particular, Γ is the special case with $k = 1$ of

$$\Gamma_k = \frac{\exp(-\max_s G_{T,k}(s)/RT)}{\exp\left(-G_{T,k}^{\ddagger}/RT\right)} \tag{14}$$

where $G_{T,k}$ is the molar generalized free energy of activation[75–77] at temperature T and location s along the minimum-energy path (MEP) through saddle point k, s is the signed distance from saddle point k along that path, $G_{T,k}^{\ddagger}$ is the value of $G_{T,k}$ at $s = 0$, and R is the gas constant.

We account for quantum mechanical effects on the reaction coordinate motion by the transmission coefficient κ. There are two kinds of primary quantum mechanical effects on reaction coordinate motion. First, systems with an energy below the effective barrier height may tunnel through the barrier. Second, systems with energies above the effective barrier, which would be transmitted with unit probability if reaction-coordinate motion were governed by classical mechanics, may be reflected by scattering off the barrier; this is a kind of diffraction, and we call it nonclassical reflection. Because tunneling is usually more important than nonclassical reflection (because the Boltzmann weighting is larger at tunneling energies than at energies above the barrier), we usually call κ the tunneling transmission coefficient. In MS-VTST and MP-VTST, we treat κ by the ground-state tunneling approximation introduced[15,75] in the single-structure version of the theory. In general notation, κ is the special case with $k = 1$ of

$$\kappa_k = \frac{\int_0^{\infty} dE P_k(E) \exp(-E/k_B T)}{\int_0^{\infty} dE P_k^{\text{QC}}(E) \exp(-E/k_B T)} \tag{15}$$

where E is the energy of reaction-coordinate motion, P_k is the quantal probability of transmission from reactants to products in the ground-vibrational state along reaction path k (which is the reaction path through saddle point k), and P_k^{QC} is the approximation to P_k implied by quasiclassical CVT. Therefore P_k^{QC} is a Heaviside step function at $E = \tilde{U}_k^{\text{VTS}}$, which yields

$$\kappa_k = \int_0^{\infty} dE P_k(E) \exp\left[-\left(E - \tilde{U}_k^{\text{VTS}}\right)/k_B T\right] \tag{16}$$

where $\tilde{U}_k^{\text{VTS}}$ is the zero-point-inclusive energy $\tilde{U}_k$ at the maximum generalized free energy of activation determined in eqn (14), and where

$$\tilde{U}_k = V_k(s) + \varepsilon_k^G(s) \tag{17}$$

where V_k and ε_k^G are respectively the potential energy and zero-point vibrational energy along path k. Note that $\tilde{U}_k^{\text{VTS}}$ depends on T because the location of the CVT transition state depends on T. Let V_k^{AG} denote the maximum zero-point-inclusive energy along reaction path k. Then we can note that κ_k differs from unity for three reasons: (i) because P_k is larger than zero for at $E_0 \leq E \leq V_k^{\text{AG}}$ (where E_0 is the quantum threshold energy), which is called tunneling, (ii) because P_k is smaller than unity for $E > V_k^{\text{AG}}$, which is called nonclassical reflection, and (iii) because sometimes $\tilde{U}_k^{\text{VTS}} \neq V_k^{\text{AG}}$, which causes vibrationally adiabatic classical reflection for $\tilde{U}_k^{\text{VTS}} < E < V_k^{\text{AG}}$. Of these three, tunneling is usually the most significant, and so we often call κ_k a tunneling transmission coefficient. It will be convenient for the discussion in Section II. D. to label the lowest-energy V_k^{AG} as V_a^{AG}. Note that the quantum threshold energy E_0 is the lowest energy at which it is possible to have tunneling, and it is $\max[\tilde{U}_1(s = -\infty), \tilde{U}_1(s = \infty)]$ for a bimolecular reaction.

We can combine the recrossing and tunneling effects by defining a generalized transmission coefficient as

$$\gamma_k = \kappa_k \Gamma_k \tag{18}$$

where Γ_k is a recrossing transmission coefficient given by the ratio of the flux through the dynamical bottleneck along path k and flux through the conventional transition state dividing surface at saddle point k.

Next we consider multi-path variational transition state theory with multidimensional tunneling (MT) for calculating the reaction rate constants. The general idea is illustrated in Fig. 1. The facets of a multi-faceted dividing surface are orthogonal the reaction paths through each of the conformations of the transition state. For a bimolecular reaction, the full MP-VTST rate constant at temperature T is

$$k^{\text{MP-VTST/MT}} = \frac{k_B T}{h} \frac{Q_{\text{elec}}^{\ddagger}}{\Phi^{\text{MS-T,R}}} \exp\left(-V^{\ddagger}/k_B T\right) \sum_{k=1}^{K} \kappa_k \Gamma_k Q_k^{\text{S},\ddagger} \tag{19}$$

where $\Phi^{\text{MS-T,R}}$ is the reactant partition function per unit volume in the center-of-mass frame given by

$$\Phi^{\text{MS-T,R}} = \Phi_{\text{trans}}^{\text{R}} \prod_{i=1}^{2} Q_{\text{elec}}^{\text{R}_i} Q_{\text{con-rovib}}^{\text{MS-T,R}_i} \tag{20}$$

Without any additional approximations, we can rearrange eqn (19) as follows:

$$k^{\text{MP-VTST/MT}} = \langle \gamma \rangle F^{\text{MS-T}} k^{\text{SS-TST}} \tag{21}$$

where $k^{\text{SS-TST}}$ is the single-structure conventional transition state rate constant in the harmonic approximation:

$$k^{\text{SS-TST}} = \frac{k_B T}{h} \frac{Q_{\text{elec}}^{\ddagger} Q_{\text{rovib},1}^{\text{SS-HO},\ddagger}}{\Phi^{\text{SS-HO,R}}} \exp\left(-V^{\ddagger}/k_B T\right) \tag{22}$$

and where we have defined

$$\Phi^{\text{SS-HO,R}} = \Phi_{\text{trans}}^{\text{R}} \prod_{i=1}^{2} Q_{\text{elec}}^{\text{R}_i} Q_{\text{rovib},1}^{\text{SS-HO,R}_i} \tag{23}$$

and

$$\langle \gamma \rangle = \frac{\displaystyle\sum_{k=1}^{K} \kappa_k \Gamma_k Q_k^{\text{S},\ddagger}}{\displaystyle\sum_{k=1}^{K} Q_k^{\text{S},\ddagger}} \tag{24}$$

where the latter may be called an averaged generalized transmission coefficient. In our original application[28] we made the approximation of replacing $Q_k^{S,\ddagger}$ by $Q_{\text{rovib},k}^{SS-OH,\ddagger}$ in eqn (24), but there is no good reason to do that, so here and in the future we will use eqn (24). We also noted that one could replace the full average of eqn (24) by a partial average:

$$\langle\gamma\rangle_P = \frac{\sum_{k=1}^{P} \kappa_k \Gamma_k Q_k^{S,\ddagger}}{\sum_{k=1}^{P} Q_k^{S,\ddagger}} \qquad (25)$$

where $P < K$.

Note that we have factored the rate constant into a single-structure local harmonic rate constant times a correction, where we will substitute a quasiharmonic treatment for the harmonic one. Alternatively, although not done here, one could factor the rate constant into a single-structure rate constant that takes some account of torsions times a correction.

When we apply MP-VTST with all κ_k equal to unity, we call the result MP-CVT. When κ_k is included, the result is called MP-CVT/MT, for example, MP-CVT/SCT.

If we set the averaged generalized transmission coefficient to unity, eqn (19) is reduced to the multi-structural conventional transition state theory (MS-TST) rate constant without tunneling:

$$k^{\text{MS-TST}} = F^{\text{MS-T}} \frac{k_B T}{h} \frac{Q_{\text{elec}}^{\ddagger} Q_{\text{rovib},1}^{SS-HO,\ddagger}}{\Phi^{\text{MS-T,R}}} \exp\left(-V^{\ddagger}/k_B T\right) \qquad (26)$$

The single-structure canonical variational theory rate constant is

$$k^{\text{SS-CVT}} = \Gamma_1 k^{\text{SS-TST}} \qquad (27)$$

In some cases, it is a good approximation to approximate $\langle\gamma\rangle$ using a single reaction path. This would be the case, for example, if all the Γ_k and κ_k have similar values and therefore one path can be used to represent all the paths. Another case where this approximation would be good is when $Q_1^{S,\ddagger}$ is much larger than all the other $Q_k^{S,\ddagger}$ such that flux along the lowest-energy path dominates the reaction. The lowest-energy saddle point usually has a larger $Q_k^{S,\ddagger}$ than any of the other transition state structures, especially at low temperature where the κ_k differ most significantly from unity. If we set $P = 1$ in eqn (25), the MP-VTST/MT rate constant is reduced to the MS-VTST/MT rate constant[27]

$$k^{\text{MS-VTST}} = \Gamma_1(T)\kappa_1(T) \frac{k_B T}{h} \frac{Q_{\text{elec}}^{\ddagger} Q_{\text{con-rovib}}^{\text{MS-T},\ddagger}}{\Phi^{\text{MS-T,R}}} \exp\left(-V^{\ddagger}/k_B T\right) \qquad (28)$$

We previously pointed out[27,28] that when one takes $P = 1$, one can use any representative structure of the transition state, not necessarily the lowest-energy one (as done here); the changes to the formulas are straightforward. If we set $\Gamma_1 = \kappa_1 = 1$ and evaluate the conformational-rovibrational partition functions by the local harmonic[28,65] approximation, then eqn (28) reduces to eqn (17) of an earlier paper.[78]

Finally we consider the calculation of Γ_k and κ_k. The calculation of these quantities requires the calculation of a reaction path through saddle point k. We calculate these quantities using the methods previously developed[79] for SS-CVT/MT calculations, in particular, using the quasiharmonic approximation defined in subsection II. A.

In the present work, we calculated κ_k using the small-curvature tunneling[68] (SCT) approximation. The calculation of Γ_k, κ_k, and $k^{\text{SS-TST}}$ are carried out using the *POLYRATE*[80] program, and the calculation of $Q_k^{S,\ddagger}$ was carried out with the *MSTor* program. In the direct dynamics calculations, MEPs in isoinertial coordinates are calculated using a reorientation of dividing surface (RODS) algorithm.[81]

It should be noted that transition state theory assumes that the internal states of the reactants are in local equilibrium, where the word "local" in this context means[82] that the products need not be present in their equilibrium population. Therefore, for unimolecular reactions in a liquid, transition state theory assumes that the coupling of the solvent to the reacting solute is strong enough to maintain an equilibrium population of reactants,[10,83] and for bimolecular reactions it also assumes that diffusion is fast enough to maintain a local equilibrium concentration of reacting partners in the vicinity of each reactant molecule. For unimolecular reactions in the gas phase, transition state theory yields what is conventionally called the high-pressure limit (in which the pressure is high enough to maintain local equilibrium of reactant states), although it is actually a plateau reached before one enters the suprahigh-pressure regime.[22] For gas-phase bimolecular reactions, the rate constant is only well defined if[84] the time for reaction is much greater then the time for establishment of local equilibrium among the reactant states, which in turn is much higher than the time for passage of a system through the transition state region of phase space; that is, the system must be in the high-pressure limit but not the suprahigh pressure limit. The rate evaluated by transition state theory corresponds to a high enough pressure that energy transfer collisions repopulate the reactive states of the reactants to maintain a thermal distribution, but a more general statement about the pressure regime to which the TST rate constant applies requires a consideration of possible intermediates, and we will defer this discussion to subsection II. D.

It is important to note that the validity of the equilibrium assumption of transition state theory is a function of the more than just the total pressure. Consider again that the basic assumption in transition state theory is that the conformational states of the reactants are in local equilibrium with each other, and this equilibrated pool of states of the reactants and the transition state are in quasiequilibrium during the reaction. When this assumption is applied to MS-VTST and MP-VTST, it means that the interconversion between the conformations of reactant is much faster than the chemical reaction, and one sometimes morphs this into a statement that torsional barriers of reactants are much smaller than the barriers of chemical reaction. However, that is an oversimplification, as seen by considering the reaction of ethanol with hydroxyl radical. The torsional barriers of ethanol are as high as $\sim$3 kcal mol^{-1}, which makes them similar to or even larger than the reaction barriers studied in the present work. But under the experimental conditions where the reaction has been studied, the concentration of hydroxyl radical is much lower than that of ethanol, and it is reasonable to assume that the conformations of ethanol are in quasiequilibrium during the reaction. Therefore MS-VTST and MP-VTST are still applicable for these reactions.

In general we label the reactant conformations by $j = 1, 2,..., J$ and we label the transition state conformations by $k = 1, 2,..., K$, and we note that for a unimolecular reaction, one can rewrite the rate constant of eqn (17) as

$$k^{\text{MP-VTST}} = \frac{\sum_{k=1}^{K} s_k}{R} \tag{29}$$

where

$$R = \sum_{j=1}^{J} r_j \tag{30}$$

$$r_j = Q_{\text{elec}}^{\text{R}} Q_j^{\text{S,R}} \tag{31}$$

and

$$s_k = \frac{k_B T}{h} Q^{\ddagger}_{\text{elec}} \exp\left(-V^{\ddagger}/k_B T\right) \kappa_k \Gamma_k Q^{S,\ddagger}_k \tag{32}$$

Note that the barrier $V^{\ddagger}$ in eqn (32) is the energy difference between the lowest-energy saddle point (*i.e.*, lowest-energy conformation of the transition state) and the lowest-energy conformation of the reactant, and the relative energy of conformation k is included in $Q^{S,\ddagger}_k$ *via* the Boltzmann factor of eqn (5).

It is interesting to compare eqn (29) to the so called generalized Winstein–Holness (GWH) equation,[85-87] which is given by

$$k^{\text{GWH}} = \sum_{j=1}^{J} P_j k^{[j]} \tag{33}$$

where we have denoted the equilibrium population of conformer j by

$$P_j = \frac{r_j}{R} \tag{34}$$

and we have associated a rate constant, $k^{[j]}$, with each conformation of reactants. (When there are two conformers of the reactant and two transition-state structures, the GWH equation reduces to the Winstein–Holness (WH) equation.) Eqn (33) can be derived from eqn (29) if we set

$$k^{[j]} = \sum_{k \subset j(k)} \frac{k_B T}{h} \frac{Q^{\ddagger}_{\text{elec}}}{r_j} \exp\left(-V^{\ddagger}/k_B T\right) \kappa_k \Gamma_k Q^{S,\ddagger}_k \tag{35}$$

where the sum is over those k that appear in a $j(k)$ for the j on the left hand side, and if we assume that two different conformers of the reactant never lead to the same conformer of the transition state; then we may associate a unique reactant conformer $j(k)$ with each transition state conformer k. (If more than one reactant conformer reacts through the same transition state conformation, then more than one $j(k)$ has the same k.) But this assumption—made to obtain eqn (33)—is not consistent with TST or with the local equilibrium condition of eqn (34). Transition state theory assumes that the rate constants for interconverting conformers of the reactant are larger than the rate constants for chemical reaction—in order to assure local equilibrium. But consider a case where local equilibrium is maintained because the barriers for interconversion of reactant conformers are lower than the barriers to reaction. Since the reaction in such a case is dominated by states with energies higher than the conformational barriers, one cannot associate a given transition state conformer with a unique path originating from a specific reactant conformer. If the barriers between reactant structures are high, they should be considered as two different reactants, not as conformers of a single reactant. Furthermore, eqn (33) is wrong in the general case. Consider, for example, a case with two similar conformations of the reactants and one conformation of the transition state. Then $k^{[2]} \approx k^{[1]}$, and eqn (33) predicts a rate constant equal to $\sim k^{[1]}$, but the correct answer is $\sim k^{[1]}/2$. (Note, as an aside, that the Curtin–Hammett principle does not suffer from this error, although some derivations[88] of it do suffer from it because they are based on the GH equation, which is inconsistent, *i.e.*, wrong.)

However, transition state theory can be formulated to handle multiple conformers properly, and that is what we have done in MS-VTST and MP-VTST. Unlike the GWH equation, we do not assume that one can identify one-to-one connections between transition state conformations and reactant conformations. Furthermore our treatment remains valid even in the regime that torsions are better descried as hindered rotors than as a collection of individual conformational states. The fact that the MS-VTST and MP-VTST methods do not require one to map the routes between the conformations of the saddle point and the conformations of the reactant or reactants is a key advantage of the correct formulation of transition state theory. The MS-VTST and MP-VTST methods fully account for the very likely reactive events in which the molecular system crosses from one reaction valley to another on the route to the transition state, contrary to the claim[87] that TST cannot handle this or the assumption[88] that this does not occur.

II. D. The treatment of intermediates

Consider a bimolecular gas-phase reaction of A with B, with a pre-reactive intermediate:

$$A + B \rightarrow AB \rightarrow AB^{\ddagger} \rightarrow P \tag{36}$$

where AB is an intermediate complex, $AB^{\ddagger}$ is the transition state, and P is the product. The reaction rate may be written as

$$\frac{d[P]}{dt} = k_{bimol}[A][B] \tag{37}$$

where, as usual, brackets denote concentrations. If the reactants A and B are in quasiequilibrium with the transition state (where quasiequilibrium is the same as equilibrium except that products and transition state species arising from products are missing), then it may be reasonable to assume that they are also in local equilibrium with AB. In that case, we can also write

$$\frac{d[P]}{dt} = k_{unimol}[AB] \tag{38}$$

where

$$k_{unimol} = k_{bimol}/K \tag{39}$$

and K is the equilibrium constant for $A + B \rightarrow AB$. Because TST assumes local equilibrium on the reactant side of the transition state, the reaction rates computed by TST are consistent with eqn (39). Because of eqn (39), the VTST rate constant with $\kappa = 1$ is independent of whether we use eqn (38) or (37). (A related but somewhat different issue is whether the canonical treatment is adequate. We note that in the case that $\tilde{U}^{VTS}$ is below the reactant ZPE (or in fact whenever it is below V_a^{AG}), the improved canonical variational transition state theory[75,89–91] (ICVT) can be used to exclude the contribution of the states with total energies lower than the V_a^{AG} in the partition function.) However, when we incorporate tunneling, the ground-state tunneling approximation depends on whether or not we assume that the complex is equilibrated, so further considerations are needed, as discussed next.

TST has the requirement that the pressure must be high enough to establish local equilibrium among the states of A and the states of B even if the concentration of AB is not at local equilibrium. Notice that local equilibrium among the states of A and the states of B is easily achieved in the presence of an inert gas. Inert gas collisions can equilibrate A and B by bimolecular collisions, but thermalization of AB complexes requires termolecular collisions involving at least one A and at least one B.

Notice that the local equilibrium for configurations (such as those corresponding to AB) between $A + B$ and $AB^{\ddagger}$ is generated in two ways. First is by nonreactive collisions with other constituents of the gas. Second, even without collisions with other bodies, is by uninterrupted evolution of the reactant local equilibrium distribution toward the transition state as governed by the Liouville equation.[84] There is, however, an important difference between these two mechanisms. The former populates all states between $A + B$ and $AB^{\ddagger}$. The latter populates only those states that can be reached by conservation of energy and angular momentum. We will focus on the energetic criterion. If the complex has states of lower energy than the lowest-energy state of A + B, then those states cannot be populated without termolecular collisions. If termolecular collisions are insufficient to populate the low-energy states of AB, then those states will be missing.

For the reactions studied in this article, all the $\tilde{U}_k^{\mathrm{VTS}}$ are higher than the reactant ZPE. In this article, we consider the low-pressure limit in which the system reaches local equilibrium only by the Liouville equation. Therefore the states with total energies lower than the lowest-energy state of ethanol + OH are not used for tunneling calculations because of the ground-state tunneling approximation (which deserves re-examination in this light, as discussed below). However, if one calculates k_{unimol} followed by setting k_{bimol} equal to Kk_{unimol}, which is the high-pressure limit, then those states will be present in the tunneling calculation, and the calculated tunneling contribution will be larger.

We evaluated the termolecular collision rate k_{termol} by using the method in ref. 92 and the method in ref. 93 and 94. The two methods predict values of k_{termol}, the rate constant for termolecular collision of ethanol, OH, and helium, that differ by two orders of magnitude. If we assume the third body in termolecular collision in ethanol + OH reaction is helium and its partial pressure is 40 torr, our calculated bimolecular rate constant at 298 K is in the middle of the values of $k_{\mathrm{termol}}[\mathrm{He}]$ evaluated by the two methods. These rough estimates are insufficient to assess the role of pre-reactive collision complexes under the experimental conditions that have been used to study the reaction of ethanol with hydroxyl. A more reliable treatment of the role of the complex AB under experimental conditions and the fall off in rate constant from the high-pressure plateau is to solve the master equation,[22] which is beyond the scope of the present article. Therefore we simply report calculations in the low-pressure limit.

II. E.   Dual-Level strategy

The computation of κ_k and Γ_k can be expensive because it requires the calculation of a reaction path and the Hessians required for generalized normal mode analyses along that path. In a direct dynamics calculation, a reaction path is usually calculated with a density functional that gives a barrier height close to the accurate (or best estimate) barrier height. However, the difference, $\delta V_{\mathrm{corr}}^{\ddagger}$, between the barrier height $V_{\mathrm{DFT}}^{\ddagger}$ calculated by this density functional and the best estimated barrier height $V_{\mathrm{BE}}^{\ddagger}$ is usually nonzero. To account for this difference, we multiply the computed MP-VTST and MS-VTST rate constants by $\exp(-\delta V_{\mathrm{corr}}^{\ddagger}/k_{\mathrm{B}}T)$ where

$$\delta V_{\mathrm{corr}}^{\ddagger} = V_{\mathrm{BE}}^{\ddagger} - V_{\mathrm{DFT}}^{\ddagger} \qquad (40)$$

This kind of correction should be used only when $\delta V_{\mathrm{corr}}^{\ddagger}$ is quite small (for instance, less than a few tenths of a kcal mol^{-1}). When $\delta V_{\mathrm{corr}}^{\ddagger}$ is larger, although the barrier is corrected, the reaction path used to calculate tunneling and recrossing effects could be quite inaccurate.

III.   Results and discussion

III. A.   Geometries and energies of saddle points

We calculated barrier heights of saddle points 1 and 2 of reaction R1a using the M08-SO/ma-TZVP and M08-HX/ma-TZVP methods. These methods respectively yield -0.13 and 1.00 kcal mol^{-1} for the barrier of reaction R1a. The former is in good agreement with our best estimate (which is 0.05 kcal mol^{-1}, as shown in Table 1). To check the geometry effect on calculated barrier heights, we calculated the barrier height by the M08-SO/ma-TZVP//M08-HX/6-31+G(d,p) method, which gives a barrier -0.10 kcal mol^{-1}; this value with the double zeta geometry is in good agreement with the value of -0.13 kcal mol^{-1} obtained with the consistently

Table 1 Energies[a] (kcal mol⁻¹) of ethanol and saddle points at geometries optimized by M08-HX/6-31+G(d,p) method

Species	Structure	CCSD(T)		BE		$T_1{}^f$
		TZ[b]	QZ[c]	DZ[d]	maug-DZ[e]	
ethanol + OH	*trans*	0.00	0.00	0.00	0.00	0.010
	g+/g⁻ [g]	0.12	0.11			
R1a saddle point	1, 2	0.21	0.35	0.12	0.05	0.026
	3, 4	0.23	0.39			0.026
R1b saddle point	1, 2	2.16	2.29	2.03	2.02	0.026
	3, 4	2.23	2.38	2.13	2.10	0.025
	5, 6	2.51	2.68			0.025
	7, 8	2.61	2.74			0.026
	9, 10	4.70	4.83			0.026
	11, 12	5.00	5.12			0.025
	13, 14	5.52	5.63			0.026
R1c saddle point	1, 2	3.31	3.60	2.42	2.74	0.042
	3, 4	3.51	3.79			0.046
	5, 6	3.93	4.19			0.039

[a] All relative energies in this article include a spin–orbit energy −0.20 kcal mol⁻¹ for OH radical, and all energies in this table are zero-point exclusive. [b] The CCSD(T)-F12a/aug-cc-pVTZ method. [c] The CCSD(T)-F12b/may-cc-pVQZ method. [d] The cc-pVDZ basis set is used in eqn (2). [e] The maug-cc-pVDZ basis set is used in eqn (2). [f] Calculated by UCCSD/may-cc-pVQZ method. [g] These are the two *gauche* structures.

optimized triple zeta geometry optimization mentioned above. Therefore we conclude, as expected, that no significant error is incurred by using the geometries optimized by the smaller basis set. We therefore performed the rest of the geometry optimizations and all the reaction-path calcuations with the smaller basis set.

We performed an exhaustive conformational structure search for the saddle points of the three reactions by generating guessed conformations based on a set of grids of torsional angles, and we optimized these conformational structures using the M08-HX/6-31+G(d,p) method. Four distinguishable structures (2 pairs of mirror images) were found for the transition state of reaction R1a. Reaction R1b has 14 saddle points (seven pairs of mirror images), and reaction R1c has six saddle points (three pairs of mirror images). Ethanol has three conformational structures. The structures of the saddle points are shown in Fig. 2–4. Previous theoretical studies[40,42,43] only considered one saddle point for each reaction in their calculations, and torsions were modeled by a one-dimensional approximation[40,43] or approximated as harmonic oscillators.[42] Because the Cartesian coordinates are not available in previous reports,[40,42,43] we do not compare our optimized geometries with previous work, and the comparison of our calculated barrier heights with other previous work leaves some questions unanswered, especially for R1b since it has several saddle points with similar energies and conformations.

Single-point energies of all the conformational structures (ethanol, hydroxyl radical, and saddle points) optimized by the M08-HX/6-31+G(d,p) method were calculated by CCSD(T)-F12a/aug-cc-pVTZ and CCSD(T)-F12b/may-cc-pVQZ, respectively. The total energy of the all-*trans* structure of ethanol infinitely separated from hydroxyl radical is taken as the zero of energy, and the relative energies of all the conformational structures are given in Table 1. The relative energies include a −0.20 kcal mol⁻¹ spin-orbit energy for the hydroxyl radical. We corrected the barrier heights calculated by the CCSD(T)-F12b/may-cc-pVQZ method using eqn (1) and (2). By applying two basis sets, cc-pVDZ and maug-cc-pVDZ, in the FBS corrections, we checked the convergence of the high-level correction with respect to adding diffuse basis functions.

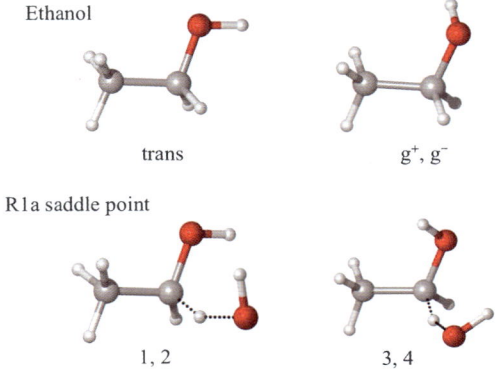

Fig. 2 Conformations of ethanol and saddle points of reaction R1a. Only one structure of each pair of mirror images is shown.

In Table 1, the CCSD(T)-F12b/may-cc-pVQZ energies are considered to be the highest quality CCSD(T)/CBS method, and they are used as the CBS result in eqn (1). As shown in Table 1, for R1a and R1b the energy differences between saddle points calculated by aug-cc-pVTZ and may-cc-pVQZ are under 0.2 kcal mol^{-1}, whereas those for R1c saddle point energies are under 0.3 kcal mol^{-1}, and the differences for reactants are 0.01 kcal mol^{-1}. We only performed the CCSDT(2)$_Q$ calculations for a few low-energy conformations of ethanol and the saddle points due to the very large computational cost. The high-level FBS correction (eqn (2)) with the larger (maug-cc-pVDZ) basis set lowers the CCSD(T) barrier heights of the three reactions by ~0.3 kcal mol^{-1} for R1a and R1b and by 0.9 kcal mol^{-1} for R1c. Furthermore, for R1c the two high-level FBS corrections differ by 0.3 kcal mol^{-1}, which indicates that a larger basis set needs to be used in eqn (2) which is not affordable with our available computational resources. The large size of the CCSDT(2)$_Q$ correction and the large T_1 diagnostic values (see Table 1) indicate that the R1c saddle points have significant multi-reference character. Our values for T_1 for the saddle points of R1c are consistent with the value of 0.044 reported by Galano et al.[95] with a different basis set.

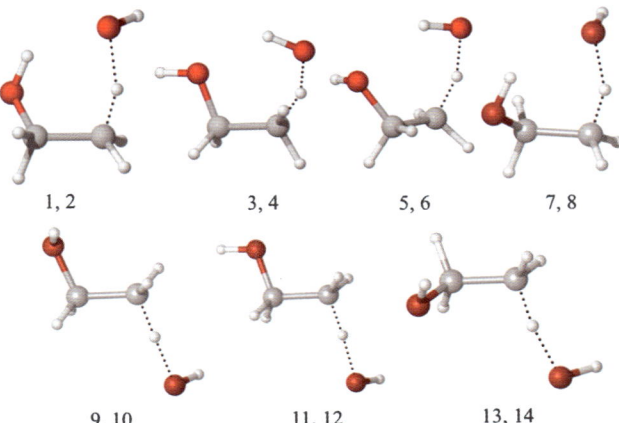

Fig. 3 Saddle points of reaction R1b. Only one structure of each pair of mirror images is shown.

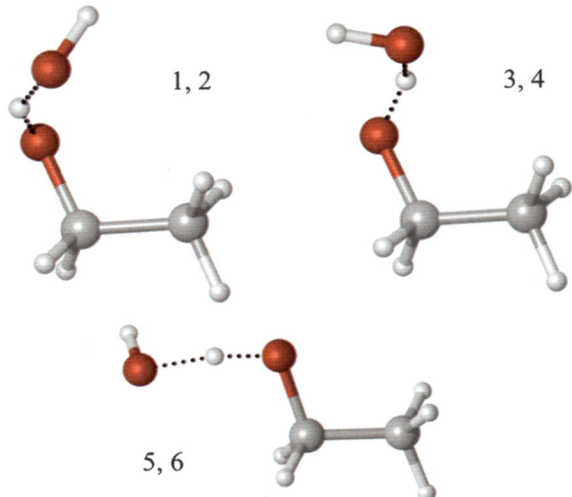

Fig. 4 Saddle points of reaction R1c. Only one structure of each pair of mirror images is shown.

According to the figures in the papers, the studies by Galano *et al.*[42] and Xu *et al.*[43] both used saddle point 1 or 2 (shown in Fig. 2.) for reaction R1a and used saddle point 5 or 6 (shown in Fig. 4) for reaction R1c. Saddle point 1 or 2 of reaction R1a is the lowest-energy saddle point of that reaction according to the CCSD(T)-F12b/may-cc-pVQZ method. However, saddle point 5 or 6 of reaction R1c is the highest-energy conformation, and it is higher than the lowest-energy saddle point of R1c by 0.6 kcal mol^{-1}. The saddle points of reaction R1b can be categorized into two groups, *i.e.*, hydrogen-bonded structures (structures 1–8 in Fig. 3) and non-hydrogen-bonded structures (structures 9–14 in Fig. 3). The structures in each group are similar in energy. In general, the non-hydrogen-bonded structures are higher than the hydrogen-bonded structures by 2–3 kcal mol^{-1}. The structures

Table 2 Calculated zero-point exclusive barrier heights^a (kcal mol^{-1}) with various density functionals using the 6-31+G(d,p) basis set

Saddle point	M08-HX	M08-SO	M06-2X	M05-2X
R1a (1, 2)	0.17	−0.87	−0.51	−0.13
R1b (3, 4)	0.31	−0.81	−0.47	−0.18
R1b (1, 2)	1.86	0.24	1.03	1.32
R1b (3, 4)	1.57	0.26	0.85	1.13
R1b (5, 6)	2.03	0.67	1.22	1.44
R1b (7, 8)	2.37	0.80	1.56	1.88
R1b (9, 10)	5.42	4.05	4.51	4.63
R1b (11, 12)	5.55	4.15	4.68	4.86
R1b(13, 14)	6.24	4.95	5.37	5.64
R1c (1, 2)	2.55	0.69	0.55	1.24
R1c (3, 4)	2.66	0.85	0.70	1.39
R1c (5, 6)	3.47	1.45	1.27	1.85

^{a} The energy of *trans*-ethanol plus hydroxyl radical is taken as zero of energy.

for R1b used in the works by Galano *et al.*[42] and Xu *et al.*[43] are one of the hydrogen-bonded structures according to the hydroxyl group orientation shown in their figures. The structures of the saddle points used in the work by Sivaramakrishnan *et al.*[40] are not clear since neither figures nor coordinates are provided in the paper. However they[40] do remark about R1b saddle points that "Our calculations find that the H-bonded saddle point lies 1.4 kcal mol^{-1} higher in energy than the geometry of Xu and Lin and also has significantly less entropy." This remark is contradictory to our findings and also is contradictory with the common expectation that an H bond usually lowers conformational energy. Even if the previous studies used the lowest-energy saddle point, they would still incur error from using only one structure[42] or from including other structures only by uncoupled one-dimensional treatments of torsions.[40,43]

Table 2 lists the calculated energies of the saddle points with various density functionals using the 6-31+G(d,p) basis set. The purpose of using such a small basis set with density functional theory is to find an efficient and affordable method for direct

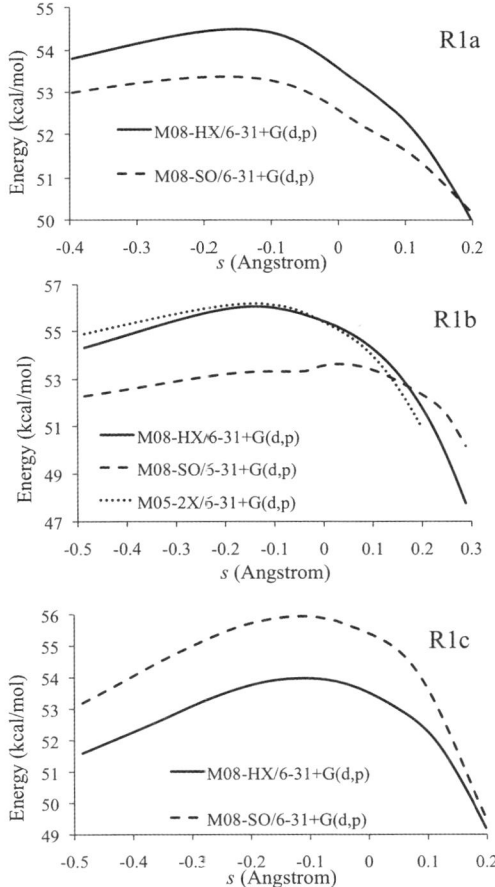

Fig. 5 The lowest-energy vibrationally adiabatic ground-state potential energy $\tilde{U}_1$ curves for each of reactions R1a–R1c. The $\tilde{U}_1$ curves of reaction R1b are calculated by using density functional integration grids that have 96 radial shells around each atom and a spherical product angular grid having 32 θ points and 64 φ points in each shell in the integrations. The $\tilde{U}_1$ curves of R1a and R1c are calculated by using density functional integration grids that are pruned from grids having 99 radial shells around each atom and 974 angular points in each shell.

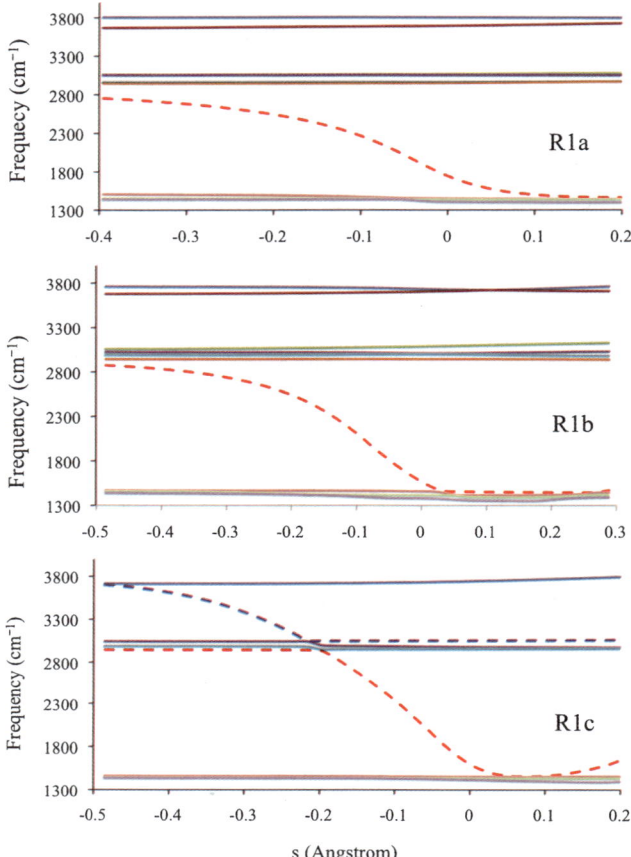

Fig. 6 The highest ten frequencies along the reaction paths of R1a–R1c calculated by the M08-HX/6-31+G(d,p) method. The frequencies are scaled by a factor of 0.972 by scaling all Hessian elements as explained in the text. These paths are the lowest-energy paths of each reaction. For the R1c reaction, two frequencies that remain almost constant between 2800 cm^{-1} and 3300 cm^{-1} are removed for better visualization.

dynamics calculations. The barrier heights by the M08-HX/6-31+G(d,p) method are in agreement with the best estimates of barriers within 0.2 kcal mol^{-1}. Therefore the M08-HX/6-31+G(d,p) method is used for straight direct dynamics calculations except where specified otherwise.

III. B. Generalized normal modes

In order to calculate variational effects, we need to calculate $G_{T,k}$ as a function of s, and in order to calculate tunneling probabilities we need to calculate $\tilde{U}_k$ as a function of s. These are calculated in the quasiharmonic approximation from generalized normal mode frequencies calculated in curvilinear coordinates[17,96] (which means that the facets of Fig. 1 are actually curved).

In our generalized normal mode analyses, a set of non-redundant internal coordinates is used, in particular, 10 bond stretches, 13 bond angles, and 4 dihedral angles (each dihedral angle corresponds to a torsion mode).

One interesting question is how many torsions there are in the saddle points of R1a–R1c. In this work, four internal rotations, namely those around the C1–O and C1–C2

Table 3 Torsional anharmonicity factors^a of ethanol and three transition states

T/K	$F^{MS,E}$	$F^{T,E}$	$F^{MS-T,E}$	$F^{MS,R1}$	$F^{T,R1}$	$F^{MS-T,R1}$	$F^{MS,R2}$	$F^{T,R2}$	$F^{MS-T,R2}$	$F^{MS,R3}$	$F^{T,R3}$	$F^{MS-T,R3}$
200	2.6	1.2	3.1	4.4	1.1	4.8	6.7	1.1	7.4	4.4	1.2	5.1
250	2.7	1.2	3.3	4.4	1.1	4.9	7.4	1.1	8.4	4.9	1.2	5.9
298	2.7	1.2	3.4	4.4	1.1	5.0	8.4	1.2	9.8	5.3	1.2	6.6
300	2.7	1.2	3.4	4.4	1.1	5.0	8.5	1.2	9.9	5.3	1.2	6.6
400	2.8	1.3	3.6	4.4	1.2	5.3	11.9	1.2	13.9	5.9	1.3	7.8
500	2.9	1.3	3.7	4.4	1.3	5.7	16.4	1.1	18.2	6.3	1.4	8.7
600	2.9	1.3	3.7	4.4	1.3	6.0	21.3	1.0	22.2	6.7	1.4	9.4
700	2.9	1.3	3.7	4.5	1.4	6.2	26.4	1.0	25.6	6.9	1.4	10.0
800	2.9	1.2	3.6	4.5	1.5	6.5	31.5	0.9	28.3	7.1	1.5	10.4
900	3.0	1.2	3.6	4.5	1.5	6.7	36.4	0.8	30.4	7.3	1.5	10.7
1000	3.0	1.2	3.5	4.5	1.5	6.9	41.0	0.8	31.9	7.4	1.5	10.9
1500	3.0	1.0	3.0	4.5	1.7	7.4	60.3	0.6	34.0	7.9	1.4	10.8
1800	3.0	0.9	2.8	4.5	1.7	7.5	69.0	0.5	32.8	8.1	1.3	10.3
2000	3.0	0.9	2.7	4.5	1.6	7.4	74.0	0.4	31.7	8.1	1.2	9.9
2400	3.0	0.8	2.4	4.5	1.6	7.1	82.2	0.4	29.0	8.3	1.1	9.0

^{a} Geometries and Hessians are calculated by the M08-HX/6-31+G(d,p) method and energies are calculated by the CCSD(T)-F12b/may-cc-pVQZ method.

Table 4 Reaction torsional anharmonicity factors calculated by eqn (10)

T/K	R1a	R1b	R1c
200	1.6	2.4	1.7
250	1.5	2.6	1.8
298	1.5	2.9	1.9
300	1.5	2.9	1.9
400	1.5	3.9	2.2
500	1.5	5.0	2.4
600	1.6	6.0	2.6
700	1.7	7.0	2.7
800	1.8	7.8	2.9
900	1.9	8.6	3.0
1000	2.0	9.2	3.1
1500	2.4	11.1	3.5
1800	2.7	11.7	3.7
2000	2.8	11.9	3.7
2400	3.0	12.1	3.8

bonds of ethanol and those around the C···H (or O···H in R1c) and O···H partial bonds, are treated with torsional anharmonicity. The bond angle between the two partial bonds involving the transferring H atom is not linear; the C···H···O bond angles of saddle points of R1a and R1b are in the range of 152 deg to 172 deg, and the O···H··· O bond angles of saddle points of R1c are 144° and 145°. A linear bend would correspond to two degenerate or nearly degenerate vibrational modes and one less torsion, but no such degenerate or nearly degenerate modes are observed in the normal mode analyses. For example, the C···H···O bending motion in the structure in which this angle is 172 deg is mostly distributed over two normal modes, in each of which it is mixed with other motions, and these modes have frequencies of 1161 and 984 cm^{-1}.

The three reactions considered here, especially R1a, have quite low barriers so that the potential energy curves in the vicinity of saddle points change quite slowly, and the locations of the maxima of the vibrationally adiabatic ground state potential curves or of the low-temperature generalized free energy of activation profiles depend sensitively on converging the frequency calculations. Therefore, to calculate a converged vibrationally adiabatic ground-state potential energy curve, it is essential to use a sufficiently fine grid for integration in the density functional calculations. The vibrationally adiabatic ground-state potential curves for the lowest-energy saddle points ($k = 1$) of the three reactions are shown in Fig. 5, as calculated with the grids specified in subsection II. A. These curves show that the maxima are located at a significant distance away from the conventional transition state ($s = 0$) in all three reactions. For example, the maximum of the $\tilde{U}_1$ curve for R1a is at $s = -0.14$ Å and is higher than the conventional transition state by 0.9 kcal mol^{-1}. The variational transition states of R1a are located between -0.18 Å and -0.15 Å for the temperatures 200–2400 K. Fig. 6 shows the ten largest frequencies calculated by the M08-HX/6-31+G(d,p) method along the lowest-energy reaction paths of reaction R1a–R1c. One frequency (dashed line in red in Fig. 6) changes dramatically in the vicinity of saddle point, in particular, this frequency decreases by 800 cm^{-1} from $s = -0.2$ Å to $s = 0$. This frequency is the vibration of the breaking bond that is turning into the vibration of the making bond.

III. C. Torsional anharmonicity

The MS-T method is applied to ethanol and the transition states of the three reactions to account the torsional anharmonicity. In the MS-T calculations, the geometries and Hessians are calculated by the M08-HX/6-31+G(d,p) method and energies are calculated by the CCSD(T)-F12b/may-cc-pVQZ method. Table 3 lists torsional anharmonicity factors calculated by eqn (6), (8), and (9) for ethanol and the three transition states. Table 4 lists the reaction torsional anharmonicity factors calculated by eqn (10) for the three reactions.

The torsional anharmonicity factors F^{MS-T} for rate calculations are based on M08-HX/6-31+G(d,p) geometries and Hessians even though κ_k and Γ_k are calculated by using other potential energy surfaces in some cases. Note that Hessians obtained by these two methods are scaled by their scaling factors that were optimized[36] for obtaining accurate zero-point energy. We compared the F^{MS-T} factors obtained by using the CCSD(T)-F12a/may-cc-pVQZ//M08-HX/6-31+G(d,p) to those obtained by using the CCSD(T)-F12a/may-cc-pVQZ//M05-2X/6-31+G(d,p) potential energy surfaces and found that they differ by less than or equal to 12%, which is similar to or smaller than the uncertainties of the dynamics methods. Therefore we conclude that it is an acceptable approximation to use slightly different potential energy surfaces for anharmonicity and for dynamics calculations.

In our previous study,[64] we calculated partition functions of ethanol by using the one-dimensional (1-D) torsional eigenvalue summation (TES) method for its torsional modes. In ethanol, the two internal rotations are nearly separable in internal coordinates (note that the normal modes of the two torsions are mixtures of the two torsions), therefore the 1-D TES method in internal coordinates is applicable to ethanol. The partition functions calculated by the MS-T method in this work and the TES method used in previous work[30] differ by about 22% at 200 K, 7% at 1000 K, and 3% at 2000 K. These differences are acceptable for treatment of the torsions.

The internal rotations of the saddle points are strongly coupled together except for the methyl group rotation of the R1a and R1c saddle points. The local periodicities of the strongly coupled torsions are calculated by Voronoi tessellation.[64,65] Fig. 7 shows a contour plot of the two-dimensional torsional potential energy surface of the R1a transition state. The two dimensions are the H–O–C1–C2 dihedral angle and H–O–H–C1 dihedral angle. Potential energies are calculated by the M08-HX/6-31+G(d,p) method. The other geometrical parameters are fixed at saddle point 1 of this reaction. There are two minima on the potential energy surface. Note that C1 is a chiral center, and the mirror images of these two minima cannot be generated

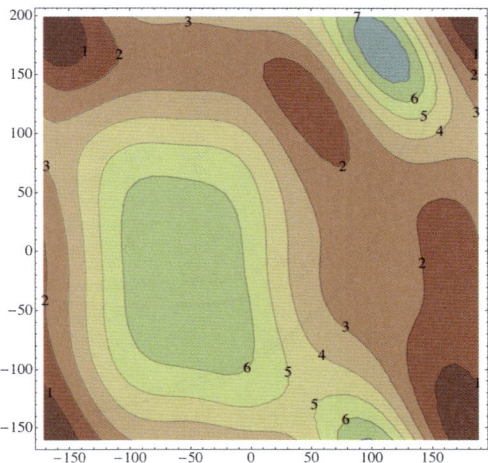

Fig. 7 Two-dimensional contour plot of the torsional potential energy surface of the R1a transition state. The abscissa is the H–O–C1–C2 dihedral angle (in degrees) and the ordinate is the H–O–H–C1 dihedral angle (in degrees). The potential energy (in kcal mol^{-1}) is calculated by the M08-HX/6-31+G(d,p) method. The other geometrical parameters are fixed at their values for saddle point 1.

by internal rotations. Fig. 7 shows that two separate 1-D rigid scans cannot yield the correct number of minima. Even if a relaxed scan can follow the minimum-energy path between the two minima, two relaxed scans will give the same information, and the partition function will be overestimated. Therefore the transition state of R1a gives a clear example of the inapplicability of a one-dimensional separable approximation. Nevertheless this approximation is widely used in the literature and has also been used for this reaction.[40]

If we let LH-SS-CVT denote the single-structure results in the quasiharmonic approximation, then the factors in Table 4 give the ratio of MS-CVT to LH-SS-CVT, and they also give the ratio of MS-CVT/SCT to LH-SS-CVT/SCT. These ratios range from 1.5 to more than an order of magnitude (12.1) with a median value of 2.7. Furthermore the factor is significantly different for each reaction and therefore has an important effect on product ratios. Thus multi-structural torsional anharmonicity is a significant factor that should not be neglected even for this small-molecule reaction.

III. D. MS-VTST reaction rate

The rate constants for the three reactions are calculated by using M08-HX/6-31+G(d,p) and M08-SO/6-31+G(d,p) potential energy surfaces, respectively. The calculated total rate constants of the three reactions are plotted in Fig. 8 together with experimental data. We also calculated the branching fraction of the three reactions and plotted them in Fig. 9. The calculated rate constants are fitted to a physically motivated four-parameter expression. This expression and its corresponding activation energy are

$$k = A\left(\frac{T + T_0}{300}\right)^n \exp\left(-\frac{E(T + T_0)}{R(T^2 + T_0^2)}\right) \tag{41}$$

$$E_a = E\frac{T^4 + 2T_0T^3 - T_0^2T^2}{(T^2 + T_0^2)^2} + nR\frac{T^2}{T + T_0} \tag{42}$$

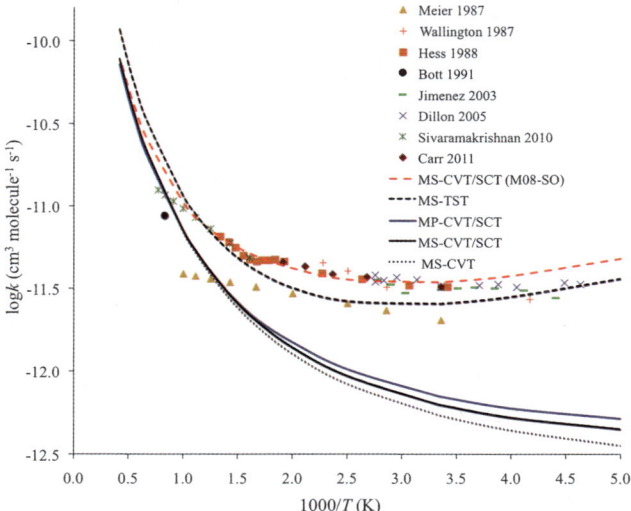

Fig. 8 Total reaction rates of R1a–R1c. The rate constants shown as black or blue solid, dotted, and dashed lines are calculated using M08-HX/6-31+G(d,p) potential energy surfaces. The rate constants shown as a red dashed line are calculated using M08-SO/6-31+G(d,p) potential energy surfaces.

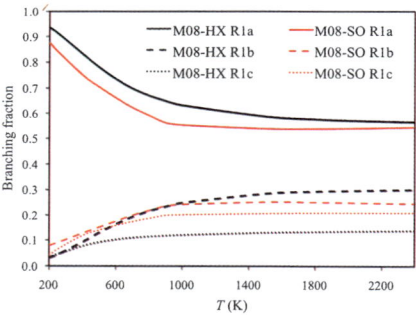

Fig. 9 Calculated branching fractions by M08-HX/6-31+G(d,p) (in black) and M08-SO/6-31+G(d,p) (in red) methods.

where A, n, E, and T_0 are fitting parameters. Eqn (41) is very similar to the eqn (8) in our previous work,[97] but it is more physically meaningful for an exoergic or ergoneutral reaction because it leads to a finite rate constant when T approaches 0 K, as it should in those cases.[98] Furthermore the activation energy becomes 0 at $T = 0$ K. The rate constants obtained by using the M08-HX/6-31+G(d,p) potential energy surfaces are

$$k_{1a} = 7.033 \times 10^{-14} \left(\frac{T + 664.9}{300} \right)^{2.931} \exp \left[-\frac{1.470(T + 664.9)}{R(T^2 + 4.421 \times 10^5)} \right] \qquad (43)$$

$$k_{1b} = 9.273 \times 10^{-14} \left(\frac{T + 299.5}{300} \right)^{2.702} \exp \left[-\frac{1.682(T + 299.5)}{R(T^2 + 8.969 \times 10^4)} \right] \qquad (44)$$

$$k_{1c} = 1.278 \times 10^{-14} \left(\frac{T + 205.5}{300} \right)^{3.160} \exp\left[-\frac{0.3278(T + 205.5)}{R(T^2 + 4.224 \times 10^4)} \right] \quad (45)$$

All rate constants are in units of cm³ molecule⁻¹ s⁻¹. The rate constants calculated by using the M08-SO/6-31+G(d,p) potential energy surfaces are

$$k_{1a} = 3.996 \times 10^{-10} \left(\frac{T + 1035}{300} \right)^{-0.07942} \exp\left[-\frac{8.209(T + 1035)}{R(T^2 + 1.072 \times 10^6)} \right] \quad (46)$$

$$k_{1b} = 2.910 \times 10^{-12} \left(\frac{T + 650.3}{300} \right)^{1.185} \exp\left[-\frac{3.632(T + 650.3)}{R(T^2 + 4.228 \times 10^5)} \right] \quad (47)$$

$$k_{1c} = 1.576 \times 10^{-14} \left(\frac{T + 575.4}{300} \right)^{3.075} \exp\left[-\frac{0.3118(T + 575.4)}{R(T^2 + 3.311 \times 10^5)} \right] \quad (48)$$

The M08-HX/6-31+G(d,p) potential energy surfaces are chosen for rate calculations because—among the small-basis-set calculations—they give barrier heights closest to our best estimates (see Tables 1 and 2). The MS-TST rate constants shown in Fig. 6 agree very well with the experimental data. The MS-TST rate constants include the multi-structural torsional anharmonicity, but they locate the reaction bottlenecks at the conventional transition states that are often not the true dynamical bottlenecks along the minimum energy paths, and also they do not include the tunneling contributions. The MS-CVT/SCT rate constants are much lower than the MS-TST rate constants, and they are very close to the MS-CVT rate constants. Therefore the difference between the MS-CVT/SCT and MS-TST rate constants is mainly caused by the variational effect. The combination of a dramatic increase of one frequency and slow changes of potential energies in the vicinity of saddle point causes significant variational effects, e.g., for reaction R1a, CVT rates are lower than TST rates by a factor 10, 5, and 2 at $T = 200$, 300, and 2400 K, respectively. The rate constants calculated by the M08-SO/6-31+G(d,p) potential energy surface have some uncertainties due to the sensitivity of the frequencies to grids, as discussed in Section III B. Using a sufficiently large grid in density functional frequency calculations is especially important for large systems because some small uncertainty in each vibrational frequency can lead to large accumulated errors in large systems that involve a large number of vibrational modes.

Tunneling contributions are not large for these hydrogen-transfer reactions because they all have quite low barrier heights, especially for reaction R1a. Note that the zero-point inclusive barrier of reaction R1a at the conventional transition state (saddle point) is lower than the reactant zero-point inclusive energy. If only the ground-state reaction and the low-pressure plateau (see subsection II. D) were considered, there would be no tunneling because it is reasonable to assume that reactants are not stabilized to a weakly bound reactant complex well at low or medium pressure (e.g., a few hundred torr[35,39] or lower[40] for most experiments that have been conducted). However, the variational transition states of reaction R1a are higher in energy than the reactant by about 0.5 kcal mol⁻¹; therefore a small amount of tunneling is obtained by the SCT method, e.g., tunneling transmission coefficients are 1.2 at 200 K and 1.1 at 300 K, respectively. We note that if the reactant complex were equilibrated with reactants, the energy levels that are below the zero-point energy of the reactants would also need to be considered in the tunneling calculations, which would lead to larger tunneling contributions. This is a good place to insert a caution about the ground-state tunneling approximation that has usually been used for the transmission coefficient in past work and that is also used here. The ground-state tunneling approximation assumes that the ground-state transmission coefficient is typical of all the transmission coefficients that make a significant contribution to the reaction rate at temperatures low enough for tunneling to be significant.[15,90] However, for a case like the present one, this cuts off tunneling due to the energetic threshold more severely than would be the case for higher-energy states, and so actually one of those higher-energy states might be more typical. This would not make a significant difference for reactions with high barriers, but it is less valid for reactions with low barriers (small positive barrier or negative barrier), such as the present one, for which using a higher-energy state to compute a representative transmission coefficient would increase the calculated rate constant a little at low temperature. We calculated the tunneling with the low energies present and found that the MS-CVT/SCT rate constant for R1a would be increased by factors of 1.8 and 1.2 at 298 and 500 K, respectively.

The MS-CVT/SCT rate constants by the M08-HX/6-31+G(d,p) method are much lower than the experimental ones, although the barrier heights calculated by the M08-HX/6-31+G(d,p) method agree very well with our best estimated values, and important effects are taken into account in the dynamics calculations, in particular, multi-structural torsional anharmonicity, the variational effect, and multi-dimensional tunneling. What could cause this discrepancy between theory and experiment? One possible cause is that our best estimates of barrier heights by coupled cluster theory may be too high. Coupled cluster theory at the CCSD(T) level is usually considered to have chemical accuracy, better than 1 kcal mol^{-1}. In the present case, the FBS correction using the CCSDT(2)$_Q$ method lowers the barrier heights calculated by the CCSD(T) method, but the best estimates of the barriers may be still higher than the accurate ones. Therefore we performed rate calculations using the M08-SO/6-31+G(d,p) method, which gives 1 kcal mol^{-1} lower barrier than the M08-HX/6-31+G(d,p) method. As shown in Fig. 8, the rate constants obtained by the M08-SO/6-31+G(d,p) method agree with experimental data very well. We conclude that the true barrier heights are between the values calculated by the M08-HX/6-31+G(d,p) and M08-SO/6-31+G(d,p) methods.

Some other reasons for the discrepancy between theory and experiment could be the presence of some stabilized pre-reaction complexes or imperfection of the dynamics and the statistical methods. For example, vibrational modes except torsional modes are still treated using a quasiharmonic approximation (harmonic oscillator formulas with frequencies scaled to account for anharmonicity), torsional barrier heights are obtained from local periodicities rather than calculated directly, mode–mode coupling is not fully taken into account for nontorsional modes, and tunneling contributions could be underestimated or overestimated by the SCT method. Furthermore we assume that the torsional anharmonicity factor remains the same at the variational transition state as at the conventional transition state. The last-named issue is particularly troublesome in the present cases because we observe a large variational effect for these reactions that is sensitive to the convergence of the frequencies, but the reaction-path independence of the torsional anharmonicity factor has not been tested.

III. E.   MP-VTST reaction rate

In the MP-VTST calculation of each reaction, the M08-HX/6-31+G(d,p) potential energy surfaces are used and the recrossing and tunneling transmission coefficients of all reaction paths are calculated explicitly. For comparison purposes, we also calculate an averaged generalized transmission coefficient $\langle \gamma \rangle_P$ using eqn (25) by considering some low-energy reaction paths for R1b. The calculated MP-CVT/SCT rate constants for the three reactions respectively are

$$k_{1a} = 8.265 \times 10^{-14} \left(\frac{T + 663.9}{300} \right)^{2.869} \exp\left(-\frac{1.396(T + 663.9)}{(T^2 + 4.407 \times 10^5)} \right) \qquad (49)$$

$$k_{1b} = 6.979 \times 10^{-14} \left(\frac{T + 327.6}{300} \right)^{2.694} \exp\left(-\frac{1.890(T + 327.6)}{(T^2 + 1.073 \times 10^5)} \right) \qquad (50)$$

$$k_{1c} = 1.285 \times 10^{-14} \left(\frac{T + 292.6}{300} \right)^{3.123} \exp\left(-\frac{0.6030(T + 292.6)}{(T^2 + 8.562 \times 10^4)} \right) \qquad (51)$$

The total MP-CVT/SCT rate constants of the three reactions are also plotted in Fig. 8. The difference between MP-CVT/SCT rates and MS-CVT/SCT rates are not noticeable at high temperatures, and they are about 15% at $T = 200$ K.

Fig. 10 shows the recrossing transmission coefficients Γ_k, tunneling transmission coefficients κ_k, and generalized transmission coefficients γ_k or $\langle \gamma \rangle$ as calculated by the M08-HX/6-31+G(d,p) method.

Reaction R1a has two pairs of saddle points that are close in energy with each other. We find that $\langle \gamma \rangle$ and γ_1 are very similar, and therefore a single reaction path can represent the whole reaction very well for this reaction.

The saddle points are more diverse in energy and in geometry for reaction R1b than for R1a and R1c. The corresponding reaction paths have quite different transmission coefficients as shown in Fig. 10. We calculated the generalized transmission coefficient γ_1 (corresponding to the reaction path corresponding the lowest-energy saddle point), $\langle \gamma \rangle_8$ (averaged over the eight reaction paths corresponding the eight

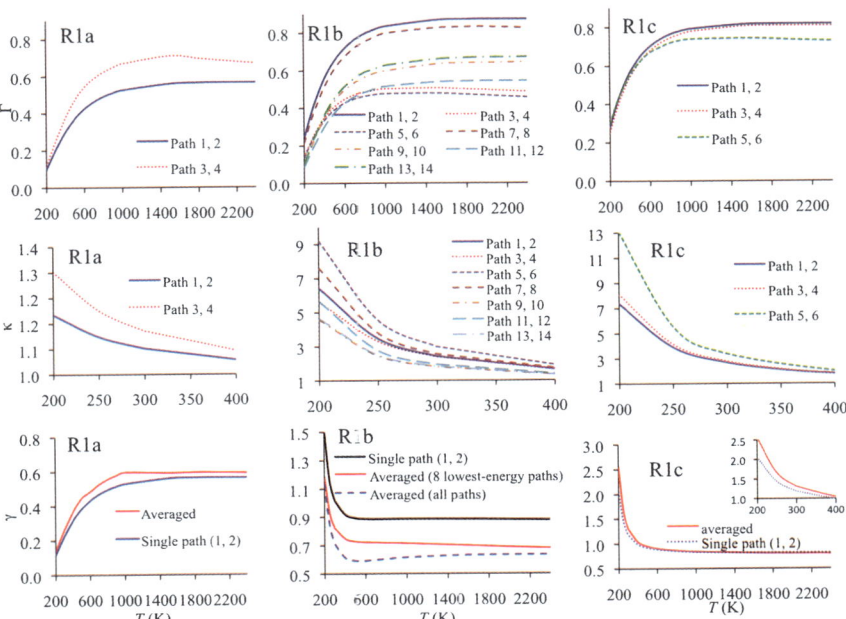

Fig. 10 Recrossing transmission coefficient Γ_k, tunneling transmission coefficient κ_k, and generalized transmission coefficient γ_k or $\langle \gamma \rangle$ of each reaction that are calculated by the M08-HX/6-31+G(d,p) method.

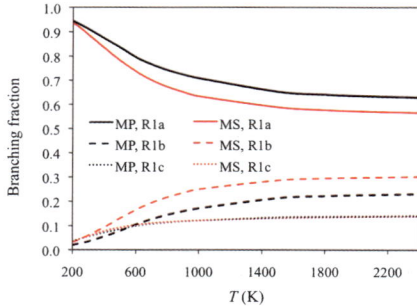

Fig. 11 Calculated branching fractions by the MP-VTST method (in black) and the MS-VTST method (in red). In the MP-VTST calculations, transmission coefficients of all reaction paths are explicitly calculated and the M08-HX/6-31+G(d,p) potential energy surfaces are used. Notice that the two sets of results for R1c are not distinguishable on the plot because they are in excellent agreement.

lowest-energy saddle points), and $\langle\gamma\rangle$ (averaged over all 14 reaction paths). At the low temperature of 200 K, $\langle\gamma\rangle_8$ and $\langle\gamma\rangle$ are very close because the sum of the weights of the eight lowest-energy paths is 94%, and the reaction is dominated by the reactive flux through these eight saddle points. The difference between $\langle\gamma\rangle_8$ and $\langle\gamma\rangle$ becomes larger at medium or high temperatures because the weight of those six highest-energy paths increase rapidly when temperature increases. For instance, the weight of the six highest-energy paths is 26% at 300 K, but it is 60% at 600 K. If we take the full path-averaged MP-VTST rate constants ($T = 200$–2400 K) as a benchmark for reaction R1b, the MS-VTST (based on γ_1 calculated using the lowest-energy path) rate constants have errors between 34% and 51%, and the MP-VTST rate constants using $\langle\gamma\rangle_8$ have errors between 4% and 22%.

The generalized transmission coefficients of the lowest-energy path of reaction R1c are very close to the averaged ones except at low temperatures. The difference between γ_1 and $\langle\gamma\rangle$ is about 20% to 13% in the range of temperature from 200 to 250 K because the two highest-energy paths have relatively large tunneling transmission coefficients, and the sum of the weights of these saddle points is 28%.

We plot the calculated branching fraction by the MS-VTST and the MP-VTST methods in Fig. 11. The R1a and R1b branching fractions obtained by the MS-VTST and the MP-VTST methods, respectively, differ by about 10%. The branching fractions for R1c are almost the same by the two methods. The branching fractions for R1a and R1b are in good agreement with a recent experiment[41] in both magnitude and temperature dependence over the entire 300–600 K temperature range of the measurement. The branching fraction for R1c has not been measured.[41]

IV. Concluding remarks

At this point we can recognize a hierarchy in the way that multiple structures and variational transition states are treated in various formulations of transition state theory. The most recent formulations, MS-VTST and MP-VTST (the latter formulated previously for unimolecular reactions and here for bimolecular ones), occupy an intermediate position between single-structure variational transition state theory[15,79,99], (called VTST, or, to emphasize the distinction, SS-VTST), which has been used very successfully for small molecules, and ensemble-averaged variational transition state theory[100–102] (EA-VTST), which has been used successfully for enzyme kinetics. EA-VTST is a multi-path method that could also be used for nonenzymatic reactions in solution, which can also be treated by single-path methods based on a potential of mean force.[19,25]

Based on the reactions studied in this work, the MS-VTST method that uses a transmission coefficient of a single reaction path to represent all reaction paths is sometimes a good approximation to the MP-VTST method in which the transmission coefficients of all reaction paths or of the most important paths are calculated explicitly. However, when a reaction has a set of saddle points that have diverse energies and geometries, the MP-VTST method with transmission coefficients averaged (with appropriate Boltzmann weighting) over all reaction paths or over the lowest-energy paths provides a better approach. More general conclusions on choosing the MS-VTST or the MP-VTST and on the strategies of choosing reaction paths will emerge as more diverse types of reactions are studied.

In general, one should be careful in viewing the agreement between experiment and theory in reaction rate calculations. The present study shows an example where agreement between theory and experimental data can be obtained by low-level calculations, while state-of-the-art theoretical methods lead to discrepancies. These discrepancies show that further development of both electronic structure theory and dynamics methods are needed to make reaction rate calculations more predictive.

Acknowledgements

The authors are grateful to Rubén Meana-Pañeda, Antonio Fernández-Ramos, and John Alecu for constructive comments on a draft of the manuscript. This work was supported by the U. S. Department of Energy, Office of Basic Energy Sciences under grant no. DE-FG02-86ER13579 and by the Combustion Energy Frontier Research Center under award no. DE-SC0001198. Some of the computations were performed as part of a Computational Grand Challenge grant at the Molecular Science Computing Facility (MSCF) in the William R. Wiley Environmental Molecular Sciences Laboratory, a national scientific user facility sponsored by the U.S. Department of Energy's Office of Biological and Environmental Research and located at the Pacific Northwest National Laboratory, operated for the Department of Energy by Battelle.

References

1 J. Zheng, Y. Zhao and D. G. Truhlar, *J. Chem. Theory Comput.*, 2009, **5**, 808.
2 X. Xu, I. M. Alecu and D. G. Truhlar, *J. Chem. Theory Comput.*, 2011, **7**, 1667.
3 I. M. Alecu and D. G. Truhlar, *J. Phys. Chem. A*, 2011, **115**, 2811.
4 E. Wigner, *Trans. Faraday Soc.*, 1938, **34**, 29.
5 H. Eyring, *Trans. Faraday Soc.*, 1938, **34**, 41.
6 M. G. Evans, *Trans. Faraday Soc.*, 1938, **34**, 49.
7 M. Polanyi, *Trans. Faraday Soc.*, 1938, **34**, 75.
8 K. J. Laidler and M. J. King, *J. Phys. Chem.*, 1983, **87**, 2657.
9 D. G. Truhlar, W. L. Hase and J. T. Hynes, *J. Phys. Chem.*, 1983, **87**, 2664; D. G. Truhlar, W. L. Hase and J. T. Hynes, *J. Phys. Chem.*, 1983, **87**, 5523(E).
10 D. G. Truhlar, B. C. Garrett and S. J. Klippenstein, *J. Phys. Chem.*, 1996, **100**, 12771.
11 D. G. Truhlar and B. C. Garrett, *Annu. Rev. Phys. Chem.*, 1984, **35**, 159.
12 T. C. Allison and D. G. Truhlar, in *Modern Methods for Multidimensional Dynamics Computations in Chemistry*, ed. D. L. Thompson, World Scientific, Singapore, 1998, p. 618.
13 D. G. Truhlar, in *Isotope Effects in Chemistry and Biology*, edited by A. Kohen and H.-H. Limbach, Marcel Dekker, Inc., New York, 2006, p. 579.
14 B. C. Garrett and D. G. Truhlar, *J. Chem. Phys.*, 1979, **70**, 1593.
15 D. G. Truhlar and B. C. Garrett, *Acc. Chem. Res.*, 1980, **13**, 440.
16 D. G. Truhlar, A. D. Isaacson, R. T. Skodje and B. C. Garrett, *J. Phys. Chem.*, 1982, **86**, 2252; D. G. Truhlar, A. D. Isaacson, R. T. Skodje and B. C. Garrett, *J. Phys. Chem.*, 1983, **87**, 4554(E).
17 C. F. Jackels, Z. Gu and D. G. Truhlar, *J. Chem. Phys.*, 1995, **102**, 3188.
18 J. Villà and D. G. Truhlar, *Theor. Chem. Acc.*, 1997, **97**, 317.
19 Y.-Y. Chuang, C. J. Cramer and D. G. Truhlar, *Int. J. Quantum Chem.*, 1998, **70**, 887.
20 Y. Georgievskii and S. J. Klippenstein, *J. Chem. Phys.*, 2003, **118**, 5442.

21 Y. Georgievskii and S. J. Klippenstein, *J. Phys. Chem.*, 2003, **107**, 9776.
22 A. Fernández-Ramos, J. A. Miller, S. J. Klippenstein and D. G. Truhlar, *Chem. Rev.*, 2006, **106**, 4518.
23 J. Zheng, S. Zhang and D. G. Truhlar, *J. Phys. Chem. A*, 2008, **112**, 11509.
24 C. Alhambra, J. Corchado, M. L. Sánchez, M. Garcia-Viloca, J. Gao and D. G. Truhlar, *J. Phys. Chem. B*, 2001, **105**, 11326.
25 Y. Kim, J. R. Mohrig and D. G. Truhlar, *J. Am. Chem. Soc.*, 2010, **131**, 11071.
26 K. J. Laidler, *Chemical Kinetics*, 2nd ed., McGraw-Hill, New York, 1965, p. 72.
27 T. Yu, J. Zheng and D. G. Truhlar, *Chem. Sci.*, 2011, **2**, 2199.
28 T. Yu, J. Zheng and D. G. Truhlar, *J. Phys. Chem. A*, 2012, **116**, 297.
29 P. Seal, E. Papajak and D. G. Truhlar, *J. Phys. Chem. Lett.*, 2012, **3**, 264.
30 I. M. Alecu and D. G. Truhlar, *J. Phys. Chem. A*, 2011, **115**, 14599.
31 J. Zheng, R. J. Rocha, M. Pelegrini, L. F. A. Ferrão, E. F. V. Carvalho, O. Roberto-Neto, F. B. C. Machado and D. G. Truhlar, *J. Chem. Phys.*, 2012, **136**, 184310.
32 U. Meier, H. H. Grotheer, G. Riekert and T. Just, *Chem. Phys. Lett.*, 1985, **115**, 221.
33 U. Meier, H. H. Grotheer, G. Riekert and T. Just, *Chem. Phys. Lett.*, 1987, **133**, 162.
34 T. J. Wallington and M. J. Kurylo, *Int. J. Chem. Kinet.*, 1987, **19**, 1015.
35 W. P. Hess and F. P. Tully, *Chem. Phys. Lett.*, 1988, **152**, 183.
36 J. F. Bott and N. Cohen, *Int. J. Chem. Kinet.*, 1991, **23**, 1075.
37 E. Jimenez, M. K. Gilles and A. R. Ravishankara, *J. Photochem. Photobiol., A*, 2003, **157**, 237.
38 T. J. Dillon, D. Holscher, V. Sivakumaran, A. Horowitz and J. N. Crowley, *Phys. Chem. Chem. Phys.*, 2005, **7**, 349.
39 S. A. Carr, M. T. Baeza-Romero, M. A. Blitz, B. J. S. Price and P. W. Seakins, *Int. J. Chem. Kinet.*, 2008, **40**, 504.
40 R. Sivaramakrishnan, M. C. Su, J. V. Michael, S. J. Klippenstein, L. B. Harding and B. Ruscic, *J. Phys. Chem. A*, 2010, **114**, 9425.
41 S. A. Carr, M. A. Blitz and P. W. Seakins, *J. Phys. Chem. A*, 2011, **115**, 3335.
42 A. Galano, J. R. Alvarez-Idaboy, G. Bravo-Perez and M. E. Ruiz-Santoyo, *Phys. Chem. Chem. Phys.*, 2002, **4**, 4648.
43 S. Xu and M. C. Lin, *Proc. Combust. Inst.*, 2007, **31**, 159.
44 R. Atkinson, D. L. Baulch, R. A. Cox, J. N. Crowley, R. F. Hampson, R. G. Hynes, M. E. Jenkin, M. J. Rossi, J. Troe and IUPAC Subcommittee, *Atmos. Chem. Phys.*, 2006, **6**, 3625.
45 Y. Zhao and D. G. Truhlar, *J. Chem. Theory Comput.*, 2008, **4**, 1849.
46 Y. Zhao, N. E. Schultz and D. G. Truhlar, *J. Chem. Phys.*, 2005, **123**, 161103.
47 Y. Zhao and D. G. Truhlar, *Theor. Chem. Acc.*, 2008, **120**, 215.
48 J. Zheng, X. Xu and D. G. Truhlar, *Theor. Chem. Acc.*, 2011, **128**, 295.
49 K. Raghavachari, G. W. Trucks, J. A. Pople and M. Head-Gordon, *Chem. Phys. Lett.*, 1989, **157**, 479.
50 T. B. Adler, G. Knizia and H.-J. Werner, *J. Chem. Phys.*, 2007, **127**, 221106.
51 G. Knizia, T. B. Adler and H.-J. Werner, *J. Chem. Phys.*, 2009, **130**, 054104.
52 E. Papajak and D. G. Truhlar, *J. Chem. Theory Comput.*, 2011, **7**, 10.
53 S. Hirata, P.-D. Fan, A. A. Auer, M. Nooijen and P. Piecuch, *J. Chem. Phys.*, 2004, **121**, 12197.
54 E. Papajak, H. Leverentz, J. Zheng and G. T. Donald, *J. Chem. Theory Comput.*, 2009, **5**, 1197.
55 T. J. Lee and P. R. Taylor, *Int. J. Quantum Chem. Symp.*, 1989, **23**, 199.
56 J. C. Rienstra-Kiracofe, W. D. Allen and H. F. Schaefer III, *J. Phys. Chem. A*, 2000, **104**, 9823.
57 J. Peiró-García, I. Nebot-Gil and M. Merchán, *ChemPhysChem*, 2003, **4**, 366.
58 N. Lambert, N. Kaltsoyannis, S. D. Price, J. Žabka and Z. Herman, *J. Phys. Chem. A*, 2006, **110**, 2898.
59 I. M. Alecu, J. Zheng, Y. Zhao and D. G. Truhlar, *J. Chem. Theory Comput.*, 2010, **6**, 2872.
60 M. J. Frisch, G. W. Trucks, H. B. Schlegel, G. E. Scuseria, M. A. Robb, J. R. Cheeseman, G. Scalmani, V. Barone, B. Mennucci, G. A. Petersson, H. Nakatsuji, M. Caricato, X. Li, H. P. Hratchian, A. F. Izmaylov, J. Bloino, G. Zheng, J. L. Sonnenberg, M. Hada, M. Ehara, K. Toyota, R. Fukuda, J. Hasegawa, M. Ishida, T. Nakajima, Y. Honda, O. Kitao, H. Nakai, T. Vreven, J. A. Montgomery, Jr., J. E. Peralta, F. Ogliaro, M. Bearpark, J. J. Heyd, E. Brothers, K. N. Kudin, V. N. Staroverov, R. Kobayashi, J. Normand, K. Raghavachari, A. Rendell, J. C. Burant, S. S. Iyengar, J. Tomasi, M. Cossi, N. Rega, J. M. Millam, M. Klene, J. E. Knox, J. B. Cross, V. Bakken, C. Adamo, J. Jaramillo, R. Gomperts, R. E. Stratmann, O. Yazyev, A. J. Austin, R. Cammi, C. Pomelli, J. Ochterski, R. L. Martin, K. Morokuma, V. G. Zakrzewski,

G. A. Voth, P. Salvador, J. J. Dannenberg, S. Dapprich, A. D. Daniels, O. Farkas, J. B. Foresman, J. V. Ortiz, J. Cioslowski and D. J. Fox, *GAUSSIAN 09 (Revision A.2)*, Gaussian, Inc., Wallingford, CT, 2009.

61 Y. Zhao, R. Peverati, K. Yang and D. G. Truhlar, *MN-GFM, version 5.1*, University of Minnesota, Minneapolis, 2011.

62 H.-J. Werner, P. J. Knowles, F. R. Manby, M. Schütz, P. Celani, G. Knizia, T. Korona, R. Lindh, A. Mitrushenkov, G. Rauhut, T. B. Adler, R. D. Amos, A. Bernhardsson, A. Berning, D. L. Cooper, M. J. O. Deegan, A. J. Dobbyn, F. Eckert, E. Goll, C. Hampel, A. Hesselmann, G. Hetzer, T. Hrenar, G. Jansen, C. Köppl, Y. Liu, A. W. Lloyd, R. A. Mata, A. J. May, S. J. McNicholas, W. Meyer, M. E. Mura, A. Nicklaß, P. Palmieri, K. Pflüger, R. Pitzer, M. Reiher, T. Shiozaki, H. Stoll, A. J. Stone, R. Tarroni, T. Thorsteinsson, M. Wang and A. Wolf, *MOLPRO*, University of Birmingham, Birmingham, *version 2010.1*, 2010.

63 M. Valiev, E. J. Bylaska, N. Govind, K. Kowalski, T. P. Straatsma, H. J. J. Van Dam, D. Wang, J. Nieplocha, E. Apra, T. L. Windus and W. A. de Jong, *Comput. Phys. Commun.*, 2010, **181**, 1477.

64 J. Zheng, T. Yu, E. Papajak, I. M. Alecu, S. L. Mielke and D. G. Truhlar, *Phys. Chem. Chem. Phys.*, 2011, **13**, 10885.

65 J. Zheng, S. L. Mielke, K. L. Clarkson and D. G. Truhlar, *Computer Phys. Comm.*, 2012, **183**, 1803.

66 J. Zheng, S. L. Mielke, K. L. Clarkson and D. G. Truhlar, *MSTor, version 2011–2*, University of Minnesota, Minneapolis, 2011.

67 D. G. Truhlar and A. Kuppermann. *J. Am. Chem. Soc.*, 1971, **93**, 1840.

68 Y.-P. Liu, G. C. Lynch, T. N. Truong, D.-h. Lu, D. G. Truhlar and B. C. Garrett, *J. Am. Chem. Soc.*, 1993, **115**, 2408.

69 D. K. Bondi, J. N. L. Connor, B. C. Garrett and D. G. Truhlar, *J. Chem. Phys.*, 1983, **78**, 5981.

70 M. M. Kreevoy, D. Ostović, D. G. Truhlar and B. C. Garrett, *J. Phys. Chem.*, 1986, **90**, 3766.

71 Y.-P. Liu, D.-h. Lu, A. Gonzàlez-Lafont, D. G. Truhlar and B. C. Garrett, *J. Am. Chem. Soc.*, 1993, **115**, 7806.

72 A. Fernandez-Ramos and D. G. Truhlar, *J. Chem. Phys.*, 2001, **114**, 1491.

73 B. C. Garrett and D. G. Truhlar, *J. Chem. Phys.*, 1983, **79**, 4931.

74 R. Meana Pañeda, D. G. Truhlar and A. Fernández-Ramos, *J. Chem. Theory Comput.*, 2010, **6**, 6.

75 B. C. Garrett, D. G. Truhlar, R. S. Grev and A. W. Magnuson, *J. Phys. Chem.*, 1983, **84**, 1730; B. C. Garrett, D. G. Truhlar, R. S. Grev and A. W. Magnuson, *J. Phys. Chem.*, 1983, **87**, 4554 (E).

76 D. G. Truhlar, A. D. Isaacson and B. C. Garrett, in *Theory of Chemical Reaction Dynamics*, ed. M. Baer (CRC Press. Boca Raton, FL, 1985), Vol. 4, pp. 65–137.

77 A. Fernández-Ramos, B. A. Ellingson, B. C. Garrett and D. G. Truhlar, *Rev. Comput. Chem.*, 2007, **23**, 125.

78 A. Fernández-Ramos, B. A. Ellingson, R. Meana-Pañeda, J. M. C. Marques and D. G. Truhlar, *Theor. Chem. Acc.*, 2007, **118**, 813.

79 A. Fernandez-Ramos, B. A. Ellingson, B. C. Garrett and D. G. Truhlar, *Rev. Comput. Chem.*, 2007, **23**, 125.

80 J. Zheng, S. Zhang, B. J. Lynch, J. C. Corchado, Y.-Y. Chuang, P. L. Fast, W.-P. Hu, Y.-P. Liu, G. C. Lynch, K. A. Nguyen, C. F. Jackels, A. F. Ramos, B. A. Ellingson, V. S. Melissas, J. Villà, I. Rossi, E. L. Coitino, J. Pu, T. V. Albu, R. Steckler, B. C. Garrett, A. D. Isaacson and D. G. Truhlar, *POLYRATE, version 2010-A*, University of Minnesota, Minneapolis, 2010.

81 J. Villà and D. G. Truhlar, *Theor. Chem. Acc.*, 1997, **97**, 317.

82 R. K. Boyd, *Chem. Rev.*, 1977, **77**, 93.

83 M. M. Kreevoy and D. G. Truhlar, *Tech. Chem. (N.Y.)*, 4th edn, 1986, **6/Pt. 1**, 13.

84 J. B. Anderson, *Adv. Chem. Phys.*, 1995, **91**, 381.

85 S. Winstein and N. J. Holness, *J. Am. Chem. Soc.*, 1955, **77**, 5562.

86 J. E. Baldwin, A. S. Raghavan, B. A. Hess and L. Smentek, *J. Am. Chem. Soc.*, 2006, **128**, 14854.

87 R. Meana-Pañeda and A. Fernández-Ramos, *J. Am. Chem. Soc.*, 2012, **134**, 346.

88 J. E. Seeman, *J. Chem. Educ.*, 1986, **63**, 42.

89 B. C. Garrett and D. G. Truhlar, *J. Phys. Chem.*, 1980, **84**, 805.

90 P. Pechukas, *Annu. Rev. Phys. Chem.*, 1981, **32**, 159.

91 A. D. Isaacson, M. T. Sund, S. N. Rai and D. G. Truhlar, *J. Chem. Phys.*, 1985, **82**, 1338.

92 L. F. Phillips, *J. Chem. Phys.*, 1990, **92**, 6523.

93 F. T. Smith, *Discuss. Faraday Soc.*, 1962, **33**, 183.

94 V. Bernshtein and I. Oref, *J. Phys. Chem. A*, 2004, **108**, 8131.
95 A. Galano, M. Francisco-Marquez and J. R. Alvarez-Idaboy, *J. Phys. Chem. B*, 2011, **115**, 8590.
96 K. A. Nguyen, C. F. Jackels and D. G. Truhlar, *J. Chem. Phys.*, 1996, **104**, 6491.
97 J. Zheng and D. G. Truhlar, *Phys. Chem. Chem. Phys.*, 2010, **12**, 7782.
98 P. S. Zuev, R. S. Sheridan, T. V. Albu, D. G. Truhlar, D. A. Hrovat and W. T. Borden, *Science*, 2003, **299**, 867.
99 B. C. Garrett and D. G. Truhlar, *J. Chem. Phys.*, 1979, **70**, 1593.
100 C. Alhambra, J. Corchado, M. L. Sánchez, M. Garcia-Viloca, J. Gao and D. G. Truhlar, *J. Phys. Chem. B*, 2001, **105**, 11326.
101 D. G. Truhlar, J. Gao, C. Alhambra, M. Garcia-Viloca, J. Corchado, M. L. Sánchez and J. Villà, *Acc. Chem. Res.*, 2002, **35**, 341.
102 D. G. Truhlar, J. Gao, M. Garcia-Viloca, C. Alhambra, J. Corchado, M. L. Sanchez and T. D. Poulsen, *Int. J. Quantum Chem.*, 2004, **100**, 1136.

Imaging the effects of the antisymmetric stretch excitation of CH_4 in the reaction with F atom

Hiroshi Kawamata,[†][a] Weiqing Zhang[‡][a] and Kopin Liu[*][b]

Received 11th January 2012, Accepted 31st January 2012

DOI: 10.1039/c2fd20004j

One quantum excitation of methane to $CH_4(v_3 = 1)$ was found to prohibit the CH bond rupture, thus slowing down the total reactivity toward F atoms, and when the reaction does occur, only stretch-excited methyl radical $CH_3(v_1/v_3 = 1)$ products have significant yields.

I. Introduction

Exciting a stretching mode of a chemical bond should increase the likelihood of the (excited) bond breaking during a chemical reaction or, at the very least, exert little impact on the reaction, according to conventional wisdom. In a recent study of the F + $CHD_3(v_1 = 1)$ reaction, counterintuitively, we found that the induced reactant vibration actually inhibits the bond rupture and slows the reaction down.[1] We posited that the unexpected result is due to the deflection of the trajectory, upon CH stretching excitation, away from the transition state by the long range anisotropic interactions in the entry valley of the reaction. Shortly after, Czako and Bowman[2,3] performed a quasiclassical trajectory study of this reaction on a highly accurate *ab initio* potential energy surface (PES),[4,5] and their results confirmed the experimental conjecture. The implication of this surprising finding lays open a vast new territory that links the mode- and bond-selective chemistry to stereo-specific reactivity.[6,7]

Upon CH stretching excitation of CHD_3 the energy deposited is primarily localized in the CH bond. It is easier to visualize the trajectory deflection with respect to this bond (or local mode) excitation. When exciting a delocalized mode, such as the antisymmetric stretch of $CH_4(v_3 = 1)$, all four CH bonds are coherently excited. It becomes unclear *a priori* how the stereo-specific effect will impact on the dynamics in the analogous F + $CH_4(v_3 = 1)$ reaction. For example, as the trajectory of the approaching F atom is steered away from the initially attacked H atom, will it end up reacting with the adjacent (also excited) H atoms? How different will the reactive outcomes be, compared to the F + $CHD_3(v_1 = 1)$ reaction? *etc.* These are the questions we try to address in this work.

II. Experimental

The rotating-source, crossed beam apparatus used in this work has been described in detail previously.[1,8,9] In brief, a supersonic expansion of a mixture of 5% F_2 in Ne at

[a]*Institute of Atomic and Molecular Sciences (IAMS), Academia Sinica, P. O. Box 23-166, Taipei, Taiwan 10617*

[b]*Department of Physics, National Taiwan University, Taipei, Taiwan 10617. E-mail: kliu@po. iams.sinica.edu.tw*

† Current address: Advanced Science Institute, RIKEN Sendai, Aramki-Aoba 519-1399, Aoba-ku Sendai, 980-0845, Japan

‡ Current address: Fritz-Haber-Institut der Max-Planck-Gesellschaft, Faradayweg 4-6, 14195 Berlin, Germany

80 psi was used to generate the F atom beam by pulsed, high-voltage discharge, and an expansion of neat CH_4 at 80 psi from another pulsed valve released the target beam. Both beams were double-skimmed before crossing in the scattering chamber. The initial collision energy E_c was controlled by varying the intersection angle of the two molecular beams. An infrared optical parametric oscillator/amplifier (OPO/A) was tuned to the R(1) transition at 3038.50 cm^{-1} to excite the CH_4 reactants to the ($v_3 = 1, j = 2$) state through a multipass ring reflector mounted in front of the first skimmer.[10] Thanks to the ring reflector, the fraction of CH_4 molecules being pumped can be accurately measured, with which the relative reactivity can then be determined.[11] The methyl radical products were probed by a (2 + 1) resonance-enhanced multiphoton ionization (REMPI) detection near 331 nm.[12] The recoil velocities of the REMPI-selected ions were then mapped by a time-sliced ion imaging technique.[8] By virtue of energy and momentum conservation, the images revealed the concomitant vibration state distribution of the HF coproducts.[9,13,14] To interrogate the effects of reactant vibrational excitation, the product images at each E_c were acquired with IR-on and IR-off alternatively to minimize the long-term drifts and other possible systematic errors. The image accumulation time was typically 0.5–1.5 h depending on the individual signal strength. For clarity, Fig. 1 depicts some of the product pairs probed in this work, along with the relevant energetics.

III.   Results and discussion

A.   REMPI spectra and IR excitation efficiency

Fig. 2 shows a portion of REMPI spectra, IR-on and IR-off, of the CH_3 products at $E_c = 3.4$ kcal mol^{-1}. Both spectra are dominated by an intense Q-head of the 0_0^0 band, whose intensity gets attenuated with IR-on, similar behaviors are observed for the weaker peaks of the rotational lines. The only exception is a vibronic band at 59883 cm^{-1} (near the P(4) line), labelled $1_1^1/3_1^1$ Q that appears enhanced with IR-on. Using an IR-UV double-resonance spectroscopic technique, this band has previously been assigned to the 3_1^1Q head of the CH_3 radical.[15] On the other hand, using the threshold method a previous study of the F + $CH_4(v = 0) \rightarrow$ HF + CH_3 reaction assigned this band as $CH_3(1_1^1Q)$ on energetic grounds.[12] The latter assignment is entirely consistent with the location of the $CH_3(1_0^1)$ band at 62887 cm^{-1} and the ground state v_1-mode vibrational frequency of 3004 cm^{-1}.[16] Hence, both spectral assignments are backed up by the rigorous optical selection rules and should be unambiguous. Several other REMPI bands, such as 2_1^1 and 2_2^0, were also probed in

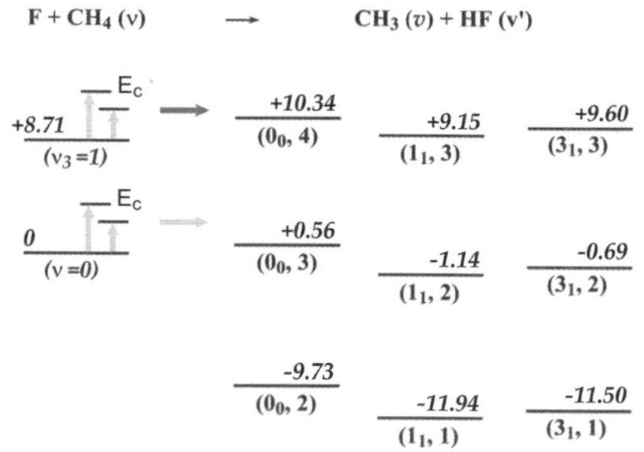

Fig. 1 Energetics of the F + $CH_4(v)$ reaction, in kcal mol^{-1}, and the product pairs probed in this work. The product-pair notation denotes (v_{CH_3}, v_{HF}) and the heat of reaction is $\Delta H_{rx} = -31.85$ kcal mol^{-1}.

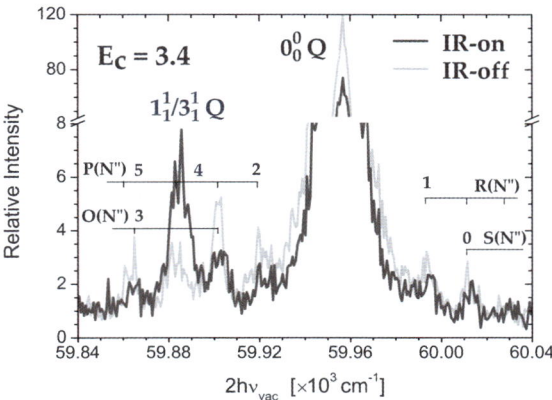

Fig. 2 Two REMPI spectra of the CH_3 product with IR-on (black) and IR-off (grey) at $E_c = 3.4$ kcal mol^{-1}. The very intense 0_0^0 Q-head peaks are truncated in order to display the weaker features. All spectral peaks are attenuated with IR-on, except the $1_1^1/3_1^1$ Q-head that overlaps with P(4) line of the 0_0^0 transition. The rotational energy of $N'' = 4$ is 0.55 kcal mol^{-1}.

this study; all indicated significant depletions upon IR excitation—albeit to a slightly lesser extent than the 0_0^0 band, suggesting lower yields than that from the corresponding ground state reaction. We then encountered a situation where the reaction of F + $CH_4(v_3 = 1)$ produces two spectroscopically indistinguishable CH_3 product states, $v_1 = 1$ and $v_3 = 1$, while suppressing the formation of all other CH_3 product states.

Integrating the 0_0^0 band gave ~40% depletion of REMPI intensity with IR-on (black) from that with IR-off (grey). To investigate the underlying process that caused the signal depletion with IR-on, we fixed the probe laser frequency at the peak of the 0_0^0 Q head and acquired the product images as shown in Fig. 3 (left). Product state speed distributions of the two images, Fig. 3 (right), proved identical—no additional features are discernible with IR-on. This demonstrated that F

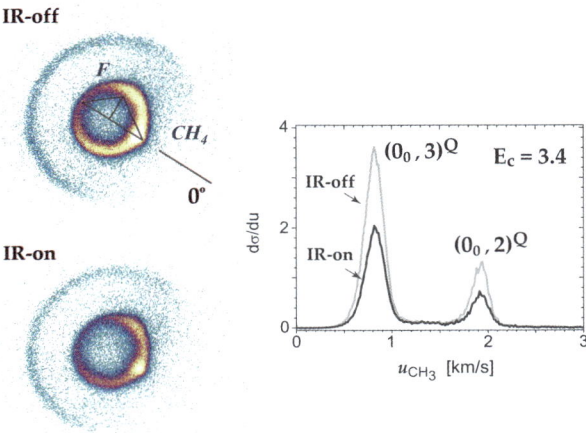

Fig. 3 Two raw images when the $CH_3(v = 0)$ products were probed at the peak of the 0_0^0 Q-head. The IR-on image is clearly weaker than the IR-off image, and no new features can be detected upon IR laser illumination. The product speed distributions are shown on the right and the extent of signal depletion yields the IR excitation efficiency, $n^*/n_0 \sim 40\%$ in this case.

$+ CH_4(v_3 = 1)$ indeed produces negligible amounts of $CH_3(0_0)$, as implied by the REMPI spectra (Fig. 2), and that the attenuation of the 0_0^0 Q signals upon IR irradiation is due solely to the depletion of ground state reactants. We therefore conclude an IR pumping efficiency of $n^*/n_0 = 40\%$, with n^* and n_0 denoting the IR-excited and the initial ground-state CH_4 molecules in the beam, respectively.

B. Product state assignment of the $1_1^1/3_1^1$ image

As mentioned earlier, the two REMPI bands, 1_1^1 and 3_1^1, overlap spectroscopically. The formation of $CH_3(v_1 = 1)$ and $CH_3(v_3 = 1)$ products, however, differs in energy by 0.45 kcal mol^{-1} (Fig. 1). To uncover the identity of the overlapped $1_1^1/3_1^1$ bands, we then adopted the threshold approach.[12,17] A pair of images acquired at the $1_1^1/3_1^1$ band is shown in Fig. 4 at $E_c = 0.9$ kcal mol^{-1}, for which the formation of $(3_1, 3)^*$ pair is energetically barely open. The IR-off image is attributed to the ground state reaction. As reported and assigned previously,[12] the product image is dominated by the resonance-mediated product pairs $(1_1, 2)$ and $(1_1, 1)$, in addition to the contribution of $(0_0, 2)^P$ from the spectrally overlapped P(4) transition. The IR-on image looks similar, but small differences are noticed upon closer inspection. The differences are more readily observed by the $P(u)$ analysis of the two images shown in the lower-left panel. Both distributions constitute two structures: a slow-velocity peak centered around 0.5 km s^{-1} and a nearly flat feature from 1 to 2 km s^{-1}. Also marked as the vertical dashed lines are the energetic limits of several product pairs; for clarity, only those from the ground state reaction are shown. Interestingly, the slower peak shows attenuation upon IR excitation, whereas the faster one gets enhanced. Based on the depletion measurement of the 0_0^0 band (Fig. 3), we scaled down the IR-off

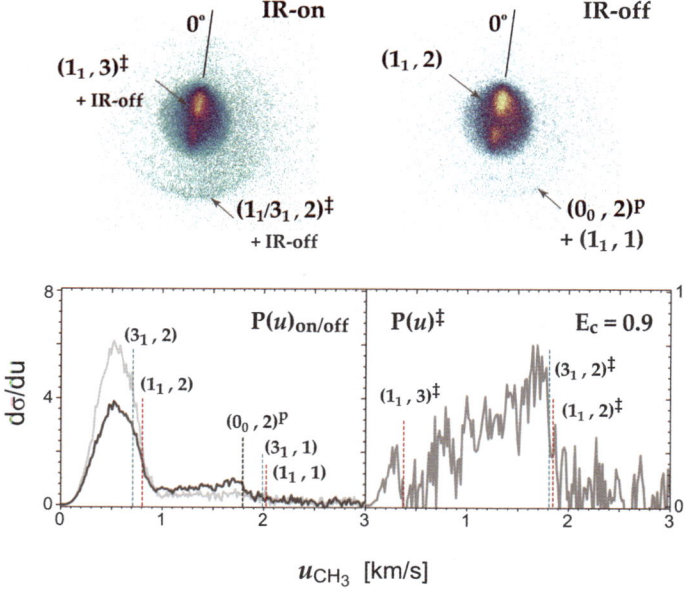

Fig. 4 Two raw images for $E_c = 0.9$ kcal mol^{-1} when the CH_3 products were probed at the peak of the $1_1^1/3_1^1$ Q-head. The product speed distributions are shown in the lower panels. The left presents the two distributions with IR-on (black) and IR-off (grey). The energetic limits of the product pairs are marked, and for clarity only those for IR-off are shown. The right panel presents the product speed distribution in the stretch-excited reaction, which was obtained by scaling the IR-off distribution by $(1 - n^*/n_0) = 0.6$ and subtracting the result from the IR-on distribution. The product-pair labels now refer to the stretch-excited reaction, as indicated by the superscript "$\neq$".

distribution by $(1 - n^{\neq}/n_0) = 0.6$ to account for the $CH_4(v = 0)$ reactants that were not exited by the IR laser and subtracted the result from the IR-on distribution. The difference is shown in the lower-right panel, which represents the signals from the $F + CH_4(v_3 = 1)$ reaction, with the energetic limits now indicated for the corresponding product pairs. A small signal labelled $(1_1, 3)^{\neq}$ is clearly seen and a broad distribution can be attributed to $(1_1, 2)^{\neq}$ and/or $(3_1, 2)^{\neq}$.

To ensure the assignments of the product pairs and to investigate the E_c-dependence of dynamical attributes, similar measurements were performed at E_c values ranging from 0.4 kcal mol^{-1} to 4 kcal mol^{-1}. Two of them are exemplified in Fig. 5, and a few representative $P(u)$ analyses are illustrated in Fig. 6. At $E_c = 0.6$ kcal mol^{-1} all observed signals from the stretch-excited reaction are ascribed to the $(1_1/3_1, 2)^{\neq}$ product pair. Despite being energetically feasible, no signal for the $(1_1, 3)^{\neq}$ pair is discernible, indicative of a dynamical threshold. At $E_c = 1.1$ kcal mol^{-1}, both $(1_1, 3)^{\neq}$ and $(3_1, 3)^{\neq}$ pairs are open. Judging from their energetic limits, however, the slow-velocity peak is more likely dominated by the $(1_1, 3)^{\neq}$ pair. As E_c increases further, the image resolution does not allow us to distinguish the $(1_1, 3)^{\neq}$ and $(3_1, 3)^{\neq}$ pairs unambiguously. Hence, we label the slow-velocity peak as $(1_1/3_1, 3)^{\neq}$, similarly for the high velocity feature $(1_1/3_1, 2)^{\neq}$. Also worth noting is the shape of the $(1_1/3_1, 2)^{\neq}$, which reflects a relatively hot rotational distribution of $HF(v = 2)$ at low E_c and a colder one at higher E_c.

C. Correlated vibrational branching ratios and excitation functions

With the product speed distribution $P(u)^{\neq}$ analyzed at each E_c, Fig. 7 summarizes the E_c-dependence of the correlated vibrational branching ratio in $F + CH_4(v_3 = 1)$. A previously reported branching ratio in $F + CH_4(v = 0)$ is reproduced in the lower panel for comparison.[18] The results of the two reactions are similar— favoring $HF(v = 2)$ at low E_c and switching to $HF(v = 3)$ at higher E_c till around 4 kcal mol^{-1}. It is worth noting that the switch-over of the vibrational-state

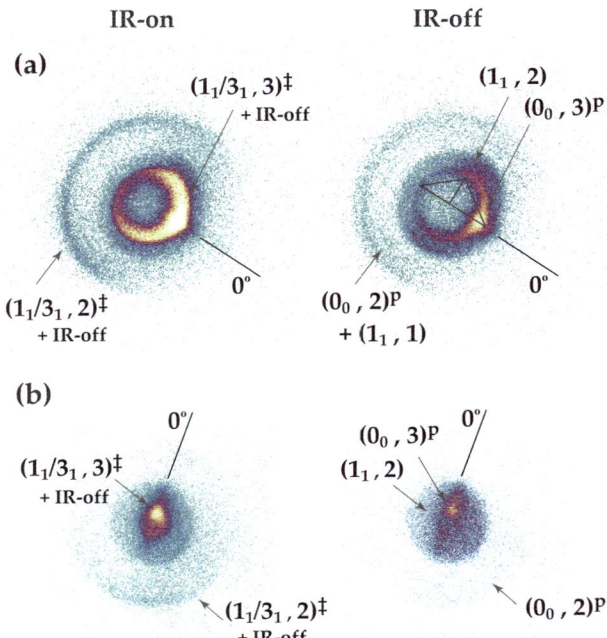

Fig. 5 Pairs of raw images of the $1_1^1/3_1^1$ band at $E_c = 3.4$ kcal mol^{-1} (a), and $E_c = 1.3$ kcal mol^{-1} (b) are shown, demonstrating the impact of the IR illumination on the product images.

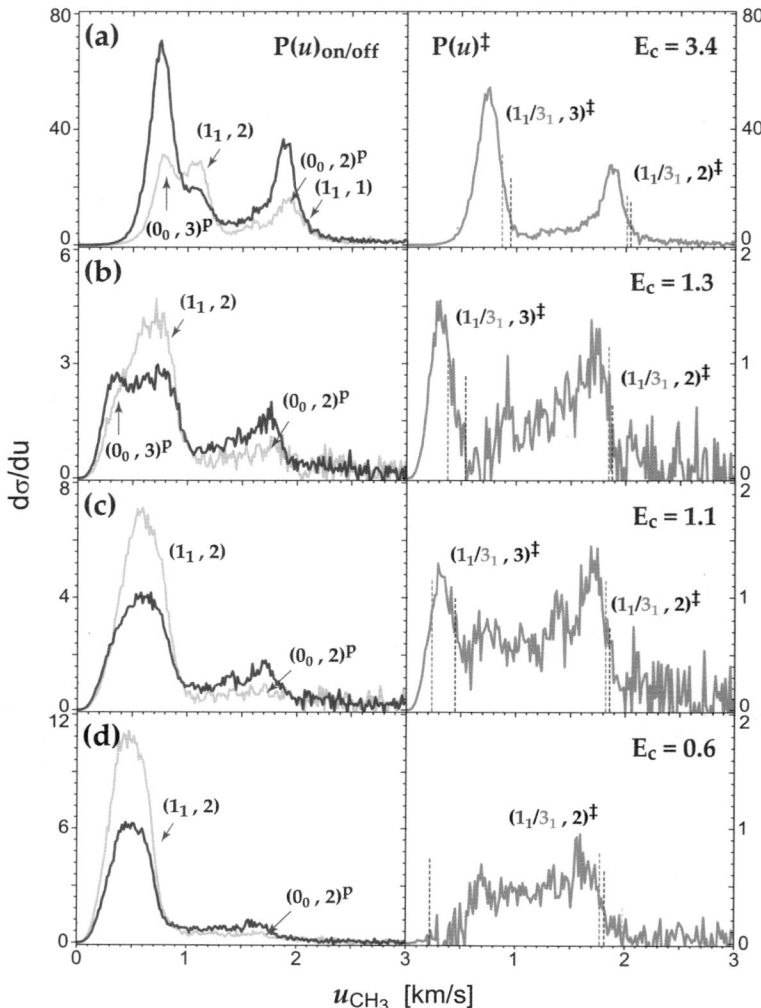

Fig. 6 Product speed distributions at four different E_c values. The left panels present the IR-on (black) and IR-off (grey) distributions, and the right panels give the corresponding distribution for the stretch-excited reaction.

propensity occurs at a branching ratio of $\sim$0.5 in both cases. Similar results were found in the previous study of the F + CHD$_3$(v_1 = 1) → CHD$_2$(v_1 = 1) + DF(v = 2–4) reaction.[1] We also note that the switch of the preferred HF vibrational branching in F + CH$_4$(v_3 = 1) occurs at higher E_c than that in F + CH$_4$(v = 0), 1.5 kcal mol^{-1} *versus* 0.75 kcal mol^{-1}. The slight difference in energetic thresholds, as indicated in the figures, can not be the sole reason. It implies that although the initial stretching excitation of CH$_4$ transforms into the stretching motion of the CH$_3$ product, it also influences the vibrational branching of the HF coproducts. Similar conclusion was drawn previously for the F + CHD$_3$(v_1 = 1) → CHD$_2$(v_1 = 1) + DF(v) reaction.[1]

The depletion of the 0_0^0 band was routinely measured to obtain the IR excitation efficiency in every experiment, which varied daily from $\sim$40% to 48%. The pairs of IR-on and IR-off images of the $1_1^1/3_1^1$ band at each E_c were experimentally normalized. Thus, it is straightforward to scale the weak $1_1^1/3_1^1$ signals to the 0_0^0 intensity

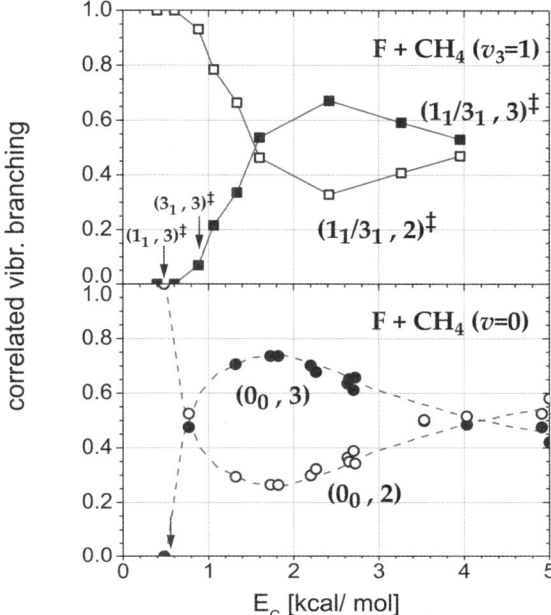

Fig. 7 Comparison of the correlated HF vibrational branching ratios for the stretch-excited reaction (upper) and the ground state reaction (lower).[18] The energetic thresholds for forming the $(1_1, 3)^{\neq}$ and $(3_1, 3)^{\neq}$ pairs are indicated. Aside from the subtle energetic difference, note the similarity of the two branching ratios.

(IR-off) at each E_c. By normalizing the relative cross section of the $CH_3(v = 0)$ products of the ground state reaction to the previously reported excitation function,[18] the excitation functions for the stretch-excited $1_1/3_1$ product state can then be deduced. However, the relative REMPI detection sensitivity has only been calibrated for the 3_1^1 band;[15] the relative sensitivity of 1_1^1 to 0_0^0 is yet to be determined. Hence the relative excitation functions for the stretch-excited products in Fig. 8 also contain the REMPI detection sensitivity factors S and S'. Compared to the previously reported $\sigma_0(0_0)$,[18] the yield of the stretch-excited $CH_3(1_1/3_1)$ products from $F + CH_4(v_3 = 1)$ appears small and its excitation function exhibits typical behavior for an activated process. The relative cross section $\sigma_0(1_1)$ from $F + CH_4(v = 0)$ seems even smaller, except for $E_c \leq 1$ kcal mol^{-1}.

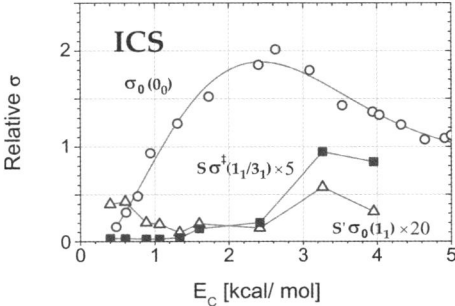

Fig. 8 Excitation functions for the reaction of $F + CH_4(v = 0) \rightarrow HF + CH_3(0_0)$, $\sigma_0(0_0)$,[18] $F + CH_4(v = 0) \rightarrow HF + CH_3(1_1)$, $\sigma_0(1_1)$, and $F + CH_4(v_3 = 1) \rightarrow HF + CH_3(1_1/3_1)$, $\sigma^{\neq}(1_1/3_1)$. The S and S' are the scaling factors to account for the relative detection sensitivity of different REMPI bands, $(1_1^1/3_1^1 Q)/(0_0^0 Q)$ and $(1_1^1 Q)/(0_0^0 Q)$.

To have a better perspective of the excitation functions, Fig. 9 presents the cross section ratios of the three reactions. As is seen in (a), the stretch-excited $\sigma^*(1_1/3_1)/\sigma_0(0_0)$ displays a peculiar E_c-dependence. It declines rapidly at low E_c, followed by a gradual rise with increasing E_c. As indicated by the branching ratio in Fig. 7, the $(1_1/3_1, 3)^*$ product pair makes a significant contribution to $\sigma^*(1_1/3_1)$ when it is energetically accessible. The thresholds for the $(1_1, 3)^*$ and $(3_1, 3)^*$ pairs are 0.44 kcal mol^{-1} and 0.89 kcal mol^{-1}, respectively. Thus, it is tempting to ascribe the low energy part (<1 kcal mol^{-1}) of Fig. 9(a) to the contribution of the $(1_1, 2/3)^*$ pair and the higher energy rise to the formation of the $(3_1, 3)^*$ pair. Note the smallness of the ordinate scale. It is unlikely that the relative sensitivity S for the $1_1^1/3_1^1$ band can be an order-of-magnitude larger. [Previous IR-UV double resonance experiment led to $S = 3.5 \pm 1.2$, favoring the 3_1^1 band over the 0_0^0 band.[15]] Since only $CH_3(1_1/3_1)$ states show signal enhancement upon IR excitation, we conclude that exciting the antisymmetric-stretch mode of methane $CH_4(v_3 = 1)$ actually inhibits the breaking of the CH bond and slows the reaction down, which conforms to the previous finding in F + $CHD_3(v_1 = 1)$ where the local CH bond was excited.[1] On the other hand, the ratios of $\sigma_0(1_1)/\sigma_0(0_0)$, Fig. 9(b), show a distinct decrease with increasing E_c, which corroborates well with previous report.[12] [Plotted in Fig. 2 of Ref. 12 are the REMPI signal ratios, whereas Fig. 9(b) depicts the cross section ratios.] Shown in Fig. 9 (c) is the ratio of the stretch-excited products $\sigma^*(1_1/3_1)/\sigma_0(1_1)$ from the two reactions. Interestingly, the ratios display a monotonic increase with E_c and the rapid declines at low E_c shown in (a) and (b) are completely cancelled out. This result seems to support the above assignment that the peculiar dependence in (a) is due to two distinct product pairs, $(1_1, 2/3)^*$ at low E_c and $(3_1, 3)^*$ at higher E_c.

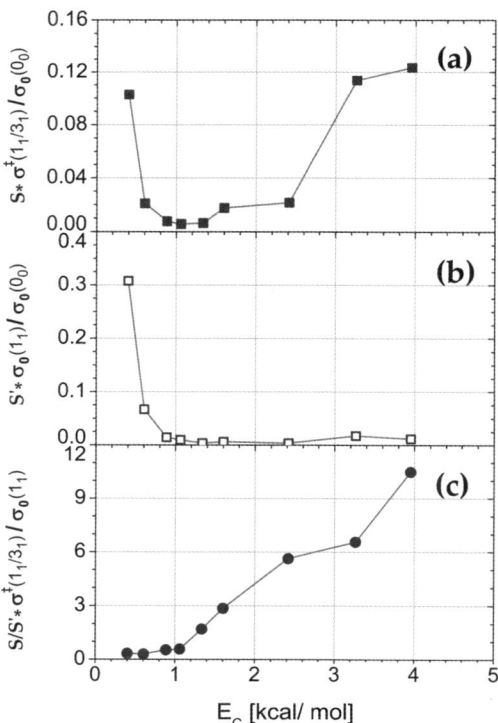

Fig. 9 Ratios of integral cross sections as a function of collision energies.

D. Correlated product angular distributions

Fig. 10 presents the product angular distributions analyzed from the four images shown in Fig. 6 for the reactions of F + $CH_4(v = 0)$ (left, DCS) and F + $CH_4(v_3 = 1)$ (right, DCS*). For the ground state reaction, only two product channels are of concern at low E_c. The situation becomes a bit more complicated at higher E_c. Generally speaking, the distributions for the $(0_0, 2)^P$ pair are broad and tend to favor the backward hemisphere at higher E_c. On the other hand, the $(0_0, 3)^P$ prefers forward scattering. Both distributions are in excellent agreement with previous results measured at 0_0^0 Q-head.[12] At lower E_c, the $(1_1, 2)$ pair displays a distinct forward-backward peaking distribution, favoring the forward direction. At $E_c = 3.4$ kcal mol^{-1}, an additional sideways-scattered component becomes apparent. Its nature is unclear: either being from a rainbow-like scattering of the $(1_1, 2)$ pair or possibly being ascribed to the formation of the $(3_1, 2)$ pair. Our

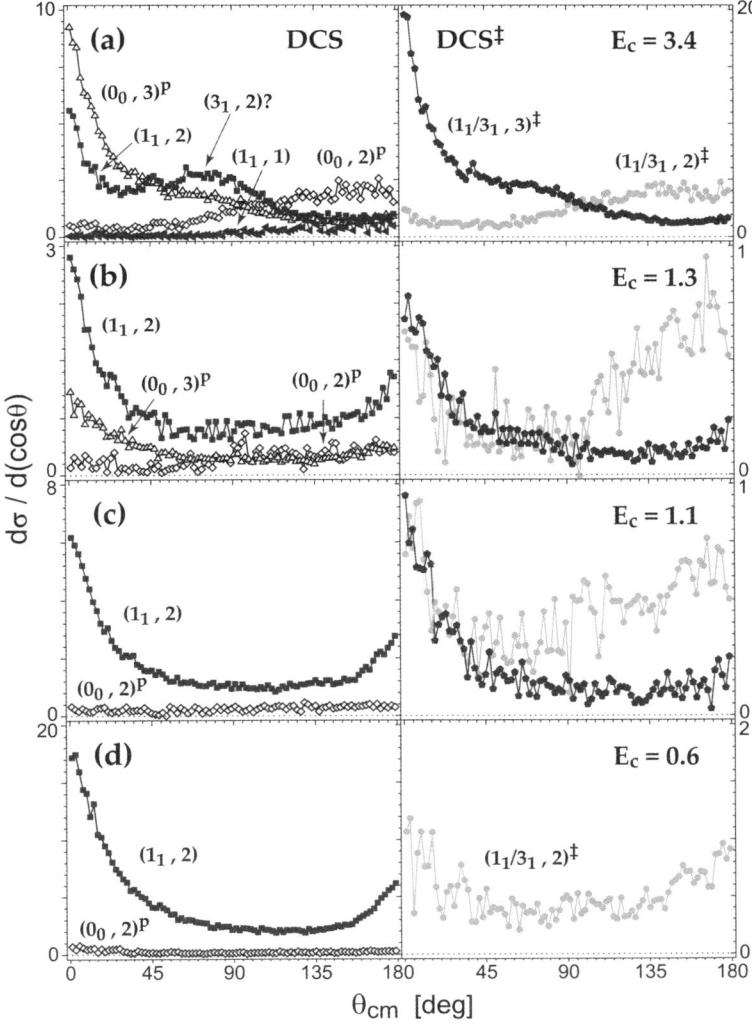

Fig. 10 Pair-correlated angular distributions for the ground state reaction (left panels) and the stretch-excited reaction (right panels).

previous assignment of $(1_1, 2)$ was for $E_c < 2.6$ kcal mol^{-1};[12] at higher E_c the contribution of the $(3_1, 2)$ pair may come into play, as hinted by the rise of the excitation function (Fig. 8).

As for the stretch-excited reaction (DCS$^\neq$, right panels), the $(1_1/3_1, 3)^\neq$ pair is always forward scattering, superimposed over a nearly isotopic component. At first

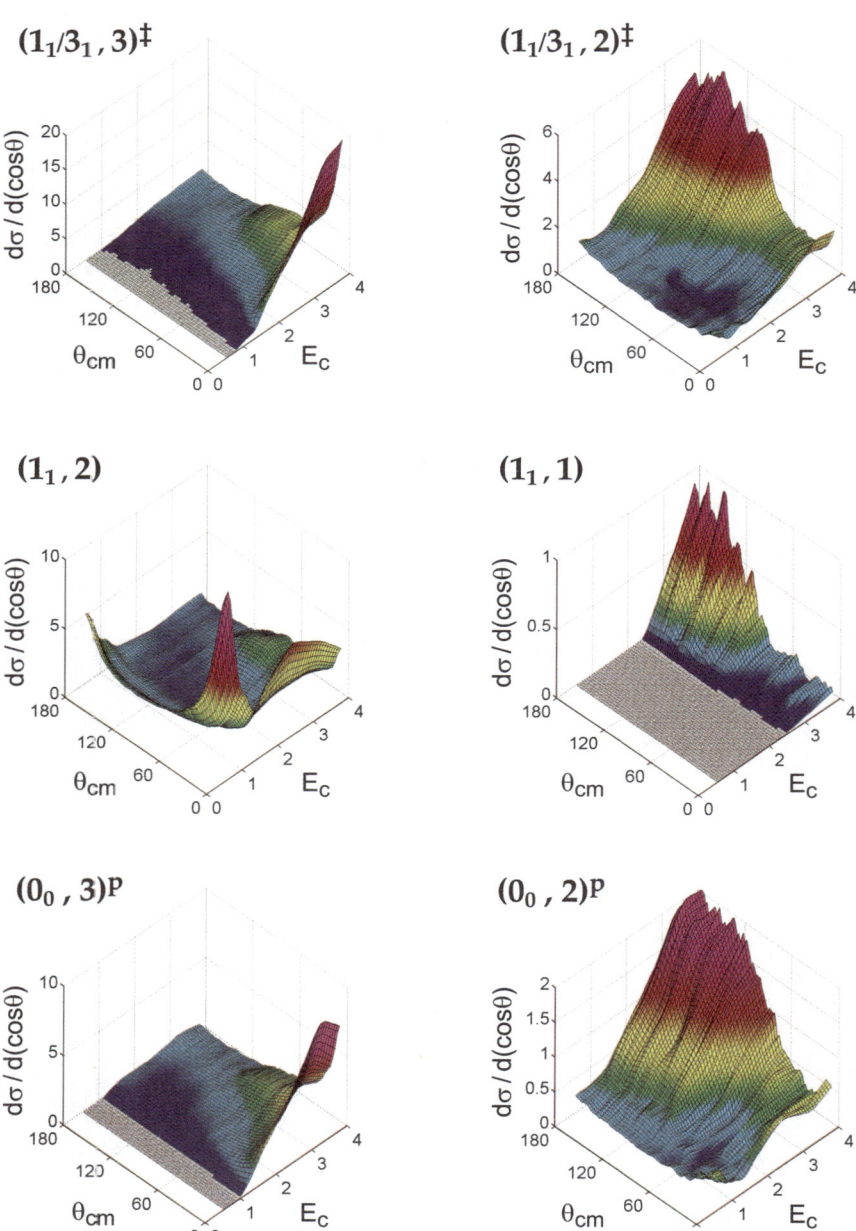

Fig. 11 3D plots of $d\sigma/d(\cos\theta)$ *versus* θ and E_c for the product pairs probed in this work. The top two are from the F + CH$_4$($v_3 = 1$) reaction. The middle two are for F + CH$_4$($v = 0$) → HF + CH$_3$(1_1) [and possibly some CH$_3$(3_1) products at higher E_c, see text], and the bottom two are for F + CH$_4$($v = 0$) → HF(v) + CH$_3$($v = 0$, N$'' = 4$).

sight, the distributions for $(1_1/3_1, 3)^{\neq}$ are not much different from that for $(1_1, 2)$, except for $E_c = 3.4$ kcal mol^{-1}. The $(1_1/3_1, 2)^{\neq}$ pair displays a forward-backward symmetric distribution at low E_c. As E_c increases, the distribution becomes dominated by a broad backward component.

A more illuminating way to summarize the evolution of those correlated angular distributions with E_c is, as presented in Fig. 11, to combine the angular distribution results (Fig. 10) with the excitation function data (Fig. 7 and 8) to generate a 3D plot of $d\sigma/d(\cos\theta)$ as a function of θ and E_c. The results are arranged as a pair according to the CH$_3$ product state, $(1_1/3_1, v)^{\neq}$, $(1_1, v)$ and $(0_0, v)^P$. As is seen, the more exothermic pairs (right column) are all characterized by the broad swaths in the backward hemisphere, indicative of the dominance of a direct rebound mechanism. The least energetic pairs (left column), on the other hand, display distinct patterns. Both $(1_1/3_1, 3)^{\neq}$ and $(0_0, 3)^P$ pairs are alike, featuring a prominent forward-peaking pattern, whereas the $(1_1, 2)$ pair displays an oscillating pattern totally different from all others. By way of contrast to the pairs on the right, we assert that these least exothermic pairs proceed through different reaction pathways, likely a time-delay or the one mediated by reactive resonances as proposed previously.[12,19–23] Note that a weaker forward-scattered component can also be seen for the $(1_1/3_1, 2)^{\neq}$ and $(0_0, 2)^P$ pairs. It is tempting to ascribe them to the resonance-mediated pathway, as the corresponding pairs on the left panel for HF($v = 3$). In that regard, the pattern of the $(1_1, 1)$ pair in the forward direction shows no trace of features like those of $(1_1, 2)$ at all. One may wonder if that part of the image signals actually arises from the spectroscopically and energetically unresolvable $(3_1, 1)$ pair. Further work will be needed to sort it out. Recently, several reduced dimensionality quantum scattering calculations for the F + CH$_4$ reaction also found resonance signature in the integral cross section.[24–26] It will be extremely interesting to extend such calculations to both the ground state and stretch-excited CH$_4$ in the DCS level of detail.

IV Conclusions

The title reaction was studied over a collisional energy range from 0.4 kcal mol^{-1} to 4 kcal mol^{-1}. We found that exciting the antisymmetric stretch vibration of CH$_4(v_3 = 1)$ suppresses the total reaction rate, and when it does react the only detectable methyl products are the stretch-excited CH$_3(1_1)$ state at low E_c and likely the CH$_3(3_1)$ state at higher E_c. In conjunction with the previous study of F + CHD$_3(v_1 = 1)$, we surmise that the reactivity suppression has little to do with the localization or not of the initial vibrational stretching excitation. Rather, we posit the mechanistic origin arises from the location of the barrier. Being an early barrier reaction, the transition-state structure is reactant-like. Thus, a stretched or compressed CH bond tends to deviate from the transition-state structure, which does not favor the reaction.[1,6] Stated in a different way, vibrationally exciting the methane reactant alters the long-range anisotropic interactions by steering the approaching F atom away from the minimum energy pathway. In an enlightening review,[27] Levine introduced the concepts of "physical shape" that governs the ability of the reactants to reach the barrier, and of "chemical shape" that is the barrier location as a function of distance and orientation, thus referring to the bending potential in the vicinity of the transition state region. In his language, the observation of vibrationally induced reactivity suppression of the title reaction is the result of a prolate physical shape of PES. A recent QCT calculation[2,3] on a highly accurate *ab intio* PES[4,5] corroborated well such type of potential landscape.

It is instructive to view the present finding of suppressing reactivity upon reactant vibrational excitation within the context of Polanyi's rules, which concern how the barrier location influences the energy requirement and energy disposal in a direct A + BC reaction.[28–30] For an exothermic reaction, according to Polanyi's rules, reactant translational energy should be more effective than vibration to surmount the barrier to reaction, thus, accelerating the reaction rate. This statement of the relative efficacy of different modes of reactant energy in promoting the reactivity is on the basis of equivalent amount of total energy. In other words, the rules say nothing about whether the reaction rate, at a fixed initial translational energy, will be enhanced or not by exciting a stretching mode of a bond in a molecule. Hence, the present stereo-specific view of the effects on reactivity by reactant vibration does not violate the venerable Polanyi's rules, rather enriches the content of the rules from a different perspective.

Acknowledgements

This work was supported by National Science Council of Taiwan, Academia Sinica, and the Air Force Office of Scientific research (grant No. AOARD-12-4020).

References

1 W. Zhang, H. Kawamata and K. Liu, *Science*, 2009, **325**, 303.
2 G. Czako and J. M. Bowman, *J. Am. Chem. Soc.*, 2009, **131**, 17534.
3 G. Czako and J. M. Bowman, *J. Chem. Phys.*, 2009, **131**, 244302.
4 G. Czako, B. C. Shepler, B. J. Braams and J. M. Bowman, *J. Chem. Phys.*, 2009, **130**, 084301.
5 G. Czako and J. M. Bowman, *Phys. Chem. Chem. Phys.*, 2011, **13**, 8306.
6 F. Wang, J.-S. Lin and K. Liu, *Science*, 2011, **331**, 900.
7 F. Wang and K. Liu, *Chem. Sci.*, 2010, **1**, 126.
8 J. J. Lin, J. Zhou, W. Shiu and K. Liu, *Rev. Sci. Instrum.*, 2003, **74**, 2495.
9 J. J. Lin, J. Zhou, W. Shiu and K. Liu, *Science*, 2003, **300**, 966.
10 J. Riedel, S. Yan, H. Kawamata and K. Liu, *Rev. Sci. Instrum.*, 2008, **79**, 033105.
11 S. Yan, Y.-T. Wu and K. Liu, *Phys. Chem. Chem. Phys.*, 2007, **9**, 250.
12 W. Shiu, J. J. Lin and K. Liu, *Phys. Rev. Lett.*, 2004, **92**, 103201.
13 K. Liu, *Phys. Chem. Chem. Phys.*, 2007, **9**, 17.
14 J. Zhou, J. J. Lin, W. Shiu and K. Liu, *Phys. Chem. Chem. Phys.*, 2006, **8**, 3000.
15 W. Zhang, H. Kawamata, A. J. Merer and K. Liu, *J. Phys. Chem. A*, 2009, **113**, 13133.
16 M. E. Jacox, *J. Phys. Chem. Ref. Data*, 2003, **32**, 1.
17 B. Zhang, J. Zhou and K. Liu, *J. Chem. Phys.*, 2005, **122**, 104310.
18 W. Shiu, J. J. Lin, K. Liu, M. Wu and D. H. Parker, *J. Chem. Phys.*, 2004, **120**, 117.
19 J. Zhou, J. J. Lin and K. Liu, *J. Chem. Phys.*, 2004, **121**, 813.
20 J. Zhou, J. J. Lin, W. Shiu and K. Liu, *J. Chem. Phys.*, 2003, **119**, 4997.
21 J. Zhou, J. J. Lin and K. Liu, *Mol. Phys.*, 2010, **108**, 957.
22 G. Czako, Q. Shuai, K. Liu and J. M. Bowman, *J. Chem. Phys.*, 2010, **133**, 131101.
23 F. Wang and K. Liu, *J. Phys. Chem. Lett.*, 2011, **2**, 1421.
24 T. Chu, K. Han and J. Espinosa-Garcia, *J. Chem. Phys.*, 2009, **131**, 244303.
25 T. Chu, X. Zhang, L. Ju, L. Yao, L. Han, M. Wang and J. Z. H. Zhang, *Chem. Phys. Lett.*, 2006, **424**, 243.
26 G. Nyman and J. Espinosa-Garcia, *J. Phys. Chem. A*, 2007, **111**, 11943.
27 R. D. Levine, *J. Phys. Chem.*, 1990, **94**, 8872.
28 J. C. Polanyi and W. H. Wang, *J. Chem. Phys.*, 1969, **51**, 1439.
29 J. C. Polanyi, *Acc. Chem. Res.*, 1972, **5**, 161.
30 J. C. Polanyi, *Science*, 1987, **236**, 680.

The dynamics of the D_2 + OH → HOD + D reaction: A combined theoretical and experimental study

Shu Liu,† Chunlei Xiao,† Tao Wang, Jun Chen, Tiangang Yang, Xin Xu, Dong H. Zhang* and Xueming Yang*

Received 10th February 2012, Accepted 19th March 2012
DOI: 10.1039/c2fd20018j

A combined theoretical and experimental study has been carried out to show the current status of comparison between experiment and theory on the title reaction. Differential cross sections and product relative translational energy distributions at collision energies of 0.25 and 0.34 eV, as well as the collision energy dependence of differential cross section in the backward direction have been measured by using crossed molecular beam experiment with D-atom Rydberg tagging technique. Theoretically, the time-dependent wave packet method has been employed to calculate state-to-state differential cross sections for the title reaction in full dimension. It is found that the experimental observations are in good accord with those of Davis and coworkers at the collision energy of 0.28 eV [Science, **290**, 958 (2000)]. The overall agreement between theory and experiment on this benchmark four-atom reaction is good, but not perfect. Further studies, both theoretical and experimental, are called to bring a complete agreement between theory and experiment on the reaction.

1   Introduction

Significant progress has been achieved in the past decades in rigorous quantum reactive scattering studies for four-atom chemical reactions.[1–24] Starting from time-independent reduced dimensionality approaches,[25,26] the development of the time-dependent wave packet method has made it possible to carry out quantum dynamics studies for four-atom reactions in full dimensions, with total reaction probabilities, integral cross sections,[13,18] state-to-state reaction probabilities, as well as state-to-state integral cross sections[20,21] calculated for some benchmark four-atom reactions. Very recently, full-dimensional quantum differential cross sections for the HD + OH → H_2O + D reaction were reported with excellent agreement between theory and experiment achieved for the first time at the state-to-state level,[23] indicating it is feasible now to calculate complete dynamical information for some simple four-atom reactions, as have been done for three-atom reactions in the past decades.

The reaction of OH with H_2 is a prototype reaction for H atom abstraction by an OH radical to form water as a product. It is important in combustion, especially in that of hydrogen, as well as in the chemistry of the atmosphere.[27,28] The reverse reaction has been studied widely as a prototype system for mode-specific chemistry.[29–32]

State Key Laboratory of Molecular Reaction Dynamics and Center for Theoretical Computational Chemistry, Dalian Institute of Chemical Physics, Chinese Academy of Sciences, Dalian, P.R. China 116023. E-mail: zhangdh@dicp.ac.cn; xmyang@dicp.ac.cn

† These authors contributed equally to this work.

Extensive experiments have been carried out to measure the thermal rate constants of these reactions, and to study the influence of initial vibrational excitation, translational excitation, and isotopic substitution on the reaction dynamics.[33] About 20 years ago, Casavecchia and coworkers carried out the first molecular beam experiment on the D_2 + OH reaction at a collision energy of about 6.3 kcal mol^{-1}.[34,35] Although they were unable to resolve the HOD vibrational levels under their experimental conditions, their experiments discovered that the HOD product is strongly backward scattered, with about 64% of the total available energy channelled into HOD internal excitation. A more recent study of OH + D_2 by Davis and coworkers using crossed molecular beams with D-atom Rydberg tagging technique[36] confirmed the earlier results, showing both strongly backward scattered products and a large fraction of energy deposited into HOD internal motion.[37] Furthermore, their experiment was able to resolve product HOD vibrational states and revealed that the newly formed OD bond in the HOD molecule preferentially excited to the $\nu = 2$ state. These experiments have presented challenges for theory, including both *ab initio* potential energy surface and reactive scattering calculation. Many quasiclassical trajectory (QCT) studies have been carried out to directly compare with the experiments since then.[33,38–44]

Theoretically, because three of the four atoms in the reaction system are hydrogens, the OH + H_2 reaction and its isotopically substituted analogs are ideal candidates for high quality *ab initio* quantum chemistry calculation of the potential energy surface as well as for accurate quantum reactive scattering study. Since Schatz and Elgersma fitted an analytic potential energy surface (PES), known as the WDSE PES,[45] numerous theoretical studies have been carried out for the reaction on the PES to study the dynamics of the reaction, and more importantly to develop new theoretical methodologies for reactive scattering involving more than three atoms.[4–5] The system has become the prototype for tetra-atomic reactions, in much the same way that the H + H_2 reaction served as the prototype for triatomics. In the past decades, a number of new potential energy surfaces, including OC,[46 47] WSLFH,[48] YZCL2,[49] and XXZ[23] PES, have been constructed for the reaction system. Extensive quantum reactive scattering calculations have been performed on these PESs to study the dynamics of this benchmark four-atom system.[14–16,19,22–24,35,47,50,51]

In the paper, we report a combined theoretical and experimental study on the title reaction. Experimentally, we performed a crossed-beam scattering study of the title reaction by using the high-resolution and highly sensitive D-atom Rydberg tagging method at collision energies of 0.25 and 0.34 eV. Theoretically, we carried out full dimensional quantum scattering calculations on the title reaction to obtain the DCS for the collision energy up to 0.4 eV. Close comparisons are made between theory and experiment, both with the current experiment and the experiment of Davis and coworkers.[37] The rest of the paper is organized as follows: In Section 2, we outline both the experimental and theoretical methods used in the present study. Section 3 gives the experimental and theoretical results for the title reaction, as well as the comparisons between theory and experiment. The conclusions follow in Section 4.

2 Methods

2.1 Experimental methods

The crossed beam scattering experiment on the OH + D_2 reaction has been revisited in this work for comparison with theoretical results. The experimental method used here is essentially the same as that used in Ref. 37 and our previous experiment on the OH + HD reaction.[23] In this work, the OH radical beam was generated by nozzle tip 193 nm photolysis of a HNO_3/H_2 mixed sample at a stagnation pressure of about 100 psi. The velocity of the OH beam was measured to be about 2.78 km s^{-1} with a speed ratio of 30. The OH beam then crosses the D_2 beam in the crossed beam region after it passes through a skimmer. The D_2 beam was generated by expanding a pure normal D_2 sample through another pulsed valve. The stagnation pressure behind the nozzle for the D_2 beam is around 250 psi. The beam velocity of D_2 was determined to be about 2.2 km s^{-1} with a speed ratio of about 10. The OH beam was fixed horizontally, while the D_2 beam was fixed in the vertical plane on a rotating mechanism so that the crossing angle between the OH and D_2 beams can be varied. In this way, the collision energy of the reaction can be changed in this experiment, in an effort to compare with the theoretical calculations to gain information on the barrier for the title reaction. In addition, we have also made efforts to determine the collision energy of the reaction more accurately so that we can make better comparison with the quantum dynamical results.

The reaction product, D-atom, was detected using the Rydberg tagging time-of-flight (TOF) technique.[36] The Rydberg tagging of the D-atom was made in the crossing region by excitation of the ground state D-atom product ($n = 1$) to a high n ($\sim n = 50$) Rydberg state via $n = 2$ state in two steps:

$$D(n = 1) + h\nu(121.6nm) \rightarrow D(n = 2),$$

and

$$D(n = 2) + h\nu(365nm) \rightarrow D^*(n \sim 50).$$

The 121.6 nm light used in the first step excitation is generated using a difference frequency four wave mixing scheme in the Kr gas cell using 212.5 nm and 845 nm. Two photons of 212.5 nm is resonant with a Kr atomic transition. About 2 mJ of 212.5 nm and 5 mJ of 845 nm laser light are generally used. Following the first step VUV excitation, the D-atom product is then sequentially excited to a high Rydberg state with $n \sim 50$ using a 365 nm light in near saturation, which is generated by frequency doubling of a dye laser pumped by the same YAG laser. This excitation scheme allows us to pump the D-atom to the high Rydberg state in high efficiency. The neutral Rydberg tagged D-atom product then flies a distance of about 350 mm to reach a Z-stack multichannel plate (MCP) detector with a grounded fine metal grid in the front of the first MCP. After passing through the grid, the Rydberg tagged D-atoms were then field-ionized by the electric field between the front plate of the MCP detector and the grid. The TOF signal detected by the MCP is then amplified by a fast pre-amplifier, and counted by a multichannel scaler (MCS). In this experiment, angular resolved TOF spectra of the D-atom product were measured at a range of laboratory angles in 5° intervals at the collision energies of 0.25 and 0.34 eV.

2.2 Theoretical methods

The potential energy surface (PES) employed in the calculation was newly constructed with the neural network (NN) method[52–54] based on $\sim$20,000 ab initio points calculated at UCCSD(T)-F12/AVTZ level of theory. It was found that the new PES, denoted as CXZ PES, is more accurate and smooth than the previous PES constructed in this group, i.e. YZCL2 and XXZ PES. The details of construction method and the PES will be reported somewhere else.[55]

The time-dependent wave packet method was employed to calculate differential cross sections for four-atom reactions in full dimension, utilizing an improved version of the reactant–product–decoupling scheme.[56] In that approach, we propagate an initial wave packet first in the reactant diatom–diatom coordinates as we did for the total reaction probability calculations by using the split-operator method. Absorption of the wave packet is carried out at the edge of the grid for the breaking

D_2 bond and saved in a hard disk for every 8 propagation steps. After finishing the propagation of the wave packet in reactant coordinates, we transform the absorbed wave functions from diatom–diatom coordinates to atom–triatom coordinates, and continue to propagate them in the product atom–triatom coordinates as we did for atom–triatom scattering calculations except only with a small basis function for the newly formed OD bond in the HOD molecule. Finally an asymptotic analysis is carried out to extract the scattering matrix elements as we did for triatomic reactions.

We carry out state-to-state calculations for all the total angular momentum J up to 45 to converge differential cross sections of the title reaction for collision energies up to 0.4 eV. The numerical parameters used in the wave packet propagation in diatom–diatom coordinates are as follows. A total number of 115 sine functions (among them, 24 for the interaction region) are employed for the translational coordinate R in a range of $[2.3, 14.6]a_0$. A total of 33 vibrational functions are employed for r in the range of $[0.7, 5.0]$ for the reagent D_2 in the interaction region. We used 3 vibrational basis functions for OH in full dimensional calculations. For the rotational motion, we used $j_{1max} = 38$ for D_2 and $j_{2max} = 18$ for OH, which results in approximately 3200 rotational basis functions for $K = 0$ and even parity, and 6000 for $K > 0$. The number of K blocks used in the calculation increases with the total angular momentum, from 1 for $J = 0$, up to 7 for $J = 45$. Thus, the maximum number of the total rotational basis reaches $\sim$38,000 functions in the 7 K-block case. The initial wave packet located at $R_0 = 11$ a.u. For lower total angular momentum, J, we propagated the wave packets for 8500 a.u. of time with a time step of 15 to converge the low energy reaction probability. For $J > 15$, we propagated the wave packets for a shorter time because the reaction probability in the low energy region is negligible.

In atom–triatom coordinates, we used a total number of 120 sine functions for the translational coordinate R' in a range of $[3.2, 14.0]a_0$, 9 vibrational basis functions for r_1', and the same 3 vibrational basis functions for r_2' as for r_2 in diatom–diatom coordinates. We used $j_{1max} = 28$, $j_{2max} = 20$, resulting in a total of $\sim$5,000 coupled rotational basis functions for $K = 0$ with even parity, and $\sim$10,000 coupled rotational basis functions for $K > 0$. The number of K blocks used in the calculation increases with the total angular momentum, from 1 for $J = 0$, up to 15 for $J = 45$. Thus, the maximum number of the total rotational basis reaches $\sim$120,000 functions in the 15 K-block case. A dividing surface is placed at $R' = 10.0$ a.u. to extract S matrix elements.

3   Results

Fig. 1 presents some time-of-flight (TOF) spectra of the D-atom products from the title reaction measured for collision energies of 0.25 and 0.34 eV. The TOF spectra for these two collision energies are both dominated by two peaks corresponding to the first and second OD stretching excitations. There is a shoulder-like structure between these two main peaks in all these spectra, corresponding to the HOD product with one OD stretching and one bending excitation. As can be seen these TOD spectra have the same structures as those obtained by Davis and coworkers at a collision energy of 0.28 eV.[37]

Four components, that correspond to the individual vibrational features of the HOD product, were used to simulate each TOF spectra of the D-atom product in Fig. 1. By adjusting the contributions of the four components, a best fit to each TOF spectrum was obtained. The simulated TOF spectra obtained were then converted to the product translational energy distributions in the center of mass (COM) frame using the laboratory to COM conversion procedure. The D-atom product flux contour maps, constructed from the individual product translational energy distributions obtained above are displayed in Fig. 2(a). The HOD product is strongly backward scattered with two main peaks and should-like structure

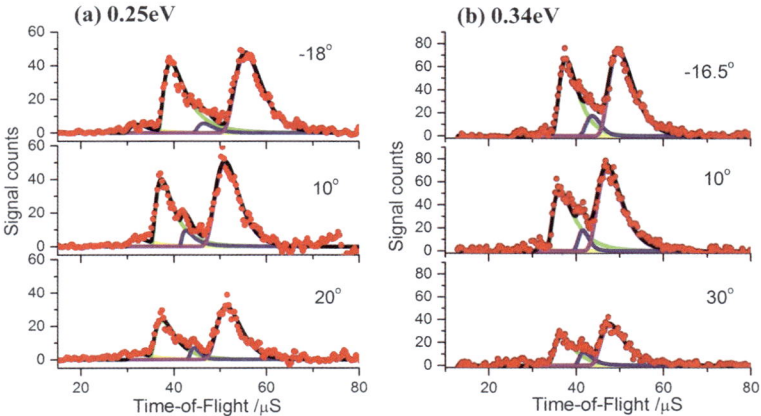

Fig. 1 The D-atom TOD spectra at three selected laboratory angles relative to the OH initial velocity at collision energies of 0.25 and 0.34 eV. Solid lines represent simulations based on best-fit translational energy and angular distributions. Four components, which correspond to four vibrational features in the HOD products, were used to fit these TOF spectra.

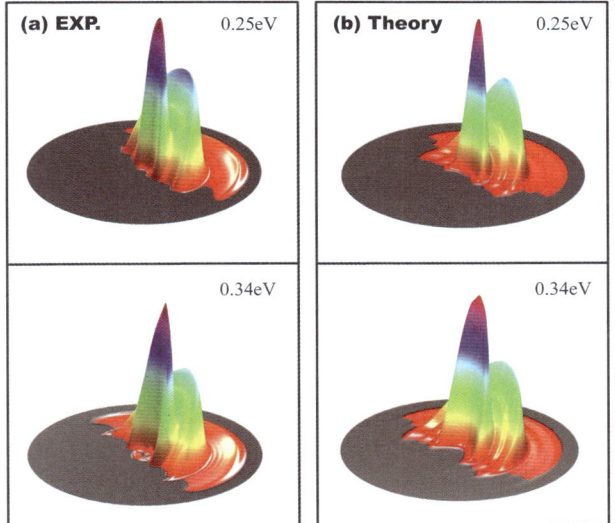

Fig. 2 Experimental (a) and theoretical (b) surface plots for product translational energy and angle distributions for the reaction at the collision energies of 0.25 and 0.34 eV.

between them, corresponding to the HOD product in (01), (02), and (11) vibrational states. Here we use (m,n) to label a vibrational state, where m is the quantum number for the bending mode and n for the OD stretching mode as in Refs. 23 and 56. As seen the D-atom product flux contour maps obtained in the current experiment have the same appearance as that obtained by Davis and coworkers at E = 0.28 eV for the reaction.[37]

Fig. 2(b) shows the theoretical DCS for the same collision energies. Overall, the agreement in the DCS between experiment and theory is good, in terms of both scattering angle and HOD product state. However, small differences between theory and experiment still can be observed. The peak heights for the (11) state given by theory are slightly lower than those of experiment at both collision energies. The experimental angle distributions are somewhat broader than the theory, in particular at a collision energy of 0.25 eV. A possible reason for the discrepancies may lie in the fact that the theoretical DCS is for the $D_2(j = 0) + OH(j = 0)$ reaction, while in the experiment both D_2 and OH are rotationally excited, in particular with normal D_2 used. In addition, although the CXZ PES used in the present study is expected to be very accurate, we still cannot guarantee that it is sufficiently accurate to provide DCS identical to the experiment.

Fig. 3 compares the theoretical energy dependence of DCS in the backward direction with the experimental measurement. Here the experimental data, measured in relative intensities, are scaled to the theoretical DCS value accordingly. As seen the experimental energy dependence of DCS at the backward direction shows a typical threshold behavior as for the HD + OH → H_2O + D reaction.[23] And the agreement between theory and experiment is excellent, indicating the theoretical reaction barrier of 0.235 eV of the present PES, calculated at CCSD(T)-F12/ AVTZ level of *ab initio* theory, is accurate.

Fig. 4 compares the theoretical and experimental product relative translational energy distribution at both collision energies. The experimental distributions show clearly two main peaks and a shoulder-like structure as in the TOF and D-atom product flux contour maps. Again, the agreement in the product translational energy distribution between experiment and theory is satisfactory, despite some small differences. As in DCS, theory underestimates the population of the (11) state for both collision energies. And at $E_c = 0.25$ eV, the relative population of the (01) and (02) states predicted by theory is also off a little bit.

Also shown in Fig. 4 are the potential-averaged five dimensional (PA5D) quantum scattering results with the OH bond fixed in its ground vibrational state.[4] The PA5D translational energy distributions are essentially identical as the full 6D results, except for a tiny energy shift. In fact, it is found that the PA5D DCS are also identical to the 6D ones for all the collision energies considered here. Therefore, the OH bond is indeed a good spectator in the dynamics of the reaction, and can be fixed in its initial vibrational state without losing any accuracy of calculations. This also means that it is sufficient for us to use HOD rovibrational states with the OH bonds fixed to expand the final product states, making the HOD rovibrational state assignment simpler.

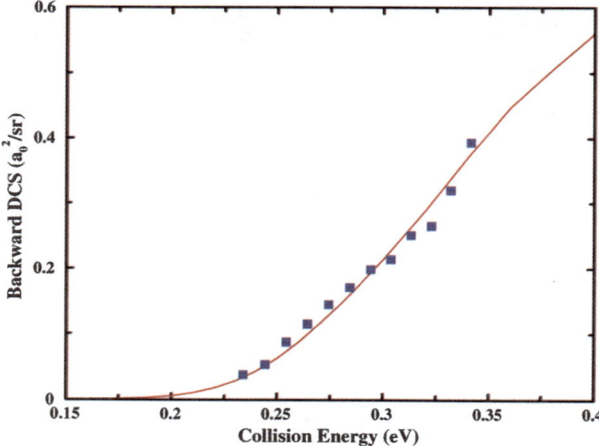

Fig. 3 Experimental and theoretical DCSs for the reaction in the backward direction as a function of collision energy.

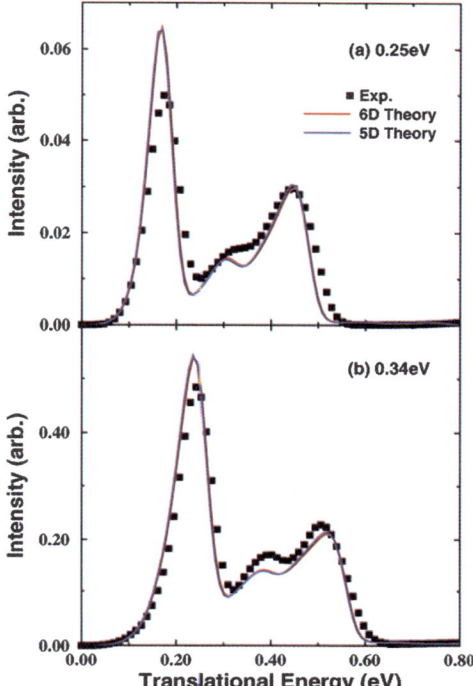

Fig. 4 Experimental and theoretical product relative translational energy distributions at the collision energies of 0.25 and 0.34 eV. The theoretical distributions were obtained from both full dimensional 6D and PA5D calculations.

Fig. 5 shows the theoretical vibrational state populations as a function of collision energy based on the PA5D calculation. Due to the large number of rovibrational levels of HOD even with the OH bond fixed and Fermi resonances between the OD stretching and bending excitations, it is difficult to assign all these states precisely. As a result, these vibrational state distributions may not be as accurate as what we have been calculating for triatomic reactions. From Fig. 5(a) one can see that the populations for (00), (10), (20), and (30) states are small, and only slightly change with the collision energy. The population for the (11) state increases slightly with the increase of collision energy in the low energy region, then oscillates around 8%. In contrast, the populations for the (01) and (02) states change substantially in the low collision energy region as shown in Fig. 5(b). As the collision energy increases from 0.15 to 0.25 eV, the (01) population drops from 80% to ~40%, while the (02) population rises from 5% to 45%. With the further increase of collision energy, the relative populations for these two states do not change too much any more. Therefore, for the title reaction, the (01) state is the dominant channel in the low collision energy region, and the (02) state becomes more important with collision energy higher than 0.24 eV. Interestingly, the total population for the (01) and (02) states only slightly fluctuates around 85% in the entire collision energy region as can be seen from Fig. 5(b). The substantial change of populations for these two states in the low collision energy region is just a shift of population from (01) to (02) states.

It is worthwhile to point out that due to Fermi resonances between the OD stretching and bending excitations it is impossible to clearly distinguish (20) and (01) states, (30) and (11) states, therefore one should not treat the populations shown in Fig. 5(a) for the (20) and (30) states too seriously, and it is better to use the

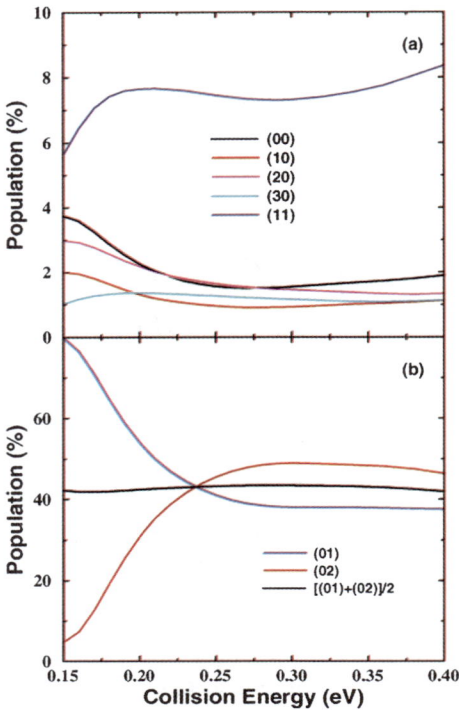

Fig. 5 The HOD vibrational state populations as a function of collision energy based on the PA5D calculation.

combined population of (20) and (01) states, and of (30) and (11) states, in order to compare with experimental results. At Ec = 0.28 eV, Davis and coworkers measured the ratio of population in the (00), (01), (11), and (02) states to be about 3 : 30 : 11 : 56,[37] in comparison with 1.5 : 0.9 : 39.9(=1.5 + 38.4) : 8.5(=1.2 + 7.3) : 48.4 of (00) : (10) : [(20) + (01)] : [(30) + (11)] : (02). Hence theory gives overestimation on the populations of (01) and underestimation on the population of the (11) and (02) states.

Fig. 6(a) presents the product HOD rovibrational energy measured from ground rovibrational level of HOD, together with the rotational energy, as a function of collision energy based on the PA5D calculation. As can be seen the product HOD rovibrational energy increases steadily with the increase of collision energy, but with two rather different increasing rates. It increases much faster in the low collision energy region between 0.15 to 0.2 eV than in the high collision energy region. In contrast, the HOD rotational energy first decreases from 0.07 eV (1.6 kcal mol^{-1}) to 0.055 eV (1.3 kcal mol^{-1}) as the collision energy increases from 0.15 eV to 0.24 eV, then increases slowly to 0.10 eV (2.3 kcal mol^{-1}) at E_c = 0.4 eV. At E_c = 0.28 eV, the current theoretical HOD rotational energy is 1.4 kcal mol^{-1}, which is considerably smaller than that of 2.4 kcal mol^{-1} given by Davis and coworkers (this average rotational energy is calculated from values given in Table I of Ref. 37, and accordingly the fraction of total energy deposited into HOD rotation should be changed to 11% instead of 5% given in the paper).

The different behavior of the product HOD rovibrational and rotational energies in low collision energy region compared to the high collision energy region is apparently related to the fast change of the populations for the (01) and (02) states. The shift of population from the (01) to the (02) state increases the overall rovibrational

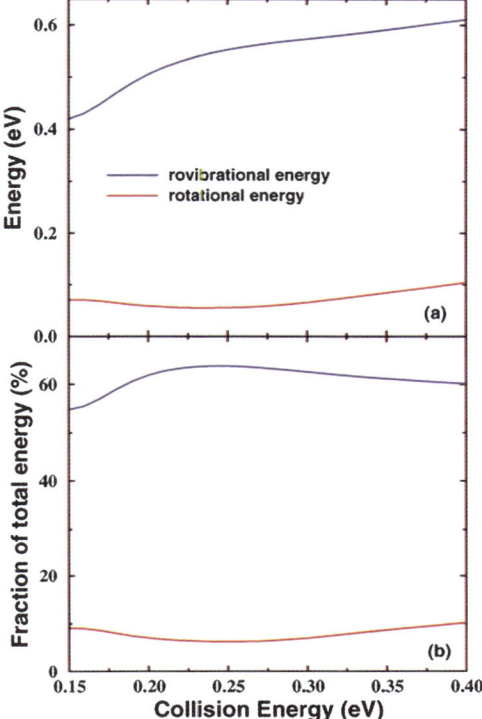

Fig. 6 (a) The product HOD rovibrational energy measured from ground rovibrational level of HOD, together with the rotational energy, as a function of collision energy based on the PA5D calculations; (b) The fraction of total available energy for the reaction deposited in the rovibrational and rotational motion of the HOD products.

energy of HOD, but decreases the rotational energy because of the lower rotational energy of the (02) state in that collision energy region.

Fig. 6(b) shows the fraction of total available energy for the reaction deposited in the rovibrational and rotational motion of the HOD product. Although the rovibrational energy of HOD increases monotonically with the collision energy, the fraction only increases in low collision energy region, from 55% at 0.15 eV to 64% at 0.25 eV, then decreases slowly with the further increase of collision energy. The fraction of total available energy deposited in the rotational motion of HOD product first decreases from 9.2% at 0.15 eV to 6.4% at 0.25 eV, then increases to 10.2% at 0.4 eV. At E_c = 0.28 eV, the fraction of total available energy deposited into HOD rovibrational excitation and rotation motion is 63% and 6.6%, respectively, compared with 69% and 11% measured from the experiment.[37]

4 Conclusions

A crossed molecular beam experiment has been carried out to study the $D_2 + OH \rightarrow HOD + D$ reaction at collision energies of 0.25 and 0.34 eV. The HOD product is found to be strongly backward scattered. The DCS and the relative translational energy distributions exhibit two main peaks and shoulder-like structure between them corresponding to the HOD product in (01), (02), and (11) vibrational states, in good agreement with the experiment of Davis and coworkers.[37] The measured energy dependence of DCS at the backward direction shows a typical threshold behavior as for the $HD + OH \rightarrow H_2O + D$ reaction.[23]

The time-dependent wave packet method has been employed to calculate the differential cross section for the $D_2(j = 0) + OH(j = 0)$ reactions in full dimensions, utilizing an improved version of the reactant–product–decoupling scheme. It is found that the agreement between theory and experiment on the DCS and the product relative translational distributions are satisfactory, and that on the energy dependence of the DCS at the backward direction is excellent. We believe the small discrepancies between theory and experiment very likely comes from the difference in rotational states of the reagents used in experimental measurements and theoretical simulations.

The PA5D quantum scattering calculations with the OH bond fixed in the ground vibrational state is able to provide essentially identical results to the full 6D calculations. This not only reveals that the OH bond is a good spectator in the reaction, but also makes it simpler to obtain vibrational distributions of the product. The vibrational distributions reveal that in the low collision region the (01) state is the dominate channel. With the rise of the collision energy, the population of the (01) state decreases and that for the (02) state increases. The (02) state becomes the dominate channel with collision energy higher than 0.24 eV. It is found that the theoretical HOD vibrational distribution at $E_c = 0.28$ eV agrees reasonably well with the experimental results of Davis and coworkers. Theory also reveals that with the increase of collision energy, the product HOD rovibrational energy increases steadily, while the HOD rotational energy first decreases, then increases. At $E_c = 0.28$ eV, the fraction of total available energy deposited into HOD rovibrational excitation and rotation motion is 63% and 6.6%, respectively, compared with 69% and 11% given by the experiment of Davis and coworkers.[37]

Overall, the agreement between theory and experiment on this benchmark four-atom reaction is good, but not perfect. To achieve better agreement, the experiment should be carried out with well-defined initial rotational distribution of the reagents. On the theoretical side, quantum dynamics for rotationally excited initial states should be performed to bring quantitative agreement between theory and experiment, as has been done for the simple triatomic reactions.

Acknowledgements

This work was supported by the National Natural Science Foundation of China, the Chinese Academy of Sciences, and the Ministry of Science and Technology of China.

References

1 D. H. Zhang and J. Z. H. Zhang, *J. Chem. Phys.*, 1993, **99**, 5615.
2 D. Neuhauser, *J. Chem. Phys.*, 1994, **100**, 9272.
3 U. Manthe, T. Seideman and W. H. Miller, *J. Chem. Phys.*, 1993, **99**, 10078.
4 D. H. Zhang and J. Z. H. Zhang, *J. Chem. Phys.*, 1994, **101**, 1146.
5 D. H. Zhang and J. C. Light, *J. Chem. Phys.*, 1996, **104**, 4544.
6 D. H. Zhang and J. C. Light, *J. Chem. Phys.*, 1996, **105**, 1291.
7 W. Zhu, J. Dai, J. Z. H. Zhang and D. H. Zhang, *J. Chem. Phys.*, 1996, **105**, 4881.
8 J. Dai, W. Zhu and J. Z. H. Zhang, *J. Phys. Chem.*, 1996, **100**, 13901.
9 S. K. Pogrebnya, J. Echave and D. C. Clary, *J. Chem. Phys.*, 1997, **107**, 8975.
10 D. H. Zhang and J. C. Light, *J. Chem. Soc., Faraday Trans.*, 1997, **93**, 691.
11 F. Matzkies and U. Manthe, *J. Chem. Phys.*, 1998, **108**, 4828.
12 D. H. Zhang, J. C. Light and S.-Y. Lee, *J. Chem. Phys.*, 1998, **109**, 79.
13 D. H. Zhang and S.-Y. Lee, *J. Chem. Phys.*, 1999, **110**, 4435.
14 D. H. Zhang, M. A. Collins and S.-Y. Lee, *Science*, 2000, **290**, 961.
15 D. H. Zhang, M. Yang and S.-Y. Lee, *J. Chem. Phys.*, 2001, **114**, 8733.
16 D. H. Zhang, M. Yang and S.-Y. Lee, *J. Chem. Phys.*, 2002, **116**, 2388.
17 D. H. Zhang, M. Yang and S.-Y. Lee, *Phys. Rev. Lett.*, 2002, **89**, 103201.
18 D. H. Zhang, M. Yang and S.-Y. Lee, *J. Chem. Phys.*, 2002, **117**, 10067.
19 E. M. Goldfield and S. K. Gray, *J. Chem. Phys.*, 2002, **117**, 1604.
20 D. H. Zhang, D. Xie, M. Yang and S. Y. Lee, *Phys. Rev. Lett.*, 2002, **89**, 283203.
21 D. H. Zhang, *J. Chem. Phys.*, 2006, **125**, 133102.

22 M. T. Cvitas and S. C. Althorpe, *J. Chem. Phys.*, 2011, **134**, 024309.
23 C. Xiao, X. Xu, S. Liu, T. Wang, W. Dong, T. Yang, Z. Sun, D. Dai, D. H. Zhang and X. Yang, *Science*, 2011, **333**, 440.
24 B. Fu, Y. Zhou and D. H. Zhang, *Chem. Sci.*, 2012, **3**, 270.
25 D. C. Clary, *J. Phys. Chem.*, 1994, **98**, 10678.
26 J. M. Bowman and G. C. Schatz, *Annu. Rev. Phys. Chem.*, 1995, **46**, 169–196.
27 J. Warnatz, *Combustion Chemistry*, edited by W. C. Gardiner, Springer-Verlag, New York, 1984.
28 T. J. Millar and D. A. Williams, Rate Coefficients in Atmospheric Chemistry.
29 A. Sinha, M. C. Hsiao and F. F. Crim, *J. Chem. Phys.*, 1990, **92**, 6333.
30 F. F. Crim, *Acc. Chem. Res.*, 1999, **32**, 877.
31 M. J. Bronikowski, W. R. Simpson, B. Girard and R. N. Zare, *J. Chem. Phys.*, 1991, **95**, 8647.
32 R. N. Zare, *Science*, 1998, **279**, 1875.
33 I. W. M. Smith and F. F. Crim, *Phys. Chem. Chem. Phys.*, 2002, **4**, 3543–3551.
34 M. Alagia, N. Balucani, P. Casavecchia, D. Stranges and G. G. Volpi, *J. Chem. Phys.*, 1993, **98**, 8341.
35 M. Alagia, N. Balucani, P. Casavecchia, D. Stranger, G. G. Volpi, D. C. Clary, A. Kliesch and H. J. Werner, *Chem. Phys.*, 1996, **207**, 389–409.
36 L. Schnieder, K. Seekamp-Rahn, E. Wrede and K. H. Welge, *J. Chem. Phys.*, 1997, **107**, 6175.
37 B. R. Strazisar, C. Lin and H. Floyd Davis, *Science*, 2000, **290**, 958.
38 A. Rodriguez, E. Garcia, J. M. Alvarino and A. Laganà, *Chem. Phys. Lett.*, 2001, **345**, 219–227.
39 M. J. Lakin, D. Troya, G. Lendvay, M. González and G. C. Schatz, *J. Chem. Phys.*, 2001, **115**, 5160.
40 J. D. Sierra, P. A. Enri'quez, D. Troya and M. Gonzalez, *Chem. Phys. Lett.*, 2004, **399**, 527–533.
41 D. Sierra, R. Martínez, J. Hernando and M. González, *Phys. Chem. Chem. Phys.*, 2009, **11**, 11520–11527.
42 J. Espinosa-Garcia, L. Bonnet and J. C. Corchado, *Phys. Chem. Chem. Phys.*, 2010, **12**, 3873–3877.
43 L. Bonnet, J. Espinosa-García, J. Corchado, S. Liu and S.-Y. Lee, *Chem. Phys. Lett.*, 2011, **516**, 137–140.
44 J. D. Sierra, L. Bonnet and M. Gonzalez, *J. Phys. Chem. A*, 2011, **115**, 7413–7417.
45 G. C. Schatz and H. Elgersman, *Chem. Phys. Lett.*, 1980, **73**, 21.
46 G. O. de Aspuru and D. C. Clary, *J. Phys. Chem. A*, 1998, **102**, 9631–9637.
47 S. K. Pogrebnya, J. Palma, D. C. Clary and J. Echave, *Phys. Chem. Chem. Phys.*, 2000, **2**, 693–700.
48 G. Wu, G. C. Schatz, G. Lendvay, D. C. Fang and L. B. Harding, *J. Chem. Phys.*, 2000, **113**, 3150.
49 M. Yang, D. H. Zhang, M. A. Collins and S. Y. Lee, *J. Chem. Phys.*, 2001, **115**, 174.
50 P. Defazio and S. K. Gray, *J. Phys. Chem. A*, 2003, **107**, 7132.
51 M. Yang, D. H. Zhang, M. A. Collins and S.-Y. Lee, *J. Chem. Phys.*, 2001, **114**, 4759.
52 M. T. Hagan and M. B. Menhaj, *IEEE Trans. Neural Networks*, 1994, **5**, 989–993.
53 W. S. Sarle, *Proceedings of the 27th Symposium on the Interface*, 1995.
54 L. M. Raff, M. Malshe, M. Hagan, D. I. Doughan, M. G. Rockley and R. Komanduri, *J. Chem. Phys.*, 2005, **122**, 084104.
55 J. Chen, X. Xu and D. H. Zhang, in preparation.
56 S. Liu, X. Xu and D. H. Zhang, *J. Chem. Phys.*, 2012, **136**, 144302.

General discussion

Professor Truhlar opened the discussion of the paper by Professor Crim: Professor Crim made an interesting analogy between intersections of potential curves in electronically nonadiabatic collisions and the intersection of two potential curves for molecular collisions with surfaces of solids: one curve for physisorption of AB at a surface and one curve for simultaneous chemisorption of A and B at the surface. On this basis the process of AB colliding with the surface and producing chemisorbed A and chemisorbed B was described as electronically nonadiabatic. However, the curve for physisorption of AB at a surface and that for simultaneous chemisorption of A and B at a surface do not represent different electronic states, but rather two different cuts through a single potential energy surface for a given electronic state, and thus this process is electronically adiabatic, not electronically nonadiabatic.

Professor Sibener remarked: I would like to make a general comment about Fleming Crim's Introductory Lecture. During this presentation there was a comprehensive and very clear discussion of how reaction dynamics are influenced by the global energetics as well as specific details of the potential energy surface(s) for a given reaction. I would like to note that both in rarefied phases as well as condensed systems, effects due to alignment and orientation between reaction partners are also, obviously, of paramount importance in influencing the evolution of a given encounter. At lower energies "adiabatic" approaches can occur during which the colliding reagents in either gas phase or gas–surface encounters can reorient themselves given the presence of anisotropic molecular polarizability; this is in contrast to more sudden collisions that do not significantly reorient at higher collision energies. Moreover, reactive gas–surface encounters can be very different than in the gas phase given that the surface-bound reaction partner may even be pre-oriented due to surface bonding that presents a non-rotating molecule to the incident reagent. Gas–surface reactions offer a further level of complexity due to the presence of the extended surface, which offers a local chemical environment that can adsorb one or more molecular fragments coming from a reaction, significantly changing the energetic and pathway options that are available for such systems.

Professor Polanyi addressed Professor Sibener: In the discussion of Fleming Crim's introductory overview, Steven Sibener remarked, correctly, that different modes of binding to a surface (*e.g.* to form one bond, as against two) can have very different rates. It is interesting to note, additionally, that even a single reaction pathway, yielding a single set of products, can have differing rates depending on the physisorption geometry through which the reaction proceeds. An example is the case of 1-bromopentane reacting to brominate Si(111) ,which has been shown by both experiment and theory to have a lower activation energy for reaction with the alkane tail vertical in the physisorbed state, than in the alternate stable physisorbed configuration with the alkane tail horizontal.[1] The reason, briefly stated, is that for the reaction to occur the hydrocarbon tail of the horizontal molecule must be lifted to bring C···Br···Si more nearly into alignment; this requires work to be done, and elevates the activation energy for reaction by way of the horizontal physisorbed state. The identification and characterisation of reactions involving differently physisorbed reagents would seem to be a topic ripe for exploration.

1 K. Huang *et al.*, *Angew. Chem., Int. Ed.*, 2012, **51**, DOI: 10.1002/anie.201202772, in press.

Professor Sibener replied: I agree with John's comments. Another important experimental extension of reactivity studies that is now getting underway, in our laboratory and perhaps elsewhere, seeks to use surface-adsorbed molecules to provide geometry-restricted (*i.e.*, highly aligned and oriented) molecular targets for attack by incident gas-phase species. Direct access to reactive scattering products and their associated dynamical properties coupled with on-surface infrared spectroscopy and scanning probe imaging will, hopefully, shed further light on reaction mechanisms and the potential energy surfaces that govern such encounters. The synergy between rarefied gas-phase reaction studies and those occurring at surfaces as herein discussed remain only partially exploited at the time of this Discussion; we should be hopeful that further advances in our knowledge of reactive encounters will be forthcoming as gas–surface scattering and surface-charaterization methods continue to evolve and be utilized for studies in chemical reaction dynamics.

Professor Nesbitt opened the discussion of the paper by Professor Bowman: In the fragmentation channel to form HCO + CH_3, there is also sufficient energy to form the less stable isomer HOC. Did you find any evidence for this contributing in your trajectory calculations?

Professor Bowman answered: HOC is an interesting, so-called "hanging" high energy isomer of HCO. That is, it's minimum energy is above the energy to form H + CO. It was first described I believe in a potential energy surface in the paper '*Ab initio* calculations of electronic and vibrational energies of HCO and HOC'.[1] We do not see evidence of this isomer in our calculations, most likely because it cannot be formed directly in the dissociation of CH_3CHO to the products HCO + CH_3.

The general question about the role of high-energy isomers in dissociation is certainly a good one. For example, we do see the high energy isomer of acetylene, vinylidene, in the dissociation of C_3H_5 as reported in the paper 'Evidence for vinylidene production in the photodissociation of the allyl radical'.[2]

1 J. M. Bowman, J. S. Bittman and L. B. Harding, *J. Chem. Phys.*, 1986, **85**, 911.
2 C. Chen, B. Braams, D. Y. Lee, J. M. Bowman, P. L. Houston and D. Stranges, *J. Phys. Chem. Lett.*, 2010, **1**, 1875–1880.

Professor Neumark asked: In your paper, you discuss both "hard" and "soft" constraints on zero-point energy in your calculations. The conclusion seems to be that the soft constraint works better. Can you explain why that might be the case? Also, in Fig. 7 of your paper, many of the trajectories have either the CH_3 and CD_3 vibrational energy below the zero point energy of the fragment. How does one interpret this in light of your overall emphasis on the importance of zero-point energy in trajectory calculations?

Professor Bowman answered: You raise an important question. First, let me note that the soft constraint does respect the correct, *i.e.*, quantum, threshold condition based on the zero-point energies (ZPEs) of the polyatomic fragments. In our experience with QCT calculations, we find that each fragment rarely comes out with vibrational energies much less than their correct ZPE. More typically, we find that one fragment satisfies its ZPE condition while the other one may not. So the hard constraint, which would discard all such trajectories, often "overcorrects" by discarding far too many trajectories. The soft constraint keeps such trajectories. I also note that the soft constraint is consistent with what is now termed "1GB", where GB refers to "Gaussian Binning". This method is described in the paper by Czako and Bowman.[1]

1 G. Czako and J. M. Bowman, *J. Chem. Phys.*, 2009 **131**, 244302.

Professor Truhlar commented: The paper mentions the problem of enforcing zero point energy and cites, among other work, my paper with Bonhommeau. Our method[1] modifies the dynamics so that the local zero point energy is maintained in addition to

This journal is © The Royal Society of Chemistry 2012

insuring that no products are formed with less than zero point energy. The local constraint is not an exact requirement of quantum mechanics (as is the product constraint), but the local constraint is important for getting through bottleneck regions. If one does not do this, a system can pass through a bottleneck region with less than local zero point energy, but extensive experience with variational transition state theory (and conventional transition state theory when there are no variational effects) and with quantized dynamical bottlenecks shows that such a procedure can severely underestimate reaction thresholds for many kinds of reaction.[2,3] In general, it is even more important to maintain zero point constraints at dynamical bottlenecks than at products, and I recommend that this should be done in the general case.

1 D. Bonhommeau and D. G. Truhlar, *J. Chem. Phys.*, 2008, **129**, 14302.
2 B. C. Garrett, D. G. Truhlar, J. M. Bowman and A. F. Wagner, J. Phys. Chem., 1986, **90**, 4305.
3 D. C. Chatfield, R. S. Friedman, S. L. Mielke, G. C. Lynch, T. C. Allison, D. G. Truhlar and D. W. Schwenke, in *Dynamics of Molecules and Chemical Reactions*, ed. R. E. Wyatt and J. Z. H. Zhang, Marcel Dekker, New York, 1996, pp. 323–386.

Professor Bowman answered: In general, it is good to have several ways to apply a zero-point energy (ZPE) constraint. As I noted in my paper, we applied "hard" and "soft constraints" to the products of photodissociation. In a general sense, the "soft constraint" is something like the constraint that Truhlar describes since is applies to the total ZPE of the products. Also, we have introduced a new Gaussian Binning method, that is now referred to as 1GB[1] which is based on the ZPE of the molecule instead of the quantum numbers of each mode being zero, for the ground vibrational state. This is also in a sense a more "coarse grained" way to deal with quantizing states in classical trajectory calculations.

1 G. Czakó and J. M. Bowman, Quasiclassical trajectory calculations of correlated product distributions for the $F+CHD_3$ ($v_1 = 0,1$) reactions using an *ab initio* potential energy surface, *J. Chem. Phys.*, 2009, **131**, 244302.

Professor Truhlar remarked: Our method constrains the local vibrational energy to be greater than or equal to the local zero point energy at each point along the trajectory, but it doesn't place a separate constraint on each mode. We found that constraining each mode led to less physical results.

Professor Bowman responded: First, there are several ways to apply a zero-point energy constraint, based on how one defines the modes. Second, the choice of constraint depends on the application. In our recent work, we do not use "instantaneous normal modes" which can have multiple imaginary frequency modes. At every time when we wish to check the zero-point energy we stop the trajectory and then locate the nearest local minimum and use the associated normal modes for analysis. We did this for the study of water dimer and water trimer. Then we did constrain the high frequency OH-stretches. This was important as we were interested in the OH radial distribution functions, which depend sensitively on the zero-point motion in those modes. This was shown in the paper 'Zero-point energy constrained quasiclassical, classical, and exact quantum simulations of isomerizations and radial distribution functions of the water trimer using an *ab initio* potential energy surface'[1]. If we used a total zero-point energy constraint in this application I believe we would not have obtained results in as good agreement with the exact quantum ones, which were also reported.

1 G. Czakó, A. L. Kaledin and J. M. Bowman, *Chem. Phys. Lett.*, 2010, **500**, 217–222.

Professor Kable remarked: We recently obtained experimental data on the photodissociation of formaldehyde at different photolysis energies. Using ion imaging of the H atom fragment we have measured the recoil kinetic energy distribution at higher energies than previously measured. These data are unpublished, but I would like to share them here as they are relevant to the question that I want to ask. Fig. 1 shows an image (Panel a) and kinetic energy distribution (Panel b) at an energy just at the barrier on the triplet surface (~90 kcal mol⁻¹). The distribution is skewed to high translational energy, indicative of triplet reaction, as characterized in previous work.[1,2] In panels c) and d) of the figure are an image and distribution for dissociation at much higher energy (102 kcal mol⁻¹). The distribution is now peaked at low translational energy, which indicates dissociation on a bound potential surface, which we interpret as following internal conversion to S_0. For internal conversion to be competitive we suggest that this energy is at or near an S_1/S_0 conical intersection (CI). Such a CI has been calculated by Robb and co-workers at an energy of ~107 kcal mol⁻¹.[3]

1 L. R. Valachovic, M. F. Tuchler, M. Dulligan, Th. Droz-Georget, M. Zyrianov, A. Kolessov, H. Reisler and C. Wittig, *J. Chem. Phys.*, 2000, **112**, 2752.
2 H-M. Yin, S. J. Rowling, A. Büll and S. H. Kable, *J. Chem. Phys.*, 2007, **127**, 064302.
3 M. Araujo, B. Lasorne, A. L. Magalhaes, M. J. Bearpark and M. A. Robb, *J. Chem. Phys.*, 2009, **131**, 144301.

Professor Kable remarked: In your three state model it seems that the ground state, S_0, is populated *via* sequential intersystem crossing: $S_1 \rightarrow T_1 \rightarrow S_0$. Is there any evidence of a conical intersection between S_1 and S_0 that would prepare S_0 *via* direct internal conversion?

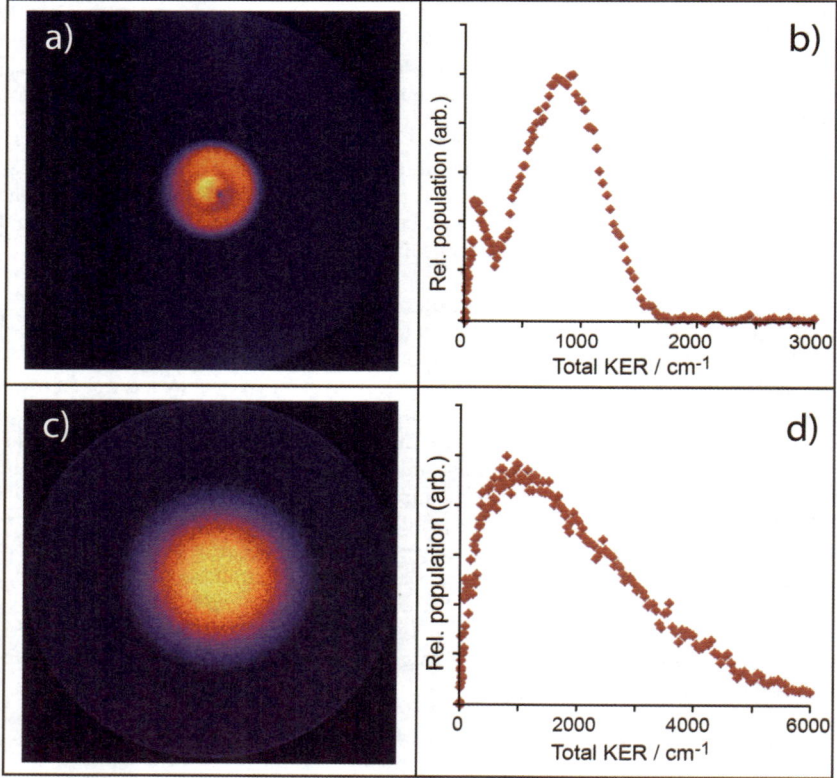

Fig. 1 Ion images and total kinetic energy distributions for $H_2CO + h\nu \rightarrow HCO + H$. In panels a) and b) total photon energy is 90 kcal mol⁻¹, while in c) and d) it is 102 kcal mol⁻¹.

Professor Bowman replied: Yes there is a conical intersection between S_1 and S_0 in acetaldehyde. This has been reported by Chen and Fang.[1] According to their limited calculations a barrier on S_1 of roughly 100 kcal mol^{-1}, relative to $CH_3CHO(S_0)$, would need to be surmounted to reach it. Further the geometry of CI is substantially "outside" the Franck–Condon region of the photoexcitation. So, we did not consider this CI in our calculations, which I should mention also took note of experiments, including your recent one, which indicate that even at photolysis energies above 100 kcal mol^{-1} the CH_3 + HCO products correlate essentially 100% from T_1.

However, it is certainly important to keep this CI in mind for future work.

1 S. Chen and Fang, *J. Chem. Phys.*, **131**, 2009, 054306.

Dr Glowacki addressed Professor Bowman and Professor Kable: I have two questions:

(1) In Fig. 5 of your paper, you show that the yield of CH_3 + HCO increases with photolysis energy. Is it best to understand these results as the T_1 channel turning on, or as the S_0 channel turning off?

(2) In your study, molecules in the S_1 state get to S_0 through a sequential mechanism, *i.e.*, $S_1 \rightarrow T_1 \rightarrow S_0$. However, I am wondering whether you reckon it is important to consider the other pathway of $S_1 \rightarrow S_0$. At the energies considered in your study, you are almost certainly below the S_1/S_0 conical intersection. However, it is often possible for tunnelling like mechanisms to occur below the intersection energy. For example, Eric Heller discussed tunnelling like mechanisms that occur below the intersection threshold.[1] I wonder if there is any experimental or theoretical evidence for this sort of behavior in your system at the energies that you are considering.

1 Heller and Brown, *J. Chem. Phys.*, 1983, **79**, 3336.

Professor Bowman responded: (1) Yes that is essentially correct. The threshold energy for the T_1 channel is the exit channel barrier on the T_1 potential energy surface.

(2) Yes tunneling is a possibility at the S_1/S_0 conical intersection (CI) even if the CI energy is above the photolysis energy. However, it also important to note that classically (and quantum mechanically too) transitions between S_1 and S_0 can occur away from the CI, depending on the magnitude of the derivative coupling, even possibly at energies below the energy of the CI. In any case, such transitions do not takes place in the current calculations, as the derivative coupling was not obtained. It might be interesting to examine that coupling in the future.

Professor Kable responded: I can comment on the experimental evidence part of Dr Glowacki's question. There is indirect evidence of direct S_1–S_0 internal conversion in acetaldehyde. The quantum yield of CH_4 + CO and CH_3CO + H increases strongly when the photolysis wavelength is shorter than 275 nm.[1] The steepness of the onset suggests a new kinetic mechanism. A reasonable interpretation of these data is that a conical intersection between S_1 and S_0 is nearby, which effectively channels the flux directly onto S_0 and away from T_1, where these chemical pathways compete with the otherwise dominant HCO + CH_3.

The evidence for direct S_1–S_0 coupling is more direct in formaldehyde, where there is a pronounced dependence of reaction on S_0 or T_1 depending on the initial H_2CO state.[1,2] This has been interpreted as weak coupling to S_0 in competition with stronger coupling to a sparse manifold of states on T_1.[2] In the case of H_2CO it is unlikely that a conical intersection between S_1 and S_0 has been reached, but that the coupling is more the tunneling type mechanism that you suggest.

1 G. K. Moortgat, H. Meyrahn and P. Warneck, *ChemPhysChem*, 2010, **11**, 3896.
2 L. R. Valachovic, M. F. Tuchler, M. Dulligan, Th. Droz-Georget, M. Zyrianov, A. Kolessov, H. Reisler and C. Wittig, *J. Chem. Phys.*, 2000, **112**, 2752.

3 W. S. Hopkins, H-P. Loock, B. Cronin, M. G. D. Nix, A. L. Devine, R. N. Dixon, M. N. R. Ashfold, H-M. Yin, S. J. Rowling, A. Büll and S. H. Kable, *J. Phys. Chem. A*, 2008, **112**, 9283.

Professor Casavecchia addressed Professor Bowman and Professor Kable:† With reference to the paper by Bowman and coworkers, where quasiclassical (QCT) three-state surface hopping calculations of acetaldehyde photodissociation to CH_3 + HCO on *ab initio* potential energy surfaces (PESs) are described, we would like to recall recent results on combined experimental/theoretical studies[1,2] of the bimolecular $O(^3P)+C_2H_4$ reaction which also exhibits a reaction channel, among several others, leading to CH_3+HCO, and ask Professor Bowman a question about the interconnection between the unimolecular and bimolecular processes leading to CH_3 + HCO formation and the role of the underlying singlet and triplet potential energy surfaces (PESs) and their nonadiabatic couplings.

The $O(^3P)$ + C_2H_4 reaction, of importance in combustion and atmospheric chemistry, stands out as paradigm reaction involving triplet and singlet state PESs interconnected by intersystem crossing (ISC) (see Fig. 2 below). Very recently, primary products from five competing channels (H + CH_2CHO, H + CH_3CO, H_2 + CH_2CO, CH_3 + HCO, CH_2 + H_2CO) and branching ratios (BRs) were determined in crossed molecular beam (CMB) experiments with soft electron-ionization mass-spectrometric detection at the collision energy, E_c, of 8.4 kcal mol^{-1}.[1] This study extends previous CMB work at higher E_c.[3,4] As some of the observed products can only be formed *via* ISC from triplet to singlet PESs (see Fig. 2), from the product BRs the extent of ISC is inferred. About 50% of the reaction is found to proceed on the triplet PES leading to H + CH_2CHO (vinoxy) and CH_2 + H_2CO (formaldehyde) products, while 50% proceeds on the singlet PES *via* $T_1 \rightarrow S_0$ ISC, leading to

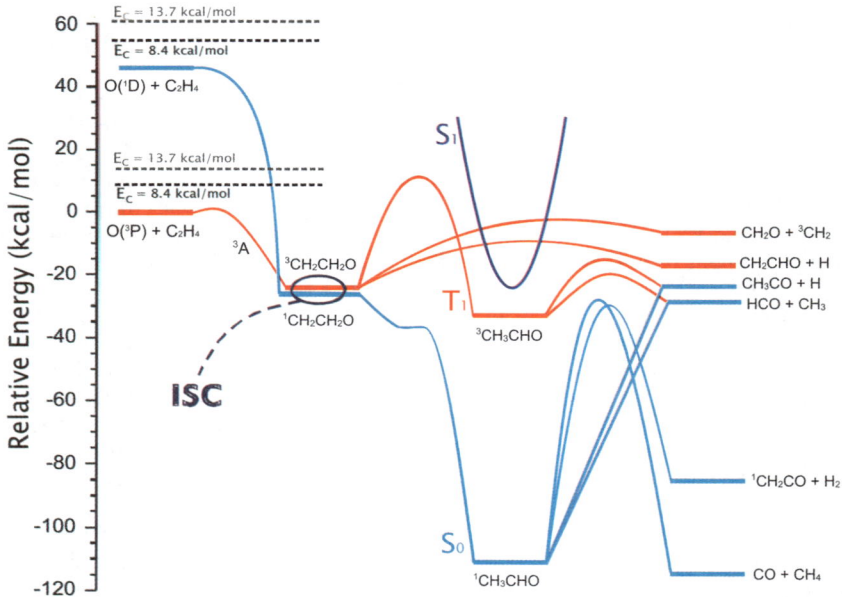

Fig. 2 Simplified schematic of triplet (red) and singlet (blue) potential energy surfaces for the $O(^3P, ^1D)+C_2H_4$ reactions. The region enclosed by the ellipse is where the intersystem crossing probability is highest between the triplet T_1 and singlet S_0 PESs.[1,6,7]

† on behalf of Dr. Francesca Leonori and Prof. Nadia Balucani

CH$_3$ + HCO, H$_2$ + CH$_2$CO (ketene) and H + CH$_3$CO (acetyl). A new full dimensional PES for the triplet state as well as spin–orbit coupling to the singlet PES were also reported by your group in a synergic fashion with our CMB experiments and roughly half a million of surface hopping trajectories were run on the coupled singlet–triplet PESs to compare with the experimental BRs and differential cross sections.[1] Both theory and experiment find almost equal contributions from the two PESs to the reaction. Detailed comparisons at the level of angular and translational energy distributions between theory and experiment were presented for the two primary channel products, CH$_3$ + HCO and H + CH$_2$CHO.[1] The agreement between experimental and theoretical functions is excellent, so implying that theory has reached the capability of describing complex multichannel nonadiabatic reactions. More recently, this was shown to be so also for the dynamics of the channel leading to CH$_2$ + H$_2$CO in another joint paper in which it was reported on a combined experimental and theoretical study of the reaction O(^{3}P)+C$_2$H$_4$ as a function of collision energy.[2]

Fig. 2 shows a schematic of the full ground state singlet (S$_0$) and triplet (T$_1$) potential energy surfaces used by Fu *et al.*[1,2] to describe theoretically and interpret the dynamics of the O(^{3}P) + C$_2$H$_4$ reaction. Indicated in Fig. 2 is also the first excited singlet surface, S$_1$, which is the state *via* excitation of which photolysis of acetaldehyde occurs in the UV region. S$_1$ acetaldehyde might convert to the T$_1$ state *via* ISC followed by C–C bond breaking to CH$_3$ + HCO if the excitation energy is greater than the dissociation barrier on the triplet PES. In turn, T$_1$ acetaldehyde may undergo ISC to S$_0$, followed by a barrierless C–C bond breaking to CH$_3$ + HCO. However, S$_0$ may also be reached by internal conversion from S$_1$. Distributions of product states in the UV photolysis showed that HCO formed on the S$_0$ PES is produced with essentially a statistical distribution (the translational energy distribution peaking at low energy values), whereas HCO formed on the T$_1$ PES is characterized by a translational energy distribution peaking at a high energy value, reflecting an impulsive energy release;[5] this is confirmed by your present calculations (see Fig. 6 in the paper).

In the O(^{3}P) + C$_2$H$_4$ reaction, a triplet biradical ^{3}CH$_2$CH$_2$O is initially formed following the initial electrophilic attack of the oxygen atom to the double bond of ethylene. Because of the large isomerization barrier (of about 12 kcal mol^{-1} above the reactant asymptote – see Fig. 2) between ^{3}CH$_2$CH$_2$O and ^{3}CH$_3$CHO (triplet acetaldehyde), formation of HCO + CH$_3$ cannot take place on the triplet PES (T$_1$ in Fig. 2) at the experimental collision energies (For the same reason this part of the PES is not accessible in photolysis studies around 300 nm). On the other hand, HCO + CH$_3$ is observed to be the most important channel (BR = 0.34 ± 0.09) at E_c=8.4 kcal mol^{-1}, which implies that ISC from T$_1$ to S$_0$ takes place very efficiently.[1] The spin–orbit induced nonadiabatic transition between T$_1$ and S$_0$ in the bimolecular O(^{3}P) + C$_2$H$_4$ reactions occurs in the PES region where the triplet and singlet biradicals are very close in energy (the energy gap being of a few kcal mol^{-1}) (see Fig. 2), in an extended region of configuration space.[1,6,7] The theoretically calculated spin–orbit T$_1$/S$_0$ coupling is about 35 cm^{-1}.[1]

It is interesting to compare the HCO + CH$_3$ product translational energy distributions P(E_T) calculated in the photolysis process at an energy of 100 kcal mol^{-1} (at which the contribution from the T$_1$ and S$_0$ PES are about the same – see Fig. 5 of the paper) with the P(E_T) measured (and also calculated) in the bimolecular reaction at E_c = 8.4 kcal mol^{-1} (ref. 1) (this corresponds to a total energy, from the singlet acetaldehyde minimum, of about 120 kcal mol^{-1}). The peaking (at about 2–3 kcal mol^{-1}) and shape of the P(E_T) for the photodissociation of ^{1}CH$_3$CHO to CH$_3$ + HCO from S$_0$ (see bottom panel of Fig. 6 in the paper) resembles very closely the peaking and shape of the P(E_T) determined experimentally and calculated theoretically for the HCO + CH$_3$ channel in the O(^{3}P) + C$_2$H$_4$ reaction (see Fig. 4B in ref. 1); this is consistent with the fact that in the bimolecular reaction the HCO + CH$_3$ products are formed solely on the S$_0$ PES, after T$_1 \rightarrow$ S$_0$ ISC has occurred. In contrast, the peaking and shape of the P(E_T) for HCO + CH$_3$ formation from T$_1$ in the photo-dissociation at a photolysis energy of 100 kcal mol^{-1} (see top panel of Fig. 6 in the paper) is quite different, rising much more slowly with respect to the P(E_T) on S$_0$, and peaking at about 7 kcal mol^{-1}, which reflects an exit potential barrier on the triplet PES (calculated to be about 6 kcal mol^{-1}) (see the paper and ref. 1). Notably, this peaking and shape of the P(E_T) function resemble closely those derived experimentally (and calculated theoretically) in the bimolecular reaction for the CH$_2$CHO + H channel, that occurs on the T$_1$ PES (see Fig. 4A in ref. 1) and is also character-ized by a significant (about 8 kcal mol^{-1}) exit potential barrier (see the paper). However, in the case of the photodissociation of acetaldehyde, the spin–orbit calcu-lations reported in your paper at this Discussion give a S$_1$/T$_1$ coupling of 5-10 cm^{-1} and roughly 65–70 cm^{-1} between T$_1$ and S$_0$. This seems to indicate that the T$_1$ → S$_0$ ISC in the UV photolysis of ^{1}CH$_3$CHO (via the S$_1$ state) in this different region (that of the acetaldehyde minimum) of the PES is even more probable than the T$_1$ → S$_0$ ISC in the O(^{3}P) + C$_2$H$_4$ reaction in the region of the CH$_2$CH$_2$O biradical. My ques-tion is: does all this suggest that, following initial UV excitation to the S$_1$ state, S$_1$ acetaldehyde first undergoes ISC to T$_1$ followed then by rapid ISC to S$_0$ (or, compet-itively, dissociation to CH$_3$ + HCO if the available energy is above the exit barrier to these products), or is S$_0$ acetaldehyde mainly produced via internal S$_1$ → S$_0$ conver-sion? I understand that there is a wavelength dependence of these inelastic processes, but are your three-state surface hopping calculations able to provide information on the relative importance of these two different pathways of formation of S$_0$ on which CH$_3$ + HCO are formed in a barrierless dissociation? Do you expect the CH$_3$ + HCO formation dynamics be different for these two different pathways? And, a question for Professor Kable, can this be elucidated experimentally in photolysis experiments as a function of wavelength?

A last question for Professor Bowman concerns the possible extension of QCT surface hopping calculations on full-dimensional triplet and singlet ab initio PESs to systems more complex than O(^{3}P) + C$_2$H$_4$ (where the extent of ISC has been experimentally and theoretically found to be of about 50%),[1] such as O(^{3}P) + CH$_2$CCH$_2$ (allene), where recently[8] ISC has been found experimentally to be as large as 90%, and its isomeric reaction O(^{3}P) + CH$_3$CCH (propyne), where very recently[9] ISC has been observed to be significantly lower than 90%. Are these calculations feasible with your current computational "technology"?

1 B. Fu, Y.-C. Han, J. M. Bowman, L. Angelucci, N. Balucani, F. Leonori, P. Casavecchia, *Proc. Natl. Acad. Sci. U. S. A.*, 2012, **109**, 9733.
2 B. Fu, Y.-C. Han, J. M. Bowman, F. Leonori, L. Angelucci, N. Balucani, A. Occhiogrosso, R. Petrucci, P. Casavecchia, *J. Chem. Phys.*, 2012, **137**, 22A532-1.
3 P. Casavecchia, G. Capozza, E. Segoloni, F. Leonori, N. Balucani and G. G. Volpi, *J. Phys. Chem. A*, 2005, **109**, 3527.
4 P. Casavecchia, F. Leonori, N. Balucani, R. Petrucci, G. Capozza and E. Segoloni. *Phys. Chem. Chem. Phys.*, 2009, **11**, 46.
5 B. R. Heazlewood, S. J. Rowling, A. T. Maccarone, M. J. T. Jordan and S. H. Kable, *J. Chem. Phys.*, 2009, **130**, 054310.
6 T. L. Nguyen, L. Vereecken, X. J. Hou, M. T. Nguyen and J. Peeters, *J. Phys. Chem. A*, 2005, **109** 7489.
7 W. Hu, G. Lendvay, B. Maiti and G. C. Schatz, *J. Phys. Chem. A*, 2008, **112**, 2093.
8 F. Leonori, A. Occhiogrosso, N. Balucani, A. Bucci, R. Petruccia and P. Casavecchia, *J. Phys. Chem. Lett.*, 2012, **3**, 75.
9 P. Casavecchia, N. Balucani, F. Leonori, V. Nevrly, S. Falcinelli and D. Stranges, poster abstract at this Discussion and work in progress.

Professor Bowman answered: Professor Casavecchia *et al.* raise an intriguing question about the relationship between the bimolecular reaction O(^{3}P) + C$_2$H$_4$ and current photodissociation studies of unimolecular dissociation of CH$_3$CHO. These two processes involve the same electronic states and yet their dynamics is quite different; more products are found in the bimolecular reaction than in most photodissociation ones. The origin of these differences is essentially in the different total energies; the bimolecular reaction occurs at higher energies than most photodissociation ones. However, there is some indication that at higher energies, photodissociation does indeed lead to many more products. This can be found in the theoretical work described in 'Quasiclassical Trajectory Calculations of the Dissociation Dynamics of CH_3CHO at High Energy Yield Many Products',[1] and also in photolysis experiments at 157 nm reported in 'Dynamics of Multidissociation Paths of Acetaldehyde Photoexcited at 157 nm: Branching Ratios, Distributions of Kinetic Energy, and Angular Anisotropies of Products'.[2] Nevertheless, differences, even at the same total energy, are to be expected owing to the different dynamical conditions. These would be fascinating to explore in detail in the future.

1 Y.-C. Han, B. C. Shepler and J. M. Bowman, *J. Phys. Chem. Lett.*,2011, **2**, 1715–1719.
2 S.-H. Lee, *J. Chem. Phys.*, 2009, **131**, 174312/1–10

Professor Kable responded: I think it is worth paraphrasing the experimental question raised by Professor Casavecchia since it is readily lost amongst the more extensive questions about the theory, which will be addressed by Professor Bowman. As I understand the question, the issue is whether the following two processes can be distinguished experimentally:
$$CH_3CHO\ (S_1) \rightarrow CH_3CHO\ (T_1) \rightarrow CH_3CHO\ (S_0) \rightarrow CH_3 + HCO\ (R1)$$
$$CH_3CHO\ (S_1) \rightarrow CH_3CHO\ (S_0) \rightarrow CH_3 + HCO\ (R2)$$
The distinguishing feature here is whether the triplet state is formed; the question is whether it subsequently decays to S_0, and if experimental measurements can distinguish the two mechanisms?

The existence of the first half of mechanism R1, the production of triplet acetaldehyde is unequivocal. There are many experiments and theory on the dynamics of $CH_3 + HCO$ on T_1 showing the involvement of T_1 in the photochemistry of acetaldehyde.[1] The production of $CH_3 + HCO$ on S_0 is also certain. Ion imaging[2] and laser induced fluorescence[3] experiments have identified the dynamical signature of reaction on S_0.[2,3]. Both sets of experiments show that the relative S_0 yield drops to essentially zero once the barrier to $CH_3 + HCO$ on T_1 is exceeded. This is strong evidence that nearly all the observed production of $HCO + CH_3$ occurs *via* the T_1 state and therefore that mechanism R1 is the predominant process, at least at relatively low photolysis energy (290 nm).

The presence of mechanism R2 is less clear. I have somewhat addressed this question in my answer to Dr Glowacki above. There is at least indirect evidence of direct S_1–S_0 internal conversion in classic photochemistry experiments,[4] under collisional conditions, at higher photolysis energy (>105 kcal mol^{-1} or λ < 275 nm). The presence of collisions might obfuscate the primary dynamics, however, they provide at least indirect evidence of S_1–S_0 internal conversion. I am not aware of any collision-free experiments that have probed this issue.

1 K. C. Thompson, D. L. Crittenden, S. H. Kable, M. J. T. Jordan, *J. Chem. Phys.*, 2006, **124**, 044302, and references therein.
2 G. A. Amaral, A. Arregui, L. Rubio-Lago, J. D. Rodríguez, L. Bañares, *J. Chem. Phys.*, 2010, **133**, 064303.
3 B. R. Heazlewood, S. J. Rowling, A. T. Maccarone, M. J. T. Jordan, S. H. Kable, *J. Chem Phys.*, 2009, **130**, 054310.
4 G. K. Moortgat, H. Meyrahn, P. Warneck, *ChemPhysChem*, 2010, **11**, 3896.

Professor Bowman opened the discussion of the paper by Professor Wester: You mentioned an interesting temperature dependence for the reaction with $(H_2O)OH^-$. I wonder if there is any relationship to the fact that the this cluster, viewed as (OH^-) $H^+(OH^-)$, has the proton delocalized at low temperature but at least somewhat localized at higher temperatures, *e.g.*, above roughly 400 K?

Professor Wester responded: This is a very interesting supposition that may well explain the observed strong temperature dependence in the proton transfer reaction of the water cluster anion with methyl iodide. At low temperature, where the $H_3O_2^-$ cluster is best described by an inversion-symmetric structure, the negative charge is delocalized over both OH groups and is likely to lead to a weaker interaction when the anion cluster approaches the methyl iodide reactant. If the increase of the internal temperature breaks this symmetry and localizes the charge on one of the OH ends of the cluster the ion–dipole attraction becomes stronger. This can increase the reaction rate, *e.g.* by increasing the range of impact parameters over which proton transfer occurs, and may thereby explain the measured temperature-dependence of the corresponding product branching ratio. To carefully test this possible explanation, more experimental and theoretical studies will be needed.

Professor Truhlar commented: I have a comment and a question. The comment: Table 1 below contains calculations of two energies of reaction by eight methods (four approximations, each with two basis sets). The best of the eight has an average error in the heat of reaction of 1.70 kcal mol^{-1}. We have recently proposed improved density functionals that give more accurate predictions for electron affinities than previous functionals, and these reactions provide a test of whether the newer functions can provide more accurate potential energy surfaces for reactions involving anions. Roberto Peverati and I have applied the new functionals, SOGGA11-X[1] and M11,[2] to the two reactions with the ma-TZVP basis set,[3] which is quite a bit smaller than the larger (aug-cc-pVTZ) basis set used in the paper. The results are in Table 1, and they show that newer density functionals would allow one to make more accurate surfaces for direct dynamics at an affordable cost.

The question: Although it is not explicitly mentioned in the article, you are assuming that only singlet CH_2 is produced in reaction 7. When only light atoms are involved, conservation of spin is often a good approximation, but iodine is heavy enough that one can wonder whether triplet CH_2 should be considered as well. Would you please comment on this?

1 R. Peverati and D. G. Truhlar, *J. Chem. Phys.*, 2011, **135**, 191102.
2 R. Peverati and D. G. Truhlar, *J. Phys. Chem. Lett.*, 2011, **2**, 2810.
3 J. Zheng, X. Xu and D. G. Truhlar, *Theor. Chem. Acc.*, 2011, **128**, 295–305.

Professor Wester answered: Regarding the comment, the new functionals and their ability to give accurate energies with a relatively small basis set seem quite promising. It will be of interest to investigate them more completely to assure that they give accurate potential barriers and potential energy minima for the unsolvated and solvated OH$^-$ + CH$_3$I reactions. They would allow direct dynamics simulations of this reactive system with more water molecules and also allow the investigation of low collision energies, both of which increase the computational demand for the simulations. Regarding the question, the simulations assume the reaction remains on the singlet potential energy surface. Triplet CH_2 is lower in energy than singlet

Table 1 Energy of reaction (kcal mol^{-1}) at 0 K for OH$^-$ + CH$_3$I → singlet products and mean unsigned error (MUE)

Method	CH$_3$OH + I$^-$	CH$_2$ + H$_2$O + I$^-$	MUE
B97-1/ECP/d^a	-64.87	28.14	1.70
SOGGA11-X	-67.34	26.74	0.66
M11	-67.11	25.66	0.66
Experiment	-66.44	26.32	—

^{a} Best of eight methods in article

CH$_2$,[1] and a singlet–triplet electronic non-adiabatic transition is possible as a result of spin–orbit coupling. It would be of interest to identify the crossing point (or points) for the singlet and triplet potential energy surfaces, and ascertain whether they can be accessed during the course of Reaction (7). It is possible that the I-atom may enhance the spin–orbit coupling and promote this electronic transition. An approach to investigate this question experimentally would be to measure the cross section for Reaction (7) *versus* collision energy and determine if the threshold is consistent with the energy of ^{1}CH$_2$ or with the energy of ^{3}CH$_2$.

1 W. L. Hase, R. J. Phillips and J. W. Simons, *Chem. Phys. Lett.*, 1971, **12**, 161.

Professor Nesbitt asked: I presume that the trajectory calculations provide most of the support for the claim that your ion molecule reactions really do exclusively proceed *via* a S$_N$2 inversion around the central C atom. Is there any direct or indirect experimental verification of this and what might the prospects be for getting such information experimentally?

Professor Wester replied: Experiment and simulation may both provide support whether an ion–molecule substitution reaction occurs *via* an inversion mechanism. In the simulation it becomes evident upon inspection of trajectories, in the experiment direct, back-scattered product ions are a signature of the nucleophilic substitution inversion mechanism,[1] which we also refer to as the direct rebound mechanism.[2] In the present study not all reactions proceed *via* this mechanism. The simulations show that reaction (6) and (11), which form I$^-$ products, may proceed *via* inversion at the carbon centre. Two-dimensional ion imaging scattering results show that additionally also a direct stripping mechanism with forward and sideways scattered products occurs for this product channel,[3] which may also occur without inversion at the carbon centre. For the highly exoergic reaction (6) for OH$^-$ + CH$_3$I the stripping mechanism is even the dominant one.

1 J. Mikosch, S. Trippel, C. Eichhorn, R. Otto, U. Lourderaj, J. X. Zhang, W. L. Hase, M. Weidemüller and R. Wester, *Science*, 2008, **319**, 183–186.
2 J. X. Zhang, J. Mikosch, S. Trippel, R. Otto, M. Weidemüller, R. Wester and W. L. Hase, *J. Phys. Chem. Lett.*, 2010, **1**, 2747–2752.
3 R. Otto, J. Brox, S. Trippel, M. Stei, T. Best and R. Wester, *Nat. Chem.*, 2012, **4**, 534.

Professor Orr-Ewing remarked: The observation from the trajectory calculations of a maximum reaction probability at an impact parameter of 4 Angstroms for reaction (6) involving substitution of an I$^-$ by OH$^-$ seems surprisingly large if the mechanism involves the classic Walden inversion. Is there significant steering of these large impact parameter trajectories as the reagents approach? Has a similar analysis of impact parameter dependent reactivity been carried out for reactions of OH$^-$ clustered with a single water molecule?

Professor Wester answered: The OH$^-$ + CH$_3$I long-range potential is highly attractive and the earliest chemical dynamics simulations for X$^-$ + CH$_3$Y S$_N$2 reactions illustrated that this potential pulls the ion and molecule together for large impact parameter collisions, and may lead to reaction.[1] For the F$^-$ + CH$_3$I S$_N$2 reaction it was found that the largest impact parameter b_{max} leading to reaction is quite accurately given by a model that assumes the orbital angular momentum is conserved as the reactants collide and the orbital angular momentum that gives a rotational barrier at the central TS which matches the collision energy identifies b_{max}.[2] Similar dynamics are found for the OH$^-$ + CH$_3$I S$_N$2 reaction. The value of b_{max} for the monosolvated OH$^-$(H$_2$O) + CH$_3$I reaction is about ten percent smaller than that for the unsolvated reaction, which is consistent with the higher central barrier and larger reactant reduced mass for the monosolvated reaction.

1. S. R. Vande Linde and W. L. Hase, *J. Chem. Phys.*, 1990, **93**,7962.
2. J. Zhang, J. Mikosch, S. Trippel, R. Otto, M. Weidemüller, R. Wester, and W. L. Hase, *J. Phys. Chem. Lett.*, 2010, **1**, 2747.

Professor Neumark remarked: Your temperature-dependent results in Fig. 8 in the paper show a remarkable enhancement in reaction rate for only a small temperature rise. You attribute this effect, quite reasonably, to a small population at the higher temperature of vibrationally excited reactants that are highly reactive. If this is the case, it would be very interesting to perform these experiments on $OD^-(D_2O)$. Is this something that you have considered or tried?

Professor Wester replied: The octupole ion trap is a versatile source for the crossed-beam spectrometer and is well suited to prepare cold deuterated water clusters. We intend to use this to carry out experiments with deuterated water cluster anions in the future. We also plan to extend our studies to a larger range of temperatures, in particular to temperatures between 100 and 300 Kelvin and to temperatures above room temperature. This will yield quite different internal state populations for the reactant cluster ions and may help to clarify which vibrations are responsible for the observed change in reactivity.

Dr Jordan asked: In Fig. 10 of your paper you note that almost 10% of the CH_2I^- velocities lie outside the kinematic cut-off and you attribute this to OH^- arising from the fragmentation in the parent OH^-/H_2O beam. What temperature was image taken at? Do you see pressure or temperature effects for this component of the image?

Professor Wester replied: The data in Fig. 10 have been obtained for $OH^-(H_2O)$ clusters trapped at 100 Kelvin in the octupole ion trap prior to the collision and we have not yet looked into an influence of temperature on these ion images. The chamber pressure is kept at a value of around 10^{-7} mbar and has not been changed during the experiment. Performing a background pressure-dependent measurement of the differential scattering cross-sections would certainly be a very suitable test of our explanation of cluster fragmentation as the origin for the products with high kinetic energy.

Professor Neumark commented: With respect to Joel Bowman's comment about the shared proton in $OH^-(H_2O)$, I would like to point out that there should be considerably less proton delocalization in $OH^-(H_2O)_2$. The OH^- anion in that cluster should be more of a distinct entity. In order to assess the role of proton delocalization on the reactivity of $OH^-(H_2O)$, it would be of interest to examine the reactivity of the larger cluster. It is, however, quite possible that energetic effects owing to additional hydration of the OH^- will dominate the reactivity of this species.

Professor Nesbitt asked: The point raised by Professor Bowman concerning the double minimum potential for the $[OH–H–OH]^-$ anion and therefore the highly delocalized nature of the wavefunction is interesting. I am wondering if the asymmetric presence of the dipolar CH_3I reagent would already be sufficient to break this symmetry and therefore localize the wavefunction into a more classical configuration?

Professor Wester responded: At short distances the interaction of the $H_3O_2^-$ cluster with the polar CH_3I will break the symmetry of the anion cluster. However, we found that even in the entrance channel complex of the reaction of $OH^-(H_2O)$ with methyl iodide, the structure of the $H_3O_2^-$ cluster is only weakly perturbed. This complex has a compact geometry where one OH group is located almost co-linearly with the C–I bond while the O–H–O axis is approximately perpendicular to the C–I axis.[1] Our calculations suggest that despite the interaction with the methyl iodide the charge is still mostly delocalized over the cluster. Thus, the ion–dipole interaction seems not to be strong enough to severely break the symmetry. However, this may change when increasing the internal temperature, because with a larger density of populated states at higher temperatures the mixing of states with different inversion symmetry due to the interaction with methyl iodide should occur more easily.

1 R. Otto, J. Brox, S. Trippel, M. Stei, T. Best and R. Wester, *Nat. Chem.*, 2012, 4, 534.

Dr Falcinelli asked: Relating to the results presented in the paper by Professor Wester, I would to point out that, as shown for example by J. Bentley in 1980, the electronic structure of the water molecule in the gas phase is different from that is usually described in undergraduate chemistry textbooks. It is usually asserted that H_2O is characterized by an sp^3 distorted hybridization with the two lone pairs degenerate. Actually, this is the case of water in liquid or solid state. In the gas phase, when the molecule is isolated, the two O–H bonding orbitals form an angle of about 104.5° and the two lone pairs are not degenerate, being one aligned along the C_{2v} molecular axis ($3a_1$, the sp^2 lone pair orbital) and the other one perpendicular to the molecular plane ($1b_1$, the $2p\pi$ oxygen non bonding orbital).[1] This is well confirmed by recent measurements from my laboratory in a Penning ionization electron spectroscopy study of water molecules by neon metastable atoms. In our experiment,[2] due to the available energy of $Ne^*(^3P_{2,0})$ atoms, the two not degenerate lone pair orbitals of the H_2O molecule are selectively involved in the ionization process. The non-degeneracy of the two lone pairs is also well confirmed by the presence of two separated bands in the electron energy spectra of H_2O, obtained by photo- or Penning ionization, assigned to the two different approach directions of the Ne^* atom to the molecule (both towards the O atom: one along the C_{2v} molecular axis and the other one along the perpendicular direction of the molecular plane).[2] In my opinion, the results presented in the paper appear to be very interesting from that point of view because in their experiment the authors should be able to obtain important information about the role of the electronic configuration change of the water molecule on the reactivity of CH_3I with anion water clusters, passing from OH^- and $OH^-(H_2O)$ towards $OH^-(H_2O)_n$ with $n \geq 2$. This should be very important also to better understand the role of the hydrogen bond interaction on reaction dynamics under micro-solvation with size-selected clusters containing water.

1 J. Bentley, *J. Chem. Phys.* 1980, **73**(4), 1805–1813.
2 B. Brunetti, P. Candori, D. Cappelletti, S. Falcinelli, F. Pirani and D. Stranges, *Chem. Phys. Lett.* 2012, **539–540**, 19-23.

Professor Crim opened the discussion of the paper by Professor Truhlar: Could you comment on the increase in complexity in going from reactions with ethanol to ones with alcohols having more sites and conformers. In particular, is analyzing experiments such as those of Suits on butanol in the offing?

Professor Truhlar replied: In our paper at this discussion we calculated thermal rate constants for the reaction of OH with ethanol. We have already applied this method to calculate thermal rate constants for more complex alcohols, in particular for reactions of HO_2 and OH with various butanols. These calculations require knowledge of the potential energy surfaces in valleys surrounding the minimum energy paths. Analyzing the scattering experiments of Suits is more complicated because we require global potential energy surfaces for that. We are currently developing global potential energy surfaces for reactions of Cl with butanols, and we plan to use them for quasiclassical trajectory calculations to simulate his experiments.

Professor Vancik commented: For all three reactions studied the structure of the transition state can be related with the kinetic isotope effect, both primary and secondary. It would be especially interesting to see the effect for the TS structures in which the hydrogen bonds are included. Since in your previous papers in which the enzymatic transition states were studied you have calculated and discussed the KIE's, it would in principle be important to also do it for the reported reactions. Is there special reason why you have not calculated the isotope effects?

Professor Truhlar answered: I agree that kinetic isotope effects (KIEs) are an important testing ground for theories of chemical kinetics, and variational transition state theory has been widely and successfully applied to both primary and secondary KIEs.[1] The only reason that MS-VTST and MP-VTST have not yet been applied to KIEs is that these formulations are relatively new, and we simply haven't gotten to that yet.

I agree that one should expect qualitatively different KIEs for transition states with hydrogen-bonded structures. The question of hydrogen bonding in transition state structures is very relevant since—in addition to the example in the paper presented at this discussion—we have also found in some other cases that some transition states have hydrogen bonds, and some do not.[2] Therefore it would be very interesting to see how the KIEs changes as functions of temperature due to the changes in the relative Boltzmann weightings of the two kinds of structures.

1 D. G. Truhlar, Variational Transition State Theory and Multidimensional Tunneling for Simple and Complex Reactions in the Gas Phase, Solids, Liquids, and Enzymes, in Isotope Effects in Chemistry and Biology, ed. A. Kohen and H.-H. Limbach, Marcel Dekker, Inc., New York, 2006, pp. 579–619.
2 O. Tishchenko, S. Ilieva and D. G. Truhlar, J. Chem. Phys., 2010, **133**, 021102.
3 P. Seal, E. Papajak and D. G. Truhlar, J. Phys. Chem. Lett., 2012, **3**, 264–271.

Dr Glowacki remarked: Your reported rate coefficients are in the high pressure limit; however, many of data points in Fig. 8 in the paper have been obtained at considerably lower pressures – e.g. the data reported by Carr et al.[1] I wonder if you have any feel for whether the agreement between experiment and theory might be improved if theoretical and experimental results at the same pressures were being compared.

1 S. A. Carr, D. R. Glowacki, C. H. Liang, M. T. Baeza-Romero, M. A. Blitz, M. J. Pilling and P. W. Seakins, Experimental and modelling studies of the pressure and temperature dependences of the kinetics and the OH yields in the acetyl + O_2 reaction, J. Phys. Chem. A, 2011, **115**, 1069–1085.

Professor Truhlar answered: The question is very relevant, and it identifies an issue with which we have been very concerned but cannot make a definitive statement. Pressure effects on unimolecular reactions have been widely studied since the early days of chemical kinetics, and they are universally agreed to be of central importance. When bimolecular reactions have a chemically bound chemical intermediate, that intermediate should usually be included in a mechanism that can then be treated by a master equation, including pressure effects on the stabilization of the intermediate and its energy distribution.[1] Pressure effects on a bimolecular reaction exhibiting only van der Waals intermediates—such as the reaction under consideration in our paper—are more subtle. Usually pressure effects on such reactions are assumed to be small, and the rate constant is assumed to be independent of pressure; this is probably a good assumption when the reaction barrier is high. The questions of high- and low-pressure limits and various pressure-regime plateaus for bimolecular reactions with a van der Waals intermediate and a low barrier are discussed briefly in the article, and we identify an issue, namely tunneling at low energies accessible by populating the low-energy states of the van der Waals complex, that may lead to a larger dependence on pressure in some cases, especially at low temperature. (One conclusion from this analysis is that it is oversimplified to say that we have calculated the high-pressure limit.) However, it is difficult to test theory for this effect simply by comparison to experiment since the pressure dependence of the rate constants may be smaller than the experimental uncertainty and since at low temperature our calculations are very sensitive to the potential energy surface, which might not be converged well enough for a comparison to experiment not to be clouded by uncertainties in the potential energy surface.

1 A. Fernández-Ramos, J. A. Miller, S. J. Klippenstein and D. G. Truhlar, *Chem. Rev.*, 2006, **106**, 4518–4584.

Dr Glowacki said: I have a couple of questions:

(1) The methods outlined in this paper clearly make a comprehensive treatment of torsional anharmonicity in bimolecular reactions. However, angular anharmonicity is also often important in the transition states for bimolecular association reaction – in particular for non-conserved modes. This is effectively the logic behind the variational reaction coordinate transition state theory (VRC-TST) developed by Klippenstein and co-workers (see, *e.g.*, ref. 1) in which one performs Monte Carlo integration over the phase space of all the non-conserved modes. I'm curious as to how the methods outlined in this paper map onto the VRC-TST type approaches, and how important you think it is to include the angular anharmonicity for these sorts of chemical systems.

(2) In your paper, you consider tens and even hundreds of closely linked saddle points. Is it necessary to include all these isomers? Or does one find that the rate coefficients converge rather quickly as a function of the number of saddle points one includes – *i.e.*, do you find that consideration of *n* structures brings you to within a few percent of the rate coefficient that you calculate using all the structures?

1 Y. Georgievskii and S. J. Klippenstein, *J. Chem. Phys.*, 2003, **118**, 5442.

Professor Truhlar responded: Thank you for asking these questions, which get to the very core of what we are trying to accomplish. In response to the first question, we note that there are two practical types of approaches for defining variational transition state dividing surfaces for gas-phase reactions. The first type, which we call reaction-path variational transition state theory (RP-VTST), is based on finding dividing surfaces orthogonal to reaction paths. This approach was initially based on hyperplanes orthogonal to rectilinear reaction coordinates[1,2] but was later generalized to curvilinear dividing surfaces orthogonal to the minimum-energy-path[3–5] and eventually to surfaces orthogonal to other reaction paths and at optimized angles with respect to reaction paths.[6–8] RP-VTST is designed for tight transition states such as one encounters both in bimolecular reactions leading to bimolecular products and unimolecular reactions leading to unimolecular products and in association and dissociation reactions with saddle points. It may be applied to loose transition states such as encountered in barrierless association reactions and their reverse dissociation reactions, but a much better strategy for loose transition states is variable-reaction-coordinate VTST (VRC-VTST) with multi-faceted dividing surfaces, as developed by Georgievskii and Klippenstein,[9,10] as an improvement of earlier efforts to define this kind of dividing surface by Wardlaw, Marcus, and Klippenstein (reviews are available (refs 11 and 12)). The POLYRATE computer program contains both kinds of VTST.

In RP-VTST, one at least initially separates vibrations from rotations and makes the quadratic approximation for vibrational potentials, yielding a generalized normal mode approximation similar to conventional normal mode analysis; corrections can then be applied for stretching anharmonicity,[2,13,14] bending anharmonicity,[3,15,16] and separable torsional anharmonicity.[17,18] These anharmonicity effects can be significant,[2,13–18] but their proper treatment requires mode coupling, and the effort to include this anharmonicity is only now becoming more important as the increased accuracy of affordable electronic structure methods now yields more accurate absolute reaction rate constants for polyatomic reactions that are no longer necessarily dominated by errors in the potential energy surfaces. One should also keep in mind that at a variational transition state, even the harmonic frequencies depend on the coordinate system, and it is important to get physical harmonic frequencies before one considers anharmonicity.[4,5] Some simple ways to correct for anharmonicity in single-structure calculations on both reactants and tight transition states are scaling the frequencies to improve the zero point energy[14] and raising all frequencies below 100 cm^{-1} to 100 cm^{-1} as a simple but highly approximate way to correct for torsional anharmonicity.[19-21]

In contrast, VRC-VTST involves a different kind of separability as the initial step; in particular one separates rotational and orbital motions of the bimolecular reactants for associations (or of products for dissociations), called transitional modes, from vibrations of the bimolecular reactants (or products—rather than separating rotations from all vibrations of the variational transition state), where the latter are called conserved modes. Note that transitional modes are those whose frequencies vanish in the limit of bimolecular reactants (or products). VRC-VTST treats anharmonicity of the transitional modes accurately (in the classical limit) by Monte Carlo integration, as you said, but it is less accurate for the conserved modes.

With that background, we can place MS-VTST and MP-VTST in perspective. MS-VTST generalizes RP-VTST to the case where reactants and products have multiple conformations generated by coupled torsions, but the tunneling (and other quantum effects on the reaction coordinate motion) and variational effects are still based on a single reaction path, assumed typical. MP-VTST generalizes MS-VTST so that all (or at least more than one) reaction paths are used to calculate the tunneling and variational effects. (Note that "variational effects" is a term we often use to denote the deviation of quasiclassical VTST from conventional TST based on a dividing surface through the saddle point and perpendicular to its imaginary frequency normal mode; thus variational effects correct for recrossing of that conventional dividing surface. "Quasiclassical" VTST is a term we often use to denote VTST without tunneling or other quantum effects on the reaction coordinate motion.) MS-VTST and MP-VTST provide (or at least attempt to provide) a more accurate treatment of torsions, but in the initial formulation they still treat other vibrational modes as harmonic and uncoupled to the torsions. Corrections for this can be very important, just as already mentioned for single-structure VTST. For example, we have already included the scaling of frequencies to improve the zero point energy, and sometimes we label the local harmonic approximation with scaled frequencies as a local quasiharmonic approximation.

In response to the second question, we find that the rate constant does not necessarily converge rapidly with respect to adding more structures. This is especially true at combustion temperatures. For example, isoheptane has 37 structures, and 17 and 31 of them of them are within 2 and 3 kcal mol^{-1}, respectively, of the lowest-energy structure.[22] Another example is the transition state for abstraction of H from carbon-3 of 1-butanol; including local zero point energy, we find that there are 32 structures below 3 kcal mol^{-1}, 80 below 4 kcal mol^{-1}, 140 below 5 kcal mol^{-1}, and 248 below 6 kcal mol^{-1}.[23] Another relevant point here is that when one searches to find all the structures, one does not necessarily find the lower-energy structures first, and if one uses the Voronoi method to determine the local periodicities, one needs all the structures. Once one has found and characterized all the local minima of the reactants and transition states (local minima of the transition states are first order saddle points), there would be little savings in not including all of them in the partition function sum.

1 B. C. Garrett and D. G. Truhlar, *J. Chem. Phys.*, 1979, **70**, 1593–1598.
2 A. D. Isaacson and D. G. Truhlar, *J. Chem. Phys.*, 1982, **76**, 1380–1391.
3 B. C. Garrett and D. G. Truhlar, *J. Chem. Phys.*, 1980, **72**, 3460–3471.

4 G. A. Natanson, B. C. Garrett, T. N. Truong, T. Joseph and D. G. Truhlar, *J. Chem. Phys.*, 1991, **94**, 7875–7892.

5 C. F. Jackels, Z. Gu and D. G. Truhlar, *J. Chem. Phys.*, 1995, **102**, 3188–3201.

6 J. Villà and D. G. Truhlar, *Theor. Chem. Acc.*, 1997, **97**, 317–323.

7 A. González-Lafont, J. Villà, J. M. Lluch, J. Bertrán, R. Steckler and D. G. Truhlar, *J. Phys. Chem. A*, 1998, **102**, 3420–3428.

8 P. L. Fast and D. G. Truhlar, *J. Chem. Phys.*, 1998, **109**, 3721–3729.

9 Y. Georgievskii and S. J. Klippenstein, *J. Phys. Chem. A*, 2003, **107**, 9776–9781.

10 J. Zheng, S. Zhang and D. G. Truhlar, *J. Phys. Chem. A*, 2008, **112**, 11509–11513.

11 D. G. Truhlar, B. C. Garrett and S. J. Klippenstein, *J. Phys. Chem.*, 1996, **100**, 12771–12800.

12 A. Fernández-Ramos, J. A. Miller, S. J. Klippenstein and D. G. Truhlar, *Chem. Rev.*, 2006, **106**, 4518–4584.

13 B. C. Garrett and D. G. Truhlar, *J. Chem. Phys.*, 1984, **81**, 309–317.

14 I. M. Alecu, J. Zheng, Y. Zhao and D. G. Truhlar, *J. Chem. Theory Comput.*, 2010, **6**, 2872–2887.

15 B. C. Garrett and D. G. Truhlar, *J. Phys. Chem.*, 1979, **83**, 1915–1924.

16 B. C. Garrett and D. G. Truhlar, *J. Phys. Chem.*, 1991, **95**, 10374–10379.

17 D. G. Truhlar, *J. Comput. Chem.*, 1991, **12**, 266–270.

18 B. A. Ellingson, V. A. Lynch, S. L. Mielke and D. G. Truhlar, *J. Chem. Phys.*, 2006, **125**, 84305.

19. L. M. Pratt, D. G. Truhlar, C. J. Cramer, S. R. Kass, J. D. Thompson and J. Xidos, *J. Org. Chem.* 2007, **72**, 2962–2966.

20 Y. Zhao and D. G. Truhlar, *Phys. Chem. Chem. Phys.*, 2008, **10**, 2813–2818.

21 R. F. Ribeiro, A. V. Marenich, C. J. Cramer and D. G. Truhlar, *J. Phys. Chem. B*, 2011, **115**, 14556–14562.

22 T. Yu, J. Zheng and D. G. Truhlar, *Phys. Chem. Chem. Phys.*, 2011, **14**, 482–494.

23 P. Seal, E. Papajak and D. G. Truhlar, *J. Phys. Chem. Lett.*, 2012, **3**, 264–271.

Professor Crim opened the discussion of the paper by Dr Liu: One of the key issues in the reaction dynamics of vibrationally excited molecules is the evolution of the initially prepared eigenstate during the perturbation of the collision. For example, calculations by Kosloff and Nesbitt[1] and by Halonen and Nesbitt[2] show that some vibrations evolve to localize the energy in a favorable bond for reaction and others to localize it in an unfavorable bond during the collision. We also observed similar behavior in calculations on CH_3D reaction with Cl.[3] Do you have any thoughts about the evolution of the initial eigenstate in the systems you study?

1 J. R. Fair, D. Schaefer, R. Kosloff and D. J. Nesbitt, *J. Chem. Phys.*, 2002, **116**, 1406.

2 L. Halonen, S. L. Bernasek and D. J. Nesbitt, *J. Chem. Phys.*, 2001, **115**, 5611.

3 S. Yoon, R. J. Holiday, E. L. Sibert and F. F. Crim, *J. Chem. Phys.*, 2003, **119**, 9568.

Dr Liu responded: Professor Crim raised a very important question. Intramolecular vibrational energy redistribution (IVR) plays a vital role in understanding the mode-specific or bond-selective chemistry. In the high-energy excitation regime, the initially prepared eigenstates of a polyatomic molecule can rapidly lose the mode-specific characters *via* IVR prior to reactions. This occurs even under collisionless conditions and often leads to the familiar RRKM behaviors. In the low vibrational excitation regime, where the state densities are sparse and mode mixings are weak, the initially prepared eigenstates can also be coupled to other states due to perturbation of the approaching reactants, and thus the initial eigenstate characters will evolve as the reaction proceeds. This is of a collision-induced type.

One way to think of the collision-induced IVR effects is to start with the concept of vibrational adiabaticity, as the works you mentioned. We also used a similar approach in the studies of the Cl + $CHD_3(v_1 = 1)$ reaction.[1,2] Based on the reaction path Hamiltonian approach[3] and the previous *ab initio* calculations,[4,5] we first constructed the vibrationally adiabatic curves that correlate the reactant vibrational states to the product pairs. However, one would not expect that vibrational adiabaticity will hold throughout a chemical reaction, and thus vibrationally non-adiabatic processes (*i.e.*, the collision-induced mode-mixings) must occur. The questions are: which modes are actively involved and to what extent, and where the nonadiabatic transitions will take place along the reaction path. Fortunately, the previous *ab initio* calculation[5] also computed the necessary non-adiabatic couplings within the reaction path Hamiltonian framework. These non-adiabatic couplings comprise the curvature coupling (*i.e.*, coupling induced by the curvature of the reaction path) of the prepared reactant mode to the reaction coordinate, and the Coriolis coupling that describes the intermode mixings induced by the twisting of the transverse vibrations about the curved reaction path as the reaction proceeds.[3,5] We then rationalized our experimental observations in the Cl + $CHD_3(\nu_1 = 1)$ reaction by taking into account of those non-adiabatic transitions.[1,2] Similar arguments have been extended to the Cl + $CH_4(\nu_3 = 1)$[6,7] and Cl + $CH_2D_2(\nu_1,\nu_6 = 1)$[8] systems to elucidate their effects on the reactive outcomes.

1 S. Yan, Y.-T. Wu, B. Zhang, X.-F. Yue and K. Liu, *Science*, 2007, **316**, 1723.
2 S. Yan, Y.-T. Wu and K. Liu, *Proc. Natl. Acad. Sci. U. S. A.*, 2008, **105**, 12667.
3 W. H. Miller, N. C. Handy and J. E. Adams, *J. Chem. Phys.*, 1980, **72**, 99.
4 W. T. Duncan and T. N. Truong, *J. Chem. Phys.*, 1995, **103**, 9642.
5 J. C. Corchado, D. G. Truhlar and J. Espinosa-Garcia, *J. Chem. Phys.*, 2000, **112**, 9375.
6 B. Zhang and K. Liu, *J. Chem. Phys.*, 2005, **122**, 101102.
7 H. Kawamata and K. Liu, *J. Chem. Phys.*, 2010, **133**, 124304.
8 J. Riedel, S. Yan and K. Liu, *J. Phys. Chem. A*, 2009, **113**, 14270.

Professor Beck asked: You have previously reported steric control of the reaction of CH stretch excited CHD_3 with Chlorine atoms[1] where you found that C–H stretch excitation increases the reaction cross section in a stereo-specific manner and also affects the product states and angular distribution. Did you also investigate possible alignment effects for the reaction of fluorine atoms with CHD_3 where C–H stretch excitation decreases the reactivity and if so what where the findings?

1 F. Wang, J.-S. Lin, K. Liu, *Science*, 2011, **331**, 900.

Dr Liu responded on behalf of Jia-Yue Yang, Dr. Fengyan Wang and Jui-San Lin: Yes, similar crossed-beam experiments have been performed for the reaction of fluorine atoms with the CH stretch-excited $CHD_3(\nu_1 = 1)$ reactants. At a fixed collision energy, the $CHD_2(\nu_1 = 1)$ product images were recorded under various directions of the linear IR laser polarization axis relative to the initial relative velocity vector (k) of the collision system. The full analysis,[1] which can reveal true 3-dimensional view of the steric effects, is currently in progress. Here, we will present the preliminary results of the two special cases at $E_c = 3.1$ kcal mol^{-1} with the IR laser polarization axis being either parallel (//) or perpendicular ($\perp$) to the k vector, which corresponds to an end-on or a side-on collisional geometry. Thus, the results can directly compared to those reported in ref. 2 for the analogous reaction with Cl-atom.

Fig. 3 below shows the normalized $CHD_2(\nu_1 = 1)$ product speed distributions from the reactions of F + aligned $CHD_3(\nu_1 = 1)$ under the //- and $\perp$-geometries. The IR laser was tuned to the R(0) transition, and thus the vibrationally excited $CHD_3(\nu_1 = 1)$ reactants are prepared in the $|JK\rangle = |10\rangle$ rotational state. As is seen, both //- and $\perp$-distributions feature two prominent peaks, corresponding to the correlated $DF(\nu' = 4)$ and ($\nu' = 3$) co-products, and small yields of $DF(\nu' = 2)$ that center around $u(CHD_3) = 2$ km s^{-1}. In terms of reactant alignment effects, while the total reactivity clearly favors the side-on approach ($\sigma_\perp/\sigma_{//} \sim 1.2$–1.3), the product DF vibration branching ratios show little alignment-dependence. Both of these results are in sharp contrast to our previous report on the reaction of Cl + aligned $CHD_3(\nu_1 = 1, |JK\rangle = |10\rangle)$.[2] In the latter case, the total reactivity is strongly in favor of the end-on attack ($\sigma_{//}/\sigma_\perp = 2.7$), and the $HCl(\nu')$ vibrational branching is also sensitively dependent on the IR polarization directions, $\sigma(\nu' = 1)/\sigma(\nu' = 0) = 0.31$ for //-geometry and 2.71 for $\perp$-geometry (see Fig. 4S and Table S2 of ref. 2). Fig. 4 below shows the polarization-dependent angular distributions of the two dominant $DF(\nu' = 3)$ and $DF(\nu' = 4)$ product channels. For $DF(\nu'=3)$ both

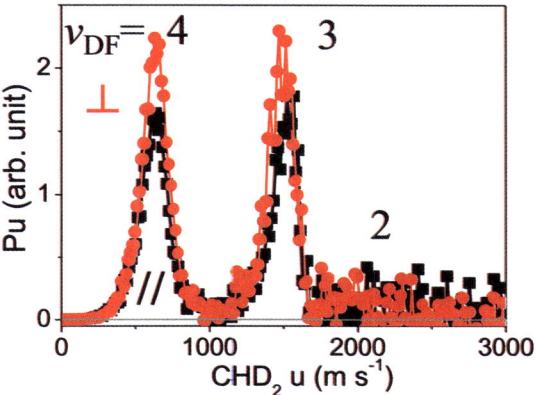

Fig. 3 Polarization-dependent product speed distributions from the F + CHD$_3$($v_1 = 1$) → DF(v) + CHD$_2$(1_1) reaction. The // (black) and ⊥ (red) symbols correspond, respectively, to parallel and perpendicular alignments of the IR polarization axis relative to the initial relative velocity vector of the collision system.

polarizations yield backward angular distributions, and the slightly preferred side-on attack (Fig. 3) manifest itself mainly in the backward direction. On the other hand, DF($v'=4$) products favor the forward hemisphere. The end-on attack (//) yields a very sharp forward peak for $\theta \leq 30°$, whereas the side-on attack (⊥) shows somewhat higher intensity over the remaining scattering angles. Hence, the effects of the reactant alignments on the product angular distributions are noticeable, but not very significant. These results, once again, contrast sharply to the Cl + aligned CHD$_3$($v_1 = 1$) reaction, where the initial polarization of the CHD$_3$($v_1 = 1$) reactants exerts huge influence on the product angular distributions (Fig. 4 (Top) of ref. 2).

It is worth noting that the pre-aligned CHD$_3$($v_1 = 1$, $|JK\rangle = |10\rangle$) reactants were prepared in the same manner, yet behave quite differently when reacting with F- or Cl-atoms: small steric effects are observed in F + CHD$_3$($v_1 = 1$), while an efficacious stereo-control of the Cl + CHD$_3$($v_1 = 1$) reaction can be achieved at will simply by rotating the direction of the IR pumping laser. Such a striking contrast illustrates clearly the balance between the anisotropic interactions in the entry valley and the pre-alignment (also the pre-orientation) of the reactants in determining the reactive outcomes. It also conforms to our previous proposition that only reactions with

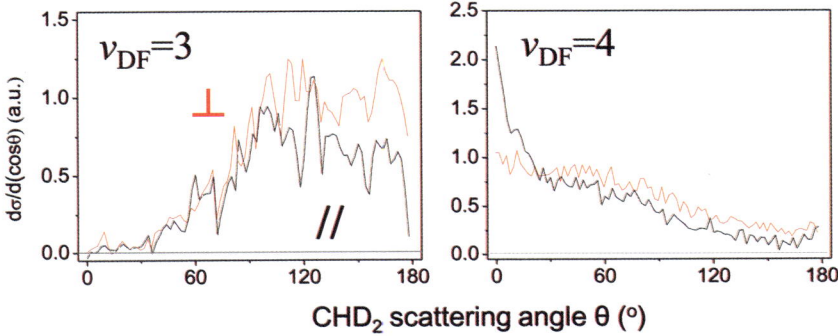

Fig. 4 Polarization-dependent angular distributions of DF($v = 3$) and DF($v = 4$) states formed coincidently with the probed CHD$_2$(1_1) products. Small differences between the two collision geometries, // (black) and ⊥ (red), are discernible in both product channels.

weak steering interactions are viable candidates for steric control (active influence over the reactive outcome by pre-alignment).[2]

1 F. Wang, K. Liu and T. P. Rakitzis, *Nat. Chem.*, 2012, **4**, 636.
2 F. Wang, J.-S. Lin and K. Liu, *Science*, 2011, **331**, 900.

Professor Polanyi remarked: Kopin Liu and his group are pushing the frontier of work on the molecular dynamics of atomic reaction with a polyatomic collision partner, specifically atomic fluorine plus methane. There will no doubt be many novelties given the complexity of methane compared with a simple diatomic, but they have undertaken to examine an aspect of the dynamics that does have a counterpart in the simple triatomic case, A + BC = AB + C, namely the effect on the reaction cross-section of increased vibration in the BC bond under attack. Their most striking finding is that increased vibration in the C–H bond of CHD_3 slows the reaction. They conclude in Section IV that this exothermic reaction has an early barrier, and that the effect of vibration, whether it compresses C–H or extends it, will be to deflect the trajectory up-hill to either side of the barrier, decreasing the reaction-probability. This same effect was clearly evident in the case of A + BC;[2] for an early-barrier energy-surface trajectories having double the activation energy vested in vibration stopped far short of the barrier-crest, failing to react. Kopin Liu and his co-workers have shown, strikingly, that the simple considerations governing reactions A + BC have relevance to their more complex case – something not previously known.

1 J. C. Polanyi, *Acc. Chem. Res.*, 1972, **5**, 161.

Professor Suits replied: I think one thing to bear in mind concerning Kopin's beautiful experiments is that the local F-D-C impact parameter is still the global impact parameter for the system. In a system as simple as Cl + ethylene, these two quantities are distinct. This is even vastly more complex in our Cl + butanol reaction, where Professor Truhlar has identified 29 distinct transition states. This is the kind of complexity we must be willing to face if we are to move these A + BC insights into the world of chemistry at large.

Dr Liu responded: Professor Suits called our attention to an important concept of the local impact parameter, which could be particularly valuable in advancing our understanding the dynamics of the site-specific reactivity in complex systems. In 1995, Levine and coworkers reported a QCT study of the Cl + CH_4 reaction, and found that many of dynamical attributes can be captured by the so-called peripheral reaction mechanism,[1] for which the concept of local impact parameter forms the backbone. In this mechanism, the large (global) impact parameter collisions can be regarded as small (local) impact parameter collisions between the approaching atom and the abstracted H atoms, because the H-atoms are sprouting away from the center-of-mass of methane. Valentini explicitly discussed the distinction between the conventional (global) impact parameter and the local impact parameter, and exploited the kinematic constraints thus derived to model the observed low rotational distributions of the H_2 products in a series of H + hydrocarbon reactions.[2] We also incorporated that concept into the line-of-centers model to examine the evolution of the product angular distributions as a function of the initial collision energy in the reactions of Cl + CH_4 and CD_4.[3]

1 X. Wang, M. Ben-Nun and R. D. Levine, *Chem. Phys.*, 1995, **197**, 1.
2 J. J. Valentini, *J. Phys. Chem. A*, 2002, **106**, 5745.
3 J. Zhou, B. Zhang, J. J. Lin and K. Liu, *Mol. Phys.*, 2005, **103**, 1757.

Professor Bowman said: In the paper 'Dynamics of the Reaction of Methane with Chlorine Atom on an Accurate Potential Energy Surface',[1] we showed, in qualitative agreement with your experiments, that at low collision energies excitation of the CH-stretch in Cl + CD$_3$H was actually less effective in promoting reaction than an equivalent amount of translational energy. This was in contradiction, as you pointed out, to the "Polanyi rule" for this somewhat late-barrier reaction. At higher collision energies, however, this rule does indeed hold. Our analysis of tens of thousands of trajectories indicates that the entrance-channel van der Waals (vdW) well is responsible (at least in some measure) for the departure from the Polanyi rule at low collision energies. This well favors an orientation of the Cl–CH$_4$ that is not the same as the reaction saddle-point geometry. Thus, at low collision energies the vdW well "steers" the reactants away from the saddle point.

Reduced dimensionality quantum calculations are underway, using our potential energy surface, by Dong-Hui Zhang and independently Dunyou Wang.

1 G. Czakó and J. M. Bowman, *Science*, 2011, **334**, 343–346.

Dr Liu replied: For the Cl + CHD$_3$($v_1 = 1$) → HCl + CD$_3$($v = 0$) reaction, our crossed-beam studies showed that the vibrational enhancement factors σ_v/σ_g, based on an equivalent amount of total energy, are less than one at low translational energies (E_c) and then become nearly unity for E_c 4 kcal mol^{-1}.[1] Your QCT calculations[2] showed a good agreement with our finding. However, when all vibrationally excited CD$_3$ products are included, the theoretical vibrational enhancement factors increase rapidly from less than one at low energies to about three at $E_c \sim 8$ kcal mol^{-1} (Fig. S1 in ref. 2). The theoretical results of all CD$_3$ states at high E_c can also be compared with the experimental finding of $\sigma_v/\sigma_g \sim 1.4$ at $E_c \sim 8$ kcal mol^{-1},[3] *i.e.*, theory and experiment disagree by a factor of two. This disagreement can be traced to the discrepancies on the vibrational excitation of the CD$_3$ product states. Experimentally, we found CD$_3$($v = 0$) accounting for about 63% of all CD$_3$ products in the Cl + CHD$_3$($v_1 = 1$) reaction at $E_c \sim 8$ kcal mol^{-1},[3] whereas theory predicted about 30%.[2] It is known that the experimental detection sensitivities of various REMPI bands can in general exhibit substantial variations. Despite our efforts in calibrating the REMPI detection sensitivities,[3,4] the cumulated errors in estimating the CD$_3$ product vibrational distribution remain significant. Given that, the present theory–experiment discrepancy on the product vibrational disposals, 30% *vs.* 63% for CD$_3$($v = 0$)/CD$_3$(all v), is perhaps acceptable, though not satisfactory. Further works in both experiment and theory will be needed.

As you correctly pointed out, at present both theory and experiment indicate that at low total energies, translational energy is more effective than vibration in promoting the reaction of Cl + CHD$_3$ → HCl + CD$_3$, in contradiction to the Polanyi's rule.[5] On the other hand, at higher total energies, both theoretical and experimental results conform to the rule, despite the quantitative differences in σ_v/σ_g.

I agree with you that the van der Waals wells or the long range anisotropic interactions in the entry valley play a key role for the departure of the Cl + CHD$_3$($v_1 = 1$) reaction from Polanyi's rule at low E_c. Compared to the analogous reactions of CHD$_3$($v_1 = 1$) with F and O(^{3}P) reactions,[6,7] however, we believe that the steering effect, exerted by the anisotropic forces in the entry valley, should be less in the Cl + CHD$_3$($v_1 = 1$) reaction.[8] Similar deviations from Polanyi's rule at low collision energies have also been observed in Cl + CH$_4$($v_3 = 1$),[9] Cl + CH$_2$D$_2$($v_1, v_6 = 1$),[10] Cl + CHD$_3$($v_b = 1$),[11] Cl + CD$_4$($v_b = 1$),[12] and O(^{3}P) + CHD$_3$($v_1 = 1$)[7] reactions. The last reaction is of particular interest in that it possesses a barrier greater than 8 kcal mol^{-1}. Yet, the additional vibrational energies (8.6 kcal mol^{-1}), upon CH stretching excitation, merely lower the reaction threshold down to 4.2 kcal mol^{-1}.[7] In other words, only about 40% of the initially deposited vibrational energy can be used to help surmount the barrier to reaction, which seems to offer an alternative view, on an equivalent amount of total energy, for understanding $\sigma_v/\sigma_g < 1$ at low E_c. Mechanically, the extent of threshold-lowering by the vibrational excitation of reactants may reflect the strength and the location (relative to the barrier) of the curvature coupling, induced by the curvature of the reaction path, of the CH stretching (v_1) motion to the reaction coordinate. Recently we exploited the polarization properties of the IR excitation laser, thereby aligning the C–H bond with respect to the approaching Cl atom, to study the stereo-specific reactivity of the Cl + $CHD_3(v_1 = 1)$ reaction.[13] The results might provide deeper insights into the precise role of the van der Waals wells in the entrance channel.

While both theory and experiment for the HCl + CD_3 product channel indicated a reactivity propensity toward CH stretching excitation (over the translation energy) at higher E_c, in accord with Polanyi's rule, the subtlety of how to extend the rule to polyatomic reactions remains. Unlike the vibrational excitation of a BC diatom in an A + BC reaction, where both the B and C atoms under attack are excited, the CH stretching excitation of CHD_3 reactants is mainly localized in the C–H bond with the three D-atoms largely unexcited, which behave as spectators in the HCl + CD_3 channel. As a Cl atom attacks the D-atoms to form DCl + CHD_2, we found the initial translation energy is more effective than vibration by a factor of ~2.6 in promoting the reactivity at a total energy of 16 kcal mol^{-1}.[3] As a result, when both isotopic product channels are considered, both vibrational and translation are nearly equivalent in their capacities to accelerate the total reaction rates.[3]

1 S. Yan, Y.-T. Wu, B. Zhang, X.-F. Yue and K. Liu, *Science*, 2007, **316**, 1723.
2 G. Czakó and J. M. Bowman, *Science*, 2011, **334**, 343.
3 S. Yan, Y.-T. Wu and K. Liu, *Proc. Natl. Acad. Sci. U. S. A.*, 2008, **105**, 12667.
4 J. Zhou, J. J. Lin, W. Shiu, S.-C. Pu and K. Liu, *J. Chem. Phys.*, 2003, **119**, 2538.
5 J. C. Polanyi and W. H. Wang, *J. Chem. Phys.*, 1969, **51**, 1439.
6 W. Zhang, H. Kawamata and K. Liu, *Science*, 2009, **325**, 303.
7 F. Wang and K. Liu, *Chem. Sci.*, 2010, **1**, 126.
8 F. Wang, J.-S. Lin and K. Liu, *Science*, 2011, **31**, 900.
9 H. Kawamata and K. Liu, *J. Chem. Phys.*, 2010, **133**, 124304.
10 J. Riedel, S. Yan and K. Liu, *J. Phys. Chem. A*, 2009, **113**, 14270.
11 G. Czakó, Q. Shuai, K. Liu and J. M. Bowman, *J. Chem. Phys.*, 2010, **133**, 131101.
12 F. Wang and K. Liu, *J. Phys. Chem. Lett.*, 2011, **2**, 1421.
13 F. Wang, K. Liu and T. P. Rakitzis, *Nat. Chem.*, 2012, **4**, 636.

Dr Chandler said: Kopin, would you discuss why you feel the laser based alignment of the reactants did not work?

Dr Liu replied: My statement referred to the reaction of an IR-laser aligned $CH_4(v_3 = 1)$ with an F atom. In general, whether the laser-based alignment of the reactants can yield large steric effects on chemical reactivity or not will depend on how strong the anisotropic interactions in the reaction entry valley are, *i.e.*, it is system dependent. As detailed in our response to Professor Beck, we found small steric effects in the F + $CHD_3(v_1 = 1)$ reaction, but a huge effect in Cl + $CHD_3(v_1 = 1)$.[1,2] The $CHD_3(v_1 = 1)$ reactants in both reactions were laser-aligned in the same way, it is the tug-of-war between the steering forces (exerted by the long-range anisotropic interactions) and the reactant pre-alignment that dictates how well the initially prepared collision geometry can be retained as the reactants approach towards the transition state.

1 F. Wang, J.-S. Lin and K. Liu, *Science*, 2011, **331**, 900.
2 F. Wang, K. Liu and T. P. Rakitzis, *Nat. Chem.*, 2012, **4**, 636.

Professor Orr-Ewing addressed Dr Liu and Professor Polanyi: The paper by Dr Liu and the subsequent discussion considered the value of ideas such as the Polanyi rules, which were developed for atom + diatomic molecule reactions, in guiding our thinking about more complicated reactions involving polyatomic molecules. Our recent work has illustrated how these ideas can still be useful in yet more complicated environments than isolated gas phase collisions. In recent experimental and computational studies, we have explored exothermic reactions of CN radicals with organic molecules in solution in liquids. The experiments use UV photolytic initiation of reaction and product detection with broadband IR absorption spectroscopy with picosecond time resolution.[1-6] These HCN-forming reactions have an early barrier on the potential energy surface for isolated reactions in the gas phase, and are known to produce vibrationally excited products. As we have shown, this dynamical behaviour persists for the reactions in solution in liquids such as chloroform and dichloromethane, with the HCN formed preferentially with 1 quantum of C–H stretch vibrational excitation, consistent with expectations from the Polanyi rules for a reaction with an early barrier. The HCN is also formed with up to two quanta of bending excitation, which is a consequence of a transition state with a flat bending potential. More recently, and inspired by the many beautiful gas phase experiments on F atom reactions with H_2 and methane that produce vibrationally excited HF, we have generated F atoms in liquid CD_3CN and observed the picosecond resolved production of DF. We see clear evidence for vibrationally hot DF products at short times followed by vibrational relaxation as the DF couples to the solvent bath modes, again consistent with expectations from the Polanyi rules despite the presence of a liquid solvent.

1 S. J. Greaves, R. A. Rose, T. A. A. Oliver, D. R. Glowacki, M. N. R. Ashfold, J. N. Harvey, I. P. Clark, G. M. Greetham, A. W. Parker, M. Towrie and A. J. Orr-Ewing, *Science*, 2011, **331**, 1423.
2 A. J. Orr-Ewing, D. R. Glowacki, S. J. Greaves and R. A. Rose, *J. Phys. Chem. Lett*, 2011, **2**, 1139.
3 D. R. Glowacki, A. J. Orr-Ewing and J. N. Harvey, *J. Chem. Phys.*, 2011, **134**, 314508.
4 R. A. Rose, S. J. Greaves, T. A. A. Oliver, I. P. Clark, G. M. Greetham, A. W. Parker, M. Towrie, and A. J. Orr-Ewing, *J. Chem. Phys.*, 2011, **134** 244503.
5 D. R. Glowacki, R. A. Rose, S. J. Greaves, A. J. Orr-Ewing and J. N. Harvey, *Nat. Chem.*, 2011, **3**, 850.
6 R. A. Rose, S. J. Greaves, F. Abou-Chahine, D. R. Glowacki, T. A. A. Oliver, M. N. R. Ashfold, I. P. Clark, G. M. Greetham, M. Towrie and A. J. Orr-Ewing, *Phys. Chem. Chem. Phys.* 2012, **14**, 10424.

Dr Liu replied: The recent work by Professor Orr-Ewing and coworkers represents one of the most beautiful examples of how a simple concept, developed for atom + diatom reactions in the gas phase, can be extended to enhance our understandings of polyatomic reactions in the condense phase. The solvents in these solution-phase reactions are sort of weaker perturbers, and thus it might be interesting to extend similar studies to solvents that act as strong perturbers.

I would like to call attention to the other aspect of Polanyi's rules, the energy requirement. We are more familiar with the concept of energy disposal because of numerous examples, both experiment and theory. Although one can infer the energy requirement of the reverse reaction from the energy disposal of the forward reaction by virtue of the microscopic reversibility arguments, caution should be exercised because of the subtleties in relating the initial and the final conditions. For example, if an exothermic reaction is studied in a state-selected manner at low collision energies, the results of the microscopic reversibility arguments will only be applicable to reaction in the endothermic direction, which leads to that state at the same total energy.

The energy requirement in a polyatomic reaction can become even more subtle. In the $Cl + CHD_3 \rightarrow HCl + CD_3$ reaction, we found the vibrational enhancement factor for both CH-stretching excitation and the bending excitation of the CHD_3 reactant are nearly the same.[1] With an equivalent amount of total energy, the stretching vibration is more effective than pure translation energy by a factor of ~1.4 in yielding $HCl + CD_3$. This finding seems in accord with the expectation of Polanyi's rules for a (slightly) late barrier reaction, as well as with the chemical intuition that the vibrational energy is directly deposited into the bond (C–H) to be broken, However, a similar enhancement factor (~1.4) was also obtained for the bend-excited

$CHD_3(v_5 = 1)$,[2] for which, unlike the C–H stretch excitation, there is no analogous counter part in the atom + diatom reaction. Intuitive understanding of this bending enhancement factor is also less obvious, because the bending mode excitation always involves non-localized, concerted motions of three or more atoms and it does not obviously map onto the reaction coordinate.

1 S. Yan, Y.-T. Wu and K. Liu, Proc. *Proc. Natl. Acad. Sci. U. S. A.*, 2008, **105**, 12667.
2 S. Yan, Y.-T. Wu, B. Zhang, X.-F. Yue and K. Liu, *Science*, 2007, **316**, 1723.

Professor Aquilanti opened the discussion of the paper by Professor Zhang:‡ The paper by Professor Zhang authoritatively discusses the case study of a four-body chemical reaction, facing the key issue of the status of comparison between theory and experiment: the authors point out that the situation is necessarily less satisfactory than for three-body processes, thus making timely a comment on the status for the latter, which continues to be crucial and presents important questions that are still open (see *e.g.* Professor Polanyi's remark earlier). Our comments are illustrated by reference to our recent work beyond the H_3 system: for the benchmark ionic He^+ + H_2 and neutral $F + H_2$ reactions, the full itinerary from PES to cross sections and rates has been made presently viable through exact time independent quantum mechanics. In Professor Zhang's paper, time dependent techniques are shown to permit progress in four-body processes, but even for three-body cases, they still need validation against time dependent ones, because of difficulties such as those encountered in going down enough in energy to cover the deep tunneling low temperature range[1] (this is of currently increasing interest in the cold and ultracold regimes). Also, time dependent methods hardly provide full scattering matrices, as needed for assessing roles of resonances[2] and for allowing complete studies of stereo-dynamical issues:[3]this will be particularly demanding when we will have to face the arguably fundamental role of chirality – perspectively emerging when we move from three to four bodies (and beyond). In our community, customarily we put the blame of disagreements between experiments and dynamical simulations on inadequacies of the potential energy surfaces provided by current quantum chemistry. Indeed, while for $He^+ + H_2$ we are close to the full configuration interaction and basis set limits,[4] even for the intensely investigated $F + H_2$ reaction (and isotopic variants) uncertainty remains on key kinetic features, such as the activation energy and branching ratios.[1,5] Under trial here are (i) difficulties of quantum chemistry for long range and spin orbit forces, (ii) the remaining uncertainties on breakdown of the Born-Oppenheimer approximation, and in general (iii) the role of excited states. The conviction can be extracted from several examples in these discussion days that, for systems of a greater complexity than those alluded at in this comment, no firm conclusions can be drawn on intriguing alternative mechanisms beyond the venerable transition state ideas: deeper understanding is needed from the quantum chemistry and the dynamical treatment of excited states (such as the pervasive conical intersections *etc.*, practically inaccessible to DFT techniques) – see *e.g.* the paper by Fu *et al.* discussed earlier;[6] a similar conclusion can be extended to the invoked role of triple fragmentations,[7] demanding basic understanding of the three-body break-up channel, still elusive to benchmark quantum mechanical investigations, even at the level of a triatomic system

1 V. Aquilanti, K. C. Mundim, S. Cavalli, D. De Fazio, A. Aguilar and J. M. Lucas, Exact activation energies and phenomenological description of quantum tunneling for model potential energy surfaces. The F + H₂ reaction at low temperature, *Chem. Phys.*, 2011, **398**, 186.
2 D. De Fazio, S. Cavalli, V. Aquilanti, A. Buchachenko and T. V. Tscherbul, On the Role of Scattering resonances in the F + HD Reaction Dynamics, *J. Phys. Chem A*, 2007, **111**, 12538–12549.

‡ This question was posed following discussion with Professor Simonetta Cavalli and Dr Dario De Fazio.

3 D. Skouteris, D. De Fazio, S. Cavalli and V. Aquilanti, Quantum stereodynamics for the two product channels of the F + HD reaction from the complete scattering matrix in the stereo-directed representation, *J. Phys. Chem A*, 2009, **113**, 14807–14812.
4 C. N. Ramachandran, D. De Fazio, S. Cavalli, F. Tarantelli, V. Aquilanti, Revisiting the Potential Energy Surface for the He + H_2^+ → HeH^+ + H Reaction at the Full Configuration Interaction Level, *Chem. Phys. Lett.*, 2009, **469**, 26–30.
5 D. De Fazio, J. M. Lucas, V. Aquilanti, S. Cavalli, Exploring the accuracy level of new potential energy surfaces for the F + HD reaction: from exact quantum rate constants to the state-to-state reaction dynamics, *Phys. Chem. Chem. Phys.*, 2011, **13**, 8571–8582.
6 B. Fu, Y. Han and J. M. Bowman, *Faraday Discuss.*, 2012, **157**, DOI: 10.1039/C2FD20010D.
7 G. de Wit, B. R. Heazlewood, M. S. Quinn, A. T. Maccarone, K. Nauta, S. A. Reid, M. J. T. Jordan and S. H. Kable, *Faraday Discuss.*, 2012, **157**, DOI: 10.1039/C2FD20015E.

Professor Bowman asked: You mentioned significant errors in doing quasiclassical trajectory (QCT) calculations and I just wish to point out that there are many instances where QCT calculations actually produce quite accurate results, provided an accurate potential energy surface is used. For example, consider the Cl + CH_4 reaction to give CH_3 + HCl. We showed[1] quantitative agreement for the HCl rotational distribution with experiments of Orr-Ewing and co-workers.[2]

1 G. Czakó and J. M. Bowman, Accurate *ab initio* potential energy surface, thermochemistry, and dynamics of the $Cl(^2P, ^2P_{3/2})$ + CH_4 → HCl + CH_3 and H + CH_3Cl reactions", *J. Chem. Phys.*, 2012, **136**, 044307.
2 C. Murray, B. Retail and A. J. Orr-Ewing, *Chem. Phys.*, 2004, **301**, 239.

Professor Truhlar asked: The OH radical has a low-lying excited electronic state only 79 cm^{-1} above the ground state, but this state was not included in the wave packet calculations. What role is this expected to play in the dynamics?

Professor Zhang replied: Under the Born–Oppenheimer approximation, that electronic state does not react, therefore was not included in our calculation. As the next step for our quantum scattering study of the reaction, we will include that state in our wave packet calculations with non-adiabatic effects properly added as we did for Cl + H_2 reaction.[1]

1 Zhigang Sun, Dong H. Zhang, and Millard H. Alexander, *J. Chem. Phys.*, 2010, **132**, 034308.

Dr Jordan commented: The agreement you have between theory and experiment seems remarkable. What part or parts of the potential energy surface do you think are responsible for the small remaining differences in the product state distributions?

Professor Zhang replied: We think the small remaining differences in the product state distributions very likely come from the difference in rotational states of the reagents used in experimental measurements and theoretical simulations, rather than from the defects in the potential energy surface. The theoretical result is only for H_2 and OH in their ground rovibrational state, while the experiment certainly involves rotationally excited reagents.

Professor Balucani commented: In the paper you have presented in this Discussion, the isotopic variant OH + D_2 → HOD + D has been investigated experimentally and theoretically. This isotopic variant is the same already studied by Alagia *et al.*,[1,2] in crossed molecular beam experiments with mass spectrometric detection and later by Davis and coworkers[3] using the same experimental technique employed in your study. For this specific isotopic variant, it has been demonstrated[2] and confirmed also by your calculation that the old O–H bond acts as a spectator and most of the available energy which is not channeled into product translation remains confined in the new O-D bond. This is also confirmed by the fact that your 5-dimensional quantum scattering results do not differ much from the full 6D results.

In your group, the isotopic variant OH + HD → H_2O + D has also been recently investigated with the same experimental and theoretical approach.[4] Quite interestingly, in that case an H_2O molecule with two equivalent bonds is formed and it is not possible to distinguish the new O–H bond from the old one because of the symmetry of the molecule. Certainly, the preexisting bond is not a simple spectator, at least as far as vibrational energy redistribution is concerned. How did you treat this specific case? What was the effect of removing one degree of freedom in your calculations compared to the fully 6-dimensional treatment? And is there any evidence in the experimental distributions that highlights the difference between the excitation of one O–H stretch rather than the vibrational excitation of the entire water molecule in its normal symmetric, antisymmetric and bending modes? Is the formalism of local modes still appropriate in the case of H_2O formation?

1 M. Alagia, N. Balucani, P. Casavecchia, D. Stranges and G. G. Volpi, *J. Chem. Phys.*, 1993, **98**, 2459.
2 M. Alagia, N. Balucani, P. Casavecchia, D. Stranges, G. G. Volpi, D. C. Clary, A. Kliesch and H.-J. Werner, *Chem. Phys.*, 1996, **207**, 389.
3 B. R. Strazisar, C. Lin, H. F. Davis, *Science*, 2000, **290**, 958
4 C. Xiao, X. Xu, S. Liu, T. Wang, W. Dong, T. Yang, Z. Sun, D. Dai, X. Xu, D.H. Zhang and X. Yang, *Science*, 2011, **333**, 440.

Professor Zhang answered: In principle, it is not possible to distinguish the new O–H bond from the old one because of the symmetry of the H_2O molecule in the OH + HD → H_2O + D reaction and one should symmetrize the final scattering wave functions properly with the nuclear spin statistics taking into account which will change the final rovibrational distribution of H_2O. Actually we did perform such a kind of procedure in our calculation, but found it only introduced tiny differences in differential cross sections and translational energy distributions. Furthermore, we found even if we treat the old O–H bond as a spectator, the final differential cross sections and translational energy distributions are very close to the fully 6-dimensional results except some small energy shift due to different vibrational energy levels with/without the old O–H bond length fixed.

Dr Glowacki enquired: A lot of the discussion regarding this paper has focused on possible inadequacies in the potential energy surface. However, your paper indicates that you performed UCCSD(T)-F12 calculations with an augmented triple zeta basis. In fact, it would be difficult to do much better at present, considering that you're calculating thousands of points. In any case, it's a pretty good surface, and so I wonder if the errors might lie in the fitting methodology rather than in the electronic structure theory. Do you have any comments on this? How prone to error is the neural network fitting methodology that you used? Do you expect the errors to be randomly distributed, or are there particular regions of configuration space that may include more fitting error than others?

Professor Zhang replied: We feel the potential energy surface used in the calculation is sufficiently good to yield a better agreement between theory and experiment on the reaction because both the *ab initio* calculations and the neural network fitting are very accurate. We believe the small differences between theory and experiment are caused by the difference in rotational states of the reagents used in experimental measurements and theoretical simulations.

Dr Skouteris asked: Last year, Cvitas and Althorpe published the results of time-dependent wavepacket calculations on the OH + H_2 isotopic variant of the same reaction, for a total angular momentum $J = 0$, using the same principle of partitioning through the reactant-product decoupling (RPD) scheme.[1] Could you please comment on the similarities and differences between your approach and theirs?

1 M. T. Cvitas and S. C. Althorpe, *J. Chem. Phys.*, 2011, **134**, 024309.

Professor Zhang answered: The approach we used in the present calculation is very similar to that used by Cvitas and Althorpe. We note their calculation is only for the total angular momentum $J = 0$ as we did many years ago.[1] It is very difficult to extend the calculation to $J > 0$, in particular to obtain differential cross section.

1 D. H. Zhang and J. C. Light, *J. Chem. Phys.*, 1996, **105**

Dr Costes remarked: The accuracy of electronic structure calculations has improved markedly in recent years and quantum-mechanical scattering calculations can now reproduce the experimental differential cross sections for prototypical 4-atom reactions obtained in high-resolution crossed-beam experiments. This is nicely exemplified in Professor Zhang's paper on $OH + D_2 \rightarrow HOD + D$ reaction dynamics.

Such a reaction possesses a significant energy barrier and experimental and theoretical studies are consequently performed in the threshold region or at higher collision energies. One important question arises: can a similar agreement between theory and experiment be expected for collision processes occurring at very low energies? In other words is the current chemical accuracy of the determination of potential energy surfaces (PESs) sufficient to describe these processes?

In Bordeaux, we have a crossed-beam set-up using cryogenically cooled pulsed-beam sources and a variable beam intersection angle which allows us to reach collision energies approaching the cold regime.[1,2] Very recently, we studied the $CO(j = 0) + H_2(j = 0) \rightarrow CO(j = 1) + H_2(j = 0)$ inelastic collision. Integral cross sections obtained in the post-threshold region of the $CO(j = 0 \rightarrow j = 1)$ transition at $E_T = 46$ J mol^{-1} (the resolution at this energy is $\delta E_T = 3$ J mol^{-1}) show structures which are identified as shape (orbiting) resonances by comparison with the results of close coupling quantum calculations performed concomitantly.[3] However, to obtain a reasonable agreement between theory and experiment, it is necessary to rescale the PES by a factor $f = 1.05$ despite the original PES being capable to reproduce the IR transitions of the CO–*para*-H_2 van der Waals complex with 0.1 cm^{-1} accuracy.[4] Clearly, very low energy scattering experiments require PESs calculated with an extreme precision.

1 C. Berteloite, M. Lara, A. Bergeat, S. D. Le Picard, F. Dayou, K. M. Hickson, A. Canosa, C. Naulin, J.-M. Launay, I. R. Sims and M. Costes, *Phys. Rev. Lett.*, 2010, **105**, 203201.
2 M. Lara, F. Dayou, J.-M. Launay, A. Bergeat, K. M. Hickson, C. Naulin and M. Costes, *Phys. Chem. Chem. Phys.*, 2011, **13**, 8127.
3 S. Chefdeville, T. Stoecklin, A. Bergeat, K. M. Hickson, C. Naulin and M. Costes, *Phys. Rev. Lett.*, 2012, **109**, 023201.
4 P. Jankowski and K. Szalewicz, *J. Chem. Phys.*, 2005, **123**, 104301.

Professor Neumark asked: You mentioned earlier that the three-body problem in quantum reaction dynamics is "solved". One very important such problem with implications outside our community is the origin of isotope enrichment in ozone. Many explanations for this effect have been proposed but I think none is entirely convincing. Is this a problem you can address with your scattering methodology?

Professor Zhang replied: The isotope enrichment in ozone is an extremely interesting problem. Unfortunately, it involves the collision induced stabilization of O_3 complex formed in the triatomic $O + O_2$ exchange reaction. At present, we can deal with the exchange reaction on a given potential energy surface, but are not able to accurately study the collision induced stabilization process yet.

Professor Truhlar remarked: Based on some of the comments, I think it is relevant to consider the accuracy of CCSD(T) results with large basis sets and whether they are definitive.

• In a recent study of C–H and O–H bond energies in hydrogen peroxide, methanol, and methane, we found a mean unsigned error (compared to experiment) of CCSD(T) extrapolated to the complete basis set limit of 0.6 kcal mol^{-1} for extrapolation from cc-pVDZ and cc-pVTZ and a mean unsigned error of 0.2 kcal mol^{-1} for extrapolation from aug-cc-pVTZ and aug-cc-pVQZ.[1]

• Extrapolated barrier heights for abstraction of H from methanol by hydroperoxy radical and methyl were somewhat more accurate (compared to best estimates) for these levels of theory.[1]

• Another recent study, in this case of 48 reaction energies and barrier heights of 16 isogyric reactions, showed a mean unsigned error of about 0.4 kcal mol^{-1} (compared to Weizmann-3.2, Weizmann-4, and CCSDT(2)$_Q$/CBS(T,Q-extrapolation)+CV+R calculations) for CCSD(T)/CBS results.[2]

• In an earlier study of 24 barrier heights for 12 diverse isogyric reactions, CCSD(T)/cc-pVTZ had a mean unsigned error of 2.3 kcal mol^{-1}, much larger than that of the best density functional approximations (0.9 kcal mol^{-1} without using unoccupied orbitals, and 0.7 kcal mol^{-1} using unoccupied orbitals), but these density functional methods are not better than CCSD(T)/aug-cc-pVTZ, which also has an MUE of 0.7 kcal mol^{-1}.[3]

Depending on one's point of view, these mean unsigned errors might be viewed as alarming, sobering, enlightening, or encouraging, but they do indicate that one should be cautious about assuming that any given potential energy surface obtained by CCSD(T), CCSD(T)-F12a, or CCSD(T)-F12b methods is accurate enough for a given purpose.

1 I. M. Alecu, J. Zheng, Y. Zhao and D. G. Truhlar, *J. Chem. Theory Comput.*, 2010, **6**, 2872–2887.
2 E. Papajak and D. G. Truhlar, *J. Chem. Phys.*, to be published.
3 J. Zheng, Y. Zhao and D. G. Truhlar, *J. Chem. Theory Comput.*, 2009, **5**, 808-821.

Contrasting the excited state reaction pathways of phenol and *para*-methylthiophenol in the gas and liquid phases†

Yuyuan Zhang,§‡[a] Thomas A. A. Oliver,§¶[b] Michael N. R. Ashfold[b] and Stephen E. Bradforth*[a]

Received 3rd March 2012, Accepted 23rd March 2012
DOI: 10.1039/c2fd20043k

To explore how the solvent influences primary aspects of bond breaking, the gas and solution phase photochemistries of phenol and of *para*-methylthiophenol are directly compared using, respectively, H (Rydberg) atom photofragment translation spectroscopy and femtosecond transient absorption spectroscopy. Approaches are demonstrated that allow explicit comparisons of the nascent product energy disposals and dissociation mechanisms in the two phases. It is found, at least for the case of the weakly perturbing cyclohexane environment, that most aspects of the primary reaction dynamics of the isolated molecule are reproduced in solution. Specifically, in the gas phase, both molecules can undergo fast X–H (X=O, S) bond dissociation upon excitation with short wavelengths ($193 < \lambda_{pump} < 216$ nm), following population of the dissociative S_2 ($1^1\pi\sigma^*$) state. Product electronic branching, vibrational and translational energy disposals are determined. Photolysis of phenol and *para*-methylthiophenol in solution at 200 nm results in formation of vibrationally excited radicals on a timescale shorter than 200 fs. Excitation of *para*-methylthiophenol at 267 nm reaches close to the S_1 ($1^1\pi\pi^*$)/S_2 ($1^1\pi\sigma^*$) conical intersection (CI): ultrafast dissociation is observed in both the isolated and solution systems—again indicating direct dissociation on the S_2 potential energy surface. Comparing results for this precursor at different excitation energies, the extent of geminate recombination and the derived H-atom ejection lengths in the condensed phase photolyses are in qualitative agreement with the translational energy release measured in the gas phase studies. Conversely, excitation of phenol at 267 nm prepares the system in its S_1 state at an energy well below its S_1/S_2 CI; the slow O–H bond fission inferred in the gas phase experiments is observed directly in the time-resolved studies in cyclohexane solution *via* the appearance of phenoxyl radical absorption after ~1 ns, with only S_1 excited state absorption discernible at earlier delay times. The slow O–H bond fission in solution provides additional evidence for a tunnelling dissociation mechanism, where the H atom tunnels beneath the lower diabats of the S_2/S_1 CI. Finally, the photodissociation of phenol clusters in solution is considered, where evidence is presented that the O–H dissociation coordinate is impeded in H-bonded dimers.

[a]*Department of Chemistry, University of Southern California, Los Angeles, CA 90089, USA.*
E-mail: stephen.bradforth@usc.edu; Fax: (+213) 740-3972; Tel: (+213) 740-0461
[b]*School of Chemistry, Cantocks Close, University of Bristol, BS8 1TS, UK*

† Electronic supplementary information (ESI) available. See DOI: 10.1039/c2fd20043k
‡ Present address: Department of Chemistry and Biochemistry, Montana State University, Bozeman, Montana 59715, USA
§ These authors contributed equally to this work
¶ Present address: Department of Chemistry, University of California, Berkeley, CA 94720, USA

I. Introduction

Weakly interacting solvents, like rare gas matrices, are generally considered excellent mimics of a gas phase environment in terms of their effect on the shapes of molecular potential energy surfaces (PESs).[1] Thus exploring the photochemistry of solutes in inert solvents such as cyclohexane provides an instructive approach to see how dissociative reaction dynamics map into the condensed phase. However, such solvents still provide some dynamical complexities when compared to the collision-free environment. For example, collision between the nascent reaction products and the solvent molecules can alter the vibrational energy disposal, as solvent molecules provide an effective sink for vibrational energy relaxation (VER).[1-4] Similarly, the translational motion of products will be eventually stopped due to the friction exerted by the solvent.[5-7] Subsequent diffusion of the incompletely separated reaction products in the solvent can lead to geminate recombination.[3,8] Although interactions with solvent molecules have modest changes on the PES landscape, differential solvation of electronic states can lead to larger changes in the location of the conical intersections (CIs), and affect how the excited state molecules approach (*i.e.*, the velocity and the angle) and thus branch through CIs.[9]

In the collision-free, gas phase environment, the X–H (X=N, O, S) bond fission in heteroaromatic molecules has been determined to be an important non-radiative deactivation pathway following UV photoexcitation.[10] In such systems, the $^1\pi\sigma^*$ state is the PES responsible for X–H bond fission, as predicted by pioneering *ab initio* calculations[11] and demonstrated by gas phase photodissociation experiments such as H(Rydberg) Atom Photofragment Translational Spectroscopy (HRA-PTS)[10,12-15] and velocity map ion-imaging (VMI).[16,17] The $^1\pi\sigma^*$ state is formed initially by 3s ← π electronic promotion (in the case that X=N, O) in the vertical Franck–Condon region. Upon X–H bond extension, the Rydberg 3s orbital evolves into the anti-bonding σ* orbital localised across the X–H bond. For simplicity, we henceforth refer to such mixed Rydberg-valence dissociative states as $^1\pi\sigma^*$ states.

In the case of phenol (PhOH), the diabatic S_2 ($1^1\pi\sigma^*$) PES is repulsive with respect to the O–H bond extension and intersects with the diabatically bound S_1 ($1^1\pi\pi^*$) state to form the S_1 ($1^1\pi\pi^*$)/S_2 ($1^1\pi\sigma^*$) CI as displayed in Fig. 1(a). Previous HRA-PTS experiments clearly demonstrated two distinct time scales for H atom generation: slow, when PhOH is excited into vibrational levels of the S_1 state at energies below the S_1/S_2 CI, and fast when the $^1\pi\sigma^*$ state is populated directly which can occur for excitation with $\lambda_{pump} < 248$ nm.[13,18-20]

The highest occupied molecular orbital (HOMO) of phenol is largely comprised of the benzene π orbital with some contribution from the oxygen $2p_x$ lone pair as shown in Fig. 1(a) whereas, in thiophenol (formed by substituting the O atom in phenol with an S atom), the S $3p_x$ lone pair makes a much greater contribution. Further, as Fig. 1(b) shows, the S_2 potential energy curve (PEC) in *para*-methylthiophenol (*p*-MePhSH) shows negligible evidence of a "shelf" in the vertical Franck–Condon (vFC) region (*cf.* phenol in Fig. 1(a)), presumably reflecting the much reduced overlap of the 4s and σ* orbitals.[21] Fig. 1(b) also shows that the dissociative S_2 and diabatically bound S_1 PECs of *p*-MePhSH intersect, but the resulting CI is very close to the S_1 potential minimum, enabling prompt formation of *para*-methylthiophenoxyl (*p*-MePhS) and H atom products even when exciting at the S_1 "origin" ($\lambda_{pump} = 295$ nm).[21,22] This is evident from the HRA-PTS experiments, which show that the H atoms generated by $\lambda_{pump} \leq 295$ nm have an anisotropic recoil velocity distribution, implying that S–H bond fission occurs on a time scale that is shorter than the parent rotational period.[21,22] Further, and in contrast to PhOH, S–H bond fission on the $1^1\pi\sigma^*$ PES yields a significant yield of both *p*-MePhS($\tilde{X}^2B_1$) and *p*-MePhS($\tilde{A}^2B_2$) radical products, as a result of branching at the S_2/S_0 ($1^1\pi\sigma^*$/S_0) CI. Branching for this family of molecules, but not in PhOH, has been rationalised by Landau–Zener arguments;[14,21,23] the energy difference between the ground and excited radical is much smaller in the case of *p*-MePhS and the CI occurs

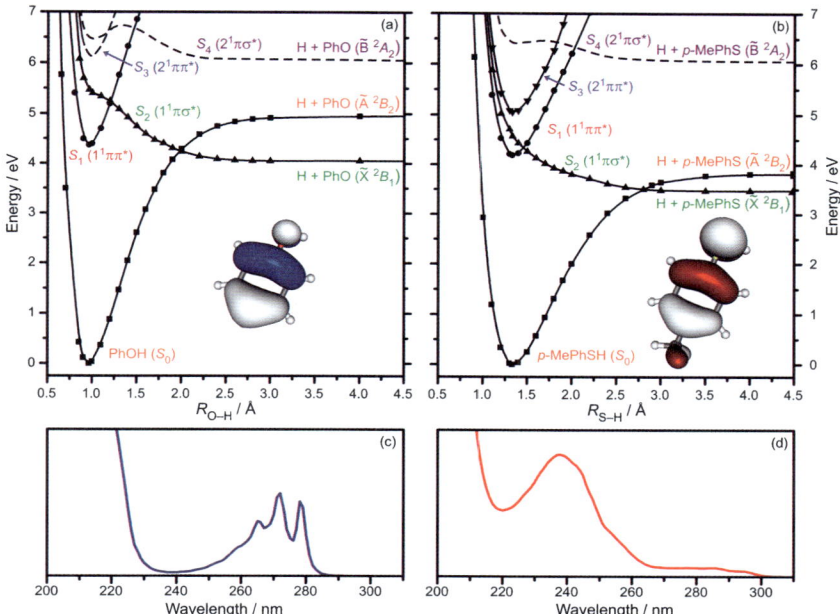

Fig. 1 (a) Optimized PECs along R_{O-H} for PhOH, calculated at the CASPT2(10/10)/aug(O)-AVTZ level for the ground state (S_0), S_1 ($1^1\pi\pi^*$) and S_2 ($1^1\pi\sigma^*$) excited states (reproduced from ref. 19); (b) CASPT2(10/10)/aug(S)-cc-pVTZ PECs in R_{S-H} for p-MePhSH for the same electronic states as in (a) plus the S_3 ($2^1\pi\pi^*$) state.[77] The dashed curves in (a) and (b) depict our best estimate of the minimum energy pathway along the given states, based on prior experimental and theoretical studies. The HOMO for each molecule is displayed in the lower right hand corner of each panel. UV electronic absorption spectra of (c) PhOH and (d) p-MePhSH in cyclohexane were taken at room temperature.

at noticeably longer R_{X-H} bond distances (see Fig. 1). Out-of-plane (a'') modes facilitate the transfer of flux between the S_2 ($^1A''$) and S_0 ($^1A'$) PECs and thus modulate the $\tilde{A}/\tilde{X}$ state radical branching. The p-MePhS products carry excitation in ring breathing and C–S wagging modes—reminiscent of the vibrational energy disposal in the PhO products from photolysis of PhOH—but the majority of the excess energy is partitioned into H atom translational excitation, especially when exciting at $\lambda_{pump} < \sim 266$ nm.[21]

We have previously studied the photodissociation of p-MePhSH in the gas phase and in one solvent, liquid ethanol.[21,22] S–H bond fission motifs characteristic of the gas phase dynamics were found to transfer to the condensed phase, and no evidence was found for other competing pathways unique to the condensed phase, such as photoionisation, excited state proton transfer or proton-coupled electron transfer. From our femtosecond transient absorption (TA) experiments, dissociation in ethanol solution was established to occur within the instrument response time (~ 50 fs). p-MePhS($\tilde{X}$) radical is clearly discernable as a result of this ultrafast dissociation, but we cannot rule out the generation of p-MePhS($\tilde{A}$) radicals.[22] The formation of an adduct was also discerned on longer timescales, and assigned to a diffusive re-encounter of the primary H atom attacking the ring of the p-MePhS radical,[22,24] its geminate partner from the initial photodissociation event.

The current study aims to investigate a broader range of energy disposals, varying the solute (PhOH *vs. p*-MePhSH), the photolysis wavelength (and thus different X–H bond fission pathways) and the environment (isolated gas phase *vs.* solution). In this way, we can further explore similarities and differences in the gas and solution photochemistry. In this report we will consider a solution environment, liquid cyclohexane, in which solute–solvent interactions will be relatively weak and thus the gas phase PESs might reasonably be expected to apply, at least to first order. Although the PESs might be little changed, the solvent may influence the location of, and dynamic approach to, individual CIs. We will therefore explore to what extent the dynamic observables such as the translational recoil motion of the H atom, the energy disposal patterns into the various electronic and vibrational degrees of freedom in the radical fragments can be deduced, and then connected back to the gas phase behaviour. Cyclohexane also provides an environment where hydrogen-bonded solute dimers can be studied; we provide a first exploration of this topic by carrying out photolysis experiments on phenol clusters.

II. Experimental and theoretical methods

The HRA-PTS experiment at Bristol has been described previously.[13,18,21,22] p-MePhSH (Sigma Aldrich, 98%) and PhOH (Fluka, 99.5%) are solids at room temperature and were placed in an inline filter and resistively heated to 70 and 50 °C, respectively, in order to generate sufficient vapour pressure. The resulting vapour pressures were each seeded in ~700 Torr of Ar and pulsed through a solenoid valve creating a supersonic beam. This was subsequently skimmed and intersected by the frequency doubled output of a tuneable nanosecond pulsed laser ($266 \geq \lambda_{pump} \geq 206$ nm) or an ArF excimer laser ($\lambda_{pump} = 193$ nm). H atom photoproducts formed in the interaction region were double-resonantly excited after a time delay $\delta t \sim 10$ ns, via the 2p state by 121.6 nm (Lyman-α) radiation, and then to a high Rydberg state ($n \sim 80$) using a second ~366 nm photon. Any prompt ions formed within the interaction region were extracted using a biased deflector plate that straddles the interaction region. Rydberg-tagged H atoms that fly the known distance, d, to the detector are field ionized upon passing through a grounded mesh and detected. H atom time-of-flight (TOF) spectra are then converted into total kinetic energy release (TKER) spectra using eqn (1):

$$\text{TKER} = \frac{1}{2} m_H \left(1 + \frac{m_H}{m_R} \right) \left(\frac{d}{t} \right)^2 \tag{1}$$

where m_H is the mass of the hydrogen atom (1.00794 u), m_R is the mass of the assumed co-fragment, (p-MePhS = 123.20 u, PhO = 93.11 u), and t is the time taken for the H atoms to reach the detector. A t^{-3} Jacobian was applied when transforming the measured TOF signal intensities into TKER space.

The TA experiments at the University of Southern California were achieved by exciting the solution of interest with UV pulses ($\lambda_{pump} = 267$ and 200 nm) and probing the transient species with a broadband super continuum pulse (typically $310 \leq \lambda_{probe} \leq 650$ nm). To make ultrashort 267 nm, the 800 nm fundamental (1 kHz, Coherent Legend) was converted to deep ultraviolet by four-wave mixing (4WM) in an argon-filled hollow core fibre.[25] The resulting pulses centred at 267 nm had a bandwidth of 6 nm, and were subsequently compressed by a pair of Gires–Tournois Interferometer negative dispersion mirrors,[26] generating a 60 ± 10 fs (FWHM) pulse with very little higher order dispersion, as determined by pulse cross-correlation by two-photon absorption in ethanol. Photolysis was also carried out at 267 and 200 nm using longer pump pulses obtained by 3rd and 4th harmonic generation by sum-frequency mixing in a BBO crystal; these pulses had bandwidths of 2 nm and 1 nm and pulse widths of 100 fs and 130 fs, respectively, as determined by cross-correlation. All experiments employed pump fluences in the range $3.2 \leq F \leq 9.6$ mJ cm^{-2}. The transient signal of the solute was linearly dependent on the pump fluence, as determined by a power dependence study in which the 267 nm pump fluence was varied across the range $0.4 \leq F \leq 13.6$ mJ cm^{-2}. Alternate pump pulses were blocked by a chopper operating at 500 Hz. The weak super continuum probe was generated by focusing a small portion of the 800 nm fundamental into a rotating CaF_2 disc. A pair of off-axis aluminium-coated parabolic mirrors was used to collimate and then focus the super continuum into the sample. The transmitted super continuum was dispersed onto a 256-pixel silicon diode array, where signals with and without the pump beam present were recorded to determine the transient absorption at a given time delay. The relative polarisation of the pump and probe radiation was controlled by rotating a zero-order half wave plate in the 800 nm beam driving the continuum generation. The temporal chirp of the continuum was corrected mathematically by setting the coherent pump + probe two-photon absorption peak from solvent alone at each probe wavelength to zero delay.[27] The sample was delivered by a wire guided gravity jet, forming a thin film of thickness 50–80 μm at the detection region. The flow rate of the liquid ensures that a fresh sample was interrogated every pump pulse.[28] For the TA experiments, various concentrations (10, 45 and 90 mM) of PhOH (99.5%, Avocado Research Chemicals) and p-MePhSH (97%, TCI America) were used in cyclohexane (>99.0%, HPLC and UV-Spectrophotometric grade, EMD or Mallinckrodt) without further purification.

Time-resolved emission from PhOH was obtained via time-correlated single photon counting (TCSPC), where phenol was excited at 267 nm and the decay of the fluorescence maximum (at 300 nm, determined by steady sate emission) was monitored. ~0.1 mM phenol solutions in cyclohexane were used to produce an optical density (O.D.) = ~0.1 in the 1 cm quartz cell.

The ground state geometries of PhOH and p-MePhSH were optimized with Møller–Plesset second order perturbation theory (MP2) and an aug-cc-pVTZ basis set. For the sulphur atom, an aug-cc-pV(T + d)Z basis was used for these and all subsequent calculations. To ensure the geometry optimization converged to a true minimum, the gradient convergence criteria were tightened to 1×10^{-6} a.u. Vertical excited state absorption (ESA) spectra for the S_1 states of PhOH and p-MePhSH were calculated using the equation of motion coupled cluster with single and double excitations (EOM-CCSD) methodology and an aug-cc-pVTZ basis set. Transition dipole moments (TDMs) for excitations from the S_1 state to higher excited singlet (S_n) states were calculated for both molecules for energies up to $E < 4.28$ eV (i.e., $\lambda > \sim290$ nm).[29]

Relaxed potential energy cuts (PECs) for PhOH in the O–H stretch coordinate, R_{O-H}, are reproduced from ref. 19. The corresponding PECs for p-MePhSH in the R_{S-H} coordinate were calculated in this study for the ground, $1^1\pi\pi^*$, $1^1\pi\sigma^*$ and $2^1\pi\pi^*$ states, which correspond to the diabatic S_0, S_1, S_2 and S_3 electronic states. These calculations were performed using complete active self consistent field (CASSCF) theory and a set of ten electrons in ten orbitals active space (10/10). The active space was comprised from the 3 benzene ring centred π, S–H σ and the S($3p_x$) conjugated lone pair occupied orbitals, three ring centred π^*, the S–H anti-bonding σ^* and S(4s) Rydberg virtual orbitals. The basis set described above was used, but with additional diffuse functions on the sulphur atom in order to allow for a better description of any Rydberg-valence mixing; henceforth this is termed the aug(S)-cc-pVTZ basis set. CASSCF calculations for each excited state were state-averaged with the S_0 wavefunction. At each given S–H bond extension, the geometry was allowed to relax on each PES, thus providing the minimum energy pathway or curve. Complete active space with second order perturbation theory (CASPT2) single point energies were then calculated at these optimized geometries, which required an imaginary level shift to avoid intruder state problems. All ab initio calculations were performed in Molpro.[30]

III. Results

i. Photolysis in the gas phase

Total kinetic energy release (TKER) spectra for the H + radical products from these photodissociation reactions obtained at many, particularly longer, photolysis wavelengths have been reported in earlier publications.[13,18,21,22,31] These provide important information about the product energy release and branching in the "unperturbed" dissociation reaction. To aid comparison with the solution studies presented here, additional TKER spectra of the products from *p*-MePhSH and PhOH photolysis at wavelengths ∼200 nm are displayed in Fig. 2(a) and (b), respectively. It is immediately observed that photolysis at 193 < λ_{pump} < 216 nm yields broad, featureless TKER spectra, indicative of highly vibrationally excited photofragments in both cases. This should be compared with behaviour found at longer photolysis wavelengths, where vibrationally structured spectra are observed. In addition, the energy partitioned into H atom translation at ∼200 nm is now substantially larger (up to 22 000 cm^{-1})—not unsurprisingly, given the much higher photoexcitation energies and the fact that most of the radical products are still formed in the same electronic states as at lower excitation energies.

Both $\tilde{X}$ and $\tilde{A}$ state *p*-MePhS radicals are observed in the photolysis of *p*-MePhSH.[21,22] These two states are separated by 3320 ± 50 cm^{-1} as shown in Fig. 1(a).[21,22] It is notable, however, that on reducing the photolysis wavelength from 216 nm to 193 nm, and thereby increasing the available energy by 5 500 cm^{-1}, very little of the additional energy appears as translation. Most of the additional energy is channelled into vibrational excitation of the nascent radical. $\tilde{X}$ state PhO products are dominant in the UV photolysis of phenol at all photolysis wavelengths so far studied, although evidence from 230 nm photolysis of phenol-d_5 suggests that some small fraction of the $\tilde{A}$ state radicals can be formed.[18] Comparing the $\lambda = 193$ nm dataset with $\lambda = 206$ nm in Fig. 2, an additional structured feature is observed centred at TKER ∼5 000 cm^{-1}. This feature is attributable to the PhO($\tilde{B}$) radical.[18] Stavros and co-workers were unable to identify PhO($\tilde{B}$) radicals in an ultrafast VMI study of phenol photolysis at 200 nm[32] and suggested, consistent with the calculated PECs presented here (Fig. 1(b)), and in ref. 18, that the H + PhO($\tilde{B}$) product channel remains closed at $\lambda = 200$ nm. Again, reducing the photolysis wavelength from 206 to 193 nm increases the total available energy by 3 200 cm^{-1} but, as Fig. 2 shows, very little of this additional energy is released as H atom translational recoil.

Table 1 compares the average TKERs for both dissociation reactions in the gas phase at these short photolysis wavelengths with results at longer wavelengths. Most of the energy release at $\lambda = 266$ nm is partitioned into translation.

ii. Solution photodissociation of *p*-MePhSH

Fig. 3(a) displays representative TA spectra of *p*-MePhSH in cyclohexane solution following excitation with a 60 fs 267 nm pump pulse. Features attributable to

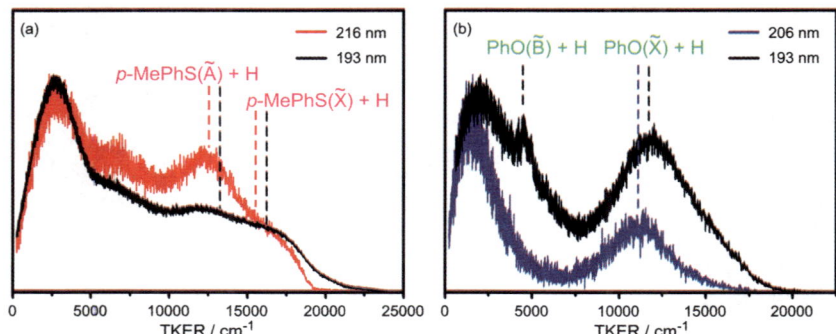

Fig. 2 TKER spectra derived from HRA-PTS studies of (a) *p*-MePhSH and (b) PhOH at photolysis wavelengths shown in legend. The combs above the datasets in each panel represent the estimated average TKER for the respective product channels as given in Table 1.

Table 1 Comparison of maximum possible (TKER$_{max}$) and average measured TKERs of products arising from excited state X–H bond fission in p-MePhSH and PhOH at selected photolysis wavelengths

λ_{phot}/nm	193	206	216	266
Excitation Energy/cm^{-1}	51 813	48 544	46 296	37 594
p-MePhSH[b]				
TKER$_{max}$ ($\tilde{X}$)/cm^{-1}	24 380	—	18 870	10 160
TKER$_{max}$ ($\tilde{A}$)/cm^{-1}	21 060	—	15 550	6 840
Average TKER($\tilde{X}$)/cm^{-1}	~16 500[c]	—	~16 000[d]	~10 000[c]
Average TKER($\tilde{A}$)/cm^{-1}	~13 200[c]	—	~12 700[d]	~6 700[c]
PhOH[a]				
TKER$_{max}$/cm^{-1}	21 800	18 530	—	7 580
Average TKER/cm^{-1}	~12 000[e]	~11 500[f]	—	~5 600[g]

[a] D_0(PhO–H) $= 30015 \pm 40$ cm^{-1} from ref. 13. [b] D_0(p-MePhS–H) $= 27430 \pm 50$ cm^{-1}, $\Delta E(\tilde{A} - \tilde{X}) = 3320 \pm 50$ cm^{-1} from ref. 21, 22. [c] ref. 21, 22. [d] ref. 21. [e] ref. 18. [f] Current work. [g] Data for 275.1 nm, the S_1 origin from ref. 13. Unpublished data at 266 nm shows a negligible difference in the average TKER.

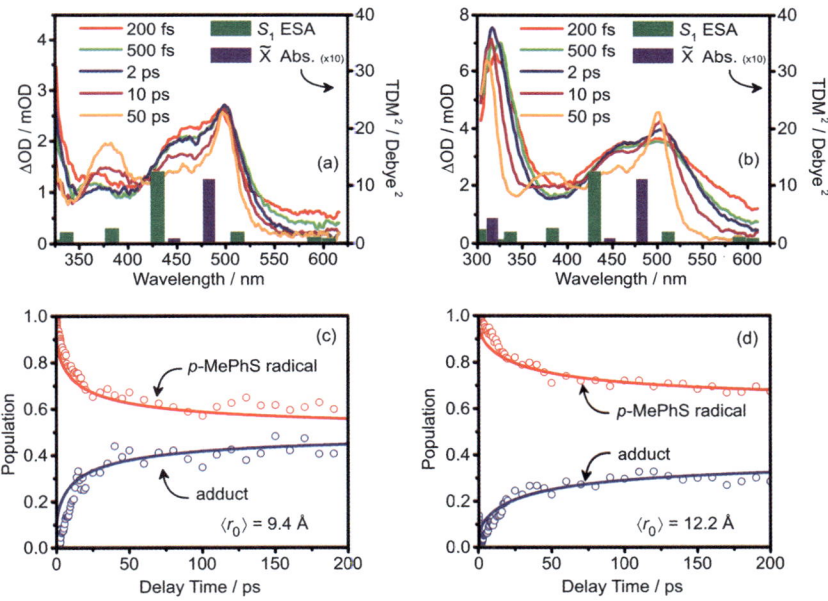

Fig. 3 (a) TA spectra measured at selected delay times following 267 nm photolysis of p-MePhSH in cyclohexane, with the pump–probe polarisation set to magic angle. (b) TA spectra measured following 200 nm photolysis of p-MePhSH in cyclohexane (magic angle polarisation). The squares of the calculated transition dipole moments (TDM^2) for transitions originating from S_1 (ESA) and from p-MePhS radicals are also shown (as bars in each panel). The radical transitions are from CASPT2/CASSCF calculations of ref. 22 and the S_1 band predictions are from the current EOM-CCSD calculations. (c, d) Geminate recombination analysis from the radical (red) and adduct (blue) kinetics after 267 nm and 200 nm photodissociation, respectively. The lines represent the survival probabilities, $\Omega(t)$, of the radical and the population rise of the adduct, as predicted by a radical–radical geminate recombination model and using physical constants described in ref. 22. From this analysis, the average ejection lengths $\langle r_0 \rangle$ are shown.

p-MePhS($\tilde{X}$) radicals, centred at $\lambda_{probe} = 500$, 450 and <320 nm,[22,33] are clearly evident within the overall instrument response time ($\sim$100 fs) and are consistent with prompt S–H bond dissociation. The band assignments and amplitudes are generally in accord with previously computed CASPT2 energies and CASSCF transition dipole moments for the gas phase radical (purple bars shown at base of Fig. 3).[22] The radical electronic absorption bands measured in cyclohexane at any given pump–probe time delay are sharper than in the equivalent TA spectrum in ethanol measured previously.[22,24] Such differences reflect the weaker solute–solvent interactions in the non-polar cyclohexane.[34,35]

Over longer time delays ($t > 5$ ps), the absorptions assigned to the p-MePhS($\tilde{X}$) radical decay and a transient centred at $\sim$380 nm appears. The latter has previously been assigned to adduct formation.[22,24] The appearance kinetics of this feature, as well as the decay kinetics at 500 nm, are insensitive to the p-MePhSH concentration (data not shown)—consistent with a product formed from geminate recombination of the primary radical pair.

We have previously assigned the 450 nm shoulder to the p-MePhS($\tilde{C} \leftarrow \tilde{X}$) transition, which in the previous CASPT2(9/8)/aug-cc-pVTZ calculations, is predicted to have weak absorption strength, arguing that it gained intensity by vibronic mixing with the stronger $\tilde{B} \leftarrow \tilde{X}$ transition centred at $\sim$500 nm.[22] In cyclohexane, unlike in ethanol, it is apparent that the 450 nm side of the band decays to about 50% of the height of the 500 nm band by $\sim$50 ps. Thus, the different kinetics at the two marker wavelengths suggests that they cannot trace the same population and that another shorter lived species must contribute to the absorption measured at 450 nm. The new electronic structure calculations presented here are helpful in understanding this aspect of the spectral evolution. The EOM-CCSD/aug-cc-pVTZ calculations suggest a series of very bright transitions originating in the excited S_1 state within the broadband continuum range (green bars shown at base of figures), consistent with the idea of a "broad underlying absorption" assigned to S_1 ESA earlier,[22] but highlight an S_1 transition centred at $\sim$430 nm with substantial TDM (green bars at base of Fig. 3). This transition is close to the 450 nm shoulder observed in our TA experiments, and it is likely therefore that the latter contains some contribution from S_1 ESA (in addition to p-MePhS($\tilde{C} \leftarrow \tilde{X}$) absorption)—thus explaining the complicated evolution of the spectrum in this region. The only part of the transient spectrum that perhaps can be assigned cleanly to parent ESA is at wavelengths $\geq$600 nm. Most of this signal decays rapidly (within 1 ps); signal to noise precludes assessing whether any small part of this absorption is long-lived (hundreds of picoseconds), as seen in our earlier studies in ethanol. The order-of-magnitude difference in the calculated TDMs for ESA compared to radical absorption suggest that little population is actually created or trapped on S_1 and its contribution appears 'exaggerated' by virtue of its much larger oscillator strength ($cf.$ the p-MePhS radical absorptions).

Fig. 3(b) displays representative TA spectra when p-MePhSH is photoexcited at 200 nm. The spectral signatures for the p-MePhS($\tilde{X}$) radical at 425–500 nm are readily observed within the 180 fs instrument response time, again implying ultrafast S–H bond fission. The initial contour of this band is much broader than that observed at 267 nm excitation, as expected if the p-MePhS($\tilde{X}$) radicals are formed with significant vibrational excitation, which subsequently cools on a $\sim$5 ps time scale. The shape of the strongest p-MePhS($\tilde{E} \leftarrow \tilde{X}$) band at $\lambda \sim$310 nm[22,33] is also better seen in this dataset.

Pulsed radiolysis experiments[36] reveal that the p-MePhSH cation has a distinct absorption feature centred at 450 nm, with an estimated maximum extinction coefficient of $\sim$1800 M^{-1} cm^{-1},[37] and a lifetime of >100 ns in cyclohexane. No such feature is observed in the present TA spectra at any delay time, indicating that photoionisation does not occur to any significant extent in these experiments.

The TA signal intensity ratio of the p-MePhS($\tilde{X}$) radical absorption at 500 nm to the adduct absorption previously noted at $\sim$380 nm is larger with 200 nm excitation, (*e.g.*, compare the 50 ps TA spectra in Fig. 3(a) and (b)). This suggests that the geminate recombination noted at 267 nm is much less efficient in the case of higher energy excitation. A more quantitative analysis of geminate recombination is presented in Fig. 3(c) and (d). The integrated area of the *p*-MePhS $\tilde{B} \leftarrow \tilde{X}$ absorption band (integrating over its half peak width $500 \leq \lambda_{probe} \leq 550$ nm to account for vibrational cooling) is plotted against the delay time. It is clear that, consistent with the initial observation, the survival probability of the *p*-MePhS radical is ~60% at 200 ps upon 267 nm photolysis, whereas a 70% survival probability can be obtained for 200 nm photolysis. The radical decay kinetics is in good accord with a full time-dependent diffusion recombination model,[22] where r_{xn} and $\langle r_0 \rangle$ are the only adjustable parameters. The former is the reaction radius—the distance between geminate partners where recombination is assumed to occur instantaneously. The latter is the average separation distance of the photofragments. Fitting the experimental data in Fig. 3 (solid lines) returns average ejection lengths of 9.4 and 12.2 Å at 267 and 200 nm photolysis, respectively, and a reaction radius $r_{xn} = 4.2$ Å. (The diffusion coefficients required in this model, D_H and $D_{p\text{-MePhS}}$ in cyclohexane, are estimated to be 7.67×10^{-4} Å^2 fs^{-1}, using D_H in water[38] and scaling for the liquid viscosities *via* the Stokes–Einstein equation, and $D_{p\text{-MePhS}} \sim D_{benzene} = 2.26 \times 10^{-4}$ Å^2 fs^{-1}, which is estimated from $D_{benzene}$ in ethanol[39] scaled by the appropriate liquid viscosity difference).

iii. Solution photodissociation of PhOH

For phenol, we first describe results at 200 nm where, on the basis of prior gas phase studies, we expect prompt O–H bond dissociation. Fig. 4 displays representative TA spectra of 10 mM phenol in cyclohexane. Although cyclohexane itself produces a moderately strong TA signal in the long-wavelength region (due to two-photon ionisation induced by the pump pulse[40–42]), fortunately it does not contribute significantly in the region where ground state phenoxyl radicals absorb.[43–46] Fig. 4 shows, similar to the ultrafast generation of *p*-MePhS radicals from *p*-MePhSH, PhO($\tilde{X}$) radicals are observed within the instrument response time (~180 fs), consistent with direct dissociation on the repulsive $^1\pi\sigma^*$ surface. Moreover, the radicals generated are clearly highly vibrationally excited, signified by the initially broad and structureless band centred at ~390 nm. This observation is also reminiscent to that of *p*-MePhSH upon 200 nm excitation. The contour of the PhO($\tilde{X}$) band contracts

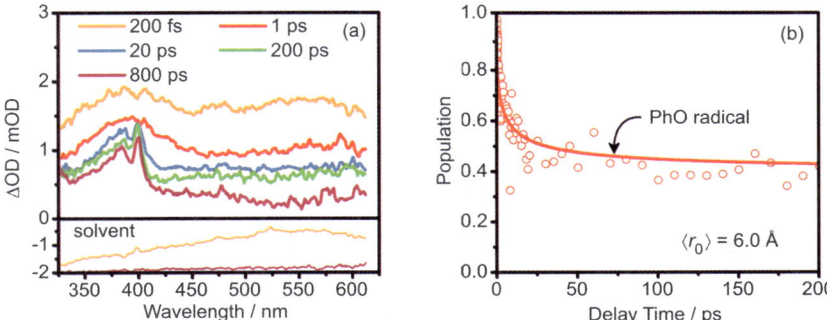

Fig. 4 (a) TA spectra measured at selected time delays following 200 nm excitation of a 10 mM phenol in cyclohexane solution. The TA spectra from the pure cyclohexane solvent, obtained with identical experimental conditions, are shown in the bottom of (a), after subtracting 2 mOD for ease of view. The polarisation of the super continuum probe pulse was set to perpendicular with respect to that of the excitation pulse. (b) Geminate recombination analysis of the radical population after photodissociation (circles) derived from dataset (a). The solid line represents the survival probability, $\Omega(t)$, as predicted by a radical–radical geminate recombination model (see text). From this analysis, $\langle r_0 \rangle = 6.0$ Å.

on a time scale of ∼5 ps, thereafter the spectrum of vibrationally "cold" phenoxyl radicals is observed. The latter is known to exhibit a characteristic vibrationally structured absorption band at 380–400 nm,[43–46] with a very similar shape to that observed here. The vibrational relaxation timescale is similar for both PhO and p-MePhS radicals; as expected, since this is primarily a function of the solvent.[47]

The decay of the PhO($\tilde{X}$) band is interpreted in terms of geminate recombination with the H atom—an average ejection length of 6 Å can be obtained from the full geminate recombination model (Fig. 4(b)). The constants used in the fitting are identical to those for p-MePhSH, with the exception of r_{xn}. The latter is found to be 3.5 Å, obtained from the best fit of the experimental data. Unlike the case of recombination after dissociating p-MePhSH, no adduct is observed (at least with an electronic absorption in our spectral window). We note that, as for the thiol system, the data also provide no evidence to support a photoionisation pathway at 200 nm as there is no spectral signature of the phenol cation (PhOH$^+$)—which has an absorption centred at 430 nm and (if it is the final product in the reaction) a life time of hundreds of nanoseconds.[48]

Now we consider the situation where photoexcitation promotes phenol to the S_1 state at energies below the S_1/S_2 CI. The TA spectra of 10 mM phenol in cyclohexane solution following excitation with a 100 fs 267 nm pulse are shown in Fig. 5(a). Within the time window shown, there is no signal attributable to the phenoxyl radical, suggesting no photodissociation. Note that at early time delays a signal due to the cyclohexane solvent also contributes to the total TA, but this makes negligible contribution after 800 ps. The TA spectra before 1 ns are in fact dominated by

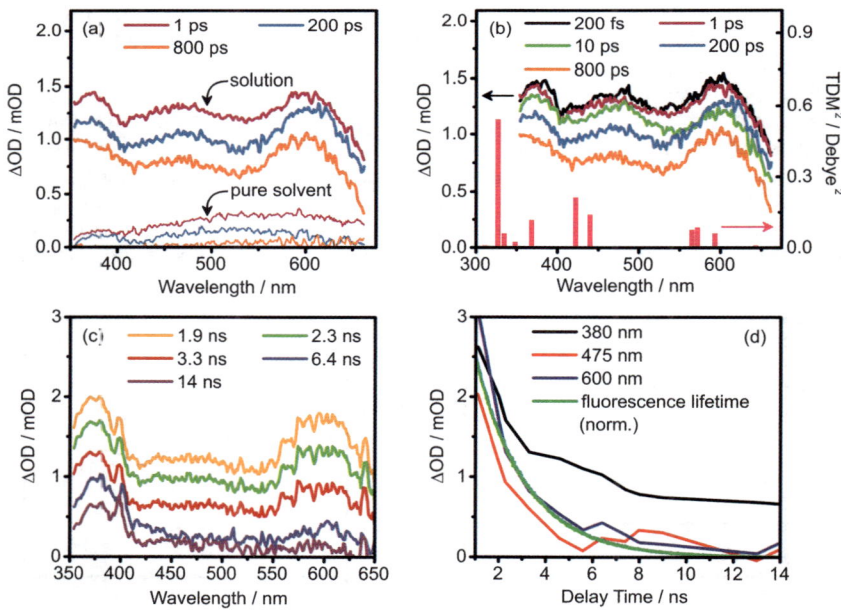

Fig. 5 (a) TA spectra measured at selected delay times following 267 nm photolysis of a 10 mM phenol in cyclohexane solution. The polarisation of the super continuum is set to magic angle with respect to the pump. Thick lines: TA signals from the phenol solution; thin lines: TA signals from pure cyclohexane. (b) TA spectra measured following 267 nm photolysis of 10 mM phenol in cyclohexane, superposed over the calculated squares of the transition dipole moments for excitations originating from S_1. (c) TA spectra measured at nanosecond delay times after 267 nm excitation of 10 mM phenol in cyclohexane. (d) Kinetics at selected wavelengths obtained from the TA experiment and S_1 fluorescence lifetime obtained from TCSPC experiment (green line).

the ESA of the S_1 state, which is populated by absorption of one 267 nm photon. This ESA is evident within our best instrument response time 52 fs (data not shown) and matches well with the calculated $S_n \leftarrow S_1$ EOM-CCSD/aug-cc-pVTZ energies and the square of the TDMs (Fig. 5(b)). Further, it has a similar shape to that obtained by Hermann *et al.*[49] in cyclohexane at 100 ps. The centres of the three major features (at 380, 475 and 600 nm) display identical kinetics, indicating that they originate from the same state.

Although there is no sign of dissociation products within the first nanosecond, we do see the appearance of the phenoxyl radical on longer time scales. Fig. 5(c) presents spectra measured at various delays in the range 1.9–14 ns, which clearly reveal the progressive appearance of the spectral signature of the phenoxyl radical. Fig. 5(d) shows that the kinetics of the 475 and 600 nm ESA bands are consistent with the fluorescence lifetime over the extended time scale (1–14 ns). The 2.1 ns fluorescence lifetime is consistent with that measured by Berlman[50] and by Bent and Hayon.[51] These are further confirmations that the spectral signature observed in 10 mM phenol in cyclohexane at $t < 1$ ns indeed originates from $S_n \leftarrow S_1$ ESA transitions. The kinetics of the 380 nm TA deviate from the S_1 lifetime, however, due to the generation of the PhO radical. The observation of PhO radical absorption is consistent with the earlier flash photolysis study of Hermann *et al.*, but no time scale for the bond fission was reported in that work as a result of the limited available time resolution.[49]

iv. Photoinduced dynamics of PhOH clusters

Both UV-visible[52,53] and IR spectroscopy[54] have demonstrated the possibility of forming phenol clusters in more concentrated cyclohexane solutions (see Fig. 6). As for gas phase clusters,[55] H-bonding between two or more phenol molecules is the main driving force for clustering. We have reproduced the literature results by monitoring the absorption strength of free O–H stretch at 3617 cm^{-1} by FT-IR spectroscopy—the results are shown in the Supplementary Information.† It is estimated from this data that only ~70% of phenol molecules in a 90 mM PhOH in cyclohexane solution exist as isolated monomers, with the remainder present as dimers and higher cluster chains.

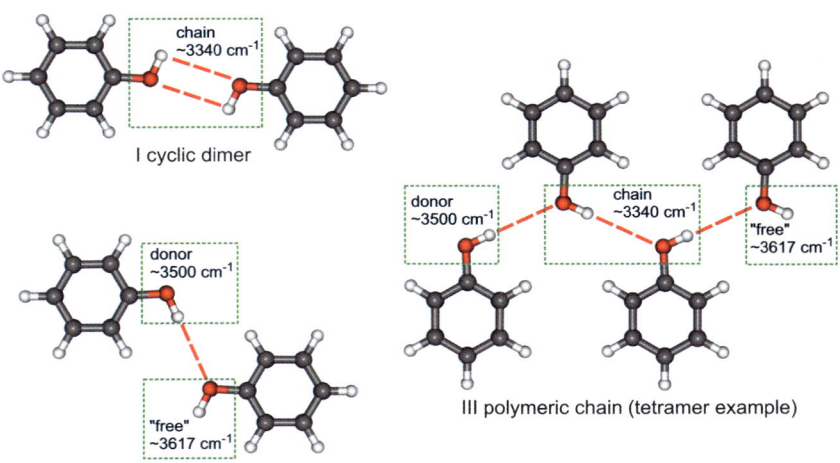

Fig. 6 Illustration of different structures of dimeric and polymeric phenol complexes and the associated OH stretch wavenumbers.

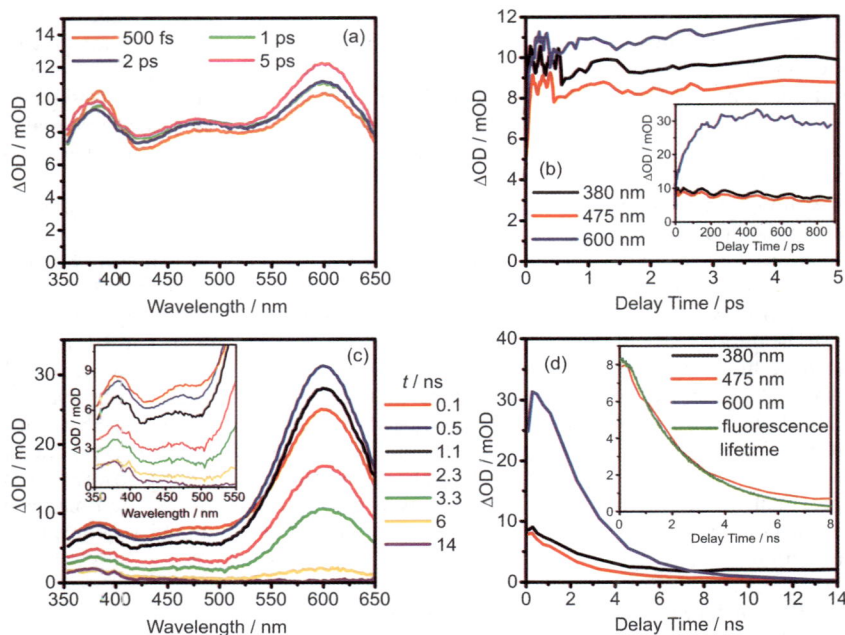

Fig. 7 (a) TA spectra measured at $t < 5$ ps following 267 nm photoexcitation of a 90 mM phenol in cyclohexane solution, with the polarisation of the super continuum probe pulse set to magic angle (54.7°) with respect to that of the excitation pulse. (b) Kinetics of selected wavelengths obtained from the TA experiment at $t < 5$ ps. Inset: $t < 900$ ps. (c) TA spectra measured at delay times in the range 100 ps to 14 ns following 267 nm photoexcitation of 90 mM phenol in cyclohexane. Inset: phenoxyl radical region displayed on an expanded scale. (d) Kinetics of selected wavelengths obtained from the TA experiment. Inset: Comparison of the kinetics at 475 nm and the fluorescence lifetime obtained from the TCSPC experiment (same data as shown in Fig. 5(d) but displayed with a different normalization factor).

Fig. 7 presents the result of femtosecond TA experiments following 267 nm excitation of a 90 mM PhOH solution. The transient spectra observed at early delay times ($t < 2$ ps) display three distinct features at 380, 475 and 600 nm, with approximately equal intensities (see Fig. 7(a)), reminiscent to those observed at low concentrations where clustering is relatively much less important. Further, the kinetics of the $\lambda = 475$ nm feature is again in good accord with the 2.1 ns fluorescence lifetime recorded at low phenol concentration (Fig. 7(d) inset). Because these kinetics are identical to those observed in 10 mM phenol solutions, and are consistent with the computed ESA spectrum for the gas phase monomer, we assign these ESA signatures to monomer (or 'monomer-like') S_1 excited phenol. However, the 600 nm peak behaves differently in these high concentration phenol solutions, growing strongly at later times. The original 'monomer-like' 600 nm ESA peak is clearly supplemented by a new transient, such that it ends up with a different set of associated kinetics *cf.* the monomer kinetics at 380 and 475 nm after $t > 2$ ps (see Fig. 7(b) and (d)). The appearance of a new transient, at concentrations where ground state phenol clustering takes off, hints that the new feature may be attributable to phenol excimers. In fact, this broad feature is very similar to the known spectrum of the benzene excimer,[56] where there is an excited state attractive interaction between two stacked π systems. Here, however, the ~100 ps appearance time is much faster than the diffusion limit (*e.g.*, the benzene excimer formation in cyclohexane has a diffusion-limited rate of $(2 \pm 1) \times 10^{10}$ M^{-1}s^{-1}[56]). Phenol is different from benzene in that it forms a stable dimer in the ground state, which already places two or more aromatic rings primed in close proximity. Therefore, the rise time seen here represents the conformational rearrangement of the cluster instead of a diffusional encounter time. We will return to discuss this feature in section IV.

By comparing Fig. 7(a) and (c) we conclude that no appreciable amount of phenoxyl radical is generated within 1 ns, but that a clearly measurable amount is produced on a longer (ns) time scale. These findings are qualitatively consistent with the observations in the 10 mM PhOH solution, but since there is a large error associated with the PhO radical yield measurement it is difficult to determine whether the radical is generated exclusively from the 70% monomer, or if it contains some contribution from the clustered form as well.

We are now poised to explore whether for phenol molecules tied up in clusters, prompt dissociation induced by exciting above the S_1/S_2 CI will turn off dimer conformational rearrangement—thereby eliminating the spectral signature of excimers. Such experiments, using 200 nm excitation, are presented in Fig. 8. We note that Fig. 4, a "control" experiment for this section which should involve negligible clustering, shows no appreciable signal attributable to PhOH S_1 ESA at longer delay times, the signature for which is displayed in Fig. 5. Although we cannot quantify yields accurately, this observation suggests that most monomeric phenol dissociates at 200 nm, or at least that very little undergoes internal conversion to get trapped on S_1.

Due to the large extinction coefficient of phenol at 200 nm, 45 mM phenol solution was used instead of 90 mM to maintain reasonable optical density samples. This diminishes the cluster fraction to ~15%. For the isolated phenol molecules that make up the majority of the solution, vibrationally excited radicals are formed within the instrument response time just as in Fig. 4, indicating direct O–H bond fission as before. Vibrational cooling occurs on a similar timescale to that observed in the low concentration experiment. However, the peak we have assigned to excimer absorption is still observed, suggesting that a significant fraction of the phenol clusters present do not directly dissociate upon 200 nm excitation. Interestingly, an underlying absorption spreading across the whole spectral window is more pronounced compared to the 10 mM solution (*cf.* Fig. 4), probably due to a lesser relative contribution from the solvent TA signal. Single wavelength analysis of 475 nm, where the monomer S_1 ESA exhibits a weak peak (*cf.* Fig. 5) but is otherwise free of contributions from PhO radical and excimer absorption, shows kinetics very similar to that of the 10 mM solution at 267 nm excitation—a single exponential decay which can be best described by a time constant of 2.1 ns. This could imply that those phenol clusters that are unable to dissociate undergo radiationless decay to S_1, where the 'monomer-like' and excimer forms are in equilibrium with the dimer exploring conformational space.

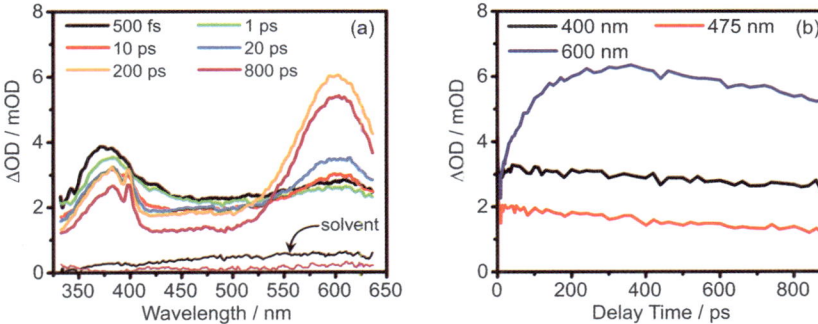

Fig. 8 (a) TA spectra measured at selected time delays following 200 nm photolysis of 45 mM phenol in cyclohexane, with the super continuum probe pulse polarisation aligned parallel to that of the excitation pulse. The thin lines are the TA signal for pure cyclohexane at 500 fs and 800 ps (same colour code as the solution) (b) Kinetics at selected wavelengths.

IV. Discussion

Let us first summarize the major results here in the context of other experimental work. Gas phase photolysis of p-MePhSH at wavelengths near 200 nm yield TKER spectra that are consistent with all so far reported,[21,22] in that they indicate formation of $\tilde{X}$ and $\tilde{A}$ state p-MePhS radical products. The recoil velocity distribution of the fast H atom products is anisotropic at all wavelengths, implying that dissociation occurs on a timescale faster than that for excited parent molecular rotation.[21,22] When p-MePhSH is photoexcited in cyclohexane, at either 267 or 200 nm, a TA attributable to p-MePhS($\tilde{X}$) radicals is observed within the instrument response time, confirming H atom loss by prompt S–H bond fission at both wavelengths. The p-MePhS($\tilde{X}$) radicals are formed vibrationally excited, most especially at λ_{pump} = 200 nm, where a broad contour is observed initially and subsequently sharpens within $\sim$5 ps. The translational energy release is reflected in the solution data by the extent of geminate recombination, $i.e.$, by the decay of the p-MePhS($\tilde{X}$) TA at 500 nm and the coincident rise of the adduct TA at 380 nm. From the ratio of radical to adduct, it is evident that geminate recombination is less efficient in the case of 200 nm excitation, which suggests a longer H atom ejection length consistent with the greater TKER observed in vacuum.

Excitation of gas phase PhOH at λ > 200 nm yields predominantly, if not exclusively, PhO($\tilde{X}$) radicals but an additional small yield of PhO($\tilde{B}$) radicals is observed at λ = 193 nm.[18] Interpreting product recoil anisotropies measured at such short wavelengths is complicated[13,18] but, based on the PESs, we expect PhOH to dissociate promptly at such short wavelengths.[32] This is in marked contrast with the HRA-PTS and time-resolved VMI results at 266 nm and neighbouring wavelengths, which show isotropic product recoil distributions and nanosecond appearance of H atoms.[13,19,20] Consistently, we find that the photochemistry of PhOH in solution is also very different at 'long' and 'short' wavelengths. As in the gas phase,[13,19] 267 nm photoexcitation of PhOH in cyclohexane must excite the former to levels of the S_1 state lying below the S_1/S_2 CI—as shown by the fact that the TA spectrum at early times (t <1 ns) is dominated by S_1 ESA. PhO($\tilde{X}$) radical absorption becomes increasingly apparent at t >1 ns. Conversely, a 200 nm photon excites PhOH well above the S_1/S_2 CI. In this case, PhO($\tilde{X}$) radicals are observed within the 180 fs instrument response time, with significant vibrational excitation, which subsequently relaxes on a time scale of $\sim$5 ps. No new transient absorptions are observed that might be attributed to PhO($\tilde{A}$) or PhO($\tilde{B}$) radicals, at least within our spectral window.

Finally, when PhOH hydrogen-bonded clusters are prepared surrounded by the cyclohexane solvent, a new TA at 600 nm rises in a time scale of $\sim$100 ps which is attributed to the formation of the excimer. While this is an expected aspect of aromatic photochemistry in the condensed phase for excited states with nanosecond lifetimes, we also see the same band when exciting PhOH at 200 nm, suggesting that the direct O–H bond fission pathway is closed for at least some of the clustered species. Further quantification of this last effect would require careful dissociation yield measurements for the monomeric species at low concentration, knowledge of the monomeric and excimeric S_1 ESA molar extinction coefficients and then conversion of the transient spectra into relative populations for each species.

i. Dissociation mechanisms

The $S_2 \leftarrow S_0$ oscillator strength in isolated PhOH is small,[10,19] and there is no evidence supporting a dramatic increase in the solution phase—for example, the absorption cross section at $\sim$240 nm is negligible in cyclohexane (Fig. 1). Therefore, upon 267 nm excitation, the initially populated excited state of phenol is unambiguously S_1, as predicted by ab $initio$ calculations, indicated in the steady state absorption spectrum, and clearly observed (both spectrally and kinetically) in our TA spectrum. The photon energy provided by 267 nm is higher than the $S_{1(v=0)} \leftarrow S_0$ transition energy, but is lower than the location of the S_1/S_2 CI. Intuitively, the deactivation pathways of excited state PhOH in this region include transfer to S_0 *via* internal conversion (IC) or radiative emission, and to T_n *via* intersystem crossing (ISC). Early studies[55,57] suggest that these three channels are responsible for >80% of the excited state deactivation for isolated PhOH molecules, with IC being the major channel. However, O–H bond fission still occurs at energies beneath the S_1/S_2 CI, on a ns timescale, and with a mean TKER of ~5600 cm^{-1} observed in the HRA-PTS experiments.[13]

In light of the large predicted IC yield and slow dissociation time scale, the O–H bond fission process was first (incorrectly) thought to occur *via* coupling at the S_0/S_2 CI after population (by IC) of high vibrational levels of S_0 with substantial O–H stretch character (see Fig.1).[10,13,31,55] This hypothesis has since been reanalyzed and refuted by new experiments (see below), and the present condensed phase experiment provides further evidence for rejecting this mechanism. The current results show a pattern of reactivity in liquid cyclohexane similar to that observed in the gas phase, namely long-lived S_1 and appearance of PhO($\tilde{X}$) radicals but only on a nanosecond timescale. Since the solvent provides an effective "sink" to dissipate excess vibrational energy,[1–4] a highly excited O–H vibration could not survive on S_0 without vibrational energy transfer to adjacent cyclohexane molecules. For example, for dilute methanol in carbon tetrachloride solution where H-bonding is absent, the relaxation time of the O–H($v = 1$) stretch level was determined to be <10 ps.[58] Therefore, the phenoxyl radicals produced on such a nanosecond time scale are unlikely attributable to this ground state dissociation mechanism.

So what accounts for the slow dissociation of phenol? The gas phase data has recently been re-interpreted in terms of H atom tunnelling from the S_1 state under the lower diabats of the S_1/S_2 CI,[19,59] as originally proposed by Sobolewski, Domcke and co-workers.[60] Evidence for this idea comes from HRA-PTS experiments which revealed that the geminate partner of the H atom, the phenoxyl($\tilde{X}^2B_1$) radical, is formed in a very limited subset of the available vibrational state density, predominantly an odd progression of v_{16a}, the nuclear motion advanced as the coupling mode facilitating the non-radiative $S_1 \rightarrow S_2$ transfer beneath the S_1/S_2 CI.[19] Pino *et al.* measured an S_1 state lifetime of ~2.4 ns using pump–probe ion time of flight measurements[59] and, very recently, Roberts *et al.* have observed the H atom elimination process directly by time-resolved VMI and deduced an appearance time $\geq$1.2 ns (limited by the maximum delay measurable in their apparatus) for translationally fast H atoms.[20] Both authors appeal to a simple 1D tunnelling model in just the O–H stretch dimension to justify a nanosecond tunnelling lifetime. This tunnelling would only become more significant with a 2D model because of the lower and narrower barrier due to the coupling mode (v_{16a}) in the branching plane; Dixon *et al.* estimated a 40% greater tunnelling using a 2D wavepacket model compared to the 1D picture.[19]

Roberts *et al.*[20] found that the measured H atom appearance timescale was seemingly invariant as the excitation wavelength was tuned between 275 and 255 nm, and argued that this was strong evidence for a tunnelling mechanism, since IC to S_0 would be expected to show a strong excitation energy dependence on account of the increasing density of vibrational states. Frequency resolved HRA-PTS experiments had previously shown that excited state O–H bond fission occurred irrespective of the initially photoprepared S_1 vibrational state.[13,19,31] The product TKER spectra reveal a marked propensity for S_1 parent vibration to carry through into the phenoxyl radical products—*i.e.* to act as a "spectator" to dissociation. Thus the departing H atom experiences essentially the same tunnelling barrier, irrespective of the particular $S_1(v)$ state excited. However, the parent lifetimes measured by Roberts *et al.* do decrease with increasing photon energy, which is hard to bring into accord with an explanation based on tunnelling in competition with IC. Simple branched kinetics dictates that if parallel processes deactivate the parent, even if the rate linking parent to product is constant, the appearance time of the product must match the decay of the parent. Thus, while the gas phase ultrafast VMI results provide some support for the tunnelling model, they are not definitive and could also be rationalised with the earlier ground state dissociation mechanism.[13,31]

It appears that the surrounding cyclohexane environment provides little change to the S_1 PES at small O–H bond extensions, which preserves the dynamical behaviour of gas phase PhOH. The dense solvent environment might be expected to push up the Rydberg states of a molecule, however, due to Pauli repulsion of solvent electron density associated with the diffuse solute Rydberg orbital, thereby changing the shape of the S_2 PES—particularly at short R_{O-H} where the Rydberg character is significant. Given such notions, we might expect some possible change in the shape of the tunnelling barrier, but the current experiments suggest that the tunnelling rate cannot be changed substantially from the gas phase otherwise either the S_1 lifetime would be reduced in solution or a vanishingly small phenoxyl radical yield would be difficult to detect.

For gas phase HRA-PTS experiments at $\lambda_{pump} < 248$ nm, the transition to S_2 ($1^1\pi\sigma^*$) state is deduced to gain oscillator strength by intensity borrowing from the S_3 ($2^1\pi\pi^*$) $\leftarrow$ S_0 transition.[19] Thus, the dissociative state can be populated directly and yield H atoms on an ultrafast time scale.[13,31,32] The state that is initially populated at 200 nm is undetermined, due to the high density of states at this excitation energy, but Fig. 2(a) confirms formation of translationally excited PhO products carrying substantial vibrational excitation. In cyclohexane solution, the observation of vibrationally hot PhO($\tilde{X}$) radicals within the instrument response time (<180 fs) following PhOH excitation at 200 nm is similarly consistent with a direct dissociation mechanism, and clearly distinct from the tunnelling mechanism observed when exciting at much longer wavelengths.

The photochemistry of p-MePhSH in cyclohexane also shows strong similarities with the gas phase behaviour. Ultrafast S–H bond fission, implied by the anisotropic H atom recoil velocity distributions in the gas phase experiments, is observed directly in the TA experiments in cyclohexane. A 267 nm photon (4.6 eV) likely excites p-MePhSH to a region above, or at least very close to, the S_1/S_2 CI (see Fig. 1). The generation of p-MePhS radicals within the first 100 fs is not surprising, as they could arise *either* by efficient coupling at the S_1/S_2 CI (by an analogous tunnelling mechanism to that in PhOH but involving a very much smaller barrier) or by direct S_2 $\leftarrow$ S_0 excitation since, unlike in phenol, this transition has reasonable oscillator strength.[21,22] Further details on the dissociation behaviour of p-MePhSH and related p-substituted thiophenols following excitation at energies close to the S_1/S_2 CI will be discussed in ref. 21.

ii. Product energy disposal

At long R_{X-H} (X=O, S) bond extension, the S_2 and S_0 PESs intersect to form the S_2/S_0 CI, which determines the eventual branching between the $\tilde{X}$ and $\tilde{A}$ states of the radical.[61] Out-of-plane motions like ring torsion and the O–H torsion encourage population transfer between the S_2 and S_0 PESs, and encourage formation of $\tilde{A}$ state radicals. Both p-MePhS($\tilde{X}$) and p-MePhS($\tilde{A}$) radicals are clearly observed in the gas phase.[21,22] Evidence for $\tilde{A}$ state radical formation in cyclohexane solution within the monitored spectral window and out instrument response time is not conclusive at this stage; this will be commented on in a future paper.[24] Calculations at the CASPT2(9/8)/aug-cc-pVTZ level predicted that the $\tilde{D}$ $\leftarrow$ $\tilde{A}$ transition, the strongest p-MePhS($\tilde{A}$) transition within our continuum probe, should lie at $\lambda \sim 363$ nm, but with a TDM less than one tenth that of the $\tilde{B}$ $\leftarrow$ $\tilde{X}$ transition at 500 nm.[22] Thus, if the $\tilde{A}/\tilde{X}$ product branching ratio in cyclohexane solution is similar to that in the gas phase, observing clear evidence of p-MePhS($\tilde{A}$) radicals is challenging.

Within the spectral range of our continuum probe, the PhO($\tilde{A}$) radical is predicted to have prominent absorption at $\sim$613 ($\tilde{C}$ $\leftarrow$ $\tilde{A}$) and $\sim$400 nm ($\tilde{E}$ $\leftarrow$ $\tilde{A}$);[62] the latter would essentially overlap with the ground state PhO radical absorption. Further, PhO(Ã) radical production is unlikely to be feasible, energetically, at 267 nm (4.65 eV) excitation (see Fig. 1). As mentioned earlier, there is some hint of some flux branching to the PhO(Ã) channel in the gas phase but only when $\lambda_{pump} \leq 230$ nm.[18] As in the thiophenols, the Ã/X̃ product branching ratio is determined by the planarity (or otherwise) of the parent molecule while in the vicinity of the S_2/S_0 CI.[61] The evidence for PhO(Ã) radical production in the gas phase is much less clear than in the case of p-MePhSH. This can be reconciled by considering the high barrier for the O–H torsion with respect to the ring. As Fig. 1 shows, the O $2p_x$ lone pair orbital aligns parallel with the π electrons of the benzene ring and forms a favourable configuration, with a torsional barrier height of 1215 cm^{-1} in the ground state (4710 cm^{-1} in the S_1 state).[63] This rigid planar geometry ensures that the dissociating flux follows the diabat, encouraging formation of PhO(X̃) radicals. In light of this, the failure to observe the PhO(Ã) radical in cyclohexane solution is consistent with the gas phase results. However, any excited state radicals formed would likely undergo fast electronic quenching to the ground state via the same S_2/S_0 CI. Such processes (see for example the possible Ã → X̃ quenching of the p-MePhS radical[22,24]) would be expected to complete within the first 500 fs, and thus be very difficult to observe the excited state radical with the 180 fs instrument response time afforded by the 200 nm pulse.

Unlike Ã state radicals, where an obvious quenching pathway can be enumerated, PhO(B̃) state radicals have no such efficient relaxation pathway (see PES in Fig. 1(a)). Therefore the absence of any new spectral signature in the 200 nm photolysis data, while not conclusive, is supportive of the gas phase observation that this channel remains closed at this photolysis wavelength.[32] Calculations for the absorption bands of PhO(B̃) would help strengthen this conclusion.

iii. Translational energy release

After X–H (X=O, S) bond fission, due to the kinematics of both molecular systems, H atoms depart with substantial kinetic energy. Of course, under a collision free environment, this information is preserved and enables detection of the product TKER. In the condensed phase, the high recoil velocity and small size of the H atom means that most can be expected to require several collisions with solvent molecules prior to translational thermalisation within the liquid. This stopping of the ballistically ejected H atoms is expected to be very fast and is not time-resolved in the present experiments, but a measure of the stopping distance (usually called the H atom ejection length) can be determined in the longer timescale diffusive geminate recombination of the radical pair.

In the spherically symmetric limit and under field-free diffusion of the geminate pair,[64,65] the long time survival probability, $\Omega(\infty)$, of the radical pair is a simple function of ejection length (r_0) and an encounter radius to which the two partners must return in order to recombine, namely $\Omega(\infty) = (1 - r_{xn}/r_0)$. The encounter radius r_{xn} is expected to be different for H recombining with PhO or with p-MePhS, but for a given radical partner should not vary with photolysis wavelength; the latter simply determines the ejection length via how much translational energy the H atom receives. In principle, with two distinct photolysis systems ejecting H atoms into cyclohexane and different photoexcitation energies we should be able to infer the stopping power of cyclohexane to H atoms of several different kinetic energies.[6] We note that, unlike photodissociation reactions producing slower, heavier fragments, we see no evidence for fast cage recombination[7,9] in the time dynamics of the X̃ state PhO or p-MePhS radicals.

The radical survival probabilities after S–H bond cleavage of p-MePhSH at 267 and 200 nm, and the ejection lengths obtained from a full time-dependent geminate recombination model,[22,64] are in qualitative agreement with the increase in the average TKER determined in the gas phase experiments. The increase in average

TKER from $\sim$8 500 to $\sim$14 000 cm^{-1} determined in the gas phase studies (Fig. 2(a) and Table 1) correlates with an increase in average ejection length from 9.4 to 12.2 Å (Fig. 3), if we assume that the energy partitioning established in the gas phase maps directly into the cyclohexane solution study. Moreover, the kinetics of the p-MePhS($\tilde{X}$) radical and the adduct as deduced by single wavelength analysis at 380 nm show a decay and rise commensurate with one another. This indicates that addition of the departing H atom to the ring, or to the sulphur atom so as to reform the parent molecule, is indeed a plausible geminate recombination channel.[24] We cannot say what fraction recombines at the S atom compared to the ring, but the relatively large r_{xn} value we derive could indicate that the ring is a major site for recombination. As can be seen in the fit, the early time kinetics of both features are not described perfectly, because absorption at these probe wavelengths is contaminated by S_1 ESA. Fortunately, the majority of the geminate recombination kinetics is not affected because of the very short S_1 state lifetime of p-MePhSH, thus allowing extraction of ejection lengths.

Comparing the H atom ejected from PhOH at 193 and 206 nm, where we expect an average TKER $\sim$12 000 cm^{-1} (see Fig. 2(b) and Table 1), the geminate recombination observed in cyclohexane appears to be best described by $\langle r_0 \rangle = 6$ Å and $r_{xn} = 3.5$ Å. Note that the deduced reaction radius is larger than the value for H + OH recombination in water (2.7 Å)[66] but, as for p-MePhSH, a larger r_{xn} suggests that PhO + H may also recombine at the ring position instead of the O atom. However, in the range of our broadband continuum there is no discernible spectral feature assignable to an H-PhO type adduct—it is possible that the electronic absorption of such species is either too weak, located outside of our probing region, or energetically unfavourable. An ejection length cannot be established from the 267 nm photodissociation, since geminate recombination in this case is faster than the nanosecond generation of the PhO radical itself.

An average ejection length of 6 Å for the H atoms formed by PhOH photolysis at 200 nm appears contradictory, given the $\sim$12 000 cm^{-1} average translational energy release in the photodissociation. In comparison, the deduced H atom ejection lengths from p-MePhSH photolysis is $\sim$9 Å (for TKER $\sim$8 500 cm^{-1} at $\lambda = 267$ nm) and $\sim$12 Å (for TKER $\sim$14 000 cm^{-1} and $\lambda = 200$ nm). Let us consider one extreme scenario—namely that the presence of solvent guides all the reactive trajectories along the diabat from p-MePhSH(S_2) to ground state radicals (i.e., the branching between p-MePhS($\tilde{X}$) and p-MePhS($\tilde{A}$) products is dramatically changed)—to see if it could provide a rationale for the apparent contradiction in ejection lengths. Even if we were to further assume that little energy is channelled into product vibrations—a result at odds with the observed narrowing of the product TA narrows with increasing time delay and allow all the reaction exoergicity to be available as product kinetic energy, the TKER$_{max}$ at 267 nm would be 10 000 cm^{-1} (Table 1), still less than the average energy release from PhOH photolysis at 200 nm. As this possibility fails to explain the discrepancy, we revisit one other possibility, that the H atom ejection length from phenol is short because the solvent induces more branching to the PhO $\tilde{A}$ state, an outcome that would substantially lower the initial average TKER. Recall that there was some evidence for PhO-d_5($\tilde{A}$) in TKER spectra for 230 nm photolysis.[18] As explained above, for this scenario to be favoured the O–H bond would need to lie out of plane as the dissociating PhOH molecule, surrounded by cyclohexane, approaches the S_2/S_0 CI. One possible scenario is that the S_2 state is prepared with the OH bond out of plane as a result of initial $S_n \rightarrow S_2$ radiationless transition. Any initial driving force toward planarity upon reaching the S_2 state could be sufficient to preserve these out of plane motions in the O–H torsion, which as the O–H bond extends upon dissociation and essentially becomes a free rotor, will remain excited on approaching the S_2/S_0 CI. Arguing against this hypothesis is the absence of any electronic absorption signature of the transient PhO($\tilde{A}$) product—though, if the oscillator strengths of the p-MePhS radical absorptions are any guide, visible transitions involving the $\tilde{A}$ state might be

This journal is © The Royal Society of Chemistry 2012

much weaker than those from the $\tilde{X}$ state.[22] Additionally, we have commented that quenching of the PhO($\tilde{A}$) radical may take place faster that the instrument response. At present, we should treat this as an interesting speculation that could possibly account for the deduced shorter H atom ejection length in phenol. However, we note that the assumptions inherent to the geminate recombination analysis are likely too simplistic when one considers (i) the distribution of H atom kinetic energies, that do not necessarily lead to a Gaussian initial radical distribution as assumed in the diffusive model, and (ii) that these molecular systems do not satisfy spherical symmetry. A full molecular dynamics (MD) simulation that includes the initial ballistic energy release would be required for a more complete investigation.[9]

The kinetic energy dependence of the H atom ejection length from p-MePhSH photolysis can be compared to that observed for the Cl and OH radicals produced by photolysis of aqueous HOCl. For this photoreaction, the majority of the excess energy after O–Cl bond fission is released as product translation, and very little O–H vibration is observed.[5,68–70] Madsen et al. have further established that the ejection length ($r_{Cl–OH}$), which ranges from 4.4 to 6.0 Å, scales linearly with initial translational energy over the range 1.6–2.8 eV (12 900–22 600 cm^{-1}).[6] This result is partially supported by MD simulations. The ejection length would be expected to be a linear function of kinetic energy if the friction, in this case from water as the environment, is velocity independent. The present results suggest that cyclohexane is much less efficient at stopping an H atom than water is in arresting heavier radicals.

iv. Excited state dynamics in phenol clusters

In phenol dimers, hydrogen-bonding is the dominant ground state interaction. Upon excitation to the S_1 state, however, π–π interactions become much stronger, and consequentially the aromatic rings move into closer proximity with each other and become aligned in a pseudo-parallel configuration.[71] The precise geometry of the excimer is determined by competing H-bonding and π-stacking interactions, as previously illustrated by rotationally resolved electronic spectroscopy.[67,72,73] The rise in the 600 nm kinetics in our 267 nm data is attributed to such conformational changes of dimeric and polymeric phenol upon excitation.

The effect of H-bonding on the $1^1\pi\sigma^*$ state and potential photodissociation has not been demonstrated in previous condensed phase studies. Intuitively, the collinear O–H$\cdots$O geometry for the H-bonded phenol dimer should significantly change the electronic structure of the $1^1\pi\sigma^*$ state at larger $R_{O–H}$ distances— becoming bound rather than repulsive. Such a situation has been described by Chipman for the excited state of the water dimer.[74] Consistent with this intuition, considering the observation of the excimer band at 600 nm, the spectra of 45 mM phenol obtained at $\lambda_{pump} = 200$ nm provide tentative evidence that at least some of the clusters do not dissociate upon photoexcitation. This 600 nm band would not appear if all photoexcited phenol molecules dissociated.

Although there have been numerous gas phase phenol dimer spectroscopy studies, the majority have focused on the S_1 state. The key observation is that the S_1 state of the phenol dimer has a notably extended lifetime (16 ns, cf. 2.1 ns for phenol monomer).[55] This has been ascribed to elimination of the otherwise important IC channel back to S_0 as the H-bonding between the two phenol molecules lowers the vibrational frequency of the O–H stretch, and in turn makes the latter a less efficient acceptor mode.[55] Such an argument implies that the electronic excitation is localised on the H-bond donor phenol which is consistent with theoretical calculations on the dimer.[67] The excited state lifetime is similarly extended in gas phase phenol/water dimers,[55,75,76] where the phenol O–H bond also acts as the H-bond donor.[76] However, the possible effects of clustering on the tunnelling dissociation under the S_1/S_2 CI on the excited state lifetime were not considered, and this process should be strongly affected. In fact, Brause suggested that in the dimer, the donor O–H bond length is shorter in S_1 compared to monomeric phenol[67] which would result in making the tunnelling barrier higher and wider (see Fig. 1(a)). As mentioned earlier, the fact that the $1^1\pi\sigma^*$ surface may become bounded at longer R_{O-H} will further impact the likelihood of dissociation.[74] Combined, this suggests that tunnelling dissociation will be a less competitive non-radiative relaxation pathway in the dimer in comparison to the monomer. In our experimental TA data, it is difficult to quantify a reduced phenoxyl yield because of the coexistence of monomer and clustered species in the solutions; the latter being a minor component. However, it is clear that the excimer absorption band at 600 nm has a lifetime longer than 2.1 ns (see Fig. 7). Additionally, the monomer S_1 absorption at 475 nm can be fitted with a bi-exponential decay function—at $t < 4$ ns, the decay is in excellent agreement with the single exponential exhibited by low concentration phenol ($\tau = 2.1$ ns), but deviates from a mono-exponential decay at longer time delays. Given that the π stacked excimer is in conformational equilibrium with an open excited state dimer structure that has little π stacking, the observation of a longer tail to the "monomer" ESA band could then also be explained.

These preliminary experiments on the photochemistry of hydrogen bonded cluster species within inert environments like cyclohexane demonstrate that this approach could provide a rich new avenue of research in comparing reaction dynamics in the two phases.

V. Conclusions

In this study, direct comparisons of the primary energy disposal and the pathways for X–H bond scission have been made for two related photolysis systems either isolated in vacuum or embedded in a weakly interacting room temperature liquid. With the energy disposal patterns revealed by HRA-PTS in the gas phase, we can interpret time-resolved liquid phase studies with a far greater level of detail. In turn, the solution femtosecond transient absorption studies provide new dynamical insights to mechanistic pathways and question the factors that influence passage through conical intersections. In particular, we find common to both phases ultrafast X–H scission (X=O, S) at short pump wavelengths near 200 nm; this is also true for p-MePhSH excitation near the S_1/S_2 CI at 267 nm. Electronic and vibrational branching patterns in the liquid are observed through characteristic bands in broadband transient spectra. These broadly mimic observations from the gas phase TKER spectra, but geminate recombination analysis suggests a smaller average translation energy release to the H atom for PhOH photolysis at 200 nm than expected from the TKER spectra. For PhOH photolysis below the S_2/S_1 CI, PhO radical formation takes place on a much longer (nanosecond) timescale in both solution and gas phase.[12,29] Because a mechanism requiring phenol having first decayed from S_1 to S_0 to remain vibrationally excited while progressing towards the S_2/S_0 CI is rather unlikely in cyclohexane, a liquid that encourages efficient (sub-nanosecond) vibrational energy relaxation, this latter observation provides new evidence against a ground state dissociation mechanism for phenol near the S_1 origin and lends further support to a tunnelling dissociation mechanism.

The use of cyclohexane as medium makes possible the measurement of excited state reaction dynamics of hydrogen bonded clusters within a liquid environment that can be explicitly compared with gas phase clusters. Our current experiments give tantalizing evidence for the shutdown of the ballistic scission of the PhO–H bond on S_2 when the excited state phenol acts as hydrogen-bond donor. This is fertile ground for dynamical simulations and new experiments that aim to explore the full nature of the solvent influence in guiding chemical dynamics.

Acknowledgements

We thank Richard Dixon for useful discussions regarding tunnelling dynamics in phenol, Mike Nix and Graeme King for help in collecting the phenol HRA-PTS spectra, and Saptaparna Das and Anirban Roy for help in collecting the phenol TCSPC data. The work at USC is supported by the US National Science Foundation under CHE-0957869 and at Bristol by EPSRC *via* a Programme Grant (EP/G00224X).

References and notes

1 C. G. Elles and F. F. Crim, *Annu. Rev. Phys. Chem.*, 2006, **57**, 273–302.
2 C. G. Elles, M. J. Cox and F. F. Crim, *J. Chem. Phys.*, 2004, **120**, 6973–6979.
3 J. Larsen, D. Madsen, J. A. Poulsen, T. D. Poulsen, S. R. Keiding and J. Thogersen, *J. Chem. Phys.*, 2002, **116**, 7997–8005.
4 C. Petersen, N. H. Dahl, S. K. Jensen, J. A. Poulsen, J. Thogersen and S. R. Keiding, *J. Phys. Chem. A*, 2008, **112**, 3339–3344.
5 C. L. Thomsen, D. Madsen, J. A. Poulsen, J. Thogersen, S. J. K. Jensen and S. R. Keiding, *J. Chem. Phys.*, 2001, **115**, 9361–9369.
6 A. Madsen, C. L. Thomsen, J. A. Poulsen, S. J. K. Jensen, J. Thogersen, S. R. Keiding and E. B. Krissinel, *J. Phys. Chem. A*, 2003, **107**, 3606–3611.
7 N. Winter and I. Benjamin, *J. Chem. Phys.*, 2004, **121**, 2253–2263.
8 C. G. Elles, M. J. Cox, G. L. Barnes and F. F. Crim, *J. Phys. Chem. A*, 2004, **108**, 10973–10979.
9 C. A. Rivera, N. Winter, R. V. Harper, I. Benjamin and S. E. Bradforth, *Phys. Chem. Chem. Phys.*, 2011, **13**, 8269–8283.
10 M. N. R. Ashfold, B. Cronin, A. L. Devine, R. N. Dixon and M. G. D. Nix, *Science*, 2006, **312**, 1637–1640.
11 A. L. Sobolewski and W. Domcke, *J. Phys. Chem. A*, 2001, **105**, 9275–9283.
12 M. G. D. Nix, A. L. Devine, B. Cronin and M. N. R. Ashfold, *Phys. Chem. Chem. Phys.*, 2006, **8**, 2610–2618.
13 M. G. D. Nix, A. L. Devine, B. Cronin, R. N. Dixon and M. N. R. Ashfold, *J. Chem. Phys.*, 2006, **125**, 133318.
14 A. L. Devine, M. G. D. Nix, R. N. Dixon and M. N. R. Ashfold, *J. Phys. Chem. A*, 2008, **112**, 9563–9574.
15 M. N. R. Ashfold, G. A. King, D. Murdock, M. G. D. Nix, T. A. A. Oliver and A. G. Sage, *Phys. Chem. Chem. Phys.*, 2010, **12**, 1218–1238.
16 M. N. R. Ashfold, N. H. Nahler, A. J. Orr-Ewing, O. P. J. Vieuxmaire, R. L. Toomes, T. N. Kitsopoulos, I. A. Garcia, D. A. Chestakov, S. M. Wu and D. H. Parker, *Phys. Chem. Chem. Phys.*, 2006, **8**, 26–53.
17 A. G. Sage, M. G. D. Nix and M. N. R. Ashfold, *Chem. Phys.*, 2008, **347**, 300–308.
18 G. A. King, T. A. A. Oliver, M. G. D. Nix and M. N. R. Ashfold, *J. Phys. Chem. A*, 2009, **113**, 7984–7993.
19 R. N. Dixon, T. A. A. Oliver and M. N. R. Ashfold, *J. Chem. Phys.*, 2011, **134**, 194303.
20 G. M. Roberts, A. S. Chatterley, J. D. Young and V. G. Stavros, *J. Phys. Chem. Lett.*, 2012, **3**, 348–352.
21 T. A. A. Oliver, G. A. King, D. P. Tew, R. N. Dixon and M. N. R. Ashfold, *in preparation*.
22 T. A. A. Oliver, Y. Zhang, M. N. R. Ashfold and S. E. Bradforth, *Faraday Discuss.*, 2011, **150**, 439–458.
23 J. S. Lim, I. S. Lim, K. S. Lee, D. S. Ahn, Y. S. Lee and S. K. Kim, *Angew. Chem., Int. Ed.*, 2006, **45**, 6290–6293.
24 Y. Zhang, T. A. A. Oliver, M. N. R. Ashfold and S. E. Bradforth, *in preparation*.
25 A. E. Jailaubekov and S. E. Bradforth, *Appl. Phys. Lett.*, 2005, **87**, 021107.
26 C. A. Rivera, S. E. Bradforth and G. Tempea, *Opt. Express*, 2010, **18**, 18615–18624.
27 C. G. Elles, C. A. Rivera, Y. Zhang, P. A. Pieniazek and S. E. Bradforth, *J. Chem. Phys.*, 2009, **130**, 084501.
28 M. J. Tauber, R. A. Mathies, X. Y. Chen and S. E. Bradforth, *Rev. Sci. Instrum.*, 2003, **74**, 4958–4960.
29 For the *p*-MePhSH S_1 ESA spectrum, $S_n \leftarrow S_1$ transitions with A'' symmetry were only calculated for $E < 3.0$ eV (*i.e.*, up to and including the 14 $^1A' \leftarrow 2 \, ^1A'$ transition) due to the large computational cost. The magnitude of A'' TDMs are low due to the orthogonal orbital overlap of the initial and final orbitals involved in the transitions. Therefore we anticipate these transitions will make only a minor contribution to the total calculated S_1 ESA spectrum.
30 MOLPRO is a package of *ab initio* programs written by H.-J. Werner, P. J. Knowles, F. R. Manby, M. Schütz, P. Celani, G. Knizia, T. Korona, R. Lindh, A. Mitrushenkov, G. Rauhut, T. B. Adler, R. D. Amos, A. Bernhardsson, A. Berning, D. L. Cooper, M. J. O. Deegan, A. J. Dobbyn, F. Eckert, E. Goll, C. Hampel, A. Hesselmann, G. Hetzer, T.

Hrenar, G. Jansen, C. Köppl, Y. Liu, A. W. Lloyd, R. A. Mata, A. J. May, S. J. McNicholas, W. Meyer, M. E. Mura, A. Nicklaß, P. Palmieri, K. Pflüger, R. Pitzer, M. Reiher, T. Shiozaki, H. Stoll, A. J. Stone, R. Tarroni, T. Thorsteinsson, M. Wang and A. Wolf, Cardiff, 2010.

31 M. N. R. Ashfold, A. L. Devine, R. N. Dixon, G. A. King, M. G. D. Nix and T. A. A. Oliver, *Proc. Natl. Acad. Sci. U. S. A.*, 2008, **105**, 12701–12706.

32 A. Iqbal, M. S. Y. Cheung, M. G. D. Nix and V. G. Stavros, *J. Phys. Chem. A*, 2009, **113**, 8157–8163.

33 Y. M. Riyad, S. Naumov, R. Hermann and O. Brede, *Phys. Chem. Chem. Phys.*, 2006, **8**, 1697–1706.

34 D. L. Gerrard and W. F. Maddams, *Spectrochim. Acta, Part A*, 1978, **34**, 1205–1211.

35 D. L. Gerrard and W. F. Maddams, *Spectrochim. Acta, Part A*, 1978, **34**, 1219–1223.

36 R. Hermann, G. R. Dey, S. Naumov and O. Brede, *Phys. Chem. Chem. Phys.*, 2000, **2**, 1213–1220.

37 Value taken from phenol cation in water, see T. N. Das, *J. Phys. Chem. A*, 2005, **109**, 3344–3351.

38 T. Ichino, PhD thesis, University of Notre Dame, 2001.

39 *CRC Handbook of Chemistry and Physics*, 91st edn, Taylor and Francis, 2009.

40 T. Shida and Y. Takemura, *Radiat. Phys. Chem.*, 1983, **21**, 157–166.

41 S. Tagawa, N. Hayashi, Y. Yoshida, M. Washio and Y. Tabata, *Radiat. Phys. Chem.*, 1989, **34**, 503–511.

42 L. D. A. Siebbeles, U. Emmerichs, A. Hummel and H. J. Bakker, *J. Chem. Phys.*, 1997, **107**, 9339–9347.

43 D. Pullin and L. Andrews, *J. Mol. Struct.*, 1982, **95**, 181–185.

44 K. Kesper, F. Diehl, J. G. G. Simon, H. Specht and A. Schweig, *Chem. Phys.*, 1991, **153**, 511–517.

45 T. N. Das, *J. Phys. Chem. A*, 2005, **109**, 3344–3351.

46 X. Y. Chen, D. S. Larsen, S. E. Bradforth and I. H. M. van Stokkum, *J. Phys. Chem. A*, 2011, **115**, 3807–3819.

47 R. M. Stratt and M. Maroncelli, *J. Phys. Chem.*, 1996, **100**, 12981–12996.

48 M. R. Ganapathi, R. Hermann, S. Naumov and O. Brede, *Phys. Chem. Chem. Phys.*, 2000, **2**, 4947–4955.

49 R. Hermann, G. R. Mahalaxmi, T. Jochum, S. Naumov and O. Brede, *J. Phys. Chem. A*, 2002, **106**, 2379–2389.

50 I. B. Berlman, *Handbook of Fluorescence Spectra of Aromatic Molecules*, Academic Press, New York, 1965.

51 D. V. Bent and E. Hayon, *J. Am. Chem. Soc.*, 1975, **97**, 2599–2606.

52 M. Ito, *J. Mol. Spectrosc.*, 1960, **4**, 125–143.

53 M. Ito, *J. Mol. Spectrosc.*, 1960, **4**, 106–124.

54 R. Mecke, *Discuss. Faraday Soc.*, 1950, 161–177.

55 A. Sur and P. M. Johnson, *J. Chem. Phys.*, 1986, **84**, 1206–1209.

56 H. Miyasaka, H. Masuhara and N. Mataga, *J. Phys. Chem.*, 1985, **89**, 1631–1636.

57 E. Pines, D. Huppert and N. Agmon, *J. Chem. Phys.*, 1988, **88**, 5620–5630.

58 E. J. Heilweil, M. P. Casassa, R. R. Cavanagh and J. C. Stephenson, *J. Chem. Phys.*, 1986, **85**, 5004–5018.

59 G. A. Pino, A. N. Oldani, E. Marceca, M. Fujii, S. I. Ishiuchi, M. Miyazaki, M. Broquier, C. Dedonder and C. Jouvet, *J. Chem. Phys.*, 2010, **133**, 124313.

60 A. L. Sobolewski, W. Domcke, C. Dedonder-Lardeux and C. Jouvet, *Phys. Chem. Chem. Phys.*, 2002, **4**, 1093–1100.

61 O. P. J. Vieuxmaire, Z. Lan, A. L. Sobolewski and W. Domcke, *J. Chem. Phys.*, 2008, **129**.

62 J. G. Radziszewski, M. Gil, A. Gorski, J. Spanget-Larsen, J. Waluk and B. J. Mroz, *J. Chem. Phys.*, 2001, **115**, 9733–9738.

63 G. Berden, W. L. Meerts, D. F. Plusquellic, I. Fujita and D. W. Pratt, *J. Chem. Phys.*, 1996, **104**, 3935–3946.

64 J. A. Kloepfer, V. H. Vilchiz, V. A. Lenchenkov, A. C. Germaine and S. E. Bradforth, *J. Chem. Phys.*, 2000, **113**, 6288–6307.

65 E. B. Krissinel and N. Agmon, *J. Comput. Chem.*, 1996, **17**, 1085–1098.

66 C. G. Elles, I. A. Shkrob, R. A. Crowell and S. E. Bradforth, *J. Chem. Phys.*, 2007, **126**, 164503.

67 R. Brause, M. Santa, M. Schmitt and K. Kleinermanns, *ChemPhysChem*, 2007, **8**, 1394–1401.

68 A. J. Bell, P. R. Pardon, C. G. Hickman and J. G. Frey, *J. Chem. Soc., Faraday Trans.*, 1990, **86**, 3831–3836.

69 C. G. Hickman, A. Brickell and J. G. Frey, *Chem. Phys. Lett.*, 1991, **185**, 101–104.

70 C. G. Hickman, N. Shaw, M. J. Crawford, A. J. Bell and J. G. Frey, *J. Chem. Soc., Faraday Trans.*, 1993, **89**, 1623–1630.

71 A. Weichert, C. Riehn and B. Brutschy, *J. Phys. Chem. A*, 2001, **105**, 5679–5691.
72 M. Schmitt, M. Boehm, C. Ratzer, D. Kruegler, K. Kleinermanns, I. Kalkman, G. Berden and W. L. Meerts, *ChemPhysChem*, 2006, **7**, 1241–1249.
73 M. Kolar and P. Hobza, *J. Phys. Chem. A*, 2007, **111**, 5851–5854.
74 D. M. Chipman, *J. Chem. Phys.*, 2006, **124**, 044305.
75 R. J. Lipert, G. Bermudez and S. D. Colson, *J. Phys. Chem.*, 1988, **92**, 3801–3805.
76 G. Berden, W. L. Meerts, M. Schmitt and K. Kleinermanns, *J. Chem. Phys.*, 1996, **104**, 972–982.
77 The $2^1\pi\sigma^*$ and $2^1\pi\pi^*$ PECs in PhOH (Fig. 1(a)) are adapted from identical calculations where the ring geometry was fixed at the S_0 minimum energy geometry (ref. 19). Here we have shifted the $2^1\pi\sigma^*$ PEC so that it corresponds to the optimized minimum energy of the H + PhO($^2A_2$) products at large R_{O-H} and to the *ab initio* $2^1\pi\sigma^* \leftarrow S_0$ excitation energy in the vFC region (ref. 19). The $2^1\pi\sigma^*$ PEC for *p*-MePhSH (Fig. 1(b)) is far more empirical and based on shifting the $1^1\pi\sigma^*$ relaxed PEC (as shown in Fig. 1(b)) to match the calculated H + *p*-MePhS($^2A_2$) product limit at large R_{S-H} (ref. 22). We have also added a slight "shelf" to this PEC in the vFC region, as we anticipate there will be low-lying Rydberg states with which it will form avoided crossings with this PES.

Ultrafast dynamics of aniline following 269–238 nm excitation and the role of the $S_2(\pi 3s/\pi\sigma^*)$ state†

Roman Spesyvtsev, Oliver M. Kirkby and Helen H. Fielding*

Received 18th April 2012, Accepted 30th April 2012
DOI: 10.1039/c2fd20076g

Femtosecond time-resolved photoelectron imaging is employed to investigate ultrafast electronic relaxation in aniline, a prototypical aromatic amine. The molecule is excited at wavelengths between 269 and 238 nm. We observe that the $S_2(\pi 3s/\pi\sigma^*)$ state is populated directly during the excitation process at all wavelengths and that the population bifurcates to two decay pathways. One of these involves ultrafast relaxation from the Rydberg component of $S_2(\pi 3s/\pi\sigma^*)$ to the $S_1(\pi\pi)^*$ state, from which it relaxes back to the electronic ground state on a much longer timescale. The other appears to involve motion along the $\pi\sigma^*$ dissociative potential energy surface. At higher excitation energies, the dominant excitation is to the $S_3(\pi\pi^*)$ state, which undergoes extremely efficient electronic relaxation back to the ground state. Our study supports some conclusions reached from H-atom photofragment translational spectroscopy measurements and pump–probe photoionization measurements and contradicts some others.

1 Introduction

Ultraviolet (UV) excitation in biological molecules can lead to excited-state photochemistry which may result in catastrophic damage.[1,2] Fortunately, many important biological chromophores, such as the DNA bases and amino acids, are remarkably photostable.[3–7] Their intrinsic photostability is achieved by efficient electronic relaxation processes that convert dangerous electronic energy into less dangerous vibrational energy, on a timescale that is faster than the time it takes for competing photochemical reactions to occur. Improving our understanding of the mechanisms of electronic relaxation processes following the absorption of UV light is thus a subject of great interest and importance.

When a molecule absorbs UV light, it is promoted to an excited electronic state in which the nuclei are no longer in their equilibrium positions. The resulting excess vibrational energy can be redistributed within the molecule in various ways: it may undergo a photochemical reaction on the excited state before relaxing back to the ground-state, or it may undergo electronic relaxation from the initially populated excited state to another state of the same multiplicity (internal conversion) or to one of different multiplicity (intersystem crossing). It is now accepted that these non-radiative processes occur via molecular configurations where two or more electronic states are degenerate, known as conical intersections.[8] These conical intersections are not isolated points but multidimensional seams and it is the topography around the seam that dictates the fate of the excited state population.

Department of Chemistry, University College London, 20 Gordon Street, London, WC1H 0AJ, UK. E-mail: h.h.fielding@ucl.ac.uk

† Electronic supplementary information (ESI) available: Fitting procedure for determining decay rates. See DOI: 10.1039/c2fd20076g

Many biological chromophores are substituted aromatics whose UV absorption spectra are dominated by strong transitions to $^1\pi\pi^*$ states. These molecules also tend to have low-lying dissociative states with $^1\pi\sigma^*$ or $^1n\sigma^*$ character,[9] which can be populated directly, albeit weakly, or indirectly by internal conversion from optically bright $^1\pi\pi^*$ states. Since Sobolewski et al.[10] suggested that these dissociative states played an important role in the non-radiative decay of various biological chromophores, there has been a surge of interest in studying the photochemistry of substituted aromatics.[11] Aromatic amines are a common structural motif in biological molecules and we are interested in the role of the $^1\pi\sigma^*$ state in aniline, a prototypical aromatic amine.

$^1\pi\sigma^*$ states have mixed Rydberg and valence antibonding character: in aniline, the N-centered $^1\pi3s$ and $^1\pi\sigma^*$ configurations form an avoided crossing at relatively modest N–H internuclear separations and the lower adiabatic potential energy surface is best described as a singlet $\pi3s/\pi\sigma^*$ state (see Fig. 1).[9] The shape of the $\pi3s/\pi\sigma^*$ potential energy surface is highly dependent on the relative stability of the Rydberg and valence configurations. If the Rydberg state is lower in energy than the dissociative valence state, as it is in aniline,[12–14] the $\pi3s/\pi\sigma^*$ potential energy surface will have a pronounced potential well in the vertical Franck–Condon region and only becomes dissociative at longer N–H bond distances. If, however, the $\pi3s$ and $\pi\sigma^*$ components are nearly degenerate in the vertical Franck–Condon region, as they are in phenol,[15] the entire potential energy surface is almost totally dissociative. Along the N–H stretch coordinate in aniline, the $\pi3s/\pi\sigma^*$ state intersects the lower-lying $^1\pi\pi^*$ state and the electronic ground state and these crossings develop into conical intersections when out of plane vibrations are introduced.[11,16–18]

Honda et al. have reported the most detailed theoretical study of the electronic structure of aniline using symmetry adapted cluster/configuration interaction.[14] They determined that the lowest singlet excited state was a $^1\pi\pi^*$ state with mainly charge-resonance character and some charge-transfer character (hereafter referred to as $S_1(\pi\pi^*)$). The vertical excitation energy of this state was calculated to be 4.20 eV. The second singlet excited state was determined as having mixed Rydberg-valence character and a calculated vertical excitation energy of around 4.53 eV (hereafter referred to as $S_2(\pi3s/\pi\sigma^*)$). The third singlet excited state was identified as another $^1\pi\pi^*$ state, resulting from local excitation on the benzene ring (hereafter referred to as $S_3(\pi\pi^*)$). The vertical excitation energy of this state

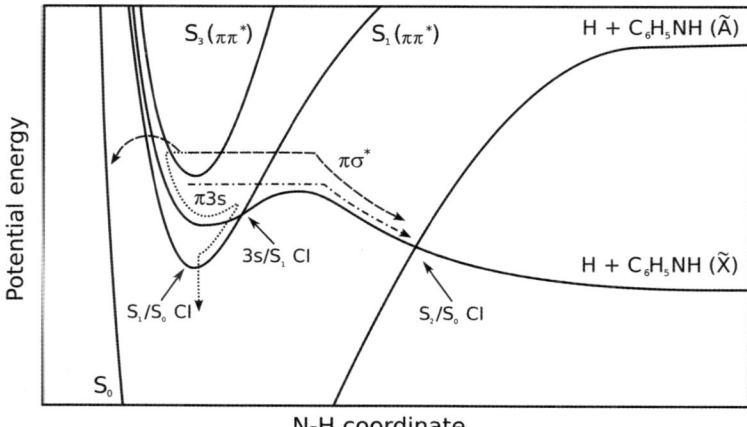

Fig. 1 Schematic potential energy curves for the four lowest singlet states of aniline along the N–H stretch coordinate. The S_0, $S_1(\pi\pi^*)$ and $S_3(\pi\pi^*)$ states are bound whereas the $S_2(\pi3s/\pi\sigma^*)$ state has mixed Rydberg and valence character and is dissociative along the N–H stretch coordinate. Various non-radiative decay mechanisms, deduced from our analysis of the time-resolved photoelectron spectra, are also illustrated.

was determined to be 5.34 eV. These calculations are consistent with the gas-phase absorption spectrum of aniline (Fig. 2 (a)) which displays absorption maxima around 4.4 eV and 5.4 eV and a fairly significant, but structureless, background between them.

There have been numerous spectroscopic investigations of the $S_1(\pi\pi^*)$ state, measuring the origin transition,[19,20] and intersystem crossing lifetimes at low excess vibrational energies.[21,22] Rice et al.[21] determined the intersystem crossing lifetime of the ground vibrational state of $S_1(\pi\pi^*)$ to be 8.5 ns and Weber et al.[22] confirmed formation of the triplet states using picosecond time-resolved photoelectron spectroscopy. Reilly et al.[19] have carried out a detailed ZEKE spectroscopy study of ground state of the cation via the $S_1(\pi\pi^*)$ state of aniline.

The $S_2(\pi 3s/\pi\sigma^*)$ state was first observed experimentally by Ebata et al.[20] They employed UV-IR double-resonance spectroscopy to study the vibrational levels of $S_1(\pi\pi^*)$; however, they observed a number of broad transitions lying at wavenumbers greater than 3500 cm^{-1} above the $S_1(\pi\pi^*)$ origin that could not be assigned to fundamental vibrations of the $S_1(\pi\pi^*)$ state. Subsequent 2 + 2 resonance-enhanced multiphoton ionisation spectroscopy found a sharp peak at 37104 cm^{-1} above the ground state, which was assigned as the $S_2(\pi 3s/\pi\sigma^*)$ origin. The observation of a short vibronic progression then led to the conclusion that the potential energy surface was not purely repulsive but had a potential minimum close to the Franck–Condon region.

Ashfold et al. have undertaken a comprehensive study of the photodissociation of aniline using H (Rydberg) atom photofragment translational spectroscopy.[12] Below the $S_2(\pi 3s/\pi\sigma^*)$ origin, they found that H atom loss occurred mainly by multiphoton processes, resonantly enhanced at the one-photon level. Decreasing the excitation wavelength to excite the first few vibrational levels of $S_2(\pi 3s/\pi\sigma^*)$ (269–260 nm)

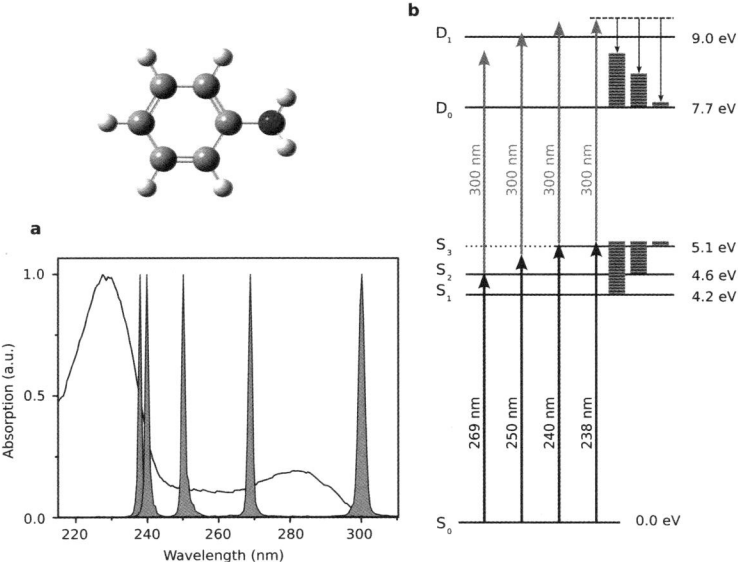

Fig. 2 (a) Gas-phase UV absorption spectrum of aniline, with pump and probe pulse profiles superimposed. (b) Excitation scheme: aniline is excited using 269 nm (4.61 eV), 250 nm (4.96 eV), 240 nm (5.17 eV) and 238 nm (5.21 eV) pump pulses. The excited state population is projected onto the photoionisation continuum using a delayed 300 nm (4.13 eV) probe pulse. Assuming a $\Delta v = 0$ propensity rule for photoionisation, the electron kinetic energy following ionisation out of the Franck–Condon region is expected to increase in the order $S_1 < S_2 < S_3$. The vibrational energies (blocks of horizontal lines) and corresponding eKEs (downward vertical arrows) are illustrated for 238 nm excitation.

was found to induce N–H bond fission, creating H atoms with anisotropic velocity distributions and anilino radicals in a few low vibrational states of the ground electronic state. These observations were explained in terms of H atom tunnelling through the barrier along the N–H bond stretch, on a timescale faster than molecular rotation, and adiabatic evolution of the vibrational mode of aniline into the corresponding vibrational mode in the anilino fragment. It should be noted that the rotation constants for aniline are 0.06–0.2 cm^{-1} in the ground and first excited singlet states,[23,24] which, assuming rotational temperatures of approximately 20 K, correspond to rotational periods of tens of picoseconds. Decreasing the excitation wavelength further led to the observation of anilino radicals with greater vibrational excitation, explained as direct excitation to $S_1(\pi\pi^*)$ followed by internal conversion through an S_1/S_2 conical intersection to the dissociative $\pi\sigma^*$ potential energy surface. At around 230 nm, changes in the profile of the kinetic energy distribution were interpreted in terms of direct excitation to $S_3(\pi\pi^*)$ followed by internal conversion through an S_3/S_2 conical intersection or through successive S_3/S_1 and S_1/S_2 conical intersections to the dissociative $\pi\sigma^*$ potential energy surface.

Castaño et al. carried out the first femtosecond time-resolved study of electronic relaxation in aniline using pump–probe photoionisation spectroscopy.[13] They observed a 165 fs timescale for all excitation wavelengths in the range 269–240 nm, which they attributed to dynamics on the $S_2(\pi3s/\pi\sigma^*)$ surface through the S_2/S_1 conical intersection. They argued that population was transferred both to the dissociative $\pi\sigma^*$ potential energy surface and to the $S_1(\pi\pi^*)$ state. They observed a long timescale (tens of picoseconds to nanoseconds) which they attributed to subsequent electronic relaxation of $S_1(\pi\pi^*)$ back to the electronic ground state. Due to the relatively minor contribution of the 165 fs component, they concluded that the dissociative $\pi\sigma^*$ potential energy surface plays a minor role in the photochemistry of aniline. Decreasing the excitation wavelength to 234 nm resulted in a very fast 21 fs timescale, which was attributed to population on $S_3(\pi\pi^*)$ undergoing extremely rapid internal conversion to $S_1(\pi\pi^*)$ which relaxes back to the ground state on a timescale of tens of picoseconds.

Stavros et al. have employed femtosecond pump–probe velocity map imaging to monitor the formation of H atoms following excitation of aniline around 240 nm and 200 nm and observe a timescale of around 155 fs for the formation of low and high kinetic energy H atoms.[25]

In this Discussion paper, we extend our recent femtosecond time-resolved photoelectron imaging (TRPEI) investigation of electronic relaxation in aniline following excitation at 236 nm,[26] to wavelengths across the range 269–238 nm. TRPEI allows the intermediate electronic states populated during electronic relaxation to be observed directly.[27–29] It involves promoting an isolated molecule to an electronically excited state with an ultrashort pump pulse. A probe pulse then ionises the excited state and the photoelectron spectrum and angular distribution are recorded as a function of the pump–probe delay. Our results confirm some of the suggestions and conclusions made from H (Rydberg) atom photofragment translational spectroscopy and pump–probe photoionisation spectroscopy, but contradict others. We observe that the $S_2(\pi3s/\pi\sigma^*)$ state is populated directly at all wavelengths, but we detect a bifurcation of the population to two decay paths. One of these is to $S_1(\pi\pi^*)$, which is subsequently observed to undergo electronic relaxation back to the electronic ground state, and the other is along the $S_2(\pi3s/\pi\sigma^*)$ potential surface to the S_2/S_0 conical intersection. At higher excitation energies, we observe direct population of $S_3(\pi\pi^*)$ but we see ultrafast internal conversion back to the electronic ground state, through what we believe is an S_3/S_0 conical intersection.

2 Methods

The excitation scheme is presented in Fig. 2. Aniline is excited above the origin of S_2 using 269–238 nm (4.61–5.21 eV) femtosecond laser pulses. The excited state population is projected onto the photoionisation continuum using a delayed 300 nm (4.13 eV) femtosecond laser pulse, selected to access as much of the ionisation continuum as possible whilst being below the onset of significant S_1–S_0 absorption. Photoelectron images are then recorded for a series of pump–probe delays.

The experimental apparatus has been described elsewhere.[30–33] Briefly, it consists of a continuous molecular beam, a 1 kHz femtosecond laser system and a velocity map imaging apparatus based on the Eppink and Parker design.[34] The molecular beam of aniline is created by passing He carrier gas through liquid aniline (Sigma-Aldrich, >99%) at 750 mbar and expanding it through a 50 μm nozzle. The beam is then collimated by a 1 mm skimmer before passing into the interaction region of the velocity map imaging spectrometer.

The cross-correlation of the pump and the probe laser pulses ranges from 150 fs to 210 fs. The pulse energies are attenuated to less than 1 μJ to minimise multiphoton processes and to keep photoelectron count-rates below 20 photoelectrons per pulse. For each pump–probe delay, we collect pump-only, probe-only and pump + probe images, in sequence, for 20 s each. This is repeated around 30 times for approximately 30 min for each pump–probe delay. The pump-only and probe-only signals are subtracted from the pump + probe signal to eliminate one-colour contributions from the images. Photoelectron spectra are recovered from the raw images using the pBasex image inversion algorithm.[35]

The energy scale of the photoelectron imaging spectrometer is calibrated by monitoring photoelectrons associated with the $^2P_{3/2}$ (12.13 eV) and $^2P_{1/2}$ (13.44 eV) spin-orbit states of Xe$^+$ following 2 + 1 resonant multiphoton ionisation (REMPI) of Xe at 249.63 nm.[36] The FWHM energy resolution of the photoelectron imaging detector is dependant on the measured energy range and for the current experiment it is around 0.06 eV at 1 eV electron kinetic energy (eKE). Photoelectron angular distributions are obtained as a result of the pBasex inversion, which has the following expression for the angular basis set:

$$I(\theta) = a[1 + \beta_2 P_2(\cos\theta)], \qquad (1)$$

where $I(\theta)$ is the probability of photoelectron emission at a particular angle θ, defined as the angle between the laser polarization and the velocity vector of the photoelectron, $P_2(\cos\theta)$ is the second order Legendre Polynomial, β_2 is the asymmetry parameter and a is a normalization constant. A raw photoelectron image recorded with a 240 nm pump pulse and a 5 fs pump–probe delay is presented in Fig. 3(a) together with the corresponding photoelectron spectrum and photoelectron angular distribution in Fig. 3(b).

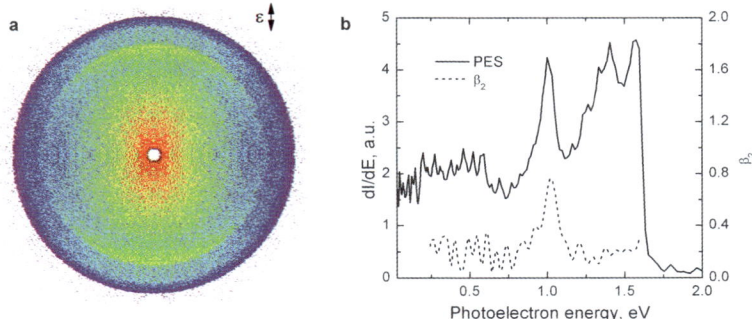

Fig. 3 (a) Velocity map image recorded with a 240 nm pump pulse, a 300 nm probe pulse and a pump–probe delay of 5 fs. (b) Corresponding photoelectron spectrum (solid line) and eKE dependence of the anisotropy parameter, β_2 (dashed line).

3 Results

3.1 Integrated photoelectron intensity

The evolution of the integrated photoelectron signal, following excitation of aniline at 269, 250, 240 and 238 nm, is plotted as a function of time in Fig. 4(a)–(d). All the measured time-profiles exhibit dynamics on more than one timescale. We fit the experimental data to a sum of exponentially decaying profiles convoluted with

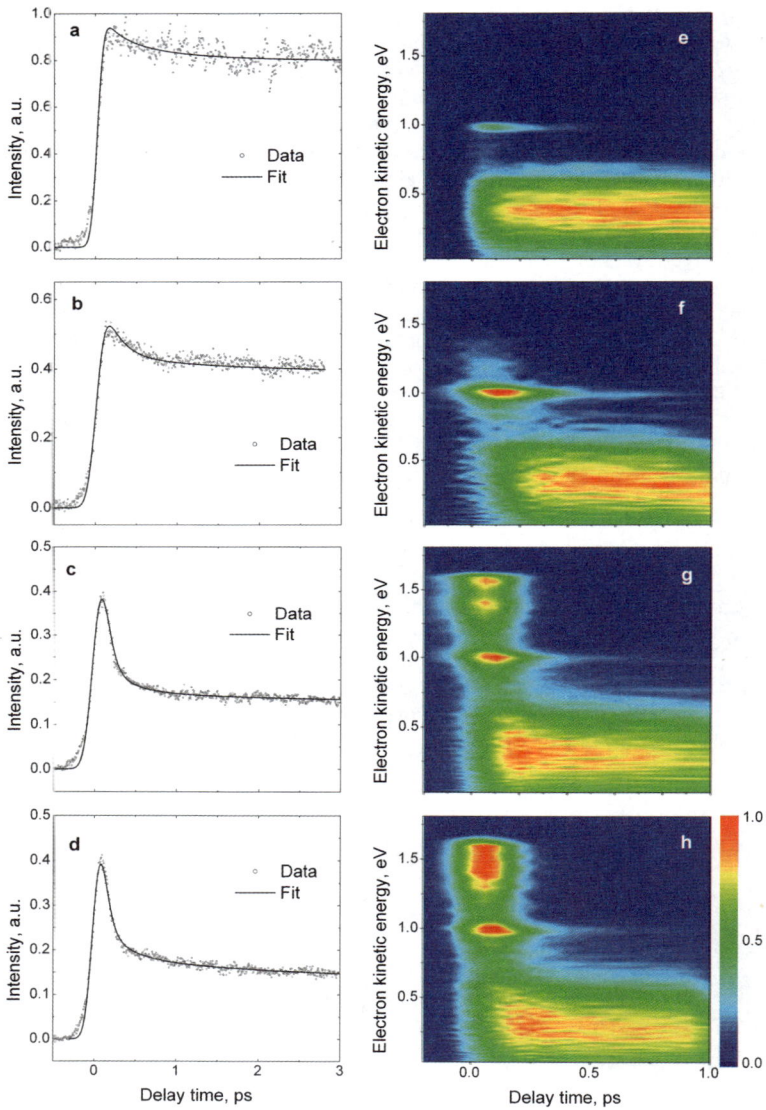

Fig. 4 Time evolution of the integrated photoelectron signal following excitation of aniline at (a) 269 nm, (b) 250 nm, (c) 240 nm and (d) 238 nm: points represent experimental data, solid lines are fits to eqn (2), vertical scales are normalised so that $\sum_i c_i = 1$ (eqn (2)). Photoelectron kinetic energy spectra as a function of pump–probe delay following excitation of aniline at (e) 269 nm, (f) 250 nm, (g) 240 nm and (h) 238 nm.

a Gaussian function representing the cross-correlation of the pump and probe laser pulses $g(t)$,

$$S(t) = \sum_i c_i e^{-t/\tau_i} \otimes g(t), \tag{2}$$

where τ_i is the ith $1/e$ decay time and c_i is its corresponding intensity, which is proportional to the product of the excited state population and the photoionisation cross-section. The decay times are those determined from the fitting procedure described in Section 3.3 below and listed in Table 1.

Following 269 nm excitation (Fig. 4(a)), approximately 20% of the integrated photoelectron signal disappears during the first picosecond and the remaining 80% has a lifetime that is longer than one nanosecond. Following 250 nm excitation (Fig. 4(b)), approximately 60% of the integrated photoelectron signal decays during the first picosecond and the remaining 40% decays on a timescale of around 600 ps. For 240 and 238 nm excitation (Fig. 4(c) and (d)), approximately 75% of the photo-electron signal decays during the few hundred femtoseconds and this increases to around 85% after one picosecond. The remaining 15% of the signal then decays on a timescale of around 200 ps (240 nm) or 100 ps (238 nm).

This energy-integrated analysis generates data similar to that reported by Castaño et al.;[13] it provides some information about timescales, but the intermediate electronic states involved in the electronic relaxation are not observed directly.

3.2 Time-resolved photoelectron spectra

The photoelectron kinetic energy spectra are plotted as a function of pump–probe delay, following excitation of aniline at 269, 250, 240 and 238 nm, in Fig. 4(e)–(h). These figures show that the fast decaying and slow decaying components of the total integrated photoelectron spectra presented in Fig. 4(a)–(d) have quite different eKE distributions.

To a first approximation, we can use the conservation of energy to assign the components of the photoelectron spectra. In rigid molecules that do not undergo large amplitude motion, there is a tendency for the excess vibrational energy in a given electronic state, $E_{vib} = h\nu_{pump} - E(S_n)$, to be approximately conserved during photoionisation:

$$eKE = h\nu_{probe} - [E(D_n) - E(S_n)] \tag{3}$$

where D_n are the electronic states of the cation and have been determined to be 7.72 eV[19] and 9.03 eV[22] for the ground and first excited electronic states, respectively. The 0–0 excitation energies of $S_1(\pi\pi^*)$ and $S_2(\pi3s/\pi\sigma^*)$ have been determined to be 4.22 eV[19] and 4.60 eV,[20] respectively. The theoretical value for the vertical excitation of the $S_3(\pi\pi^*)$ state is 5.34 eV.[14] To the best of our knowledge an accurate value for the 0–0 excitation energy of $S_3(\pi\pi^*)$ has not been reported, but we estimate it to be around 5.06 eV from the gas-phase absorption spectrum (Fig. 2). Using this information, we assign peaks in the photoelectron spectrum around 1.5 eV and 1 eV to

Table 1 Summary of $1/e$ decay lifetimes extracted from the fitting procedures described in the text

Pump wavelength (nm)	$S_3(\pi\pi^*)$ lifetime (fs)	$S_2(\pi3s)$ lifetime (fs)	$S_2(\pi3s/\pi\sigma^*)$ lifetime (ps)	$S_1(\pi\pi^*)$ lifetime (ps)	Cross correlation FWHM (fs)
269	—	230 ± 70	$0.9^{+0.9}_{-0.3}$	>1 ns	150 ± 45
250	—	200 ± 45	$1.6^{+2.0}_{-0.8}$	600 ± 35	200 ± 40
240	65 ± 20	280 ± 170	$1.8^{+4.7}_{-1.5}$	185 ± 15	210 ± 20
238	70 ± 15	330 ± 150	$2.4^{+4.8}_{-1.1}$	100 ± 40	185 ± 15

photoionisation from $S_3(\pi\pi^*)$ and the $\pi3s$ component of $S_2(\pi3s/\pi\sigma^*)$, respectively. The photoelectron spectrum at lower eKE (<0.7 eV) is broad and vibrationally unresolved. Energetically, this region is most likely to correspond to photoionisation from $S_1(\pi\pi^*)$ or $S_2(\pi3s/\pi\sigma^*)$. Ionisation from $S_3(\pi\pi^*)$ to D_1 is a possibility (eKE < 0.2 eV) but is expected to be considerably weaker than ionisation to the ground state of the cation.[22]

The most intense features in the 240 and 238 nm photoelectron spectra are the asymmetric bands visible at high eKE, which correspond to ionisation from $S_3(\pi\pi^*)$. As the pump–probe delay increases, the intensity of this band decreases very rapidly on a timescale of around one hundred femtoseconds. A short vibrational progression is visible in the 240 nm spectrum with a vibrational wavenumber of approximately 1200 cm^{-1} and remains visible throughout the decay.

The peak at 1 eV, corresponding to ionisation from the $\pi3s$ component of the $S_2(\pi3s/\pi\sigma^*)$ state, is visible in all four spectra; it is narrow and symmetric, as expected for Rydberg states that obey the $\Delta\upsilon = 0$ propensity rule. From the eKE, we determine the origin of the $S_2(\pi3s/\pi\sigma^*)$ state to be 4.6 eV, which is in excellent agreement with the more precise value of 37 104 cm^{-1}(4.6003 eV) obtained using 2 + 2 resonance enhanced multiphoton ionisation.[20] The asymmetry parameter rises to a maximum value of $\beta_2 = 1.1$ across this narrow peak, which is consistent with photoelectrons being ionised from a 3s Rydberg state. As the pump–probe delay increases, this feature decays less rapidly than $S_3(\pi\pi^*)$, but still on the timescale of hundreds of femtoseconds. The anisotropy of this band is preserved during its decay. Interestingly, the decay timescale does not appear to vary significantly between 269 nm excitation (resonant with the S_2–S_0 origin) and 238 nm (0.6 eV above the S_2 origin).

The broad band observed at low eKE has two components, with both visible in all four spectra. Following excitation at 250–238 nm, the lower eKE component has a Gaussian profile centered around 0.2–0.26 eV; it grows with a timescale of a few hundred femtoseconds and decays on a timescale that decreases with decreasing excitation wavelength (from 600 ps at 250 nm to around 100 ps at 238 nm). This band seems most likely to correspond to photoionisation from $S_1(\pi\pi^*)$. Following excitation at 269 nm, the low eKE component is asymmetric and has a maximum photoelectron intensity at 0.3 eV. The asymmetric shape arises because $S_1(\pi\pi^*)$ is also populated directly during the excitation process at this wavelength, as expected from the absorption spectrum. This band decays on a time scale >1 ns that we cannot measure in our experiment. In all four spectra, the higher eKE component of the broad photoelectron band, around 0.6 eV, decays on a picosecond timescale; we believe this component corresponds to photoionisation from $S_2(\pi3s/\pi\sigma^*)$ (see Discussion).

3.3 Timescales

To extract timescales, we fit our experimental photoelectron spectra to a sum of exponentially decaying spectral profiles convoluted with a Gaussian cross-correlation of the pump and probe laser pulses $g(t)$,

$$S(\text{eKE}, t) = \sum_i c_i(\text{eKE})e^{-t/\tau_i} \otimes g(t), \tag{4}$$

where $c_i(\text{eKE})$ is the intensity of the ith decay at a given eKE and τ_i is its corresponding $1/e$ decay time. Multidimensional fitting procedures are prone to finding local minima, especially if there are a large number of parameters to fit. We avoid this by fitting discernible portions of the photoelectron spectra in turn, rather than fitting all the parameters simultaneously;† the $1/e$ decay times obtained in this way are summarised in Table 1.

First, the lifetime of the long-lived component of the broad, Gaussian shaped band at low eKE, τ_0, is determined by fitting the integrated photoelectron counts from 0 to 400 ps to a single exponential decay.

For the 269 nm and 250 nm data, the decay time of the 3s component of $S_2(\pi 3s/\pi \sigma^*)$, τ_2, the cross-correlation, $g(t)$, and zero-time were determined by fitting the photoelectron signal integrated over the range 0.9–1.1 eV with two decays, one of which was fixed to τ_0. The decay time associated with the shorter lived 0.6 eV shoulder of the broad band at low eKE, τ_1, was then determined from the fit of the photoelectron signal integrated over the range 0.48–0.72 eV to three decays, where two of the time constants were fixed to τ_0 and τ_2.

For the 240 nm and 238 nm data, the decay time of the $S_3(\pi\pi^*)$ component, τ_3, the cross-correlation, $g(t)$, and zero-time were determined by fitting the photoelectron signal integrated over the range 1.3–1.7 eV to eqn (4) with two decays, one of which is τ_0. The decay time of the 3s component of $S_2(\pi 3s/\pi \sigma^*)$, τ_2, was then obtained by fitting the integrated photoelectron counts over the range 0.9–1.1 eV with three decays, two of which were fixed to τ_0 and τ_3. Finally, the decay time associated with the shorter lived 0.6 eV shoulder of the broad band at low eKE, τ_1, was determined from the fit of the photoelectron signal integrated over the range 0.48–0.72 eV to four decays, where three of the time constants were fixed to τ_0, τ_2 and τ_3.

3.4 Relaxation pathway

Using the $1/e$ lifetimes summarised in Table 1, the time-resolved photoelectron spectra for 269 and 250 nm excitation were fitted to eqn (4) over the entire energy range to obtain the spectra associated with each of the decay times, $c_i(eKE)$. As eqn (4) fits a series of exponentially decaying functions that have zero pump–probe as their origin, positive values of $c_i(eKE)$ correspond to exponential decay and negative values of $c_i(eKE)$ correspond to exponential growth. Such a representation has been shown to be a valuable method for unravelling relaxation mechanisms from time-resolved photoelectron spectra as long as exponential kinetics are valid.[37,38]

The results of the fits to the data are shown in Fig. 5(a) and (b). The spectra associated with the 230 fs (269 nm) and 200 fs (250 nm) timescales have positive amplitude components in the region of the photoelectron spectrum corresponding to the sharp $\pi 3s$ Rydberg component of $S_2(\pi 3s/\pi \sigma^*)$, at 1 eV, and negative amplitude components in the region of the photoelectron spectrum corresponding to photoionisation from $S_1(\pi\pi^*)$, at low eKE. This suggests that population flows from the $\pi 3s$ component of $S_2(\pi 3s/\pi \sigma^*)$ to $S_1(\pi\pi^*)$ on the timescale of a couple of hundred femtoseconds. The spectra associated with the longest lifetimes (>1 ns for 269 nm and 600 ps for 250 nm) approximately mirror the negative amplitude components of the spectra associated with the 230 fs and 200 fs lifetimes. The spectra associated with the 0.9 ps (269 nm) and 1.6 ps (250 nm) timescales have large errors associated with them, suggesting that fitting exponential kinetics is not as appropriate for this component as it is for the others. This is discussed in more detail below. The sums of the coefficients $c_i(eKE)$ represent the photoelectron spectra of the initially populated states. They suggest that for 269 nm, both $S_1(\pi\pi^*)$ and $S_2(\pi 3s/\pi \sigma^*)$ are populated during the initial excitation process but that for 250 nm, only $S_2(\pi 3s/\pi \sigma^*)$ is populated significantly, as we might expect from the absorption spectrum (Fig. 2(a)). The time-resolved photoelectron spectra can be reconstructed using the timescales τ_i and coefficients $c_i(eKE)$ obtained from the fitting procedure and are seen to reproduce all the dynamical features of the experimental spectra (compare Fig. 5(c) and (d) with Fig. 4(e) and (f), respectively).

It was not possible to determine the spectra associated with the various decay times for 240 and 238 nm excitation using the same procedure that was employed for 269 and 250 nm. The main reason for this is that there are too many exponential contributions to the low eKE part of the photoelectron spectrum (0–0.7 eV) at short delays when the photoelectron signal is low. Instead, the dynamics were modelled using the $1/e$ lifetimes summarised in Table 1, together with the fitted cross-correlations and smoothed photoelectron spectral profiles deduced from the time-resolved photoelectron spectra. The profiles for $S_3(\pi\pi^*)$ were modelled using the

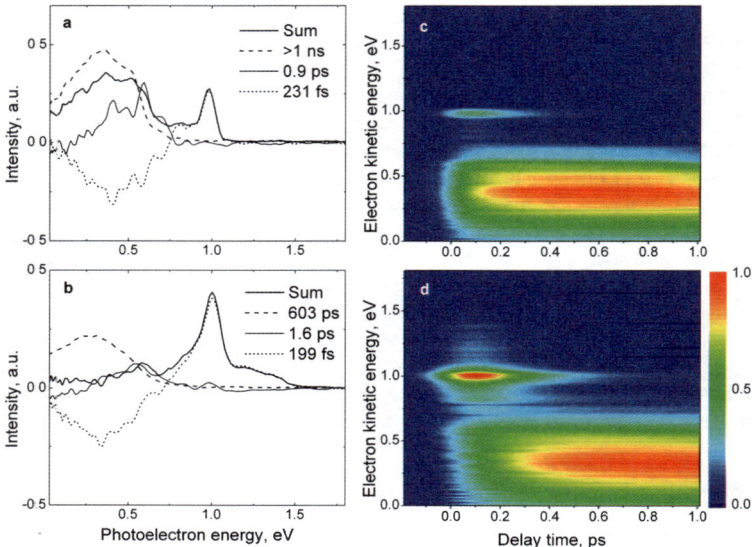

Fig. 5 Spectral components of the three decays derived using the fitting procedure described in the text for (a) 269 nm and (b) 250 nm excitation. Modelled photoelectron spectra for (c) 269 nm and (d) 250 nm excitation, using the timescales listed in Table 1 and the coefficients in (a) and (b) of this Figure.

experimental spectral profiles for the 1.2–1.7 eV range and adding a smooth monotonic tail. The $\pi 3s$ component of $S_2(\pi 3s/\pi\sigma^*)$ was modelled using a Gaussian profile (FWHM = 0.1 eV) centered at 1 eV. The $S_1(\pi\pi^*)$ shape was fixed to the long time (>7 ps) profile. The spectral shapes are presented in Fig. 6(a) and (b).

It is assumed that only $S_3(\pi\pi^*)$ and $S_2(\pi 3s/\pi\sigma^*)$ (including the $\pi 3s$ component) are populated during the initial excitation and that the rise of population in $S_1(\pi\pi^*)$ equals the decay of the population from the $\pi 3s$ component of $S_2(\pi 3s/\pi\sigma^*)$. This has been shown to be the case for 269 and 250 nm (above) and also for 236 nm;[26] although, at 269 nm a large fraction of $S_1(\pi\pi^*)$ is also populated directly. Using these assumptions, the relative amplitudes of the three spectral components were fitted to the experimental data and the resulting models are presented in Fig. 6(c) and (d). These modelled spectra were then subtracted from the experimental photoelectron spectra (Fig. 4(g) and (h)) to obtain difference spectra (Fig. 6(e) and (f)).

The difference spectra represent the photoelectron spectra of the fast component in the low eKE range (0–0.7 eV) and show a variation in lifetime across the photoelectron spectrum. The average decay times of the difference spectra are faster than the decays listed in Table 1 due to the restriction on the rise of the $S_1(\pi\pi^*)$ population. The difference spectra appear to be shifting to lower eKEs on a sub-hundred femtosecond timescale.

4 Discussion

4.1 $S_2(\pi 3s/\pi\sigma^*)$

Our experiments show that the $S_2(\pi 3s/\pi\sigma^*)$ state is populated directly at all wavelengths. The $\pi 3s$ Rydberg component of $S_2(\pi 3s/\pi\sigma^*)$ decays on the timescale of a couple of hundred femtoseconds. For 269 and 250 nm, the spectra associated with the 1/e decay of the $\pi 3s$ component of $S_2(\pi 3s/\pi\sigma^*)$ (Fig. 5(a) and (b)) have negative amplitudes in the region of the photoelectron spectrum corresponding to

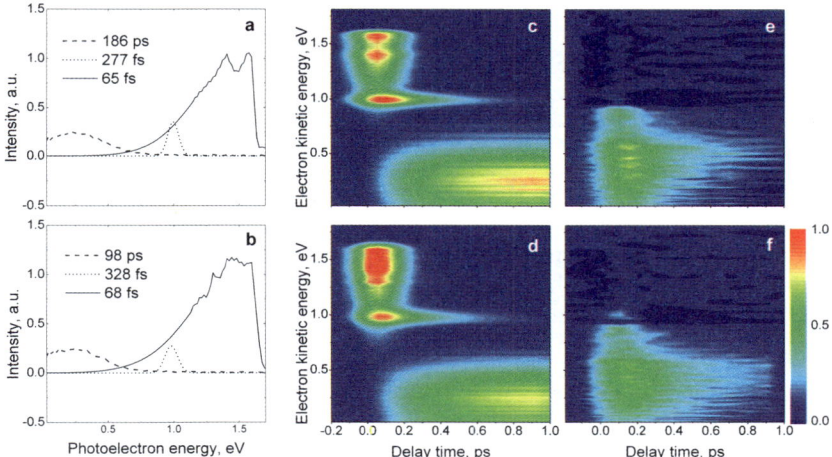

Fig. 6 Smoothed photoelectron spectral profiles for (a) 240 nm and (b) 238 nm excitation. Reconstructed photoelectron spectra for (c) 240 nm and (d) 238 nm excitation, using timescales listed in Table 1 and the smoothed profiles presented in (a) and (b) of this Figure. Difference spectra for (e) 240 nm and (f) 238 nm excitation, as described in the text.

photoionisation from $S_1(\pi\pi^*)$. This suggests that there is an efficient mechanism for transferring population from the Rydberg component in the Franck–Condon region of $S_2(\pi3s/\pi\sigma^*)$ to the $S_1(\pi\pi^*)$ state, through an S_2/S_1 conical intersection (Fig. 2). Similar behaviour was observed following excitation at 236 nm.[26] This model also fits the data well for 240 and 238 nm excitation (Fig. 6(c) and (d)). The anisotropy of the photoelectron distribution is preserved during the decay, suggesting that the S_2/S_1 conical intersection is located before the barrier in the vertical Franck–Condon region (Fig. 1). Within experimental error, the timescale of S_2/S_1 internal conversion is independent of excitation energy above the S_2 origin (in the range 0–0.6 eV), suggesting that the vibrational mode connecting the vertical Franck–Condon region with the S_2/S_1 conical intersection seam is independent of excitation energy.

Population that is transferred from the $\pi3s$ component of $S_2(\pi3s/\pi\sigma^*)$ to $S_1(\pi\pi^*)$ subsequently decays relatively slowly, on a timescale that decreases from >1 ns at the S_2 origin (269 nm), to around 100 ps at 0.6 eV above the S_2 origin (238 nm). The absence of any negative amplitude features in the spectra associated with this decay for 269 and 250 nm excitation (Fig. 5(a) and (b)), indicates that the population moves directly out of the observation window. We could not find any evidence of the lowest energy triplet, which is known to be formed by intersystem crossing from lower vibrational states of $S_1(\pi\pi^*)$.[22] Thus, it is most likely that the $S_1(\pi\pi^*)$ population decays back to the electronic ground state S_0. There is also a possibility of population passing diabatically through the S_2/S_1 conical intersection, at modest N–H bond lengths, to the $\pi\sigma^*$ component of the $S_2(3s/\pi\sigma^*)$ state to form the anilino radical and H atom; this could account for the fast H atoms observed in H-atom photofragmentation translational spectroscopy[12] and femtosecond pump–probe velocity map imaging mearsurements.[25]

The femtosecond pump–probe photoionisation experiments reported by Castaño and coworkers[13] identified a decay with a lifetime of 165 fs which, within experimental error, matches the lifetime we assign to the $\pi3s$ Rydberg component of $S_2(\pi3s/\pi\sigma^*)$. They attributed the 165 fs lifetime to dissociation along the $\pi\sigma^*$ component of $S_2(\pi3s/\pi\sigma^*)$, by reference to H atom photofragmentation studies,[12] and to internal conversion to $S_1(\pi\pi^*)$. They determined that dissociation along the $\pi\sigma^*$ potential energy surface was not a significant relaxation pathway, although it is not surprising that the 165 fs decay was a minor component of the total integrated decay profile, since both the $\pi 3s$ Rydberg component of $S_2(\pi 3s/\pi\sigma^*)$ and $S_1(\pi\pi^*)$ lie within the observation window of the photoionisation experiments.

Our experiments have also revealed another non-radiative decay channel that has a 1–2 ps lifetime and is characterised by a broad band in the photoelectron spectrum from 0–0.6 eV. In principle, this component of the photoelectron spectrum could correspond to photoionisation from $S_1(\pi\pi^*)$, but we believe that it is more likely that it corresponds to photoionisation from $S_2(\pi 3s/\pi\sigma^*)$ for the following reasons. The decay associated spectra for 269 and 250 nm excitation (Fig. 5(a) and (b)) and fits to the 240 and 238 nm spectra (Fig. 6(c) and (d)), suggest that the state with the 1–2 ps lifetime is populated directly during the excitation process. From the absorption spectrum (Fig. 2(a)), it seems unlikely that direct population of $S_1(\pi\pi^*)$ at 238 and 240 nm would be measurable in our experiments. In addition, the 269 nm decay associated spectra (Fig. 5(a)) suggest that the spectrum of the directly excited component with a 0.9 ps decay has considerable overlap with the directly excited $S_1(\pi\pi^*)$ spectrum that decays on a timescale greater than 1 ns. It would be difficult to rationalise two decay timescales for $S_1(\pi\pi^*)$ following 269 nm excitation below the barrier along the N–H bond stretch (Fig. 1) and one 1–2 ps decay timescale for $S_1(\pi\pi^*)$ following excitation above the barrier. Moreover, like the $\pi 3s$ component of $S_2(\pi 3s/\pi\sigma^*)$, this feature has a decay time that does not vary significantly with excitation energy. Thus, we believe that the 1–2 ps non-radiative decay channel is most likely to correspond to population across the $\pi 3s$ and $\pi\sigma^*$ components of $S_2(\pi 3s/\pi\sigma^*)$.

The lifetime of the $S_2(\pi 3s/\pi\sigma^*)$ population has a large uncertainty associated with it, which we attribute at least in part to a variation in the time constant across the spectral profile (Fig. 6(e) and (f)). This would be consistent with a change in the vertical detachment energy along the NH stretch coordinate,[37] or possibly with population being trapped in the upper section of a conical intersection, as has been observed in thioanisol[39] and methylamine.[40] It is worth noting that the decay time derived from our fitting procedure is similar to the timescales attributed to motion on $\pi\sigma^*$ surfaces in indole and 5-hydroxyindole.[41] The absence of any negative amplitude features in the spectra associated with this decay following 269 and 250 nm excitation (Fig. 5(a) and (b)) suggests that the population moves out of the observation window of our experiment, either passing diabatically through the S_2/S_0 conical intersection at extended N–H bond lengths to form the anilino radical and the H atom (which could explain the observation of anisotropic H atom distributions following excitation at wavelengths >260 nm in H atom photofragment translational spectroscopy experiments)[12] or nonadiabatically to form vibrationally hot S_0.

4.2 $S_3(\pi\pi^*)$

Following excitation above the origin of the $S_3 \leftarrow S_0$ transition (240 and 238 nm), our experiments show that direct excitation to the $S_3(\pi\pi^*)$ state is the dominant process, as expected from the absorption spectrum. The $S_3(\pi\pi^*)$ state decays extremely rapidly, on a timescale of <100 fs. The vibrational structure observed in the $S_3(\pi\pi^*)$ photoelectron spectrum is maintained during the decay; it is likely to correspond to a progression in a totally symmetric mode but does not provide any dynamical information. The significant decrease in the total photoionisation signal within the lifetime of the $S_3(\pi\pi^*)$ state (Fig. 4(c) and (d)) suggests that the majority of the $S_3(\pi\pi^*)$ population moves out of the observation window of our experiment, most probably to the ground state through an S_3/S_0 conical intersection.

A rapid (<100 fs) decay of total photoionisation signal was also observed in femtosecond pump–probe studies following excitation at wavelengths <240 nm,[13] but was attributed to internal conversion through an S_3/S_1 conical intersection. H atom loss was observed in photofragment translational spectroscopy studies at similar wavelengths and was proposed to arise from coupling to the $\pi\sigma^*$ component of the $S_2(\pi 3s/\pi\sigma^*)$ potential energy surface through an S_3/S_2 conical intersection or successive S_3/S_1 and S_1/S_2 conical intersections. However, we do not see any convincing evidence of significant population flow into $S_1(\pi\pi^*)$ or $S_2(\pi3s/\pi\sigma^*)$ on a <100 fs timescale. It is possible that there is some internal conversion through an S_3/S_1 conical intersection: a very small rise in $S_1(\pi\pi^*)$ population due to $S_3(\pi\pi^*)$ decay would be difficult to disentangle from the significant rise in $S_1(\pi\pi^*)$ population due to decay of the $\pi3s$ Rydberg component of $S_2(\pi3s/\pi\sigma^*)$. In principle, it is also possible that a small proportion of the population undergoes internal conversion through an S_3/S_2 conical intersection; however, we do not observe any delay in the rise of population in the $S_2(\pi3s/\pi\sigma^*)$ state, which we believe lies within our observation window. Therefore, we conclude that the dominant decay pathway for $S_3(\pi\pi^*)$ population is rapid (<100 fs) internal conversion to the electronic ground state through an S_3/S_0 conical intersection.

Theory has predicted that the $S_1 \leftarrow S_0$ transition has mainly charge-resonance and some charge-transfer character and that the $S_3 \leftarrow S_0$ transition is the result of a local excitation on the benzene ring,[14] so we might expect an S_3/S_0 conical intersection to lie along a mode similar to the benzene 'prefulvene mode' which is understood to lead to ultrafast S_1/S_0 internal conversion in the 'channel 3' region of benzene.[42–44] Once back on the ground-state surface, there would be two possible mechanisms for fragmentation. One would involve passing adiabatically through the S_0/S_2 conical intersection at extended N–H bond lengths to generate H atoms with high translational energy. The other would be by unimolecular dissociation of vibrationally hot S_0 to generate H atoms with low translational energy. These mechanisms would be consistent with the bimodal distribution observed in H (Rydberg) atom photofragment translational spectroscopy experiments.[12]

5 Conclusions

This time-resolved photoelectron spectroscopy study of aniline, following absorption of ultraviolet light in the range 269–238 nm, has provided detailed information about the intermediate electronic states involved in the electronic relaxation of this biologically important chromophore, supporting some conclusions reached following H atom photofragment translational spectroscopy measurements and pump–probe photoionisation measurements, and contradicting some others.

Our results confirm that aniline possesses a low-lying electronic state with significant Rydberg character in the Franck–Condon region and reveal that this state is populated directly over a wide range of excitation wavelengths (269–238 nm). Our analysis of the photoelectron spectra suggests that the directly excited $S_2(\pi3s/\pi\sigma^*)$ population bifurcates to two non-radiative decay channels. Interestingly, the decay times of both $S_2(\pi3s/\pi\sigma^*)$ paths are independent of excitation energy. One channel involves relaxation back to the electronic ground state by rapid internal conversion (200–300 fs) through an S_2/S_1 conical intersection located close to the Franck–Condon region, followed by much slower internal conversion (100 ps–ns) to S_0. It is possible that population also passes diabatically through the S_2/S_1 conical intersection to form the anilino radical and H atom, which would be consistent with other experimental observations of fast H atoms. The other channel involves motion on the $S_2(\pi3s/\pi\sigma^*)$ potential energy surface with a decay time of 1–2 ps, most likely towards the S_2/S_0 conical intersection at extended N–H bond lengths, where it can either dissociate or return to the electronic ground state.

Close to the S_2–S_0 origin (269 nm excitation), both $S_2(\pi3s/\pi\sigma^*)$ and $S_1(\pi\pi^*)$ are populated during the excitation process. At higher excitation energies, between the absorption maxima of $S_1(\pi\pi^*)$ and $S_3(\pi\pi^*)$ (250 nm excitation), only $S_2(\pi3s/\pi\sigma^*)$ is populated during the excitation process. Increasing the excitation energy further to just below the absorption maximum of $S_3(\pi\pi^*)$ (240 and 238 nm excitation) results in significant population of the $S_3(\pi\pi^*)$ state. Somewhat surprisingly, the dominant relaxation pathway is an extremely rapid (<100 fs) internal conversion back to the electronic ground state, most likely through an S_3/S_0 conical intersection.

The observation of these competing non-radiative relaxation channels in aniline illustrates that its photostability arises from a subtle balance between dynamics on different potential energy surfaces and between Rydberg and valence states. As well as providing the motivation to reinvestigate the electronic relaxation of other chromophores possessing excited electronic states with mixed Rydberg-valence character, this study highlights the need for time-resolved photoelectron spectroscopy with higher energy probe pulses to monitor both the formation of the products of dissociation and the return of population to the electronic ground state. Such experiments are planned by our group. It is clear that high-level quantum molecular dynamics calculations using accurate potential energy surfaces and calculations of the time-resolved photoelectron spectra are also highly desirable.

Acknowledgements

The authors acknowledge support from the European Marie Curie Initial Training Network Grant No. CA-ITN-214962-FASTQUAST for research funding and a studentship (R.S.) and the EPSRC for studentship funding (O.M.K.). The authors are grateful to Andreas Kafizas for assistance with the gas-phase UV absorption spectrum of aniline and Graham Worth and Mike Ashfold for useful discussions about the interpretation of the data.

References

1 F. de Gruijl, *Eur. J. Cancer*, 1999, **35**, 2003–2009.
2 M. Ichihashi, M. Ueda, A. Budiyanto, T. Bito, M. Oka, M. Fukunaga, K. Tsuru and T. Horikawa, *Toxicology*, 2003, **189**, 21–39.
3 C. Z. Bisgaard, H. Satzger, S. Ullrich and A. Stolow, *ChemPhysChem*, 2009, **10**, 101–110.
4 A. Iqbal and V. G. Stavros, *J. Phys. Chem. Lett.*, 2010, **1**, 2274–2278.
5 H. Satzger, D. Townsend, M. Z. Zgierski, S. Patchkovskii, S. Ullrich and A. Stolow, *Proc. Natl. Acad. Sci. U. S. A.*, 2006, **103**, 10196–10201.
6 A. L. Sobolewski, W. Domcke and C. Hattig, *Proc. Natl. Acad. Sci. U. S. A.*, 2005, **102**, 17903–17906.
7 C. T. Middleton, K. de La Harpe, C. Su, Y. K. Law, C. E. Crespo-Hernandez and B. Kohler, *Annu. Rev. Phys. Chem.*, 2009, **60**, 217–239.
8 S. Matsika and P. Krause, *Annu. Rev. Phys. Chem.*, 2011, **62**, 621–643.
9 H. Reisler and A. Krylov, *Int. Rev. Phys. Chem.*, 2009, **28**, 267–308.
10 A. L. Sobolewski, W. Domcke, C. Dedonder-Lardeux and C. Jouvet, *Phys. Chem. Chem. Phys.*, 2002, **4**, 1093–1100.
11 M. N. R. Ashfold, G. A. King, D. Murdock, M. G. Nix, T. A. A. Oliver and A. G. Sage, *Phys. Chem. Chem. Phys.*, 2010, **12**, 1218.
12 G. A. King, T. A. A. Oliver and M. N. R. Ashfold, *J. Chem. Phys.*, 2010, **132**, 214307.
13 R. Montero, A. P. Conde, V. Ovejas, R. Martinez, F. Castaño and A. Longarte, *J. Chem. Phys.*, 2011, **135**, 054308.
14 Y. Honda, M. Hada, M. Ehara and H. Nakatsuji, *J. Chem. Phys.*, 2002, **117**, 2045–2052.
15 M. N. R. Ashfold, B. Cronin, A. L. Devine, R. N. Dixon and M. G. Nix, *Science*, 2006, **312**, 1637–1640.
16 R. N. Dixon, T. A. A. Oliver and M. N. R. Ashfold, *J. Chem. Phys.*, 2011, **134**, 194303.
17 Z. Lan, W. Domcke, V. Vallet, A. L. Sobolewski and S. Mahapatra, *J. Chem. Phys.*, 2005, **122**, 224315–224315.
18 B. Chmura, M. F. Rode, A. L. Sobolewski, L. Lapinski and M. J. Nowak, *J. Phys. Chem. A*, 2008, **112**, 13655–13661.
19 X. Song, M. Yang, E. R. Davidson and J. P. Reilly, *J. Chem. Phys.*, 1993, **99**, 3224.
20 T. Ebata, C. Minejima and N. Mikami, *J. Phys. Chem. A*, 2002, **106**, 11070–11074.
21 R. Scheps, D. Florida and S. A. Rice, *J. Chem. Phys.*, 1974, **61**, 1730–1747.
22 B. Kim, C. P. Schick and P. M. Weber, *J. Chem. Phys.*, 1995, **103**, 6903.
23 W. E. Sinclair and D. W. Pratt, *J. Chem. Phys.*, 1996, **105**, 7942–7956.
24 E. R. T. Kerstel, M. Becucci, G. Pietraperzia, D. Consalvo and E. Castellucci, *J. Mol. Spectrosc.*, 1996, **177**, 74–78.

This journal is © The Royal Society of Chemistry 2012

25 G. M. Roberts, C. A. Williams, J. D. Young, S. Ullrich, M. J. Paterson and V. G. Stavros, *J. Am. Chem. Soc.*, DOI: 10.1021/ja3029729.

26 R. Spesyvtsev, O. M. Kirkby, M. Vacher and H. H. Fielding, *Phys. Chem. Chem. Phys.*, 2012, **14**, 9942–9947.

27 A. Stolow, A. E. Bragg and D. M. Neumark, *Chem. Rev.*, 2004, **104**, 1719–1757.

28 T. Suzuki, *Annu. Rev. Phys. Chem.*, 2006, **57**, 555–592.

29 A. Stolow and J. G. Underwood, in *Time-resolved photoelectron spectroscopy of nonadiabatic dynamics in polyatomic molecules*, ed. S. A. Rice, John Wiley & Sons, Inc., Hoboken, NJ, 2008, ch. 6, pp. 497–584.

30 A. D. G. Nunn, R. S. Minns, R. Spesyvtsev, M. J. Bearpark, M. A. Robb and H. H. Fielding, *Phys. Chem. Chem. Phys.*, 2010, **12**, 15751–15759.

31 R. S. Minns, D. S. N. Parker, T. J. Penfold, G. A. Worth and H. H. Fielding, *Phys. Chem. Chem. Phys.*, 2010, **12**, 15607–15615.

32 D. S. N. Parker, R. S. Minns, T. J. Penfold, G. A. Worth and H. H. Fielding, *Chem. Phys. Lett.*, 2009, **469**, 43–47.

33 G. A. Worth, R. Carley and H. H. Fielding, *Chem. Phys.*, 2007, **338**, 220–227.

34 A. T. J. B. Eppink and D. H. Parker, *Rev. Sci. Instrum.*, 1997, **68**, 3477.

35 G. A. Garcia, L. Nahon and I. Powis, *Rev. Sci. Instrum.*, 2004, **75**, 4989–4996.

36 J. E. Sansonetti and W. C. Martin, in *Handbook of Basic Atomic Spectroscopic Data*, National Institute of Standards and Technology, 2011.

37 A. M. Lee, J. D. Coe, S. Ullrich, M. L. Ho, S. J. Lee, B. M. Cheng, M. Z. Zgierski, I. C. Chen, T. J. Martinez and S.A., *J. Phys. Chem. A*, 2007, **111**, 11948–11960.

38 O. Schalk, A. E. Boguslavskiy and A. Stolow, *J. Phys. Chem. A*, 2010, **114**, 4058–4064.

39 J. S. Lim and S. K. Kim, *Nat. Chem.*, 2010, **2**, 627–632.

40 D.-S. Ahn, J. Lee, J.-M. Choi, K.-S. Lee, S. J. Baek, K. Lee, K. K. Baeck and S. K. Kim, *J. Chem. Phys.*, 2008, **128**, 224305.

41 R. Livingstone, O. Schalk, A. E. Boguslavskiy, G. Wu, B.L.T., A. Stolow, M. J. Paterson and D. Townsend, *J. Chem. Phys.*, 2011, **135**, 194307–194307.

42 T. J. Penfold and G. A. Worth, *Chem. Phys.*, 2010, **375**, 58–66.

43 D. S. N. Parker, R. S. Minns, T. J. Penfold, G. A. Worth and H. H. Fielding, *Chem. Phys. Lett.*, 2009, **469**, 43–47.

44 R. S. Minns, D. S. N. Parker, T. J. Penfold, G. A. Worth and H. H. Fielding, *Phys. Chem. Chem. Phys.*, 2010, **12**, 15607–15615.

Reaction dynamics of Cl + butanol isomers by crossed-beam sliced ion imaging

Armando D. Estillore, Laura M. Visger-Kiefer† and Arthur G. Suits*

Received 27th March 2012, Accepted 27th April 2012

DOI: 10.1039/c2fd20059g

Butanol is now prominent among the prototype renewable biofuels. We have studied oxidation of a variety of butanol isomers under single collision conditions using chlorine atom as the oxidizing agent to gain detailed insight into the energetics and dynamics of these reactions. The interaction of chlorine atom radicals with butanol isomers: n-butanol, iso-butanol, sec-butanol, and tert-butanol have been studied by crossed-beam dc slice ion imaging techniques. The hydroxybutyl radicals generated from the H-abstraction processes were probed by single photon ionization using an F_2 excimer laser. After background subtraction and density-to-flux correction of the raw images, translational energy distribution and product angular distributions were generated. At low collision energy, the hydroxyalkyl products are backscattered with respect to the alcohol beam and the scattering shifts to the forward direction as the collision energy is increased. The translational energy distributions are reminiscent to that of Cl + pentane reactions we studied earlier, i.e. a sharp forward peak ~80% of the collision energy appears at the high collision energy. Isomer-specific details of the reactions will be discussed.

Introduction

Over the years, a wide array of theory, experiments, and computation has been performed in establishing the dynamics associated with the reaction of Cl atoms with hydrocarbons (RH).[1–3] However, related studies on the dynamics of Cl atom reactions with oxygenated hydrocarbons like alcohols are relatively limited. These oxygenated hydrocarbons (ROH) systems offer special dynamical interest to both theory and experiment. In addition to its multiple reactive sites and the degree of polarity of the molecule, the complexity of the potential energy surfaces present some serious challenges on the part of the researchers in interpreting the results.

Only a few of the experiments on the dynamics of Cl + ROH reactions involve direct probing the hydroxyalkyl radical or the nascent HCl (v, J) products using various detection methods. Ahmed et al.[4,5] first combined crossed molecular beam with velocity map imaging techniques for the reaction of Cl($^2P_{3/2}$) atoms with smaller alcohols such as methanol, ethanol, and 2-propanol using single photon ionization as a product probe of the scattered hydroxylalkyl products. Results showed that 30–40% of the available energy of the reaction was deposited into product translation with the detected product angular distributions to be predominantly sideways/backward scattered with respect to the alcohol beam suggesting direct rebound dynamics. Rudić and coworkers[6] measured the rotational distribution of the HCl(v = 0) products formed from the reaction of Cl atoms with methanol, ethanol, and dimethyl

Department of Chemistry, Wayne State University, Detroit, MI 48202, USA. E-mail: asuits@chem.wayne.edu

† Present address: Department of Chemistry, University of Michigan, Ann Arbor, MI, 48105.

ether to be relatively hotter compared to that of Cl + RH reactions. These results, coupled with *ab initio* calculations,[7] were ascribed to the dipole–dipole interaction of the HCl and the radical co-product in the exit channel of the PES. In a state-to-state experiment of the Cl + $CH_3OH \rightarrow HCl(v, J) + CH_2OH$ reaction, Bechtel and coworkers[8] employed the *photoloc* technique combined with REMPI methods to investigate the nascent HCl (v, J). The probed HCl $(v = 0, 1)$ products revealed direct reaction with both stripping and rebound mechanism for $v = 0$ and stripping reactions for $v = 1$ with the CH_2OH co-products acting as spectator in direct contrast to the case for $v = 0$. In the imaging study of the differential cross sections (DCS) of Cl + CH_3OH, Murray and coworkers[9] found that the nascent HCl $(v = 0)$ products are forward scattered for $J = 2$ and backwards for $J = 5$.

Recently, there has been a growing interest in the chemistry of butanol isomers primarily because of the possibility to use these alcohols as an alternative transportation fuel to address the problem of diminished natural oil and gas resources. A significant number of detailed kinetic investigations[10–16] and theoretical calculations[17–19] have been performed to establish an accurate kinetic model in the combustion of butanol. These alcohols are produced from fermentation and non-fermentative biosynthesis and possess energy densities closer to gasoline and do not readily absorb moisture from the atmosphere.[20–23] In addition, branched alcohols have octane numbers higher than the straight chain counterparts. Substantial kinetic databases have been established for the reaction of Cl with smaller alcohols such as methanol, ethanol, and propanol.[24–29] The few kinetic studies of Cl + butanol reactions are limited to determining the rate constant and mechanism of the reaction.[30–34] However, the associated dynamics of the reaction has not been touched upon. The kinetics of these reactions has been investigated in a variety of independent studies. Kinetic measurements have shown that the room temperature rate constant for Cl + *n*-butanol[31] is $(1.96 \pm 0.19) \times 10^{-10}$ cm³ molec⁻¹ s⁻¹, Cl + *sec*-butanol[30] is $(1.32 \pm 0.14) \times 10^{-10}$ cm³ molec⁻¹ s⁻¹, Cl + iso-butanol[32] is $(1.82 \pm 0.14) \times 10^{-10}$ cm³ molec⁻¹ s⁻¹ and Cl + *tert*-butanol[32] is $(3.26 \pm 0.19) \times 10^{-11}$ cm³ molec⁻¹ s⁻¹.

In this paper, we present a dynamics study on the non-state selective H-abstraction reaction of $Cl(^2P_{3/2})$ with butanol isomers (*n*-butanol, *sec*-butanol, iso-butanol, and *tert*-butanol) by crossed-beam dc slice imaging. The product hydroxyalkyl radicals were probed *via* single photon ionization at 157 nm. As has been seen in the past, this product probe is a sensitive and systematic approach to studying the underlying dynamics of complicated polyatomic chemical reactions.[4,5,35–44]

Experiment

The scattering experiments were performed using a crossed-beam dc slice ion-imaging set-up described elsewhere.[5,45] Briefly, both beams were produced in separate supersonic expansion in two source chambers ($\sim 10^{-7}$ torr base pressure and $\sim 10^{-5}$ torr operational pressure) fixed at 90° to each other and collimated by a skimmer before entering the differentially pumped interaction region. The Cl atom beam was generated from the photolysis of oxalyl chloride, $(COCl)_2$ using the 193 nm output of an ArF excimer laser (60 mJ, 10 Hz) at the nozzle of a piezoelectric pulsed valve. The dissociation dynamics of $(COCl)_2$ was known to produce Cl, $Cl^*(^2P_{1/2})$, and CO as reported by Ahmed and coworkers[46] at 230 nm and further investigated by Hemmi and Suits[47] at 193 nm. The excited spin–orbit $Cl(^2P_{1/2})$ produced at 230 nm is anticipated at 193 nm. However, after entrainment in the supersonic expansion, most of the $Cl(^2P_{1/2})$ are likely to be quenched to the ground state.[48]

The alcohols (*n*-butanol, *sec*-butanol, iso-butanol, and *tert*-butanol), seeded 5% in He or H_2, were expanded from another piezielectric pulsed valve in another source chamber with a total pressure of 4 bar, and crossed the Cl beam at 90° in the interaction region. An F_2 excimer laser (GAM EX-10, ~ 0.5 mJ, 10 Hz) was used to effect ionization of the hydroxyalkyl radical products $(m/z = 73)$ of the reactions. The probe laser was loosely focused using an MgF_2 lens ($f = 135$ cm) into the collision region of the two crossed molecular beams. The ions were accelerated *via* a four-electrode dc slice ion optics assembly[49] to impact on a 75 mm diameter dual microchannel plate (MCP) detector coupled to a fast phosphor screen held at 5kV (Burle Electro-Optics). The front of the MCP assembly was held at ground potential and the back plate was pulsed to "gate" the central slice of the reaction products at a specific mass by application of a high voltage pulse (+2.2 kV/+1 kV bias, 70 ns width) using a commercial pulse generator (DEI PVX-4140). The timing of the pulsed molecular beam nozzles, firing of the photolysis and probe lasers, and detector gate pulse were controlled using a delay generator (BNC 555). The resulting image was recorded using a charged coupled device camera (Mintron 2821e, 512 × 480 pixels). The dc slice imaging detection scheme and megapixel acquisition program IMACQ were used to accumulate the raw images containing centroided data.[50] Image accumulation to reach a satisfactory signal to noise ratio took 1–3 h. The conditions of the experiment are illustrated in an accompanying video shown in our previous publication.[38,39]

Results

The reactively scattered hydroxyalkyl radical from the Cl atom reactions with butanol isomers are shown in Fig. 1 with the nominal Newton diagram overlaid in the images. These images are background subtracted and density-to-flux corrected from the raw images obtained in the experiment as described in our previous reports. The improved instrumental resolution and the ability to produce an intense source of Cl radicals enabled us to obtain considerably more detailed images of the hydroxyalkyl radical than was previously possible. The spread in the collision energy is roughly 25% full-width half maximum.

All the butanol isomers have two fundamentally different abstraction sites: One at the hydroxyl end of the molecule and the other involving primary or secondary C–H groups. Thermochemical data indicate the bond strength to be 104 and 94 kcal mol^{-1} for H abstraction in the O–H and C–H sites, respectively. From this data alone, the C–H bond abstraction site is seen to be more favorable.

Moreover, a kinetics investigation on the specificity of H atom abstraction showed preference of H-atom abstraction on the alkyl backbone with $\sim$3% contribution from the hydroxyl end.[26,27] A dynamical study on the reaction of Cl atom with deuterated methanol: CH_3OD and CD_3OH showed no DCl products detected for the former and measured only the nascent DCl for the latter reaction.[8]

Fig. 2A shows the total translational energy distributions, integrated over all angles. The distributions peak at lower energy and grow with collision energy. Fig. 2B are the corresponding angular distributions, integrated over all recoil directions derived from the hydroxyalkyl images after background subtraction and density-to-flux correction, described in detail in our previous reports.[38,39] The angular distributions at low collision energy (E_c) show flux in all direction with preference in the backward direction with respect to the alcohol beam. Increasing the E_c shows a shift to a more forward peaking trend.

To gain further insight into the dynamics, we also present the translational energy distributions for the forward (0°–60°), sideways (60°–120°), and backward (120°–180°) regions of the distributions as shown in Fig. 3 and tabulated in Table 1.

Discussion

A surprising result here is the similarity of the dynamics of the title reaction with that of Cl + pentanes reaction as will be discussed here. One of the significant differences of Cl atom reactions with alcohols compared to alkanes is the degree of rotational excitation of the nascent HCl (v, J) products. It is rotationally cold for the alkanes while it is modestly rotationally excited in alcohols.[1,6] This warmer HCl rotational

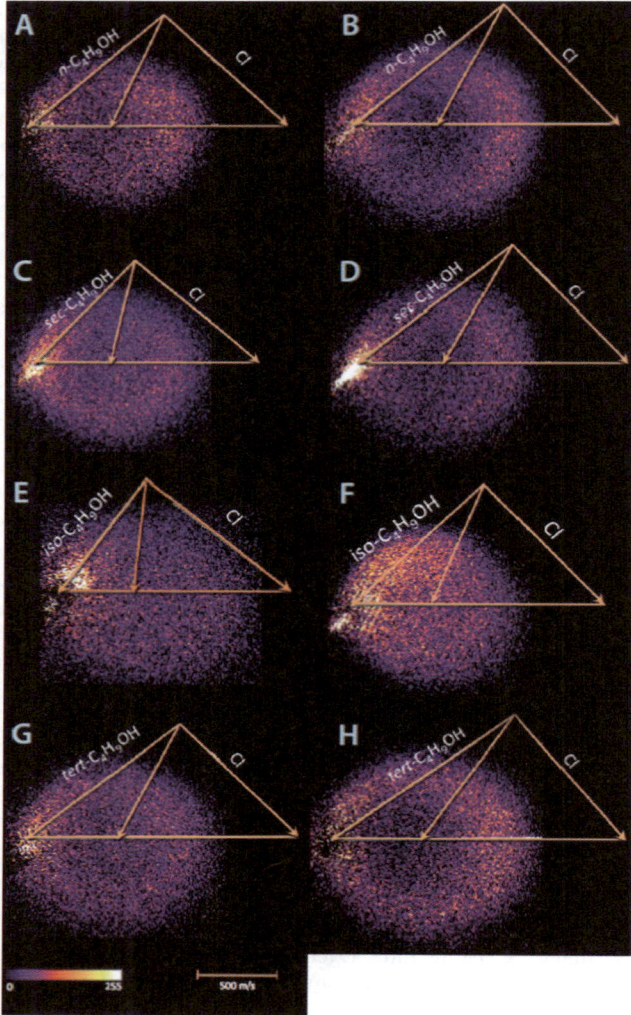

Fig. 1 Sliced scattering data for reaction of Cl with butanol isomers. A,B: *n*-butanol at 6.8 and 8.0 kcal mol⁻¹; C,D:2-butanol at 5.7 and 8.0 kcal mol⁻¹; E, F: isobutanol at 5.3 and 7.0 kcal mol⁻¹, and G,H:*t*-butanol at 7.0 and 9.2 kcal mol⁻¹, respectively.

distribution is ascribed to a dipolar interaction between HCl and the hydroxyalkyl moieties in the exit channel of the potential energy surface.[6] The difference in rotational distributions may suggest a marked contrast in overall dynamics between saturated and oxygenated hydrocarbons towards reactions with chlorine atoms but the extent of this distinction remains an interesting question. We first turn to the angular distributions shown in Fig. 2B. At lower collision energy, the product flux is scattered in the backward direction with respect to the direction of the alcohol beam (0°). The product angular distribution shifts to the forward distribution as the collision energy is increased. The absence of symmetric angular distributions support the hypothesis of a direct or rebound dynamics at low collision energy as a result of low impact parameter collisions and stripping for high collision energy coupled to high impact parameter collisions. However, in the *ab initio* MP2/6-311G** calculation of Garzon and coworkers,[31] it was found out that the reaction

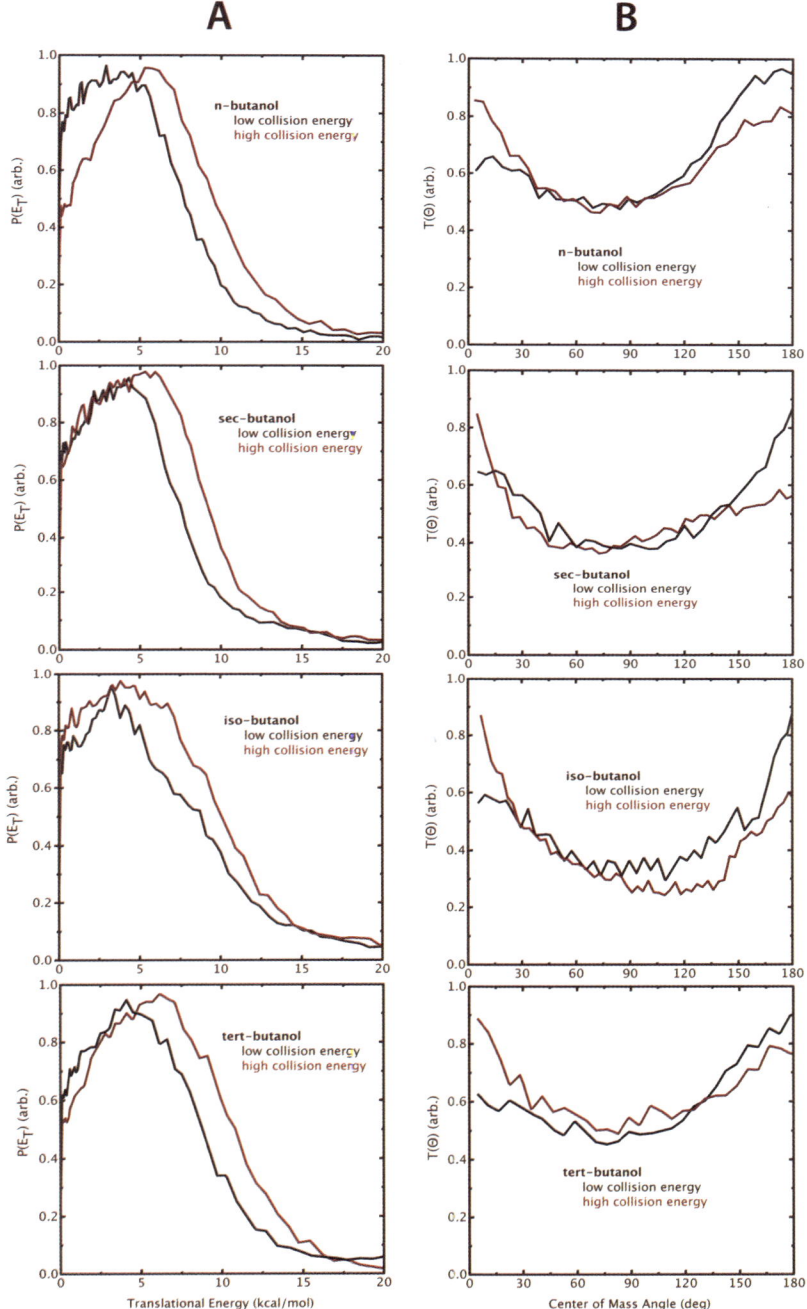

Fig. 2 Global translational energy and angular distributions extracted from Fig. 1.

of Cl + butanol proceed *via* molecular complexes appearing along the reaction pathway. Conventional wisdom informs us that for complex mediated reaction, the angular distribution is expected to be highly symmetric in the forward-backward regions or a highly isotropic distribution.[51] Although our current results show

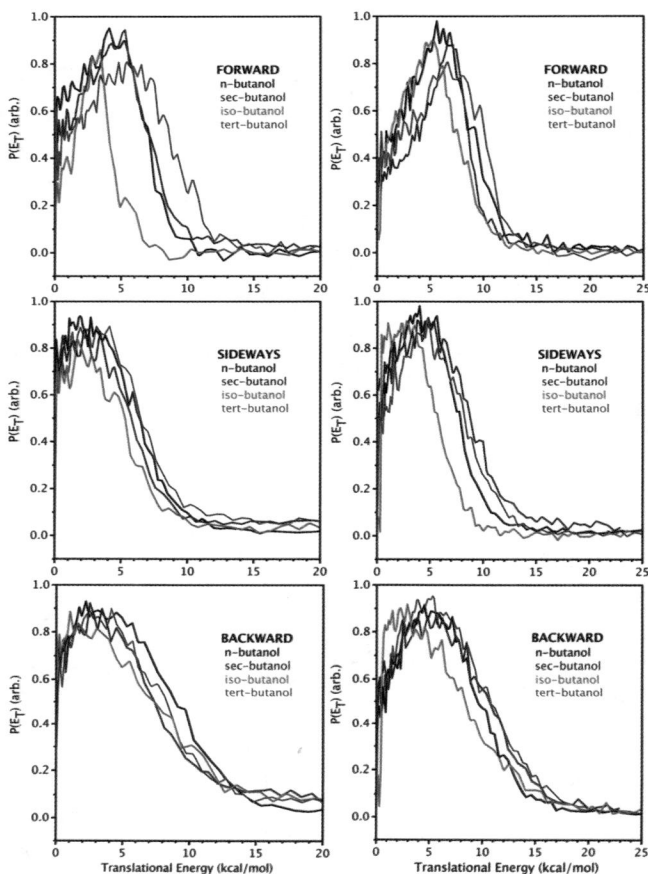

Fig. 3 Angle-dependent translational energy distributions from more detailed analysis of images in Fig. 1. Low collision energy series at left.

Table 1 Most probable collision energy, E_C (kcal mol^{-1}); average translational energy release, $\langle E_T \rangle$ (kcal mol^{-1}), average reduced translational energy, $\langle E_T \rangle^* = \langle E_T \rangle / E_C$

	E_C	$\langle E_T \rangle$ total	$\langle E_T \rangle$ fwd	$\langle E_T \rangle$ sdw	$\langle E_T \rangle$ bwd	$\langle E_T \rangle^*$ total	$\langle E_T \rangle^*$ fwd	$\langle E_T \rangle^*$ sdw	$\langle E_T \rangle^*$ bwd
n-	6.8	5.8	4.0	4.1	5.5	0.86	0.59	0.60	0.81
butanol	8.0	5.4	6.3	5.4	6.6	0.68	0.79	0.68	0.83
sec-	5.7	4.7	4.9	4.7	5.2	0.82	0.86	0.82	0.91
butanol	8.0	5.2	6.6	6.1	6.8	0.65	0.83	0.76	0.85
iso-	5.3	4.2	4.0	3.1	4.8	0.79	0.75	0.58	0.91
butanol	7.0	4.3	5.3	4.0	6.0	0.61	0.76	0.57	0.86
tert-	7.0	5.8	6.2	6.0	6.3	0.83	0.86	0.86	0.90
butanol	9.2	6.4	6.7	4.9	7.4	0.70	0.73	0.53	0.80

otherwise, a critical parameter influencing the dynamics is, of course, the collision energy. Global statements such as "the reaction is complex-mediated (or direct)" must be associated with a particular collision energy to be meaningful. Unfortunately we have not yet examined these reactions at thermal energies or below.

This subtle difference in behavior of Cl + butanol isomers is reminiscent of the previously observed distribution in Cl + smaller alcohols. *Ab initio* calculations for Cl + $CH_3OH \rightarrow HCl + CH_2OH$ identified minima in the potential energy surface suggesting and argued that the reaction was complex-mediated,[52] however, these van der Walls and weak H-bonds wells are too shallow to allow reaction complexes to exist at the collision energies employed.[6,7] In separate experimental investigations, Ahmed and co-workers[4,5] found the reaction to be backward-sideways scattered suggesting direct rebound dynamics using single photon ionization as a product probe for the hydroxyalkyl radical at 8.7 kcal mol^{-1} collision energy. Bechtel and coworkers[8] revisited the reaction using co-expansion technique probing the nascent HCl (v, J) by 2 + 1 REMPI. Their findings suggest stripping mechanism for HCl $(v = 1)$ and both stripping and rebound mechanism for HCl $(v = 0)$ products. At 5.6 kcal mol^{-1} mean collision energy of Cl + CH_3OH, a state-to-state probing of HCl products showed propensity for forward scattering for $J = 2$, and more pronounced backward scattering for $J = 5$. These results provide additional support to our experimental findings for butanol isomers.

The question now is how do we reconcile this experiment–theory discrepancy? One of the factors to consider is the issue of the stability of the complex formed in the reaction. The 'osculating' complex model by Herschbach[53] postulated that the collision complex lifetime τ, has to survive of at least ~5 times its rotational period τ_r, to qualify a reaction to be a long-lived complex-mediated mechanism. For a short-lived complex, the initial approach of the reactants is not scrambled causing the angular distribution, $T(\theta)$ to be asymmetric. In addition to the collision energy contribution, one of the major problems in dealing with large polyatomic systems such as the case herein is the presence of conformational isomers that potentially blur the angular distributions. Although great success has been achieved in investigating behaviour of different conformers in unimolecular dissociation,[54,55] studying their effects in full collision reactions remain a challenge.[40] For butanol alone, the presence of 14 stereoisomers with small differences in energetics,[56,57] in addition to the anisotropic shape of the molecule, foil the distribution as Cl atom approaches the butanol at all possible angles. It is however, noteworthy to mention that in the present experiment we are detecting all rotational and vibrational levels of the hydroxyalkyl radical, *i. e.*, achieving near-universal detection. The angular distribution may show a forward-backward symmetry in a state-to-state probing the nascent HCl (v, J) products at collision http://www.thunder-ranch.com/356.html energy values near or at thermal energies. As mentioned, one of the shortcomings in the current report is our inability to go to low enough collision energies especially for the *n*-butanol case. This arises from background interference from the beams. We hope in the future to achieve lower collision energies to examine these effects and determine where we enter a complex-mediated regime.

A closer inspection of the translational energy distribution provides additional information in the reaction dynamics. Fig. 2A shows the total translational energy integrated over all angles. The distributions show a tendency to preserve the collision energy into product recoil as expected for a heavy–light–heavy reaction, where a heavy atom abstracts a light atom attached to another heavy atom.[58,59] We also scale the translational energy distribution by dividing it by the collision energy and the plots are shown in Fig. 4 and summarized in Table 1. At high collision energy, $\sim70\%$ of the product have $P(E_T)$ equal to or less than the collision energy. This fraction is smaller in the low collision energy. The translational energy for the forward scattered product can be predicted using the spectator-stripping model[60] where the departing HCl product has the same momentum as the approaching Cl atom: $(35/36)^2 \times E_c$. The values reported in Table 1 are significantly lower than the predicted value, *i.e.* 4.0 kcal mol^{-1} (experiment) against 6.5 kcal mol^{-1} (model) for the 6.8 kcal mol^{-1} collision energy for the Cl + *n*-butanol reaction, suggesting likely rotational excitation of the hydroxyalkyl radical. The forward distributions

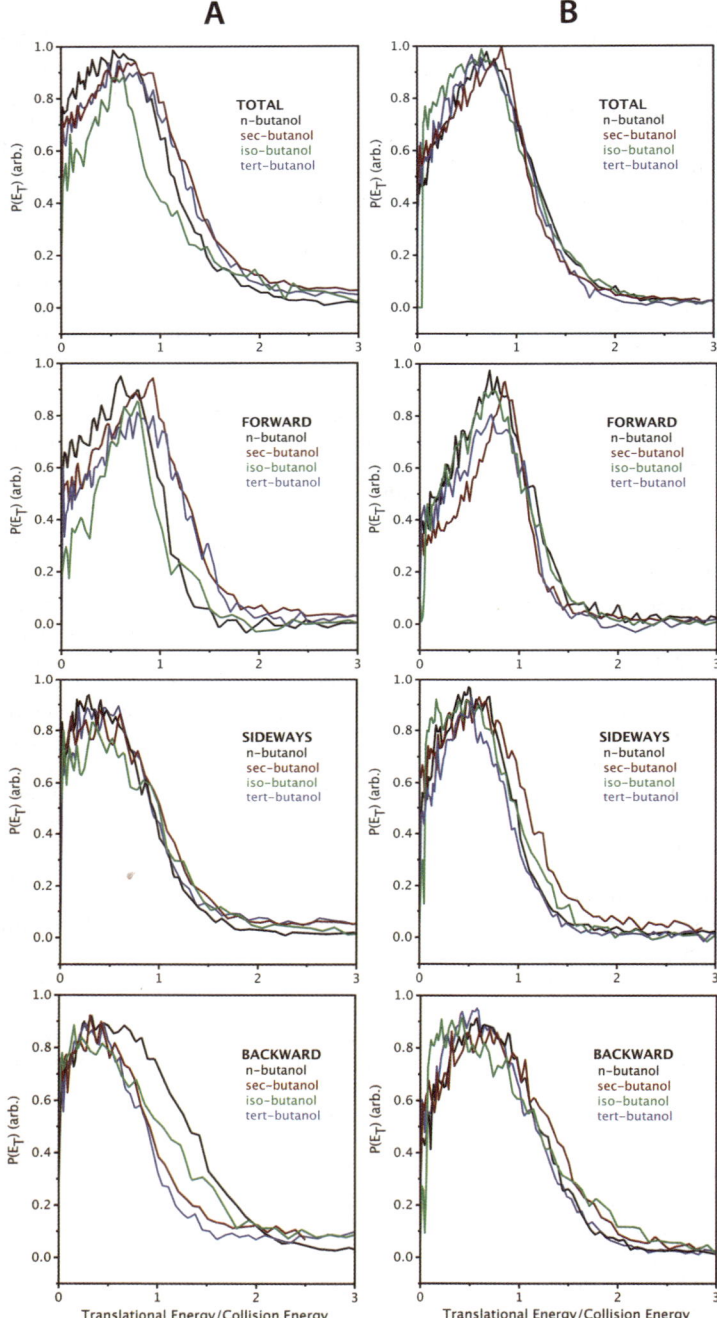

Fig. 4 "Reduced" translational energy distributions highlighting tendency for translation to follow collision energy.

peak at ~80% of the collision energy in sharp similarity with Cl + pentanes reaction.[38] The backward scattered peak, which results from the low-impact parameter collisions and collinear collisions, can be modeled by:

$$\langle E_T \rangle = E_C \cos^2 \beta + E_R \sin^2 \beta$$

where β is the skew angle for the reaction, E_C is the collision energy, and E_R is the energy release in the reaction. Since the reaction considered here is a heavy-light-heavy reaction, the second term of the equation above vanishes because the skew angle is highly acute, $\beta \sim 11.5°$. In all cases, the observed translational energy release for the backward scattered products are lower than the predicted value from the model of Evans et al.[61] The disagreement can be due to the fact that this model works well for triatomic systems and not for polyatomic molecules. Polyatomic systems possess several repositories for the excess energies: vibrations or rotation of the alkyl radical product are two of the possibilities. A Franck–Condon argument by Liu and Suits for the reaction of O(^{3}P) + butane, showed 4–5 kcal mol^{-1} for the relaxation energy of the alkyl radical from the transition state geometry is not available for product recoil.[43] The sideways-scattered product resulting from intermediate impact parameter collisions, showed with consistency, the lowest average fraction of energy in translation (see Table 1). It is likely that in this case, there is efficient rotational excitation of the hydroxyalkyl radicals but with no compensating coupling of the exoergicity into recoil. We do not report the fraction of available energy appearing in translation because of the lack of scientific reports on the enthalpy of reactions for the different butanol isomers with chlorine atom.

As mentioned at the outset of this section, one puzzling aspect of the dynamics of the title reaction is its similarity to the pentane reaction with chlorine atom. In the reaction of Cl atoms with pentanes studied at $\sim$5 kcal mol^{-1} and $\sim$9 kcal mol^{-1} collision energies, the angular distributions showed backward scattering and enhanced forward scattering for the two collision energies, respectively. The energies for the C–H bonds in pentane and butanol are very similar and the structures resemble each other with mass difference of only 2 amu. One key difference between the two is in the potential energy surface: Cl + pentanes appear to be barrierless reactions and modestly exoergic. For Cl + butanol reactions, a classical barrier of 1.34 kcal mol^{-1} has been calculated for the most energetically favored pathway, i. e. H abstraction in the β-C position.[31] Even with the presence of potential well in the Cl + butanol reaction, it is not "deep enough to cause long-lived complex to survive,"[6,8] thus the direct reaction mechanism observed in our crossed-beam experiments.

Conclusion

Understanding the chemistry of butanol and its isomers has attracted attention from the fields of theory and computation and experimental chemical kinetics following its importance in combustion. We present here the first dynamics approach to this important molecular system by crossed beam imaging of the hydroxyalkyl products generated from the reaction of chlorine atoms with butanol isomers. While the HCl product distributions for oxygenated hydrocarbons are markedly different from the HCl product distribution of saturated hydrocarbons, the results presented here for Cl + butanol isomers reaction showed similarity with that of Cl + pentanes. The product angular distribution showed backward scattering at low collision energy with enhanced forward scattering with respect to the alcohol beam with increased collision energy. The product translational energy distributions further support the similarity of these reactions. At high collision energy, a sharp peak of $\sim$80% of the collision energy is seen in the forward scattered products. The sideways-scattered product showed the lowest fraction of energy appearing in translation.

Acknowledgements

This work was supported by the Director, Office of Science, Office of Basic Energy Science, Division of Chemical Science, Geoscience and Bioscience, of the

U. S. Department of Energy under the Contract No. DE-FG02-04ER15593. A. D. E. acknowledges the support from the Graduate School of Wayne State University through the Graduate Dissertation Fellowship Award. We thank Professor Don Truhlar for initially directing us to this system.

Notes and references

1 C. Murray and A. J. Orr-Ewing, *Int. Rev. Phys. Chem.*, 2004, **23**, 435–482.
2 J. J. Valentini, *Annu. Rev. Phys. Chem.*, 2001, **52**, 15–39.
3 A. Teslja and J. J. Valentini, *J. Chem. Phys.*, 2006, **125**, 132304.
4 M. Ahmed, D. S. Peterka and A. G. Suits, *Chem. Phys. Lett.*, 2000, **317**, 264.
5 M. Ahmed, D. S. Peterka and A. G. Suits, *Phys. Chem. Chem. Phys.*, 2000, **2**, 861.
6 S. Rudic, C. Murray, D. Ascenzi, H. Anderson, J. N. Harvey and A. J. Orr-Ewing, *J. Chem. Phys.*, 2002, **117**, 5692.
7 S. Rudic, C. Murray, J. N. Harvey and A. J. Orr-Ewing, *J. Chem. Phys.*, 2004, **120**, 186–198.
8 H. A. Bechtel, J. P. Camden and R. N. Zare, *J. Chem. Phys.*, 2004, **120**, 4231.
9 C. Murray, A. J. Orr-Ewing, R. L. Toomes and T. N. Kitsopoulos, *J. Chem. Phys.*, 2004, **120**, 2230.
10 S. M. Sarathy, M. J. Thomson, C. Togbe, P. Dagaut, F. Halter and C. Mounaim-Rousselle, *Combust. Flame*, 2009, **156**, 852–864.
11 P. S. Veloo, Y. L. Wang, F. N. Egolfopoulos and C. K. Westbrook, *Combust. Flame*, 2010, **157**, 1989–2004.
12 S. Vranckx, K. A. Heufer, C. Lee, H. Olivier, L. Schill, W. A. Kopp, K. Leonhard, C. A. Taatjes and R. X. Fernandes, *Combust. Flame*, 2011, **158**, 1444–1455.
13 M. R. Harper, K. M. Van Geem, S. P. Pyl, G. B. Marin and W. H. Green, *Combust. Flame*, 2011, **158**, 16–41.
14 J. T. Moss, A. M. Berkowitz, M. A. Oehlschlaeger, J. Biet, V. Warth, P. A. Glaude and F. Battin-Leclerc, *J. Phys. Chem. A*, 2008, **112**, 10843.
15 G. Black, H. J. Curran, S. Pichon, J. M. Simmie and V. Zhukov, *Combust. Flame*, 2010, **157**, 363–373.
16 R. Grana, A. Frassoldati, T. Faravelli, U. Niemann, E. Ranzi, R. Seiser, R. Cattolica and K. Seshadri, *Combust. Flame*, 2010, **157**, 2137–2154.
17 P. Seal, E. Papajak, T. Yu and D. G. Truhlar, *J. Chem. Phys.*, 2012, **136**, 034306–034310.
18 X. Xu, E. Papajak, J. Zheng and D. G. Truhlar, *Phys. Chem. Chem. Phys.*, 2012, **14**, 4204–4216.
19 J. Zheng and D. G. Truhlar, *Phys. Chem. Chem. Phys.*, 2010, **12**, 7782–7793.
20 N. Qureshi and T. C. Ezeji, *Biofuels, Bioprod. Biorefin.*, 2008, **2**, 319.
21 N. Qureshi, B. C. Saha and M. A. Cotta, *Biomass Bioenergy*, 2008, **32**, 176.
22 N. Qureshi, B. C. Saha, B. Dien, R. E. Hector and M. A. Cotta, *Biomass Bioenergy*, 2010, **34**, 559.
23 N. Qureshi, B. C. Sahaa, R. E. Hector, S. R. Hughes and M. A. Cotta, *Biomass Bioenergy*, 2008, **32**, 168.
24 T. Yamanaka, M. Kawasaki, M. D. Hurley, T. J. Wallington, W. F. Schneider and J. Bruce, *Phys. Chem. Chem. Phys.*, 2007, **9**, 4211–4217.
25 F. Taketani, K. Takahashi, Y. Matsumi and T. J. Wallington, *J. Phys. Chem. A*, 2005, **109**, 3935–3940.
26 M. A. Crawford, Z. Li, H. A. Heuerman and D. Kinscherff, *Int. J. Chem. Kinet.*, 2004, **36**, 584–590.
27 C. A. Taatjes, L. K. Christensen, M. D. Hurley and T. J. Wallington, *J. Phys. Chem. A*, 1999, **103**, 9805–9814.
28 T. J. Wallington, L. M. Skewes, W. O. Siegl, C.-H. Wu and S. M. Japar, *Int. J. Chem. Kinet.*, 1988, **20**, 867–875.
29 S. A. Cheema, K. A. Holbrook, G. A. Oldershaw and R. W. Walker, *Int. J. Chem. Kinet.*, 2002, **34**, 110–121.
30 B. Ballesteros, A. Garzon, E. Jimenez, A. Notario and J. Albaladejo, *Phys. Chem. Chem. Phys.*, 2007, **9**, 1210.
31 A. Garzon, C. A. Cuevas, A. A. Ceacero, A. Notario, J. Albaladejo and M. Fernandez-Gomez, *J. Chem. Phys.*, 2006, 125.
32 H. Wu, Y. Mu, X. Zhang and G. Jiang, *Int. J. Chem. Kinet.*, 2003, **35**, 81–87.
33 M. D. Hurley, T. J. Wallington, L. Laursen, M. S. Javadi, O. J. Nielsen, T. Yamanaka and M. Kawasaki, *J. Phys. Chem. A*, 2009, **113**, 7011.
34 F. Wicktor, A. Donati, H. Herrmann and R. Zellner, *Phys. Chem. Chem. Phys.*, 2003, **5**, 2562–2572.

35 H. U. Stauffer, R. Z. Hinrichs, P. A. Willis and H. F. Davis, *J. Chem. Phys.*, 1999, **111**, 4101–4112.
36 P. A. Willis, H. U. Stauffer, R. Z. Hinrichs and H. F. Davis, *J. Chem. Phys.*, 1998, **108**, 2665–2668.
37 A. D. Estillore, L. M. Visger, R. I. Kaiser and A. G. Suits, *J. Phys. Chem. Lett.*, 2010, **1**, 2417–2421.
38 A. D. Estillore, L. M. Visger and A. G. Suits, *J. Chem. Phys.*, 2010, **132**, 164313.
39 A. D. Estillore, L. M. Visger and A. G. Suits, *J. Chem. Phys.*, 2010, **133**, 074306.
40 A. D. Estillore, L. M. Visger-Kiefer, T. A. Ghani and A. G. Suits, *Phys. Chem. Chem. Phys.*, 2011, **13**, 8433–8440.
41 R. L. Gross, X. Liu and A. G. Suits, *Chem. Phys. Lett.*, 2003, **376**, 710–716.
42 C. Huang, W. Li, A. D. Estillore and A. G. Suits, *J. Chem. Phys.*, 2008, **129**, 074301.
43 X. Liu, R. L. Gross, G. E. Hall, J. T. Muckerman and A. G. Suits, *J. Chem. Phys.*, 2002, **117**, 7947–7959.
44 X. Liu, R. L. Gross and A. G. Suits, *J. Chem. Phys.*, 2002, **116**, 5341–5344.
45 C. Huang, W. Li and A. G. Suits, *J. Chem. Phys.*, 2006, **125**, 133107.
46 M. Ahmed, D. Blunt, D. Chen and A. G. Suits, *J. Chem. Phys.*, 1997, **106**, 7617.
47 N. Hemmi and A. G. Suits, *J. Phys. Chem. A*, 1997, **101**, 6633–6637.
48 S. Y. T. van de Meerakker, H. L. Bethlem and G. Meijer, *Nat. Phys.*, 2008, **4**, 595–602.
49 D. Townsend, M. P. Minitti and A. G. Suits, *Rev. Sci. Instrum.*, 2003, **74**, 2530–2539.
50 W. Li, S. D. Chambreau, S. A. Lahankar and A. G. Suits, *Rev. Sci. Instrum.*, 2005, **76**, 063106.
51 W. B. Miller, S. A. Safron and D. R. Herschbach, *Discuss. Faraday Soc.*, 1967, **44**, 108–122.
52 J. T. Jodkowski, M.-T. Rayez, J.-C. Rayez, T. Berces and S. Dobe, *J. Phys. Chem. A*, 1998, **102**, 9230–9243.
53 G. A. Fisk, J. D. McDonald and D. R. Herschbach, *Discuss. Faraday Soc.*, 1967, **44**, 228.
54 F. Filsinger, J. Küpper, G. Meijer, J. L. Hansen, J. Maurer, J. H. Nielsen, L. Holmegaard and H. Stapelfeldt, *Angew. Chem., Int. Ed.*, 2009, **48**, 6900–6902.
55 M. H. Kim, L. Shen, H. Tao, T. J. Martinez and A. G. Suits, *Science*, 2007, **315**, 1561–1565.
56 K. Ohno, H. Yoshida, H. Watanabe, T. Fujita and H. Matsuura, *J. Phys. Chem.*, 1994, **98**, 6924–6930.
57 J. Moc, J. M. Simmie and H. J. Curran, *J. Mol. Struct.*, 2009, **928**, 149–157.
58 R. D. Levine, *Molecular Reaction Dynamics*, Cambridge University Press, Cambridge, 2005.
59 P. Casavecchia, K. Liu and X. Yang, in *Tutorials in Molecular Reaction Dynamics*, ed. M. Brouard and C. Vallance, RSC Publishing, Cambridge, 2010.
60 J. C. Polanyi, *Discuss. Faraday Soc.*, 1967, **44**, 293–307.
61 G. T. Evans, E. van Kleef and S. Stolte, *J. Chem. Phys.*, 1990, **93**, 4874–4883.

Between ethylene and polyenes - the non-adiabatic dynamics of *cis*-dienes†

Thomas S. Kuhlman,[a] **William J. Glover,**[bc] **Toshifumi Mori,**[bc] **Klaus B. Møller**[a] **and Todd J. Martínez**[*bc]

Received 23rd March 2012, Accepted 30th April 2012
DOI: 10.1039/c2fd20055d

Using Ab Initio Multiple Spawning (AIMS) with a Multi-State Multi-Reference Perturbation theory (MS-MR-CASPT2) treatment of the electronic structure, we have simulated the non-adiabatic excited state dynamics of cyclopentadiene (CPD) and 1,2,3,4-tetramethyl-cyclopentadiene (Me$_4$-CPD) following excitation to S$_1$. It is observed that torsion around the carbon–carbon double bonds is essential in reaching a conical intersection seam connecting S$_1$ and S$_0$. We identify two timescales; the induction time from excitation to the onset of population transfer back to S$_0$ (CPD: ∼25 fs, Me$_4$-CPD: ∼71 fs) and the half-life of the subsequent population transfer (CPD: ∼28 fs, Me$_4$-CPD: ∼48 fs). The longer timescales for Me$_4$-CPD are a kinematic consequence of the inertia of the substituents impeding the essential out-of-plane motion that leads to the conical intersection seam. A bifurcation is observed on S$_1$ leading to population transfer being attributable, in a 5 : 2 ratio for CPD and 7 : 2 ratio for Me$_4$-CPD, to two closely related conical intersections. Calculated time-resolved photoelectron spectra are in excellent agreement with experimental spectra validating the simulation results.

1 Introduction

Molecules possessing π-electrons play an essential role in organic photochemistry and photophysics.[1] In particular molecules possessing conjugated carbon–carbon double bonds participate in a plethora of reactions induced by light such as photo-isomerisation,[2–11] electrocyclic ring-opening and closing,[12–20] sigmatropic rearrangement[21–23] and cycloaddition.[12,24] Polyenes are prominent examples of such systems *e.g.* retinal, the photoinduced *cis–trans* isomerization of which, initiates the vision process.[25–28] Polyenes also exhibit ultrafast photophysics for example in the light-harvesting and energy-transfer process by carotenoids in the photosynthetic reaction center[29,30] and the non-adiabatic transitions of isolated, longer polyenes such as *all-trans*-2,4,6,8-decatetraene.[31] Thus, fully understanding the intricate nature of the excited state dynamics of polyenes has far reaching implications in photochemistry, photophysics, and photobiology.

[a]*Department of Chemistry, Technical University of Denmark, Kemitorvet 207, DK-2800 Kg, Lyngby, Denmark*
[b]*Department of Chemistry and PULSE Institute, Stanford University, Stanford, California 94305, USA*
[c]*SLAC National Accelerator Laboratory, Menlo Park, California 94309, USA. E-mail: todd. martinez@stanford.edu.*

† Supporting information: Vibrational frequencies of CPD and Me$_4$-CPD and coordinates of optimized geometries for the ground state and MECIs as well as complete ref. 65.

The rich set of photoinduced phenomena exhibited by polyenes is a consequence of the complex nature of the excited states of π-electron systems and conjugated systems in particular. This is evident even in the simplest π-electron system, ethylene, where the lowest $\pi \rightarrow \pi^*$ excited state has zwitterionic character at the Franck–Condon point but diradical character at twisted geometries.[32] For polyenes the picture is even more complicated due to the presence of a low lying (optically dark) electronic state with a large doubly excited character.[33] Due to the dark nature of this doubly excited state it often eludes direct observation, however, it is well established that it plays a significant role in the photochemistry of longer polyenes with more than three conjugated double bonds.[33]

In between the distinct ethylene and the longer polyenes, the role of the doubly excited state in the dienes is more subtle. In the case of s-trans-dienes, where the two double bonds are in a trans-configuration with respect to the single bond, numerous studies have investigated the state-ordering of the bright (π,π^*)-state and the doubly excited state.[34–40] Both experimental and theoretical studies suggest that the doubly excited state plays a significant role in the initial dynamics following excitation to the bright state giving the s-trans-dienes some of the characteristics of the longer polyenes. However, the longer time dynamics resemble that of ethylene.[41–43] In the case of (acyclic) s-cis-dienes, fewer studies exist, which is primarily a consequence of the predominance of the s-trans-conformation under ambient conditions. For butadiene, for example, the equilibrium concentration at room temperature of the s-cis-conformation is only ~3%,[44,45] necessitating very specific experimental conditions for its investigation.[46–48] On the other hand, if the molecule is locked in the s-cis-configuration, such as in cyclopentadiene (CPD), experimental investigations are feasible. CPD exhibits intriguing photochemistry such as electrocyclic ring-closure to bicyclo[2.1.0]pent-2-ene and tricyclo[2.1.0.0^{2,5}]pentane[49–51] as well as sigmatropic hydrogen shifts.[23] Furthermore, CPD is a prototype molecule for many other five-membered rings containing two double bonds, including heterocycles.

Numerous experimental studies have investigated the low-lying valence states in CPD[52–57] but only few have discussed the spectral position of the doubly excited state.[54,57] Despite the inability to directly locate this dark state in absorption spectroscopy, several time-resolved studies have invoked non-adiabatic transitions involving it to explain the sub 200 fs dynamics observed in ion- and photoelectron signals subsequent to excitation to the bright (π,π^*)-state.[23,58,59] The short timescales involved can also be inferred from the lack of detectable fluorescence of CPD,[60] further hinting at the involvement of a doubly excited state. The timescales of the observed dynamics have furthermore been suggested to be dependent on specific nuclear motion as exemplified in the slow-down of the dynamics from CPD to the substituted 1,2,3,4-tetramethylcyclopentadiene (Me$_4$-CPD).[59]

In this paper, we uncover the specific nuclear dynamics leading to non-adiabatic transitions as well as the connection between static structure and electronic and nuclear dynamics, thereby furthering an understanding of the rules governing excited state dynamics in molecular systems in general. We employ first-principles quantum dynamical methods to investigate the ultrafast excited state dynamics in CPD and Me$_4$-CPD following excitation to the bright (π,π^*)-state. We use the Ab Initio Multiple Spawning (AIMS) method,[61] which can model non-adiabatic effects in a framework that allows simultaneous solution of the electronic structure alongside the nuclear dynamics. We are able to carry out these calculations using an electronic structure method (multi-reference perturbation theory) that can treat both the static and dynamic electron correlation effects that are important in unsaturated and conjugated hydrocarbons.

In the following, we unravel the non-adiabatic nuclear dynamics of CPD and Me$_4$-CPD following excitation to the (π,π^*)-state. The observed dynamics is discussed in light of the dynamics of ethylene and other polyenes, in particular the similar s-trans-butadiene. Finally, we relate the results to experimental data from time-resolved photoelectron spectroscopy by direct simulation of the experimentally observed signals.

2 Theoretical methods

2.1 Electronic structure, geometry optimizations and dynamics

Multi-state multi-reference complete active space second-order perturbation theory (MS-MR-CASPT2)[62–64] calculations were performed using the MOLPRO 2006.2 molecular electronic structure package[65] without the use of symmetry, the 6-31G** basis set and a level shift of 0.2 Hartrees. The active space used in all calculations of neutral species consisted of four electrons distributed in four orbitals - the two π-orbitals and the two lowest lying π^*-orbitals, with state-averaging over the lowest three singlet states (SA-3-CAS(4,4)). Optimization of stationary points was performed using built-in routines in MOLPRO, whereas optimization of degeneracy points between potential energy surfaces was performed using either the CIOpt code,[66] which uses a penalty function method, or a locally modified version of MOLPRO. Relaxed pathways were calculated using a Nudged Elastic Band (NEB) method.[67,68] The NEB method is conventionally used for finding the minimum energy path between a pair of local minima or stable states, however, here it is used where at least one of the endpoints is not such a configuration. This use of NEB can lead to ambiguity of the path around the endpoint, however, the method is only invoked to demonstrate that barriers present on interpolated paths might not be present if the path is relaxed and as such whether the NEB method converges to the actual minimum energy path is not of importance here.

AIMS calculations were performed using the in-house code combining AIMS dynamics with electronic structure calculations performed in MOLPRO 2006.2[69] at the MS-MR-CASPT2/6-31G** level of theory with state-averaging over three states including analytic MS-MR-CASPT2 non-adiabatic coupling matrix elements.[70,71] The initial positions and momenta of trajectory basis functions (TBFs) were sampled from the 0 K Wigner distribution of the harmonic approximation to the vibrational ground state.[72] The geometry and vibrational frequencies used to obtain the Wigner distribution were calculated at the MP2/6-31G** level of theory in MOLPRO. The initial 40 (24 for Me$_4$-CPD) TBFs were placed on S$_1$ and propagated for 193.5 fs (8000 au) with a time step of 0.39 fs (16 au) using the independent first-generation approximation.[69,73] The final time was chosen long enough to capture the essential dynamics on the excited states.[59] In cases where all population (>99%) had been transferred to the ground state before the final time was reached, the calculation was stopped as our focus here is on the excited state dynamics and timescale of non-adiabatic transfer and not on possible thermal reactions taking place on the vibrationally hot ground electronic state. The dynamics of Me$_4$-CPD was simulated in separate calculations by scaling the mass of H$_{(6)}$–H$_{(9)}$ to match the mass of a methyl group (see Fig. 1). This approximation allows us to

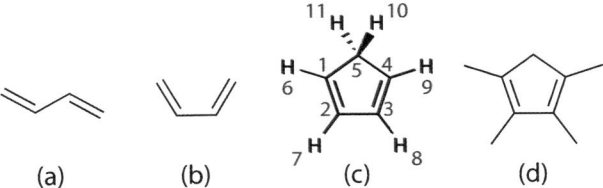

(a)   (b)   (c)   (d)

Fig. 1 Molecular structures of (a) *s-trans*-butadiene, (b) *s-cis*-butadiene, (c) cyclopentadiene (CPD) with numbering, and (d) 1,2,3,4-tetramethylcyclopentadiene (Me$_4$-CPD). Whereas *s-trans*-butadiene belongs to the C_{2h} point group, the three molecules where the double bonds are in an *s-cis*-conformation belong to C_{2v}.

separate the effects of changes in mass from changes in the electronic properties upon methylating CPD and is motivated by the experimental observation that the increased density of states due to the methyl groups is not reflected in the dynamics.[59] Nuclear densities were evaluated by Monte Carlo sampling of the pertinent integrals for the desired internal coordinate.[74]

2.2 Time-resolved photoelectron spectra

In order to calculate the time-resolved photoelectron spectrum it is necessary to calculate the matrix element connecting the initial N-electron neutral state with the final state corresponding to the combination of an $(N-1)$-electron cation state and a continuum electron. In the sudden approximation[75,76] the latter two can be assumed independent, whereby the final state is the product of the cationic and continuum wave functions. In the dipole approximation the matrix element of interest is

$$\left\langle \Psi_I^N(\mathbf{r}_1,...,\mathbf{r}_N)\left|\hat{\mu}(\mathbf{r}_1,...,\mathbf{r}_N)\right|\Psi_F^{N-1}(\mathbf{r}_2,...,\mathbf{r}_N)\Psi_{kf}^{el}(\mathbf{r}_1)\right\rangle = \left\langle\phi_{IF}^D(\mathbf{r}_1)\left|\hat{\mu}(\mathbf{r}_1)\right|\Psi_{kf}^{el}(\mathbf{r}_1)\right\rangle_{\mathbf{r}_1}, \tag{1}$$

where the equality follows from the strong orthogonality condition between the cation and the ionized electron.[77] k and f label the continuum electron momentum and angular momentum respectively. The Dyson orbital ϕ_{IF}^D is given by

$$\left\langle\phi_{IF}^D(\mathbf{r}_1)\right| = \sqrt{N}\left\langle\Psi_I^N(\mathbf{r}_1,...,\mathbf{r}_N)\left|\Psi_F^{N-1}(\mathbf{r}_2,...,\mathbf{r}_N)\right\rangle_{\mathbf{r}_2,...,\mathbf{r}_N}, \tag{2}$$

where the integration is over all electronic coordinates except for the ones of the ejected photoelectron, $\mathbf{r}_1$. Assuming the electric field of the probe pulse, ε_ω, to be a δ-function in time and energy (the Bersohn-Zewail model[78]), the time-resolved photoelectron spectrum can be given as a discrete distribution according to

$$\sigma_{stick}(E_k,\Delta t) \propto \sum_{I,F}\sum_f \left|\left\langle\phi_{IF}^D(\mathbf{r}_1,\Delta t)\left|\hat{\mu}(\mathbf{r}_1)\right|\Psi_{kf}^{el}(\mathbf{r}_1)\right\rangle\right|^2 \delta(\hbar\omega - \Delta E(\Delta t) - E_k), \tag{3}$$

where E_k is the kinetic energy of the ejected photoelectron, Δt is the time between pump and probe pulses, $\hbar\omega$ is the energy of the probe photon, and ΔE is the vertical ionization potential from the neutral state I to the cationic state F, i.e. the difference in the potential energies V_I and V_F at the center of the neutral state wave packet. The polarization vector of the probe pulse is suppressed in eqn (3) because we compute an isotropic average over all possible relative orientations of the molecule and probe pulse. The above equation is in essence a stick-spectrum. To include the finite extent in time and energy of the probe pulse the stick spectrum is convoluted with a Gaussian χ in time and energy

$$\sigma(E_k,\Delta t) = \chi(E_{width},t_{width}) \otimes \sigma_{stick}(E_k,\Delta t). \tag{4}$$

The calculations of the photoelectron spectra include the three neutral states employed in the dynamics calculations, as well as two cationic states calculated from MS-MR-CAS(3/4)-PT2 with molecular orbitals taken from the CASSCF solution of the neutral states. The Dyson orbitals were calculated from the neutral and cationic MS-MR-CASPT2 mixing coefficients and CI vectors according to ref. 71, and the transition matrix element was evaluated numerically using the ezDyson code.[79] The latter calculations employed a grid of $192 \times 192 \times 192$ points with a size of $12 \times 12 \times 12$ au^3, a maximum angular momentum $f = 7$, analytical isotropic averaging and a Coulomb radial function for the free electron.

Although a high-level, correlated electronic structure method is used in the simulations, inevitably some discrepancy between the energies obtained from calculations and those from experiment is to be expected. Therefore, in order to be able to compare the calculated spectra with those from experiment a correction was employed for ΔE to match to the experimental value of the photoelectron kinetic energy, E_k, at the Franck–Condon (FC) point such that $\Delta E \rightarrow \Delta E - \Delta$ with

$$\Delta = [\delta E_{S_1 \leftarrow S_0}^{\text{exp.,vert.}} - \delta E_{S_1 \leftarrow S_0}^{\text{CASPT2}}(\mathbf{Q}_{\text{FC}})] + [\text{IP}_{D_0 \leftarrow S_0}^{\text{CASPT2}}(\mathbf{Q}_{\text{FC}}) - \text{IP}_{D_0 \leftarrow S_0}^{\text{exp.,vert.}}] \qquad (5)$$

In eqn (5), δE indicates an energy difference between two neutral states, whereas the vertical ionization potential indicated by IP is the energy difference between the neutral ground state and the cationic ground state and $\mathbf{Q}_{FC}$ indicates the FC geometry. This correction ensures that at least the predicted kinetic energy of the photoelectrons ejected from S_1 close to time zero matches the experimental value. To determine Δ, the experimental vertical IP of 8.57 eV is used for CPD, and the corresponding value of 7.52 eV for Me_4-CPD[80,81] ($\text{IP}_{D_0 \leftarrow S_0}^{\text{CASPT2}} = 8.41$ eV). Furthermore, it is assumed that the vertical excitation energy corresponds to the spectral position of the band maximum[59] giving $\delta E_{S_1 \leftarrow S_0}^{\text{exp.,vert.}} = 5.17$ eV for CPD and 4.96 eV for Me_4-CPD ($\delta E_{S_1 \leftarrow S_0}^{\text{CASPT2}} = 5.46$ eV).

The experimental spectra also exhibit features associated with two-photon ionization, however, the present scheme does not allow for the much more difficult calculation of the two-photon ionization cross-section. As a consequence the probability of two-photon ionization was assumed to be unity for all trajectory basis functions at all times. In other words, we neglect any geometry dependence of the two-photon ionization cross-section and the relative intensity of the one-photon and two-photon ionization spectra are undetermined by our calculations.

Another important factor in simulating the time-resolved photoelectron spectrum is the full-width-at-half-maximum (FWHM) of the experimental cross-correlation (XC) between the pump and probe pulses, which for a one probe and one pump pulse ionization scheme $(1 + 1')$ was determined to be $\text{FWHM}_{(1+1')} \sim 160$ fs.[59] This value was used for t_{width} in eqn (4) when calculating the one-photon time-resolved photoelectron spectra. Assuming equal duration of the pump and probe pulse and a Gaussian pulse shape, the FWHM of the XC in a $(1 + 2')$ ionization scheme is given by $\text{FWHM}_{(1+2')} = \sqrt{\dfrac{3}{4}} \times \text{FWHM}_{(1+1')} \sim 139$ fs, which was used for t_{width} when calculating the two-photon spectra. The factor of $\sqrt{\dfrac{3}{4}}$ results from $\text{XC}_{(1+2')}$ being the outcome of a convolution between a Gaussian and a product of two Gaussians, whereas $\text{XC}_{(1+1')}$ is just the outcome of a convolution between two Gaussians. The spectral bandwidth of the laser pulses was approximately 25 meV and furthermore a detection resolution has to be taken into account.[59] However, due to our limited sampling we used 150 meV for E_{width} in eqn (4).

3 Results and discussion

3.1 Electronic structure and potential energy surfaces

In a diene the interaction between the two ethylene units gives rise to two bonding π-orbitals and two anti-bonding π^*-orbitals. The ground state configuration, $\pi_1^2 \pi_2^2$, is termed N, whereas the four possible one-electron promotions give rise to four excited states termed $V_1 - V_4$ by Mulliken (i.e. $V_1 - V_4$ are the (π, π^*)-states).[82] For symmetric structures, V_2 and V_3 will belong to the totally symmetric representation of the pertinent point group, whereas V_1 and V_4 will belong to B_u and B_2 for C_{2h} and C_{2v} respectively (see Fig. 1). Furthermore, a totally symmetric, doubly excited configuration is possible, which will mix with V_2, lowering the energy of this state and giving it partial doubly excited character.[54] This doubly excited configuration can be understood as arising from excitation of both of the ethylene units to their lowest triplet states but coupled to an overall singlet.[83] In the following we will use N, V_1, and V_2 as diabatic labels associated with the electronic character of the state as described here, whereas S_0–S_2 will be strictly adiabatic labels (*i.e.* denoting only the order of the adiabatic electronic states and not their character).

In the equilibrium ground state geometry of cyclopentadiene (CPD), *i.e.* the Franck–Condon (FC) point, only $H_{(10)}$ and $H_{(11)}$ (see Fig. 1 for numbering) protrude from the plane defined by the five-membered carbon skeleton and the remaining hydrogens, and the molecule possesses an overall C_{2v} symmetry. From the FC geometry, we find the vertical excitation energy to S_1 to be 5.46 eV in very good agreement with the best estimate for the vertical transition energy of 5.43 ± 0.05 eV found from a combination of high-level theoretical methods and spectroscopic simulations[84] and in the range 5.19–6.46 eV determined using various high-level electronic structure methods.[85–90] The calculated vertical value is, however, slightly different from the spectral position of the band maximum found from experiment of 5.17–5.33 eV.[52–55,59] At the FC geometry, S_1 can be identified as the V_1-state characterized by the HOMO → LUMO single $\pi \to \pi^*$ excitation. Similarly, we find the vertical excitation energy to S_2 at this geometry to be 6.51 eV falling in between previous calculated values of 6.31–7.05 eV[85,86,88–90] and slightly above the value of 6.2 eV suggested on the basis of experimental results.[54] The S_2 state possesses a large doubly excited character of ∼50% primarily due to the HOMO → LUMO double $(\pi)^2 \to (\pi^*)^2$ excitation and can be identified as the V_2-state. The V_2-state has not been directly observed in absorption spectroscopy but resonance Raman depolarization ratios suggest that the minimum of this state lies below that of the singly excited V_1, and a conical intersection connecting these two states can thus be expected.[57]

Upon excitation from the ground state, the initial nuclear motion will primarily be in the direction of steepest descent along the gradient. In both S_1 and S_2 the gradient at the FC geometry corresponds closely to movement in the bond alternation coordinate, where the two double bonds contract and the connecting single bond extends (see Fig. 2). If one further examines the gradient for all starting geometries from the initial Wigner sampling an out-of-plane component of the gradient is also observed in some cases, however, the largest component is always in-plane.

Subsequent to the initial nuclear motion along the direction of steepest descent, one could expect motion towards a lower energy configuration on the excited state potential energy surface unless a large momentum leads to motion in a different direction. The minimum energy configurations on the potential energy surface of both excited states located in this work correspond to conical intersections (MECIs) connecting S_1 with S_0 and S_2 with S_1 (see Fig. 3 and 4). Two additional MECIs were also located between S_1 and S_0, which are 39 and 48 meV higher in energy than the minimum energy configuration of S_1. Two of the S_1S_0-MECIs located in this work, *eth*1-MECI and *eth*2-MECI, result primarily from torsion around a single double bond, wherefore these MECIs are termed ethylene-like, whereas the last, *dis*-MECI, results from a disrotatory mechanism where both double bonds twist to some degree. The double bond torsion might also result in the carbon backbone distortion observed in Fig. 3(a)–(c).

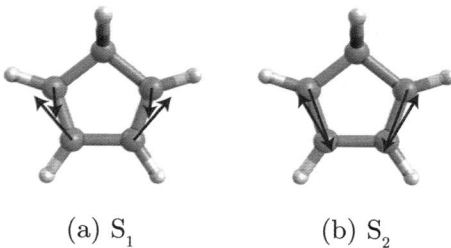

(a) S_1 (b) S_2

Fig. 2 Gradient at the FC geometry for S_1 (a) and S_2 (b). The gradient for both states points in a direction, which leads to a large change of the value of the bond-alternation coordinate defined as $R = R_{12} + R_{34} - R_{23}$ with the numbering given in Fig. 1.

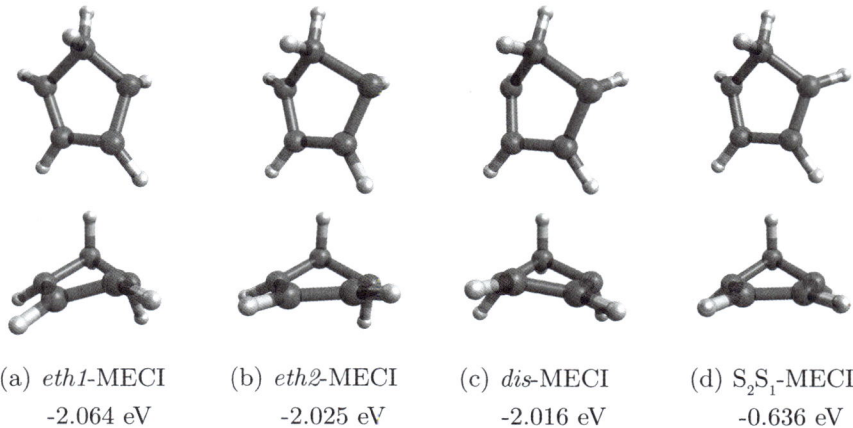

(a) *eth1*-MECI	(b) *eth2*-MECI	(c) *dis*-MECI	(d) S_2S_1-MECI
-2.064 eV	-2.025 eV	-2.016 eV	-0.636 eV

Fig. 3 Geometries at the determined minimum energy conical intersections (MECI) viewed from two different angles and relative energies of S_1 with respect to the Franck–Condon point. All four MECIs are energetically accessible after photoexcitation to S_1.

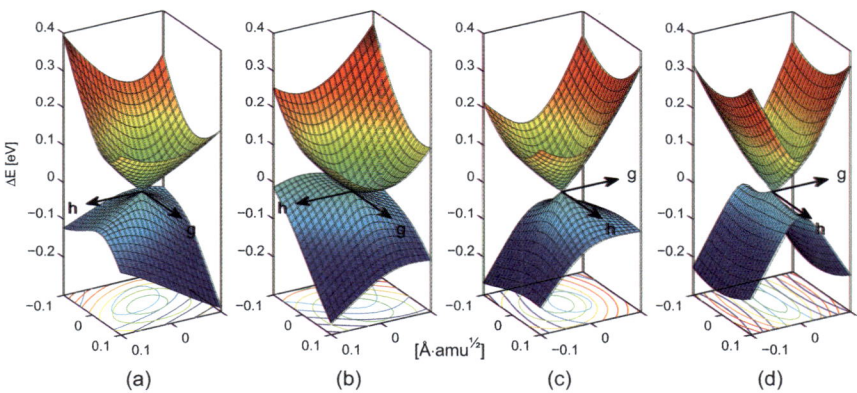

Fig. 4 Conical intersections in the branching space of the g– and the h-vector, in mass-weighted displacement, which have been orthogonalized according to the prescription of Yarkony.[92] (a) *eth1*-MECI intersection between S_1 and S_0, (b) *eth2*-MECI intersection between S_1 and S_0, (c) *dis*-MECI intersection between S_1 and S_0, and (d) intersection between S_2 and S_1.

The potential energy surfaces around the MECIs are illustrated in Fig. 4 along the g- and h-vectors defining the branching space, where the degeneracy of the two electronic states is lifted to first order. All intersections appear peaked but have different characteristics.[91] Moving away from the degeneracy point of the S_1S_0-MECIs, an inspection of the electronic structure shows that the intersections are between a state similar in character to V_1 and the ground state N, the first being ionic with charge separation between the two carbons of the double bonds (larger charge separation for the most twisted double bond), the second being of a more diradicaloid character. These two states, V_1 and N, are observed to mix to the largest extent for the *dis*-MECI. The S_2S_1-MECI is akin to a crossing of non-interacting diabatic states; one similar to V_1 and the other characterized by a large doubly excited character $\sim 60\%$ and thus similar to V_2. This latter intersection, which was expected on the basis of experimental findings in ref. 57, leads to S_1 possessing the character of V_2 in some regions of the potential energy surface.

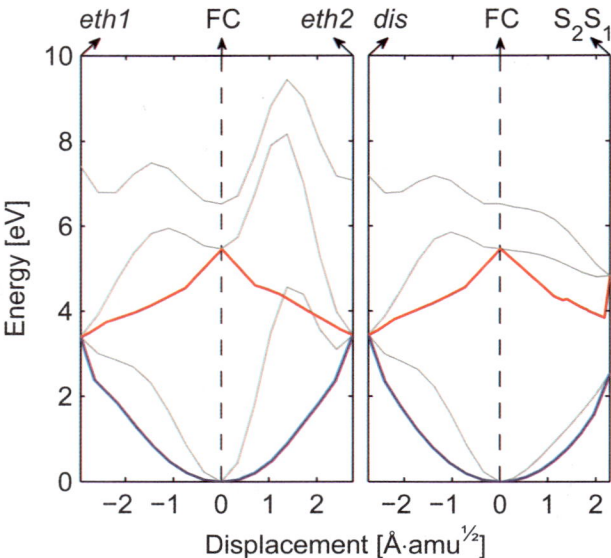

Fig. 5 Potential energy surfaces connecting the Franck–Condon (FC) geometry with the four minimum energy conical intersections (MECI) as a function of mass-weighted displacement. The thin grey lines represent linearly interpolated surfaces, whereas the thick colored lines represent relaxed paths between the two end points on the given surface from the Nudged Elastic Band algorithm.[67,68] The structures were interpolated from the FC geometry to the MECIs. Consequently, the value on the abscissa is only relevant as a distance to the FC geometry indicated by a dashed vertical line and cannot be used for comparing the distance between two geometries on either side of a vertical line.

To better visualize the connection between the initial FC geometry and the four MECIs, linearly interpolated potential energy surfaces have been calculated for geometries between these points of interest (see Fig. 5). The surfaces can also act as a guide to the intuition on the path followed subsequent to excitation to S_1. From the interpolated surfaces it appears there is a barrier on the direct path from the FC geometry to all the S_1S_0-MECIs, whereas the path towards the S_2S_1-MECI appears to be downhill. If this latter path is followed, the S_2S_1-MECI would most likely be passed on the lower half of the cone, see Fig. 4(d), although population transfer to S_2 is possible. In this scenario the electronic state character of S_1 would thus change from V_1 to V_2 over the course of the dynamics and population could be transferred back to S_0 from this part of the surface at some point along the conical intersection seam. However, if the surfaces are relaxed in directions perpendicular to the direction of the given path the picture changes somewhat (see Fig. 5). On these relaxed paths, shown in bold, the barriers in the direction of the S_1S_0-MECIs disappear. Thus, only dynamics calculations can decide whether the part of the S_1 potential energy surface of V_2-character is visited, and whether the population transfer back to S_0 occurs from these V_2 regions or if a simpler picture only involving an S_1-surface of V_1-character is adequate.

3.2 Dynamics

3.2.1 Population transfer - setting the timescales. From the AIMS calculations it is observed that both CPD and Me$_4$-CPD exhibit ultrafast population decay from the initially excited S_1 state back to S_0 on a sub 200 fs timescale (see Fig. 6). Very few spawning events from S_1 to S_2 are observed and the total population transfer to S_2 is <0.1%. For both molecules the onset of population decay is, however, preceded by a delay period, the induction time.[93,94] In the case of CPD this period

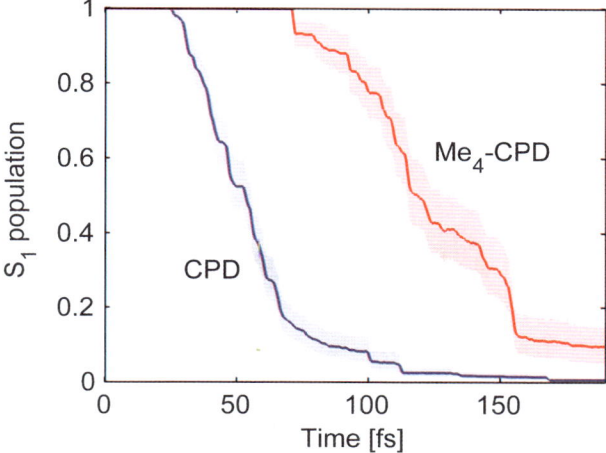

Fig. 6 Population of the S_1 state for CPD (blue) and Me$_4$-CPD (red) with standard deviations from bootstrapping indicated by the shaded regions.

is ~25 fs, while the half-life of the S_1 population decay yields a time constant of $\tau_{1/2}(\text{CPD}) \approx 28$ fs. The half-life $\tau_{1/2}$ is defined as the time it takes the S-shaped S_1 population curve to reach 0.5 (following the initial induction time).[95] Another option is to fit the population decay to a model such as

$$P(t) = H(t - t_0)\exp(-(t - t_0)/\tau) + H(t_0 - t), \tag{6}$$

where H is the Heaviside step function, t_0 is the induction time and τ is the time-constant of the exponential decay. Due to the non-exponential nature of the population decay, the model in eqn (6) primarily fits the tail of the population falloff. Nonetheless, using such a model gives $\tau(\text{CPD}) \approx 31$ fs and $\tau(\text{CPD}) \approx 25$ fs.

In contrast to CPD, the induction time for Me$_4$-CPD is ~71 fs and $\tau_{1/2}(\text{Me}_4\text{-CPD}) \approx 48$ fs, while the exponential fit yields $t_0(\text{Me}_4\text{-CPD}) \approx 88$ fs and $\tau(\text{Me}_4\text{-CPD}) \approx 44$ fs. The results are thus in line with the experimental observation of a slow down of the non-adiabatic dynamics upon methylation of CPD.[59] The following discussion will explore the background for these observed timescales and their differences.

3.2.2 In-plane motion. The initial dynamics following excitation to S_1 in both CPD and Me$_4$-CPD is characterized by significant in-plane nuclear motion, as is common in conjugated molecules such as *s-trans*-butadiene,[43] and in line with the observation that the gradient for S_1 at the FC geometry corresponds to such motion (see Fig. 2). The promotion of an electron from a bonding π-orbital to an antibonding π^*-orbital leads to a weakening of the double bonds, causing an elongation of these in conjunction with a contraction of the $C_{(2)}$–$C_{(3)}$ single bond. This nuclear motion can be quantified by the bond alternation coordinate defined by $R = R_{12} + R_{34} - R_{23}$, the time-dependence of which is given in Fig. 7. A significant in-plane distortion is observed by the rapid increase in this coordinate over the first 15 fs before oscillatory motion around the new equilibrium ensues after ~25 fs. No significant difference is observed between the two molecules as could be expected on the basis of their similar vibrational frequencies for in-plane motion (albeit for the ground state, see Supporting Information†). The mean, about which the coordinate oscillates at later times, of ~1.55 Å for CPD and ~1.54 Å for Me$_4$-CPD is close to the value of 1.54 Å found for the *eth*1-MECI geometry (the coordinate has a value of 1.58 and 1.62 Å for the *eth*2- and *dis*-MECI respectively) and the value of 1.51 Å found for the S_2S_1-MECI geometry (see Fig. 3). This indicates that the in-plane

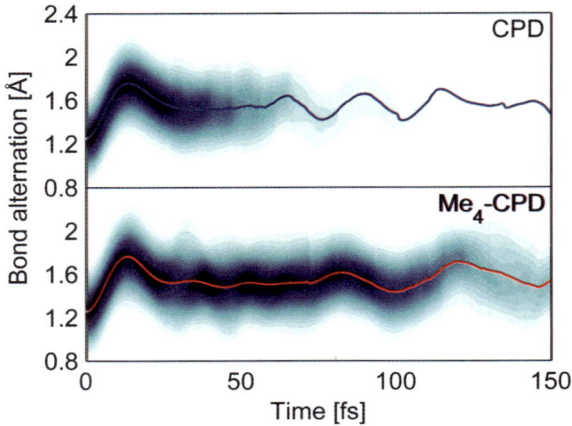

Fig. 7 Projection of the S_1 wavepacket density on the bond alternation coordinate with the expectation value indicated by the colored line for CPD (top, blue) and Me$_4$-CPD (bottom, red). The bond alternation coordinate is defined as $R = R_{12} + R_{34} - R_{23}$ with the numbering given in Fig. 1.

motion towards the intersection with S_0 is over within the first ∼25 fs and is identical in the two molecules. It is thus evident that the in-plane motion cannot be the source of the different time-scales observed for the population transfer between S_1 and S_0.

3.2.3 Out-of-plane motion. In conjunction with the in-plane nuclear motion, out-of-plane motion is also observed, although this occurs on a longer timescale due to the lower frequency modes involved and since the initial gradient is in the direction of in-plane motion (see Fig. 2). Fig. 8 shows the time-dependence of the backbone torsion given by the absolute value of the dihedral angle $\angle C_{(1)}C_{(2)}C_{(3)}C_{(4)}$. In the case of CPD a steady increase in the expectation value of the backbone torsion is observed over the course of the first ∼50 fs before an unstructured behavior takes over. In contrast, it takes ∼70 fs before the maximum of the expectation value is reached for the first time in the case of Me$_4$-CPD. Ring deformation through

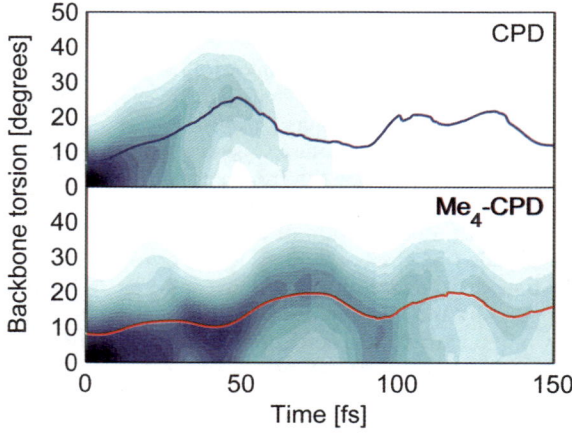

Fig. 8 Projection of the S_1 wavepacket density on the backbone torsion coordinate with the expectation value indicated by the colored line for CPD (top, blue) and Me$_4$-CPD (bottom, red). The backbone torsion is the absolute value of the dihedral angle $\angle C_{(1)}C_{(2)}C_{(3)}C_{(4)}$, with the numbering given in Fig. 1.

backbone torsion is also observed in the structures of the S_1S_0-MECIs in Fig. 3(a)–(c), for which it takes on values of 24.5°, 8.5° and 18.3° for the *eth*1-, *eth*2- and *dis*-MECI respectively. It is thus obvious that ring deformation is essential in reaching the conical intersection seam. The value of the backbone torsion of 24.5° for the *eth*1-MECI is reached within ∼50 fs for the expectation value in the case of CPD, whereas it is only reached by a very small part of the population in the case of Me_4-CPD. This difference between the two molecules does indicate that the motion in the backbone torsion coordinate is important as a cause of the different time scales for the two molecules, especially in explaining the initial delay period before significant population decay begins. However, from Fig. 8 it is also evident that this cannot be the full story as the expectation value of the backbone torsion drops significantly for CPD after peaking at ∼50 fs and the subsequent mean of the coordinate is very similar to that of Me_4-CPD and therefore cannot explain the observed slower population decay on a longer time scale.

The geometries of the S_1S_0-MECIs are also characterized by a significant torsion in one (or both in the case of the *dis*-MECI) of the double bonds leading to the out-of-plane bend of the CH_2-group and a neighboring hydrogen, see Fig. 3. We quantify this torsion as the degree of twist of the most twisted double bond, the time-dependence of which is given in Fig. 9. The value of this coordinate is 50.5°, 58.9°, and 43.4° for the *eth*1-, *eth*2- and *dis*-MECI respectively, whereas it is 31.7° at the S_2S_1-MECI. Again we observe a larger degree of out-plane nuclear motion for CPD compared to Me_4-CPD on a faster time scale. However, compared to the backbone torsion, it takes >100 fs for the expectation value for Me_4-CPD to reach its maximum at ∼35°, a value which is obtained after only ∼28 fs by CPD. It thus appears that the double bond torsion is essential in reaching the conical intersection seam and combined with distortion of the backbone (which could be a consequence of the double bond torsion) can explain the different time scales observed in the population decay for the two molecules.

From the spawning events we can assign population transfer to one of the S_1S_0-MECIs by using the spawning geometries as starting points for optimization of a S_1S_0-MECI. This procedure reveals a bifurcation on S_1 with 71% and 78% of the population transfer being attributable to the *eth*1-MECI and 27% and 22% to the *eth*2-MECI for CPD and Me_4-CPD respectively. For CPD, a very small part of the population, 2%, can be assigned to the *dis*-MECI. This bifurcation is not directly revealed by the overlapping distributions of RMSD between the spawning geometries and the three S_1S_0-MECIs (see Fig. 10). The RMSD distributions on the other

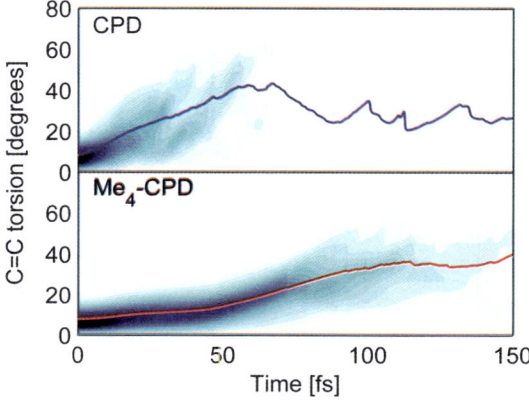

Fig. 9 Projection of the S_1 wavepacket density on the C=C torsion angle (for the most twisted ethylene unit) with the expectation value indicated by the colored line for CPD (top, blue) and Me_4-CPD (bottom, red). The torsion is given by $\max[\cos^{-1}((\hat{\mathbf{R}}_{12} \times \hat{\mathbf{R}}_{56})\cdot(\hat{\mathbf{R}}_{12} \times \hat{\mathbf{R}}_{37})),\cos^{-1}((\hat{\mathbf{R}}_{34} \times \hat{\mathbf{R}}_{28})\cdot(\hat{\mathbf{R}}_{34} \times \hat{\mathbf{R}}_{59}))]$, with the numbering given in Fig. 1.

hand do reveal the preference for the ethylene-like MECIs for Me_4-CPD, whereas this is not obvious for CPD.

3.2.4 Electronic character. Having discussed the nuclear dynamics we turn our attention to the electronic character of the states involved. For the *s-trans*-dienes the two lowest excited states are close in energy at the FC geometry, which for *s-trans*-butadiene has led to the observation of an ultrafast exchange of electronic character between S_1 and S_2 taking place within the first 5 fs subsequent to excitation to the bright state.[43] This exchange of electronic character is observed by the change in transition dipole moment to S_0 such that the initially bright state S_1 becomes dark and the initially dark state S_2 becomes bright. The change of character is unambiguous for *s-trans*-butadiene. In the case of CPD (and Me_4-CPD) the two lowest excited states are separated to a larger extent - at the FC geometry the calculated energy splitting is 1.05 eV, and labels can unambiguously be assigned as the transition dipole moment to S_1 is 2.81 D, whereas it is only 0.28 D to S_2. Fig. 11 shows the time-development of the ratio of the squared transition dipole moments $\langle|\mu_{01}|^2\rangle/\langle|\mu_{02}|^2\rangle$. For both molecules the ratio starts out >10 (the value is 100 for the FC geometry), however, it drops within the first 10 fs to ~3 and stays at that level. It is thus apparent that there is a mixing of the electronic character, and an unambiguous assignment to bright and dark (or equivalently to V_1 and V_2) of the two adiabatic states S_1 and S_2 is not possible at later times.

For *s-trans*-butadiene it is known that charge-transfer states play an essential role in the excited state dynamics.[43] Furthermore, in that molecule charge separation occurs on S_1 and is preceded by twisting of a single methylene unit akin to the twist of a single double bond in the cyclopentadienes. In the latter molecules the torsional motion is, however, frustrated due to the ring structure and the twist does not reach the extremum of 90° corresponding to complete out-of-plane twist as is observed in *s-trans*-butadiene. As a consequence a significantly smaller charge separation is observed in the cyclopentadienes.

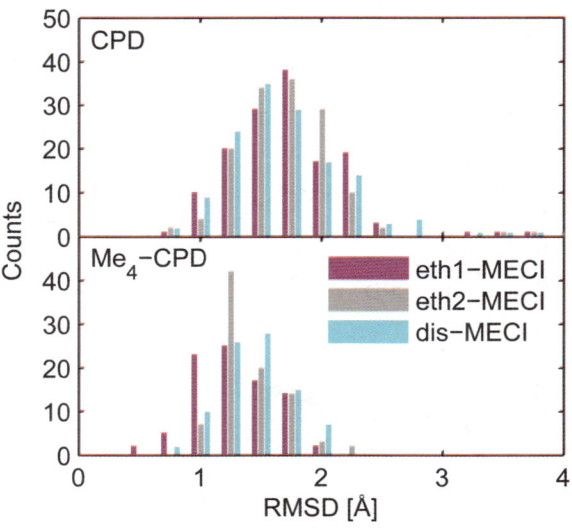

Fig. 10 Histograms of the RMSD between the spawning geometries and the geometry at the *eth*1-MECI (magenta), the *eth*2-MECI (grey), and the *dis*-MECI (cyan) for CPD (top) and Me_4-CPD (bottom).

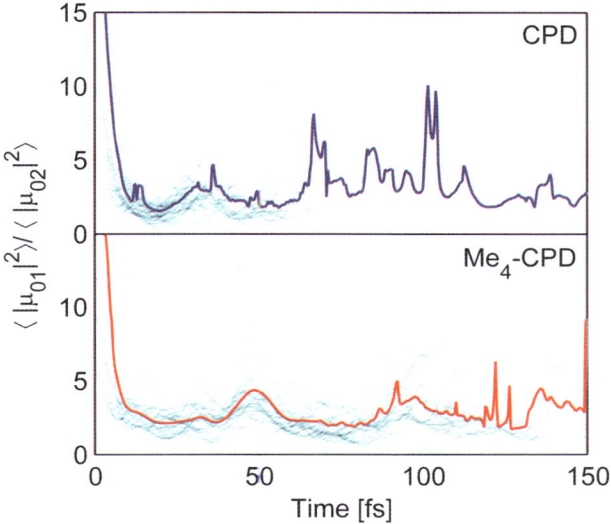

Fig. 11 Histogram of the ratio of the average of the square of the transition dipole moment between S_0 and S_1 (μ_{01}) and between S_0 and S_2 (μ_{02}) with the average value indicated by the colored line for CPD (top,blue) and Me$_4$-CPD (bottom,red). Data is from trajectories on S_1 only.

3.3 Summary of the excited state reaction mechanism for cyclopentadienes

Having established the nuclear dynamics leading to non-adiabatic transition between S_1 and S_0 and the electronic character of the states involved, a complete picture of the excited state dynamics emerges and in particular an understanding of the differences between CPD and Me$_4$-CPD and between the cyclopentadienes and *s-trans*-butadiene is established. At the FC geometry, S_1 and S_2 can clearly be identified as the V_1 and V_2 states, where the former is primarily a bright, HOMO $\rightarrow$ LUMO singly excited $\pi \rightarrow \pi^*$ state, whereas the latter is a dark state dominated by a large doubly excited character $(\pi)^2 \rightarrow (\pi^*)^2$. Excitation to S_1 creates a wavepacket which starts evolving in time. Initial motion primarily along in-plane modes, such as along the bond-alternation coordinate but for CPD also along out-of-plane modes, takes this wavepacket out of the FC region in ~25 fs (see Fig. 12). The motion in the bond-alternation coordinate is typical of conjugated molecules as exemplified by *s-trans*-butadiene.[43] As a consequence of this nuclear motion the wavepacket enters a region of the potential energy surface where the electronic state character mixes significantly and an unambiguous assignment of diabatic labels to the adiabatic states S_1 and S_2 is no longer possible.

After the initial nuclear motion, additional out-of-plane motion in Me$_4$-CPD occurs from ~25 to ~71 fs after excitation, which is not observed for CPD. This prolonged initial period is a consequence of the slow-down of motion along out-of-plane modes for Me$_4$-CPD compared to CPD due to the inertia of the substituents. This motion in particular involves torsion in the double bonds very similar to that in *s-trans*-butadiene and to the smaller ethylene. Due to the rigid ring structure such motion could have been thought to be absent in the cyclopentadienes but seems to occur nonetheless although slightly suppressed. Although the two ethylene-like S_1S_0-MECIs, to which most population transfer can be assigned, primarily result from torsion in only one double bond, the spawning geometries reveal a slight disrotatory mechanism, where torsion occurs around both double bonds. The difference in torsion of the two double bonds for most of the spawning geometries falls somewhere in between that for the ethylene-like S_1S_0-MECIs and the *dis*-MECI.

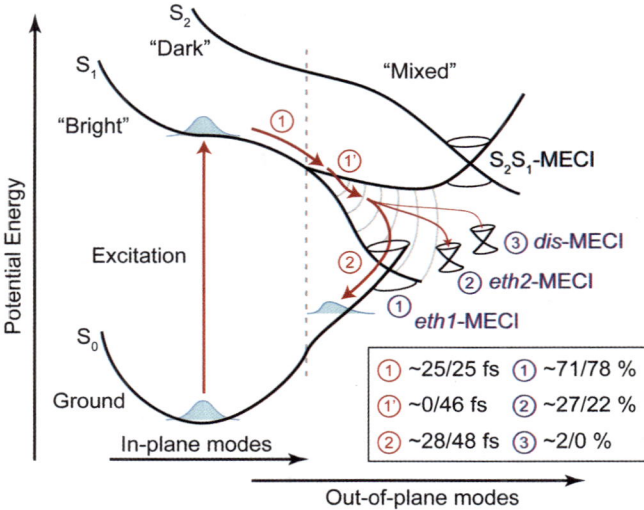

Fig. 12 Interpretation of the dynamics following excitation to S_1 with timescales and branching percentages indicated in the format CPD/Me$_4$-CPD. Initially (1), nuclear motion occurs along in-plane modes such as the bond-alternation coordinate, however, for CPD motion along out-of-plane modes also takes place during this time window. Over the course of the initial nuclear motion the electronic state character mixes making an unambiguous assignment of diabatic labels to S_1 and S_2 impossible. Following the initial nuclear motion additional out-of-plane motion takes place in Me$_4$-CPD (1'), which is absent in CPD, as a consequence of the slower motion of the former due to the heavy substituents. Finally (2), bifurcation on S_1 leads to a non-adiabatic transition primarily assignable to two ethylene-like MECIs.

The out-of-plane motion is thus reminiscent of a reaction path towards a bicyclo [2,1,0]pentene structure and the disrotatory pathway is in accordance with the Woodward–Hoffmann rules.[12] The initial in-plane and out-of-plane motion on S_1 *i.e.* the induction time from excitation to the onset of population transfer sets the first timescale of ~25 fs for CPD and ~71 fs for Me$_4$-CPD (see Fig. 12).

After nuclear motion on S_1 during the first timescale, the second timescale is set by the non-adiabatic transition back to S_0 which takes place with half-lives of $\tau_{1/2}$(CPD) ≈ 25 fs and $\tau_{1/2}$(Me$_4$-CPD) ≈ 48 fs (see Fig. 12). The transition can primarily by assigned to the *eth*1-MECI, which accounts for 71 and 78% of the population transfer for CPD and Me$_4$-CPD respectively, whereas the *eth*2-MECI accounts for 27 and 22% respectively. In the case of CPD, a small population transfer of 2% can be assigned to the *dis*-MECI. An important observation which can be drawn from the determined time-scales is that the slow-down of the non-adiabatic dynamics *i.e.* the longer time-scale of non-adiabatic transition in Me$_4$-CPD compared to CPD is largely accounted for on the basis of the inertia of the substituents and is thus a consequence of a kinematic effect and not due to a difference in the final vibrational density of states (DOS) of S_0. From Fermi's Golden rule one would expect a faster transition in the case of Me$_4$-CPD due to a higher DOS, however, recent studies have shown that this can lead to erroneous conclusions regarding the relative rate of transition, and these findings are thus in line with that conclusion.[96] Also, the differences cannot be due to differences in electronic structure, since these are not included in the present simulation.

3.4 Femtosecond time-resolved photoelectron spectra

From the nuclear dynamics two time-scales of importance have been identified - one time-scale during which nuclear motion takes place only on the initially excited state with the wavepacket moving away from the FC region, and one time-scale for the non-adiabatic transfer back to S_0. Using time-resolved photoelectron spectroscopy two time-scales have also been identified experimentally and a comparison between the experiment and the present simulations are found in Table 1. The assignment of the first timescale differs between simulation and experiment and, furthermore, it is observed that the timescales determined from the present simulations are slightly shorter than the ones determined experimentally. The discrepancies in the timescales could be the consequence of approximating the exciting laser pulse by a δ-function in time and energy and the use of the harmonic approximation for constructing the Wigner distribution of the ground state vibrational wave function in the simulations. Both these approximations can lead to an initial wavepacket in the electronic excited state slightly different from the one prepared in the experiments. However, the discrepancies could also partly arise from differences in the method of determining timescales between simulation and experiment. Therefore, to be able to make a direct comparison between theory and experiment and validate the simulations the time-resolved photoelectron spectra are calculated on the basis of the simulated dynamics.

Fig. 13 shows the calculated one-photon spectrum (top), two-photon spectrum (center), and combined spectrum (bottom) of CPD. The combined spectrum has been constructed by assuming the maximum intensity of the one-photon spectrum to be 20 times that of the two-photon spectrum following experimental findings.[59] In accordance with the experimental spectrum,[59] the combined calculated spectrum exhibit a low energy band at $E < 0.5$ eV due to one-photon ionization centered at $t = 0$ and a delayed, broad band due to two-photon ionization (see Fig. 13 (bottom)). Thus, it is evident that the present simulation is able to reproduce the experimental data satisfactorily.

Both bands of the calculated spectrum are observed to originate from ionization out of the same adiabatic state, S_1‡. The disappearance of the low-energy, one-photon band is a consequence of a fast increase in ionization potential from S_1 to D_0, the ground state of the cation, when the wavepacket leaves the FC region and slides down the potential energy surface. It is thus the energetic factor in the last term in eqn (3) that leads to the decay of the low energy band by effectively closing the one-photon probe window. Through two-photon ionization a new probe window is opened further down the potential energy surface resulting in the band centered at a kinetic energy of 1.9 eV. This window stays open longer than the

Table 1 Comparison between timescales from experimental time-resolved photoelectron spectroscopy[59] and from the present simulations

| Molecule | Experimental[a] | | Simulation[b] | |
	τ/fs	Interpretation	τ/fs	Interpretation
CPD	39	$S_2 \rightarrow S_1$ transition	25(31)	Nuclear dynamics on S_1
	51	$S_1 \rightarrow S_0$ transition	28(25)	$S_1 \rightarrow S_0$ transition
Me$_4$-CPD	68	$S_2 \rightarrow S_1$ transition	71(88)	Nuclear dynamics on S_1
	76	$S_1 \rightarrow S_0$ transition	48(44)	$S_1 \rightarrow S_0$ transition

[a] The definition of the S_1 and S_2 labels in ref. 59 is not identical to the one used throughout this work. [b] The values in parentheses are determined from fitting the S_1 population decays using eqn (6).

‡ There is a contribution to the two-photon spectrum <0.75 eV from ionization out of S_0 and S_2, however, this is hidden below the much stronger one-photon band in the combined spectrum and does not play a significant role when comparing calculation to experiment.

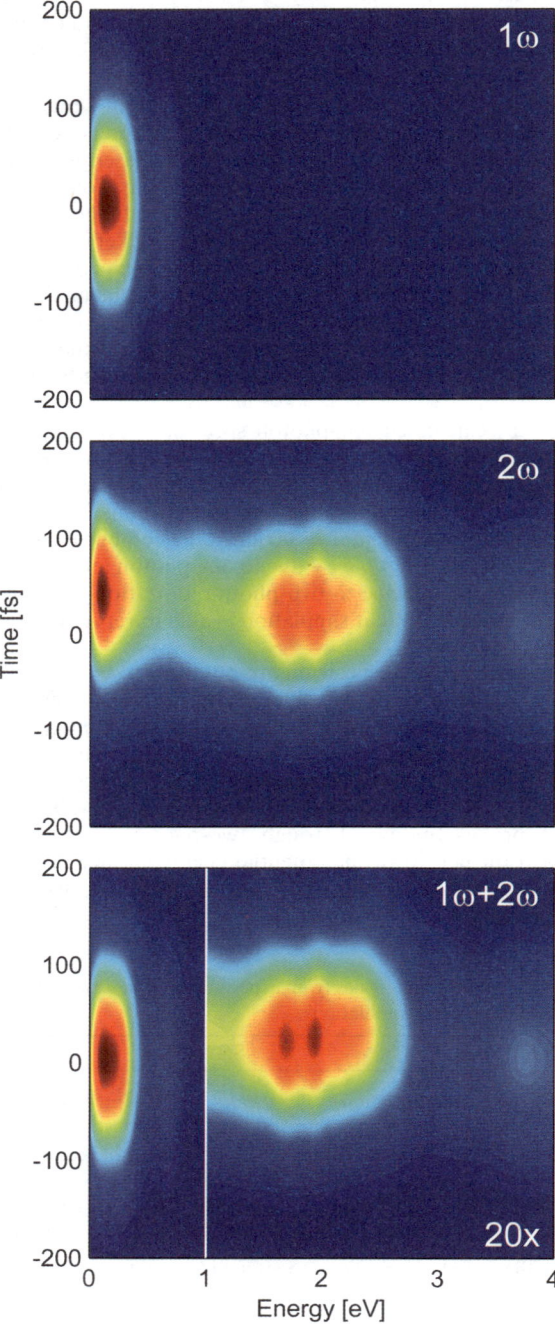

Fig. 13 Time-resolved photoelectron spectra of CPD. TOP: normalized one-photon spectrum (1ω), center: normalized two-photon spectrum (2ω) and bottom: combined spectrum where the region above 1 eV has been multiplied by a factor of 20 to resemble the presentation of the experimental spectrum in ref. 59. It has furthermore been assumed that the maximum intensity of the one-photon spectrum was 20 times that of the two-photon spectrum in accordance with experimental results.[59]

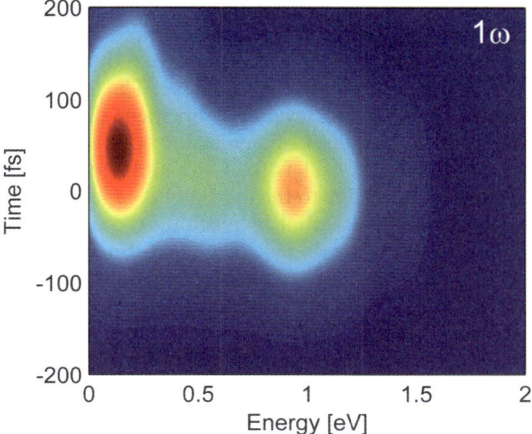

Fig. 14 One-photon time-resolved photoelectron spectrum of Me$_4$-CPD.

one-photon window until population decay back to the ground state finally leads to the decay of the band.

Fig. 14 shows the calculated one-photon spectrum of Me$_4$-CPD. The spectrum of Me$_4$-CPD was not presented in the experimental work but it was mentioned that the spectrum is similar to that of 5-propyl-cyclopentadiene but red-shifted by 0.1 eV, and this spectrum will therefore be used for comparison. The calculated spectrum in Fig. 14 exhibits two bands, a high energy band centered around 0.95 eV and $t = 0$ and a delayed, lower energy band below 0.3 eV, which is in accordance with what is observed experimentally although the bands are shifted slightly from the experimental values of 1.1 and 0.3 eV§. However, again the present simulation is able to reproduce the experimental data satisfactorily.

Both bands of the calculated spectrum originate from ionization out of the same adiabatic state S$_1$, and one could thus interpret the two bands as a single band moving down in kinetic energy as a function of time as the molecule distorts away from the FC geometry and slides down the potential energy surface. This nuclear motion would lead to an increase in the ionization potential and thus a decrease in kinetic energy of the photoelectrons. By comparing a photoelectron spectrum calculated assuming unit ionization probability with the one in Fig. 14, one can somewhat deduce the cause of the band splitting observed. As the band splitting is apparent in both spectra, the dip separating the band into two is seemingly not due to a change in ionization probability as a consequence of changing electronic character of S$_1$ but most likely a consequence of nuclear dynamics leading to a sudden change in the ionization potential.

4 Conclusion

Using Ab Initio Multiple Spawning with electronic structure at the MS-MR-CASPT2 level of theory, we have simulated the excited state dynamics following excitation to S$_1$ in cyclopentadiene (CPD) and 1,2,3,4-tetramethylcyclopentadiene (Me$_4$-CPD). At the Franck–Condon (FC) geometry, S$_1$ is easily identified as resulting from a HOMO → LUMO, $\pi \to \pi^*$ excitation, whereas S$_2$ has a pronounced doubly excited character of ~50%. However, the electronic character mixes significantly as the dynamics unfold on the S$_1$ surface, thus the adiabatic S$_1$ label is not

§ *I.e.* red-shifted by 0.1 eV from the values of 1.2 and 0.4 eV found for 5-propyl-cyclopentadiene.

synonymous with either of the diabatic V_1 and V_2 labels. Subsequent to excitation, initial motion along the bond-alternation coordinate takes the wavepacket on S_1 out of the FC region, whereafter out-of-plane motion ensues due to torsion in the double bonds similar to that of ethylene or s-trans-butadiene. The motion is reminiscent of a disrotatory mechanism towards the bicyclo[2,1,0]pentene photoproduct. The induction time from excitation to the onset of significant population transfer back to S_0 was determined to be $\sim$25 fs and $\sim$71 fs for CPD and Me_4-CPD respectively. The longer timescale for Me_4-CPD is due to the inertia of the methyl substituents slowing down the out-of-plane motion essential in reaching the conical intersection seam between S_1 and S_0 and this slow-down also leads to a longer timescale of non-adiabatic transition. The timescale of non-adiabatic transitions, given by the half-life of the population decay from S_1 to S_0, were determined to be $\sim$28 fs and $\sim$48 fs for CPD and Me_4-CPD respectively.

To make direct connection to experimental observables the time-resolved photoelectron spectra were calculated on the basis of the simulations and were seen to be in correspondence with the experimental spectra. The bands observed in the spectra mainly derive from one- or two-photon ionization out of S_1. In the case of CPD the decay of the one-photon band is due to an increasing ionization potential effectively closing the probe window, whereas the two-photon band decays due to population transfer back to S_0. As a consequence of the substantially lower ionization potential of Me_4-CPD compared to CPD, the one-photon probe window allows for probing of the wavepacket over a longer timescale resulting in the observation of a band shifting down in energy as function of time due to the wavepacket sliding down the S_1 potential energy surface. The final decay of the band results from a combination of the probe window closing due to energetic factors as well as population decay back to S_0.

Acknowledgements

This work was supported by the AMOS program within the Chemical Sciences, Geosciences and Biosciences Division of the Office of Basic Energy Sciences, Office of Science, US Department of Energy. T. S. Kuhlman acknowledges financial help from The Danish Ministry of Science, Technology and Innovation through a Elite Research Scholarship during his stay at Stanford University and SLAC National Accelerator Laboratory.

References

1 N. J. Turro, J. C. Scaiano and V. Ramamurthy, Modern Molecular Photochemistry of Organic Molecules, University Science Books, Sausalito, California, 2010.
2 D. H. Waldeck, Chem. Rev., 1991, 91, 415–436.
3 B. E. Kohler, Chem. Rev., 1993, 93, 41–54.
4 M. Ben-Nun and T. J. Martinez, Chem. Phys. Lett., 1998, 298, 57–65.
5 J. Quenneville and T. J. Martinez, J. Phys. Chem. A, 2003, 107, 829–837.
6 T. Schultz, J. Quenneville, B. Levine, A. Toniolo, T. J. Martinez, S. Lochbrunner, M. Schmitt, J. P. Shaffer, M. Z. Zgierski and A. Stolow, J. Am. Chem. Soc., 2003, 125, 8098–8099.
7 W. Fuß, C. Kosmidis, W. E. Schmid and S. A. Trushin, Angew. Chem., Int. Ed., 2004, 43, 4178–4182.
8 C. Dugave and L. Demange, Chem. Rev., 2003, 103, 2475–2532.
9 B. G. Levine and T. J. Martinez, Annu. Rev. Phys. Chem., 2007, 58, 613–634.
10 L. M. Frutos, T. Andruniow, F. Santoro, N. Ferre and M. Olivucci, Proc. Natl. Acad. Sci. U. S. A., 2007, 104, 7764–7769.
11 M. Barbatti, M. Ruckenbauer, J. J. Szymczak, A. J. A. Aquino and H. Lischka, Phys. Chem. Chem. Phys., 2008, 10, 482–494.
12 R. B. Woodward and R. Hoffmann, The Conservation of Orbital Symmetry, Verlag Chemie, Weinheim, 1970.
13 W. Fuß, W. E. Schmid and S. A. Trushin, J. Chem. Phys., 2000, 112, 8347–8362.
14 M. Ben-Nun and T. J. Martinez, J. Am. Chem. Soc., 2000, 122, 6299–6300.

This journal is © The Royal Society of Chemistry 2012

15 A. Hofmann and R. de Vivie-Riedle, *J. Chem. Phys.*, 2000, **112**, 5054–5059.
16 M. Garavelli, C. S. Page, P. Celani, M. Olivucci, W. E. Schmid, S. A. Trushin and W. Fuß, *J. Phys. Chem. A*, 2001, **105**, 4458–4469.
17 R. C. Dudek and P. M. Weber, *J. Phys. Chem. A*, 2001, **105**, 4167–4171.
18 F. Rudakov and P. M. Weber, *Chem. Phys. Lett.*, 2009, **470**, 187–190.
19 J. Bao, M. P. Minitti and P. M. Weber, *J. Phys. Chem. A*, 2011, **115**, 1508–1515.
20 S. Deb and P. Weber, *Annu. Rev. Phys. Chem.*, 2011, **62**, 19–39.
21 S. A. Trushin, S. Diemer, W. Fuß, K. L. Kompa and W. E. Schmid, *Phys. Chem. Chem. Phys.*, 1999, **1**, 1431–1440.
22 W. Fuß, W. E. Schmid and S. A. Trushin, *J. Am. Chem. Soc.*, 2001, **123**, 7101–7108.
23 W. Fuß, W. E. Schmid and S. A. Trushin, *Chem. Phys.*, 2005, **316**, 225–234.
24 W. L. Dilling, *Chem. Rev.*, 1969, **69**, 845–877.
25 Q. Wang, R. W. Schoenlein, L. A. Peteanu, R. A. Mathies and C. V. Shank, *Science*, 1994, **266**, 422–424.
26 H. Kandori, Y. Shichida and T. Yoshizawa, *Biochemistry (Moscow)*, 2001, **66**, 1197–1209.
27 D. Polli, P. Altoè, O. Weingart, K. M. Spillane, C. Manzoni, D. Brida, G. Tomasello, G. Orlandi, P. Kukura, R. A. Mathies, M. Garavelli and G. Cerullo, *Nature*, 2010, **467**, 440–U88.
28 T. J. Martinez, *Nature*, 2010, **467**, 412–413.
29 B. P. Krueger, G. D. Scholes, R. Jimenez and G. R. Fleming, *J. Phys. Chem. B*, 1998, **102**, 2284–2292.
30 P. J. Walla, P. A. Linden, C. P. Hsu, G. D. Scholes and G. R. Fleming, *Proc. Natl. Acad. Sci. U.S.A*, 2000, **97**, 10808–10813.
31 V. Blanchet, M. Z. Zgierski, T. Seideman and A. Stolow, *Nature*, 1999, **401**, 52–54.
32 L. Salem, *Science*, 1976, **191**, 822–830.
33 B. S. Hudson, B. E. Kohler and K. Schulten, in *Linear Polyene Electronic Structure and Potential Surfaces*, ed. E. C. Lim, Academic Press, New York, 1982, vol. 6, pp. 1–95.
34 V. Vaida, R. E. Turner, J. L. Casey and S. D. Colson, *Chem. Phys. Lett.*, 1978, **54**, 25–29.
35 L. J. Rothberg, D. P. Gerrity and V. Vaida, *J. Chem. Phys.*, 1980, **73**, 5508–5513.
36 J. P. Doering and R. McDiarmid, *J. Chem. Phys.*, 1980, **73**, 3617–3624.
37 J. P. Doering and R. McDiarmid, *J. Chem. Phys.*, 1981, **75**, 2477–2478.
38 R. McDiarmid and J. P. Doering, *Chem. Phys. Lett.*, 1982, **88**, 602–606.
39 R. R. Chadwick, D. P. Gerrity and B. S. Hudson, *Chem. Phys. Lett.*, 1985, **115**, 24–28.
40 R. R. Chadwick, M. Z. Zgierski and B. S. Hudson, *J. Chem. Phys.*, 1991, **95**, 7204–7211.
41 W. Fuß, W. E. Schmid and S. A. Trushin, *Chem. Phys. Lett.*, 2001, **342**, 91–98.
42 F. Assenmacher, M. Gutmann, G. Hohlneicher, V. Stert and W. Radloff, *Phys. Chem. Chem. Phys.*, 2001, **3**, 2981–2982.
43 B. G. Levine and T. J. Martinez, *J. Phys. Chem. A*, 2009, **113**, 12815–12824.
44 J. G. Aston, G. Szasz, H. W. Woolley and F. G. Brickwedde, *J. Chem. Phys.*, 1946, **14**, 67–79.
45 B. R. Arnold, V. Balaji and J. Michl, *J. Am. Chem. Soc.*, 1990, **112**, 1808–1812.
46 M. E. Squillacote, R. S. Sheridan, O. L. Chapman and F. A. L. Anet, *J. Am. Chem. Soc.*, 1979, **101**, 3657–3659.
47 P. Huber-Wälchli and H. H. Günthard, *Spectrochim. Acta, Part A*, 1981, **37**, 285–304.
48 J. J. Fisher and J. Michl, *J. Am. Chem. Soc.*, 1987, **109**, 1056–1059.
49 J. I. Brauman, L. E. Ellis and E. E. van Tamelen, *J. Am. Chem. Soc.*, 1966, **88**, 846–848.
50 E. E. van Tamelen, J. I. Brauman and L. E. Ellis, *J. Am. Chem. Soc.*, 1971, **93**, 6145–6151.
51 G. D. Andrews and J. E. Baldwin, *J. Am. Chem. Soc.*, 1977, **99**, 4851–4853.
52 L. Pickett, E. Paddock and E. Sackter, *J. Am. Chem. Soc.*, 1941, **63**, 1073–1077.
53 R. P. Frueholz, W. M. Flicker, O. A. Mosher and A. Kuppermann, *J. Chem. Phys.*, 1979, **70**, 2003–2013.
54 R. McDiarmid, A. Sabljic and J. P. Doering, *J. Chem. Phys.*, 1985, **83**, 2147–2152.
55 A. Sabljic and R. McDiarmid, *J. Chem. Phys.*, 1990, **93**, 3850–3855.
56 R. McDiarmid and A. Gedanken, *J. Chem. Phys.*, 1991, **95**, 2220–2221.
57 Q.-Y. Shang and B. S. Hudson, *Chem. Phys. Lett.*, 1991, **183**, 63–68.
58 F. Rudakov and P. M. Weber, *J. Phys. Chem. A*, 2010, **114**, 4501–4506.
59 O. Schalk, A. E. Boguslavskiy and A. Stolow, *J. Phys. Chem. A*, 2010, **114**, 4058–4064.
60 N. Nakashima, S. R. Meech, A. R. Auty, A. C. Jones and D. Phillips, *J. Photochem.*, 1985, **30**, 207–214.
61 M. Ben-Nun and T. J. Martinez, *Adv. Chem. Phys.*, 2002, **121**, 439–512.
62 H.-J. Werner, *Mol. Phys.*, 1996, **89**, 645–661.
63 J. Finley, P. Malmqvist, B. Roos and L. Serrano-Andres, *Chem. Phys. Lett.*, 1998, **288**, 299–306.
64 T. Shiozaki, W. Gyorffy, P. Celani and H.-J. Werner, *J. Chem. Phys.*, 2011, **135**, 081106.

65 H.-J. Werner, P. J. Knowles, R. Lindh, F. R. Manby, M. S., *et al.* MOLPRO, version 2006.1,† a package of ab initio programs, see http://www.molpro.net.
66 B. G. Levine, J. D. Coe and T. J. Martinez, *J. Phys. Chem. B*, 2008, **112**, 405–413.
67 H. Jónsson, G. Mills and K. W. Jacobsen, in *Nudged elastic band method for finding minimum energy paths of transitions*, ed. B. J. Berne, G. Ciccotti and D. F. Coker, World Scientific, 1998, ch. 16, pp. 385–404.
68 D. Sheppard, R. Terrell and G. Henkelman, *J. Chem. Phys.*, 2008, **128**, 134106.
69 B. G. Levine, J. D. Coe, A. M. Virshup and T. J. Martinez, *Chem. Phys.*, 2008, **347**, 3–16.
70 T. Mori and S. Kato, *Chem. Phys. Lett.*, 2009, **476**, 97–100.
71 T. Mori, W. J. Glover, M. Schuurman and T. J. Martinez, *J. Phys. Chem. A*, 2011, **116**, 2808–2818.
72 M. J. Davis and E. J. Heller, *J. Chem. Phys.*, 1984, **80**, 5036–5048.
73 M. D. Hack, A. M. Wensmann, D. G. Truhlar, M. Ben-Nun and T. J. Martinez, *J. Chem. Phys.*, 2001, **115**, 1172–1186.
74 J. D. Coe, B. G. Levine and T. J. Martinez, *J. Phys. Chem. A*, 2007, **111**, 11302–11310.
75 B. T. Pickup and O. Goscinski, *Mol. Phys.*, 1973, **26**, 1013–1035.
76 B. T. Pickup, *Chem. Phys.*, 1977, **19**, 193–208.
77 S. Patchkovskii, Z. Zhao, T. Brabec and D. M. Villeneuve, *Phys. Rev. Lett.*, 2006, **97**, 123003.
78 R. Bersohn and A. H. Zewail, *Ber. Bunsenges. Phys. Chem.*, 1988, **92**, 373–378; J. Petersen, N. E. Henriksen and K. B Møller, *Chem. Phys. Lett.*, 2012, **539-540**, 234–238.
79 L. Tac, C. M. Oana, V. A. Mozhayskiy and A. I. Krylov, *ezDyson*, http://iopenshell.usc.edu/downloads/ezdyson.
80 P. Derrick, L. Asbrink, O. Edqvist, B.-O. Jonsson and E. Lindholm, *Int. J. Mass Spectrom. Ion Phys.*, 1971, **6**, 203–215.
81 V. Kiselev, A. Sakhabutdinov, I. Shakirov, V. Zverev and A. Konovalov, *Zh. Org. Khim.*, 1992, **28**, 2244–2252.
82 R. S. Mulliken, *Rev. Mod. Phys.*, 1942, **14**, 265–274.
83 M. B. Robin, *Higher Excited States of Polyatomic Molecules*, Academic Press, New York, 1975, vol. II.
84 Y. J. Bomble, K. W. Sattelmeyer, J. F. Stanton and J. Gauss, *J. Chem. Phys.*, 2004, **121**, 5236–5240.
85 H. Nakatsuji, O. Kitao and T. Yonezawa, *J. Chem. Phys.*, 1985, **83**, 723–743.
86 L. Serrano-Andrés, M. Merchan, I. Nebot-Gil, B. O. Roos and M. Fülscher, *J. Am. Chem. Soc.*, 1993, **115**, 6184–6197.
87 H. Nakano, T. Tsuneda, T. Hashimoto and K. Hirao, *J. Chem. Phys.*, 1996, **104**, 2312–2320.
88 J. D. Watts, S. R. Gwaltney and R. J. Bartlett, *J. Chem. Phys.*, 1996, **105**, 6979–6988.
89 M. Schreiber, M. R. J. Silva, S. P. A. Sauer and W. Thiel, *J. Chem. Phys.*, 2008, **128**, 134110.
90 J. Shen and S. Li, *J. Chem. Phys.*, 2009, **131**, 174101.
91 D. R. Yarkony, *J. Phys. Chem. A*, 2001, **105**, 6277–6293.
92 D. R. Yarkony, *J. Chem. Phys.*, 2000, **112**, 2111–2120.
93 K. B. Møller and A. H. Zewail, *Chem. Phys. Lett.*, 1998, **295**, 1–10.
94 K. B. Møller, N. E. Henriksen and A. H. Zewail, *J. Chem. Phys.*, 2000, **113**, 10477–10485.
95 K. B. Møller and A. H. Zewail, *Chem. Phys. Lett.*, 2002, **351**, 281–288.
96 T. S. Kuhlman, T. I. Sølling and K. B. Møller, *ChemPhysChem*, 2012, **13**, 820–827.

Infrared driven CO oxidation reactions on isolated platinum cluster oxides, $Pt_nO_m^+$†

Alexander C. Hermes,[a] Suzanne M. Hamilton,[a] Graham A. Cooper,[a] Christian Kerpal,[b] Dan J. Harding,[b] Gerard Meijer,[b] André Fielicke*[b] and Stuart R. Mackenzie*[a]

Received 10th February 2012, Accepted 24th February 2012
DOI: 10.1039/c2fd20019h

This collaboration has recently shown that infrared excitation can drive decomposition reactions of molecules on the surface of gas-phase transition metal clusters. We describe here a significant extension of this work to the study of bimolecular reactions initiated in a similar manner. Specifically, we have observed the infrared activated CO oxidation reaction ($CO_{(ads)}$ + $O_{(ads)} \rightarrow CO_{2(g)}$) on isolated platinum oxide cations, $Pt_nO_m^+$. Small platinum cluster oxides $Pt_nO_m^+$ ($n = 3$–7, $m = 2$, 4), have been decorated with CO molecules and subjected to multiple photon infrared excitation in the range 400–2200 cm^{-1} using the Free Electron Laser for Infrared eXperiments (FELIX). The $Pt_nO_mCO^+$ clusters have been characterised by infrared multiple photon dissociation spectroscopy using messenger atom tagging. Evidence is observed for isomers involving both dissociatively and molecularly adsorbed oxygen on the cluster surface. Further information is obtained on the evolution of the cluster structure with number of platinum atoms and CO coverage. In separate experiments, $Pt_nO_mCO^+$ clusters have been subjected to infrared heating *via* the CO stretch around 2100 cm^{-1}. On all clusters investigated, the CO oxidation reaction, indicated by CO_2 loss and production of $Pt_nO_{m-1}^+$, is found to compete effectively with the CO desorption channel. The experimental observations are compared with the results of preliminary DFT calculations in order to identify both cluster structures and plausible mechanisms for the surface reaction.

1 Introduction

One of the major driving forces behind research into isolated metal clusters is the prospect of developing a better understanding of the chemistry of defect sites which play a key role in practical heterogeneous catalysis.[1,2] Very small clusters (less than 10 metal atoms) with no long-range order can have a variety of different coordination sites and, in many respects, represent experimentally and computationally tractable model catalytic systems.[3–5] To this end the full range of modern gas-phase spectroscopic and mass spectrometric techniques have been brought to bear on a central aim in cluster science – understanding the evolution of electronic and

[a]Department of Chemistry, University of Oxford, Physical and Theoretical Chemistry Laboratory, South Parks Road, Oxford, OX1 3QZ, UK. E-mail: stuart.mackenzie@chem.ox.ac.uk; Fax: +44 (0)1865 275410; Tel: +44 (0)1865 275156
[b]Fritz-Haber-Institut der Max-Planck-Gesellschaft, Faradayweg 4-6, D-14195 Berlin, Germany. E-mail: fielicke@fhi-berlin.mpg.de; Fax: +49 (0)30 8413 5603; Tel: +49 (0)30 8413 5622

† Electronic supplementary information (ESI) available: Relative energies and structures of the calculated lowest-lying $Pt_4O_2CO^+$ structures with dissociatively adsorbed O_2 and the lowest-lying structures with O_2 in a peroxo motif. See DOI: 10.1039/c2fd20019h

geometrical structure with cluster size. Reflecting their central role in catalysis, much of this effort has been directed at clusters of transition metal atoms. Our understanding of these systems has improved enormously over recent years with experiments finally providing reliable spectroscopic data against which to benchmark computational studies.

One major advance in the spectroscopy of small clusters has been the recent emergence of infrared multiple photon dissociation (IR-MPD) spectroscopy as a powerful method for investigating the vibrational structures of isolated metal clusters. The use of inert messenger tagging,[6,7] in which absorption of IR photons is registered by the concomitant loss of, for example, a physisorbed rare-gas atom, has extended the range of the method beyond cluster adsorbate systems[8] to naked metal clusters.[9–12] These action spectroscopy techniques couple the sensitivity of mass spectrometric detection with the high fluence and wide wavelength tunability of modern laser sources and thus permit the study of isolated clusters in molecular beams or ion traps.

We have recently demonstrated that, in addition to driving the desorption of weakly-bound species, direct infrared pumping may be used to trigger interesting chemistry on isolated clusters in a manner similar to temperature programmed reaction on extended surfaces.[13,14] IR excitation of nitrous oxide molecularly bound on cationic rhodium clusters was shown to cause efficient decomposition of the N_2O moiety resulting in the generation of rhodium oxide clusters and the loss of molecular nitrogen. All indications are that the reaction is thermal with the same processes occurring with similar efficiency following pumping of the Rh–O mode in $Rh_nON_2O^+$. These studies yield a wealth of valuable and unique information on desorption and reaction barriers as a function of cluster size and structure. This was exemplified recently when we demonstrated that the marked difference in the efficiency of the surface reaction on $Rh_5N_2O^+$ and $Rh_5ON_2O^+$ clusters arises as a result of subtle variations in the barrier heights for reaction and N_2O desorption between the two.[15] A second example is the size-dependence of the IR-induced dehydrogenation of methane initially molecularly bound to cationic platinum clusters.[16] These experiments are only possible using an isotopically enriched Pt sample, which has also been used in the studies described in the following.

Here, we report a significant extension of this infrared driven cluster surface reactivity to the study of bimolecular reactions. Specifically, we have studied the CO oxidation reaction on platinum cluster oxides which can be summarised as

$$CO_{(ads)} + O_{(ads)} \rightarrow CO_{2(g)}. \tag{1}$$

CO-oxidation over transition metal catalysts is one of the best studied reactions of its type.[17] It is believed to proceed via a classic Langmuir–Hinshelwood mechanism following adsorption of CO and dissociative adsorption of O_2 to the catalytic surface. Temperature programmed reaction studies on Pt(111) as well as polycrystalline platinum surfaces indicate at least two CO_2 formation mechanisms on extended surfaces. α–CO_2, produced around 160 K, results from the reaction of hot oxygen atoms produced upon O_2 dissociation and thus depends critically on the presence of co-adsorbed O_2 molecules. By contrast β–CO_2, whose production exhibits a pronounced maximum at 260 K, results from reaction of CO with co-adsorbed oxygen adatoms and is thus highly sensitive to the local environment.[18] The ratio of the two components is strongly dependent on both the O_2 and CO coverage.

In pioneering studies of their type, reaction (1) has been studied by Heiz et al. on monodisperse platinum clusters deposited on MgO(100) films in the size regime Pt_n ($n \leq 20$).[19] Strong variations in the number of CO_2 molecules produced per platinum atom were observed as a function of cluster size. In particular, step increases in efficiency around $n = 8$ and $n = 15$ were interpreted as resulting from the change from pseudo-two to three–dimensional structures.

On isolated gas-phase clusters, Shi and Ervin were the first to establish a full catalytic cycle involving the oxidation of CO on Pt_n^- ($n = 2$–6) anionic clusters under thermal conditions.[20] Using guided ion beam techniques, Pt_n^- clusters were oxidised by collision with O_2 or N_2O to produce $Pt_nO_m^-$ and subsequently reacted with CO to regenerate the original clusters. The same N_2O/CO scheme was used by Balaj *et al.* in Fourier transform ion cyclotron resonance (ICR) experiments to study reactions of cationic clusters, Pt_n^+ ($n = 6$–8).[21] The long storage times in the ICR experiments permitted the establishment of steady state conditions which were poisoned by a high surface CO coverage.

Our approach represents something of a hybrid method for studying reactions on clusters. First we decorate isolated clusters with reactive precursors (O_2 and CO) under thermal conditions (~300 K), allowing the stabilisation of intermediate species not usually observed under single collision reaction conditions. Then, *via* IR pumping of an allowed vibrational transition, we raise the internal energy of the cluster complex to initiate surface processes such as desorption, reaction-desorption, *etc.* Simultaneous mass spectrometric monitoring of all mass channels permits depletions in one channel to be matched with enhancements in others providing a detailed picture of any reactive dynamics resulting in moiety loss from the cluster.

2 Experimental

All experiments were carried out at the Free Electron Laser for Infrared eXperiments (FELIX)[22] facility in the Netherlands. The basic experimental setup has been described previously in detail[10,23] and is shown schematically in Fig. 1. The essential features are as follows. $^{194}Pt_n^+$ cluster cations are generated *via* 532 nm pulsed laser ablation of an isotopically enriched ^{194}Pt foil wrapped around a target rod. The ablation products are entrained within a short, intense (8 bar backing pressure) pulse of helium gas seeded with argon (~1%). Cooling and clustering occurs as the gas pulse travels down a narrow channel held at ambient temperature.

In a development from previous experiments, two pulsed valves allow controlled injection of molecular oxygen, O_2 (typical backing pressure 500 mbar), and/or carbon monoxide, CO (1.2 bar), respectively, into the reaction channel. These gases can react with the preformed clusters generating a range of $Pt_nO_mCO^+$ species allowing the user a degree of control over the product distribution *via* the partial pressures of the respective reactants. This approach was found to provide more flexibility than could be obtained with a single valve and premixed O_2/CO.

After exiting the cluster channel, the molecular beam passes through a 2 mm diameter skimmer and a 1 mm diameter aperture into the extraction region of a reflectron time-of-flight mass spectrometer. Here it is met by the counter-propagating

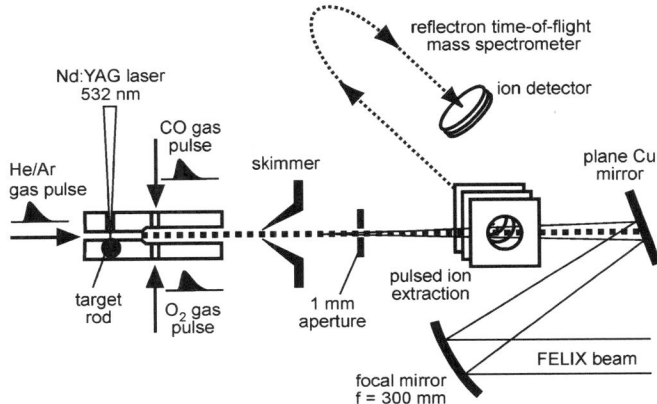

Fig. 1 Scheme of the experiment used for IR multiple photon dissociation of metal-cluster rare-gas complexes.

IR beam delivered by FELIX (*ca.* 20 mJ/pulse) which is loosely focused onto the aperture and synchronized with the molecular beam ensuring good spatial and temporal overlap of the IR and cluster beams.

Mass spectra are collected at a rate of 10 Hz with the FELIX beam enabled on alternate shots to collect 'on' and 'off' spectra. From these, a difference mass spectrum is obtained at each wavelength. This difference, as a function of wavelength, yields the IR-MPD spectrum for each point in the mass spectrum. Depletion of a parent ion peak is indicative of a resonant loss process, whose nature can be determined by looking for concomitant enhancements in product species. Similarly, the relative cross sections can be calculated from the relative depletion and normalization by the photon fluence.

All IR spectra are recorded at 5 cm^{-1} intervals and several scans are averaged to increase the spectral signal to noise ratio. The IR-MPD spectra cover the range 400–2200 cm^{-1} encompassing both metal oxide (Pt–O) vibrations as well as the CO stretching region. The reactivity studies concentrated primarily on processes induced following absorption in the CO stretch region around 2100 cm^{-1}.

Two conceptually different experiments are performed: Firstly, conventional IR-MPD spectra of Ar-tagged clusters (which are formed efficiently even at 300K), in tandem with quantum-chemical spectral simulations, provide important information on the cluster structures. In this case, whenever IR radiation is absorbed by the cluster, rapid intramolecular redistribution of energy leads to heating of the cluster and the subsequent desorption of the physisorbed Ar atom:

$$Pt_nO_mCOAr^+ + IR \rightarrow Pt_nO_mCO^+ + Ar \qquad (2)$$

Secondly, infrared pumping is used to heat the clusters in order to trigger chemical reaction between moieties co-adsorbed on the cluster surface:

$$Pt_nO_mCO^+ + IR \rightarrow Pt_nO_{m-1}^+ + CO_2 \qquad (3)$$

In these experiments, there is deliberately no mass selection in the cluster source. This permits the spectra of all species generated in the beam to be recorded simultaneously. However, it does present potential problems in correlating unambiguously enhancements in one channel with depletions elsewhere. To this end, and in order to maximise our chances of observing IR-induced CO oxidation reactions, O_2 was used as the source of oxygen atoms. In this way complexes with *even* numbers of O atoms *i.e.*, $Pt_nO_m^+$ ($m = 2, 4,...$) and $Pt_nO_mCO^+$ ($m = 2, 4,...$) are generated preferentially. The products of any CO oxidation reaction (3) are consequently observed as enhancements in the *odd* (*i.e.*, $m - 1$) oxide mass channels against a nominally zero background.

3 Computational

In support of the experiments described above we have undertaken a computational study of low-lying $Pt_nO_2CO^+$ ($n = 3$–5) structures and plausible CO oxidation pathways on these clusters.

All calculations were carried out using Density Functional Theory (DFT) as implemented in the Turbomole 6.0 package.[24] In common with our previous findings, no single exchange–correlation functional provides the best performance for every type of calculation that is required.[14] However, the TPSS functional[25,26] has previously yielded good agreement with experiments of this type[27] (including Pt_n^+ spectra)[28] and provides a good compromise in performance for each type of calculation. In our previous work with high-spin rhodium clusters, we used the hybrid variant of this functional (TPSSh).[26] However, in the current work, in which low spin states dominate, we find both functionals favour the same low-energy structures, including the putative global minimum, with only small differences in the relative energies. As a result, for the current study, the TPSS functional has been used throughout. Triple-ζ valence basis sets (def2-TZVP)[29,30] were used for all atoms, with def2–ecp effective core potentials representing the core electrons of the platinum atoms.

Initial structures for the clusters were derived as follows: A range of $Pt_nO_2^+$ structures was generated *via* a combination of DFT-based basin hopping[31] and by addition of O_2 to the experimentally determined Pt_n^+ structures. CO was then added to these structures in a variety of likely binding geometries to provide the starting points of the calculations. A range of spin states were explored for each cluster and normal mode analyses confirmed the identity of minima or transition states as well as providing the spectral simulations. In all of the low-lying geometries calculated, the doublet spin state ($2S + 1 = 2$) is predicted to lie significantly lower in energy than higher spin states. As a result, all of the computational data presented henceforth refer to the doublet unless stated otherwise. Spin–orbit coupling was not included in the calculations here. The results of previous computational studies which have included spin–orbit coupling are inconclusive with several reporting no effect on the relative isomer ordering.[32]

For ease of comparison with the experimental data, the calculated stick spectra are broadened with a Gaussian line shape (18 cm^{-1} line width).

4 Results and discussion

4.1 IR-MPD spectroscopy of $Pt_nO_mCO^+$

A typical mass spectrum recorded in the region of the Pt_4^+ cluster is shown in Fig. 2 illustrating the range of complexes produced in the source. The dominant species present, as designed, are complexes with a single CO and/or O_2 molecule (note that the mass spectrum yields no conclusive information regarding the molecular/dissociative nature of the adsorption). Although there is evidence of monoxide clusters, these have low intensity compared with the even oxides. All intense peaks in the mass spectrum have an Ar-tagged counterpart 40 u higher in mass resulting from the 1% Ar used in the carrier gas. These are used for the IR-MPD spectra in order to gain structural information. Interestingly, secondary clustering (with O_2, CO) was

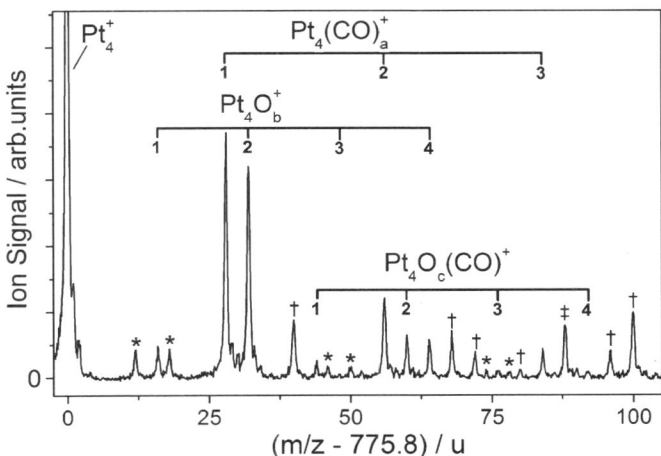

Fig. 2 Time-of-flight mass spectrum recorded in the region of Pt_4^+ complexes in the absence of IR radiation. (*) Minor Pt_4X^+ species (X = N, H_2O, *etc.*). † Ar tagged species. (‡) $Pt_4O_2(CO)_2$. The spectrum is dominated by species with even numbers of oxygen atoms as expected when using molecular O_2 as the oxygen source.

observed to be significantly less efficient without the presence of the Ar suggesting that it plays a role in the clustering process itself.

Fig. 3 (a) shows the full IR-MPD spectrum of the $Pt_4O_2COAr^+$ cluster from 400–2200 cm^{-1}. This region encompasses the fundamental vibrational transitions from Pt–O stretches/bends (400–700 cm^{-1}) to the CO stretch (2125 cm^{-1}) and a number of strong, well-resolved, bands are observed. Aid in the assignment of the spectrum comes from the simulated spectra provided by the DFT calculations. The simulated IR spectrum for the lowest energy structure obtained (the putative global minimum) is shown in Fig. 3 (b). This is a buckled two dimensional Pt_4 structure with the CO bound atop and with the oxygen dissociatively adsorbed in bridge sites. The calculated spectrum shows excellent agreement with the experimental spectrum below 750 cm^{-1} and the CO stretching band observed at 2125 cm^{-1} is similarly well–reproduced.

One important spectral feature (*ca.* 940 cm^{-1}) cannot, however, be explained by the lowest energy calculated structure. On the basis of comparison with simulated spectra of calculated structures two possible carriers for this band have been identified. One involves isomers with atop bound oxygen atoms. These typically lie >1 eV above the putative global minimum structure and have calculated Pt–O stretches up to 900 cm^{-1}. They also have additional features in the low frequency range (400–600 cm^{-1}) for which, however, there is little evidence in the experimental spectrum. An intriguing alternative, whose agreement with the experimental spectrum is marginally better, is the presence of peroxo-type isomers of which one example is shown in Fig. 3 (c). The structure shown in Fig. 3 (c) has been selected for its match with the experimental band but is not the lowest energy peroxo-structure calculated (see ESI†). Given the potential need to scale the calculated frequencies, as well as possible spectral shifts introduced by the presence of co-adsorbed Ar atoms, it is impossible to assign this single spectral feature to a particular peroxo–structure. All such peroxo–structures, however, are calculated to lie >1.4 eV above the lowest energy isomer (see ESI†) and have peroxo bands in the range 900 ± 50 cm^{-1}. It is noteworthy that the spectral position of the peroxo band is unusually sensitive to the exchange correlation functional used in the simulation, appearing, on average,

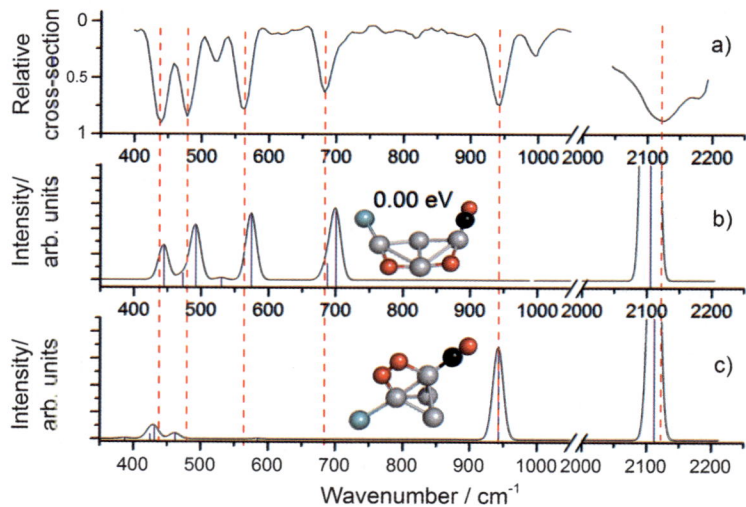

Fig. 3 (a) Experimental IR–MPD spectrum of $Pt_4O_2COAr^+$ (note the spectrum is shown inverted, *i.e.*, cross section increasing down to reflect the fact that this spectrum is recorded by observing the depletion of $Pt_4O_2COAr^+$). (b) Calculated spectrum (TPSS/def2-TZVP) of the calculated lowest lying isomer identified. (c) Calculated spectrum of a low-lying peroxo-type isomer (the Ar tagged structure 12, see ESI†).

100 cm^{-1} further to the blue in calculations using TPSSh compared to the TPSS spectrum shown in Fig. 3.

On extended platinum surfaces, temperature programmed desorption experiments provide good evidence of both molecular and dissociative binding of O_2, showing O_2 evolution at between 150–200 K (molecularly adsorbed) and >700 K (dissociatively adsorbed).[33]

Whether the band at 940 cm^{-1} is assigned to atop–bound oxygen atoms or peroxo–structures, in order to account for all features in the observed spectrum it is necessary to invoke the presence of multiple isomers within our molecular beam. The possible presence of the peroxo–structures is particularly unexpected given their calculated high energy, especially given the apparent absence of other lower-lying isomeric forms. It is, however, possible that the peroxo–structures become trapped in a deep local minimum during the clustering process, perhaps as a result of boiling off pre-adsorbed Ar atoms. This raises the possibility that the structures of the Ar-tagged clusters are not representative of the same species without the Ar present.

Ar–tagging is not necessary in recording IR–MPD spectra in the CO stretching region, with CO loss itself being an effective signature of IR absorption. Fig. 4 shows the CO–loss IR-MPD spectrum of $Pt_nO_mCO^+$ clusters recorded in the region 2050–2200 cm^{-1}. The observed peak centres deduced from attenuated spectra are presented in Table 1. Two trends are clear; firstly, for otherwise identical species, there is a small red-shift in the CO stretch, ν_{co}, with increasing number of platinum atoms, n. This has been observed previously for CO binding on bare clusters by Gruene et al.[34] and can be understood in terms of a simple back-bonding model of CO chemisorption in which additional electron density increasingly weakens the CO bond by

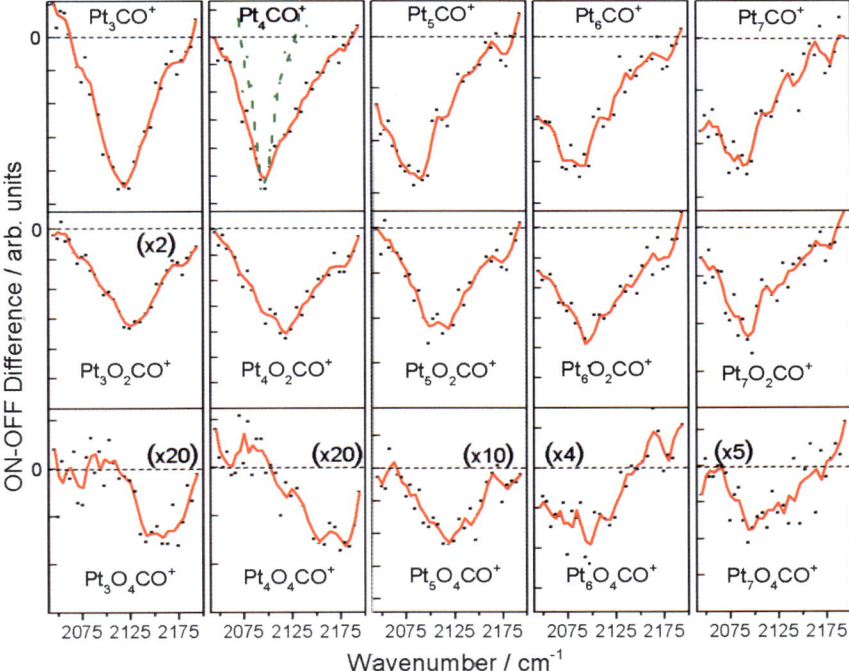

Fig. 4 IR–MPD spectra of $Pt_nO_mCO^+$ ($n = 3$–7, $m = 0, 2, 4$) clusters in the CO stretch region (black dots: raw data, red line: 5 point adjacent average). The green dashed line shows the spectrum recorded for Pt_4CO^+ with 5 dB attenuation of the IR beam illustrating the degree of saturation achieved. Note the slight red shift of the ν_{CO} peak as the number of platinum atoms increases and the blue-shift with increasing oxygen coverage.

Table 1 CO stretching band centres for $Pt_nO_mCO^+$ ($n = 3$–7, $m = 0, 2, 4$) extracted from Gaussian fits of the IR-MPD spectra. Where the signal to noise was acceptable ($n = 4$–6) the IR-attenuated data was used otherwise the full power spectra were used. For species common to both studies, these values are in good agreement with those of the earlier study by Gruene et al.[34]

n	m	Peak centre, ν_{CO}/cm^{-1}	Band shift $\Delta\nu/cm^{-1}$
3	0	2120 ± 5^a	0
3	2	2127 ± 8^a	7
3	4	2156 ± 8^a	46
4	0	2097 ± 2.5	0
4	2	2125 ± 2.5	28
4	4	2164 ± 5	67
5	0	2080 ± 2.5	0
5	2	2112 ± 5	32
5	4	2118 ± 5	37
6	0	2078 ± 5	0
6	2	2092 ± 5	14
6	4	2087 ± 10^a	9
7	0	2072 ± 8^a	0
7	2	2090 ± 5^a	18
7	4	2109 ± 10^a	37

^{a} Full power spectrum used.

π–backdonation. By way of comparison, the vibrational frequency of CO atop–bound on extended Pt(111) surfaces is 2095 ± 10 cm^{-1}.[35,36] Secondly, there is a marked blue-shift in ν_{co} with increasing oxygen coverage (i.e., down Fig. 4). This effect is largest on the very smallest clusters. This, too can be understood qualitatively on the basis of back-bonding with oxygen adatoms locally fixing electron density and reducing the scope for π–backdonation to CO. A similar effect was observed for H-atoms co-adsorbed with CO on cobalt cluster cations.[37]

4.2 IR-driven CO oxidation reactions

While spectral characterisation is performed using the Ar messenger with attenuated IR intensity as described above, in other experiments the $Pt_nO_mCO^+$ clusters were subjected to intense infrared excitation via the CO stretch in order to initiate surface reaction processes and/or desorption of the CO. The results for a typical cluster, $Pt_4O_2CO^+$, are shown in Fig. 5. The top spectrum shows the depletion in the parent ion ($Pt_4O_2CO^+$) signal as a function of excitation wavenumber. This spectrum has been corrected for the component which depletes into this mass channel, e.g., Ar loss from $Pt_4O_2COAr^+$, and CO loss from $Pt_4O_2(CO)_z$ ($z \geq 2$). The loss in the parent signal is mirrored by a concomitant increase in the signal observed in a variety of lower mass channels. Perhaps unsurprisingly, the largest of the enhancements observed in the same spectral region is in the $Pt_4O_2^+$ channel and corresponds to IR–MPD resulting in simple CO loss:

$$Pt_4O_2CO^+ + IR\ (\sim 2120\ cm^{-1}) \rightarrow Pt_4O_2^+ + CO. \tag{4}$$

More interestingly from a reactivity perspective, the same spectral feature is also observed in the Pt_4O^+ mass channel. In principle, this signal could arise from CO loss from the monoxide cluster,

$$Pt_4OCO^+ + IR\ (\sim 2120\ cm^{-1}) \rightarrow Pt_4O^+ + CO. \tag{5}$$

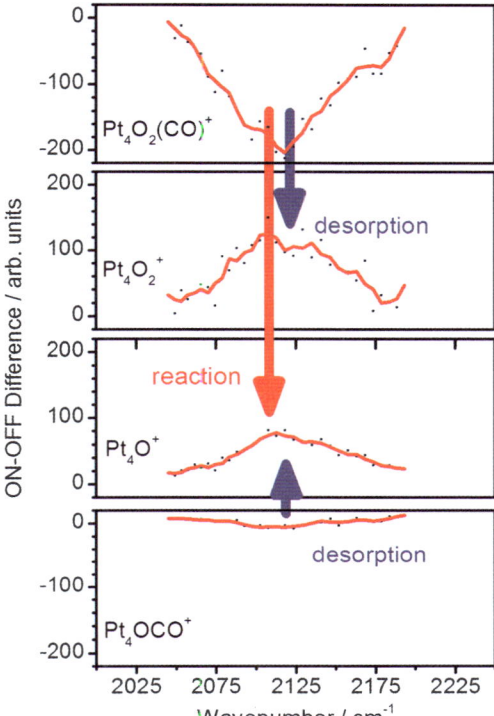

Fig. 5 IR-MPD spectrum of $Pt_4O_2CO^+$ showing a strong signal depletion upon excitation in the region of the CO stretching frequency at 2120 cm^{-1} (black dots: raw data, red line: 5 point adjacent average). Accompanying the above depletion, enhancements are observed at the same spectral position in the $Pt_4O_2^+$ and Pt_4O^+ mass channels. The depletion in the Pt_4OCO^+ channel (bottom), whilst observed, is much too small to account for the increase in Pt_4O^+ by CO loss.

However, as the intensity difference spectrum at the bottom of Fig. 5 shows, the depletion in the Pt_4OCO^+ channel is much too small to account for the increase in the Pt_4O^+ signal observed. Indeed, the observed Pt_4O^+ enhancement is larger than the total Pt_4OCO^+ signal (see Fig. 2) due to the use of O_2 in the production of the clusters.

Instead, the Pt_4O^+ signal is interpreted as the signature of cluster surface reactivity induced by infrared heating of the parent $Pt_4O_2CO^+$ cluster complex, in this case CO oxidation generating CO_2:

$$Pt_4O_2CO^+ + IR\ (\sim 2120\ cm^{-1}) \rightarrow Pt_4O^+ + CO_2. \tag{6}$$

This is an important finding, as it indicates that infrared driven cluster chemistry is more general than the surface decomposition we have studied previously.[13,14]

In these experiments, it is possible to quantify the change in signal intensity (*i.e.*, the depletions and enhancements) in each channel and, within mutual experimental uncertainties, the sum of the $Pt_4O_2^+$ and Pt_4O^+ enhancements is equal to the total depletion observed in the parent ion channel. This indicates that CO loss and the $CO + O \rightarrow CO_2$ reaction are the only significant processes which result in parent ion signal loss.

Similar CO oxidation reactions are observed on all platinum cluster sizes studied here (*i.e.*, $Pt_nO_2CO^+$, $n = 3–7$) as shown in Fig. 6. By comparing (integrated) intensity differences in the different product channels across the spectral region explored it is possible to extract representative branching ratios for the reaction *vs.* CO desorption channels and these are shown in Table 2. These branching ratios show little size dependence in the cluster range studied here with the reactive channel accounting for approximately 35–45% of the parent signal loss. It would appear that there are no inert clusters in this size range, nor any with anomalously high reactivity. It is difficult to compare these numbers directly with the kinetic studies on isolated clusters such as those of Shi *et al.*[20] and Balaj *et al.*[21] as mass spectrometric collision experiments are, by their nature, insensitive to ineffective collisions (*e.g.*, $CO + Pt_n^- \rightarrow Pt_n^- + CO$). However, Balaj *et al.* noted that, unlike $n = 6$–8, Pt_5^+ failed to establish an efficient catalytic cycle due to CO binding unreactively to the oxide and dioxide clusters.

The experimental data regarding the CO oxidation reaction following pumping of the peroxo band at 940 cm^{-1} are inconclusive. Any depletion in the $Pt_4O_2CO^+$ signal that is observed in this spectral region is, unfortunately, obscured by a larger depletion into this mass channel by Ar loss from $Pt_4O_2COAr^+$ (see Fig. 3). In addition, there is no evidence of enhancement in this region in the Pt_4O^+ channel which would signify CO_2 loss. The spectrum in the CO loss channel is further obscured by the fact that $Pt_4O_2^+$ and its Ar– tagged variant have strong absorptions of their own in this region.

4.3 Calculated reaction pathways

We have undertaken preliminary calculations in an attempt to identify a plausible mechanism for the CO oxidation reaction on the $Pt_4O_2CO^+$ cluster. For the example of the lowest energy structure, key stationary points in the pathway are shown in Fig. 7. The pathways were identified by identifying likely transition states (in this

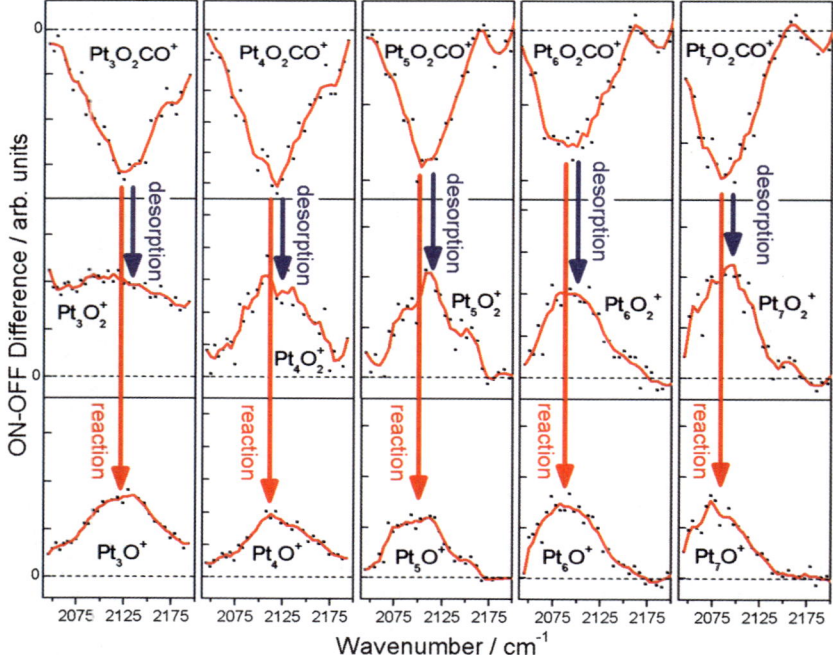

Fig. 6 Infrared driven CO oxidation $Pt_nO_2CO^+$ ($n = 3$–7) following excitation in the CO stretch region around 2100 cm^{-1} (black dots: raw data, red line: 5 point adjacent average). Both CO desorption (leading to $Pt_nO_2^+$) and CO oxidation (generating Pt_nO^+) represent important channels across the range of cluster size studied.

Table 2 Branching ratios for the CO oxidation reaction and CO loss following IR heating of $Pt_nO_2CO^+$ clusters

n	Spectral range/cm^{-1}	% Reaction[a]	% Desorption[a]
3	2088–2169	43	57
4	2083–2153	38	62
5	2079–2134	40	60
6	2055–2134	46	54
7	2055–2123	38	62

[a] Estimated uncertainty = ±5%.

case by analogy with those calculated for the same reaction on extended surfaces by Eichler[38]) and then using the eigenvector following method[39] to link these with local minima.

These calculations proved considerably more challenging than similar pathways for the unimolecular decomposition of N_2O on rhodium clusters which we have studied previously. In the present case the metal atom framework is much more flexible than was the case for rhodium and the platinum atoms are more intimately involved in the reaction. Indeed, the metal atom framework itself is highly fluxional and we have identified several mechanisms involving apparent migration of Pt atoms. Furthermore, the addition of even a single additional oxygen atom to a bare Pt_n^+ cluster is enough to change its structure markedly. We calculate the CO binding energy to the lowest energy structure to be 2.71 eV which thus represents the minimum energy required to desorb the CO. This compares well with the CO binding to Pt_n^- clusters experimentally determined by Shi *et al.* to be 220–250 kJ mol^{-1} (2.3–2.6 eV) by collisional induced dissociation and the calculated ~2.6 eV binding energy on neutral Pt_4 determined by Xu *et al.*[40] The presence of

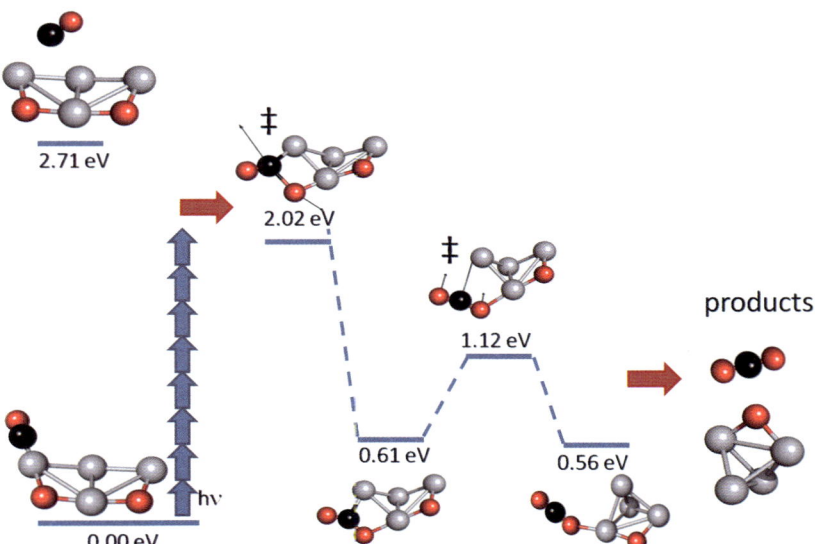

Fig. 7 Key stationary points calculated on the CO oxidation reaction pathway on $Pt_4O_2CO^+$ heated by infrared multiple-photon excitation *via* the CO stretching vibration. Structures shown result from vector following from the transition states and do not necessarily represent the lowest energy structures.[40] The arrows shown represent the approximate energy of each IR photon.

oxygen adatoms, can, of course, profoundly affect the binding energies of co-adsorbates and we recently demonstrated such an effect in infrared driven reactivity.[15]

The activation barrier for $CO_{(ads)} + O_{(ads)} \rightarrow CO_{2(g)}$ on Pt(111) extended surfaces is strongly coverage-dependent[41] but at low coverage has variously been determined to be from 100 kJ mol^{-1} (1 eV) by molecular beam scattering,[42] to 166 kJ mol^{-1} (1.7 eV) by temperature programmed desorption.[43] In this context the reaction barrier in Fig. 7 is markedly higher than might be expected. It is also significantly higher than the 0.74 eV barrier identified in the DFT study on extended Pt(111).[38] These reaction barriers, however, and indeed the geometries of the transition states themselves, are highly dependent on the exchange–correlation functional used in the calculations.

5 Summary and conclusions

Infrared multiple photon absorption spectroscopy has been applied to CO decorated platinum cluster oxides in the gas-phase. Well-resolved vibrational structure in both the Pt–O stretch and CO stretch regions permit structural assignments by comparison with DFT simulations. Additional features around 940 cm^{-1} provide evidence of higher-lying isomeric forms, possibly peroxo-structures.

In a significant extension of the technique over the unimolecular decompositions to which it has been applied to date, efficient CO oxidation has been observed following IR excitation of $Pt_nO_2CO^+$ in the region of the CO stretch. Preliminary calculations of the reaction pathway have identified plausible transition states below the CO desorption energy consistent with the experimental findings. These observations lend further support to the thermal nature of the reactivity observed and offer the prospect of studying important catalytic reactions as a function of both cluster size and isomeric structure. As we have shown previously, this cluster analogue of temperature programmed reactions is exquisitely sensitive to the relative barriers to reaction and/or desorption.

Acknowledgements

We gratefully acknowledge the support of the Stichting voor Fundamenteel Onderzoek der Materie (FOM) for providing FELIX beam time and the FELIX staff, in particular Dr B. Redlich and Dr A.F.G. van der Meer, for their skilful assistance. This work is supported by the Deutsche Forschungsgemeinschaft through research grant FI893/3-1. DJH thanks the Alexander-von-Humboldt-Stiftung for support. ACH and GAC are grateful to the EPSRC for support. We are grateful to the Oxford Supercomputing Centre and the EPSRC UK National Service for Computational Chemistry Software (NSCCS) at Imperial College London for the use of their computing resources in performing the computational aspects of this work.

References

1 G. A. Somorjai, *Chemistry in two dimensions: surfaces*, Cornell University Press, Ithaca, 1981.
2 A. Cho, *Science*, 2003, **299**, 1684–1685.
3 P. B. Armentrout, *Annu. Rev. Phys. Chem.*, 2001, **52**, 423–461.
4 P. B. Armentrout, *Eur. J. Mass Spectrom.*, 2003, **9**, 531–538.
5 M. B. Knickelbein, *Annu. Rev. Phys. Chem.*, 1999, **50**, 79–115.
6 M. Okumura, L. I. Yeh, J. D. Myers and Y. T. Lee, *J. Chem. Phys.*, 1986, **85**, 2328–2329.
7 G. Gregoire and M. A. Duncan, *J. Chem. Phys.*, 2002, **117**, 2120–2130.
8 A. Fielicke, G. von Helden, G. Meijer, D. B. Pedersen, B. Simard and D. M. Rayner, *J. Am. Chem. Soc.*, 2005, **127**, 8416–8423.
9 P. Gruene, D. M. Rayner, B. Redlich, A. F. G. van der Meer, J. T. Lyon, G. Meijer and A. Fielicke, *Science*, 2008, **321**, 674–676.
10 A. Fielicke, G. von Helden and G. Meijer, *Eur. Phys. J. D*, 2005, **34**, 83–88.

11 D. J. Harding, T. R. Walsh, S. M. Hamilton, W. S. Hopkins, S. R. Mackenzie, P. Gruene, M. Haertelt, G. Meijer and A. Fielicke, *J. Chem. Phys.*, 2010, **132**, 011101.

12 D. J. Harding, P. Gruene, M. Haertelt, G. Meijer, A. Fielicke, S. M. Hamilton, W. S. Hopkins, S. R. Mackenzie, S. P. Neville and T. R. Walsh, *J. Chem. Phys.*, 2010, **133**, 214304.

13 S. M. Hamilton, W. S. Hopkins, D. J. Harding, T. R. Walsh, P. Gruene, M. Haertelt, A. Fielicke, G. Meijer and S. R. Mackenzie, *J. Am. Chem. Soc.*, 2010, **132**, 1448–1449.

14 S. M. Hamilton, W. S. Hopkins, D. J. Harding, T. R. Walsh, M. Haertelt, C. Kerpal, P. Gruene, G. Meijer, A. Fielicke and S. R. Mackenzie, *J. Phys. Chem. A*, 2011, **115**, 2489–2497.

15 A. C. Hermes, S. M. Hamilton, W. S. Hopkins, D. J. Harding, C. Kerpal, G. Meijer, A. Fielicke and S. R. Mackenzie, *J. Phys. Chem. Lett.*, 2011, **2**, 3053–3057.

16 D. J. Harding, C. Kerpal, G. Meijer and A. Fielicke, *Angew. Chem., Int. Ed.*, 2012, **51**, 817–819.

17 G. Ertl, *Angew. Chem., Int. Ed.*, 2008, **47**, 3524–3535.

18 T. Matsushima, *Surf. Sci.*, 1983, **127**, 403–423.

19 U. Heiz, A. Sanchez, S. Abbet and W. D. Schneider, *J. Am. Chem. Soc.*, 1999, **121**, 3214–3217.

20 Y. Shi and K. M. Ervin, *J. Chem. Phys.*, 1998, **108**, 1757–1760.

21 O. P. Balaj, I. Balteanu, T. T. J. Rossteuscher, M. K. Beyer and V. E. Bondybey, *Angew. Chem., Int. Ed.*, 2004, **43**, 6519–6522.

22 D. Oepts, A. F. G. van der Meer and P. W. van Amersfoort, *Infrared Phys. Technol.*, 1995, **36**, 297–308.

23 A. Fielicke, A. Kirilyuk, C. Ratsch, J. Behler, M. Scheffler, G. von Helden and G. Meijer, *Phys. Rev. Lett.*, 2004, **93**, 023401.

24 Turbomole v6.0 2009, a development of University of Karlsruhe and Forschungszentrum Karlsruhe GmbH, 1989–2007, TURBOMOLE GmbH, since 2007; available from http://www.turbomole.com, (2009).

25 J. M. Tao, J. P. Perdew, V. N. Staroverov and G. E. Scuseria, *Phys. Rev. Lett.*, 2003, **91**, 146401.

26 V. N. Staroverov, G. E. Scuseria, J. M. Tao and J. P. Perdew, *J. Chem. Phys.*, 2003, **119**, 12129–12137.

27 A. Fielicke, P. Gruene, M. Haertelt, D. J. Harding and G. Meijer, *J. Phys. Chem. A*, 2010, **114**, 9755–9761.

28 D. J. Harding, C. Kerpal, D. M. Rayner and A. Fielicke, *J. Chem. Phys.*, 2012, **136**, 211103.

29 F. Weigend, M. Haser, H. Patzelt and R. Ahlrichs, *Chem. Phys. Lett.*, 1998, **294**, 143–152.

30 F. Weigend and R. Ahlrichs, *Phys. Chem. Chem. Phys.*, 2005, **7**, 3297–3305.

31 D. J. Wales and J. P. K. Doye, *J. Phys. Chem. A*, 1997, **101**, 5111–5116.

32 L. Xiao and L. C. Wang, *J. Phys. Chem. A*, 2004, **108**, 8605–8614.

33 Y. Ohno and T. Matsushima, *Surf. Sci.*, 1991, **241**, 47–53.

34 P. Gruene, A. Fielicke, G. Meijer and D. M. Rayner, *Phys. Chem. Chem. Phys.*, 2008, **10**, 6144–6149.

35 J. Yoshinobu and M. Kawai, *Surf. Sci.*, 1996, **363**, 105–111.

36 J. V. Nekrylova and I. Harrison, *Chem. Phys.*, 1996, **205**, 37–46.

37 I. Swart, A. Fielicke, D. M. Rayner, G. Meijer, B. M. Weckhuysen and F. M. F. de Groot, *Angew. Chem., Int. Ed.*, 2007, **46**, 5317–5320.

38 A. Eichler, *Surf. Sci.*, 2002, **498**, 314–320.

39 J. Simons, P. Jorgensen, H. Taylor and J. Ozment, *J. Phys. Chem.*, 1983, **87**, 2745–2753.

40 Y. Xu, R. B. Getman, W. A. Shelton and W. F. Schneider, *Phys. Chem. Chem. Phys.*, 2008, **10**, 6009–6018.

41 J. Wintterlin, S. Volkening, T. V. W. Janssens, T. Zambelli and G. Ertl, *Science*, 1997, **278**, 1931–1934.

42 C. T. Campbell, G. Ertl, H. Kuipers and J. Segner, *J. Chem. Phys.*, 1980, **73**, 5862–5873.

43 J. L. Gland and E. B. Kollin, *J. Chem. Phys.*, 1983, **78**, 963–974.

Product state and speed distributions in photochemical triple fragmentations

G. de Wit, B. R. Heazlewood, M. S. Quinn, A. T. Maccarone,†
K. Nauta, S. A. Reid,‡ M. J. T. Jordan* and S. H. Kable*

Received 8th February 2012, Accepted 17th April 2012
DOI: 10.1039/c2fd20015e

The clearest dynamical signature of a roaming reaction is a very cold distribution of energy into the rotational and translational degrees of freedom of the roaming donor fragment (e.g. CO) and an exceptionally hot vibrational distribution in the roaming acceptor fragment (e.g. H_2, CH_4). These signatures were initially identified in joint experimental/theoretical investigations of roaming in H_2CO and CH_3CHO and are now used to infer the presence of roaming mechanisms in other photodissociation reactions. In this paper we construct a phase space theory (PST) model of triple fragmentation (3F) and show that the dynamical signature of 3F is similar to that of the roaming donor fragment. The PST model starts with a calculation of two-body fragmentation (2F) of a generic molecule, ABC into AB + C. Every AB fragment with sufficient energy to undergo subsequence spontaneous dissociation is allowed to dissociate and the PST distribution of energy into A + B products is calculated for every initial AB state. Using $CH_3CHO \rightarrow HCO + CH_3 \rightarrow H + CO + CH_3$ as an example, we calculate that the energy disposal into the rotational and translational degrees of freedom of the 3F products is very low, and is similar to the dynamical signature expected for production of CO via a roaming mechanism. We compare the 3F PST model with published experimental data for photodissociation of CH_3CHO and CH_3OCHO at energies above the 3F threshold.

I. Introduction

Since the first report of a "roaming" reaction in 2004[1] there has been a flurry of experiments and theory concerning this mechanism. Roaming is associated with a simple bond fission reaction, for example, a molecule dissociating to two radical fragments. If the pair of incipient fragments have enough translational energy to reach the van der Waals (vdW) region of configuration space, but not sufficient to reach the final product asymptote, they can "roam" in the vdW region, accessing regions of configuration space that can be quite removed from the original structure.[2,3] From these distant configurations the system can either reform the original reactant molecule, or a chemical reaction can occur in the vdW region. The term "roaming reaction" has been used to describe a self-abstraction reaction in the vdW region that leads to molecular rather than radical products. Indeed a consensus

School of Chemistry, University of Sydney, Sydney, NSW, 2006, Australia. E-mail: scott. kable@sydney.edu.au; m.jordan@chem.usyd.edu.au

† Present address: Department of Chemistry, University of Wollongong, Wollongong, NSW, 2522, Australia
‡ Permanent address: Department of Chemistry, Marquette University, Milwaukee, WI, 53233, USA.

is emerging that whenever there is a simple, barrierless, bond-cleavage reaction, there will be a concomitant roaming process.[2,3]

The earliest reports of roaming reactions, in formaldehyde, H_2CO,[1,4–6] and acetaldehyde, CH_3CHO,[7–9] were accompanied by extensive theoretical calculations to support and explain the experimental findings. In these two simplest aldehydes, there is a "conventional" reaction pathway *via* a tight transition state (TS), as well as a "roaming pathway" to produce the molecular products RH + CO (R = H, CH_3). The TS path has a significant barrier with an early, three-centre TS. Products that form *via* this mechanism recoil down the very high exit channel barrier, resulting in substantial translational energy release. The roaming products, however, are born at large separation producing vibrationally-hot RH, much less translational and rotational energy. Indeed, it was the observation of different rotational distributions in H_2CO photolysis that provided the first tantalizing glimpse of roaming.[10] However, it was the power of imaging experiments on the CO fragment that showed not only the two different translational energy distributions, but also yielded correlated information about the H_2 co-fragment.[1,4–6] Quasi-classical trajectory (QCT) calculations on a global H_2CO potential energy surface (PES) provided the explanation of the experimental distributions and led to our current concept of roaming.[1,11,12]

Later experiments on CH_3CHO, again accompanied by QCT calculations, showed similar characteristics in the roaming products – considerable vibrational excitation in CH_4, with much less rotational and translational energy in the CO co-fragment.[7–9] These experiments also showed that there was no correlation between the CO velocity and angular momentum vectors (**v**–**J** correlation), while CO formed *via* the TS mechanism showed a preference for **v** and **J** to be perpendicular.[7] This experimental finding was supported by QCT theory,[9] which demonstrated that TS trajectories were strongly influenced by the planar nature of the three-centred TS, leading to the **v**–**J** correlation, while the roaming trajectories sampled a wide variety of angles, which destroyed any such correlation. The dynamical signatures of the roaming mechanism in H_2CO and CH_3CHO dissociation are thus (i) rotationally and translationally cold CO, (ii) vibrationally hot but translationally cold RH and (iii) no CO **v**–**J** correlation.

Since these early studies, roaming has been established or inferred, experimentally, in a variety of molecular photodissociations, including $(CH_3)_2CO$,[13] CH_3OOCH_3,[14] CH_3OCHO[15] and NO_3.[16,17] Roaming has also been identified in the photoisomerisation of $C_6H_5NO_2$.[18] Very recently, roaming has been demonstrated experimentally and theoretically to be present in photodissociation of NO_3 on both ground and excited states with roaming implicated in the non-radiative curve-crossing between these states.[19]

The onus of proof for roaming in H_2CO and CH_3CHO was held very high because roaming was an unanticipated, and perhaps controversial, mechanism. Experiments were accompanied by a significant theoretical effort, and the photon energy was kept low to limit the access to competing photochemical pathways. As roaming becomes a more accepted mechanism in photodissociation dynamics and its characteristics better understood, the burden of proof is becoming more relaxed. If the dynamical signatures of roaming, learned from H_2CO and CH_3CHO, prove to be both reliable and unique, then this will lead to more rapid experimental recognition of roaming in a wider variety of systems and environments.

The aim of this paper is to explore the uniqueness of the roaming dynamical signatures, particularly the low velocity and rotational angular momentum that characterised the roaming CO distribution in the two smallest aldehydes. In particular, we use phase space theory (PST)[20–22] to explore these distributions when there is sufficient photon energy to produce three fragments; so-called triple fragmentation (3F). Previous photodissociation experiments on CH_3CHO and CH_3OCHO, at sufficient energy for 3F,[15,23–25] have implicated roaming, but the interpretation has not yet been supported by QCT calculations at these energies. We explore whether

3F provides a plausible alternate explanation for the observed product state distributions.

II. Theoretical considerations

A. General considerations of triple fragmentation

Triple photofragmentation of molecules is a well-studied field. The dynamics of breaking two bonds is often defined as being sequential or concerted, and within the concerted framework, there is often discussion of whether the two bonds break synchronously or asynchronously. There are a couple of excellent reviews on this topic that summarise the field, up to the year 2000.[26,27] There have been many more subsequent experimental and theoretical investigations, but the general ideas have remained consistent. In this work we consider only sequential three-body dissociation.

Fig. 1 is a sketch of the energetics of a generic ABC molecule, where A, B, and C represent any photochemical fragment. We consider the connectivity of the molecule to be A–B–C (that is, not a ring structure). There are three, two-body fragmentation (2F) processes; two of these are simple bond cleavage processes, breaking the A–B or B–C bond. Frequently, the production of AB + C and A + BC will be barrierless, with bond cleavage correlating directly with the ground state radical fragments. The other 2F process, producing AC + B will invariably have a high barrier, involving a tight transition state for making the new A–C bond and breaking the A–B and B–C bonds. This reaction often produces closed shell products, and hence is often at a much lower final energy than the radical products. It is this set of products that is also involved with the roaming mechanism, where A roams about BC to abstract C, or C roams about AB to abstract A. We will not explicitly consider roaming pathways, but return to the inclusion of roaming once the dynamical signatures of the conventional pathways have been considered in more detail.

Fig. 1 also shows the 3F energy, which, of course, correlates with all three initial 2F processes. The breaking of the A–B or B–C bonds in the AB or BC radicals often correlates, diabatically, with excited state A + B or B + C products, which leads to a curve crossing with the adiabatic ground state surface. Whether the adiabatic potential energy surface has a barrier or not depends on the strength of the coupling between the diabatic curves. For example, there is a ∼24 kJ mol⁻¹ exit channel barrier in the C–C bond cleavage of CH_3CO radical.[28] Breaking the C–H bond in the formyl radical has a very small exit channel barrier of <10 kJ mol⁻¹,[29] while the dissociation of many alkylperoxy radicals is barrierless.[30]

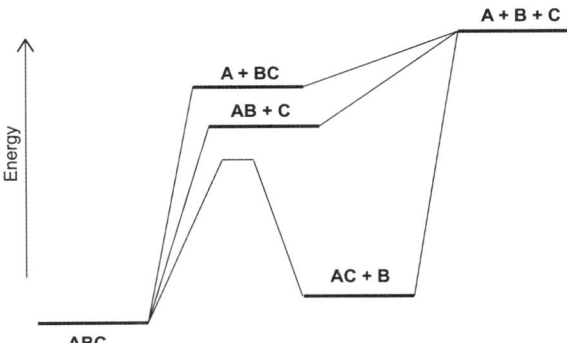

Fig. 1 Generic energy level diagram of triple fragmentation of ABC into A + B + C *via* three two-body pathways, two of which are barrierless bond cleavage and the third *via* a tight transition state.

B. Phase space theory (PST) model of triple fragmentation (3F)

Initially, we assess the product state distributions in the case where both primary and secondary dissociations are barrierless. The dynamics of dissociation *via* barrierless potential energy surfaces (PESs) are generally modeled reasonably well by statistical theories.[22,31-35] These models assume that reaction is slow, intramolecular vibrational redistribution complete, and that configuration space, or phase space, is well-sampled. As a consequence, statistical models predict that all possible product states are produced in proportion to their degeneracy, limited only by physical conservation laws.

Our theoretical model to assess 3F starts with a PST calculation for the disposal of energy into the first two fragments, AB + C. Hunter *et al.* provide a clear and concise description of the implementation of PST in photodissociation dynamics and we adopt their approach.[22] We take each fragment to be a rigid rotor, harmonic oscillator and count all such states up to the limit of available energy. The harmonic oscillator density of states will be an underestimate at high available energy, but we retain this model because of its simplicity. Each of the fragments that we consider is treated as a linear or symmetric (prolate or oblate) rotor, which, for the fragments we investigate here (HCO, CO, CH_3, CH_3O) is either exact or reasonably accurate.

Every state of AB is assessed for whether the AB internal energy, $E_{int}(AB)$, is above the AB thermochemical dissociation energy, $E_{diss}(AB)$. If so, the available energy for the second step, $E_{avail}(2)$ becomes

$$E_{avail}(2) = E_{int}(AB) - E_{diss}(AB) \qquad (1)$$

A second PST calculation is then performed for every individual (AB,C) state for the secondary reaction $AB \rightarrow A + B$, using $E_{avail}(2)$, $J(AB)$ as J_{parent}, and $E_{trans}(AB)$ to provide the initial velocity components to the A and B fragments.

The final velocity of A and B is determined from the velocity of the AB recoil from step 1, $ABC \rightarrow AB + C$, convolved with the velocity imparted in step 2, $AB \rightarrow A + B$. We assume that the two velocity vectors are independent, in which case the final velocity is $v_{tot} = (v_1^2 + v_2^2)^{1/2}$. The final angular momentum is evaluated explicitly in the second PST calculation, using the angular momentum of AB, determined from the first step, to determine the available phase space, and hence the magnitude and degeneracy of the final rotational states in the second step.[22] The final vibrational energy is also determined directly from the second PST calculation as the initial vibration of the AB fragment is assumed to be statistically partitioned (that is, intramolecular vibrational redistribution, IVR, is complete).

It is important to account correctly for the probability of the A + B states in the second (inner) PST calculation. Each unique, dissociative (AB, C) state, i, will spawn a number of final (A + B) states, j, each with their own probability, $P_j(A + B)$ which is proportional to the PST degeneracy $g_j(A + B)$. The partition function requires that the sum of $P_j(A + B)$ is unity:

$$\sum_j P_j(A + B) = 1 \qquad (2a)$$

If, however, the degeneracy of the (AB, C) state, $g_i(AB,C) > 1$, then we require:

$$\sum_j P_j(A + B) = g_i(AB, C) \qquad (2b)$$

This sum rule ensures that the correct number of total states are counted. That is, irrespective of how many of (A + B) states are possible, they can only arise from the $g_i(AB,C)$ states that dissociate.

Calculations were performed below the triple fragmentation energy to inspect and calibrate the two-body results. The available energy was then raised above the AB dissociation energy and representative calculations were performed at 1000 and

3000 cm^{-1} above this energy as well as at energies corresponding to the experimental wavelengths of 248, 243 and 234 nm.

Quadruple precision code for the PST calculations at a total energy, E_{avail}, above the AB + C two-body fragmentation channel was written in-house in Fortran-90 and implemented at the Australian National Computational Infrastructure (NCI) facility. The data required for these calculations is given in the Appendix. Every possible vibrational and rotational state of both AB and C fragments with energy, $E < E_{avail}$, was counted explicitly. The state count was binned for the translational and vibrational degrees of freedom, using 100 bins over the available energy range. Angular momentum states were binned in units of $J\hbar$.

III. Results

A. Benchmarking studies: Two-body fragmentation in acetaldehyde

Our PST results are initially benchmarked against experimental photodissociation results for CH_3CHO. There are many photodissociation pathways available at modest photolysis energy, as shown below and in Fig. 2:

$$CH_3CHO + hv \rightarrow CH_3 + HCO \quad \text{(R1)}$$

$$CH_3CHO + hv \rightarrow CH_4 + CO \ (\textit{via} \ TS) \quad \text{(R2a)}$$

$$CH_3CHO + hv \rightarrow CH_4 + CO \ (\textit{via} \ \text{roaming}) \quad \text{(R2b)}$$

$$CH_3CHO + hv \rightarrow CH_3CO + H \quad \text{(R3)}$$

$$CH_3CHO + hv \rightarrow CH_2CO + H_2 \quad \text{(R4)}$$

$$CH_3CHO + hv \rightarrow CH_2CHO + H \quad \text{(R5)}$$

$$CH_3CHO + hv \rightarrow C_2H_2 + H_2O \quad \text{(R6)}$$

$$CH_3CHO + hv \rightarrow CH_3 + CO + H \quad \text{(R7)}$$

The lowest energy 3F pathway (R7) produces $CH_3 + CO + H$, which correlates with the 2F reactions (R1)–(R3). The 3F pathways to $CH_3 + CO + H$ from each of reactions (R1)–(R3) are shown in Fig. 2, including small exit channel barriers in the secondary decomposition of the radicals, which are ignored in the model. Reactions (R4)–(R6) also have their own 3F pathways, but these lead to different sets of products, at higher energy.

Due to the importance of carbonyl chemistry in the atmosphere, the 2F reactions (R1)–(R3) have been investigated by a wide range of techniques under different conditions. Quantum yields of each channel have been reported as a function of wavelength, temperature and pressure in the energy range relevant here.[36,37] The two simple bond cleavage channels are barrierless, while the formation of molecular products occurs over a high barrier or *via* roaming.[7,8] The dynamics of the HCO + CH_3 channel have been investigated *via* state-resolved measurements of the HCO vibrational and rotational states and translational energy, *via* laser induced fluorescence spectroscopy and have been previously reported to be described well by PST.[35]

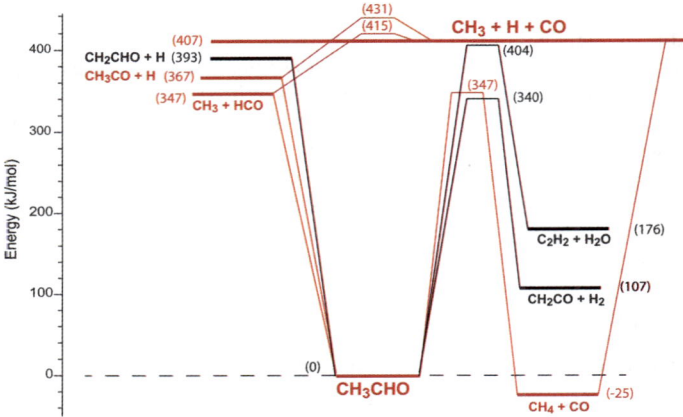

Fig. 2 Energy level diagram for the two body and three body dissociation of CH_3CHO. Only lowest energy triple fragmentation to $CH_3 + CO + H$ is shown here in red, but other three body channels occur at higher energy.

The translational energy of the CH_3 and HCO fragments have also been measured by ion imaging using state-specific REMPI of each fragment,[23,24] or by one-photon ionization using vacuum ultraviolet radiation.[25]

Acetaldehyde photochemistry *via* (R1) is complicated by the presence of a triplet channel, which is active for $\lambda < 318.5$ nm.[24,35] The triplet channel correlates with the same products – ground state $CH_3 + HCO$, however an exit channel barrier of ~24 kJ mol^{-1} changes the dynamics on the triplet surface in favour of greater release of translational energy and less of internal energy of the fragments.[28] For our initial benchmarking we choose experimental data that clearly correspond to (R1) on the S_0 state of acetaldehyde.[24,35]

We have shown previously that PST provides a good model for the HCO rotational state distributions obtained from laser induced fluorescence experiments at a photolysis wavelength of 323 nm (see, for example, Fig. 4 in ref. 35). Fig. 3 shows the translational energy release in the HCO fragment, in the ground vibrational state, obtained from ion imaging experiments after 325 nm dissociation of CH_3CHO,[24] together with our 2F PST results. At this wavelength only reaction on S_0 is possible, and the experimental PST distributions are in reasonable agreement. The HCO ($v = 0$) fragment was imaged following HCO 2 + 1 REMPI, at approximately 239 nm. Bañares and co-workers also report the CH_3 translational energy distribution, and although the PST results are not shown, the agreement is similar to that seen in Fig. 3.

The PST results are likely to be worst for low available energies; relatively few vibrational states are populated and the validity of PST requires fast IVR. The agreement between PST and the HCO distributions obtained following 325 and 323 nm photolysis of CH_3CHO suggest that IVR is indeed fast and that PST will provide a good model for product state distributions at shorter photolysis wavelengths.

B. Predictions of triple fragmentation in CH_3CHO

Despite the wealth of studies on CH_3CHO photochemistry, to our knowledge there has been no explicit study of 3F dynamics in this system. Thermochemically, the lowest energy CH_3CHO 3F products are $CH_3 + H + CO$, which lie 34 037 cm^{-1} above CH_3CHO. [see Appendix]. This channel will be open for photolysis wavelengths shorter than 294 nm. Fig. 4 shows the PST triple fragmentation translational energy distributions for CO and H, for several energies above the triple fragmentation asymptote.

Translational energy distributions for both fragments peak at low energies, with the distribution broadening and shifting to higher energy as the available energy

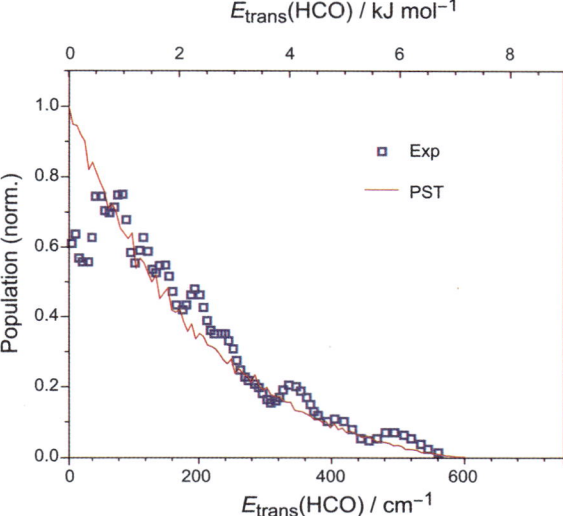

Fig. 3 The HCO ($v = 0$) translational energy distribution from photodissociation of CH_3CHO at 325 nm. Experimental data adapted from ref. 24.

to triple fragmentation, $E_{avail}(3F)$, increases. These distributions are considerably colder than what would be expected from CO arising from the 2F TS mechanism, (R2a). For example, $E_{avail}(3F) = 1000$ cm^{-1} corresponds to an energy more than 37 000 cm^{-1} above the CH_4 + CO asymptote. Even if only the energy corresponding to the exit channel barrier (31 100 cm^{-1}) were distributed to relative product translational energy, the component carried by the CO fragment would be greater than 11 000 cm^{-1}; the CO translational energy shown in Fig. 4(a) accounts for less than 2% of this value.

It is also worth noting that the CO and H translational energy distributions shown in Fig. 4 are not momentum matched. In a 2F process, the H atom recoil energy would be 28 times that of CO, as shown by the dashed line in Fig. 4(b). In the 3F process the H atom carries away only 3 times the energy of the CO fragment. The reason is apparent in Fig. 5, which shows the HCO speed distribution from the initial

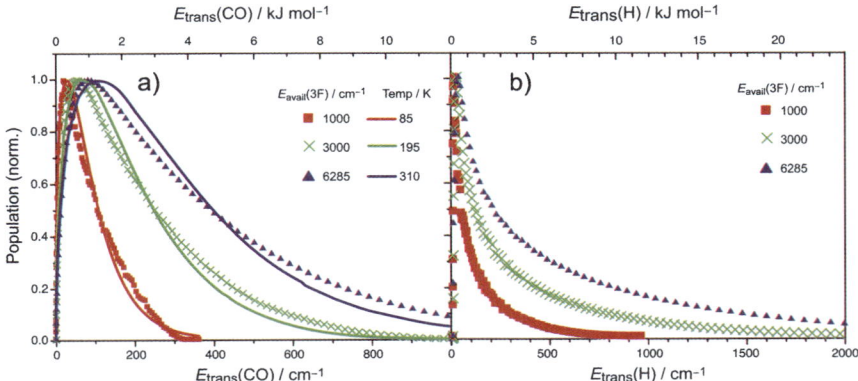

Fig. 4 Triple fragment PST translational energy distributions, (a) CO, (b) H-atom, for various energies above the triple fragmentation energy in CH_3CHO. The CO translational distributions can be fit reasonably to Boltzmann distributions (solid lines) with temperatures as indicated in panel (a). The H-atom distributions are clearly non-Boltzmann.

2F process for fragments with enough vibrational energy to dissociate, along with the velocity of both the CO and H 3F fragments, all calculated at $E_{\text{avail}}(3F) = 6285$ cm^{-1}. The CO obtains the vast majority of its speed from the HCO parent radical, formed in the 2F dissociation to CH$_3$ + HCO. The H-atom, however, gains the majority of its speed in the second step, where HCO dissociates to H + CO. The CO distributions also resemble Boltzmann distributions, as shown in the figure, because the speed/energy is dominated by the 2F step; 2F-PST distributions resemble Boltzmann distributions at high energy. The H-atom speed/energy distribution, however, cannot be fit to a Boltzmann distribution.

Three predicted 3F angular momentum distributions for the CO fragment are shown in Fig. 6. Similarly to Fig. 4, the CO angular momentum distributions broaden as $E_{\text{avail}}(3F)$ increases and the maxima shift to higher $J(CO)$. The $J(CO)$ distributions are also relatively cold, with the most probable values of $J(CO) \leq 12$ for all energies considered. Again this rotational distribution is significantly colder than would be expected from CO formed *via* the tight TS mechanism, reaction (2a). For example, for photolysis at 308 nm (corresponding to 3471 cm^{-1} above the CH$_3$ + HCO asymptote, and 1570 cm^{-1} below the triple fragmentation energy) the experimental $J(CO)$ distributions identified as from the tight TS channel peak at approximately $J(CO) = 30$,[7] with trajectory simulations predicting a maximum value of $J(CO) \approx 60$.[3,38] Our PST simulations for triple fragmentation (Fig. 6) predict, at 1000 cm^{-1} *above* the asymptote, a maximum at $J(CO) = 5$. The $J(CO)$ distributions can be fit approximately to Boltzmann distributions, with temperatures as indicated in Fig. 6.

The product state distributions for H and CO, born from triple fragmentation of CH$_3$CHO, have similar dynamical signatures to those associated with roaming reactions, *viz.* they are rotationally and translationally cold, especially when compared to the relevant 2F available energy for CO. In the next section we explore whether 3F distributions might provide an alternative explanation for the dynamical features recently attributed to roaming reactions in both CH$_3$CHO[23] and CH$_3$OCHO[15] at photolysis energies above the 3F energy threshold.

IV. Discussion

A. Photolysis of CH$_3$CHO above the triple fragmentation energy

Bañares and co-workers used a 2 + 1 REMPI scheme to image the CO($v = 0$) arising from photolysis of CH$_3$CHO at 248 nm.[23] They saw clear evidence of two different

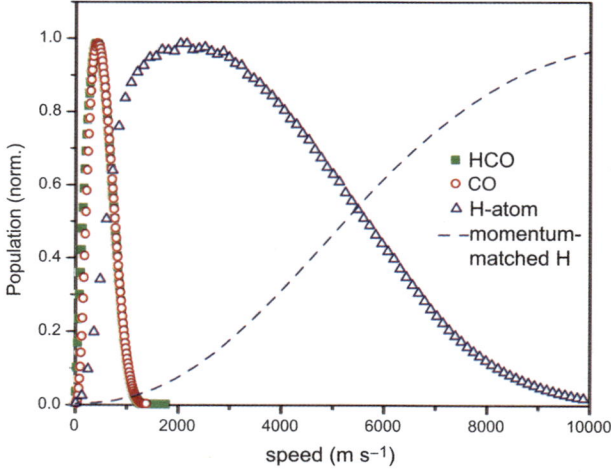

Fig. 5 Speed distributions for parent HCO (above dissociation limit) and triple fragments H and CO for $E_{\text{avail}}(3F) = 6285$ cm^{-1} in CH$_3$CHO.

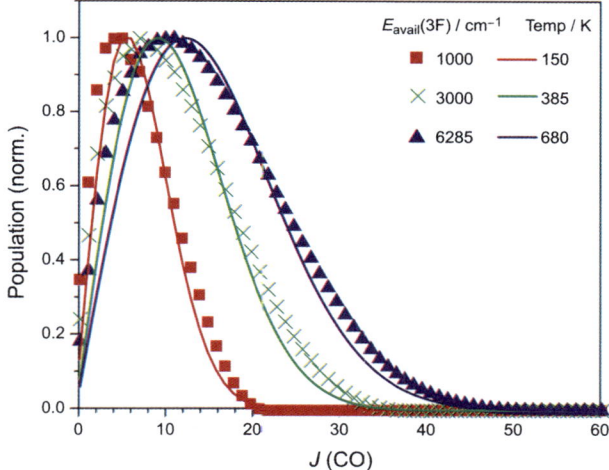

Fig. 6 The triple fragment CO PST angular momentum, $J(CO)$, distribution as a function of energy above the CH_3CHO triple fragmentation energy. The $J(CO)$ distributions can be fit reasonably to Boltzmann distributions (solid lines) with temperatures as indicated.

components in the CO velocity and angular momentum distributions. One component, with translationally and rotationally excited CO, appears consistent with the formation of $CH_4 + CO$ via a tight TS, (R2a). The second, with translationally and rotationally cold CO, was attributed to formation of $CH_4 + CO$ via a roaming mechanism, (R2b).

Photolysis at 248 nm is equivalent to $E_{avail}(3F) = 6285$ cm^{-1}. The full PST $CO(v = 0)$ translational energy distribution from 3F is shown in Fig. 7, together with data from ref. 23, obtained at $J(CO) = 7$. In this figure, and in all figures where we compare the 3F PST distributions with experimental data we use wavenumber as the common unit of energy on the lower abscissa, but show the units of the original experimental data on the upper axis.

The triple fragment $CO(v = 0)$ PST translational energy distribution shown in Fig. 7 is significantly colder than the experimental $CO(v = 0)$ distribution, although it is possible that the experimental distribution includes a very cold component, as a shoulder at low energy. This very cold component is only seen in the lowest $J(CO) = 7$ slice.[23] The discrepancy between the PST triple fragmentation CO translational distribution and the long tail in the "cold" experimental distribution reported by Bañares and co-workers suggests that this distribution is not due to triple fragmentation and may indeed be due to a roaming reaction (R2b).

In a further attempt to confirm that triple fragmentation was unimportant, Bañares and co-workers performed separate ion imaging experiments on H-atoms produced following photolysis of CH_3CHO at 248 nm. These results are shown in Fig. 8, together with the triple fragmentation PST H-atom translational energy distributions. Bañares and co-workers suggest that these data arise from CH_3CHO dissociation to acetyl, $CH_3CO + H$ (R3). Our 2F PST calculation via (R3) supports this suggestion, as shown in Fig. 8. The PST and experimental data peak at the same energy and have a similar width and shape, although the experimental tail is a little longer than the PST calculation. Indeed, the experimental tail slightly exceeds the 2F available energy, which indicates, perhaps, a small contribution from multiphoton absorption. Nonetheless, the PST and experimental data peak are in good agreement.

The 3F H-atom translational energy distribution is extremely cold and not consistent with the observed distribution, which extends significantly further than $E_{avail}(3F) = 6285$ cm^{-1} at 248 nm.

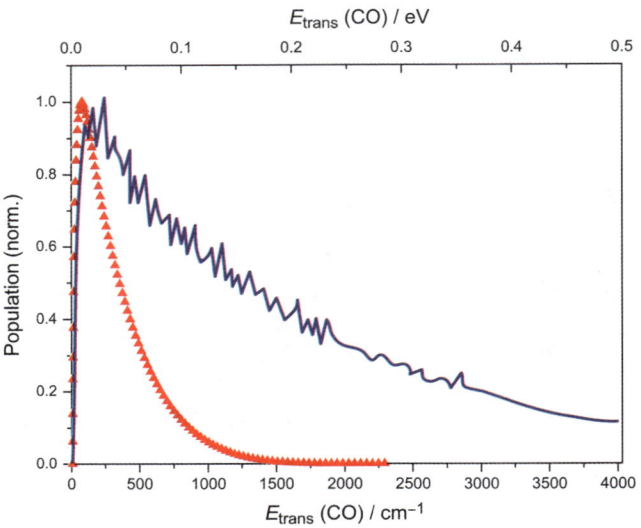

Fig. 7 The triple fragment $CO(v = 0)$ PST translational energy distribution at $E_{avail}(3F) = 6285$ cm^{-1} (red triangles) and the $CO(v = 0)$ translational energy distribution obtained from ion imaging at $J(CO) = 7$ after photolysis of CH_3CHO at 248 nm (blue line) (adapted from ref. 23).

B. Photolysis of CH$_3$OCHO above the triple fragmentation energy

Lin and coworkers have recently used a (2 + 1) REMPI scheme and ion imaging to measure $CO(v = 0)$ fragments following the photodissociation of CH_3OCHO at 234 nm.[15] They observed two distinct $CO(v = 0)$ distributions, and inferred two different reaction mechanisms.

Similar to CH_3CHO, there are a number of possible photodissociation pathways available to CH_3OCHO at modest photolysis energy, as shown below and in Fig. 9:

$$CH_3OCHO + hv \rightarrow CH_3O + HCO \tag{R8}$$

$$CH_3OCHO + hv \rightarrow CH_3OH + CO \ (via \ TS) \tag{R9a}$$

$$CH_3OCHO + hv \rightarrow CH_3OH + CO \ (via \ roaming) \tag{R9b}$$

$$CH_3OCHO + hv \rightarrow CH_3COCO + H \tag{R10}$$

$$CH_3OCHO + hv \rightarrow CH_2O + HCOH \tag{R11}$$

$$CH_3OCHO + hv \rightarrow CH_2O + CH_2O \tag{R12}$$

$$CH_3OCHO + hv \rightarrow CH_4 + CO_2 \tag{R13}$$

$$CH_3OCHO + hv \rightarrow CH_3O + H + CO \tag{R14}$$

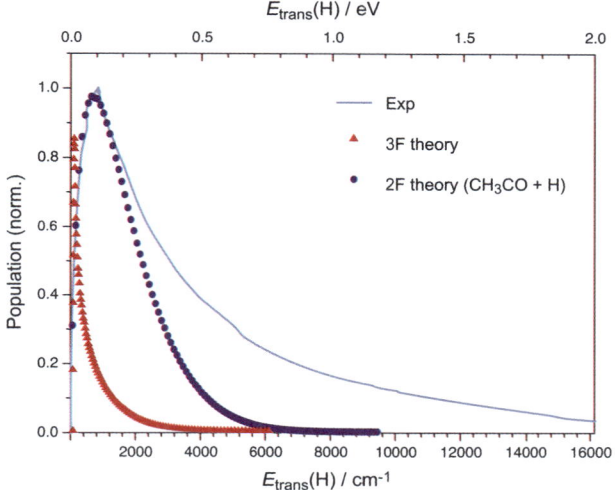

Fig. 8 Calculated 2F ($CH_3CO + H$) and 3F H-atom translational energy distributions at $E_{avail}(3F) = 6285$ cm^{-1} (red triangles). Also shown is the H-atom translational energy distribution obtained from ion imaging (adapted from ref. 23) after photolysis of CH_3CHO at 248 nm (blue line).

One experimentally observed $CO(v = 0)$ distribution is characterized by rotationally hot CO with a broad, relatively hot distribution of translational energies and was attributed to dissociation to $CH_3OH + CO$ *via* a tight TS, reaction (R9a). The second $CO(v = 0)$ distribution was rotationally and translationally colder, and was ascribed to production of $CH_3OH + CO$ *via* a roaming mechanism, reaction (R9b).[15] As seen above, rotationally and translationally cold CO also arises from triple fragmentation, in this case $CH_3OCHO \rightarrow HCO + CH_3O \rightarrow H + CO + CH_3O$; photolysis at 234 nm is equivalent to $E_{avail}(3F) = 3735$ cm^{-1}.

The full PST $CO(v = 0)$ translational energy distribution from triple fragmentation is shown in Fig. 10, together with data from ref. 15, obtained at $J(CO) = 10$. The cold experimental distribution appears to be consistent with the calculated 3F distributions. This cold distribution may arise from roaming, reaction (R9b), but CO from triple fragmentation cannot be discounted.

In an attempt to confirm that the cold CO distribution was not due to triple fragmentation, Chao *et al.* also obtained (2 + 1) REMPI ion images of the H-atom obtained following photolysis of CH_3OCHO at 243.1 nm [15]. Their experimental H-atom translational energy distribution is shown in Fig. 10(b), together with the 3F H-atom translational energy distribution determined at $E_{avail}(3F) = 2135$ cm^{-1}. As in the CO translational energy distribution, there are two distinct H-atom distributions, one at very low translational energies and the other, broader, distribution extending to larger translational energies. Chao *et al.* ascribe the hotter, broader distribution to H-atoms arising from the two-body dissociation to $CH_3OCO + H$, (R9). The cold H-atom component, however, is peaked at the same energies as H-atoms arising from 3F at 243.1 nm. Again these results do not, by themselves, preclude triple fragmentation. However, in this case, the burden of proof for claiming a roaming mechanism needs to be more than the observation of a rotationally and translationally cold fragment.

Because both roaming and 3F yield rotationally and translationally cold fragments it is challenge to determine the mechanism unambiguously from the product state distributions. We suggest a number of experimental approaches that might be used to distinguish the two possibilities.

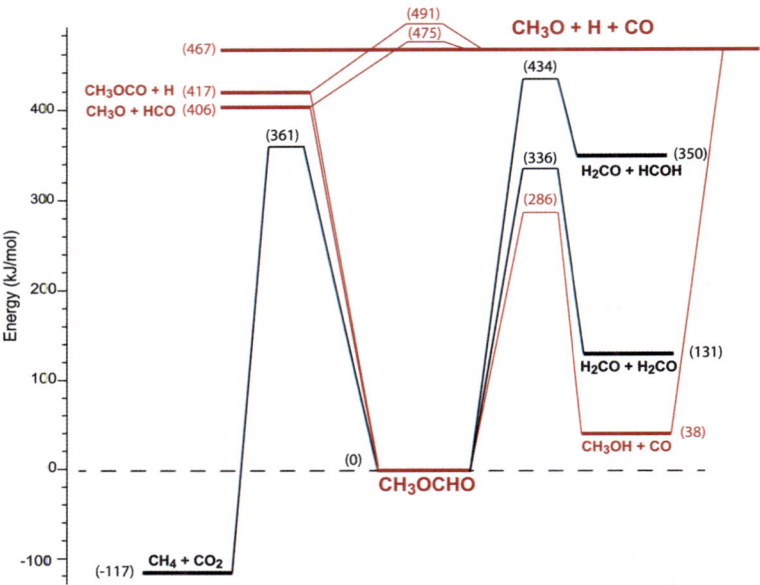

Fig. 9 Energy level diagram for the two body and three body dissociation of methyl formate, CH_3OCHO. Triple fragmentation to $CH_3O + CO + H$ is shown here in red, but other three body channels occur at other energies.

1) The simplest approach is to examine the experimental translational energy distribution of both possible secondary fragments to investigate whether either violates energy conservation for triple fragmentation. For example, Bañares and co-workers[23] point out that the broad H-atom kinetic energy distribution they report for CH_3CHO photolysis extends to energies that are not accessible to triple fragmentation products and therefore reject this mechanism.

2) As shown above, the translational energy distributions of fragments born in the second step of a triple fragmentation are not momentum matched. As a consequence, it is incorrect to take the translational energy distribution of one fragment and use it to determine the translational energy distribution in an assumed co-fragment. This may lead to an incorrect conclusion that energy conservation has been violated and triple fragmentation erroneously rejected. Indeed, we suggest that reporting of translational energy as total kinetic energy release (TKER) should be avoided when the triple fragmentation channel is open.

3) Triple fragmentation must occur at significantly higher energy than the energy required for production of the same fragment by a roaming reaction; the roaming threshold occurs just below the two-body radical dissociation threshold. The photolysis energy dependence of the two pathways will therefore be different. At the extreme, lowering the photolysis energy below the triple fragmentation threshold would provide unambiguous evidence for the presence or absence of triple fragmentation.

The PST model presented here for product state distributions arising from 3F is a simple approach. There are a number of improvements that can be made to it, none of which, however, will change the overall conclusions that we have drawn: that translational and rotational distributions arising from triple fragmentation can be very cold. An obvious addition to our model is the inclusion of a centrifugal barrier, as suggested by Reisler and co-workers.[22] This would place additional constraints on the orbital angular momentum of both the $ABC \rightarrow AB + C$ and $AB \rightarrow A + B$ reactions, limiting further the $J(CO)$ distribution obtained.

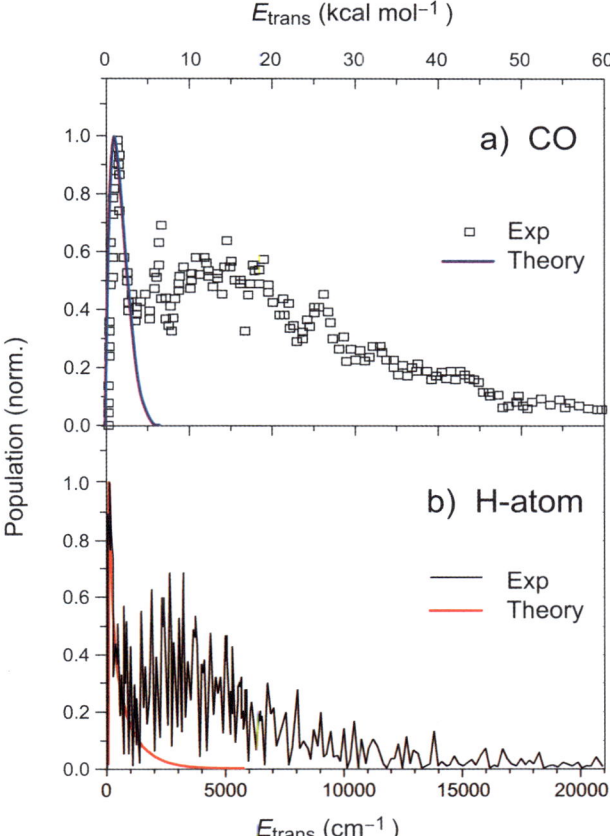

Fig. 10 PST distribution for H and CO($v = 0$) triple fragmentation products from dissociation of CH_3OCHO with $E_{avail}(3F) = 3735$ cm^{-1} for CO and 2135 cm^{-1} for H-atoms. Also shown are experimental distributions at the same available energy, corresponding to photolysis of CH_3O-CHO at 234 nm (for CO) and 243 nm (for H). Data adapted from ref. 15.

A second improvement is to incorporate the effects of an exit channel barrier to triple fragmentation. In the two cases investigated here there is a small exit channel barrier of approximately 8.4 kJ mol^{-1} (~700 cm^{-1}) for HCO $\rightarrow$ H + CO.[39] In general, an exit channel barrier will partition more energy into translation and less into internal degrees of freedom, however, the effect on the distributions reported above would be small. In effect this would shift the CO and H-atom and translational energy distributions in Fig.7, Fig. 8 and Fig. 10 to slightly higher energy, but not change any of the conclusions. Regardless of these possible improvements to the model, the conclusion that triple fragmentation leads to translationally and rotationally cold fragments is robust.

V.   Summary

We have developed and used phase space theory to predict the nature of product state distributions arising from sequential triple fragmentation reactions, showing that the secondary fragmentation products are extremely translationally and rotationally cold. This is a similar dynamical signature to that identified in roaming reactions[2,3] and our analysis shows that, when triple fragmentation is possible, the presence of such distributions is not a sufficient condition to conclude a roaming reaction mechanism.

Table 1 Constants (cm^{-1}) required for phase space theory calculations of sequential triple fragmentation of CH_3CHO and CH_3OCHO

CH$_3$O		CH$_3$		HCO		CO
Mode (sym)	Vib. Frequency	Mode (sym)	Vib. Frequency	Mode (sym)	Vib. Frequency	Vib. Frequency
1 (a_1)	2840	1 (a_1')	3004.43	1 (a')	2434.48	2169.8136
2 (a_1)	1412	2 (a_2'')	606.453	2 (a')	1080.76	
3 (a_1)	1047	3 (e')	3160.821	3 (a')	1868.17	
4 (e)	2778	4 (e')	1396			
5 (a_1)^a	914					
5 (e)^a	1465					
6 (a_1)^a	651.5					
6 (e)^a	1210					
Rotational Constants:						
A	5.2097		9.57789		24.329602	
B	0.931682		9.57789		1.4939571	1.931281
C	0.931682		4.74202		1.3986631	

^{a} Jahn–Teller split components of the CH$_3$O nominally degenerate vibrational mode.

We compared our triple fragmentation PST distributions to experimental data for CH_3CHO and CH_3OCHO photolysis at energies above the triple fragmentation energy.[23,15] We find that, for photolysis of CH_3CHO at 248 nm, the experimentally observed "cold" distributions are not consistent with triple fragmentation, supporting the hypothesis that they arise from a roaming reaction.[23] In CH_3OCHO photolysis at 234 and 243 nm, however, the "cold" distributions observed experimentally are consistent with triple fragmentation and, although roaming may indeed be present, triple fragmentation cannot be ruled out.[15] The burden of proof for claiming a roaming mechanism therefore needs to be more stringent.

We suggest a number of experimental checks that may be used to determine whether triple fragmentation is important, including a test of energy conservation, checks of momentum matching and examining the product state distributions as a function of available energy above and below the triple fragmentation threshold.

Appendix

The vibrational frequencies (cm^{-1}) and rotational constants (cm^{-1}) utilised in the PST calculations are given in Table 1. All vibrational frequencies are as reported in ref. 40. The degenerate e symmetry vibrations of the methoxy radical, CH_3O, were treated as their experimentally observed Jahn–Teller split components and the degenerate e' symmetry CH$_3$ vibrations were treated as two distinct vibrations with identical frequencies. The rotational constants for CH$_3$ and CH$_3$O are from ref. 41, CO from ref. 40 and HCO from ref. 42

The energetic parameters required for the PST calculations are shown in Fig. 2 and Fig. 9. The data for Fig. 2 were adopted from ref. 43, while energies in Fig. 9 were adopted from ref. 15. Additional energies for the barriers for HCO → H + CO and CH$_3$CO → CH$_3$ + CO are from ref. 39 and ref. 44, respectively.

Acknowledgements

This work was funded by the Australian Research Council (grant DP1094559). Parts of this research were undertaken on the NCI National Computational Infrastructure

Facility in Canberra, Australia, which is supported by the Australian Commonwealth Government.

References

1 D. Townsend, S. A. Lahankar, S. K. Lee, S. D. Chambreau, A. G. Suits, X. Zhang, J. Rheinecker, L. B. Harding and J. M. Bowman, *Science*, 2004, **306**, 1158.
2 A. G. Suits, *Acc. Chem. Res.*, 2008, **41**, 873.
3 J. M. Bowman and B. C. Shepler, *Annu. Rev. Phys. Chem.*, 2011, **62**, 531.
4 S. D. Chambreau, S. A. Lahankar and A. G. Suits, *J. Chem. Phys.*, 2006, **125**, 044302.
5 S. A. Lahankar, S. D. Chambreau and A. G. Suits, *J. Chem. Phys.*, 2006, **125**, 044303.
6 S. A. Lahankar, S. D. Chambreau, X. Zhang, J. M. Bowman and A. G. Suits, *J. Chem. Phys.*, 2007, **126**, 044314.
7 P. L. Houston and S. H. Kable, *Proc. Natl. Acad. Sci. U. S. A.*, 2006, **103**, 16079.
8 B. R. Heazlewood, M. J. T. Jordan, S. H. Kable, T. M. Selby, D. L. Osborn, B. C. Shepler, B. J. Braams and J. M. Bowman, *Proc. Natl. Acad. Sci. U. S. A.*, 2008, **105**, 12719.
9 B. C. Shepler, B. J. Braams and J. M. Bowman, *J. Phys. Chem. A*, 2007, **111**, 8282.
10 R. D. van Zee, M. F. Foltz and C. B. Moore, *J. Chem. Phys.*, 1993, **99**, 1664.
11 X. Zhang, S. Zou, L. B. Harding and J. M. Bowman, *J. Phys. Chem. A*, 2004, **108**, 8980.
12 J. M. Bowman and X. Zhang, *Phys. Chem. Chem. Phys.*, 2006, **8**.
13 V. Goncharov, N. Herath and A. G. Suits, *J. Phys. Chem. A*, 2008, **112**, 9423.
14 R. Sivaramakrishnan, J. V. Michael, A. F. Wagner, R. Dawes, A. W. Jasper, L. B. Harding, Y. Georgievskii and S. J. Klippenstein, *Combust. Flame*, 2011, **158**, 618.
15 M.-H. Chao, P.-Y. Tsai and K.-C. Lin, *Phys. Chem. Chem. Phys.*, 2011, **13**, 7154.
16 M. P. Grubb, M. L. Warter, A. G. Suits and S. W. North, *J. Phys. Chem. Lett.*, 2010, **1**, 2455.
17 M. P. Grubb, M. L. Warter, K. M. Johnson and S. W. North, *J. Phys. Chem. A*, 2011, **115**, 3218.
18 M. Hause, N. Herath, R. Zhu, M. C. Lin and A. G. Suits, *Nat. Chem.*, 2011, **3**, 932.
19 A very recent paper on excited state roaming in NO₃ appeared after this paper was submitted: North *et al.*, Science (2012), ref to be added in proof.
20 W. J. Chesnavich and M. T. Bowers, *J. Chem. Phys.*, 1977, **66**, 2306.
21 P. Pechukas, J. C. Light and J. Rankin, *J. Chem. Phys.*, 1966, **44**, 794.
22 M. Hunter, S. A. Reid, D. C. Robie and H. Reisler, *J. Chem. Phys.*, 1993, **99**, 1093.
23 L. Rubio-Lago, G. A. Amaral, A. Arregui, J. G. Izquierdo, F. Wang, D. Zaouris, T. N. Kitsopoulos and L. Bañares, *Phys. Chem. Chem. Phys.*, 2007, **9**, 6123.
24 G. A. Amaral, A. Arregui, L. Rubio-Lago, J. D. Rodríguez and L. Bañares, *J. Chem. Phys.*, 2010, **133**, 064303.
25 V. A. Schubert and S. T. Pratt, *J. Phys. Chem. A*, 2010, **114**, 11238.
26 C. Maul and K.-H. Gericke, *Int. Rev. Phys. Chem.*, 1997, **16**, 1.
27 C. Maul and K.-H. Gericke, *J. Phys. Chem. A*, 2000, **104**, 2531.
28 K. C. Thompson, D. L. Crittenden, S. H. Kable and M. J. T. Jordan, *J. Chem. Phys.*, 2006, **124**, 044302.
29 H. Y. Wang, J. A. Eyre and L. M. Dorfman, *J. Chem. Phys.*, 1973, **59**, 5199.
30 A Miyoshi, *Int. J. Chem. Kin.*, 2012, **44**, 59.
31 S. W. North and G. E. Hall, *J. Chem. Phys.*, 1996, **104**, 1864.
32 L. Bonnet, P. Larregaray and J. C. Rayez, *Phys. Chem. Chem. Phys.*, 2005, **7**, 3540.
33 H.-M. Yin, K. Nauta and S. H. Kable, *J. Chem. Phys.*, 2005, **122**, 194312.
34 H.-M. Yin, S. J. Rowling, A. Büll and S. H. Kable, *J. Chem. Phys.*, 2007, **127**, 064302.
35 B. R. Heazlewood, S. J. Rowling, A. T. Maccarone, M. J. T. Jordan and S. H. Kable, *J. Chem. Phys.*, 2009, **130**, 054310.
36 G. K. Moortgat, H. Meyrahn and P. Warneck, *ChemPhysChem*, 2010, **11**, 3896.
37 A. Horowitz and J. G. Calvert, *J. Phys. Chem.*, 1982, **86**, 3105.
38 B. C. Shepler, B. J. Braams and J. M. Bowman, *J. Phys. Chem. A*, 2008, **112**, 9344.
39 H. Y. Wang, J. A. Eyre and L. M. Dorfman, *J. Chem. Phys.*, 1973, **59**, 5199.
40 M. E. Jacox, Vibrational and Electronic Energy Levels of Polyatomic Transient Molecules in NIST Chemistry WebBook, NIST Standard Reference Database Number 69, ed. P. J. Linstrom and W. G. Mallard, National Institute of Standards and Technology, Gaithersburg MD, 20899, http://webbook.nist.gov, (retrieved Jan 20, 2012).
41 B. Ruscic, *et al.*, *J. Phys. Chem. Ref. Data*, 2005, **34**, 573.
42 S.-H. Lee and I.-C. Chen, *J. Chem. Phys.*, 1995, **103**, 104.
43 B. R. Heazlewood, A. T. Maccarone, D. U. Andrews, D. L. Osborn, L. B. Harding, S. J. Klippenstein, M. J. T. Jordan and S. H. Kable, *Nat. Chem.*, 2011, **3**, 443.
44 J. S. Francisco, *Chem. Phys.*, 1998, **237**, 1.

General discussion

Professor Gruebele opened the discussion of the paper by Professor Bradforth: You mentioned solid evidence for a tunneling pathway below the conical intersection exists, based on isotopic effects. In polyatomic systems, isotope effects are more subtle than in one dimension, even if an effective one-dimensional representation is possible. In addition, the reaction involves shifts between surfaces with different dipoles, so solvent electrostatics could play an important role by dephasing a coherent tunneling process. In particular, inert solvents that introduce some electrostatic fluctuations could lead to damping of the tunneling dynamics.

Rather than jumping from cyclohexane to water, could you use other reasonably inexpensive solvents with greater electrostatic fluctuations, such as ethyl-propyl ether, methyl furan or cyclohexylether, to look at damping of the tunneling contribution? These would introduce small electrostatic fluctuations without severely shifting the electronic energy levels, and it seems that residual energy shift could be compensated for by tuning the excitation wavelength appropriately in the experiment.

Small amounts of polar trace solvent (*e.g.* water) could also have a deleterious effect *via* formation of hydrogen-bonded clusters within the apolar solvent. Have you seen any evidence of this when using not perfectly 'dry' cyclohexane?

Professor Bradforth replied: The result of the isotopic substitution experiment is not described in our manuscript in this volume and it is still preliminary data. What I can say is that when we photoexcite PhOD at 267 nm, within our signal to noise, we do not observe a spectral signature of phenoxyl radicals within the 14 ns time window used. This suggests a sizeable kinetic isotope effect for dissociation and is consistent with tunneling. However, as you rightly point out, these experiments are complicated by trace amounts of water in the cyclohexane solvent and also from absorbed water in the starting phenol, which together can amount to 5–10% of H_2O as a co-solute in the cyclohexane compared to the 10 mM phenol concentration. As suggested from the data presented for phenol dimers, phenol–water hydrogen-bonded clusters would likely have different excited state dynamics from isolated phenol. We do not see evidence for phenol–water hydrogen-bonded clusters in the infrared spectrum but this is something we are currently exploring in greater detail.

Your observations are absolutely correct that one needs to consider the tunneling process in higher-dimensionality and the effects of dephasing by the solvent. This is our first result on H-atom tunneling in solution and we indeed plan to explore some of the extensions you suggest. To do so, we will need the capability to measure out longer into the nanosecond timescale using broadband transient absorption. Such an instrument is currently under construction.

Professor Neumark said: In gas phase photodissociation experiments, a key question is whether a photoexcited molecule dissociates on an excited state or instead undergoes internal conversion to the ground state prior to dissociation. In your condensed phase experiments on phenol, one would expect that once internal conversion occurs, the molecule relaxes quickly and does not dissociate. However, in the gas phase dissociation on the ground state to PhO + H can in principle occur. Do your experiments point to excited state dissociation being the dominant mechanism in the gas phase and solution at all wavelengths, even when the appearance time of the PhO radical is very long?

Professor Bradforth replied: At higher energies ($E_{exc} > 5.0$ eV), the rapid fragment appearance time in ultrafast experiments and anisotropic H-atom distribution in polarized photolysis experiments certainly points to excited state dissociation in both gas and condensed phases. We have tried to address this issue in phenol specifically at excitation energies below the S_1/S_2 conical intersection, but so far just at one wavelength, 267 nm. Because the photofragment appearance times are so long in the gas phase[1,2] we were interested to see if the dissociation channel would be shut off in the liquid exactly because of the solvent's ability to very efficiently vibrationally relax molecules reaching the ground electronic state with high ν(O–H) after internal conversion. Instead, as described in our manuscript, we see production of phenoxyl radicals in cyclohexane and with similar nanosecond rise times as seen in the gas phase. This is strongly supportive of dissociation occurring on the excited state by a tunneling mechanism. We note however that in both the isolated molecule and in cyclohexane, dissociation is the minor pathway. Internal conversion and intersystem crossing are dominant.[3,4]

1 G. A. Piro, A. N. Oldani, E. Marceca, M. Fujii, S. I. Ishiuchi, M. Miyazaki, M. Broquier, C. Dedonder and C. Jouvet, *J. Chem. Phys.*, 2010, **133**, 124313.
2 G. M. Roberts, A. S. Chatterley, J. D. Young and V. G. Stavros, *J. Phys. Chem. Lett.*, 2012, **3**, 348–352.
3 A. Sur and P. M. Johnson, *J. Chem. Phys.*, 1986, **84**, 1206–1209.
4 E. Pines, D. Huppert and N. Agmon, *J. Chem. Phys.*, 1988, **88**, 5620–5630.

Professor Truhlar said: In order to calculate the solvatochromic shifts, Alek Marenich and I performed calculations on the excitation energies in cyclohexane solution with the M05 density functional,[1] the corrected linear response (cLR) model for vertical excitation energies in liquid solutions,[2–4] and the 6-311+G(2df,2p) basis set. We found that S_1 has a very small solvatochromic shift (13 cm^{-1}), but S_2 is red-shifted by 763 cm^{-1}. Furthermore, as the O–H bond is stretched from its equilibrium value of 0.96 Å, the solvatochromic lowering of the S_2–S_0 energy gap increases, for example to 1296 cm^{-1} at 1.2 Å, then decreases again, but not to zero. This contrasts with the behavior speculated in the article, where the Rydberg character of the S_2 state is anticipated to lead to a larger S_2–S_0 gap. (Even at an O–H distance of 0.80 Å, where the Rydberg character of S_2 is even larger, the cLR solvatochromic effect is to lower the gap by 279 cm^{-1}.) This difference in our result from that anticipated may be attributed to a difference in the polarity of the S_2 and S_0 states. At O–H distances of 0.96, 1.2, and 1.6 Å, the Mulliken charge on the OH group in the S_2 state is -0.33, -0.56, and -0.45, respectively, whereas in the ground state one finds, -0.07, -0.06, and -0.09, respectively. Thus the excited state is better solvated by solvent polarization. However, the cLR model does not explicitly take account of the solute–solvent repulsion that is postulated to raise the energy of S_2 in solution. Can experimental methods be used to resolve whether the electrostatic effect or the repulsion effect dominates?

1 Y. Zhao, N. E. Schultz and D. G. Truhlar, *J. Chem. Phys.*, 2005, **123**, 161103.
2 M. Caricato, M. Mennucci, J. Tomasi, F. Ingrosso, R. Cammi, S. Corni and G. Scalmani, *J. Chem. Phys.*, 2006, **124**, 124520.
3 M. Mennucci, C. Capelli, C. A. Guido, R. Cammi and J. Tomasi, *J. Phys. Chem. A*, 2009, **113**, 3009.
4 A. V. Marenich, C. J. Cramer, D. G. Truhlar, C. A. Guido, B. Mennucci, G. Scalmani and M. J. Frisch, *Chem. Sci.*, 2011, **2**, 2143–2161.

Professor Bradforth responded: First, we appreciate the input of detailed calculations on how the various excited state potential energy surfaces shift on solvation by cyclohexane. Our zero-order picture, that the interaction of the solute with solvent is weak, is of course oversimplified but we feel a useful starting point is to expect that the potential energy surface shapes, if not the absolute gaps between surfaces, are only modestly changed from vacuum. Clearly it is the shapes of the surfaces, particularly at the locus of conical intersections that govern the crude reaction dynamics. The R(O–H) dependent solvation effect you calculate for the S_2–S_0 gap is important in this regard, particularly if it is predominantly an effect due to the stabilization of the S_2 state. As suggested by your question, it is known that the Rydberg character of this state decreases as the bond length becomes more extended as does the dipole moment.[1] Also, as you state, there are various factors that contribute to stabilization or destabilization of Rydberg states in the condensed phase. Your calculations have captured the electrostatic stabilization of the excited state dipole but the spatial extent of the Rydberg orbital leads to Pauli repulsion with the adjacent cyclohexane molecules which will lead to a blue shift on the S_2–S_0 gap, decreasing on lengthening the O–H bond.

The absorption spectrum tells us about the stabilization in the Franck–Condon region only and, in principle, an experiment that captures the time-evolving emission spectrum for a molecule promptly dissociating on S_2 could give us the energy gap along the hydrogen atom dissociation coordinate. However, that experiment would also require time resolution of a few femtoseconds, as that is the expected timescale for direct dissociation, and uncertainty broadening would overwhelm the relatively small gap changes we are discussing.

1 A. L. Sobolewski and W. Domcke, *J Phys Chem A*, 2001, **105**, 9275–9283.

Professor Truhlar asked: Regarding the promoting mode in phenol, Xuefei Xu and I carried out calculations by CASSCF, MR-QDPT, and the fourfold way[1] for three modes, namely the OH torsion, mode 16a, and mode 16b, and we found large diabatic coupling in each of them, especially the OH torsion, whose role you find to be missing, as you explain by a G_4 symmetry argument.[2] As you pointed out, imposing correct quantum mechanical symmetry arguments on classical or quasi-classical dynamics may require considering additional quantum mechanical aspects; and I note that one requires consideration of quantum mechanical phases to impose even–odd interference effects associated with identical-particle symmetry,[3] to enforce symmetry constraints on internal-state distributions associated with electronic angular momentum,[4] or to quantize tunneling-split states.[5] The role of symmetry is also important for linking gas-phase processes with the solution phase for substituted phenols, as presented in your paper and your group's poster at this conference.[6] In light of this intriguing symmetry effect, would you summarize the main experimental findings that show a symmetry difference between phenol and symmetrically substituted phenols on the one hand and asymmetrically substituted phenols on the other?

1 H. Nakamura and D. G. Truhlar, *J. Chem. Phys.*, 2002, **117**, 5576–5593.
2 R. N. Dixon, T. A. A. Oliver and M. N. R. Ashfold, *J. Chem. Phys.*, 2011, **134**, 194303.
3 C. W. McCurdy and W. H. Miller, *J. Chem. Phys.*, 1977, **67**, 463–468.
4 M. D. Hack and D. G. Truhlar, *J. Chem. Phys.*, 1999, **110**, 4315-4337.
5 W. H. Miller, *J. Phys. Chem.*, 1979, **83**, 960-963.
6 D. Murdock, S. Harris, T. Karsili, I. Clark, G. Greetham, M. Towrie and M. Ashfold, poster P05 at Faraday Discussion 157: Molecular Reaction Dynamics in Gases, Liquids, and Interfaces, Assisi, Italy 25–27 June 2012.

Professor Ashfold replied: The translational energy distribution of the H atoms formed following excitation to the S_1 state of phenol displays a structured component centered at a total kinetic energy release (TKER) $\sim$5500 cm^{-1} and an underlying background that peaks at much lower TKER. The structure is attributable to formation of phenoxyl radical products, in their ground ($\tilde{X}^2B_1$) state, following H atom loss by tunneling under the $S_1(^1\pi\pi^*)/S_2(^1\pi\sigma^*)$ conical intersection (CI).[1] $C_6H_5O(\tilde{X})$ products are also formed when exciting at energies above the S_1/S_2 CI, but these show a very different TKER distribution. Illustrative TKER spectra obtained following photolysis of C_6D_5OH at 273.815 nm (the S_1–S_0 origin) and at 230 nm (*i.e.* at an energy well above the S_1/S_2 CI) are shown in the right inset within Fig. 1.[2] The C_6D_5O radicals formed at long wavelengths populate a very limited sub-set of the many vibrational levels that would be accessible simply on energetic grounds. The observed activity in ν_{16a} (an out-of-plane ring puckering vibration) is understandable on Franck–Condon grounds: the frequency of the ν_{16a} mode drops considerably, then recovers, during the evolution from the S_0 to S_1 to S_2 states of the molecule and ultimately to the ground state radical. However, all attempts to assign the fastest peak in these TKER spectra to formation of $C_6D_5O(\tilde{X})$ fragments in their $v = 0$ vibrational level leads to contradictory values for the O–H bond dissociation energy $D_0(C_6D_5O–H)$. This is illustrated by the red data in Fig. 1, in which we plot the TKER of the fastest peak (TKER(max)) measured at many different photolysis energies. Data taken at 'long' and 'short' wavelengths (i.e. at energies that are, respectively, below and above the S_1/S_2 CI) both fall on straight lines with unit gradient, but the two lines are offset. Internal consistency can be achieved, however, by associating the fastest peak in spectra recorded at long wavelength to $C_6D_5O(\tilde{X})$ radicals with $\nu_{16a} = 1$ (and, by implication, assuming that all populated product states involve an odd number of quanta in ν_{16a}). This finding has been rationalized by recognizing that a symmetric phenol is best described within the non-rigid molecular group G_4 (since it exists in two equivalent configurations (rotamers), the degeneracies of which are lifted by torsional tunneling) and that, within G_4, parent mode ν_{16a} (an out-of-plane ring puckering vibration) is the lowest frequency motion of the appropriate (a_2) symmetry to promote non-adiabatic coupling between the $S_1(^1\pi\pi^*)$ and $S_2(^1\pi\sigma^*)$ potentials.[3] Entirely analogous behavior has been identified in phenol itself,[1] and in 'symmetrically' substituted phenols like 4-fluorophenol.[4]

'Asymmetrically' substituted phenols like 3-fluorophenol and 3-chlorophenol, in contrast, are not constrained by G_4 symmetry. They exist in distinguishable cis- and trans-conformers, and – in principle at least – any nuclear motion of a'' symmetry could promote non-adiabatic coupling (tunneling) between the S_1 and S_2 potentials. Given this relaxed symmetry restriction, OH torsion (τ_{OH}) is deduced to have the largest interstate coupling constant – as found in the (non-G_4 symmetry restricted) theoretical treatments of S_1/S_2 coupling in bare phenol reported here by Truhlar[5] and in the earlier studies from Domcke's group.[6] τ_{OH} is a 'disappearing mode' upon O–H bond fission, but its participation would satisfy the symmetry requirements for coupling between the $S_1(^1A')$ and $S_2(^1A'')$ states and thus allow formation of $v = 0$ radical products. Thorough studies of 3-fluorophenol photolysis, for example, at numerous different photolysis energies, above and below the S_1/S_2 CI, return TKER(max) values that when plotted against the photolysis energy fall on a common line of unit gradient – consistent with some formation of $v = 0$ radical products at all wavelengths.

The black data in Fig. 1 are from photolysis studies of 4-methoxyphenol (also known as mequinol).[7] TKER spectra obtained at 'long' and 'short' wavelengths (such as those shown in the left inset) are qualitatively similar to those from bare phenol but, in contrast to phenol, the plot of TKER(max) versus photolysis energy yields a single straight line of unit gradient. Such a finding is consistent with the foregoing discussion since methoxy substitution – even in the 4-position – lowers the symmetry (4-methoxyphenol exhibits resolvable cis- and trans-isomers[8]) and thus allows OH torsion to promote tunneling under the S_1/S_2 CI.

In summary, we concur that OH torsion is likely to provide the dominant S_1/S_2 coupling in all 'asymmetric' phenols (i.e. phenols wherein substitution has lifted the degeneracy of the two rotamers) and that, in such cases, some of any radical products formed by tunneling from the S_1 state will appear in their $v = 0$ level. The 'symmetric' phenols are subject to G_4 symmetry constraints, however. Now τ_{OH} has the wrong symmetry to promote S_1/S_2 coupling, and the out-of-plane ring puckering motion ν_{16a} is deduced to be the dominant coupling mode. This response has focused on the way substitution can affect the molecular symmetry, and thus the S_1/S_2 coupling and the detailed vibrational energy disposal within the radical products formed following O–H bond fission by tunneling from the S_1

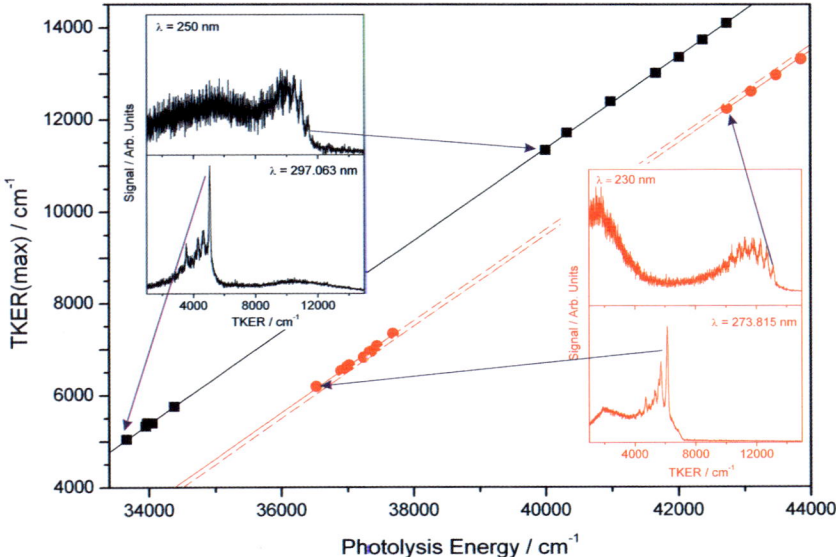

Fig. 1 Plots showing TKER(max) values obtained following photolysis of C_6D_5OH (red points) and 4-methoxyphenol (black points) at many different photolysis energies along with lines of best fit (and gradient of unity). Illustrative TKER spectra from which these data derive are shown in the insets. The lower panel in each inset shows the spectrum obtained when exciting at the respective S_1–S_0 origins, while the upper panels show spectra recorded at energies well above the respective S_1/S_2 CIs. The arrows indicate the data points which derive from each of the displayed TKER(max) peaks.

state. Of course, the substituent can also perturb (stabilize/destabilize) the ring π-system and thereby profoundly influence the barrier under S_1/S_2 CI and thus the actual rate of H atom loss by tunneling from the S_1 state.[9]

1 M. G. D. Nix, A. L. Devine, B. Cronin, R. N. Dixon and M. N. R. Ashfold, *J. Chem. Phys.*, 2006, **125**, 133318.
2 G. A. King, T. A. A. Oliver, M. G. D. Nix and M. N. R. Ashfold, *J. Phys. Chem. A*, 2009, **113**, 7984.
3 R. N. Dixon, T. A .A. Oliver and M. N. R. Ashfold, *J. Chem. Phys.*, 2011, **134**, 194303.
4 A. L. Devine, M. G. D. Nix, B. Cronin and M. N. R. Ashfold, *Phys. Chem. Chem. Phys.*, 2007, **9**, 3749.
5 D. G. Truhlar, reported in his question
6 O. P. J. Vieuxmaire, Z. Lan, A.L. Sobolewski and W. Domcke, *J. Chem. Phys.*, 2008, **129**, 224307.
7 T. N. V. Karsili, A. M. Wenge and M. N. R. Ashfold, (unpublished).
8 G. N. Patwari, S. Doraiswamy and S. Wategaonkar, *J. Phys. Chem. A*, 2000, **104**, 8466.
9 T. N. V. Karsili, A. M. Wenge, D. Murdock, S. J. Harris, J. N. Harvey, R. N. Dixon and M. N. R. Ashfold, (awaiting submission).

Professor Nesbitt asked: I was quite intrigued by your comment that the rather slow 2.4 ns scale lifetimes for H atom ejection in S_1 phenol were thought to be dominated by tunneling at energies below the conical intersection point, and that, furthermore, quantitative support for these time scales had been previously offered based on a 1D tunneling model involving only the O–H stretch coordinate. At least from a simple organic chemistry perspective, one might imagine some stabilization in phenoxy radical arising from a "quinone" like resonance delocalization of the radical center onto the aromatic ring and partial double bond character in the CO bond. I am therefore wondering why there might not be more collective nuclear motion taking place with OH bond breaking, that might require consideration of tunneling in much higher dimensionality and potentially effective tunneling masses differing substantially from that of an H atom. Malonaldehyde is the classic example of this, where essentially all atoms in the molecule rearrange substantially throughout the proton transfer event. Given the exponential sensitivity of tunneling to this effective mass, what is the gold standard for such calculations and how conclusively should we take such an agreement with 1D model predictions?

Professor Bradforth responded: The HRA-PTS experimental data published by the Ashfold group provides clear evidence for excitation in odd quanta of v_{16a} in the phenoxyl radical – a key signature that this mode is involved in coupling the diabatic S_1 and S_2 states, *en route* to H atom tunnelling and eventual O–H bond cleavage.[1,2] Your comment about the possible quinoidal nature of phenol is pertinent, and something that has been addressed previously.[1] Upon S_1–S_2 excitation, the vibrational wavenumbers associated with v_4, v_{16a} and O–H torsional modes halve in magnitude,[3] the latter mode vanishes upon dissociation, whereas the wavenumber values of former modes in the phenoxyl radical are similar to those in the S_0 phenol molecule.[1] The change in the frequencies in S_1 is a consequence of the slightly quinoidal nature of this state; the C–O bond contracts and the C–C bond increase in length.[4] If we then turn to symmetry considerations, phenol can be viewed in the G_4 permutation point group; v_{16a} is the lowest frequency vibration of the correct symmetry able to couple the S_1 and S_2 states. In 2D dynamic simulations, that include both the O–H stretch and v_{16a} ring-twisting co-ordinates, the effective barrier to tunnelling was found to be reduced by 40% compared to an earlier 1D model which only included O–H stretch.[5] As you correctly point out in the case of malonaldehyde, framework re-arrangement can assist in the tunnelling process. Dixon *et al.* performed identical 2D dynamic simulations but added ring relaxation as a function of O–H stretch; the excited state structures were optimised at the CASPT2(10/10)//CASSCF(10/10)/aug(O)-cc-pVTZ level. The resulting phenoxyl product populations from wavepacket calculations revealed very similar population distributions of v_{16a} in phenoxyl products.[1]

Phenol is not a small molecule, and reliable on-the-fly trajectory surface-hopping calculations, that for this particular problem would require nanosecond simulation times, would be far too computationally expensive. Extensive conical intersection searches in the C_s point group by Domcke and co-workers[6] have revealed prefulvenic structures for the S_1/S_0 CI, but have not yielded any further geometric details about the CI of interest compared to the previous study.[2] At the present time, the 2D dynamic simulations are the "gold standard", especially compared to previous 1D efforts.[5]

1 M. G. D. Nix, A. L. Devine, B. Cronin, R. N. Dixon and M. N. R. Ashfold, *J. Chem. Phys.*, 2006, **125**, 133318.
2 R. N. Dixon, T. A. A. Oliver and M. N. R. Ashfold, *J. Chem. Phys.*, 2011, **134**, 194303.
3 H. D Bist, J. C. D. Brand and D. R. Williams, *J. Mol. Spect.*, 1966, **21**, 76; *ibid.*, 1967, **24**, 402, 413.
4 C. Ratzer, J. Küpper, D. Spandenberg and M. Schmitt, *Chem. Phys.*, 2002, **283**, 153.
5 G. A. Pino, A. N. Olfdani, E. Marceca, M. Fujii, S.-I. Ishiuchi, M. Miyazaki, M. Broquier, C. Dedonder and C. Jouvet, *J. Chem. Phys.*, 2010, **133**, 124313.
6 O. P. J. Vieuxmaire, Z. Lan, A. L. Sobolewski and W. Domcke, *J. Chem. Phys*, 2008, **129**, 224307.

Miss Harris addressed Professor Bradforth: As Professor Bradforth is aware, the Bristol group have undertaken complementary studies to those presented in this paper at the ULTRA facility at the Rutherford Appleton Laboratory.[1] Rather than probing the electronic transitions of the species produced following UV excitation *via* a transient UV/visible probe, however, these studies have used a broadband transient infrared pulse to identify the products created (and depleted) by photo-excitation, and to elucidate the extent of the product vibrational excitation.

This journal is © The Royal Society of Chemistry 2012

Firstly, we have studied the dynamics of the processes following UV (267 nm) excitation of 4-methylthiophenol in cyclohexane. There are two main features of this spectrum as shown in Fig. 2; a bleach signal, reflecting the proportion of ground state molecules that have been excited by the UV pump pulse, and a transient gain signal attributable to a ring breathing vibrational mode of the radical produced following S–H bond fission. In accord with the UV/visible probe experiment presented by Professor Bradforth, the radical signal is apparent immediately, though is initially broad and centred at lower wavenumber than at later delay times, δt – suggesting that the radical is formed vibrationally hot and subsequently cools. After prompt formation, the total area of the radical signal decreases as shown in the kinetic trace inset within Fig. 2, with loss of population complete after ∼60 ps, in agreement with the data obtained using the UV/visible probe.[2] The observed bleach signal is seen to recover on a similar timescale, with partial recovery complete after ∼60 ps, indicating repopulation of the ground electronic state of the 4-methylthiophenol molecule. The UV/visible probe experiment reveals that geminate recombination of the produced radical pairs yields a new product, 4-MePh(H)S, where the H atom has returned to a position on the benzene ring.[3] We have not been able to identify any signature of this adduct in the IR probe experiments, but the latter studies do confirm recreation of the ground electronic state by radical recombination; comparing the extents to which the radical population decays and the ground state parent reforms suggests that this is, in fact, the dominant recombination process. This example serves to highlight the complementarity of the two techniques; in this case the UV/visible probe has allowed

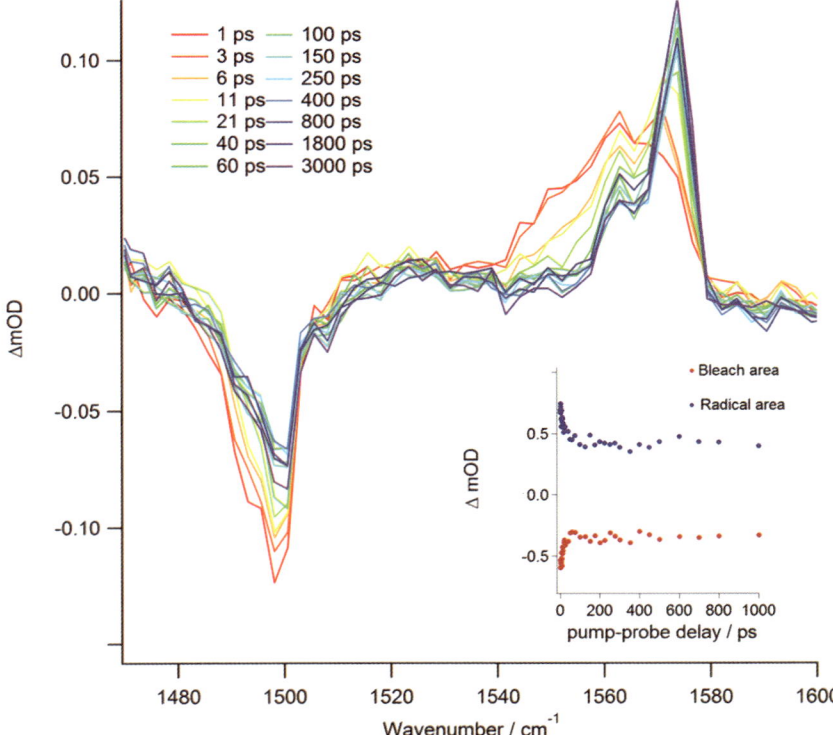

Fig. 2 Transient spectra obtained following 267 nm excitation of a 45 mM solution of 4-methylthiophenol in cyclohexane at varying pump-probe delays probing in the range of 1470–1600 cm⁻¹. Kinetic traces corresponding to the change in the total areas of the parent bleach (1480–1506 cm⁻¹) and radical (1539–1582 cm⁻¹) product peaks are shown in the inset.

observation of a relatively minor product, the 4-MePh(H)S adduct, presumably on account of the large oscillator strength of the observed electronic transition, whereas the IR probe experiment allows monitoring of the recovery of the ground state population by radical recombination (information that is not accessible using the UV/visible probe).

Professor Bradforth also suggests that studying solvated molecular clusters in an inert solvent such as cyclohexane may provide an interesting link between the photo-dissociation dynamics of a molecular species as an isolated gas phase molecule through to those of the solvated molecule in a highly interacting solvent environment – complementing studies in intermediate environments such as gas phase clusters, and monomers solvated in an inert solvent – and presents some preliminary studies in this area. Previous IR studies have shown that the O–H stretch frequency of phenols in weakly interacting solvents is significantly red-shifted at higher concentrations (through hydrogen bonding interactions with other phenol molecules, where dimers and other oligomeric species may be formed), and in more polar solvents.[4] This suggests that IR probing of the O–H stretch region of solutions of phenol at different concentrations (*i.e.* with differing amounts of clustering) may offer a route to exploring the varying dissociation dynamics of different cluster sizes. Fig. 3 shows recent results obtained following UV excitation (267 nm) of phenol solutions of different concentrations with subsequent IR probing in the range 3700–3200 cm^{-1} at a fixed pump–probe delay of $\delta t = 10$ ps. These spectra are dominated by bleach signals corresponding to the O–H stretching wavenumbers of the phenol monomer (the sharp band at 3616 cm^{-1}) and oligomers (a broad band in the 3550–3200 cm^{-1} region). The transient IR probe spectra obtained at $\delta t = 10$ ps show another clear absorption feature at $\sim$3540 cm^{-1}, which is tentatively assigned to the "free" O–H stretch vibration of phenol molecules excited to the S_1 state. No analogous gain feature corresponding to population of the S_1 states of the dimer or of the oligomeric species of phenol are observed. We can envisage at least two possible explanations for this. First, we may expect that the band associated with these O–H stretching

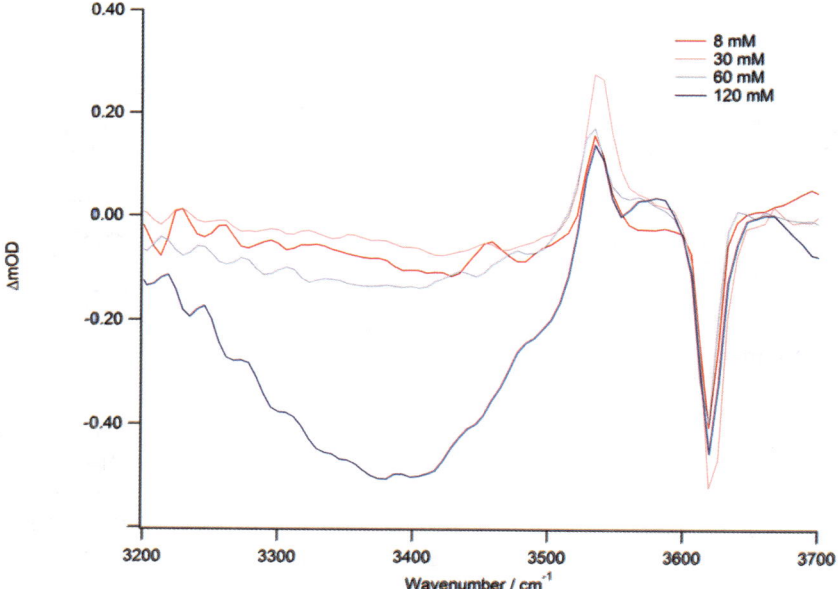

Fig. 3 Transient IR spectra obtained following 267 nm excitation of solutions comprising differing concentrations of phenol in cyclohexane at a pump–probe delay of 10 ps, probing in the range of 3200–3700 cm^{-1}.

motions in the excited state will be broad (as is the case for the ground state molecules) and shifted only slightly from the ground state bleach feature. In such a case, the excited state absorption may simply be hidden beneath the ground state bleach feature. Second, UV excitation may cause the dimeric and oligomeric species to break apart or rearrange – thereby resulting in a complete absence of, or a greatly diminished contribution from, the excited state feature corresponding to the hydrogen bonded species. Elucidation of the exact processes occurring in this case and detailed explanation of these results is the subject of on-going work but serves to illustrate that, as suggested by Professor Bradforth, investigation of the photodissociation dynamics of UV excited clusters in solution is an interesting and relatively unexplored area of research. These results also demonstrate the usefulness of employing both transient UV/visible and transient IR probing methods to gain a more comprehensive picture of the products created and the processes occurring following UV excitation of aromatic and heteroaromatic molecules in the solution phase.

1 S. J. Harris, D. Murdock, M. N. R. Ashfold, I. P. Clark, G. M. Greetham and M. Towrie, unpublished results, 2012.
2 Y. Zhang, T. A. A. Oliver, M. N. R. Ashfold and S. E. Bradforth, *Faraday Discuss.*, 2012, **157**, DOI: 10.1039/C2FD20043K.
3 T. A. A. Oliver, Y. Zhang, M. N. R. Ashfold and S. E. Bradforth, *Faraday Discuss.*, 2011, **150**, 439.
4 G. Keresztury, F. Billes, M. Kubinyi and T. Sundius, *J. Phys. Chem. A*, 1998, **102**, 1371–1380.

Professor Huse added: The work by the groups of Ashfold and Bradforth proposes to explore hydrogen bonding between solvated phenol molecules. In this context, S. Harris commented on the paper, showing in Fig. 3 in her question the infrared response to electronic excitations with clear ground-state bleaching signals of the free (3620 cm^{-1}) and hydrogen-bonded (<3520 cm^{-1}) OH stretching vibration (ν_{OH}). A clear induced absorption of the product free ν_{OH} vibration at 3532cm^{-1} is visible in the same figure at all concentrations with a similar Lorentzian linewidth (~18cm^{-1}) as the bleached ground state signal of the free ν_{OH}.

The strong bleaching signal of the hydrogen-bonded ν_{OH} vibration is reminiscent of the same signal from acetic acid dimers (formed in aprotic solvents). In particular, the broad bleaching signal of the hydrogen-bonded ν_{OH} vibration has very likely the same origin as found in acetic acid dimers which the group of Thomas Elsaesser extensively studied using various third-order femtosecond infrared spectroscopies (pump–probe, transient grating, 2-pulse-photon echo, 3-pulse-photon echo peak shift, and Fourier-transform 2-dimensional spectroscopy).[1-6] While the dimer is a stable structure in aprotic solvents with very little free ν_{OH} vibrations present[7] (see also Fig. 4.2 of ref. 8 and data in Fig. 4. in this response). Phenol in cyclohexane can form a multitude of hydrogen-bonded structures (*cf.* Fig. 6 of the paper), which is probably the reason for the less structured bleaching signal as compared to ν_{OH} bleaching in acetic acid dimers. However, the dominant underlying coupling mechanisms should still exist, namely (i) Fermi-resonances of combination and overtones with the first excited state of the ν_{OH} mode and (ii) Franck–Condon progressions of hydrogen-bond modes. These two coupling mechanisms (along with Davydov coupling) had been discussed for many decades,[9] showing very similar coupling strength in the acetic acid dimer.[10] Interestingly, the dominant spectral features arise from Fermi resonances of various combination and overtones (2 δ_{OH} = 2, ν_{C-O} = 2, ν_{C-O} = 1 + ν_{C-O} = 1, ...) with the ν_{OH} mode. Franck–Condon progressions of (low-frequency) hydrogen-bond modes manifest prominently in coherent oscillations in any of the third-order spectroscopies of the ν_{OH} mode employed.

It would indeed be interesting to study how hydrogen bonding influences the chemistry as studied in the paper. Hydrogen bonding could provide an efficient hydrogen transfer mechanism similar to intramolecular H-transfer (*e.g.*

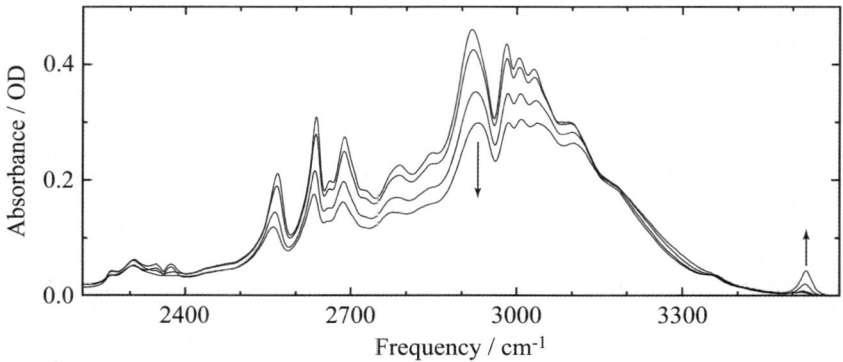

Fig. 4 Temperature-dependent absorption in the region of the OH stretching vibration of 0.5 M acetic acid dissolved in CCl₄. The spectra were recorded at −25, −10, 30, and 70 °C with the absorption of the free OH group at 3538cm⁻¹ increasing with temperature.

Rini *et al.*[11,12]) and resulting phenoxyl carbonyl groups (instead of phenoxyl radicals) could act as hydrogen bond acceptors (red-shifting the $\nu_{C=O}$ 0→1 transition). Different formation and relaxation times of the possible products should provide additional information to identify the various chemical reaction pathways.

1 K. Heyne, N. Huse, E. T. J. Nibbering and T. Elsaesser, Coherent vibrational dynamics of intermolecular hydrogen bonds in acetic acid dimers studied by ultrafast mid-infrared spectroscopy, *J. Phys.: Condens. Mat.*, 2003, **15**, 129.

2 K. Heyne, N. Huse, E. T. J. Nibbering and T. Elsaesser, Ultrafast coherent nuclear motions of hydrogen bonded carboxylic acid dimers, *Chem. Phys. Lett.*, 2003, **369**, 591.

3 K. Heyne, N. Huse, E. T. J. Nibbering and T. Elsaesser, Ultrafast relaxation and anharmonic coupling of O–H stretching and bending excitations in cyclic acetic acid dimers, *Chem. Phys. Lett.*, 2003, **382**, 19.

4 N. Huse, K. Heyne, J. Dreyer, E. T. J Nibbering and T. Elsaesser, Vibrational Multilevel Quantum Coherence due to Anharmonic Couplings in Intermolecular Hydrogen Bonds, *Phys. Rev. Lett.*, 2003, **91**, 197401.

5 K. Heyne, N. Huse, J. Dreyer, E. T. J. Nibbering, T. Elsaesser and S. Mukamel, Coherent low-frequency motions of hydrogen bonded acetic acid dimers in the liquid phase, *J. Chem. Phys.*, 2004, **121**, 902.

6 N. Huse, B. D. Bruner, M. L. Cowan, J. Dreyer, E. T. J. Nibbering, R. J. D. Miller and T. Elsaesser, Anharmonic couplings underlying ultrafast vibrational dynamics of hydrogen bonds in liquids, *Phys. Rev. Lett.*, 2005, **95**, 147402.

7 Y. Fujii, H. Yamada, and M. Mizuta, Self-Association of Acetic Acid in Some Organic Solvents, *J. Phys. Chem.*, 1988, **92**, 6768.

8 Nils Huse, Multidimensional Vibrational Spectroscopy of Hydrogen-Bonded Systems in the Liquid Phase: Coupling Mechanisms and Structural Dynamics, Mensch & Buch Verlag, Berlin, 2007.

9 G. M. Florio, T. S. Zwier, E. M. Myshakin, K. D. Jordana and E. L. Sibert, Theoretical modeling of the OH stretch infrared spectrum of carboxylic acid dimers based on first-principles anharmonic couplings, *J. Chem. Phys.*, 2003, **118**, 1735 and references therein.

10 J. Dreyer, Hydrogen-bonded acetic acid dimers: Anharmonic coupling and linear infrared spectra studied with density functional theory, *J. Chem. Phys.*, 2005, **122**, 184306.

11 M. Rini, A. Kummrow, J. Dreyer, E.T.J. Nibbering and T. Elsaesser, Femtosecond mid-infrared spectroscopy of condensed phase hydrogen-bonded systems as a probe of structural dynamics, *Faraday Discuss.*, 2003, **122**, 27.

12 M. Rini, J. Dreyer, E. T. J. Nibbering and T. Elsaesser, Ultrafast vibrational relaxation processes induced by intramolecular excited state hydrogen transfer, *Chem. Phys. Lett.*, 2003, **374**, 13.

Professor Wester asked: The central chromophore anion of the Green Fluorescent Protein (GFP) is known to fluoresce efficiently inside the protein, but not isolated in solution. This is attributed to excited state quenching *via* a conical intersection, which becomes accessible due to the large amplitude motion in solution that is suppressed in the rigid protein configuration. You find a longer S_1 lifetime in phenol clusters where the hydrogen bonds stiffen the cluster structure. What are the prospects for observing a related enhancement of fluorescence in the phenol cluster with respect to the monomer due to this longer S_1 lifetime?

Professor Bradforth responded: The monomer phenol S_1 lifetimes from transient absorption in this work agree well with time-resolved fluorescence data and even the 2 ns fluorescence lifetime in the gas phase.[1,2] The same gas phase experiments report an increased excited state lifetime for phenol dimer of 16 ns.[3] Rather than a stiffening of the structure, this was attributed to elimination of internal conversion mediated by the free O-H stretch as promoting mode. But the data in our paper (inset of Fig. 7(d) compared to Fig. 5(c)) suggests that the S_1 lifetime, at least when phenol is excited at 266nm, is changed relatively little on increasing the fraction of phenol hydrogen-bonded dimers within cyclohexane. However, in considering the current data it is important to recall two factors. 70% of the equilibrium population at 90 mM is still isolated monomers. Second, dissociation is not the major decay pathway for S_1 phenol so the fact that clustering inhibits dissociation may not necessarily be reflected in much of a change in the excited state lifetime. What is surprising is we don't see the turning off of internal conversion when the cluster is surrounded by cyclohexane. This perhaps suggesting a different hydrogen-bonded dimer structure compared to the gas phase.[4]

We note that the excimer lifetime (measured at 600 nm) tracks a different geometric conformation of the dimer that has rearranged to achieve π stacking. For this reason we do not compare the lifetime of the 600 nm band to the monomer lifetime. Fluorescence from the excimer is not observable in cyclohexane, suggesting that the excimer has a stronger oscillator strength in absorption and much weaker one in emission compared to the monomer.

1 C. Ratzer, J. Küpper, D. Spangenberg, and M. Schmitt, *Chem. Phys.*, 2002, **283**, 153.
2 G. A. Pino, A. N. Oldani, E. Marceca, M. Fujii, S. I. Ishiuchi, M. Miyazaki, M. Broquier, C. Dedonder and C. Jouvet, *J. Chem. Phys.*, 2010, **133**, 124313.
3 A. Sur and P. M. Johnson, *J. Chem. Phys.*, 1986, **84**, 1206–1209.
4 V. Poterya, L. Sistik, P. Slavicek and M. Farnik, *Phys. Chem. Chem. Phys.*, 2012, **14**, 8936–8944

Dr Glowacki asked: I have recently been involved in collaborative work at Bristol looking at the details of vibrational relaxation dynamics of HCN[1-4] in solvents ranging from dichloromethane (DCM) to tetrahydrofuran (THF). In that work, we found a range of relaxation times, from ~150–200 ps in DCM[1-3] to ~10 ps in THF[4]. We have been able to effectively interpret this range of relaxation times according to overlap of solute/solvent spectral bands, and this has shed considerable microscopic information on the mechanisms for energy transfer from the vibrationally excited solute to the solvent.[2,4] You report vibrational relaxation timescales for PhO(X) of ~5 ps, which is very fast. Do you have any insight as to what sorts of solvent/spectral overlaps might account for such rapid cooling, and does this provide any qualitative insight into what modes might be most important in facilitating this rapid energy transfer from solute to solvent?

1 S.J. Greaves *et al.*, Vibrationally quantum-state-specific reaction dynamics of H atom abstraction by CN radical in solution, *Science*, 2011, **331**, 1423–1426.
2 D. R. Glowacki *et al.*, Ultrafast energy flow in the wake of solution phase bimolecular reactions, *Nat. Chem.*, 2011, **3**, 850–855.
3 D. R. Glowacki *et al.*, Product Energy Deposition for CN + alkane H abstractions in Gas and Solution Phases, *J. Chem. Phys.*, 2011, **134**, 214508.
4 R. A. Rose *et al.*, Reaction dynamics of CN radicals with tetrahydrofuran in liquid solutions, *Phys. Chem. Chem. Phys.*, 2012, **14**, 10424–10437.

Professor Bradforth replied: Your example of HCN is rather special as the C–H stretch has an unusually long vibrational lifetime due to its high frequency. Therefore, only solvents that have similar frequency modes can rapidly relax the C–H vibration; otherwise IVR to lower frequency vibrations usually takes place first.[1] The vibrations being cooled in the hot phenoxyl fragment produced by photodissociaton (ring distortions) are likely to have much better overlap with the solvent spectral density. The recent experiments carried out by the Ashfold group at Bristol, and reported by Professor Ashfold in his earlier response, are likely an excellent source of information to address mode specific relaxation dynamics after the bond scission is complete.

1 L. K. Iwaki and D. D. Dlott, *J. Phys. Chem. A.*, 2000, **104**, 9101–9112.

Professor Huse commented: you mention in the introduction of the paper that no excited state proton transfer has been observed upon photoexcitation of either MePhOH or *p*-MePhSH, neither in cyclohexane nor in ethanol. The hydrogen release is reminiscent of work by Rini *et al.*[1] in the sense that very fast hydrogen release (within 3–5 periods of the OH stretching vibration) rather proton transfer occurs. Naively, one could imagine that in solvents such as ethanol and water, hydrogen bonding could promote proton transfer to the solvent rather than hydrogen release/transfer. Is there an intuitive explanation for the absence of proton transfer beyond the theoretical calculations (Fig 1. of the paper) and the experimental findings (Fig. 3 of the paper)?

1 Rini *et al.*, *Faraday Discuss.*, 2003, **122**, 27.

Professor Bradforth responded: By way of clarification, the statement about the absence of excited state proton transfer in our paper only identified the MePhSH system. We have not yet reported on excited state dynamics for phenol in hydrogen-bonding solvents and, in fact, early results do suggest differences in reaction dynamics compared to cyclohexane, but interestingly do not support excited-state proton transfer. The work you cite describes intramolecular hydrogen atom transfer in the excited state promoting tautomerization. For such systems the H-atom acceptor is primed to receive the incoming atom and rearrange the solute electronic structure, whereas in simpler single functional group aromatics the H-atom transfer coordinate is really much better described as straight bond dissociation. When the excitation energy launches the MePhSH system above the conical intersection, prompt H-atom dissociation can occur on timescales comparable to or faster than the S–H vibrational period. Past work on photoacids with phenolic chromophores, such as β-naphthol, suggests that excited state proton transfer to the solvent is often on the tens to hundreds of picoseconds timescale (as cited by Rini *et al.*[1]). It may be for some systems like those studied here that competing kinetics disfavors excited state proton transfer in hydrogen-bonding solvents compared to other reaction channels. Clearly the thermodynamics of each solute/solvent system needs to be carefully assessed before making such generalizations.

1 Rini *et al.*, *Faraday Discuss.*, 2003, **122**, 27.

Dr Roberts remarked: The discussion paper by Bradforth and co-workers presents data which evidences H-atom tunnelling under a $^1\pi\pi^*/^1\pi\sigma^*$ conical intersection in the excited state dynamics of phenol in a cyclohexane solvent. This is analogous to the behaviour observed in their gas phase H-atom Rydberg tagging translational spectroscopy measurements. Has the influence of the solvent on the dynamics been investigated in any further detail by performing experiments in different solvents (*e.g.* polar, aprotic *etc.*)? If so, how are the excited state tunnelling dynamics affected?

Professor Bradforth replied: We have also explored the excited state dynamics for phenol in two polar solvents at the same excitation energies where tunneling is operative in vacuum and in cyclohexane. In water and ethanol, several additional complications come into play. First, hydrogen bonding of the phenolic hydrogen to the solvent can be expect to impede the dissociation pathway, as we have seen in the phenol dimer clusters. Second, the polarity of the environment stabilizes the heterolytic dissociation products, $PhO + H^+ + e^-_{(solv)}$ relative to the position of this thermodynamic asymptote in the gas phase and these products are instead observed in water. Curiously, the timescale for appearance of the $PhO + H^+ + e^-_{(aq)}$ fragments also takes place on a several nanosecond timescale, which is unusual for aqueous phase photoionization pathways. In ethanol the homolytic fission seen in cyclohexane still seems to dominate with phenoxyl again observed on a nanosecond timescale; this suggests H atom tunneling is still occurring. In some senses, the experiments reported here serve as a control experiment – gas phase reaction dynamics are by and large preserved in cyclohexane where the electrostatic effects are less strong, which establishes the viability for tunneling in this system even in a condensed environment. But once the solvent can stabilize other channels, such as charge separation, the dominant gas phase mechanism must outcompete newer channels to survive. We are currently exploring the phenol fragmentation in water in some detail, trying to establish whether the proton and electron transfer motions are coupled.

Professor Gruebele opened the discussion of the paper by Professor Fielding: The dual-path mechanism of the S_2 ($\pi 3s\ \pi\sigma^*$) is very interesting, and I wonder how general it is in related biomolecules, such as the *meta/ortho/para* nucleobase analogs C_5NH_4–NH_2 In those molecules, heteroatom substitutions, particularly ring nitrogens putting an extra lone pair of electrons into the ring system, are common. Your experiments nicely demonstrated that CASPT2 calculations (with your collaborator Guérin of Dijon) produce reasonable energy ordering in the carbon-only ring case. Do you believe the dual-path mechanism will persist in these heterocyclic systems, or will shifts due to the lone pair and more electronegative heteroatom tip the balance in favor of either S_2–S_0 or S_2–S_1–S_0?

Professor Fielding answered: This is an interesting question and one for which I do not have an immediate answer other than that it will be very interesting to investigate the effect of nitrogen substitution in the ring using time-resolved photoelectron imaging.

Professor Ashfold said: Professor Fielding has provided a beautiful illustration of the potential of time resolved photoelectron spectroscopy (TRPES) studies for unravelling the excited state photophysics of heteroaromatic molecules like aniline. The abstract of that paper concludes with a statement that the present analysis involves some agreement and some contradiction with conclusions reached in other recent studies of the UV photochemistry of aniline. Can we be clear what these contradictions are? Relative to conclusions drawn in the earlier H Rydberg atom photofragment translational spectroscopy (PTS) study,[1] I sense the main point of contention may be the relative importance of direct versus indirect population of the $S_2(1\pi(3s/\sigma^*))$ state. The present analysis implies some direct population of this state at all wavelengths studied. I do not see this as inconsistent with the earlier results – indeed we noted that the observation of H atoms with kinetic energies consistent with formation of the partner radical products in their $v = 0$ state could indicate some propensity for direct population of the S_2 state at wavelengths as short as $\lambda = 240$ nm.

1 G. A. King, T. A. A. Oliver and M. N. R. Ashfold, *J. Chem. Phys.*, 2010, **132**, 214307.

Professor Fielding answered: Our data and analysis are in agreement with previous experimental observations but our interpretation of the conclusions of earlier work was that our analysis contradicted two points.

First, following excitation to wavelengths $\lambda \leq 240$ nm, the rapid decay observed in femtosecond pump–probe photoionisation experiments[3] was attributed to internal conversion through an S_3/S_1 conical intersection. It is our understanding that H atom loss observed in photofragment translational spectroscopy studies[1] and femtosecond pump–probe velocity map imaging studies,[5] was proposed to arise predominantly from coupling to the $\pi\sigma^*$ component of $S_2(\pi 3s/\pi\sigma^*)$ through successive S_3/S_1 and S_1/S_2 conical intersections, or through an S_3/S_2 conical intersection. However, the absence of significant negative amplitude features in our spectra associated with the ~100 fs decay (Fig. 6 of our discussion paper and Fig. 4 in ref. 6) suggests that there is not any substantial population of $S_1(\pi\pi^*)$ or $S_2(\pi 3s/\pi\sigma^*)$ during relaxation from $S_3(\pi\pi^*)$, which is why we propose that an extremely efficient decay path takes the majority of the $S_3(\pi\pi^*)$ population straight back to the ground state, most probably through an S_3/S_0 conical intersection. Preliminary calculations carried out by Matthieu Sala and Stephane Guerin (Dijon) in collaboration with our group, suggest that population may pass diabatically through conical intersections with the lower singlet states located close to the Franck-Condon region before passing adiabatically through an S_3/S_0 conical intersection.

Second, as Professor Ashfold states, our data shows that the $S_2(\pi 3s/\pi\sigma^*)$ state is populated directly at all the excitation wavelengths we studied. From our understanding, this contradicts the suggestion made following other recent studies,[1,3,5] that it is $S_1(\pi\pi^*)$ that is excited directly and that subsequent H atom loss proceeds by coupling to $S_2(\pi 3s/\pi\sigma^*)$. We are happy to note that the photofragment translational spectroscopy studies[1] suggested the possibility of direct excitation of $S_2(\pi 3s/\pi\sigma^*)$.

1 G. A. King, T. A. A. Oliver and M. N. R. Ashfold, *J. Chem. Phys.*, 2010, **132**, 214307.
2 Y. Honda, M. Hada, M. Ehara and H. Nakatsuji, *J. Chem. Phys.*, 2002, **117**, 5.
3 R. Montero, A. P. Conde, V. Ovejas, R. Martinez, F. Castaño and A. Longarte, *J. Chem. Phys.*, 2011, **135**, 054308.
4 R. Spesyvtsev, O. M. Kirkby and H. H. Fielding, *Faraday Discuss.*, 2012, **157**, DOI: 10.1039/C2FD20076G.
5 G. M. Roberts, C. A. Williams, J. D. Young, S. Ullrich, M. J. Paterson and V. G. Stavros, *J. Am. Chem. Soc.*, 2012, **134**, 12578–12589.
6 R. Spesyvtsev, O. M. Kirkby, M. Vacher and H. H. Fielding, *Phys. Chem. Chem. Phys.*, 2012, **14**, 9942.

Professor Ashfold asked: I would be interested to hear Professor Fielding's views on the relative efficiencies with which the different experiments sense the various excited states of interest. The PTS experiments depend on the partial UV absorption cross-sections (and the subsequent probability for eventual N–H bond fission). The available data suggests that the oscillator strength of the $3s/\sigma^* \leftarrow \pi$ transition is at least one order of magnitude smaller than that for either $\pi^* \leftarrow \pi$ excitation.[1] The TRPES yield, in contrast, is sensitive to the initial UV absorption cross-section and that of the subsequent ionisation step. Is it reasonable to expect that the ionisation cross-section from the S_2 Rydberg state will be larger than that of the S_1 and S_3 valence states, and that the TRPES experiments are thus particularly sensitive to S_2 population?

1 Y. Honda, M. Hada, M. Ehara and H. Nakatsuji, *J. Chem. Phys.*, 2002, **117**, 5.

Professor Fielding replied: This is a good point to make. It seems perfectly reasonable to assume that the ionisation cross-section for the Rydberg state is larger than those of the $\pi\pi^*$ states. None of the experiments allow the relative yields to be determined directly, but calculations of ionisation cross-sections could allow the yields to be estimated from photoelectron spectroscopy experiments.

Professor Ashfold queried: The S_2 state lifetime is deduced to be ~200 fs,[1,2] though the present analysis suggests that another subset of this same state has a longer (ps) lifetime. How might this come about?

1 R. Montero, A. P. Conde, V. Ovejas, R. Martinez, F. Castaño and A. Longarte, *J. Chem. Phys.*, 2011, **135**, 054308.
2 R. Spesyvtsev, O. M. Kirkby and H. H. Fielding, *Faraday Discuss.*, 2012, **157**, DOI: 10.1039/C2FD20076G.

Professor Fielding responded: This might occur if the population on the S_2 excited state bifurcates to take population to both the S_2/S_1 and S_2/S_0 conical intersections.

Professor Bowman remarked: A perhaps naive question is, what role do triplet state(s) play in the aniline photochemistry?

Professor Fielding responded: Following excitation near the origin of the $S_1(\pi\pi^*)$ excited state, the triplet state is formed on a nanosecond timescale[1,2]. We also observed triplet formation on a nanosecond timescale at low excess energies in $S_1(\pi\pi^*)$ and thought that we might find some evidence for ultrafast intersystem crossing at higher excitation energies, analogous to benzene,[3,4] but we did not.

1 R. Scheps, D. Florida and S. A. Rice, *J. Chem. Phys.*, 1974, **61**, 1730–1747.
2 B. Kim, C. P. Schick and P. M. Weber, *J. Chem. Phys.*, 1995, **103**, 6903.
3 D. S. N. Parker, R. S. Minns, T. J. Penfold, G. A. Worth and H. H. Fielding, *Chem. Phys. Lett.*, 2009, **469**, 43–47.
4 R. S. Minns, D. S. N. Parker, T. J. Penfold, G. A. Worth and H. H. Fielding, *Phys. Chem. Chem. Phys.*, 2010, **12**, 15607–15615.

Professor Neumark asked: The delay time of 1-2 ps to reach the S_2/S_0 conical intersection seems rather long, given that S_2 is primarily a dissociative state (Fig 1 in your paper). What is your explanation for this? Does the Rydberg character of this state at short N–H bond distances play a role, or do the one-dimensional potential energy curves in Fig. 1 leave out too much information to enable one to draw conclusions about the dynamics?

Professor Fielding replied: The one-dimensional potential energy curves in Fig. 1 are misleading as they leave out a lot of information. The barrier on the $\pi3s/\pi\sigma^*$ surface seems likely to be important.

Domcke, Sobolewski and coworkers[1] observed that following excitation to the $\pi\sigma^*$ state of pyrrole, which also has a barrier along the N–H stretch coordinate, the high energy part of the wave packet reached the $\pi\sigma^*/S_0$ conical intersection within 10 fs but the low energy part of the wave packet took a few hundred femtoseconds to reach the conical intersection because it had to tunnel through the barrier. The barrier height for pyrrole was determined to be 0.26 eV but our preliminary estimation for aniline is around 0.3 eV, which could explain a slightly longer tunnelling time.

As mentioned in our discussion paper, it is worth noting that this 1–2 ps long timescale is similar to that observed in indole and hydroxyindole[2] where, like aniline, the $\pi\sigma^*$ and $\pi\pi^*$ states are coupled at short N–H distances.

Alternatively, as we also mention in the discussion paper, the population may become trapped in the upper cones of the conical intersection, as has been observed in thioanisole and methylamine.[3,4]

1 V. Vallet, Z. Lan, S. Mahapatra, A. L. Sobolewski and W. Domcke, *J. Chem. Phys.*, 2005 **123**, 144307.
2 R. Livingstone, O. Schalk, A. E. Boguslavskiy, G. Wu, L. Therese Bergendahl, A. Stolow, M. J. Paterson and D. Townsend, *J. Chem. Phys.*, 2011, **135**, 194307.
3 J. S. Lim and S. K. Kim, *Nat. Chem.*, 2010, **2**, 627–632.

4 D.-S. Ahn, J. Lee, J.-M. Choi, K.-S. Lee, S. J. Baek, K. Lee, K. K. Baeck and S. K. Kim, *J. Chem. Phys.*, 2008, **128**, 224305.

Dr Stavros said: Helen, You presented some really nice data on the excited state dynamics in aniline following photoexcitation at a range of wavelengths. I wanted to ask whether you carried out measurements following photoexcitation directly to the S_1 origin and if so, whether you saw any anisotropy in the photoelectron angular distributions consistent with photoelectrons being ejected from the 3s Rydberg state?

Professor Fielding responded: We have not yet looked at excitation energies below the $S_2(\pi 3s/\pi\sigma^*)$ origin (269 nm) but this is an interesting question and we will check at lower excitation energies to see if we do see evidence of photoelectrons ejected from the 3s Rydberg state.

Professor Truhlar opened the discussion of the paper by Professor Suits: Firstly, the paper raises the interesting issue of the number of isomers and cites papers by Ohno *et al.* and Moc *et al.* showing 14 isomers for butanol. Our work shows 29 structures for n-butanol, consisting of 14 pairs of mirror images and one structure that does not have a distinguishable mirror image due to a plane of symmetry.[1] For isobutanol we found nine structures including four pairs of mirror images.[1]

Secondly, the paper also points out that the transition state of the $O(^3P)$ plus butane reaction releases 4–5 kcal mol^{-1} into the butyl radical by structural relaxation of the radical from its structure at the reactant to the product radical's lowest-energy equilibrium structure. Rubén Meana Pañeda, Ewa Papajak, and I have carried out this calculation for Cl + n-butanol (using the M06-HX density functional and the 6-311+G(2df,2p) basis set). For abstraction of H at C-1, we obtain 5.85 kcal mol^{-1}, and for abstraction at C-2 we obtain 6.90. We predict that abstraction at C-1 is barrierless, but for the C-2 abstraction we also carried out the calculation for relaxation from the lowest-energy saddle point, which yields only 1.36 kcal mol^{-1}, considerably less than the quoted results for O + butane. How does this much smaller number affect the discussion of recoil energy?

1 P. Seal, E, Papajak, T. Yu, and D. G. Truhlar, Statistical Thermodynamics of 1-Butanol, 2-Methyl-1-Propanol, and Butanal, *J. Chem. Phy.*, 2012, **136**, 034306/1–10.

Professor Suits replied: Thank you for your reappraisal of the conformational isomerism in these butanol isomers. This is clearly a lesson we learn in extending the methods of reaction dynamics to larger systems: this conformational heterogeneity may obscure some of the issues we wish to investigate.

On the question of the internal energy in the radical product of these reactions you point out that the relaxation energy of the radical from the transition state geometry is significantly lower than from the parent geometry for abstraction at C-2. But I wonder how close to the transition state configuration these systems actually approach at these collision energies, and I look forward to your dynamical calculations to gain some insight into this. My intuition suggests the relevant value will be somewhere between the transition state number you mention and the larger value corresponding to relaxation from the parent geometry. It is also interesting to note, as mentioned in the paper, that close collisions are inefficient at coupling the reaction exoergicity into recoil because of the acute skew angle for the reaction, while for large impact parameter collisions we see evidence of spectator stripping dynamics, which will also preclude this. For these reasons as well, I suspect the relevant number for the vibrational excitation in the radical may be closer to the parent relaxation energy.

Professor Nesbitt commented: The CH bond strengths are generally weaker by 10–20 kcal mol^{-1} or so than OH bond strengths, with CH bond strengths themselves varying substantially between primary, secondary and tertiary hydrogens. 1) Is the channel for OH bond cleavage by Cl energetically open at your highest collision energies? 2) Could perhaps HCl J-state resolved REMPI experiments provide direct energetic evidence for the presence or absence of these various primary, secondary and tertiary H atom abstraction pathways?

Professor Kable commented: The CH bond enthalpy for α-, β-carbon atoms increases with distance from the OH group, at least for the smaller n-alcohols. For example, the CH_3CHOH radical is about 7 kcal mol^{-1} more stable that the CH_2CH_2OH radical.[1] As a consequence, the radical abstraction of a H-atom from an alcohols should favor abstraction at the α site, followed by β, *etc.* For Cl atoms reacting with n-butanal, the rate constant for Cl abstraction has been reported to drop by a factor of ~2 from α- to β-carbon and again to the γ carbon.[2] Could the different dynamics at low and high collision energy have something to do with preferential reaction at the α-C at low energy, but more a more statistical reaction with the various C–H bonds at higher energy, when the difference in reaction enthalpy has less effect?

1 Z. F. Xu, K. Xu and M. C. Lin, *ChemPhysChem.*, 2009, **10**, 972.
2 H. Wu, Y. Mu, X. Zhang, G. Jiang, *Int. J. Chem. Kinetics*, 2003, **35**, 81.

Professor Suits answered: This is indeed something we hoped to see manifested in the differential cross sections. However, all isomers gave essentially the same differential cross section, and we believe the conformational heterogeneity blurs this effect, making the DCS less sensitive. Professor Orr-Ewing has suggested the rotationally-selected DCSs as one way to probe this, and this should certainly provide additional information. I suspect the most sensitive test of the distinct energetics (and the corresponding TS structures) would be to look at the HCl vibrational distributions as a function of scattering angle. We have plans to do this.

Professor Orr-Ewing said: The differential cross sections (DCSs) and kinetic energy release (KER) distributions reported by Prof Suits for the Cl + butanol reaction are very similar to those we reported for the Cl + methanol reaction at a mean collision energy of 23.4 kJ mol^{-1}.[1] We detected the HCl product of the reaction in $v' = 0$ and rotational levels with quantum numbers $J' = 2$–5, and used velocity map imaging to obtain the CM-frame angular and KER distributions. The DCSs show both backward and forward scattering components that change their ratios as J' changes. With the aid of electronic structure calculations and direct dynamics trajectory calculations,[2,3] we drew similar conclusions to those presented in the paper by Estillore *et al.*, such as that the shallow wells on the PES do not lead to formation of a long-lived complex, but that the forward and backward scattering instead results from direct reactions over a range of impact parameters. Given the observed similarities of the scattering dynamics for the Cl + butanol and Cl + methanol systems, are we now in a position to develop a generic model for Cl + alcohol reaction dynamics, or are the DCSs not sufficiently discriminating measurements of the differing dynamics? Do we instead need to look for more subtle indicators, such as we have done by measuring the rotational excitation of HCl products and the variation of the scattering with product quantum states, as illustrated by our studies of Cl + methanol and several other reactions of Cl atoms with organic molecules?[2,4-8]

1 C. Murray, A. J. Orr-Ewing, R. L. Toomes and T. N. Kitsopoulos, *J. Chem. Phys.* 2004, **120**, 2230.
2 S. Rudic, C. Murray, D. Ascenzi, H. Anderson, J.N. Harvey and A. J. Orr-Ewing, *J. Chem. Phys*, 2002, **117**, 5692.
3 S. Rudic, C. Murray, J. N. Harvey and A. J. Orr-Ewing, *J. Chem. Phys.*, 2004, **120**, 186.
4 S. Rudic, C. Murray, J. N. Harvey and A. J. Orr-Ewing, *Phys. Chem. Chem. Phys.*, 2003, **5**, 1205.

5 C. Murray, B. Retail and A. J. Orr-Ewing, *Chem. Phys.*, 2004, **301**, 239.
6 J. K. Pearce, C. Murray, P. N. Stevens and A. J. Orr-Ewing, *Mol. Phys.*, 2005, **103**, 1785.
7 C. Murray, J. K. Pearce, S. Rudic, B. Retail and A. J. Orr-Ewing, *J. Phys. Chem. A*, 2005, **109**, 11093.
8 R. L. Toomes, A. J. van den Brom, T. N. Kitsopoulos, C. Murray and A. J. Orr-Ewing, *J. Phys. Chem. A*, 2004, **108**, 7909.

Professor Suits replied: Indeed we were quite disappointed to see so little variation in the product differential cross sections with butanol isomer, and we have ascribed this to blurring of any isomeric signature by the conformational variety presented by each isomer and the ease with which any alkyl H atom reacts at these collision energies. You showed beautifully the influence of the OH dipole on the HCl rotational distributions, but I fear as the molecules get larger and conformationally diverse these effects may be overwhelmed by reaction at sites far from the oxygen. Nevertheless, it is in part with hope of seeing such effects that were are now implementing a sensitive Doppler-free probe of quantum state-selected HCl in these and other reactions in crossed-beams. HCl vibration is another place to look for some sensitivity to the abstraction site, both in the alcohols and in particular in alkenes in which the energetics vary greatly for reaction with different H atoms. We will be investigating this as well.

Professor Bowman asked: A general question is whether there is some way to detect multiple collision of the Cl with the various reaction sites in butanol or whether the dynamics is direct and abstraction occurs at each site independently?

Professor Suits answered: For the forward scattered component it is quite clear that these are direct. For the backscattered component multiple collisions are possible, but I do not believe we can find a clear signature of this in experiment. Dynamical calculations will offer the best hope of finding how important they are in these systems.

Professor Balucani communicated:† With reference to the interesting paper by Arthur Suits and coworkers,[1] where the reaction dynamics of several structural isomers of butanol with atomic chlorine is established, we would like to report some recent results we have obtained in the Perugia laboratory regarding the reactions of ground-state atomic oxygen with two structural isomers of unsaturated hydrocarbons, namely methylacetylene (CH_3CCH) and allene (CH_2CCH_2).[2] Even though these two species are characterized by different functional groups (a $C \equiv C$ triple bond in the case of methylacetylene and two contiguous double bonds for allene), quite often in their reactions they partly access regions of the underlying potential energy surface which link one reaction to the other. For instance, in the reaction $CN + CH_3CCH$ equal amounts of cyanoacetylene and cyanoallene have been observed.[3]

We have recently published a first account on the reactive system $O + CH_2CCH_2$.[2] Amongst the possible reaction channels we have verified that five of them are actually open

$$O + C_3H_4 \rightarrow CH_2{=}C{-}CHO + H \quad \Delta_r H°_{0K} = -58.6 \text{ kJ mol}^{-1} \quad \text{(a)}$$
$$\rightarrow CH_2CO + CH_2 \quad \Delta_r H°_{0K} = -102.1 \text{ kJ mol}^{-1} \quad \text{(b)}$$
$$\rightarrow C_2H_3 + HCO \quad \Delta_r H°_{0K} = -100.4 \text{ kJ mol}^{-1} \quad \text{(c)}$$
$$\rightarrow C_2H_4 + CO \quad \Delta_r H°_{0K} = -500.4 \text{ kJ mol}^{-1} \quad \text{(d)}$$
$$\rightarrow C_2H_2 + H_2CO \quad \Delta_r H°_{0K} = -324.7 \text{ kJ mol}^{-1} \quad \text{(e)}$$

By means of the crossed molecular beam method with mass spectrometric detection empowered by soft electron impact ionization we have been able to determine

---

† On behalf of Dr. Francesca Leonori, Dr. Vaclav Nevrly, Dr. Stefano Falcinelli, Prof. Domenico Stranges, and Prof. Piergiorgio Casavecchia.

their differential cross sections and product branching ratios at a collision energy of 39.3 kJ mol^{-1}.[2] Quite surprisingly for a system like this, we have inferred that most of the reaction proceeds via intersystem crossing (ISC) to the underlying singlet potential energy surface and, in fact, the channels which require ISC to be accessed (1d and 1e) account for more than 90% of the entire reaction. In particular, the most exothermic channel leading to C_2H_4 + CO accounts for ~80%. These observations have posed the question of how important it is to consider nonadiabatic effects (ICS) for this and other similar systems involved in combustion chemistry. For instance, also in the case of the $O(^3P)$ + C_2H_4 reaction it has been recently established that 50% of the reaction proceeds via ISC to the underlying singlet potential energy surface.[4] Overall, these observations are somewhat surprising because only systems like $O(^3P)$ + CH_3I, where the heavy iodine atom favours ISC via spin-orbit coupling, have previously manifested important effects of ISC in their reaction dynamics.[5]

More recently, we have faced the study of the O + CH_3CCH reaction under the same experimental conditions used for the O + CH_2CCH_2 reaction. For both systems, the reaction starts with the addition of the O atom either to the triple bond of methylacetylene[6] or to one of the double bonds of allene.[7] Once added to the triple bond of methylacetylene, the bound intermediate formed can directly dissociate (in this case producing also a CH_3 elimination channel in addition to the H-displacement one) or rearrange. ISC is also possible for this system and indeed, once ISC has occurred, the rearrangements of the intermediates lead to the same singlet methyl ketene and acrolein which, by dissociating, produce C_2H_4 + CO and $C_2H_2+H_2CO$. As a consequence, the centre-of-mass functions (product angular and translational energy distributions) appear to be identical for these two channels to those found for the $O(^3P)+CH_2CCH_2$ reaction. From our preliminary analysis, the main difference in the case of the two reactions is the reduced extent of ISC in the case of the $O(^3P)+CH_3CCH$ reaction with respect to the CH_2CCH_2 one. On the contrary, for each reaction channel the dynamics appears to be very similar. This can be appreciated by comparing the time-of-flight spectra recorded at the same angle (the centre-of-mass angle is similar for the two systems, as well as the collision energy) for the mass-to-charge (m/z) ratio of 26 (see Fig. 5). The same four channels contribute to this mass in both reactions (in the case of the $O+CH_3CCH$ reaction also a small fraction of the H $_2+CH_2CCO$ channel appears to contribute, while the equivalent channel is not present in the $O+CH_2CCH_2$ case), as indicated with product labels of different colour.

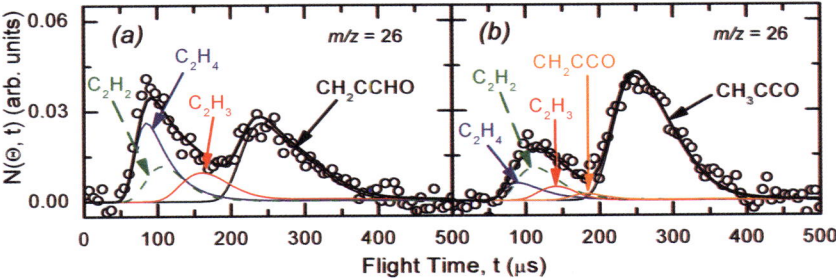

Fig. 5 Open circles: laboratory time-of-flight (TOF) distributions measured at $m/z = 26$ at the centre-of-mass angle of 37° for the reactions $O(^3P)+CH_2CCH_2$ at E_c=39.3 kJ mol^{-1} (left panel) and $O(^3P)+CH_3CCH$ at E_c=38.9 kJ mol^{-1} (right panel). The solid black curve represents the calculated total TOF distribution when using the weighted best-fit centre-of-mass functions for the various contributing channels (see ref. 2 in the paper). The separate contributions from the various channels to the calculated global TOF distributions are shown in different colours and indicated with the product formula.

As can be seen, the ratio between the slow peak centered around 250 μs, which originates from the dissociative ionization of the alkoxy radical corresponding to the H-elimination channel on the triplet surface, and the fast peak to which three different products contribute, all originating from the singlet surface (although the C_2H_3 + HCO channel can originate also from the triplet surface, statistical calculations indicate that it mostly arises from the singlet surface, at least in the case of the O + CH_2CCH_2 reaction[7]), is much smaller in the case of the O + CH_2CCH_2 reaction (left panel) with respect to the O + CH_3CCH reaction (right panel). This indicates that the ratio of triplet to singlet products increases when the hydrocarbon reactant changes from allene to methylacetylene, that is the extent of ISC decreases moving from allene to methylacetylene. This is not surprising in the light of the topology of the somewhat different potential energy surfaces for the two reactions. In particular, in the case of O + allene ISC can occur very efficiently in the region where triplet and singlet oxyallyl diradicals (CH_2CCH_2O) are very close (within 5–10 kJ mol^{-1}) to each other,[7] while in the case of O + methylacetylene ISC is expected to be significantly less efficient because the energy separation between the triplet and singlet diradical keto-carbene (CH_3CCHO) is much larger (about 40 kJ mol^{-1}).[6] A quantitative estimate of the extent of ISC in the case of the O + CH_3CCH reaction is currently underway; the results will be compared with those obtained for the related O + CH_2CCH_2 reaction and the effect of the different structure of the two C_3H_4 isomers on the extent of ISC will be discussed in detail.

1 A. D. Estillore, L. M. Visger-Kiefer and A. G. Suits, *Faraday Discuss.*, 2012, **157**, doi:10.1039/c2fd20059g.
2 F. Leonori, A. Occhiogrosso, N. Balucani, A. Bucci, R. Petrucci and P. Casavecchia, *J. Phys. Chem. Lett.*, 2012, **3**, 75–80.
3 N. Balucani, O. Asvany, R. I. Kaiser and Y. Osamura, *J. Phys. Chem. A*, 2002, **106**, 4301–4311.
4 B. Fu, Y.-C. Hana, J. M. Bowman, L. Angelucci, N. Balucani, F. Leonori and P. Casavecchia, *Proc. Natl. Acad. Sci. U. S. A.*, 2012, **109**, 9733–9738.
5 M. Alagia, N. Balucani, L. Cartechini, P. Casavecchia, M. van Beek, G. G. Volpi, L. Bonnet and J. C. Rayez, *Faraday Discuss.*, 1999, **113**, 133–150.
6 S. Zhao, W. Wu, H. Zhao, H. Wang, C. Yang, K. Liu and H. Su, *J. Phys. Chem. A*, 2009, **113**, 23–34.
7 T. L. Nguyen, L. Vereecken, X.-J. Hou, M. T. Nguyen and J. Peeters, *J. Phys. Chem. A*, 2005, **109**, 7489–7499.
8 F. Leonori, N. Balucani, V. Nevrly, S. Falcinelli, D. Stranges and P. Casavecchia, in preparation.

Professor Suits responded: I am delighted to see this rich contribution included in this session. I would like to point out some relevant papers they may wish to cite.[1,2]

1 A. M. Schmoltner, P. M. Chu, R. J. Brudzynski, Y. T. Lee, *J. Chem. Phys.*, 1989, **91**, 6926–6936.
2 A. M. Schmoltner, S. Y. Huang, R. J. Brudzynski, P. M. Chu, Y. T. Lee, Crossed molecular bean study of the reaction of (O^3P) with allene, *J. Chem. Phys.*, 1993, **99**, 1644-1653.

Professor Balucani said: I wish to thank Professor Suits for reminding us of the pioneering work of Y.T. Lee and his group on the crossed molecular beam study of the reactions of atomic oxygen with unsaturated hydrocarbons, namely O(^3P)+allene,[1] O(^3P) + ethylene[2] and O(^3P) + acetylene.[3] In particular, in the previous dynamic study of the reaction O(^3P) + allene,[1] the use of ^{18}O atoms permitted them to identify some of the open channels; however, the use of "hard" ionization detection prevented the observation of all the products and the estimate of branching ratios.

1 A. M. Schmoltner, S. Y. Huang, R. J. Brudzynski, P. M. Chu and Y. T. Lee, *J. Chem. Phys.*, 1993, **99**, 1644.
2 A. M. Schmoltner, P. M. Chu, R. J. Brudzynski, Y. T. Lee, *J. Chem. Phys.*, 1989, **91**, 6926.
3 A. M. Schmoltner, P. M. Chu, Y. T. Lee, *J. Chem. Phys.*, 1989, **91**, 5365.

Professor Nesbitt addressed Professor Suits and Professor Polanyi: Of course, CH abstraction chemistry by Cl atoms in the condensed phase can be highly selective, which is part of the great success of synthetic organic chemistry. This is basically by virtue of having relatively large activation barriers that only a small fraction of molecules can surmount for a given kT. As a result, small changes in relative barrier heights can make exponentially large differences in reaction rates. Could such a mechanism be operating here in the gas phase as well, whereby a Cl atom sticks transiently and "wanders" down the butanol "spine" to preferentially harvest the weaker CH bond?

Professor Polanyi responded: Those of us who, some time ago, were attracted to the field of reaction dynamics in gases, always dreamed that the vocabulary being developed might be useful in discussing reactions in solution (–only occasionally did we think of surface reaction, where the same applies). It is heartening, therefore, to learn of the parallels between reactions in different environments, being drawn today.

First a negative comment. A usage against which I have battled, clearly in vain, has been the habit of referring to new-born reaction products (in their ultimate quantum states) as 'nascent'. If a you visit a maternity ward and ask to see the nascent babies, provided that they don't throw you out they will take you to the delivery room, since 'nascent' means 'in the process of being born' – not new-born. It did not occur to me (though it should have) that we would eventually have legitimate need of the term 'nascent' when femtosecond lasers became available and pioneers like Ahmed Zewail and, at this meeting notably Thorsten Bernhardt's laboratory, begin studying what are truly nascent species, coming from time-resolved photodissociation of a parent molecule.

Now for the positive. In connection with the specificity that we observe in chemical reactions in gases and in solution, David Nesbitt asks whether a reagent molecule, having encountered its reaction partner at a site of low reactivity may then "wander" down its "spine" until it comes to a more reactive region. My reply is that this type of 'indirect' reaction has indeed been noted in the gas phase. It is shown in Fig. 12 of a review article[1] in which is pictured an 'abortive reaction' of H with the I end of ICl. During several H···I vibrations the H repeatedly approaches the I, constantly rebounding without reaction into the region of van der Waals attraction. Since the impact parameter of H with respect to ICl is non-zero, this extended interaction causes the H not only to vibrate against ICl but also to rotate around it. Vibration plus rotation causes the H to follow a wandering path, tracing out something like the shape of the petals of a daisy. The concurrent rotation of the HI axis along with the H···I vibration ultimately brings the attacking H-atom closer to the Cl end of the ICl molecule under attack, than to its I-end where it began. Since the formation of the HCl reaction-product is the path with the lowest activation energy, reaction at the Cl-end ensues. For obvious reasons, this type of reactive pathway was termed 'migratory'. It has recently found its analogue in photochemical events near threshold for product formation that allow the separating species to 'roam' in the van der Waals binding-region. 'Migration', like 'roaming' represents an 'indirect' mode of reaction, taking the reactive trajectory from reagents to products by a meandering path, rather than across the lowest barrier (the famous 'col' or transition state) that leads directly from reagents to products. In the early experimental examples (see for example Fig. 7 of the review cited[1]) the experimenters had the pleasure of recording the vibration, rotation and translation in precisely the same product formed by the indirect migratory pathway (leading to highly vibrationally and rotationally excited HCl with, therefore, very low translation) and also the product formed by the direct pathway (leading to low vibration and rotation, hence a high degree of translational excitation).

Does this discussion lend substance to Professor Nesbitt's dream of reagent A "wandering down the spine" of its collision-partner, B? I think so. In a favourable case, failed reactive encounters can hold a system together for long enough to permit A to wander to a more favourable site for reaction at B. This is not, however, to deny the validity of the point made by Professor Donald Truhlar, in the course of this Discussion, that selective reaction can frequently occur as a result of failed encounters followed, later and independently, by a fortunate (direct) collision with a reactive site on collision partner B. A question that remains for the future is the relative importance to chemistry of the lingering encounters involving migration, as against the (less tender) brief encounters.

1 J. C. Polanyi, *Science*, 1987, **236**, 680.

Professor Suits replied: Well, I believe in all of this we must not overlook the energetics or the time scale of the reaction. Our Cl + butanol reactions were carried out at 4 and 10 kcal mol^{-1}. At these enegiers the reactive "cone of acceptance" is large for all alkyl H atoms. There is no reason to seek a more favorable site. In a sense it is a shortcoming of the experiment. We would enjoy going to lower collision energies where our sensitivity to these effects would be greater. Another alternative might be different reactive targets such as (^{3}OP), with corresponding and revealing differences in reactivity.

Professor Gruebele asked: In these reactions, as molecular size grows, the prefactor A in the Arrhenius law $k = Ae^{-E_a/kT}$ shrinks because the probability of direct collision at the proper reaction site decreases. On the other hand, the Boltzmann factor plays a much more dramatic (exponentially scaling) role in tuning reaction probabilities: with two channels possible, the lower activation barrier channel is much more likely to win out, even if the steric factor A is larger for the other channel. Nonetheless, at very low reaction energies the attacking reagent could be trapped at the surface of the molecule (migratory reactions popular in physical organic mechanisms would be an example of this effect), and in the resulting complex the reagent then finds the reactive site with much higher probability than by single direct collision. Is there evidence in your data of such unusually large prefactors in halogen atom–alkanol reactions, particularly as the chain length of the alcohol increases?

Professor Suits replied: In fact, for these reactions, particularly at these collision energies, all alkyl sites are reactive, and the reaction cross section will simply grow with the size of the molecule. We have seen something related to what you describe in Cl reaction with alkenes, however.[1] In that case the Cl atom adds to the double bond quite readily, after which it can find the highly reactive allylic hydrogens very efficiently. At higher collision energies we could see direct H abstraction from remote alkyl sites but in this case the direct reactions were always forward scattered. We believe all the low impact parameter collisions were "captured" at the unsaturation site leading to complexes and likely resulting in formation of the resonantly stabilized allylic radicals.

This is also an opportunity to point out that although the exponential scaling of the energy factor leads this to control outcomes in general, there are many cases of dynamical control of reactions. For unimolecular reaction, simple bond fission can overcome an energetically favored molecular elimination pathway because of large differences in the pre-exponential factor. This is what drives "roaming" reactions: they are molecular pathways that benefit from the large A-factors of simple bond fission.[2]

1. A. D. Estillore, L. M. Visger and A. G. Suits, *J. Chem. Phys.*, 2010, **133**, 074306.
2. A. G. Suits, *Acc. Chem. Res.*, 2008, **41**, 873.

Professor Truhlar remarked: One issue raised in the discussion is the extent to which product branching ratios (relative yields) in the Cl + butanol reactions might be controlled by the relative bond energies. We[1] have calculated the relative bond energies for the five radicals produced by abstraction of H from 1-butanol and the four radicals produced by abstraction of H from 2-butanol. The calculations were carried out at the CCSD(T)-F12a/jul-ccpVTZ//M08-HX/MG3S level with scaled M08-HX/MG3S frequencies (the methods and basis sets are explained in the literature[1-8]). These calculations refer to the lowest energy conformers of the radicals. The relative bond energies are given in the ΔD_0 column of Table 1. We see differences of up to 9.5 kcal mol^{-1} in the relative bond energies, and even if attention is limited to the carbon radicals, we see differences of up to 6.0 kcal mol^{-1}.

The last two columns of the table give finite-temperature values of relative enthalpies and relative free energies at 298 K. These calculations include all conformers by using the MS-T method that is reviewed in our paper at this discussion. If relative free energies of activation were to correlate with thermodynamic quantities of the reactants and products, they would be more likely to correlate with relative free energies of reaction than with bond energies or relative enthalpies of reaction. We see that, even at room temperature (attending this discussion in Italy in the summer rather than the in UK provided a good reminder of why the "room temperature" language, which was established before air conditioning, refers to 298 K rather than a more normal modern laboratory temperature of 293 K), the relative free energies differ from the relative bond energies by as much as 1.4 kcal mol^{-1} (or, if we limit attention to carbon radicals, by as much as 0.6 kcal mol^{-1}). Comparison of the last two columns shows that this difference is mainly due to entropic factors, not energetic factors. Thus while the relative free energies of reaction are dominated by energetic effects, the entropic effects are by no means negligible even at room temperature, and they become even more important at higher temperature. One of advances in quantum chemical theory in recent years is that it is now being applied more often and with greater reliability to calculate free energies, not just energies.

Table 1 Relative energies, enthalpies, and free energies (kcal mol^{-1}) of radicals produced by hydrogen abstractions from 1-butanol and 2-butanol[a]

Radical	ΔD_0	ΔH_{298}	ΔG_{298}
radicals produced from 1-butanol			
$C_4H_9\dot{O}$	9.4	9.5	10.8
$C_3H_7\dot{C}HOH$	0.0	0.0	0.0
$C_2H_5\dot{C}HCH_2OH$	4.3	4.3	3.9
$CH_3\dot{C}HCH_2CH_2OH$	2.6	3.2	2.7
$\dot{C}H_2(CH_2)_2CH_2OH$	5.8	5.9	5.7
radicals produced from 2-butanol			
$(CH_3)_2CHCH_2\dot{O}$	9.5	9.0	10.3
$(CH_3)_2CH\dot{C}HOH$	0.0	0.0	0.0
$(CH_3)_2\dot{C}CH_2OH$	1.6	1.9	1.3
$\dot{C}H_2CH(CH_3)CH_2OH$	6.0	5.8	5.4

[a] For each precursor molecule, the energies are relative to the most stable radical. The finite temperature results are based on nine distinguishable conformers (four pairs of mirror images plus one symmetrical structure) for the 1-butoxyl radical, 36 conformers (18 pairs of mirror images) each for the 1-hydroxy-1-butyl and 1-hydroxy-2-butyl radicals, 38 conformers (19 pairs of mirror images) each for the 1-hydroxy-3-butyl and 1-hydroxy-4-butyl radicals, 3 conformers (1 pair of mirror images plus 1 symmetrical structure) for the 2-methyl-1-propoxyl radical, 12 conformers (6 pairs of mirror images) for the 1-hydroxy-2-methyl-1-propyl radical, 8 conformers (4 pairs of mirror images) for the 3-hydroxy-2-methyl-2-propyl radical, and 18 conformers (9 pairs of mirror images) for the 3-hydroxy-2-methyl-1-propyl radical.

1 E. Papajak, P. Seal, X. Xu and D. G. Truhlar, *J. Chem. Phys.*, in press.
2 K. Raghavachari, G. W. Trucks, J. A. Pople, M. Head-Gordon, *Chem. Phys. Lett.*, 1989, **157**, 479.
3 G. Knizia, T. B. Adler and H.-J. Werner, *J. Chem. Phys.*, 2009, **130**, 054104.
4 E. Papajak and D. G. Truhlar, *J. Chem. Theory Comput.*, 2010, **6**, 597.
5 Y. Zhao and D. G. Truhlar, *J. Chem. Theory Comput.*, 2008, **4**, 1849.
6 M. J. Frisch, J. A. Pople and J. S. Binkley, *J. Chem. Phys.*, 1984, **80**, 3265.
7 B. J. Lynch, Y. Zhao and D. G. Truhlar, *J. Phys. Chem. A*, 2003, **107**, 1384.
8 I. M. Alecu, J. Zheng, Y. Zhao and D. G. Truhlar, *J. Chem. Theory Comput.*, 2010, **6**, 2872.

Professor Bowman opened the discussion of the paper by Professor Kable: First a question and then a comment.

You noted that a cold rotational distribution in the CO fragment from CH_3CHO photodissociation is expected from the three-body fragmentation to CO + H + CH_3. And you caution therefore against inferring that the cold CO came from a roaming pathway to give CO + CH_4. This is true, however, wouldn't a simultaneous detection of CH_4 or CH_3 distinguish between these two pathways? So are you cautioning against inferring anything based just on the CO internal energy distribution?

The comment is meant to indicate that in fact roaming can also lead to the same three-body channel, *via* the following sequential mechanism. $CH_3CHO \rightarrow CH_4^* +$ CO, where CO is rotationally cold and CH_4^* is highly excited methane, formed *via* roaming. Then CH_4^* dissociates to CH_3 + H. We actually see this in QCT calculations at high photolysis energies where the three-body channel is open. However, in this case we actually see two "cold" CO rotational distributions, the colder one from CH_3 + $HCO^* \rightarrow CH_3$ + H + CO and the warmer one from the roaming pathway.[1] This is illustrated in Fig. 6.

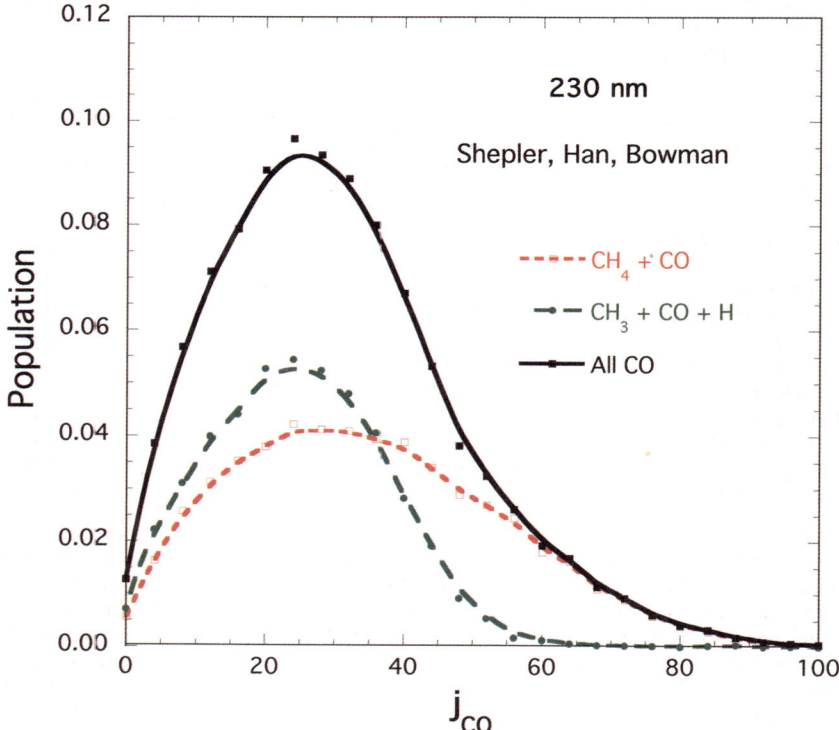

Fig. 6 CO rotational distribution from CH_3CHO photodissociation at a total energy corresponding to a photolysis wavelength of 230 nm.

1 B. C. Shepler, Y. Han, and J. M. Bowman, unpublished.

Professor Kable responded: In his question, Professor Bowman implies that obtaining more information about the other reaction products can lead to a clear distinction between roaming and 3F products. In this he is quite correct. Clearly, the detection of extremely vibrationally hot CH_4, as was the case in our experiments on CH_3CHO below the 3F threshold,[1] argues for a roaming mechanism. If hot CH_4 fragments were also detected above the 3F threshold then the same argument for a roaming mechanism would apply. The measurement of the CH_3 energy distribution is quite a different case. At first glance, the CH_3 fragment is likely born from the simple C–C bond cleavage in the primary 2F step. Subsequent dissociation of the HCO does not affect the CH_3 distribution. Therefore, one would not expect a measurement of the CH_3 distribution to be very informative about the presence or absence of triple fragmentation, or indeed roaming.

The statement above, however, has a rider that is nicely captured by your comment. Production of CH_4 + CO via a very high transition state, as shown schematically in Fig. 2 of our paper, will produce fragments with significant translational energy; your own calculations suggest ~50% of the available energy.[1] Therefore an insignificant amount of CH_4 will have sufficient internal energy to undergo secondary fragmentation. In your comment you remind us that the production of CH_4 + CO via roaming produces extremely highly vibrationally excited CH_4.[1] The distribution of translational energy more closely resembles that of the HCO + CH_3 channel from which the roaming is derived. In this case, one might expect a significant fraction of CH_4 fragments to undergo secondary dissociation to CH_3 + H. As your figure demonstrates, the distribution of energy in the CO fragment from roaming and 3F is the same, which is what is expected in the case that CO is the primary fragment. Using the same argument as for CH_3 above, the disposal of energy into the CO is therefore unaffected by the subsequent break up of CH_4.

This leads to consideration of the energy disposal in the 3F fragments, H and CH_3 following roaming. Our model uses PST and therefore cannot be used to model CO + CH_3 + H formed following roaming production of CO + CH_4 and the subsequent break up of CH_4. However, there are lessons from the modeling that can be applied. Firstly, it is only the high energy tail of the CH_4 internal energy distribution that will produce 3F fragments. Therefore the most likely available energy for the second step is zero; the 3F products will therefore have little internal energy. The translational energy in the roaming products is known to be small. Breaking the CH bond in CH_4 is barrierless, therefore the second step will also produce little translational energy. The combination of two processes that produce little translational energy will result in 3F fragments (H and CH_3) that are also very slow. Experimental detection of these 3F fragments, however, will be difficult because the 2F processes that produce H (+ CH_3CO) and CH_3 (+ HCO) on the S_0 state are also barrierless. As a consequence the distribution of translational energy from the 2F process will also be peaked at very low energy, which will mask the low energy contribution from 3F.

1 B. R. Heazlewood, M. J. T. Jordan, S. H. Kable, T. M. Selby, D. L. Osborn, B. C. Shepler, B. J. Braams and J. M. Bowman, *Proc. Natl. Acad. Sci. U. S. A.*, 2008, **105**, 12719.

Professor Suits commented: For sequential triple fragmentation of acetone, the product CO cannot in general be found at low translational energy in the center of mass frame. The reason is that the acetyl radical produced in the first step has a substantial exit barrier, so that even if there is just enough energy remaining for dissociation, the CO product will have significant rotational and translational energy.

An interesting related point concerns the production of $CO(v = 1)$ from a roaming event at an energy in the 3-body continuum, say in formaldehyde for example.

CO($v = 1$) has been observed as a roaming product from formaldehyde at lower energies. Do you suspect a drop in the yield of CO($v = 1$) once the 3 body fragmentation energy (for CO($v = 0$)) is reached?

Professor Kable replied: This is a very interesting question to consider. In my response to Professor Bowman earlier, I commented on the independence of various 2F and 3F processes (*i.e.* the primary two-body event and the subsequent dissociation of the hot product). The 3F processes we have modelled in formaldehyde, which mirror those in our paper on acetaldehyde and methylformate, have considered

$$H_2CO \rightarrow H + HCO \rightarrow H + H + CO \text{ (R1)}$$

In this case the roaming and 3F events are independent. The competition between roaming and 3F depends on the competition between roaming and H + HCO radical production. I don't see this competition changing substantially when the threshold for spontaneous HCO product fragmentation is reached.

There is more than one way to achieve triple fragmentation, as highlighted in Fig 1 of our paper. Consider the following 3F pathway:

$$H_2CO \rightarrow H_2 + CO \text{ (roaming)} \rightarrow H + H + CO \text{ (R2)}$$

Roaming and 3F are now in direct competition. Your work shows that H_2 is born with a large amount of vibrational energy, therefore pathway R2 might be significant. I haven't thought too deeply about this, but it might be that you can observe CO products associated with the H_2 continuum in your ion imaging.

Your question, of course, was more subtle than my answer thus far because you asked about the production of CO($v = 1$) when the 3F threshold for CO($v = 0$) was reached. Following my long-winded answer above, I would expect that the attainment of the 3F threshold for pathway R1 should have no effect on the roaming CO($v = 1$) yield from roaming as they are independent. *Via* pathway R2, CO is the primary fragment and its dynamics should be dictated by the roaming event itself. Therefore I do not see an obvious link between the 3F threshold (essentially the H_2 dissociation energy) and the observed yield of roaming CO($v = 1$). Therefore I do not expect a drop in CO($v = 1$) born from roaming when the CO($v = 0$) 3F threshold is reached. Indeed, you might see more low energy CO from pathway R1, once the CO($v = 1$) 3F threshold is reached.

Dr Glowacki asked: This is a very interesting paper, and an excellent example of how coupled chemical channels can result in triple fragmentations following photo-excitation. An interesting question is whether triple fragmentations observed under low pressure laboratory conditions carry over to atmospheric temperatures and pressures. Along these lines, I wanted to highlight some relevant work that I was involved with a few years ago,[1] in which we found evidence for triple fragmentations at atmospheric temperatures and pressures under thermal (*i.e.*, non-photochemical) conditions using an approach very similar to that outlined in Professor Kable's paper.

The chemical system we looked at involved the atmospheric oxidation of methylglyoxal (MGLY), which can be described using the following scheme:

$$MGLY + OH \rightarrow H_2O + CH_3COCO^* \text{ (R1)}$$
$$CH_3COCO^* \rightarrow CH_3CO^* + CO \text{ (R2)}$$
$$CH_3CO^* \rightarrow CH_3 + CO \text{ (R3a)}$$
$$CH_3CO^* + M \rightarrow CH_3CO + M \text{ (R3b)}$$

Reaction (R1) has a energy of -120 kJ mol^{-1}, producing both H_2O^* and CH_3COCO^*, whose energy distribution I modelled using a statistical prior distribution (which is closely related to the phase space theory approach). Comparing the nascent CH_3COCO^* energy distribution with RRKM dissociation $k(E)$s shows that unimolecular dissociation of CH_3COCO^* effectively competes with collisional relaxation under atmospheric conditions.

CH$_3$COCO* dissociation yields CH$_3$CO* and CO (R2) with a unity yield. The nascent CH$_3$CO* has two possible fates: it may further dissociate to CH$_3$ + CO (R3a), or it may be collisionally stabilized with atmospheric bath gases (R3b). In order to determine the branching between these channels, we performed a subsequent prior distribution calculation to determine the energy of the nascent CH$_3$CO*, using a convolution nearly identical to that described in Professor Kable's paper. Interestingly, our results showed that ~52% of the nascent CH$_3$CO distribution is above the dissociation threshold. In order to determine whether the nascent CH$_3$CO* could dissociate under atmospheric temperatures and pressures, we used the calculated CH$_3$CO energy distribution as the initial conditions vector in an energy grained master equation. It turned out that ~40% of the nascent CH$_3$CO dissociated at 760 torr and 298 K. At lower pressures, our prediction was in broad agreement with experimental observations, but it needs to be emphasized that the experimental error bars were very large, owing to the fact that the CH$_3$ + CO yield was measured using rather indirect methods.

So the MGLY oxidation system provides an excellent example of a triple fragmentation that occurs under thermal conditions and atmospheric pressures, driven by non-equilibrium kinetics. I suspect that the effects of non-equilibrium kinetics are widespread within the atmosphere over a range of thermal and photochemical conditions in which they are often neglected. In recent work I have even gone so far as to argue that non-equilibrium kinetics extend to some synthetic organic solution phase chemistry systems.[2,3] Hence, I very much look forward to seeing whether the effects highlighted in Professor Kable's paper extend to atmospheric conditions, and welcome any thoughts that he has on this issue.

1 M. Baeza-Romero, D. R. Glowacki, M. A. Blitz, D. E. Heard, M. J. Pilling, A.R. Rickard and P. W. Seakins, *Phys. Chem. Chem. Phys.*, 2007, **9**, 4114–4128.
2 D. R. Glowacki, C. H. Liang, S. Marsden, J. N. Harvey and M.J. Pilling, *J. Am. Chem. Soc.*, 2010, **132**, 13621–13623.
3 L. Goldman, D. R. Glowacki and B. K. Carpenter, *J. Am. Chem. Soc.*, 2011, **133**, 5312–5318.

Professor Kable replied: In the example by Prof Glowacki, the radical formed after the initial abstraction of a H-atom from methylglyoxal (MGLY) by OH, undergoes triple fragmentation. In a funny way, the OH radical is playing a similar role to a photon in the systems we describe in that it leaves behind an energized molecule that dissipates the energy by breaking chemical bonds. Indeed the subsequent chemistry of the CH$_3$–CO–CO radical resembles that of glyoxal itself where triple fragmentation to 2CO + H$_2$ is well known.

The use of a prior distribution instead of PST under thermal, collisional conditions is appropriate. The main difference between prior and PST is the correct accounting of angular momentum conservation in PST. This is particularly notable for low angular momentum of the parent, which imposes a severe restriction on the range of angular momentum states of the products. Under atmospheric conditions, where there is both higher, and a broader distribution of J, this restriction is significantly relaxed. Indeed, Reisler and co-workers show that the PST distribution approaches the prior distributions for increasing temperature of the parent.[1] Therefore, I think that if we were to use the PST model in our paper here on your system, it would give much the same results as you obtained using a prior model.

1 M. Hunter, S. A. Reid, D. C. Robie and H. Reisler, *J. Chem. Phys.*, 1993, **99**, 1093.

Professor Nesbitt asked: This theoretical issue of triple fragmentation (3F) processes in photchemistry is certainly an interesting one with relevance to "roaming" dynamics, with suggestions of "triple whammy" (*i.e.*, perfectly synchronous) 3-body photofragmentation dating back to at least when I was still in graduate school! I wonder if you would comment on what is now the "gold standard" experimental proof of a triple fragmentation event. In particular, would triple coincidence experiments with high internal energy neutrals be feasible, or might there be even simpler ways exploiting the REMPI/LIF capabilities for detecting correlated H, CO, and CH_3 angular distributions?

Professor Kable answered: The study of triple fragmentation has certainly reached middle age! When Professor Nesbitt was in graduate school the predominant question was whether the three fragments were formed in a single dynamical step, such as the production of $H_2 + 2CO$ from glyoxal, or in separate steps. I will leave it to Professor Neumark to comment on the triple coincidence experiments. Here, I will simply comment on the "gold standard" for distinguishing triple fragmentation from roaming, which was the focus of the paper.

All three 3F products have a commensurate 2F reaction that will produce them at lower energy, albeit three different 2F reactions as shown in Fig. 1 of our paper. The gold standard, therefore is to carry out the reaction below the 3F threshold. If this is not possible, then there are two strategies. Firstly, measurement of the roaming co-fragment, as discussed in the answer to Professor Bowman's question above should provide unambiguous evidence as to the mechanism. Secondly, the photolysis energy dependence of 3F should be different to roaming. The general form of the dependence of roaming product state distribution on photolysis energy is not clear (it has only been measured in formaldehyde to our knowledge). But, if it follows loosely that of the associated radical channel then as photolysis energy is increased, roaming will produce increasingly hotter and faster fragments. The 3F distribution, however, remains very cold.

Professor Neumark remarked: In response to David Nesbitt's question about three-body decay of CH_3CHO, I think the best way to investigate such a channel would be charge exchange of the cation *via* collisions with Cs atoms, followed by coincidence detection of all three fragments. As shown by Bob Continetti and others,[1] this process produces highly excited neutrals that, depending on geometry and energetics of the cation and neutral, can then undergo three-body decay. In our laboratory, we can investigate three-body decay of a neutral species that can be formed by photodetachment of a negative ion, but I don't believe CH_3CHO is a suitable candidate.

1 J. D. Savee, J. E. Mann, C. M. Laperle and R. E. Continetti, *Int. Rev. Phys. Chem.*, 2012, **30**, 79.

Professor Stranges commented:‡ In relation to the paper presented by Professor S.H. Kable and his co-workers, where a phase space theory model for triple fragmentation (3F) has been developed and used to indicate that the dynamical signature of 3F is similar to that of the donor fragment in the roaming mechanism, we would like to present an example where only the roaming mechanism is consistent with the observations. We have recently studied the photodissociation of 2-propyl (isopropyl), $(CH_3)_2CH$, radical at 248 nm by using the photofragment translational spectroscopy technique.[1] By absorbing a 248 nm photon the isopropyl radical is excited into the 3s Rydberg state which decays into the ground state by internal conversion forming internally hot radicals with 115 kcal mol^{-1} of vibrational excitation. We have observed three primary dissociation channels:

$(CH_3)_2CH \rightarrow CH_3CHCH_2 + H$ (1)
$(CH_3)_2CH \rightarrow C_3H_5 + H_2$ (2)
$(CH_3)_2CH \rightarrow CH_3CH_2CH_2 \rightarrow CH_2CH_2 + CH_3$ (3)

‡ On behalf of Dr. E. Ripani.

In this short communication we will focus on the H_2 elimination channel (2). In addition to experimental data, we have characterized the ground state potential energy surface (PES) of the C_3H_7 system by means of ab initio calculations using the Gaussian 03 package.[2] Energy and vibrational frequencies of the stable and transition state structures were calculated at the B3LYP/cc-pVDZ level followed by single point electronic energy calculations at the QCISD(T)/cc-pVTZ level and zero point energy ZPE corrections by B3LYP/cc-pVDZ anharmonic frequency analysis. The results of these calculations are reported in Fig.7. The H-atom loss channel, (1), is essentially a barrierless dissociation process, while in the case of the H_2 elimination channel, (2), we did not find a true transition state which was correlating to the products on one side and to the isopropyl radical on the other side.

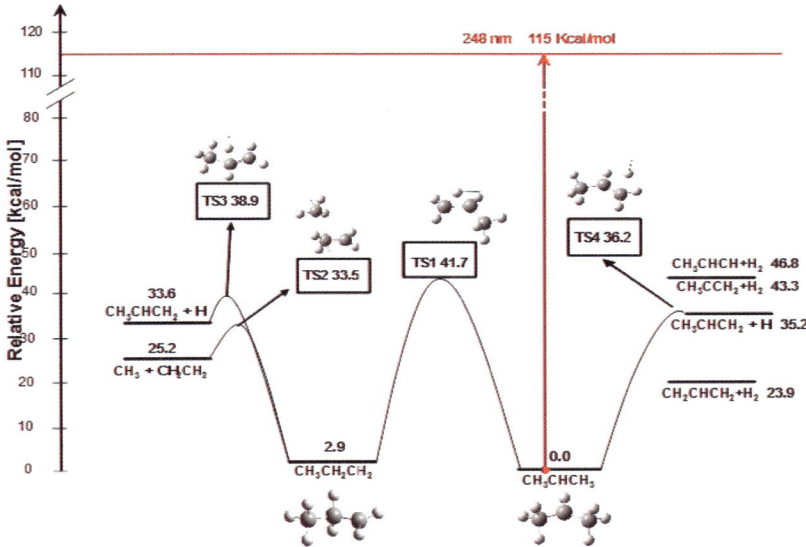

Fig. 7   Potential energy barriers for the ground state PES of the C_3H_7 system.

In Fig. 8 are reported the Time-cf-Flight (TOF) spectra recorded at $m/e = 41$, $C_3H_5^+$, at different laboratory angles. These spectra have been fit by using two different contributions: the blue line is associated to a contribution of the dissociative ionization of propene, C_3H_6, inside the electron impact ionizer, while the red line is associated to the formation of C_3H_5 in the H_2 elimination channel. The contribution from reaction (1) has been obtained, without interferences, by fitting the TOF spectra for $m/e = 42$, $C_3H_6^+$ (not reported). As it can be seen in the TOF spectra in Fig. 8 the contribution due to the dissociative ionization of propene has a different shape with respect to the global TOF spectrum, giving a clear evidence of the presence of the second contribution. This contribution can only come from the formation of C_3H_5 together with H_2, channel (2), because the available energy from reaction (1) is not high enough to open a secondary dissociation process (sequential triple fragmentation) where propyne looses a hydrogen atom, reaction (4):

$$(CH_3)_2CH \rightarrow CH_3CHCH_2 + H \rightarrow C_3H_5 + H + H \text{ (4)}$$

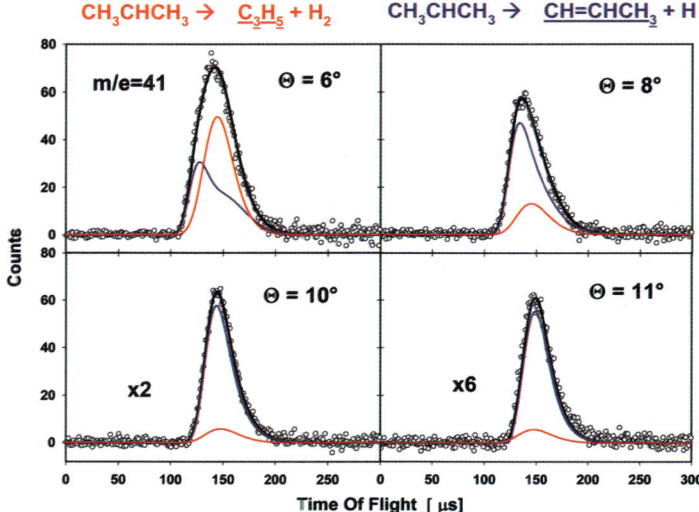

Fig. 8 TOF spectra of $m/e = 41$, $C_3H_5^-$, from the photodissociation of isopropyl radical at 248 nm at the indicated laboratory angles. The circles are experimental points, the blue line is the contribution from the H_2 elimination channel, the red line is the contribution due to the dissociative ionization of propene, C_3H_6, and the black line is the sum of the two contributions in the simulation.

In Fig. 9 is reported the translational energy distribution function, $P(E_T)$, utilized to simulate the contribution of the H_2 elimination channel. This distribution is peaked at 7 kcal mol^{-1} and the mean translational energy, $<E_T>$, is 8.3 kcal mol^{-1}. These results are quite surprising because it is well known that the concerted H_2 elimination channel is accompanied by a large exit barrier and a large fraction of the available energy is converted into translation with a $P(E_T)$ peaked at high energy. In this case if the allyl radical (CH_2CHCH_2) plus H_2 are formed, being the available energy 91 kcal mol^{-1}, less than 10% is released into product translational energy. This unusually low fraction suggests the formation of highly vibrationally excited H_2 molecules, which is compatible with a roaming mechanism. However the best way to prove the existence of this particular reaction dynamics is to measure the vibrational distribution of the H_2 molecules.

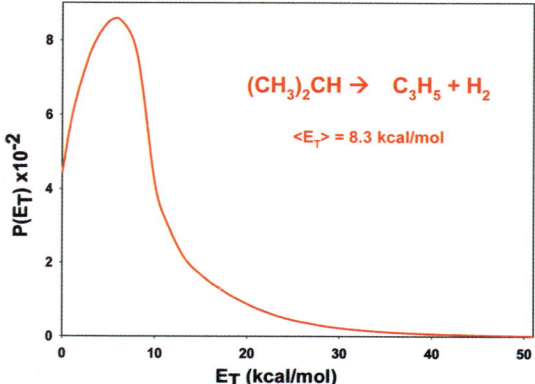

Fig. 9 Translational energy distribution function, $P(E_T)$, for $(CH_3)_2CH \rightarrow C3H5C_3H_5 + H_2$ dissociation channel used to simulate its contribution in the TOFs spectra reported in Fig. 8.

1 D. Stranges, M. Stemmler, X. Yang, J.D. Chesko, A. G. Suits and Y.T. Lee, *J. Chem. Phys.*, 1998, **109**, 5372; D. Stranges, P. O'Keeffe, G. Scotti, R. Di Santo and P.L. Huston, *J. Chem. Phys.*, 2008, **128**, 151101; J. D. Chesko, D. Stranges, A. G. Suits, Y. T. Lee, *J. Chem. Phys.*, 1995, **103**, 6290.
2 M. J. Frisch, G. W. Trucks, H. B. Schlegel *et al.*, GAUSSIAN 03, Revision B. 01, M. J. Frisch, Gaussian Inc., Pittsburgh PA 2003.

Professor Kable responded: These are very interesting results on a complex system. The problem with complex systems such as this is the number of product channels, transition states and pathways that are available. The data that you present are consistent with a roaming mechanism for H_2 production, but you are right to be cautious.

Your Fig. 7 includes isomerisation from the *i*-propyl to n-propyl radical. Did you also look for H_2 elimination from the n-propyl side? This would produce allyl + H_2, amongst other less stable radicals. It is also thermochemically feasible to produce a variety of triple fragmentation products, including propargyl + $2H_2$ and acetylene + CH_3 + H_2. However, dynamically, elimination of H_2 must surely proceed over a significant barrier. Therefore, even if the transition state is accessible at this energy, the height of the barrier from the exit channel side, coupled with the light mass of H_2 must produce fast moving, high energy H_2 fragments, unlike what is seen in your experiments. I cannot think of another likely mechanism for H_2 production that would produce products with low translational energy.

Dr Grubb commented: The paper provides a list of measureable signatures of roaming dynamics in the photoproducts of formaldehyde and acetaldehyde, the last of which is that no CO *v–j* correlation is observed. The rationale for this signature is that: "the roaming trajectories sampled a wide variety of (abstraction) angles, which destroyed any such correlation". The widespread belief that roaming dynamics lead to an absence of directed forces in the exit channel is incorrect. Recent publications on NO_3 NO + O_2 photodissociation, which has been shown to proceed entire *via* roaming pathways,[1] reveal the presence of strong perpendicular *v–j* correlations in the NO fragment.[2] The supporting *ab initio* calculations suggest that the dissociation step is confined to the NO_3 molecular plane, implying strong directed forces on the fragments and a non-statistical distribution of O–O abstraction angles. There is no reason to assume roaming abstraction angles will not be constrained. The intra-molecular abstraction step of roaming has been shown to yield identical product distributions as the corresponding bimolecular abstraction reaction. For instance, in formaldehyde, Christoffel and Bowman[3] show that the H_2CO CO + H_2 roaming channel produces identical product state distributions to the bimolecular H − HCO → CO + H_2 reaction. Vector correlations in the roaming products are therefore governed by the same dynamical constraints as the corresponding bimolecular reaction, and may or may not be statistical.

Furthermore, the statistically expected fragment *v–j* correlation resulting from an unconstrained reaction is not zero. It has been shown that a fragment *j* dependent perpendicular *v–j* correlation arises statistically from rotationally cold parent molecules.[4] This arises due to angular momentum conservation. The angular momentum of the parent molecule must equal the sum of the total angular momentum of the fragments and their orbital angular momentum, *L*. The orbital angular momentum is by definition perpendicular to the fragment velocity, and thus if a large *L* is required to conserve the total angular momentum this must be balanced by a *v* perpendicular to *j* correlation in the fragments as well. Fig. 10 shows the expected H_2 *j* dependent *v–j* correlations for formaldehyde dissociation when particular CO *j* states are probed. Significant statistical *v–j* correlations are expected from this reaction, and if none are experimentally observed, this would actually imply that dynamical constraints are present.

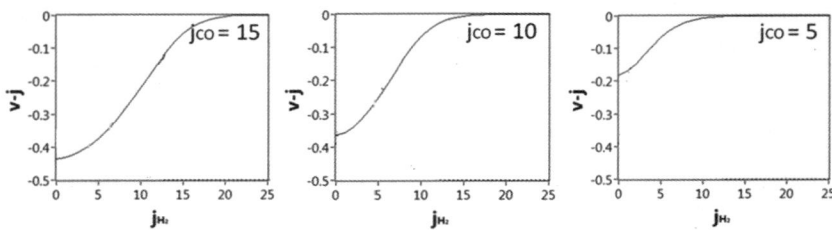

Fig. 10 Statistical vector correlations expected in the CO (j_{CO} = 5, 10, 15) fragments arising from formaldehyde (T_{rot} = 50 K) photodissociation. A v–j value of -0.5 corresponds to a limiting v perpendicular to j correlation.

1 M. Grubb *et al.*, *Science*, 2012, **355** 1075.
2 M. Grubb *et al. Phys. Chem. Chem. Phys.*, 2012, **6733**, 14.
3 K. M. Christoffel and J. M. Bowman, *J. Phys. Chem. A*, 2009, **113**, 4138.
4 S. W. North and G. E. Hall, *J. Chem. Phys.*, 1996, **104**, 1864.

Professor Kable responded: We are happy to accept that roaming reactions may produce a non-statistical v–J correlation and note that the two papers on NO_3 referred to by Mr Grubb were published after we submitted this paper. It will be very interesting, in the future, to consider whether the vector correlations in the roaming reaction do indeed match those of the relevant abstraction reaction or not. This will lead to information about the angular extent of roaming compared to the range of impact parameters in the bimolecular event.

Professor Ashfold asked: Kable and colleagues highlight the shortage of unambiguous experimental data for triple fragmentation (3F) processes. The cases considered in this paper eventually occur on the ground state potential energy surface (PES), though this of course is not a necessary requirement when exciting at higher energies – where 3F processes must surely be commonplace. One example from our own group, the photolysis of H_2S (and D_2S) at the Lyman-α wavelength,[1] exemplifies two types of 3F process. The excited state populated at this energy is dissociative with respect to extending one S–H bond, but also favours linear geometries – where it shares a conical intersection with the ground state PES at extended R_{S-H}. Dissociation thus involves simultaneous stretching of one S–H bond and opening of the bond angle. The topology of the excited state PES ensures that the dissociating molecules pass through linearity outside the region of the CI and evolve towards electronically excited H + SH(A) products. The SH(A) products are formed in a very wide range of rovibrational states, including the complete set of bound (and quasi-bound) rotational levels associated with the $v = 0$ state; indeed, some H atom products appear with kinetic energies consistent with dissociating trajectories that are heading towards yet higher angular momentum product states, which are unable to survive the associated centrifugal force. These products exemplify an asynchronous, concerted 3F process. The SH(A) radicals that are formed are all unstable with respect to predissociation, yielding ground state S + H products within a few ns – a sequential 3F route to the atomic products.

1 P. A. Cook, S. R. Langford, R. N. Dixon and M. N. R. Ashfold, *J. Chem. Phys.*, 2001, **114**, 1672, and references therein.

Professor Kable answered: There are indeed a very large number of triple fragmentation examples in the literature, and, as Professor Ashfold points out, many of these occur on the excited electronic state of the parent molecule. At photolysis energies corresponding to common wavelengths such as Lyman-α (121 nm), or the excimer laser wavelengths of 193 or 157 nm, triple fragmentation is common, and it is considered it explicitly in many papers. At such high photolysis energy, dissociation

This journal is © The Royal Society of Chemistry 2012

frequently occurs on a repulsive potential energy surface, leading to fast recoiling products. These fast products may still have enough energy to further fragment, such as in the case Professor Ashfold presents. With such a high initial "kick" the secondary products retain their high translational energy.

When the initial dissociation occurs on the ground state potential energy surface the fragments are born with much lower translational energy and commensurately higher internal energy. The take-home message from our paper is that when 3F occurs on the ground state of both parent and products, the recoil translational energy can be extremely cold, even at large excess energy above the 3F threshold.

Dr Glowacki addressed Professor Suits and Professor Kable: In all fields of chemistry, people argue all the time about concerted *vs.* stepwise, and which is the appropriate description for experimental observations. What appears as experimentally 'concerted' may actually be 'stepwise' when one carries out electronic structure theory and subsequent dynamics calculations. Ultimately, it depends on the time resolution that one has available. Any dynamical system has a range of significant eigenvalues, or timescales, but it often the case that many of the dynamical timescales are often beyond what the experiments are capable of resolving. In Professor Crim's introductory lecture, he drew attention to the idea of the 'triple whammy'. Does Professor Kable's system represent an example of the 'triple whammy'?

Professor Suits answered: You are correct that people debate the question of concerted *vs.* stepwise reactions, and it is important to be clear about definitions. Although I don't believe "triple whammy" has ever been given a precise definition, in general it was used to mean a concerted reaction giving rise to three fragments, most notably in glyoxal photodissociation. "Concerted" was operationally defined as being faster than the rotational period. In general Professor Kable was discussing step-wise dissociation, and I believe he was clear about this. If one measures the momentum distribution of only one product of a three-body decomposition process, then some modeling or assumptions about the process is necessary to infer the momentum distributions of the undetected products. The community that uses photofragment translational spectroscopy has generally employed forward-convolution fitting of the data and has for years incorporated such modeling as part of the analysis. The imaging community has not routinely addressed this but it would be a valuable addition to the resources for studying photochemistry with ion imaging.

Professor Kable responded: We deliberately avoided using the term "triple whammy" in our paper! But to answer the question, we firstly need to define what is meant by triple whammy. In our response, we assume that this means a concerted process, either synchronous or asynchronous, which underlies the influence of timescale on the explanation. Under this definition, the systems that we describe do not represent examples of triple whammies. Indeed they are as far from triple whammy as you can get in that we assume implicitly the formation two separated primary fragments before considering the secondary fragmentation step.

Dr Jordan opened the discussion of the paper by Professor Martinez: In ethylene, C_2H_2, there are a number of Rydberg states very close in energy to the V state, for which dynamic correlation energy is lower than in the V state (which has valence character). A "correct" picture for ethylene can only be obtained by allowing valence/Rydberg mixing. How important are Rydberg states in these larger dienes? Have you included Rydberg states in your treatment of their electronic energy?

Professor Martinez replied: We did not include Rydberg states in these calculations on dienes. This could be done in principle, as the multi-state multi-reference perturbation theory (MS-MRMP) we use in the "on the fly" dynamics does include the dynamic electron correlation effects that are crucial to get a balanced treatment of valence and Rydberg states. However, it would be computationally costly. We recently did look at the excited state dynamics of ethylene including the Rydberg states[1] (also using MS-MRMP with ab initio multiple spawning). Interestingly, the Rydberg states are largely spectators in ethylene dynamics –although some population is diverted to the 3s Rydberg state, it very quickly recrosses back onto the valence state.

1 T. Mori, W. J. Glover, M. S. Schuurman and T. J. Martinez, *J. Phys. Chem. A*, 2012, **116**, 2808–2818.

Professor Neumark commented: I have both a comment and a question. I found some aspects of your paper to be a bit unclear on first reading, mainly because what you call the "initial state" in your calculations of time-dependent photoelectron spectra is not the neutral ground state, but rather the non-stationary excited state created by the pump pulse. Hence, for example, eqn 3 in the paper looks to be the standard expression for a one-photon process from a stationary state; there is no indication that initial state is evolving in time. A similar issue arises in several of your figures such as Fig. 14 in which you report a "one-photon time-resolved photo-electron spectrum". My question also pertains to Fig. 14. Since this spectrum represents one-photon ionization out of the excited state, why is there a dip around 0.5 eV? Is this due to nuclear dynamics in the neutral excited state or in the cation? I presume your simulations must provide insight into the origin of this effect.

Professor Martinez responded: Thanks for your comments – we have modified eqn 3 to make it (hopefully) more clear, indicating explicitly that the Dyson orbital depends on the pump–probe time delay (since it depends on the evolving excited state wavepacket). It may be useful to also note that eqn 3 is written for a neutral excited state wavepacket described by a single Gaussian trajectory basis function. When the excited state wavepacket is represented by multiple basis functions (which can happen after spawning), these are all included by a population-weighted summation over the basis functions on the neutral excited state. By "one-photon time-resolved photoelectron spectrum," we just mean that this is a femtosecond TRPES spectrum calculated including only one-photon photoionization in the probe step, i.e. two- photon photoionization processes are excluded.

We included two cation states in the computation of the TRPES spectra. However, only the lowest (D_0) cation state contributes to the TRPES in Figure 14. Thus, the observed dip does not come from a mixing of D_0 and D_1, for example. Instead, it seems that it is due to the dynamics of the neutral excited state wave-packet, which leads to a very rapid change in the excited state ionization potential. We expect that this would fill in to some extent with more extensive statistics, i.e. sampling of more initial conditions. We plan to explore this in more detail in subsequent work where we will compare directly to the experimental TRPES spectra for CPD and Me_4-CPD. This may require a more realistic description of Me_4-CPD (including the methyl groups and their influence on the electronic structure explicitly – note that these calculations represent Me_4-CPD by adjusting the mass of the H atoms of CPD which would be methyl groups in Me_4-CPD).

Professor Crim asked: Do your studies of ethene and cyclopentadiene allow you to make predictions about the behavior of larger polyenes? For example, do you expect charge transfer contributions to be important in general?

Professor Martinez responded: Yes, we do expect that charge transfer contributions will be observed quite generally. Indeed, explicit calculations for stilbene suggest that charge transfer states are critical in the nonadiabatic transitions from the lowest bright excited state back to the ground state.[1] However, there also do exist intersections that do not involve charge transfer states – for example, the out-of-plane puckering distortion in benzene couples the S_0 and S_1 states that are best described as covalent and radicaloid.[2] Similarly, there is an S_1/S_0 intersection in butadiene that couples covalent and radical states (characterized by significant torsion about all C–C bonds).[3] The involvement of this "transoid" intersection may vary with increasing polyene length or differing substitution patterns. It is best to keep in mind that there will usually be at least two classes of mechanisms for the excited state dynamics (covalent and charge-transfer types) which may compete in the photochemistry of polyenes.

1 J. Quenneville and T. J. Martínez, *J. Phys. Chem. A*, 2003, **107**, 829–837.
2 A. L. Thompson and T. J. Martínez, *Faraday Disc.*, 2011, **150**, 293–311.
3 B. G. Levine and T. J. Martínez, *J. Phys. Chem. A*, 2009, **113**, 12815–12824.

Dr Ripani commented:§ In relation to the paper presented by Professor Martinez and his co-workers, aimed at simulating the non-adiabatic excited state dynamics of cyclopentadiene and 1,2,3,4-tetramethyl-cyclopentadiene following the excitation to the S1 state, we would like to make a short communication. We have recently studied the photodissociation of 2-methyl-propene, cis-2-butene, trans-2-butene, 2-methyl-2-butene and 2,3-dimethyl-2-butene, following the excitation with 193 nm photons using the Photofragment Translational Spectroscopy technique.[1]

All these mono olefins share the same photochemical properties and their excited state dynamic have been extensively studied[2,3] with pump-probe femtosecond spectroscopies. For all these molecules the absorption of a UV photon is followed by an ultrafast internal conversion to the ground state.

The systematic study of the effect of the methyl substitution on the double bond is an invaluable tool to underline which molecular motions are involved in the non adiabatic dynamics following the absorption of a UV photon.

For methyl substituted ethylenes, as also for the cyclopentadiene and 1,2,3,4-tetramethyl-cyclopentadiene, the effect of methyl substitution is to decrease the internal conversion rates as the result of an increasing inertia and steric hindrance of those molecular motions that drive the decay from the excited states. The effect is particularly strong for the 2,3-dimethyl-2-butene that possess an excited state lifetime of several picoseconds, much greater than all other molecules with lower degree of methylation which show a lifetime on the order of hundreds of femtoseconds or less.

The study of the photodissociation of the five molecules above may have revealed some features underlying minor processes where the methyl substitutes do not act as simple bulky groups.

In Fig. 11 below are shown Time of Flight (TOF) spectra, recorded at $m/e = 1$ and $\theta = 90°$, for the H-atom elimination channel of 2-methyl-propene, cis-2-butene, trans-2-butene and 2-methyl-2-butene while in Fig. 12 below are reported translational energy distributions functions ($P(E_T)$s), used for the simulations of the former TOFs.

The characteristics of the $P(E_T)$ are consistent with a statistical dissociation process from vibrationally hot molecules in the ground states, underlying a common dissociation dynamics for all of them.

Totally different is the TOF spectrum of the 2,3-dimethyl-2-butene reported in Fig 13(a). For the 2,3-dimethyl-2-butene, besides the main component that can be simulated with the same $P(E_T)$ used for the other molecules, a fast component can be observed at short flight times. The component makes a contribution of 10 % to the total H signal and can be simulated with the $P(E_T)$ shown in Fig 13(b).

The characteristics of this $P(E_T)$ are indicative of a non statistical dissociation dynamic where a large fraction of the available energy is channelled into translation. The high energy cut-off of this $P(E_T)$, located at energies lower than 70 kcal mol^{-1} is consistent with the production of ground state products in a one photon process.

§ On behalf of Prof. D. Stranges.

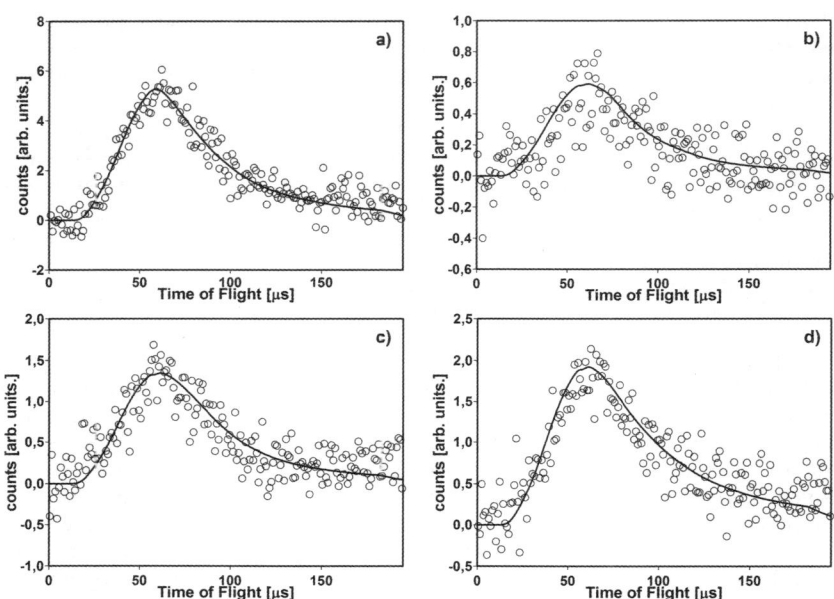

Fig. 11 H-atom TOF spectra, recorded at $\vartheta = 90°$, for: a) 2-methyl-propene, b) *cis*-2-butene, c) *trans*-2-butene and d) 2-methyl-2-butene.

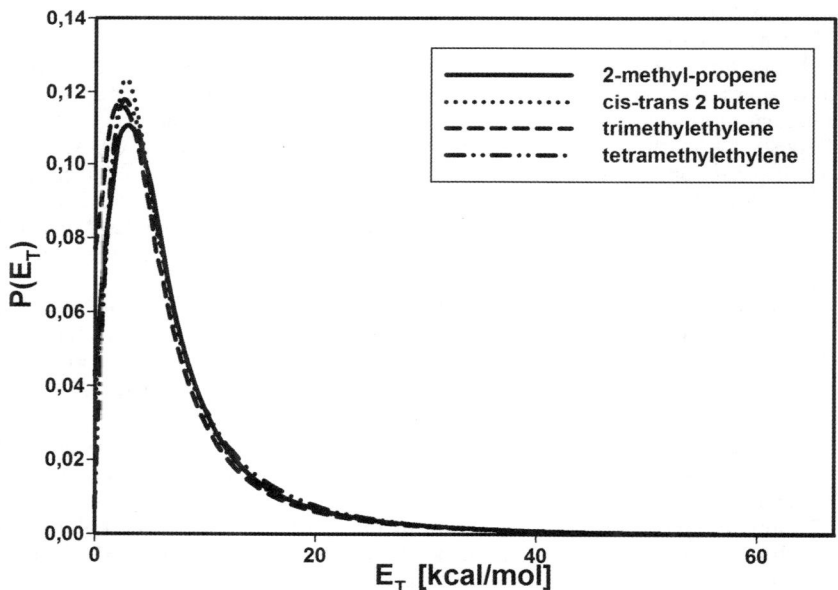

Fig. 12 Translational energy distributions, $P(E_T)$, used to simulate H-atom TOF spectra reported in Fig. 11.

Similar observations were made in the photodissociation of cyclohexene at 193 nm.[4] The presence of a non statistical H-atom elimination channel may underline the presence of other competing decay mechanisms from the excited states that involve directly the alkyl substitutes and start to give a contribution when constrained geometries or molecules with particularly long excited state lifetimes are considered.

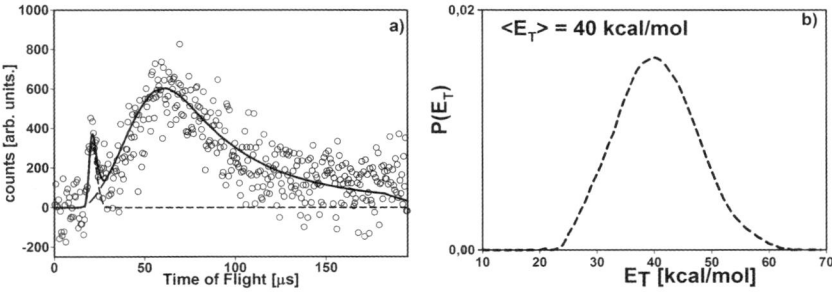

Fig. 13 a) H-atom TOF spectrum, recorded at $\theta = 90°$, for 2,3-dimethyl-2-butene. b) $P(E_T)$ used to simulate the fast component.

A possible correlation can be made with the liquid phase photochemistry of simple alkenes as it is known that upon direct irradiation with $\lambda < 200$ nm, one of the main products is the [1,3]-H sigmatropic shift (see ref. 5 below and references therein).

1 D. Stranges, M. Stemmler, X. Yang, J. D. Chesko, A. G. Suits and Y. T. Lee, *J. Chem. Phys.*, 1998, **109**, 5372; D. Stranges, P. O'Keeffe, G. Scotti, R. Di Santo and P. L. Houston, *J. Chem. Phys.*, 2008, **128**, 151101; J. D.Chesko, D. Stranges, A. G. Suits, Y. T. Lee, *J. Chem. Phys.*, 1995, **103**, 6290.
2 J. M. Mestdagh, J. P. Visticot, M. Elhanine and B. Soep, *J. Chem. Phys.*, 2000, **113**, 237.
3 G. Wu, A. E. Boguslavskiy, O. Schalk, M. S. Schuurman and A. Stolow, *J. Chem. Phys*, 2011, **135**, 164309.
4 X. Zhao, R. E. Continetti, A. Yokoyama, E. J. Hintsa, and Y. T. Lee, *J. Chem. Phys.*, 1989, **91**, 4118.
5 P. J. Kropp, Photorearrangement and Fragmentation of Alkenes in *Organic Photochemistry and Photobiology*, eds., W. Horspool and F. Lenci, 2004, pp 13–15, CRC Press Inc, Boca Raton FL.

Dr Glowacki asked: During discussion of Professor Bowman's paper, I pointed out that it is often possible for $S_0 \rightarrow S_1$ tunnelling-like mechanisms to occur despite the fact that the system is below the conical intersection energy.[1] When this sort of tunnelling occurs far from stationary points, it becomes difficult to treat with time-independent stationary phase approximations. Presumably though, tunnelling that occurs far from stationary points comes out of AIMS quite naturally. Your figure 12 qualitatively shows that the bulk of the hopping occurs at the CI. However, I'm wondering whether you find any evidence for $S_0 \rightarrow S_1$ hopping below the intersection energy in sloped regions of the potential which have a relatively constant energy gap?

1 See, *e.g.*, Heller and Brown, *J. Chem. Phys.*, 1983, **79**, 3336.

Professor Martinez replied: For this case, we did not analyze the data to characterize the sloped and peaked character of the intersections involved in the dynamics. However, we have done this recently for ethylene, where there are both sloped (ethylidene-like) and peaked (twisted-pyramidalized) intersections. Although the sloped intersections in this case are a minor channel (most of the population is transferred through the more peaked twisted-pyramidalized intersections), one does see transitions from S_1 to S_0 that occur below the intersection energy.[1]

Also relevant here is the recently introduced optimal spawning algorithm.[2] This allows one to place "spawned" basis functions in phase space according to the criteria that they maximize the coupling to the "parent" basis function subject to the constraint that the classical energy of the phase space center of the parent and spawned basis functions are the same. The biggest difference between this optimal spawning and the usual spawning algorithm is that it is more likely to lead to both position and momentum adjustments for the spawned basis function. The position adjustments in this case correspond to what would look like tunneling in surface hopping.

1 A. M. Virshup, J. Chen and T. J. Martinez, *J. Chem. Phys.*, 2012, **137**, 22A519.
2 S. Yang, J. D. Coe, B. Kaduk and T. J. Martínez, *J. Chem. Phys.*, 2009, **130**, 134113.

Professor Sibener opened the discussion of the paper by Stuart R. Mackenzie: I would like to note that there is a fundamental difference between pre-adsorbing reactants on clusters at lower temperatures followed by heating to elevated temperatures sufficient for reaction to occur as opposed to doing isothermal studies where both adsorption and reaction occur at the same temperature. Is it clear that one is probing the same part of the reaction landscape following these two kinetic procedures? For systems involving extended (large) substrates such as metallic single crystals it is likely safe to say that one reaches the same geometric configuration in most instances using either procedure. For clusters extending from a few atoms up to sizes involving nanoparticles adsorption may restructure the catalyst to a notable extent. This in turn may cause the reaction path to evolve along a different local chemical and physical environment than when adsorption and reaction occur under the same isothermal conditions. This is, potentially, of particular relevance to small systems which are themselves highly fluxional. or, again, which significantly rearrange given the presence of a strongly bound moiety.

Dr Mackenzie responded: Your first point is well taken and it is certainly true that the processes we are observing in the isolated cluster add another dimension to temperature programmed reaction studies on extended surfaces. That said there are clear similarities between the two methods. Each adsorbs precursors at low temperature and induces reactivity by means of heating the system. The spectral information we gain provides detailed information on both the cluster structure and the specific vibrational mode (or chromophore) involved in the multiple photon absorption but comes at the cost of any precise knowledge of the cluster internal energy/temperature following heating. We see our approach as complementary to both single collision reactivity studies and fully equilibrated thermal reactions of clusters as we learn directly about intermediate structures and barriers which are unobserved in other methods.

With regard to the restructuring involved during reaction we have seen examples of both extremes. Our modelling of IR-driven decomposition of nitrous oxide on rhodium clusters suggests that the metal atom framework is largely passive with the same isomeric form retained throughout. The whole process looks very much like a reaction on the metal cluster surface.[1] The current case of platinum oxide clusters is markedly different. Here, the platinum atoms are very much involved in the chemistry and whole cluster structure is indeed highly fluxional. Likewise, our recent IR-MPD studies show that the structures of platinum carbide[2] and platinum oxide clusters[3] are qualitatively different to those of naked platinum clusters.[4] Even the latter, though, may teach us some relevant lessons regarding reactions at interfaces undergoing (local) surface melting.

1 S. M. Hamilton, W. S. Hopkins, D. J. Harding, T. R. Walsh, M. Haertelt, C. Kerpal, P. Gruene, G. Meijer, A. Fielicke and S. R. Mackenzie, *J. Phys. Chem. A*, 2011, **115**, 2489–2497.
2 D. J. Harding, C. Kerpal and A. Fielicke, *manuscript in preparation*.
3 C. Kerpal, D. J. Harding, A. C. Hermes, G. Meijer, S. R. Mackenzie and A. Fielicke, *J. Phys. Chem. A*, 2012, in press, DOI: 10.1021/jp3055137.
4 D. J. Harding, C. Kerpal, D. M. Rayner and A. Fielicke, *J. Chem. Phys.*, 2012, **136**, 211103.

Professor Bernhardt asked: The idea of infrared induced reactivity in clusters is very appealing and your data very nicely confirm that it is possible to trigger the reaction of coadsorbed ligands on a metal cluster by infrared laser pulses. To follow up on this idea it would be certainly very useful to extract quantitative data from such experiments. Do you think that it would be possible to determine the internal energy of the cluster complexes by, *e.g.*, laser power dependent measurements, maybe in connection with a calibration on a well known system. The measurement of activation energy barriers of the infrared induced reactions would also be very interesting.

Dr Mackenzie replied: Having now demonstrated this conceptual approach in two qualitatively different systems, attempting to quantify the barriers involved is a major aim of current and future studies. This, of course, is the key information from a chemical perspective. To extract hard numbers will require detailed knowledge of the number of photons absorbed (and hence the energy pumped into the complex). As you suggest this information may be extracted from a power dependence studies coupled with quantitative modelling of the absorption process including single-photon absorption cross sections and anharmonicities. There are examples in the literature in which such combined experimental/modelling studies have been successful for well characterised systems,[1] but the clusters we have studied thus far would represent a step change in complexity. Finally, the detailed power dependence studies which would be required are notoriously time-consuming and perhaps better suited to laboratory laser-based experiments rather than the time-pressure environment of experimental runs at a (inter)national facility. Certainly, finding a well-known system to use as a calibration would be useful in this regard but part of the fascination of cluster science is the extreme differences exhibited by seemingly similar systems.

1 A. Bekkerman, E. Kolodney, G. von Helden, B. Sartakov, D. van Heijnsbergen and G. Meijer, *J. Chem. Phys.*, 2006, **124**, 184312.

Dr Roberts questioned: Figure 6 in the discussion paper by Mackenzie and co-workers presents evidence of IR induced CO desorption and CO oxidation on $Pt_nO_2CO^+$ (n = 3–7) clusters. For cluster sizes n = 4–7 the profiles of the IR-MPD spectra for the parent ($Pt_nO_2CO^+$) depletion, desorption product ($Pt_nO_2^+$) enhancement and oxidation product (Pt_nO^+) enhancement are comparable for a given n. However, for n = 3 the IR-MPD spectral profile for the desorption product enhancement is notably broader than the profiles of the corresponding parent depletion and oxidation product enhancement. What are the origins of this difference?

Dr Mackenzie answered: In all likelihood several factors contribute here. Firstly, conditions were optimised for slightly larger clusters and thus the signals for the Pt_3^+ clusters are typically weak and the spectra in general correspondingly noisy. Secondly, all clusters in the size range studied complex with multiple CO molecules (see Fig. 2 in the article). Adsorption of multiple CO molecules results in small spectral shifts which are largest for the smaller metal clusters resulting in broader spectra in the desorption channels. Finally, the smaller clusters also bind Ar more effectively than do the larger species and simultaneous Ar + CO desorption may also contribute to the $Pt_3O_2^+$ spectrum. The same broad spectrum in the $Pt_3O_2^+$ channel is observed in all our data sets including those recorded with attenuated IR intensity.

Professor Wester commented: You seem to not see CO oxidation following infrared absorption on frequencies associated with Pt-O transitions. What causes this and would it be possible to observe in future experiments? Could blackbody-related thermal emission as an additional energy loss channel play a role?

Dr Mackenzie responded: If, as we propose, the chemistry we observe is thermal in nature, we would expect to see it occur irrespective of the chromophore pumped. However, two factors in particular make this difficult:

1) The Pt–O stretches/bends are markedly weaker than the CO stretch. Our DFT spectral simulations suggest the oscillator strength of the CO stretch is approximately a factor of 20 greater than for the Pt–O bands. Experimentally, this is reflected in the respective widths of the spectral features in the two regions.

2) By virtue of their respective frequencies, a factor of between 3 and 5 more quanta have to be absorbed in the Pt–O bands to raise the internal energy of the cluster by the same amount as a single CO stretching quantum. In the present case, assuming the barrier to reaction is, as we calculate, ca. 2.0 eV, even accounting for the thermal energy of the cluster at 300K, approximately 5 photons at 2100 cm^{-1} are required to surmount the barrier. This rises to more than 15 or 24 photons if the Pt–O bands at 680 cm^{-1} or 440 cm^{-1}, respectively are used for excitation.

We observed similar results in our previous study of IR-driven nitrous oxide decomposition on rhodium clusters.[1,2] The $Rh_nN_2O^+$ system has much lower reaction barriers (≈ 0.75 eV) and we demonstrated that reaction was observed following pumping of any of the N_2O vibrational modes (or indeed the Rh–O stretch in $Rh_5ON_2O^+$)[3] but not when the lower frequency metal–metal modes were excited.

We can make a trivial point regarding reaction barriers and the Ar atom binding energies. The lower frequency regions of the IR spectrum (Fig. 3 in the paper) have been recorded using the Ar-messenger technique. The observation of efficient Ar loss suggests both that IVR is efficient and that the Ar binding is markedly lower than the barrier to CO oxidation.

As for radiative emission representing a significant competitive decay channel, we think this unlikely on the time scales of the experiment. The delay between IR excitation and ion extraction/detection is of the order of 30 µs for the experiments described in this paper, very much shorter than the likely black-body emission rate.

1 S. M. Hamilton, W. S. Hopkins, D. J. Harding, T. R. Walsh, P. Gruene, M. Haertelt, A. Fielicke, G. Meijer and S. R. Mackenzie, J. Am. Chem. Soc., 2010, 132, 1448–1449.
2 S. M. Hamilton, W. S. Hopkins, D. J. Harding, T. R. Walsh, M. Haertelt, C. Kerpal, P. Gruene, G. Meijer, A. Fielicke and S. R. Mackenzie, J. Phys. Chem. A, 2011, 115, 2489–2497.
3 A. C. Hermes, S. M. Hamilton, W. S. Hopkins, D. J. Harding, C. Kerpal, G. Meijer, A. Fielicke and S. R. Mackenzie, J. Phys. Chem. Lett., 2011, 2, 3053–3057.

Professor Nesbitt commented: It could be quite interesting to probe the final quantum states of the CO_2 products, which are formed vibrationally hot and would report on the exit channel dynamics for oxidation processes in your size selected Pt clusters. At the shot noise limit, direct IR laser absorption methods for molecules such as CO_2 can detect down to number densities of roughly 10^7 cm^{-1} per quantum state, possibly even lower with IR frequency comb cavity ringdown methods. I am wondering if you would provide more information on your experimental Pt cluster densities and maximal rates for CO_2 production, in order to help assess if such experiments might be feasible.

Dr Mackenzie answered: Any information on the reaction products would indeed be interesting. However, the number densities in the current molecular beam experiment are, in all probability, much too low for the experiments you are proposing. We haven't fully quantified the number densities for this system, but we detect small numbers of a particular cluster ion per duty cycle. One additional complication to the experiment you propose is that we rely on mass selective detection in these experiments to distinguish between similar processes occurring on different cluster sizes. Unless we introduced mass selectivity in the source (thereby losing the multiplex advantage we currently have) we would be unable to distinguish CO_2 molecules originating from different cluster sizes. In principle, both of these problems could be addressed if mass–selected clusters were accumulated in an ion trap provided the space charge limit is not exceeded.

An alternative approach to gain the same information may be to determine the product kinetic energy distribution using a variant of VMI. This would at least take advantage of the sensitivity of mass-spectrometric detection. However, such experiments would themselves be hampered by a lack of knowledge of the electronic/isomeric/vibrational structure of the other reaction product/recoil partner (*e.g.*, the oxide complex Pt_4O^+).

Dr Glowacki said: Dr Mackenzie, I have a few questions and a comment.

(1) Do you have any handle on the lifetime of the complex prior to IR excitation? In particular, I'm interested in knowing whether you are certain that the complex is fully thermalized prior to IR excitation?

(2) You show the branching ratios for the CO oxidation and CO loss channels, but do you have any data for the total yield? Is it unity? Is there any possibility that the complex can relax prior to dissociation, in which case the total yield would be less than unity?

(3) I am slightly confused by Fig. 7 in the paper. It seems to show that the IR pulses make it possible for the CO oxidation channel to switch on, but that there is not enough energy for the CO loss channel. So based on the figure, one might naively expect only CO oxidation. Perhaps I am missing something here - any comments would be most welcome.

Finally, it seems like this system may be particularly well suited to an RRKM-master equation type of analysis. We have recently developed a flexible, open-source master equation code called MESMER, which is freely available for download at http://sourceforge.net/projects/mesmer/. It outputs a range of potentially useful data, including microcanonical rate coefficients, thermal rate coefficients, and partition functions.

Dr Mackenzie answered: Taking these questions in turn:

(1) We get no explicit information on the lifetime of the complexes prior to irradiation except that any we detect have clearly survived the *ca.* 100 μs delay between formation and irradiation. Most of this time is spent in the "high pressure" region of the cluster source (~10 mbar He) which is thermalised to the room temperature source. That said, the fact that we need to invoke the presence of high lying isomers to fully explain the IR-PMPD spectra, implies that there may be some kinetically trapped species which have not been efficiently annealed.

(2) I think there are two related points here. Firstly, only on the very strongest transitions such as the CO stretching region, do we observe depletions in the parent ion signal approaching unity and this only with the full unattenuated IR beam. Where depletions are observed, we are confident that we capture all major channels as the sum of all daughter signals matches well the observed parent ion depletion. We are only sensitive to decay processes resulting in some form of fragmentation (Ar-loss, CO-loss, CO_2-loss, etc.). Were a complex to relax, *e.g.*, radiatively, before dissociation it would still be observed in the parent ion channel.

(3) The arrows are added to Fig. 7 in the paper by way of cartoon merely to indicate the relative photon energies by comparison with the barriers involved. We have no way of controlling (or indeed knowing) exactly how many photons are absorbed. The calculations indicate that the reaction channel opens energetically before the CO desorption channel. However, the rate of reaction at each energy is clearly also important and the desorption channel should be entropically favoured. Hence, as soon as this channel opens it probably competes effectively with the reaction.

Master equation modelling of the various processes involved would indeed be useful and such calculations are, in fact, under way for the simpler $Rh_nN_2O^+$ system we have studied previously. MESMER sounds like a useful tool in this regard. The difficulties will come from the fact that the energetics are only known crudely from the DFT simulations and we do not know exactly how many photons are absorbed and hence what the internal energy of our clusters are.

Professor Neumark asked: Can you say a bit more about the peroxo cation structure that contributes to the infrared spectrum in Fig. 3 in the paper at 940 cm⁻¹? Does it represent a strongly bound local minimum structure that is formed in a small percentage of collisions, in which case it could be a "dead end" from the perspective of CO oxidation? Or is it a precursor species to the more stable structure in Fig. 3b?

Dr Mackenzie replied: Firstly, let me reiterate that we cannot be certain that it is a peroxo structure – whilst this provides the best fit to the spectrum, the 940 cm⁻¹ band may be the signature of an atop bound O atom.

However, assuming it is the peroxo structure, it seems likely to be a kinetically trapped structure along the path to O_2 dissociation and the lower-lying structures. It is, of course, not unusual, to trap species in deep local minima in molecular beam experiments.

As to whether it is a "dead end" for reactivity, two points seem pertinent. Firstly, we do not see the signature of IR-induced reaction (Pt_4O^+), or indeed loss of the $Pt_4O_2CO^+$ parent when pumping this band for the reasons discussed in the paper (and expanded upon the response to Professor Wester's question). Absorption of several photons is probably required to trigger reaction. Were the peroxo-structure to decompose at energies below the reaction barrier (as calculations predict – see Fig. 14 below), our chromophore would be lost and no more energy could be absorbed. Secondly, TPD experiments on extended surfaces (where the previous problem doesn't arise) detect CO oxidation reactions at around 160 K which can be attributed to reactions of hot O atoms formed upon O_2 dissociation. This competes with the molecular O_2 descrption at this temperature.

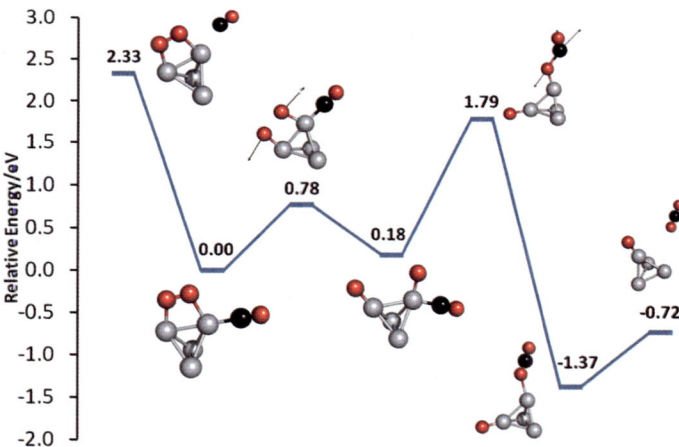

Fig. 14 Reaction profile calculated for the peroxo structure shown in Fig 3 in the paper. The peroxo "chromophore" is lost long before the reaction channel opens. *n.b.*, The starting peroxo structure structure lies 1.5 eV above the lowest energy structure observed.

Vibrationally bond-selected chemisorption of methane isotopologues on Pt(111) studied by reflection absorption infrared spectroscopy†

Li Chen,[a] Hirokazu Ueta,[a] Régis Bisson[b] and Rainer D. Beck*[a]

Received 18th January 2012, Accepted 14th February 2012
DOI: 10.1039/c2fd20007d

Reflection absorption infrared spectroscopy (RAIRS) was used to probe for vibrational bond-selectivity in the dissociative chemisorption of three partially deuterated methane isotopologues on a Pt(111) surface. While a combination of incident translational energy and thermal vibrational excitation produces a nearly statistical distribution of C–H and C–D bond cleavage products, we observe that laser excitation of an infrared active C–H stretch normal mode leads to highly selective dissociation of a C–H bond for CHD_3, CH_2D_2, and CH_3D. Our results show that vibrational energy redistribution between C–H and C–D stretch modes due to methane/surface interactions is negligible during the sub-picosecond collision time which indicates that vibrational bond-selectivity may be the rule rather than the exception in heterogeneous reactions of small polyatomic molecules.

Introduction

Laser control of chemical reactions by vibrational reactant excitation[1,2] has been sought for intensively ever since the invention of the laser because its success (or failure) provides insight into the reaction dynamics and because it offers new ways to control the outcome of a chemical reaction beyond what can be achieved by thermal activation. For bimolecular reactions in the gas phase, vibrational bond selectivity has first been demonstrated by Crim and coworkers[3] for the bond-selected H-atom abstraction from HOD controlled *via* O–H overtone excitation. A prerequisite for successful vibrational control is that the vibrational excitation stays localized in the selected bond(s) of the reactant molecule during the approach of the reaction partner. Results published during the last two decades by several groups attest that this can be achieved for gas phase reactions of small molecules[4–8] such as H_2O and CH_4 and their isotopologues.

Vibrational control in gas/surface reactions is of even greater interest due to the importance of heterogeneous catalysis in the chemical industry. However, the participation of a metallic solid in the reaction adds many degrees of freedom (and relaxation channels), which may aid in the fast delocalization of the reactant vibrations potentially preventing vibrationally bond-selected chemistry for gas/surface reactions. In fact, fast vibrational energy redistribution within a hypothetical transient molecule/surface complex is the fundamental assumption in the statistical descriptions of gas/surface reactivity developed by Harrison and coworkers.[9,10] Furthermore, clear evidence for fast energy transfer of vibrational energy between an

[a]Laboratoire de Chimie Physique Moléculaire, Ecole Polytechnique Fédérale de Lausanne, Switzerland. E-mail: rainer.beck@epfl.ch
[b]Aix-Marseille Univ, PIIM, CNRS, UMR 7345, 13397 Marseille, France

† This paper is dedicated to Prof. Fleming Crim on the occasion of his 65th birthday.

incident molecule and the electronic degrees of freedom of the metal surface has been reported for several molecule/surface systems[11,12] by Wodtke's group. On the other hand, vibrationally mode specific reactivity has been observed for the chemisorption of CH_4 and CH_2D_2 on Ni(100).[13–15] These quantum state-resolved measurements showed that excitation of different nearly isoenergetic vibrational modes of the incident reactant molecule can lead to very different reaction probabilities in disaccord with the assumption of complete vibrational energy redistribution prior to reaction. A first example of bond-selectivity in a gas/surface reaction has recently been reported by Killelea *et al.*[16] They observed that for the chemisorption of CHD_3 on Ni(111), a strong preference for C–H bond cleavage can be induced by excitation of the incident CHD_3 to its v_1 normal mode, a C–H stretch vibration localized in the unique C–H oscillator. In their work, Killelea *et al.*[16] identified the chemisorption products by mass spectrometry using temperature-programmed desorption (TPD) which is known to occur selectively upon surface heating between sub-surface D-atoms and chemisorbed methyl products on the surface of the Ni(111) crystal.[17]

In order to investigate to what extent vibrationally bond-selective dissociation can be realized with other reactant molecules as well as on different metal surfaces, we apply reflection absorption infrared spectroscopy (RAIRS)[18,19] as a more direct and generally applicable detection method of the nascent products of methane chemisorption on Pt(111) and use it to quantify the branching ratio between C–H and C–D cleavage channels for all three partially deuterated methane isotopologues. Comparison of the branching ratios observed for the dissociation of methane activated by a mix of translational and thermal vibrational energy from a hot nozzle beam ($T_n = 700$ K) with those for slower methane from a room temperature nozzle of $T_n = 294$ K excited to an infrared active C–H stretch mode provides clear evidence for bond- selective dissociation for each of the three partially deuterated methane isotopologues CHD_3, CH_2D_2, and CH_3D.

Experimental

A schematic of our experimental setup is shown in Fig. 1. The apparatus consists of a continuous molecular beam source connected to an ultra-high vacuum (UHV)

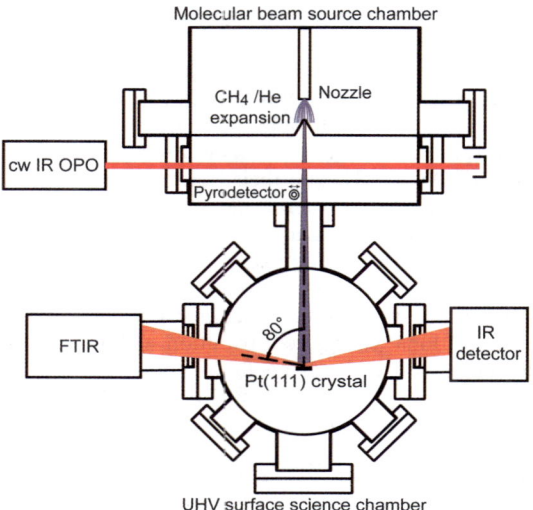

Fig. 1 Schematic of the molecular beam/surface science apparatus combining IR-pumping of the incident reactants and online RAIRS detection of adsorbates.

surface science chamber with a base pressure of 3×10^{-11} mbar. Reactant molecules in the molecular beam incident on a single crystal surface can be prepared in specific ro-vibrational eigenstates by infrared pumping with radiation from a single-mode continuous-wave optical parametric oscillator (cw-OPO). The Pt(111) target surface is mounted in the UHV chamber on a home-built sample manipulator which allows for sample heating to > 1300 K and cooling to 78 K. The chamber is also equipped with an Ar-ion sputter gun for sample cleaning and a cylindrical mirror analyzer for Auger electron spectroscopy (AES) for elemental surface analysis. An evacuated Fourier transform infrared spectrometer (FTIR) (Bruker Vertex V70) with an external InSb IR-detector is used to perform reflection absorption infrared spectroscopy (RAIRS) to obtain structural information on the chemisorption products deposited on the Pt(111) surface by the incident molecular beam. RAIRS offers several advantages over the product detection techniques used in previous state-resolved chemisorption studies such as AES,[20,21] secondary ion mass spectrometry (SIMS),[22] and temperature programmed desorption (TPD).[16] Being non-invasive, it enables on-line monitoring of the uptake of chemisorption products on the sample surface during the molecular beam deposition. In contrast to the high energy electrons or ions which bombard the sample surface during AES or SIMS analysis, the infrared radiation emitted by the thermal IR source of the FTIR neither affects the incident reactant molecules in the molecular beam nor can it lead to desorption or dissociation of the chemisorbed reaction products on the target surface. Furthermore, RAIRS is applicable to a wide range of adsorbates, providing direct structural information through the vibrational spectra of the absorbed species. In the work described here it enables the unambiguous identification of different isotopologues of chemisorbed methyl species on the Pt(111) surface with submonolayer sensitivity.

To detect the nascent products of methane chemisorption, the clean Pt(111) surface is cooled to 150 K and a background RAIR spectrum is collected by averaging 256 FTIR scans of 4 cm^{-1} resolution in 35 s. We then start to expose the surface to a molecular beam of 2–4% methane seeded in helium carrier gas, either with or without IR pumping of a C–H stretch vibration by the cw-OPO, and monitor the appearance and growth of the RAIR absorption signals in the C–H and C–D stretch region (2000–3100 cm^{-1}) due to the uptake of the surface bound reaction products by continuously recording RAIR spectra throughout the 80 min deposition time. The RAIRS absorption signal is proportional to the normal component of the vibrational transition dipole moment of the adsorbate vibration(s) and the surface coverage θ of the adsorbed species. We use Auger electron spectroscopy (AES) detection to calibrate the RAIRS signals of the different adsorbate species, CH$_3$(ads), CH$_2$D(ads), CHD$_2$(ads) and CD$_3$(ads), in terms of their surface coverage θ.

The deuterated methane isotopologues CD$_4$, CHD$_3$, CH$_2$D$_2$, and CH$_3$D with 99% isotopic purity and 98% chemical purity were obtained from Cambridge Isotope Laboratories and were further purified by passing them through an oxygen and water trap (Supelpure-O). CH$_4$ of 5.5 purity was obtained from Messer Griesheim and also purified in the same way. The translational energy E_{trans} of the incident methane, controlled via the nozzle temperature T_n, was determined by time-of-flight measurements using a fast chopper wheel in combination with an on-axis quadrupole mass filter. For the quantum state-specific preparation of ro-vibrationally excited states of methane in the molecular beam, we excite the incident methane with up to 1 Watt of cw, single mode infrared radiation produced as the idler wave of an optical parametric oscillator (OPO, Aculite, Argos 2400) tunable from 2600–3100 cm^{-1}. We stabilize the OPO idler frequency by locking it to a 1.5 MHz (FWHM) wide Doppler-free saturation hole (Lamb dip) in the Doppler broadened room-temperature absorption line of the same ro-vibrational methane transition used for state-preparation in the molecular beam. The Lamb dip was detected by retro-reflecting typically 50–100 mW of the idler output through a 1.7 m long gas cell filled with 30–90 µbar of the same methane isotopologues as used in the seeded molecular beam.

Vibrational excitation of the methane isotopologues in the molecular beam is detected and quantified by a pyroelectric detector (Eltec 406), which can be translated into the molecular beam in the third differential pumping stage (Fig. 1). Phase sensitive detection of the pyroelectric detector signal in combination with chopping of the excitation laser enables us to measure the laser fluence dependence of the vibrationally excited fraction of methane in the molecular beam. Analysis of the recorded fluence dependence of the pyroelectric detector signal indicates that the infrared pumping in the molecular beam is saturated so that we excite approximately 50% of the methane molecules to a given final rotational state in $v = 1$.

Results

Fig. 2 shows RAIR spectra of the nascent chemisorption products of five different isotopologues of methane on Pt(111) at a surface temperature of 150 K produced by an 80 min exposure to a molecular beam of $\approx 3\%$ of the indicated methane isotopologue seeded in helium using a nozzle temperature of 700 K without laser excitation. For these laser-off depositions, the chemisorption was activated by a combination of translation energy (≈ 50–55 kJ mol^{-1}) and thermal vibrational energy (≈ 5–10 kJ mol^{-1}) of the incident reactant molecules due to nozzle heating.

RAIRS detects only adsorbate vibrations with a vibrational transition dipole component along the surface normal due to the surface dipole selection rule.[18] For incident CH$_4$ on Pt(111) at $T_s = 150$ K, the only detected product is CH$_3$(ads), characterized by a strong line near 2883 cm^{-1} due to the symmetric CH$_3$ stretch and a weaker line at 2755 cm^{-1} due to a Fermi resonance with the C–H bend overtone in agreement with previous RAIRS studies[23,24] of chemisorbed CH$_3$ on Pt(111). The RAIR spectrum confirms that a single C–H bond is cleaved in CH$_4$ chemisorption to form a CH$_3$ group bound with the three-fold axis along the surface normal on a top site of the Pt(111) surface and that the methyl product is stable at $T_s = 150$ K. No RAIRS signal for the H–Pt(111) vibration is detected in our setup due to the low frequency cutoff at 1800 cm^{-1} of the InSb detector used. Similarly,

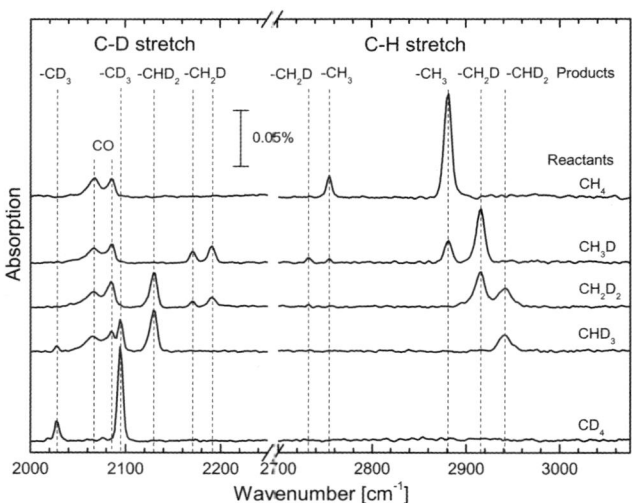

Fig. 2 RAIR spectra of nascent products of the dissociative chemisorption of five different methane isotopologues on Pt(111) at $T_s = 150$ K, activated by 50–55 kJ mol^{-1} incident translational energy and 5–10 kJ mol^{-1} of thermal vibrational excitation from nozzle heating to 700 K. Peaks near 2070 cm^{-1} (defect sites) and 2080 cm^{-1} (terrace site) are due to a small coverage (0.3% ML) of a CO present as an impurity in the molecular beam.

chemisorption of CD_4 produces CD_3(ads) which is detected by a strong line at 2095 cm^{-1} due to the symmetric CD_3 stretch vibration and a weaker line at 2028 cm^{-1} again due to a Fermi resonance with the C–D bend overtone. The RAIRS signals at 2080 cm^{-1} and 2060 cm^{-1} are due to a small coverage (<0.3% ML) of adsorbed CO on the terrace and step sites of the Pt(111) sample, respectively.[25] CO is an impurity in our molecular beam, probably resulting from the reaction of methane in the hot nozzle with traces of H_2O, and could not be avoided despite different attempts to purify the reagent gases. Chemisorbed CO is easily detected by RAIRS because of its large normal transition dipole moment, which we observe to be approximately 50 times larger than for the symmetric CH_3 stretch of CH_3(ads) on Pt(111).

The RAIR spectra following molecular beam exposure of the Pt(111) surface for the three partially deuterated methane isotopologues all show products consistent with dissociative chemisorption *via* two different reaction channels corresponding to cleavage of either a single C–H or C–D bond for activation of the reaction by incident translational and thermal vibrational energy (Table 1). We calibrate the observed RAIRS absorption signal for the four different methyl isotopologues in terms of surface coverage using calibrated Auger detection of the methyl carbon atoms. We use the measured methyl coverages to determine the branching ratios for translational and thermal vibrational activation given in Table 2. The C–H/C–D cleavage branching ratios observed for the hot nozzle beam are near the statistical limit with a small preference for C–H bond cleavage due to a kinetic isotope effect.[26]

Bond-selective chemisorption by IR pumping of C–H stretch vibrations

To probe to what extent the C–H/C–D cleavage branching ratio can be altered by bond-selective vibrational excitation of the incident methane prior to impact on the Pt(111) single crystal, we performed deposition experiments with a molecular beam of C–H stretch-excited partially deuterated methanes (CH_3D, CH_2D_2, and CHD_3) prepared by saturated infrared laser pumping. For state-prepared methane depositions, the nozzle temperature (T_n) was reduced from 700 K (laser-off

Table 1 Nascent products of the dissociative chemisorption of different methane isotopologues

$$CH_4 \longrightarrow CH_3(ads) + H(ads)$$

$$CH_3D \diagup\diagdown \begin{array}{l} CH_2D(ads) + H(ads) \\ CH_3(ads) + D(ads) \end{array}$$

$$CH_2D_2 \diagup\diagdown \begin{array}{l} CHD_2(ads) + H(ads) \\ CH_2D(ads) + D(ads) \end{array}$$

$$CHD_3 \diagup\diagdown \begin{array}{l} CD_3(ads) + H(ads) \\ CHD_2(ads) + D(ads) \end{array}$$

$$CD_4 \longrightarrow CD_3(ads) + D(ads)$$

Table 2 C–H and C–D cleavage branching ratios for state-resolved and thermal ensembles of partially deuterated methane isotopologues

Reactant 2–4% in He	Molecular beam conditions	E_{trans} (kJ mol^{-1})	E_{vib} (kJ mol^{-1})	E_{total} (kJ mol^{-1})	C–H cleavage	C–D cleavage
CHD$_3$	Laser-off, T_n = 700 K	54.6	8.3	62.9	30%	70%
	Laser-excited ν_1, T_n = 294 K	25.6	35.9	61.5	~100%	No detectable signal
CH$_2$D$_2$	Laser-off, T_n = 700 K	49.4	7.3	56.7	55%	45%
	Laser-excited ν_6, T_n = 294 K	23.2	35.9	59.1	~100%	No detectable signal
CH$_3$D	Laser-off, T_n = 700 K	49.3	6.3	55.6	78%	22%
	Laser-excited ν_4, T_n = 294 K	23.2	36.1	59.3	~100%	No detectable signal

experiments) to 294 K, so that the C–H stretch-excited methane dominates the reactivity of the molecular beam. An identical T_n = 294 K exposure of the Pt(111) surface without IR pumping yields no detectable RAIRS signal of methane chemisorption products (not shown). The reduction of T_n from 700 K to 294 K reduces the incident translational energy by approximately 25–30 kJ mol^{-1} and the thermal vibrational energy between 5–10 kJ mol^{-1} for CH$_4$ to CD$_4$. For the deposition of C–H stretch excited methanes, these amounts of translational and thermal vibrational energy are replaced by $\approx$ 36 kJ mol^{-1} of C–H stretch vibrational energy added by IR laser pumping of a specific ro-vibrational eigenstate. Fig. 3a–c present comparisons of the RAIRS detected chemisorption products with and without IR-pumping for each of the partially deuterated methane isotopologues.

The RAIR spectra recorded following the laser-on deposition of C–H stretch excited methane isotopologues are characterized by a complete absence of C–D bond cleavage products (Fig. 3a–c), providing direct evidence for strong bond-selectivity in the chemisorption of all three partially deuterated methanes on Pt(111) afforded by a single quantum of C–H stretch vibration.

Since the RAIRS detection does not interfere with the molecular beam/surface reaction, we can monitor the build-up of chemisorption products on the Pt(111) surface directly throughout the methane deposition. This enables us to convert a time series of RAIR spectra into an uptake curve of product coverage vs. reactant exposure by calibrating the RAIRS absorption signal in terms of product coverage and the molecular beam intensity in terms of reactant exposure. For a coverage of less than a monolayer (ML), the absorption signal dI/I_0 rather than the absorbance $-\log(I/I_0)$ is proportional to adsorbate coverage since all adsorbates are exposed to the same IR intensity (Lambert–Beer law does not apply). Because of the small transition dipole moment for the stretching vibrations of the adsorbed methyl species, dipole–dipole coupling is too weak to cause any nonlinearity in the RAIRS absorption signal.

Fig. 4a–c show the uptake curves of the different methyl product species resulting from the C–H and C–D cleavage reactions shown in Table 1 for the three partially deuterated methane isotopologues CH$_3$D, CH$_2$D$_2$, and CHD$_3$ obtained from the RAIRS data recorded throughout the molecular beam deposition experiments. The solid lines in Fig. 4 are the result of fitting a Langmuir site-blocking model[27] for the methyl uptake on the cold Pt(111) surface to the RAIRS data using eqn (1).

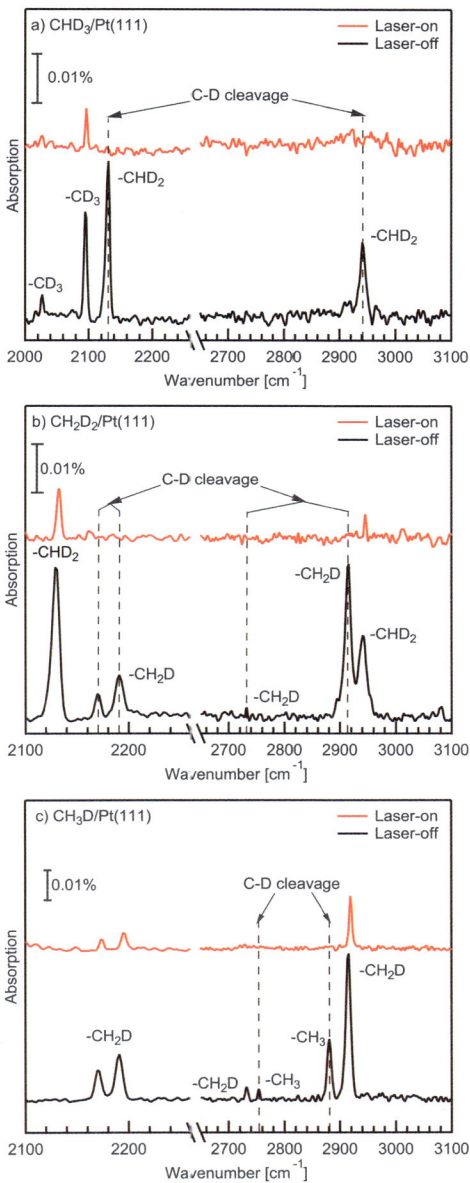

Fig. 3 a–c: Comparison of RAIR spectra of a near saturation coverage of nascent chemisorption products of (a) CHD$_3$, (b) CH$_2$D$_2$, and (c) CH$_3$D on Pt(111) at $T_s = 150$ K, for activation by translational and thermal vibrational energy supplied by nozzle heating to $T_n = 700$ K (bottom-black traces); for quantum state specific IR pumping of a C–H stretch normal mode vibration (a) CHD$_3(v_1)$, (b) CH$_2$D$_2(v_6)$, and (c) CH$_3$D(v_4) in a molecular beam using $T_n = 294$ K (top-red traces). C–D cleavage products are absent for chemisorption of the C–H stretch excited partially deuterated methane isotopologues.

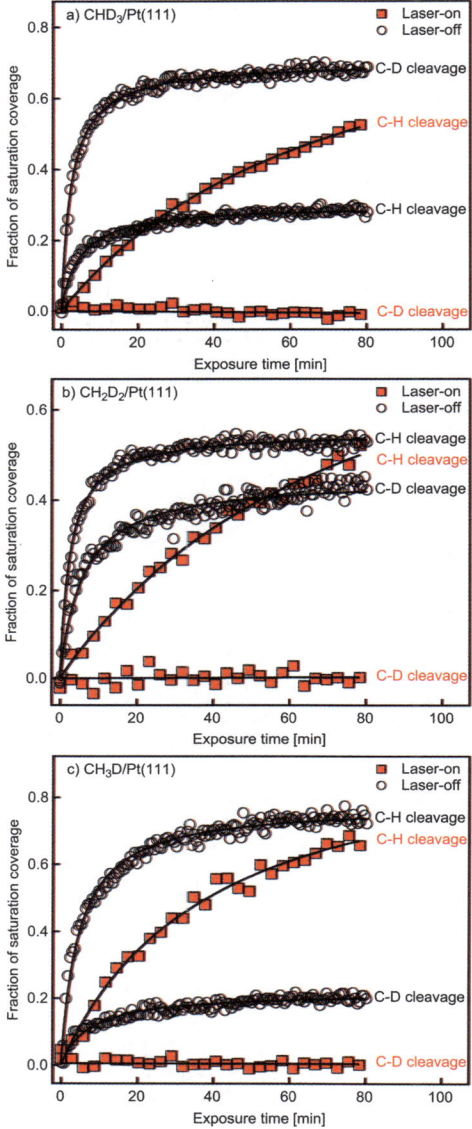

Fig. 4 a–c: Methyl product uptake curves for three partially deuterated methane isotopologues incident on Pt(111) (T_s = 150 K) for thermal activation (laser-off, T_n = 700 K) and state-specific C–H stretch excitation (laser-on, T_n = 294 K). For the laser-on deposition 25–30 kJ mol^{-1} of translational and 5–10 kJ mol^{-1} of thermal vibrational energy of the incident methane are replaced by 36 kJ mol^{-1} in C–H stretch excitation (See Table 2 for specific translational and vibrational energies for each experiment). The solid lines represent a fit of a site-blocking model to the coverage data to determine the saturation coverage used for the determination of the product branching ratios shown in Table 2. Since the excited fraction in the molecular beam is not accurately known for the laser-on depositions, we cannot interpret the slopes of the uptake curves in terms of reactivity relative to the ground state.

$$\theta(t) = A \frac{Bt}{1 + Bt} \qquad (1)$$

where $\theta(t)$ is the time dependent methyl product coverage in monolayers (ML), t the molecular beam deposition time, B a constant, and A the saturation coverage (ML). We use the saturation coverage A_{C-H} and A_{C-D} obtained from the fits shown in Fig. 4a–c to calculate the laser-off branching ratios given in Table 2. Fig. 4 shows that for deposition of C–H stretch excited methane isotopologues, only C–H cleavage products are detected, indicating complete bond-selectivity for all three partially deuterated isotopologues.

The initial slope AB of the uptake curve plotted as a function of exposure reflects the sticking coefficient S_0 of the clean surface and the curvature of the uptake curve yields the coverage dependence of the sticking coefficient $S(\theta)$. Detailed analysis of the uptake curves as a function of the incident dose for the different state-prepared methane isotopologues can be used to determine the state-resolved sticking coefficients for $CH_4(\nu_3)$, $CH_3D(\nu_4)$, $CH_2D_2(\nu_6)$, and $CHD_3(\nu_1)$ comparing the effect of a single quantum of C–H stretch excitation on the reactivity of each of the four methane isotopologues. However, a lack of precise knowledge of the excited fraction prepared by the infrared laser pumping has so far prevented us from detecting significant systematic variations in the reactivity of these four species. A preliminary analysis shows that the state-resolved reactivities of $CH_4(\nu_3)$, $CH_3D(\nu_4)$, $CH_2D_2(\nu_6)$, and $CHD_3(\nu_1)$ differ by less than a factor of 3.

Discussion and conclusions

We observe that a single quantum of C–H stretch excitation is sufficient to make the dissociative chemisorption on Pt(111) completely bond-specific for all of the three partially deuterated methane isotopologues. While the vibrational energy conferred by the infrared laser pumping is nearly identical for the three species (≈ 36 kJ mol^{-1}), the vibrational amplitude is distributed over an increasing number of C–H bonds for the excitation of $CHD_3(\nu_1)$, $CH_2D_2(\nu_6)$ and $CH_3D(\nu_4)$. The narrowband cw-infrared pumping (OPO linewidth 1 MHz) in the collision-free molecular beam prepares a single molecular eigenstate of the reactant containing stretching amplitude in either one, two, or three C–H bonds, depending on the isotopologues used. For the case of CHD_3, we prepare an eigenstate that closely resembles the ν_1-A_1 normal mode corresponding to a localized stretch excitation of the unique C–H bond. Wang and Sibert[28] used high order perturbation theory to calculate the vibrational eigenstates of several methane isotopologues up to 9000 cm^{-1} in energy. Their calculations show the $CHD_3(\nu_1)$ state prepared here via the R(1), K = 1 transition at 3005 cm^{-1} to contain 94% ν_1 and 5% $3\nu_3$ character where ν_3 is the CD_3 umbrella mode. For $CH_2D_2(\nu_6)$, the laser prepared eigenstate via the $^0Q(1)$ transition at 3004 cm^{-1} is the 100% pure antisymmetric C–H stretch normal mode ν_6-B_1. Finally for CH_3D, the $^0Q(1)$ transition at 3017 cm^{-1} prepares an eigenstate, which is predicted to be 96% ν_4 and 3% $2\nu_5$, where ν_4 is the antisymmetric C–H stretch (E-symmetry) in CH_3D and ν_5 is the C–H bend at 1471 cm^{-1}.

Single-mode IR pumping prepares an eigenstate of the isolated reactant molecule in the molecular beam which can undergo no time evolution as long as there are no perturbations due to gas-phase collisions or to interactions with the metal surface. Therefore, no intramolecular vibrational redistribution (IVR) occurs during the 100–200 microsecond time-of-flight from the laser excitation region up to a few Å from the surface impact site. When the reactant molecule is sufficiently close to the Pt(111) surface, surface-induced IVR becomes possible due to interactions with the surface. The eigenstate of the isolated molecules prepared by infrared pumping is not an eigenstate of the combined molecule/surface system but can be described in terms of a (time dependent) superposition of eigenstates of the molecule/surface system. Surface induced intramolecular vibrational energy flow has been predicted in calculations and was suggested to lead to mode specific CH_4 dissociation probabilities.[29] The fact that we observed completely bond-selective chemisorption for all three species studied here indicates that, while surface induced IVR may cause vibrational energy flow between the different C–H oscillators, energy transfer between C–H and C–D bonds is negligible during the sub-picosecond molecule/surface interaction time before impact for any of the three partially deuterated methanes.

The incomplete IVR and the ensuing bond-selectivity are in contradiction with statistical models[9,10] for the dissociative chemisorption of methane on Pt(111) which assume complete randomization of the initial vibrational state in a short-lived physisorbed complex formed by the incident molecule with several surface atoms prior to dissociation of a C–H bond. Statistical models[9,10] calculate the reaction probability from the rate of dissociation of the physisorbed complex obtained from statistical rate theory (RRKM) and the rate of intact desorption of the physisorbed molecule. A dynamically biased extension of a statistical model was recently proposed by Donald and Harrison[30] which is claimed to predict CH_4/Pt(111) sticking coefficients measured using effusive sources, supersonic beams, and eigenstate-resolved experiments. The new model treats rotation and translation parallel to the surface as spectator degrees of freedom and assumes that internal vibrational energy of the CH_4 participates in the dissociation with an average vibrational efficacy of 0.4 relative to normal translational energy. In the dynamical biased version, it is assumed that only 40% of the internal vibration energy is randomized together with normal translational energy and surface phonon energy as exchangeable energy before dissociation occurs. While it might be possible to adjust the parameters of the statistical model to fit an ensemble average of the reactivity of a statistical distribution of initial ro-vibrational states, it cannot correctly describe the bond-selective dissociation of the reactant prepared in a single ro-vibrational eigenstate and therefore fails to capture the state- and bond-specific dynamics of the methane chemisorption process.

Comparison of the laser-off branching ratios with those for state-specific C–H stretch excitation for the partially deuterated methane isotopologues shows that the control over the branching ratio increases with increasing deuteration. For CH_3D, C–H cleavage is the dominant reaction pathway for thermal activation and state specific laser excitation of a C–H vibration suppressed the minority C–D cleavage pathway. On the other hand, for the dissociation of CHD_3, the bond selectivity due to C–H stretch excitation is sufficiently strong to suppress the dominant C–D cleavage channel and make the minority C–H cleavage channel unique.

Due to limitations in the tuning range and our frequency stabilization scheme (Lamb-dip locking), we have so far only investigated the state-resolved methane dissociation reactions for excitation of the strongest infrared active C–H stretching fundamental transitions. Improvements in the experimental setup, which are currently underway, will enable us to also probe the extent of vibrational bond selectivity for other vibrational states of methane such as C–D stretch vibrations as well as overtones and combination bands including both C–H and C–D stretch and bend excitation. The results of these future experiments will provide even more detailed information about the possibilities and limitations of bond-selective control by vibrational reactant excitation and the role of surface-induced intramolecular vibrational energy transfer in chemisorption reactions of the different methane isotopologues.

Acknowledgements

We gratefully acknowledge financial support provided by the Swiss National Science Foundation (Grant No. 134709/1) and the Ecole Polytechnique Fédérale de Lausanne.

This journal is © The Royal Society of Chemistry 2012

References

1 R. N. Zare, *Science*, 1998, **279**, 1875–1879.
2 F. F. Crim, *Acc. Chem. Res.*, 1999, **32**, 877–884.
3 A. Sinha, M. C. Hsiao and F. F. Crim, *J. Chem. Phys.*, 1990, **92**, 6333–6335.
4 M. J. Bronikowski, W. R. Simpson, B. Girard and R. N. Zare, *J. Chem. Phys.*, 1991, **95**, 8647–8648.
5 J. D. Thoemke, J. M. Pfeiffer, R. B. Metz and F. F. Crim, *J. Phys. Chem.*, 1995, **99**, 13748–13754.
6 S. Yoon, R. J. Holiday and F. F. Crim, *J. Phys. Chem. B*, 2005, **109**, 8388–8392.
7 C. J. Annesley, A. E. Berke and F. F. Crim, *J. Phys. Chem. A*, 2008, **112**, 9448–9453.
8 Z. H. Kim, H. A. Bechtel and R. N. Zare, *J. Am. Chem. Soc.*, 2001, **123**, 12714–12715.
9 V. A. Ukraintsev and I. Harrison, *J. Chem. Phys.*, 1994, **101**, 1564–1581.
10 A. Bukoski, D. Blumling and I. Harrison, *J. Chem. Phys.*, 2003, **118**, 843–871.
11 J. D. White, J. Chen, D. Matsiev, D. J. Auerbach and A. M. Wodtke, *Nature*, 2005, **433**, 503–505.
12 Q. Ran, D. Matsiev, D. J. Auerbach and A. M. Wodtke, *Phys. Rev. Lett.*, 2007, **98**.
13 P. Maroni, D. C. Papageorgopoulos, M. Sacchi, T. T. Dang, R. D. Beck and T. R. Rizzo, *Phys. Rev. Lett.*, 2005, **94**, 4.
14 R. D. Beck, P. Maroni, D. C. Papageorgopoulos, T. T. Dang, M. P. Schmid and T. R. Rizzo, *Science*, 2003, **302**, 98–100.
15 L. B. F. Juurlink, R. R. Smith, D. R. Killelea and A. L. Utz, *Phys. Rev. Lett.*, 2005, **94**, 208303.
16 D. R. Killelea, V. L. Campbell, N. S. Shuman and A. L. Utz, *Science*, 2008, **319**, 790–793.
17 A. D. Johnson, S. P. Daley, A. L. Utz and S. T. Ceyer, *Science*, 1992, **257**, 223–225.
18 F. M. Hoffmann, *Surf. Sci. Rep.*, 1983, **3**, 107–192.
19 M. Trenary, *Annu. Rev. Phys. Chem.*, 2000, **51**, 381–403.
20 L. B. F. Juurlink, P. R. McCabe, R. R. Smith, C. L. DiCologero and A. L. Utz, *Phys. Rev. Lett.*, 1999, **83**, 868–871.
21 M. P. Schmid, P. Maroni, R. D. Beck and T. R. Rizzo, *J. Chem. Phys.*, 2002, **117**, 8603–8606.
22 R. Bisson, T. T. Dang, M. Sacchi and R. D. Beck, *J. Chem. Phys.*, 2008, **129**.
23 D. J. Oakes, M. R. S. Mccoustra and M. A. Chesters, *Faraday Discuss.*, 1993, **96**, 325–336.
24 D. H. Fairbrother, X. D. Peng, M. Trenary and P. C. Stair, *J. Chem. Soc., Faraday Trans.*, 1995, **91**, 3619–3625.
25 B. E. Hayden, K. Kretzschmar, A. M. Bradshaw and R. G. Greenler, *Surf. Sci.*, 1985, **149**, 394–406.
26 H. F. Winters, *J. Chem. Phys.*, 1976, **64**, 3495–3500.
27 D. R. Killelea, V. L. Campbell, N. S. Shuman, R. R. Smith and A. L. Utz, *J. Phys. Chem. C*, 2009, **113**, 20618–20622.
28 X. G. Wang and E. L. Sibert, *J. Chem. Phys.*, 1999, **111**, 4510–4522.
29 L. Halonen, S. L. Bernasek and D. J. Nesbitt, *J. Chem. Phys.*, 2001, **115**, 5611–5619.
30 S. B. Donald and I. Harrison, *Phys. Chem. Chem. Phys.*, 2012, **14**, 1784–1795.

On probing ions at the gas–liquid interface by quantum state-resolved molecular beam scattering: the curious incident of the cation in the night time

Andrew W. Gisler and David J. Nesbitt*

Received 17th February 2012, Accepted 19th March 2012
DOI: 10.1039/c2fd20026k

There has been a long standing controversy over the preferential presence or absence of cations *vs.* anions at the gas–liquid interface, dating back to early theoretical efforts by Onsager and co-workers [*J. Chem. Phys.*, 1934, **2**, 528]. In the present work, we describe our first efforts to selectively probe ions at the interface *via* a completely novel approach, based on scattering high energy, jet cooled molecular projectiles from the surface of hydrogen bonded liquids with dissolved alkali halide salts as the source of solvated charges. In particular, this work focuses on preliminary results from quantum state- resolved scattering studies as a function of anion (Cl^-, I^-), and cation identity (Li^+, Na^+, K^+) in alkali halide/glycerol solutions. By way of physical picture, a quadrupolar projectile such as CO_2 preferentially aligns parallel (anions) or perpendicular (cations) to the interface, which could therefore make rotational energy transfer sensitive to the *sign of the ion charge*. Experimentally, we find that the impulsive scattered (IS) CO_2 rotational distributions reveal a clear dependence on anion identity (*e.g.*, hotter for $[I^-]$ than $[Cl^-]$), but with essentially no corresponding sensitivity to cation species (*e.g.*, $[Li^+]$ indistinguishable from $[K^+]$). Detailed trajectory calculations are used to provide additional insight into the anticipated and observed experimental trends.

I. Introduction

The traditional view of inorganic ions at liquid interfaces is that the observation of increased surface tension suggests that ions are repelled from liquid interfaces.[1,2] Recently there has been renewed interest in the subject because of experimental evidence and theoretical support that larger, more polarizable halogens (Br^-, I^-) exist, and can even be enhanced, in the surface layer.[3-9] In addition to existing at the gas–water interface, these ions are thought to strongly disrupt hydrogen-bonded networks in the interfacial region.[5] Anions dissolved in glycerol are expected to exhibit a similar, though smaller, surface enhancement compared to aqueous solutions.[10] Glycerol proves to be an extremely useful analog for aqueous systems, due to its much lower vapor pressure and thus ease of study under vacuum conditions. Theoretical simulations of the glycerol surface suggest that hydroxyl OH comprises 60% of the surface with the remaining 40% consisting of CH groups.[11,12] Of those OH groups, none of them appear to be "free" (*i.e.*, pointing into the vacuum) and therefore available for hydrogen bonding.[13,14] Of particular relevance, the solubility

JILA, University of Colorado and National Institute of Standards and Technology, and Department of Chemistry and Biochemistry, University of Colorado, Boulder, Colorado 80309-0440. E-mail: djn@jila.colorado.edu of simple alkali halides in glycerol is quite comparable to that of water on a mole fraction basis.

Low energy molecular beam scattering at the gas–liquid interface is a powerful tool that exclusively probes the very topmost surface species.[15] In general, molecules that strike the liquid can undergo one of three pathways: they can either i) impulsively/inelastically scatter (IS) from the liquid after one or a few collisions, ii) thermally equilibrate with the surface and then desorb (trapping-desorption, TD) or iii) explicitly dissolve further into the liquid. This technique has been combined with time-of-flight mass-spectrometry to study the branching between these pathways by monitoring the translational degree of freedom.[16–18] In scattering studies of DCl from salty glycerol, Nathanson and co-workers noted a *decrease* in the fractional thermalization (*i.e.*, accommodation coefficient, α) of DCl compared to pure glycerol. This has been attributed to the ions tying up hydrogen bonds, thereby smoothing and stiffening the surface.[19] Molecular beams have also been scattered from self assembled monolayers (SAMs) which act as tunable liquid analogs.[20] Of particular interest to this work, Morris and co-workers have observed CO_2 collisions with OH terminated SAMS and see *increased* IS scattering compared to CH_3 terminated SAMS, which is ascribed to increased attraction of the quadrupolar CO_2 to the dipole of the OH group overcoming the rigidity of the surface.[21] Contributions from our group have exploited infrared laser absorption methods to monitor the quantum state-resolved rovibrational distribution of scattered molecules as well as high resolution Dopplerimetry to probe final quantum state-resolved translational distributions[22,23] In the present study, we extend these quantum state-resolved laser spectroscopy tools to explore scattering of CO_2 at hyperthermal kinetic energy (15 kcal mol^{-1}) from a series of gas–alkali halide/glycerol solutions. One main goal of this work is to exploit the quantum state-resolved scattered flux distributions as a "messenger" for the presence/absence of cations or anions in the surface layer.

II.   Experimental/theoretical background

Previous studies have described the experimental setup in detail; we present only aspects of special relevance to the current studies.[22,24] The fast incident molecular beam is formed from a mixture of 7% CO_2 reverse seeded in H_2 and supersonically expanding in a 300 μs pulse through a 500 μm pinhole, with a 3 mm diameter skimmer 1.5 cm downstream to collimate the angular divergence to < 6°. These gas molecules impinge on the liquid surface 10 cm from the pulsed valve at 45° with respect to the surface normal. The surface is generated using the method of Lednovich and Fenn,[25] based on a 12.7 cm rotating frosted glass wheel (0.2 Hz) viscously dragging a sheet of liquid from the reservoir, scraped by a stainless steel razor blade and leaving a thin (≈ 0.5 mm) freshly renewed liquid layer. The scattered molecules are detected at the 45° specular angle *via* high-resolution Beer's law absorption with a tunable narrow-linewidth PbSe diode laser beam parallel to and ≈ 1 cm above our liquid surface in a 16-pass Herriot cell configuration. Detection of the signal beam and reference on matched liquid nitrogen cooled Indium Antimonide (InSb) detectors in a fast servo-loop subtractor circuit yields absorbance sensitivities (≈ 1–2×10^{-6} Hz$^{-1/2}$), which are within a factor of ≈ 2 of the shot noise limit for ≈ 1 μW of light on each InSb. The laser is scanned in 2 MHz steps across each Doppler broadened transition, integrating over the initial 100 μs rising edge of the transient absorption, which avoids contamination from molecules thermalized with the walls of the chamber before entering the laser beam path. To greatly extend the range of rotational states accessible with our present diode laser, we have performed studies with both hyperthermal $^{12}CO_2$ and $^{13}CO_2$ molecular beams, allowing a much more detailed probe of the very hot, inelastically scattered IS molecules. Of special importance in the present work is a new capability for multiple liquid reservoirs on a rotating carousel in the vacuum chamber. This permits one to repeat experiments on different liquid solutions under identical conditions, which allows

This journal is © The Royal Society of Chemistry 2012

us to extract small changes in rotational state distributions with high reproducibility and precision.

In parallel with these experimental studies, molecular dynamics (MD) simulations have been performed to investigate the incremental influence of ions present on an interface. In previous work, such MD simulations of CO_2 scattered off of a fluorinated self-assembled monolayer (FSAM) have been shown to agree well with experimental results for CO_2 scattered from a perfluorinated liquid (PFPE),[26] providing valuable intuition into gas–liquid scattering dynamics.[27,28] As a first effort, we have modified the CO_2 PFPE potential surface of Hase and co-workers;[27] specifically, we have added electrostatic charge terms in the potential for a small fraction of the surface atoms (1 out of 9) corresponding to a 4 Å spacing and ≈ 2 M solution in glycerol. To implement this readily in the VENUS code, terms in the potential interaction are augmented to mimic Coulomb attraction/repulsion between a singly charged ion and partial charges present on each atom of CO_2. The partial charges ($Q_C = -2Q_O = 0.36\ Q_{electron}$) are chosen to match the experimental electric quadrupole moment[29,30] of CO_2 ($Q_{zz} = -2Q_{xx} = -2Q_{yy} = 4.278$ Debye Å), with simulations performed for both positively and negatively charged surface "ions." Finally, as these charges influence slightly the overall CO_2 binding energy with respect to the surface, the corresponding exponential repulsion parameters for the "ion"–CO_2 potential have been chosen to achieve physically motivated well depths ($D_0 \approx 5$–7 kcal mol^{-1}) for CO_2 + cation (collinear) and CO_2 + anion (T-shaped) interactions. Though obviously an approximate model of surface ion–CO_2 interactions, this does serve to isolate how the contribution of the quadrupole-cation *vs.* anion interaction affects the scattering dynamics for a fixed ion–CO_2 potential well depth, ion mass, *etc.* Initial translational, rotational energies of the CO_2 projectile are chosen to match our typical experiment conditions ($E_{inc} = 15$ kcal mol^{-1}, $T_{rot} = 15$ K, incident angle 45°), with ≈ 8000 trajectories (with randomly sampled initial orientation and lateral displacement) scattering from both positive and negative "ions." This yields a wealth of dynamical information; in the interest of time, however, the theoretical analysis in this work addresses only the degree of electrostatic alignment and anisotropy with which the CO_2 molecule "lands" on the surface.

III. Results and analysis

Experimental absorption profiles of CO_2 molecules scattered over a range of rotational states spanning ≈ 2000 cm^{-1} are presented in Fig. 1a, with line shapes fitted to Voigt profiles shown in red. Each transition is scanned three times for CO_2 scattering from each of the three liquid reservoirs over the course of 10 min, with the signals integrated over Doppler shift frequency to rigorously obtain column-integrated densities (#/cm^2) in each rovibrational state. These densities are proportional to the absolute density (#/cm^3) scattered into a particular quantum state and detection angle, which after correction for degeneracy/Hönl–London factors, permits a standard Boltzmann plot, as shown in Fig. 1b. The experimental data can be fit remarkably well by a two-temperature model,[22,24]

$$\frac{A_J}{S_J} = N[\alpha P_{TD}(J) + (1 - \alpha)P_{IS}(J)]$$

$$P_{TD/IS}(J) = \frac{(2J + 1)e^{-E_{rot}(J)/k_B T_{rot}(TD/IS)}}{Q_{rot}(TD/IS)}$$

even out to rotational energies > 10-fold in excess of kT for the liquid. The simplest interpretation of this is that one population (blue dashed line) of molecules *thermally* equilibrates with the surface (*i.e.*, trapping-desorption, TS) before desorbing, while a second population (red solid line) of molecules undergoes a small but finite number of interactions with the surface (*i.e.*, inelastic scattering, IS) and leaves with *hyperthermal* levels of rotational excitation. Here, A_J is the column-integrated absorbance,

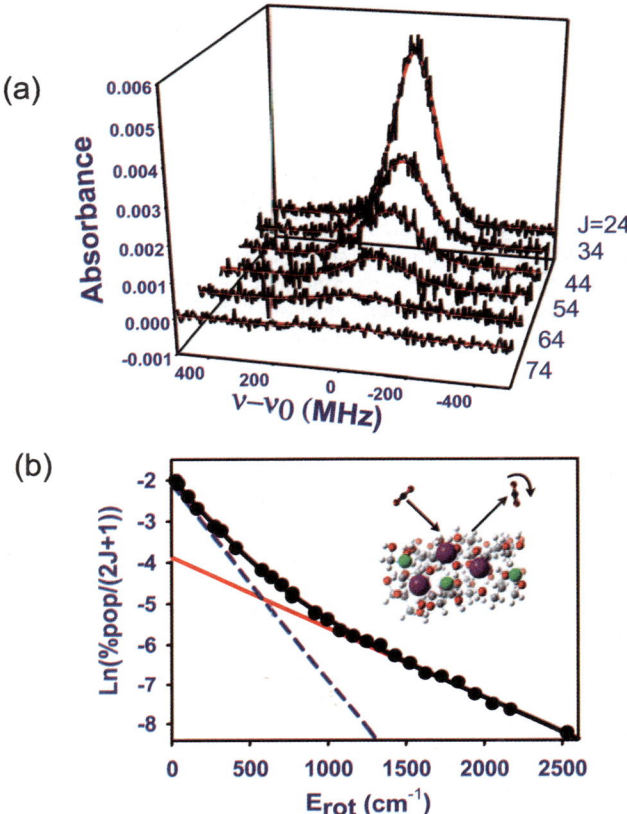

Fig. 1 (a) Doppler absorption profiles (black) for a range of rovibrational transitions of scattered CO_2 molecules out of the ground vibrational state ($00^01 \leftarrow 00^00$). Lower rotational J state is indicated to the right, with a Voigt lineshape fit for each transition in red. (b) A Boltzmann plot for CO_2 scattering from 300 K 2.5M NaI/Glycerol. The two temperature fit (see text) is shown as the black line, with contributions from thermal desorption (TD, with $T_{rot}(TD)$ fixed at 300 K) and impulsively scattered (IS, $T_{rot}(IS) = 780(20)$K) indicated by blue and red lines, respectively, with the TD fraction corresponding to $\alpha = 0.62(3)$.

S_J is the Hönl–London factor, N is a normalization factor related to the number density of the scattered flux, α is the fraction of detected CO_2 which has thermally desorbed from the surface, and $P_{TD/IS}$ are Boltzmann factors reflecting probability of the molecule being in an individual rotational (J) state. To facilitate fitting the data, the TD population is fixed at the temperature of the liquid surface, the validity of which has been shown previously.[28] The fact that this IS distribution can be so well described by a single "temperature" parameter over such a significant dynamic range is a remarkable and far from obvious result, though now observed and confirmed in numerous gas–liquid state-resolved scattering experiments.[22–24,28,31,32]

If ions are present at the very surface of the liquid and/or disrupt the surface properties, we might expect this to be evident in rotational distributions of the scattered molecules. By way of example, we compare in Fig. 2 Boltzmann plots for CO_2 at the same incident energy ($E_{inc} = 15$ kcal mol^{-1}) scattered from i) pure glycerol and ii) 5.0 M NaI in glycerol. Clearly, the addition of NaI *increases* the degree of CO_2 rotational excitation, particularly evident at the higher J levels dominated by IS scattering. Interestingly, however, we observe that the fractional thermalization (α) of the CO_2 *increases* with the addition of salt, from 0.58(1) in pure glycerol to

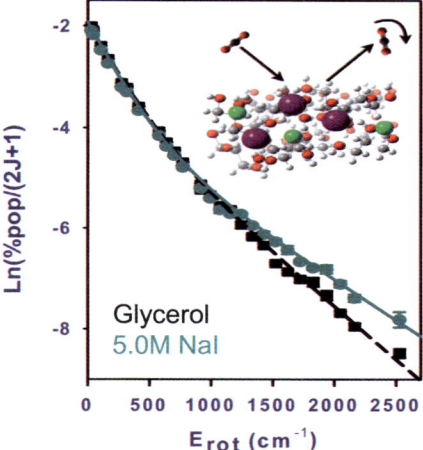

Fig. 2 A two temperature Boltzmann plot for pure glycerol (black dashed) and for 5.0M NaI/ Glycerol (green solid). Adding NaI salt to glycerol *increases* the rotational temperature of the IS molecules (T_{rot}(IS) = 690(20) K for pure glycerol and T_{rot}(IS) = 900(30) K for NaI/glycerol). The TD fraction α also increases slightly as NaI is added (α = 0.58(1) for pure glycerol and 0.65(2) for NaI/glycerol), which could reflect an increased attractive potential between CO_2 and the surface ion species.

0.65(2) for 5.0 M NaI. Stated alternatively, a greater fraction of CO_2 molecules thermalize at the gas–salty glycerol interface, but the ones that don't thermalize come off substantially hotter, with T_{rot}(IS) = 900(30) K (with NaI) *vs.* T_{rot}(IS) = 690(20) K (without NaI). The fact that more molecules come into thermal equilibrium than with pure glycerol would be consistent with the stronger binding anticipated between CO_2 and ions at the interface. This makes for interesting comparison with work by Nathanson and co-workers on collisions of dipolar DCl with salty glycerol, where the addition of salts indicate a *decrease* in α,[19] and which may reflect differences between dipolar (DCl) *vs.* quadrupolar (CO_2) scattering dynamics. One must be cautious, however, as the two studies are also probing different aspects of the scattered flux; Nathanson and co-workers observe translational kinetic energy into a narrow scattering solid angle, while we are sensitive to rotational/translational excitation in the laser probe path perpendicular to the scattering plane.[15] Analysis of the data exploiting high resolution Dopplerimetry, with additional quantum state-resolved experiments using DCl scattering are currently underway to help elucidate these issues.

Theory and experiment both suggest that small, "hard" ions, such as the alkali cations, are expected to reside a few monolayers below the surface while large polarizable ions, such as the larger halides (I^-), will preferentially reside at the glycerol surface.[2] If this is the case, then we might expect the identity of the deeply solvated cation to have little effect on the scattering dynamics, whereas the identity of the anion could play a larger role in the surface that the CO_2 molecule "sees." In order to test this prediction, we have performed scattering experiments for a series of glycerol solutions with the same halide anion but varying the alkali *cation*. Sample results are reported in Fig. 3b, which reveal essentially indistinguishable rotational distributions for both LiI and KI solutions, with the inset in Fig. 3b illustrating this comparison in greater detail. This insensitivity to cation identity would be qualitatively consistent with a simple physical picture of cation *depletion* from the uppermost surface layer. The fact that there might be relevant information hidden in the *absence of an effect* is quite noteworthy and, as any diligent reader of the Sherlock Holmes stories can readily appreciate, has also motivated our choice of title for this paper.[33]

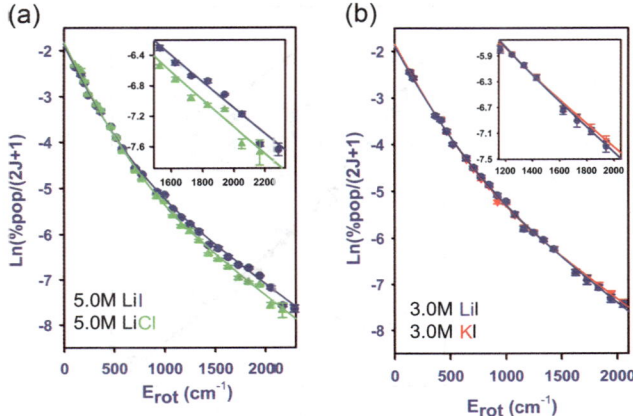

Fig. 3 Boltzmann plots demonstrating effects of varying anion (a) and cation (b) species. In (a) there is a small but quite reproducible increase in rotational excitation between LiCl (green triangles) and LiI (blue circles), particularly evident in the high rotational states corresponding to the IS pathway and indicating a sensitivity in surface properties to the anion identity. In (b), by way of contrast, there is essentially no difference in the scattered rotational populations between LiI (blue circles) and KI (red triangles) solutions with glycerol. These data would suggest that i) anion identity has a greater impact than cation identity on glycerol surface properties, and ii) the magnitude of this effect increases from Cl^- to I^-. This is also consistent with a preference for larger, more polarizable anions at the top most layer of the gas–liquid interface.

By way of a control experiment, Fig. 3a displays sample Boltzmann plots for CO_2 scattering from LiI and LiCl glycerol solutions, which show modest but nevertheless quite reproducible differences, particularly at the high rotational energies corresponding to IS scattering (see Fig. 3a inset). Specifically, I^- solutions appear to provide a larger rotational "kick" to the scattering CO_2 molecules than for Cl^- solutions. In contrast to the above cation studies, this demonstrates an unambiguous sensitivity to the anion identity, which would at least be consistent with a preferential *presence* of the more polarizable I^- *vs.* Cl^- anion in the interfacial surface layer. Furthermore, these anion effects seem to be most evident at the higher rotational energies, which preferentially impact internal state distributions for the *impulsively scattered* (IS) fraction of collisional events.

Though clearly much more experimental and theoretical work will need to be done to elucidate these intriguing effects, it is nevertheless useful and perhaps even appropriate in a Faraday Discussion to offer a simple physical picture for the putative collision dynamics. Due to its strongly quadrupolar nature, the incident CO_2 would energetically prefer to orient itself *parallel* to the interface with the C atom directly over the *anion* (i.e., "belly flop" collisions). Conversely, when the projectile approaches a *cation* at the interface, the CO_2 would energetically prefer to line up *perpendicular* to the interface with one of the O atoms attracted toward the positive charge (i.e., "nose dive" collisions). More quantitatively, potential curves as a function of polar angle are shown in Fig. 4a for CO_2 at a distance 5 Å away from a single positive (left) or negative (right) ion in the top surface layer. As expected, these reveal a non-trivial rotational barrier (≈ 4000 cm^{-1}), with a clear preference for *parallel* ($\theta = 0°$, $180°$) *vs.* *perpendicular* ($\theta = 90°$) alignment of the CO_2 projectile with respect to the surface. How such a preferential alignment would influence the asymptotic levels of CO_2 rotational excitation is not immediately obvious, particularly coming in at an incident angle away from $0°$. It is nevertheless worth noting that the differential experimental sensitivity to choice of anion *vs.* cation clearly signals an *influence* consistent with a preferential *presence* of anions *vs.* cations in

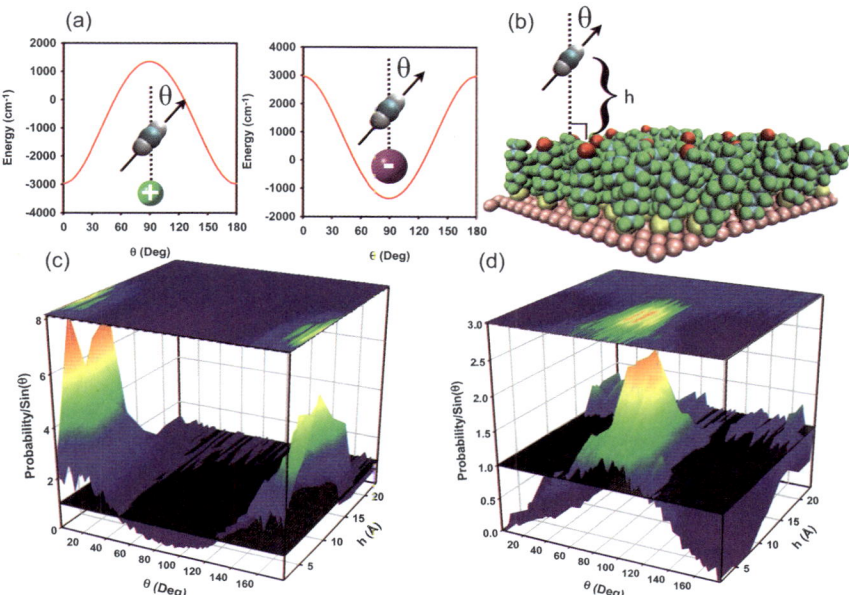

Fig. 4 Quadrupolar CO_2 molecules spatially aligning with respect to the gas–liquid interface as a function of ion charge; CO_2 approaching an ion (and associated E-field gradients) will align with respect to the surface. (a) Angular potential anisotropy between CO_2 and ions at a distance 5 Å. (b) MD simulations described in the text for CO_2 collisions with a modified FSAM, with red atoms representing anions or cations embedded in the surface layer. Even at 15 kcal mol^{-1} incident energy, CO_2 has time to align strongly (c) perpendicular or (d) parallel with respect to the surface due to cations or anions, respectively, with a maximum alignment occurring ≈ 7 Å from the FSAM surface.

the very topmost layer in the interfacial region sampled by quantum state-resolved collisional studies.

It is an interesting question as to whether these anisotropic potentials would be sufficiently strong to induce alignment on the time scale of the incoming CO_2 trajectory, particularly over our range of experimental kinetic energies. As described in Section I, we have also performed MD simulations of CO_2 colliding with FSAM surfaces distributed with anions and cations. A sample snapshot of the MD collision system is displayed in Fig. 4b, with histograms of the incoming CO_2 spatial alignment distributions shown in Fig. 4c. As the molecules approach the anion (cation) dominated surface, they clearly tend to become quite strongly aligned parallel (perpendicular) to the surface, even at relatively high incident collision energies of 15 kcal mol^{-1}. Far from the surface (*i.e.*, > 20 Å), the CO_2 feels an approximately uniform electric field with no gradient to align the quadrupole. Moving closer, CO_2 feels a strongly anisotropic field gradient and achieves a strong maximum average alignment near ≈ 7 Å. On further approach toward the minimum energy distance of ≈ 4 Å, it starts to "see" and therefore tilt the CO_2 projectile toward individual ions, which, due to random sampling of the initial CO_2 trajectory lateral coordinate, tends to diminish the average spatial alignment slightly. In any event, the histograms in Fig. 4c (for surface cations) and 4d (for surface anions) clearly testify to the presence of sufficient time at 15 kcal mol^{-1} collision energies for substantial electrostatic alignment of CO_2 prior to sampling the repulsive wall interactions with the surface.

In summary, we have reported the first results from a novel approach for probing ions at the gas–liquid interface, based on molecular beam scattering of quadrupolar

(CO$_2$) projectiles. In conjunction with quantum state-resolved infrared absorption, this permits us to study internal (rovibrational) degrees of freedom of molecules that scatter from the liquid surface, as well as external (translational) degrees of freedom by high resolution Dopplerimetry. Unlike nonlinear spectroscopies such as SHG and SFG that have a multilayer associated probe depth,[34] molecular beam scattering arguably offers an exclusive sensitivity to properties of the topmost interfacial layer. The observation that inelastically scattered (IS) distributions depend on anion but not cation identity suggests a greater impact of anions on surface scattering properties of salty glycerol, and would be consistent with increased propensity for polarizable negative ions such as I$^-$ at the gas–liquid interface.

This discussion raises and addresses some intriguing issues, but also leaves many interesting directions for future investigation. For example, one might reasonably expect that highly aligned "nose dive" or "belly flop" collisions at the interface would lead to different "signatures" of rotational excitation, at least at the initial stages of recoil. However, the electrostatic potential experienced by the recoiling CO$_2$ in the outgoing exit channel is just as strongly anisotropic as in the incoming channel. It becomes an interesting question as to whether these strong field gradients adiabatically extract energy away from rotational excitation, and thereby "erase" or at least significantly "blur" memories of the strong alignment forces felt in the interfacial region. Of equal interest would be the extent of electrostatic steering and orientation/alignment dynamics at the gas–ionic solution interface for strongly *dipolar* projectiles such as HCl, H$_2$O or NH$_3$, for which the *E-field* rather than *E-field gradient* represent the relevant term in the Hamiltonian. Finally, our present treatment has considered only monopolar fields due to isolated monovalent ions; it would also be interesting to ask if such strong field gradient and alignment effects survive averaging over the more *dipolar* field interactions due to monovalent (or multivalent) counterions nearby and/or below the surface. We are currently working to make advances in each of these areas, and would hope that the present work may stimulate complementary efforts from other research groups in similar or related directions.

Acknowledgements

This work has been supported by the Air Force Office of Scientific Research, with additional funding for the diode laser spectrometer from the National Science Foundation. We also would like to acknowledge support and helpful discussions with members of the NSF CCI Center for Energetic Non-equilibrium Chemistry at Interfaces (CENECI).

References

1 L. Onsager and N. N. T. Samaras, *J. Chem. Phys.*, 1934, **2**, 538.
2 P. Jungwirth and D. J. Tobias, *Chem. Rev.*, 2006, **106**, 1259.
3 J. H. Hu, Q. Shi, P. Davidovits, D. R. Worsnop, M. S. Zahniser and C. E. Kolb, *J. Phys. Chem.*, 1995, **99**, 8768.
4 P. Jungwirth and D. J. Tobias, *J. Phys. Chem. B*, 2001, **105**, 10468.
5 D. F. Liu, G. Ma, L. M. Levering and H. C. Allen, *J. Phys. Chem. B*, 2004, **108**, 2252.
6 P. B. Petersen, R. J. Saykally, M. Mucha and P. Jungwirth, *J. Phys. Chem. B*, 2005, **109**, 10915.
7 C. Schnitzer, S. Baldelli and M. J. Shultz, *J. Phys. Chem. B*, 2000, **104**, 585.
8 E. A. Raymond and G. L. Richmond, *J. Phys. Chem. B*, 2004, **108**, 5051.
9 P. B. Petersen and R. J. Saykally, *J. Am. Chem. Soc.*, 2005, **127**, 15446.
10 P. Jungwirth and B. Winter, *Annu. Rev. Phys. Chem.*, 2008, **59**, 343.
11 I. Chorny, I. Benjamin and G. M. Nathanson, *J. Phys. Chem. B*, 2004, **108**, 995.
12 M. Oh-E, H. Yokoyama and S. Baldelli, *Appl. Phys. Lett.*, 2004, **84**, 4965.
13 S. Baldelli, C. Schnitzer, M. J. Shultz and D. J. Campbell, *J. Phys. Chem. B*, 1997, **101**, 4607.
14 T. Krebs, G. Andersson and H. Morgner, *Chem. Phys.*, 2007, **340**, 181.
15 G. M. Nathanson, *Annu. Rev. Phys. Chem.*, 2004, **55**, 231.

16 M. E. King, K. M. Fiehrer, G. M. Nathanson and T. K. Minton, *J. Phys. Chem. A*, 1997, **101**, 6556.
17 M. E. King, G. M. Nathanson, M. A. Hanninglee and T. K. Minton, *Phys. Rev. Lett.*, 1993, **70**, 1026.
18 M. E. King, M. E. Saecker and G. M. Nathanson, *J. Chem. Phys.*, 1994, **101**, 2539.
19 A. H. Muenter, J. L. DeZwaan and G. M. Nathanson, *J. Phys. Chem. B*, 2006, **110**, 4881.
20 B. S. Day, S. F. Shuler, A. Ducre and J. R. Morris, *J. Chem. Phys.*, 2003, **119**, 8084.
21 J. W. Lu, W. A. Alexander and J. R. Morris, *Phys. Chem. Chem. Phys.*, 2010, **12**, 12533.
22 B. G. Perkins, T. Häber and D. J. Nesbitt, *J. Phys. Chem. B*, 2005, **109**, 16396.
23 B. G. Perkins and D. J. Nesbitt, *Proc. Natl. Acad. Sci. U. S. A.*, 2008, **105**, 12684.
24 B. G. Perkins and D. J. Nesbitt, *J. Phys. Chem. B*, 2006, **110**, 17126.
25 S. L. Lednovich and J. B. Fenn, *AIChE J.*, 1977, **23**, 454.
26 E. Martinez-Nunez, A. Rahaman and W. L. Hase, *J. Phys. Chem. C*, 2007, **111**, 354.
27 J. J. Nogueira, S. A. Vazquez, O. A. Mazyar, W. L. Hase, B. G. Perkins, D. J. Nesbitt and E. Martinez-Nunez, *J. Phys. Chem. A*, 2009, **113**, 3850.
28 B. G. Perkins and D. J. Nesbitt, *J. Phys. Chem. B*, 2008, **112**, 507.
29 C. Graham, J. Pierrus and R. E. Raab, *Mol. Phys.*, 1989, **67**, 939.
30 C. Graham, D. A. Imrie and R. E. Raab, *Mol. Phys.*, 1998, **93**, 49.
31 B. G. Perkins and D. J. Nesbitt, *J. Phys. Chem. A*, 2007, **111**, 7420.
32 B. G. Perkins and D. J. Nesbitt, *J. Phys. Chem. A*, 2010, **114**, 1398.
33 A. C. Doyle, The Memoirs of Sherlock Holmes: Silver Blaze, *The Strand Magazine*, 1892.
34 K. B. Eisenthal, *Chem. Rev.*, 1996, **96**, 1343.

Dynamics of molecular and polymeric interfaces probed with atomic beam scattering and scanning probe imaging

Ryan D. Brown, Qianqian Tong, James S. Becker,
Miriam A. Freedman, Nataliya A. Yufa and S. J. Sibener*

Received 8th February 2012, Accepted 20th February 2012
DOI: 10.1039/c2fd20016c

The scattering of atomic and molecular beams from well-characterized surfaces is a useful method for studying the dynamics of gas-surface interactions, providing precise information on the energy and momentum exchange which occur in such encounters. We apply this technique to new systems including disordered films of macromolecules, complex interfaces of macromolecular systems, and hybrid organic-semiconductor interfaces. Time-lapse atomic force microscopy studies of diblock copolymer structural evolution and fluctuations complement the scattering data to give a more complete understanding of dynamical processes in these complex disordered films. Our new scattering findings quantitatively characterize changes in interfacial dynamics including confinement in thin films of poly(methyl methacrylate) and changes in the physical properties of poly(ethylene terephthalate) films as they transform from the glassy to their semicrystalline phase. Further measurements on a hybrid organic-semiconductor interface, methyl-terminated silicon (111), reveal that the surface thermal motion and gas-surface energy accommodation are dominated by local molecular vibrations while the interfacial lattice dynamics remain accessible through helium scattering. High temperature atomic force microscopy allows direct, real-time visualization of structural reorganization and defect migration in poly(styrene)-*block*-poly(methyl methacrylate) films, revealing details of film reorganization and thermal annealing. Moreover, we employed lithographically created channels to guide the alignment of polymer microdomains. This, in turn, allows direct observation of the mechanisms for diffusion and annihilation of dislocation and disclination defects. In summary, this paper elaborates on the power of combining atom scattering and scanning probe microscopy to interrogate the vibrational dynamics, energy accommodation, energy flow, and structural reorganization in complex interfaces.

1. Introduction

Since the discovery of atom diffraction from alkali halide interfaces,[1] atom scattering techniques have been extended from a method of observing structure and lattice dynamics of clean ordered interfaces[2,3] to increasingly complex systems. The successful application of helium atom scattering toward the investigation of defects,[4] adsorbate decoration,[5,6] and thin film interfaces[7,8] has directed the evolution of this method as a tool for characterizing the dynamic and structural properties of increasingly complex interfaces. Herein we present the extension of atomic

The James Franck Institute and Department of Chemistry, The University of Chicago, 929 E. 57th Street, Chicago, Illinois 60637, USA. E-mail: s-sibener@uchicago.edu

scattering studies to include disordered organic films of macromolecules as well as complex ordered organic films. Our application of atom scattering to glassy and crystalline polymeric films as well as organically terminated semiconductor interfaces allows us to access interfacial vibrational dynamics and energy accommodation characteristics. Scanning probe microscopy techniques, specifically high temperature time-lapse atomic force microscopy, supplement the insight gained from atom scattering through real space visualization of the structural evolution and fluctuations in these disordered and structurally complex interfaces. The characterization of vibrational dynamics through atomic scattering combined with the real space visualization of complex polymer film evolution from atomic force microscopy gives us new insight into the vibrational dynamics and structural reorganization of disordered films on the atomic, nanoscopic, and microscopic length scales.

Neutral helium atom scattering (HAS) is a non-destructive and uniquely surface sensitive probe of surface structure and vibrations. Diffractive HAS arises due to the helium atom's de Broglie wavelength being on the order of an angstrom, which also imparts sensitivity to atomic scale fluctuations and vibrations. Helium's relatively light mass allows for the observation of small energy and momentum exchanges from collisions with a surface, and thus the low energy vibrations are accessible through time-of-flight scattering methods. Atomic scattering can be useful in characterizing interfacial vibrations through measuring the surface temperature dependence of the elastically scattered helium atoms as well as the single and multiple phonon energy exchanges between the helium atom and surface. Atomic force microscopy (AFM), specifically in an AC or tapping imaging configuration, is also a relatively non-perturbative scanning probe technique capable of visualizing real space events on the nanometer and micrometer length scales.

We begin this paper by demonstrating HAS's ability to detect subtle changes in vibrational dynamics of polymeric thin films through Debye–Waller attenuation measurements.[9] We find that the amplitudes of the surface mean-square displacements (MSD) normal to the surface plane are depressed for a 10 nm thick poly(methyl methacrylate), or PMMA, film relative to that of a thicker 50 nm film, and that this is corroborated by a suppression of annihilation events in the multiphonon lineshape. The sensitivity of the inelastic lineshape is investigated for chemical composition and molecular weight in other glassy polymers. Building upon these successes, we show that HAS is capable of examining the surface vibrational properties of poly(ethylene terephthalate) (PET) through temperature and angle sensitive Debye–Waller attenuation measurements, and that this method is capable of observing changes in surface dynamics arising due to a transition from a homogenous amorphous phase to a heterogeneous semi-crystalline phase. We observe a 15% enhancement in the amplitude of perpendicularly polarized MSDs for the semi-crystalline polymer, or a softening of these vibrational modes, yet a 25% suppression of the MSD amplitude in the surface plane. We can attribute these seemingly conflicting trends as features arising from different domains of this complex disordered surface.

We then move on to apply these Debye–Waller attenuation measurements to a methyl-terminated silicon interface in order to characterize the vibrational dynamics, energy accommodation, and gas-surface interaction potential. The CH_3–Si(111) interface exhibits thermal motion similar to a rigid semiconductor in the z direction, but in the surface plane its magnitude is closer to that of an alkanethiol SAM. We are able to use the angular dependence of the Debye–Waller factor to determine the helium-surface well depth and the surface Debye temperature. Further, we demonstrate the observation of single elastic vibrational modes of the underlying silicon lattice through single-phonon inelastic collision events for this complex surface. These measurements are capable of interrogating the dispersion relation of the underlying Si(111) lattice Rayleigh wave despite the presence of the covalently bound methyl overlayer.

Finally, we use time-lapse AFM to directly observe local structural fluctuations and reorganization in a complex diblock copolymer film. We present the observation of domain break up and reorganization and discuss how *in situ* imaging of this process lends insight into the mechanism of polymer diffusion in a poly(styrene)-*block*-poly(methyl methacrylate), PS-*b*-PMMA, film. Nanoconfinement of the diblock copolymer in channels causes alignment of the polymer domains in this film and allows direct observation of isolated defect interactions. These observations give insight into the mechanisms guiding defect migration and annealing for both dislocations and disclinations in these complex polymer films.

2. Experimental

Helium atom scattering experiments were performed using a high-energy and high-momentum resolution scattering apparatus which is described in detail elsewhere.[10] The apparatus can briefly be described as follows: A supersonic expansion of helium through a nozzle cooled by a closed cycle helium refrigerator giving a nearly mono-energetic beam of helium atoms with typical velocity distributions ($\Delta v/v$) of less than one percent. Time resolution is achieved by modulating the beam with a mechanical chopper. The beam was collimated to a 0.22° angular width and 4 mm spot size on a target mounted to a six-axis manipulator in the UHV scattering chamber, with a chopper-to-target distance of 0.5 m. The sample temperature is regulated by using a button heater to heat against cooling provided by a closed cycle helium refrigerator. The reflected atoms enter a differentially pumped rotatable detector with a target-to-ionizer flight path of 1 m for the PMMA, PS, and PB experiments and 0.5 m for the PET and methyl-terminated silicon experiments. The atoms are detected by ionization followed by mass selection.

The PMMA films were prepared by spin casting a solution of the polymer on Si(100) substrates with their native oxide. These films were annealed under argon to ensure uniformity, and film thickness was verified using ellipsometry. The PET films were prepared using a spin casting technique in which a PET solution was applied to a clean Au/Mica substrate, producing films of 80 nm depth. The film quality was verified with optical microscopy, and the film thickness was confirmed with ellipsometry and AFM. Upon completion of the scattering studies from the amorphous film, the sample was crystallized by annealing at 440 K for 30 min, which yielded films with a crystalline fraction of approximately 55% as confirmed by infrared reflection-absorption spectroscopy (IRRAS) measurements in a separate UHV chamber.[11]

The methylated silicon samples were prepared by the Lewis group at Caltech using their previously published method.[12] The functionalized wafers were transported from Pasadena to Chicago in sealed containers under inert atmosphere and loaded into the helium atom scattering apparatus. Angular scans were performed using a square wave chopping pattern with a 50% duty cycle and the rotatable detector arm (overall instrumental resolution 0.5°) was swept through the scan range under computer control. Energy-resolved time-of-flight measurements were performed using a single slit chopping pattern with a 1% duty cycle. The Debye–Waller attenuation experiments used a 46 meV beam and sample temperatures between 200 K and 500 K in 50 K steps, and the phonon spectroscopy experiments utilized beam energies of 43 to 65 meV and surface temperatures of 140 to 200 K. The Lewis group also prepared a set of deuterated methyl samples, CD_3-Si(111), so that the vibrational characteristics of this surface could be examined as well.

The atomic force microscopy experiments were performed using an Asylum MFP-3D atomic force microscopy with commercial silicon cantilevers and a polymer heater mount. We employed Olympus ACT160TS silicon cantilevers (k $\sim$ 42 N m^{-1}) for imaging domain reorganization and fluctuations in the unconfined films, and Olympus ACT240TS silicon cantilevers (k $\sim$ 2 N m^{-1}) for imaging defect annealing in the diblock films confined in channels. All imaging was performed using

AC or tapping mode in order to minimize the physical damage from the cantilever during imaging. The polymer heater mount situates the sample in direct contact with the heating element (300 to 700 K) while maintaining atmospheric control with a purging flow of room temperature argon. Drift correction allowed for imaging a single region of the film over long time periods.

The unconfined PS-b-PMMA films were prepared by spin coating a solution PS-b-PMMA (M_w = 77 kg mol^{-1}, Polymer Source) in toluene onto silicon nitride substrates, resulting in a 30 nm thick film. The scanning frame rate was between 30 to 180 s and the scan sizes ranged from 500 nm to 2 μm. For the films confined to channels, nanopatterns were introduced onto silicon nitride substrates using photoresist e-beam lithography then transferring the pattern to the substrate with a 95% CF$_4$, 5% O$_2$ plasma etch. The resulting etched channels are about 500 nm wide, 20 μm long, and 50 nm deep. After the substrates were cleaned, we spin coated a solution of PS-b-PMMA (M_w = 77 kg mol^{-1}, Polymer Source) in toluene at 4000–5500 rpm for 60 s. The films were then annealed at 523 K for 150 min under an inert argon atmosphere so that the PS-b-PMMA films formed well ordered aligned cylindrical domains with low defect density.

3. Results and discussion

3.1 Helium atom scattering as a probe of thin film polymer dynamics

The extension of HAS to the study of disordered films is complemented by improvements in the theoretical treatment of atom scattering from disordered surfaces.[13,14] In the case of a structurally disordered film, the nature of the scattering is quite different than from ordered surfaces. For structurally disordered surfaces there are no coherent diffraction channels available for scattering and thus the helium scatters diffusely. While there is no coherent scattering due to the absence of long-range order, there is still a rich palette of information available regarding the dynamical properties of such surfaces. Attenuation measurements of the elastically scattered fraction as well as analysis of the line shape of the inelastic features give insight into energy accommodation and momentum exchange in these gas-surface collisions. Furthermore, the nature of the gas-surface potential can be determined through careful analysis of the diffuse elastic and inelastically scattered components.

Time-of-flight spectroscopy was employed to observe the energy exchange characteristics of the PMMA surface, and a pseudorandom cross-correlation chopping technique was used to maximize signal-to-noise. Beam energies employed in this experiment range from 9.7 to 54 meV, with the sample temperatures ranging from 60 to 490 K, and scattering at the specular condition was performed at angles of 37.42°, 32.42°, and 24.42° from the surface normal. Fig. 1 shows an overlay of representative spectra at the specular condition as the surface temperature is varied. These spectra are comprised of a relatively sharp diffuse elastic peak (at cryogenic beam energies and sample temperatures from 60 to 120 K), and a broad inelastic, or multiphonon, feature.

The attenuation of the elastic scattering component can be described theoretically using the Debye–Waller model.[9] The model relates the degree of the attenuation of the elastic component to the magnitude of the mean-square displacement (MSD) of the surface atoms or molecules from their equilibrium position. This analysis has been applied to the cases of metals,[15] self-assembled monolayers,[7] and organic crystals,[16] and in the case of PMMA we successfully demonstrate its application to a disordered polymer surface. In case of HAS, the elastically scattered component can be described by eqn (1) below, where the term $2W(T)$ is the Debye–Waller factor, I is the observed elastic intensity, and I_0 is the theoretical elastic intensity at $T_S = 0$ K. Eqn (2) defines the perpendicular and parallel momentum exchanges for the elastically scattered helium atoms, with the Beeby correction[17] accounting for the well depth, D, of the helium-polymer interaction potential, E_i the incident

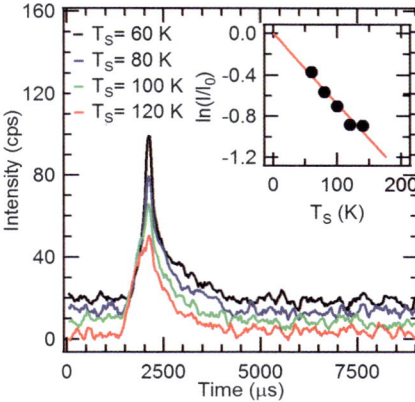

Fig. 1 Time-of-flight spectra at the specular condition from a 10 nm thick PMMA film with $E_B = 11$ meV and $\theta_i = 37.42°$ for sample temperatures ranging from 60 to 120 K. The elastic peak clearly decreases with sample temperature, and the inset shows a linear plot of the Debye–Waller factor $vs.$ T_S.

helium atom kinetic energy, k_i the magnitude of the incident helium wave vector, and θ representing the angle from the surface normal for the incident (i) and scattered (f) helium atom wave vectors. The Debye–Waller factor itself can be separated into the contribution from perpendicular (z) and in plane ($\parallel$) displacements, and thus at specular scattering conditions the ΔK^2 term disappears and the attenuation is dependent only on displacements normal to the surface.

$$I = I_0 e^{-2W(T)}, 2W = \Delta k_z^2 \langle u_z^2 \rangle + \Delta K \langle u_\parallel^2 \rangle \tag{1}$$

$$\Delta k_z = k_i \left\{ \left[\cos^2(\theta_f) + \frac{D}{E_i} \right]^{\frac{1}{2}} + \left[\cos^2(\theta_i) + \frac{D}{E_i} \right]^{\frac{1}{2}} \right\}, \Delta K = k_i \left[\sin(\theta_f) - \sin(\theta_i) \right] \tag{2}$$

As is apparent from eqn (1), the temperature dependent thermal attenuation of the elastic feature can be used to characterize the MSDs arising from the surface vibrational modes. A plot of the natural log of the elastic intensity *versus* surface temperature is linear when the surface oscillators exhibit harmonic behavior and attenuation is due to the coarsening of the surface from thermal motion. Traditionally, this analysis is applied to the introduction of dynamic disorder in an ordered system, where a coherent elastic peak attenuates due to the coarsening of the ordered lattice. In the case of disordered polymer interfaces there is no ordered lattice and no coherent scattering is observed. Instead, the dynamic disorder at the surface attenuates diffuse elastic scattering intensity at a given solid angle by opening further diffuse scattering channels. We find that the Debye–Waller model fits the observed attenuation of our diffuse elastic peak intensity quite well, as seen in Fig. 1.

The inset of Fig. 1 shows the Debye–Waller plot of the spectra shown, and the linearity indicates that the harmonic approximation holds in the temperature regime studied. The fact that we are able to perform this analysis on a structurally disordered organic surface is impressive, but more remarkable is that this technique is sensitive to subtle changes in surface vibrations due to thin-film confinement. The peak intensity was taken as the height of the diffuse elastic peak, and the helium-polymer potential well depth was assumed to be 7 meV for this system since this value was experimentally obtained for high density alkanethiol self-assembled monolayers.[7] The 10 nm PMMA samples had smaller MSDs ($6.0 \pm 0.5 \times 10^{-5}$ Å^{2}K^{-1}) than those of the thicker 50 nm films ($6.9 \pm 0.5 \times 10^{-5}$ Å^{2}K^{-1}). This reduced thermal motion due to thin film confinement has also been observed for bulk film MSDs by neutron scattering,[18] and possibly arises from strong substrate polymer interactions which can propagate to the interface in ultrathin films.

The multiphonon envelope line shape of the scattered helium also supports the Debye–Waller attenuation measurements in showing intrinsic differences in inelastic scattering profiles due to film thickness confinement effects. Since the 10 nm film is "stiffer" than the 50 nm film, it should show differences in its energy transfer profile. Fig. 2 overlays energy transfer profiles for both the 10 and 50 nm films at four different sample temperatures. It is interesting to note that the thicker film has more intensity attributed to annihilation events (the energy transfer scale is for the helium atom, thus a positive energy transfer is the atom gaining energy by annihilating surface vibrational modes) than the thinner film for all four sample temperatures. The suppression of thermal motion due to substrate interactions could account for the systematic differences observed in these energy transfer profiles.

The line shape of the multiphonon envelope was fit to the semi-classical theory developed by Manson, Celli, and Himes[13,14] to determine the nature of the gas-polymer potential. This model contains two situations for the nature of the potential, a continuum condition and a discrete scattering center condition. The continuum of atomic centers model has been applied to metal surfaces while the discrete model has been applied to an organic molecule monolayer. The differential reflection coefficient is shown in eqn (3), and the key component in our studies is the surface temperature dependent form factor prior to the exponential.

$$\frac{dR}{d\Omega_f dE_f} \propto |k_f| |\tau_{fi}|^2 \left(\frac{\hbar\pi}{\omega_0 k_B T_S} \right)^n \exp \left[-\frac{(\Delta E + \hbar\omega_0)^2}{4 k_B T_S \hbar\omega_0} \right] \tag{3}$$

The term τ_{fi} is the scattering transition matrix, which is the product of the Jackson-Mott matrix element of the surface potential and a parallel momentum cutoff term.[19] The classical recoil energy, $\hbar\omega_0$, is approximated as $\hbar^2 k^2 / 2M_{eff}$, where k is the total momentum transfer, $k_f - k_i$, and M_{eff} is the effective surface mass. The

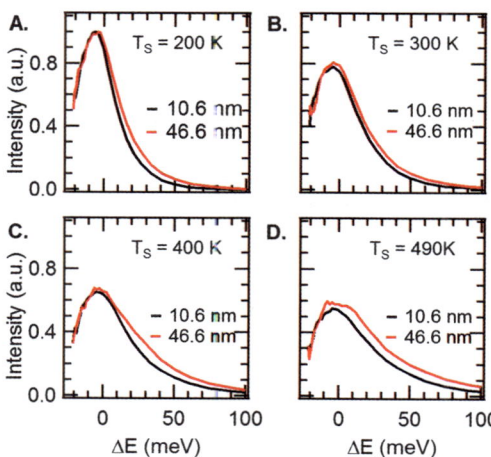

Fig. 2 Overlay of energy transfer spectra of a 31 meV beam comparing thin (black) and thick (red) PMMA films at four different sample temperatures at the specular condition $\theta_i = \theta_f = 24.42°$. As the sample temperature increases, the divergence in intensity of the annihilation (positive ΔE) side of the spectra increases.

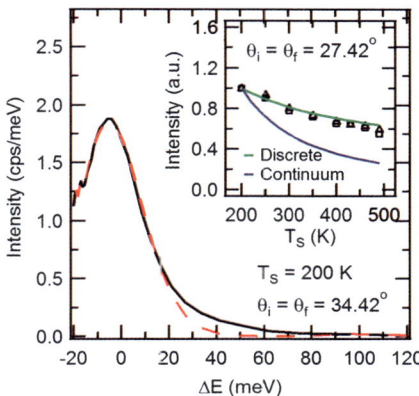

Fig. 3 An overlay of the experimental energy transfer spectrum of a 31 meV beam from the 47 nm PMMA film (black solid line) with the semiclassical discrete center model fit (red dashed line). The inset is the experimental intensity of the multiphonon feature at this beam energy for the thin (square) and thick (triangle) PMMA films compared to the semiclassical model predictions for the discrete (green) and continuum (blue) scattering potentials.

difference between the two models is in the power of the surface temperature dependence of the surface form factor, which is $n = 3/2$ for the continuum case and $n = 1/2$ for the discrete center case.

The PMMA multiphonon line shape is fit well by using the discrete model, and very poorly by the continuum model, as shown in Fig. 3. The only significant deviation from the model occurs in the high energy tail on the annihilation side of the spectra, which most likely arises from multiple scattering events not accounted for in this model. When the inelastic scattering profiles of polystyrene (PS) and polybutadiene (PB) were compared to PMMA, little difference was observed.[20] However, inherent differences in the multiphonon lineshape appear when scattering from polymer films of different molecular weight.[21]

3.2 Helium atom scattering as a probe of changes in dynamics due to phase transitions

The successful application of HAS for characterizing thin polymer film dynamics can be extended to more complex disordered systems, specifically those which undergo transitions between glassy and crystalline phases. The particular phase change polymer studied was poly(ethylene terephthalate), or PET, which undergoes a transition from a homogenous amorphous glass to a heterogeneous semi-crystalline film above its glass transition temperature (T_g), 348 K, and below its melt temperature of 523 K. The semi-crystalline film consists of crystalline lamellae comprised of stacked planes of *trans*-oriented polymer chains with amorphous domains between these lamellae. As in the case of thin polymer films, the sensitivity of helium atom scattering to surface thermal motion was employed to observe changes in surface vibrational dynamics due to a phase transition from an amorphous to a semi-crystalline PET film. The resulting Debye–Waller analysis truly demonstrates the power of helium atom scattering to detect subtle changes in interfacial dynamics.

Elastic scattering was observable at low beam energies (7.1 meV) and surface temperatures (40–120 K). The scattering profile from PET is similar to that of PMMA in that there is a sharp diffuse elastic feature and broader inelastic multiphonon component in the time-of-flight spectra. The Debye–Waller analysis was performed by fitting the elastic peak using a Gaussian function (FWHM ~1 meV) and the peak intensity was taken as the height of the elastic component fit. As in the case with PMMA, the Beeby correction was assumed to have a well depth of 7 meV due to its use in similar systems.[7,16] Four independent measurements were made at each of five specular conditions ($\theta_i = \theta_f = 37.7°$, $33.7°$, $29.7°$, $25.7°$, and $21.7°$), and the normalized natural log of the peak intensities were averaged for each sample temperature at each kinematic condition. Using eqn (1) and (2), the perpendicular MSD was found to be $2.7 \pm 0.2 \times 10^{-4}$ $Å^2K^{-1}$ for the amorphous surface and $3.1 \pm 0.1 \times 10^{-4}$ $Å^2K^{-1}$ for the semi-crystalline interface (Fig. 4).

In-plane MSDs were characterized by measuring the thermal attenuation of the diffuse elastic peak at non-specular scattering conditions. Using the same beam energy, and the most glancing scattering condition ($\theta_i = 37.7°$), thermal attenuation measurements were performed at $\theta_f = 29.7°$ and $13.7°$, which correspond to in-plane momentum exchanges of -0.43 $Å^{-1}$ and -1.38 $Å^{-1}$, respectively. The perpendicular contributions to the Debye–Waller factor, as determined from the specular decay rate, are subtracted from the off-specular decay rate, leaving the Debye–Waller factor as a function of ΔK^2. An overlay of these decay rates at different non-specular conditions is presented in Fig. 5. The linear relationship between the decay rate and ΔK^2 confirms the efficacy of this analysis, and was used to extract the values for the in-plane MSD. The parallel MSD was $2.2 \pm 0.1 \times 10^{-3}$ $Å^2K^{-1}$ for amorphous PET and $1.6 \pm 0.1 \times 10^{-3}$ $Å^2K^{-1}$ for the semi-crystalline film.

Helium atom scattering is able to resolve a relatively small ($\sim$15%) but significant softening of the surface for the vibrations polarized perpendicular to the surface while detecting a 25% stiffening for the vibrations polarized in the surface plane due to a transition from a homogeneous amorphous film to a composite semi-crystalline film. This surprising and seemingly contradictory result can be explained by invoking the heterogeneous nature of the semi-crystalline film.[22,23] A polymer system which fully crystallizes, polyethylene (PE), was studied using neutron scattering and the crystalline film MSD was suppressed to 70% of the magnitude of those in the amorphous film.[24] Since fully crystalline polymers have reduced MSDs, the apparent softening of perpendicular MSDs in PET must arise from the remaining amorphous regions. Neutron scattering studies of fullerenes embedded in amorphous PS films have observed an enhancement of bulk MSDs despite the strong attractive potential of the fullerene, and this effect is attributed to frustrated packing of polymer chains away from the C_{60} molecules.[25] In PET, the tight polymer chain packing in the crystalline lamellae should prevent efficient chain packing in the remaining amorphous domains, much like C_{60} in PS. This frustrated polymer chain packing results in

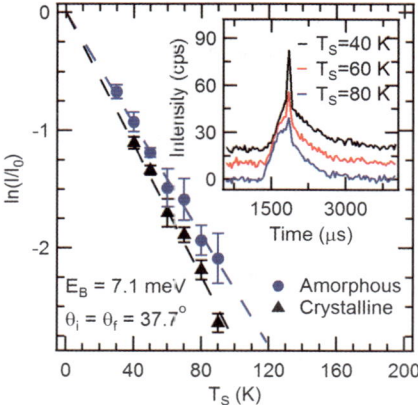

Fig. 4 Comparison of the Debye–Waller factor at the specular condition for amorphous and semicrystalline PET. The inset is an overlay of typical time-of-flight spectra at the three sample temperatures employed in this investigation.

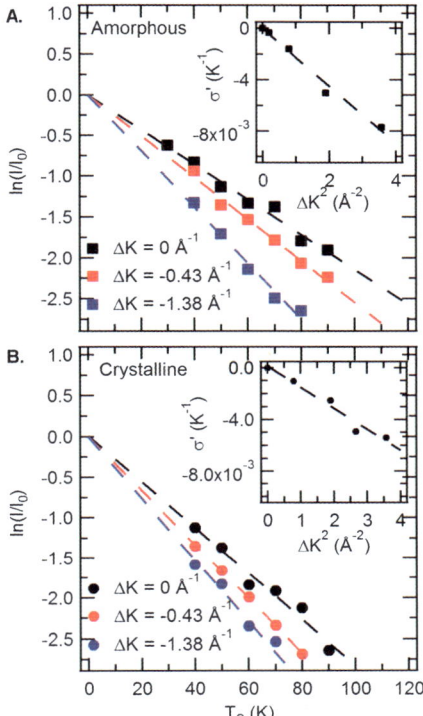

Fig. 5 Overlays of the Debye–Waller factor at specular and two non-specular conditions for amorphous (A) and semicrystalline (B) PET. The slope of the inset plots gives the in-plane mean-square displacements for the two surfaces.

a greater free volume and thus larger MSD in the interstitial amorphous regions, which explains the apparent softening of the PET interface upon crystallization.

As in the case of SAMs,[7] the parallel MSDs are an order of magnitude greater than those of the perpendicular polarization for both surfaces. The decrease of the in-plane MSD for the semi-crystalline film relative to the amorphous film originates from the crystalline lamellae. Since crystalline lamellae are more dense than the amorphous regions of the film, reduced free volume from the crystalline regions of the interface can explain this suppression of in-plane MSDs. These measurements support the presence of both crystalline lamellae and amorphous regions, distinct from those of the non-crystalline film, at the interface of this composite film.

3.3 Helium atom scattering as a probe of interfacial dynamics in a complex ordered film

The evolution of helium atom scattering as a probe of clean and adsorbate decorated surfaces continues as recently this technique has been used to study the structure, thermal motion, and specific librations of chemisorbed organic films.[7,8] We demonstrate the extension of this technique to a class of complex interfaces, best described as hybrid interfaces, in our studies of methyl-terminated silicon(111). Analysis of helium atom diffraction allows us to characterize the gas-surface potential, surface MSDs, and energy accommodation in gas-surface collisions. We also present the first observation of single phonon spectroscopy in methyl-terminated silicon, as well as the first observation of the lattice's vibrational band structure, through our inelastic HAS studies.

Methyl-terminated silicon is of great interest as it represents the next generation of passivated silicon interfaces. Since the early strategy of covalent termination by hydrogen lacked suitable resistance to oxidation, Lewis *et al.* developed a two step technique for covalently terminating the Si(111) interface with methyl units which results in high surface coverages.[12] Structural studies using low energy electron diffraction (LEED) and SPM techniques have confirmed the presence of large, pristine, highly ordered domains at this interface while transmission IR (TIR) and high resolution electron energy loss spectroscopy (HREELS) techniques were employed to characterize the molecular vibrations of the methyl terminated interface.[26-29] We present the use of Debye–Waller analysis and inelastic helium atom scattering to investigate the atomic scale librations, gas-surface collisional energy accommodation, and the vibrational dynamics of this hybrid interface.

Unlike the case for the disordered polymer films, CH_3–Si(111) gives intense, sharp diffraction peaks with minimal diffuse background. In this case, Debye–Waller formalism can be applied as a measurement of the dynamic disordering of an ordered lattice, which should arise due to the vibrations of both the methyl groups and the underlying silicon lattice. Rather than using time-of-flight spectroscopy to monitor the thermal attenuation of elastic scattering, we employed square wave chopping to make angularly resolved measurements yielding diffraction patterns at each given sample temperature. These scans show clearly resolved diffraction peaks which decay as the sample temperature increases, as seen in Fig. 6. This decay fits well with the Debye–Waller model of thermal attenuation not only in the linearity of the inset plot, but in that the diffraction peak decays at a much faster rate than the specular peak due to contributions from in-plane thermal motion. The parallel momentum exchange is defined by the elastic diffraction condition below, where h and k are integer indices for the reciprocal lattice unit vectors.

$$\Delta K = G = h\vec{b}_1 + k\vec{b}_2 \tag{4}$$

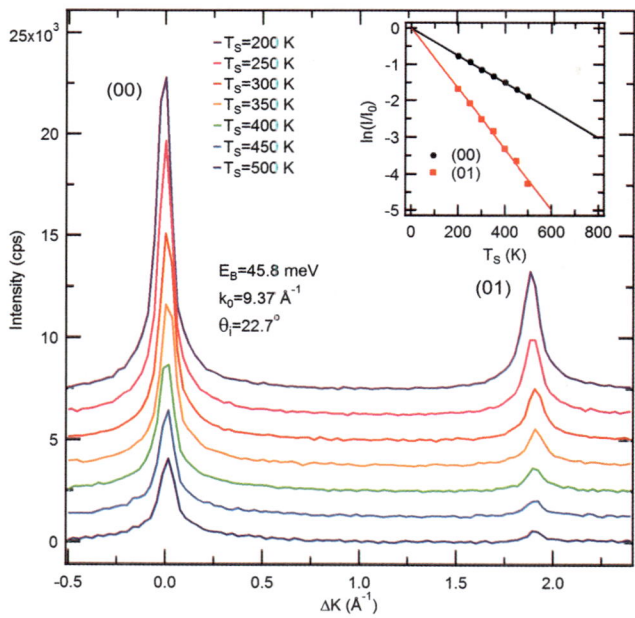

Fig. 6 Representative helium atom diffraction scans for a typical Debye–Waller attenuation experiment for CH_3–Si(111). The inset is the Debye–Waller factor *vs.* T_S for the zeroth and first order diffraction peaks.

For the CH_3–Si(111) surface the reciprocal lattice vector for the (01) diffraction peak was observed to be 1.90 Å^{-1}, which corresponds to the 3.82 Å spacing of the underlying silicon lattice.

These diffraction peaks were fitted using two Gaussian functions to account for the narrow coherent elastic component as well as the broad multiphonon and diffuse scattering contributions, and the elastic intensity was extracted as the numerical integral of the elastic component of the two Gaussians. Repeat measurements of these thermal attenuation measurements were performed at five different specular conditions in order to elucidate the value for the He–CH_3–Si(111) well depth.

$$\sigma_{hk} \equiv -\frac{d(2W(T_S))}{dT_S} = -\Delta k_z^2 \frac{d\langle u_z^2 \rangle}{dT_S} - \Delta K^2 \frac{d\langle u_\parallel^2 \rangle}{dT_S} \quad (5)$$

In the case of the specular, or zeroth order, diffraction peak, the ΔK^2 term vanishes and the perpendicular component can be simplified to the form in eqn (6).

$$\Delta k_z^2 = 4k_i^2 \left[\cos^2(\theta_i) + \frac{D}{E_i} \right] \quad (6)$$

As seen in Fig. 7D, a plot of σ_{00} versus $\cos^2(\theta_i)$ should be linear with an intercept that, when divided by the slope, gives the ratio of the Beeby well depth D to the incident beam energy. This method gave He-surface interaction potential well depths of 7.5 ± 2.6 meV and 6.0 ± 3.9 meV for the CH_3 and CD_3 terminated surfaces, respectively. It is worth noting that these values are in agreement with potentials applied to SAMs and methyl-terminated organic crystals,[7,16] and that although these values differ, they fall well within the precision of the method. We used the experimentally determined values for the well depths of each surface, but note that using the average of the two values changes our calculated MSD by less than 5% due to the large beam energy used relative to D.

Analysis of these rates of thermal attenuation of the elastic diffraction peaks allowed the quantification of the perpendicular and parallel MSDs for both surfaces. The perpendicular MSD was 1.0 ± 0.1 × 10^{-5} Å^{2}K^{-1} and 1.2 ± 0.2 × 10^{-5} Å^{2}K^{-1} for the respective CH_3 and CD_3 surfaces. The enhancement of the mean-square displacement in the CD_3 surface is in accordance with the expected impact of isotopic substitution. Eqn (7) details how the surface Debye temperature (θ_D) can be extracted from σ_{00}, given that the surface temperature is within the order of θ_D.

$$\sigma_{00} = \frac{24m_{He}[E\cos^2(\theta_i) + D]}{M_{eff}k_B\theta_D^2} \quad (7)$$

There are two unknown parameters in this equation, M_{eff} and θ_D, so the appropriate treatment of mass is necessary in order to quantify the surface Debye temperature. Scattering studies from silica glass[30] found the effective surface mass to be

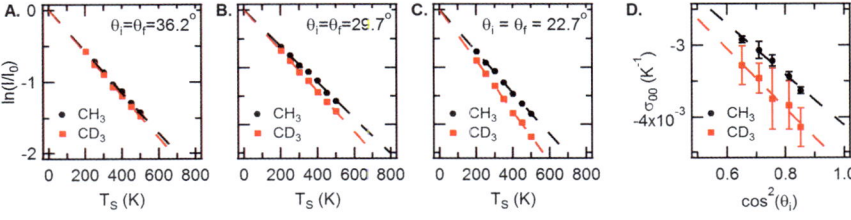

Fig. 7 Comparison of representative specular decays at three different incident angles (A–C) with fits (dashed lines) for the CH_3 and CD_3 terminated surfaces. Fig. 7D is the plot of the specular decay rate, with fits, as a function of $\cos^2(\theta_i)$ for CH_3–Si(111) and CD_3–Si(111). The intercept is used to extract the He-surface potential well depth.

18 amu, or roughly the mass of a hydroxyl unit terminating the surface, so we employed the mass of the terminal methyl group for our analysis. This assumption combined with the experimentally observed well depths resulted in a surface Debye temperature of 983 ± 31 K, or 683 cm^{-1}, for the CH$_3$ surface ($M_{eff} = 15$ amu) and 824 ± 40 K, or 573 cm^{-1}, for the CD$_3$ surface ($M_{eff} = 18$ amu). These values are both higher than the Debye temperature of bulk silicon (645 K) and that of the estimate for the Si(111) surface (476 K),[31] and much higher than those observed for organic thin films.[32,33] The HREELS measurements of the molecular vibrational modes of the CH$_3$–Si(111) surface[27,28] indicate that the Si–C stretching mode is 680 cm^{-1}, which corresponds to 980 K. While no spectroscopic data are available on the vibrational modes of the CD$_3$–Si(111) surface, comparison to spectroscopic studies of gas phase CD$_3$SiH$_3$ and CD$_3$SiD$_3$ can be helpful. The gas phase Si–C stretch is 700 cm^{-1} for CH$_3$SiH$_3$, 645 cm^{-1} for CD$_3$SiH$_3$ and 619cm^{-1} for CD$_3$SiD$_3$, so further suppression of the mode's energy due to the mass of the second layer of silicon could easily depress this value into the range of 570 cm^{-1}.[34]

Covalent methyl termination appears to have stiffened the surface in the normal polarization, and the surface Debye temperatures indicate that the Si–C stretching mode dominates the thermal motion and energy accommodation at this interface. The relatively low perpendicular MSD also implies a stiffer surface, since its magnitude is closer to that of a rigid semiconductor or metal than that of a thin organic film. Enhancement of the surface Debye temperature by the addition of an organic adlayer has been observed.[35] These results demonstrate the sensitivity of helium atom scattering to changes in the local chemical environment.

The parallel MSDs were obtained from the σ_{hk} for the first and second order diffraction peaks. These attenuation rates were averaged to obtain a parallel MSD of $7.1 \pm 5.1 \times 10^{-4}$ Å^{2}K^{-1} for CH$_3$–Si(111) and $7.2 \pm 5.3 \times 10^{-4}$ Å^{2}K^{-1} for CD$_3$–Si(111). While the perpendicular MSDs resembled those of a rigid semiconductor, these values are consistent with those observed in other thin organic films.[7,36] The addition of a single methyl functional group was sufficient to transition the interfacial dynamics from a regime mediated by substrate lattice vibrations to one in which the thermal motion and energy accommodation is dominated by local molecular modes.

The thermal motion is dominated by local molecular modes for the methyl-terminated silicon interface, but the lattice dynamics are still accessible through inelastic helium atom scattering. Since the first observation of the surface waves of an alkali halide using helium atom scattering,[3] the use of helium atom scattering to investigate the surface vibrational structure has been extended to increasingly complex interfaces.[37,38] Knowledge of the surface phonon band structure allows for the characterization of the local bonding character of the interface, so inelastic helium atom scattering enables us to investigate how the local silicon bonding is affected by the covalent attachment of a methyl terminal group. We report the first helium atom scattering measurements of lattice vibrational modes of not only CH$_3$–Si(111), but any semiconductor system with covalent termination by an organic functional group.

For this set of experiments a single shot chopping method was employed, with a duty cycle of 1% and beam energies ranging from 35 to 64 meV. Typical sample temperatures ranged from 140 K to 200 K to reduce the attenuation from diffuse scattering. A representative time-of-flight spectrum is shown in Fig. 8 with an inset of the Rayleigh wave dispersion curve along the $\bar{\Gamma} - \bar{M}$, or nearest neighbor, azimuth for the CD$_3$–Si(111) surface. This spectrum consists of a weak diffuse elastic peak (relative to the coherent diffraction peaks) and a clear inelastic peak with a longer flight time than the elastic peak, corresponding to the creation of a surface vibration during collision. Changing the kinematic conditions at a given beam energy allows us to map out the dispersion of these interfacial vibrational modes across the surface Brillouin zone. The vibrational structure resembles that observed for the H–Si(111) surface,[37] so methyl termination not only preserves the underlying

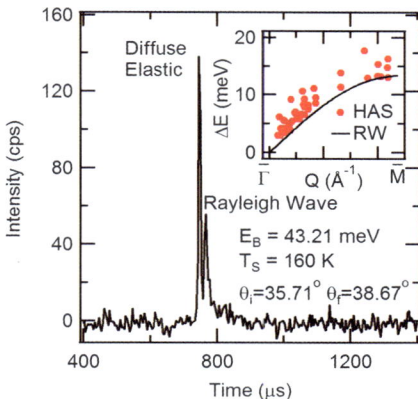

Fig. 8 A time-of-flight spectrum demonstrating single phonon inelastic helium scattering from CD_3–Si(111). The inelastic peak arises from a creation event with the surface Rayleigh wave. The inset is the Rayleigh wave dispersion along the $\bar{\Gamma} - \bar{M}$ azimuth (nearest neighbor) for the CD_3–Si(111) surface.

lattice structure but also its low energy vibrational band structure. The low frequency of the surface modes accessible by helium atom scattering means that these modes are unlikely to mix significantly with the molecular modes of the methyl groups. This does not exclude the possibility of an altered band structure for the higher frequency lattice modes, but in the case of the Rayleigh wave it is relatively unperturbed. An important clarification is that while we are probing lattice vibrational modes using helium atom scattering, we are not in fact scattering from the silicon lattice itself. The helium atom scattering is non-penetrative, so this motion is projected from the underlying lattice onto the surface methyl groups.

In summary, helium atom scattering experiments discussed herein demonstrate the power of this technique in characterizing and quantifying surface vibrations in disordered and complex systems. We have demonstrated its sensitivity to subtle changes in surface librations due to local environmental effects arising from confinement (*i.e.* thickness) and substrate interactions, phase change, and chemical functionalization. Furthermore, we have demonstrated the utility of using the temperature and angle dependent attenuation of elastically scattered atoms to gain fundamental insights on energy accommodation in gas-solid collisions for hybrid interfaces.

3.4 Mechanisms of defect migration and annihilation in aligned diblock copolymers

As a complement to our atomic scattering data, we now discuss our use of high temperature time-lapse atomic force microscopy to obtain real-space images of mesoscopic phase separation, chain diffusion, and structural reorganization in diblock copolymer films. Diblock copolymer systems are of great technological interest due to their self-organizing behavior, which has been used for a variety of applications,[39,40] and earlier atomic force microscopy investigations of the structural evolution in diblock copolymers involved *ex situ* heating and static imaging below T_g.[41–43] While these studies gave useful insight into the structural organization and defect migration of these films, they did not yet allow direct imaging of the domain fluctuations or reorganization in real time. Time-lapse AFM imaging above T_g gives *in situ* real-space observations of such behavior in diblock copolymer films, which allows a more complete characterization of the mechanisms directing structural evolution and fluctuations. Previously we used this method to demonstrate that direct visualization of PS-*b*-PMMA domain fluctuation and growth, Fig. 9, allows us to determine that domain boundaries fluctuate through polymer chain tube-like reptation and are most likely restored by a curvature minimization force.[44]

Previous studies have used time-lapse atomic force microscopy measurements to study coarsening processes in polymer films, but they were limited by complicated structure of phase-separated diblock copolymer films.[45,46] Our innovation is to use lithographically modified substrates to enforce a highly ordered cylindrical domain alignment and lower defect densities.[39] The low defect densities afford us the luxury of studying *isolated defect interactions* while the alignment enables us to study the migration of these defects both along and across domain boundaries. Our high temperature imaging studies offer insight into the energy barriers and mechanism for defect migration and annihilation in these complex polymer films.

In this paper, we focus on the interactions of isolated dislocations and disclinations, Fig. 10. These defects can migrate through a combination of climb (diffusion along a domain interface) and glide (diffusion across a domain interface) mechanisms, which have been predicted by theory but not directly observed due to the disordered nature of the previous films studied. Both mechanisms should be involved in the film reorganization, but the climb should have a much higher velocity than the glide mechanism due to the large energetic cost from repulsion of immiscible domains. The linear cylindrical alignment in this study allows for the direct observation of both the climb and glide mechanism velocities through our time-lapse AFM images.

Dislocations of opposite orientation will attract and annihilate leaving a defect-free region of the film. This annihilation proceeds with a combination of glide and climb mechanisms as seen in Fig. 11. Note that the velocity along the domain interface is much higher than across, as predicted for climb and glide mechanisms. Detailed observation of the glide and climb mechanism velocities allows for the characterization of their respective diffusion constants as well as quantification of the activation barrier to defect diffusion. Disclinations do not move with the same mechanisms as the dislocations due to the topological constraints, so they must

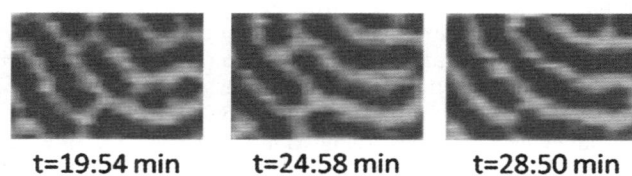

t=19:54 min   t=24:58 min   t=28:50 min

Fig. 9 Sequential AFM phase images at 448 K showing PMMA domain (dark) fluctuations within the PS matrix (light).

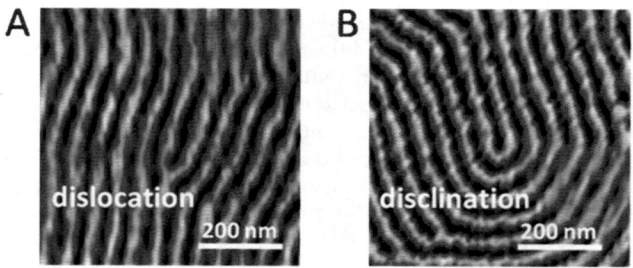

Fig. 10 Examples of topological defects which are observed in cylinder-forming PS-*b*-PMMA films. A shows a dislocation and B a disclination with a PMMA core structure. PMMA domains appear dark and PS domains appear bright in the AFM phase image.

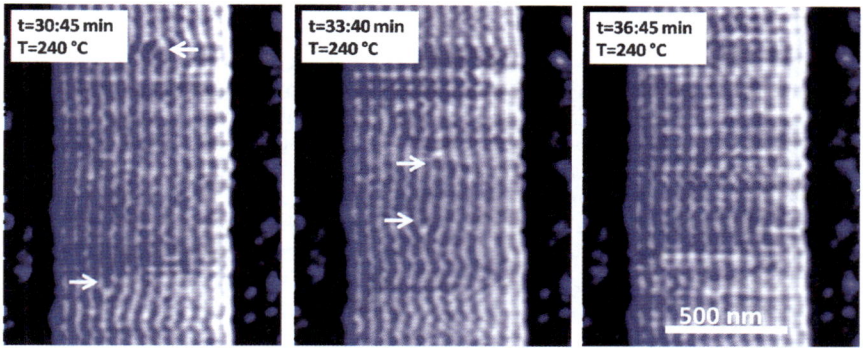

Fig. 11 Representative AFM phase images showing the annihilation of a dislocation pair. The dislocations are indicated with white arrows. The images were acquired at 513 K.

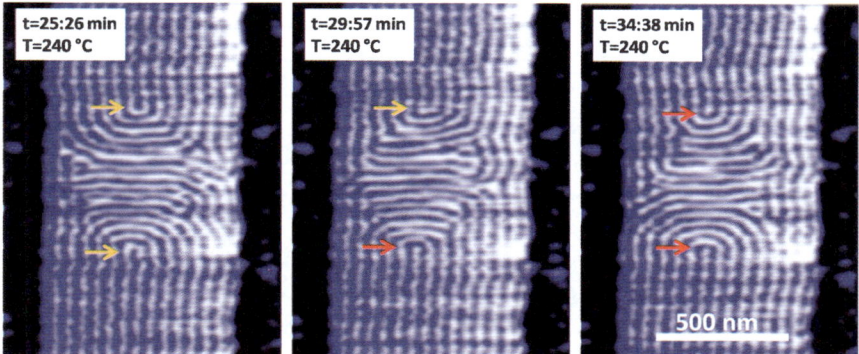

Fig. 12 AFM images showing the attraction of a disclination quadrupole. The yellow arrows correspond to the PMMA disclination cores, while the red arrows correspond to the PS disclination cores. These images were acquired at 513 K.

migrate *via* domain breaking and reorganization. Fig. 12 shows the approach of two oppositely oriented disclinations. The yellow arrows correspond to a PMMA disclination core while the red arrows correspond to a PS disclination core. The disclinations migrate by breaking the disclination lines into dislocations, which then are free to interact with the surrounding domains and annihilate, and then switching the disclination core structure between PS and PMMA. While we were not able to observe dislocation emission, which should quench rapidly, we are able to visualize the core-switching component of this mechanism quite clearly. This mechanism was predicted theoretically[47] but has not been previously observed. Our ability to make direct observations of film reorganization and defect migration has led to greater insights into the precise energetics and kinetics involved in diblock copolymer film annealing.

4. Conclusions

In this paper we have shown that helium atom scattering can be applied as a precision probe of surface vibrational dynamics and gas-surface interaction potentials to disordered and complex macromolecular films as well as to a hybrid interface. We demonstrated the ability of helium atom scattering to detect subtle changes in surface vibrational dynamics due to thin film confinement, polymer chain molecular weight, and phase through our studies of PMMA and PET films. Helium atom scattering also allowed us to characterize the surface thermal motion, gas-surface energy accommodation, and the interfacial lattice dynamics of a hybrid interface, methyl-terminated silicon. Time-lapse atomic force microscopy complemented our atom scattering studies by allowing *in situ* real space imaging of mesoscopic structural evolution in PS-*b*-PMMA, a diblock copolymer film. By employing a lithographically modified substrate, we were able to use cylindrical alignment of polymer domains to directly observe isolated defect pair migration both along and across domain boundaries, as well as the mechanism of annihilation. Helium atom scattering and time-lapse atomic force microscopy are powerful complementary tools, which together give profound insight into interfacial dynamics at atomic, nanoscopic, and microscopic length scales for complex condensed matter interfaces.

Acknowledgements

We would like to thank Nathan S. Lewis, Erik Johansson, and Leslie E. O'Leary at the California Institute of Technology for providing us with high quality methyl-terminated silicon samples. This work was supported at the University of Chicago by the Air Force Office of Scientific Research, DTRA (Grant No. HDTRA1-11-1-0001), and the NSF Materials Research Science and Engineering Center at the University of Chicago.

References

1 I. Estermann and O. Stern, *Z. Phys.*, 1930, **61**, 95.
2 M. J. Cardillo, G. E. Becker, S. J. Sibener and D. R. Miller, *Surf. Sci.*, 1981, **107**, 469.
3 G. Brusdeylins, R. B. Doak and J. P. Toennies, *Phys. Rev. Lett.*, 1980, **44**, 1417.
4 B. Poelsema and G. Comsa, *Faraday Discuss. Chem. Soc.*, 1985, **80**, 247.
5 K. D. Gibson and S. J. Sibener, *Phys. Rev. Lett.*, 1985, **55**, 1514.
6 F. Hofmann and J. P. Toennies, *Chem. Rev.*, 1996, **96**, 1307.
7 N. Camillone, C. E. D. Chidsey, G. Y. Liu, T. M. Putvinski and G. Scoles, *J. Chem. Phys.*, 1991, **94**, 8493.
8 S. B. Darling, A. M. Rosenbaum and S. J. Sibener, *Surf. Sci.*, 2001, **478**, L313.
9 D. Farias and K. H. Rieder, *Rep. Prog. Phys.*, 1998, **61**, 1575.
10 B. Gans, P. A. Knipp, D. D. Koleske and S. J. Sibener, *Surf. Sci.*, 1992, **264**, 81.
11 Y. Zhang, Y. L. Lu, Y. X. Duan, J. M. Zhang, S. K. Yan and D. Y. Shen, *J. Polym. Sci., Part B: Polym. Phys.*, 2004, **42**, 4440.
12 A. Bansal, X. L. Li, I. Lauermann, N. S. Lewis, S. I. Yi and W. H. Weinberg, *J. Am. Chem. Soc.*, 1996, **118**, 7225.
13 J. R. Manson, V. Celli and D. Himes, *Phys. Rev. B: Condens. Matter*, 1994, **49**, 2782.
14 J. R. Manson and J. G. Skofronick, *Phys. Rev. B: Condens. Matter*, 1993, **47**, 12890.
15 V. Bortolani, V. Celli, A. Franchini, J. Idiodi, G. Santoro, K. Kern, B. Poelsema and G. Comsa, *Surf. Sci.*, 1989, **208**, 1.
16 G. Bracco, J. Acker, M. D. Ward and G. Scoles, *Langmuir*, 2002, **18**, 5551.
17 J. L. Beeby, *J. Phys. C: Solid State Phys.*, 1971, **4**, L359.
18 C. L. Soles, J. F. Douglas, W. L. Wu and R. M. Dimeo, *Macromolecules*, 2003, **36**, 373.
19 M. B. Li, J. R. Manson and A. P. Graham, *Phys. Rev. B: Condens. Matter*, 2002, **65**, 195404.
20 M. A. Freedman, J. S. Becker, A. W. Rosenbaum and S. J. Sibener, *J. Chem. Phys.*, 2008, **129**, 044906.
21 M. A. Freedman, J. S. Becker and S. J. Sibener, *J. Phys. Chem. B*, 2008, **112**, 16090.
22 T. Kanaya, M. Imai and K. Kaji, *Phys. B*, 1996, **226**, 82.
23 D. A. Ivanov, T. Pop, D. Y. Yoon and A. M. Jonas, *Macromolecules*, 2002, **35**, 9813.
24 T. Kanaya, U. Buchenau, S. Koizumi, I. Tsukushi and K. Kaji, *Phys. Rev. B: Condens. Matter*, 2000, **61**, R6451.
25 A. Sanz, M. Ruppel, J. F. Douglas and J. T. Cabral, *J. Phys.: Condens. Matter*, 2008, **20**, 104209.
26 L. J. Webb, S. Rivillon, D. J. Michalak, Y. J. Chabal and N. S. Lewis, *J. Phys. Chem. B*, 2006, **110**, 7349.
27 T. Yamada, T. Inoue, K. Yamada, N. Takano, T. Osaka, H. Harada, K. Nishiyama and I. Taniguchi, *J. Am. Chem. Soc.*, 2003, **125**, 8039.

28 T. Yamada, M. Kawai, A. Wawro, S. Suto and A. Kasuya, *J. Chem. Phys.*, 2004, **121**, 10660.
29 H. B. Yu, L. J. Webb, R. S. Ries, S. D. Solares, W. A. Goddard, J. R. Heath and N. S. Lewis, *J. Phys. Chem. B*, 2005, **109**, 671.
30 W. Steurer, A. Apfolter, M. Koch, W. E. Ernst, E. Sondergard, J. R. Manson and B. Holst, *Phys. Rev. B: Condens. Matter Mater. Phys.*, 2008, **78**, 045427.
31 J. S. Ha and E. F. Greene, *J. Chem. Phys.*, 1989, **91**, 571.
32 J. J. Hernandez, J. A. Li, J. Baker, S. A. Safron and J. G. Skofronick, *J. Vac. Sci. Technol., A*, 1996, **14**, 1788.
33 C. N. Borca, S. Adenwalla, J. W. Choi, P. T. Sprunger, S. Ducharme, L. Robertson, S. P. Palto, J. L. Liu, M. Poulsen, V. M. Fridkin, H. You and P. A. Dowben, *Phys. Rev. Lett.*, 1999, **83**, 4562.
34 A. J. F. Clark and J. E. Drake, *Can. J. Spectrosc.*, 1977, **22**, 79.
35 D. Q. Feng, P. A. Dowben, R. Rajesh and J. Redepenning, *Appl. Phys. Lett.*, 2005, **87**, 181918.
36 T. Y. B. Leung, P. Schwartz, G. Scoles, F. Schreiber and A. Ulman, *Surf. Sci.*, 2000, **458**, 34.
37 R. B. Doak, Y. J. Chabal, G. S. Higashi and P. Dumas, *J. Electron Spectrosc. Relat. Phenom.*, 1990, **54**, 291.
38 P. Santini, P. Ruggerone, L. Miglio and R. B. Doak, *Phys. Rev. B: Condens. Matter*, 1992, **46**, 9865.
39 D. Sundrani, S. B. Darling and S. J. Sibener, *Nano Lett.*, 2004, **4**, 273.
40 R. A. Segalman, *Mater. Sci. Eng., R*, 2005, **48**, 191.
41 T. L. Morkved, W. A. Lopes, J. Hahm, S. J. Sibener and H. M. Jaeger, *Polymer*, 1998, **39**, 3871.
42 J. Hahm, W. A. Lopes, H. M. Jaeger and S. J. Sibener, *J. Chem. Phys.*, 1998, **109**, 10111.
43 J. Hahm and S. J. Sibener, *J. Chem. Phys.*, 2001, **114**, 4730.
44 N. A. Yufa, J. Li and S. J. Sibener, *Macromolecules*, 2009, **42**, 2667.
45 C. Harrison, D. H. Adamson, Z. D. Cheng, J. M. Sebastian, S. Sethuraman, D. A. Huse, R. A. Register and P. M. Chaikin, *Science*, 2000, **290**, 1558.
46 C. Harrison, Z. D. Cheng, S. Sethuraman, D. A. Huse, P. M. Chaikin, D. A. Vega, J. M. Sebastian, R. A. Register and D. H. Adamson, *Phys. Rev. E: Stat. Phys., Plasmas, Fluids, Relat. Interdiscip. Top.*, 2002, **66**, 011706.
47 N. M. Abukhdeir and A. D. Rey, *New J. Phys.*, 2008, **10**, 063025.

Nonadiabatic dynamics at metal surfaces: Independent electron surface hopping with phonon and electron thermostats

Neil Shenvi[a] and John C. Tully[*b]

Received 21st February 2012, Accepted 17th April 2012
DOI: 10.1039/c2fd20032e

Independent Electron Surface Hopping (IESH) is a computational method for accounting for nonadiabatic electronic transitions in simulations of molecular motion at metal surfaces. IESH is applicable in cases of strong coupling where the electronic friction model is suspect, and has been demonstrated to accurately reproduce the results of detailed molecular beam experiments on vibrationally inelastic scattering of nitric oxide from the (111) surface of gold. However, in its original form, IESH represents a closed system without energy flow outside the local region of explicitly included substrate atoms. In this paper we propose two "thermostats" for introducing energy flow to the bulk, one for phonons and the other for electronic excitations.

1. Introduction

It is now well established through both theory and experiment that the Born–Oppenheimer approximation[1] – the foundation of most theories of chemical reaction dynamics – is not generally valid for treating molecular motion at metal surfaces.[2–4] The forces experienced by a molecule interacting with a metal surface, and the resulting rates and pathways of energy exchange and chemical reaction, can be dramatically altered by nonadiabatic electronic excitations in the metal as well as by charge transfer to and from the molecule. Two methods have been employed for introducing nonadiabatic effects into classical mechanical molecular dynamics based simulations of molecule–metal interactions. The first, the *electronic friction* approach,[5–9] accounts for the dissipation of energy from adsorbate motion into electron hole pair (EHP) excitations *via* a spatially and directionally dependent friction. While the friction terms are sometimes chosen empirically, procedures have been developed to compute frictions from first principles, *e.g.*, using density functional theory (DFT).[9–13] Electronic friction theories have proved quite successful for simulating energy transfer at metal surfaces in a number of applications.[11–14] However, the friction model has limitations that arise from the weak-coupling assumption upon which it is based.[9] Specifically, the model assumes that nuclear motion evolves on a manifold of parallel, infinitesimally-spaced potential energy surfaces, with transitions among these surfaces produced by nonadiabatic EHP transitions. This is not adequate for cases where multiple non-parallel potential energy surfaces of chemically distinct character interact, such as at crossings or avoided crossings. Such behavior is exhibited quite generally in molecule–surface interactions.[15] Obvious examples are those involving electron transfer from the surface to an adsorbate species or *vice versa*, a frequent occurrence due to the image potential

[a]Department of Chemistry, Duke University, Durham, NC 27708, USA
[b]Department of Chemistry, Yale University, New Haven, CT 06520-8107, USA. E-mail: john. tully@yale.edu

lowering of the energies of ionic states near metal surfaces. The potential energy surfaces for ionic and neutral states can have very different topographies, with dramatic dynamical consequences. The electronic friction picture of losing or gaining kinetic energy in relatively small increments, such as single quantum vibrational transitions, governed by the ground state potential energy surface appears inadequate in these situations.

A well-studied example is the lifetime of a vibrationally excited ($v = 1$) CO molecule adsorbed on the (100) face of copper. Calculations based on the Born–Oppenheimer approximation predict lifetimes in the millisecond range, whereas inclusion of nonadiabatic EHP excitations reduces the lifetime by 9 orders of magnitude to a few picoseconds,[16] in agreement with experiment.[17] More recent demonstrations abound of the inadequacy of the Born–Oppenheimer approximation to describe dynamics at metal surfaces. Notably, experiments by the Wodtke,[18] McFarland,[19,20] Hasselbrink,[21] and Somorjai[22] groups demonstrate that individual electrons in a metal can receive as much as 1.5 eV or more of excitation upon molecular scattering, adsorption or chemical reaction; electrons do not play a passive role in chemistry at metal surfaces. This apparent large deposition of energy into a single electron is difficult to reconcile with the electronic friction weak coupling theory.

These results have prompted development of a theory that can apply to strong coupling situations, the *Independent Electron Surface Hopping* (IESH) theory.[23] Rather than moving on a single potential energy surface with dissipation, there are sudden hops from one potential energy surface to another, as dictated by the quantum mechanical electronic state amplitudes along the path traced out by the nuclei. IESH has been applied to only one chemical system so far, the scattering of NO molecules from the (111) face of gold.[24] For this system it has been remarkably successful in reproducing a wealth of detailed observations on de-excitation of initially highly vibrationally excited molecules[25] and vibrational excitation of molecules initially in the ground vibrational state,[26] over a wide range of incident translational energies and surface temperatures. All interactions employed in this work were derived from *ab initio* DFT calculations.[24] This success notwithstanding, there are limitations to IESH as well. First, IESH is more computationally demanding than electronic friction. Low energy EHP transitions that are described naturally by electronic friction must be represented by transitions among close-lying electronic states in IESH. Thus a relatively large number of discrete states must be invoked to adequately represent the metallic continuum. Methods have been developed for choosing the energies of these discrete states efficiently, allowing for far fewer states than if equally-spaced levels were chosen.[27] As a result, the NO/Au(111) simulations required only relatively modest computational time. Nevertheless, IESH can require significantly more CPU time than electronic friction.

A second limitation is that the IESH method addresses an isolated system of atoms and electronic energy levels. While flow of energy among atomic motions, phonons and conduction electrons occurs properly within this finite system, there is no connection to the rest of the bulk solid. Thus the total energy of the system, nuclear motion plus electronic, is conserved in IESH, in contrast to electronic friction theory which includes dissipation directly. This is often not a problem. For example, in the NO/Au(111) simulations 396 moving Au atoms were explicitly included in the dynamics,[25] and this number can be easily increased if a larger heat capacity is needed. It is more computationally challenging to increase the number of discrete electronic levels, however, and even for phonons, processes that involve large energy release may require a procedure for both phonon and electron energy dissipation.

In this paper we propose methods for introducing both phonon and electronic dissipation into the IESH theory. We then present the results of preliminary calculations of adsorbate vibrational relaxation and ultrafast laser-induced desorption, both employing the interaction potentials developed previously for NO interacting with Au(111).

2. The IESH method

Before presenting our procedures for developing phonon and electron thermostats for IESH, we must briefly outline the method. We do this with reference to the NO–Au(111) system, with the simplifying assumption that only two molecular electronic configurations, the neutral and negative ion states, need be considered. The method is based on a discretized version of the widely-used Newns–Anderson model,[28] extended to allow for interactions that vary with atomic positions, $\mathbf{R}$. A critical feature of the Newns–Anderson model is that electrons are non-interacting, so that the many-electron Hamiltonian can be expressed as a sum of one-electron terms:

$$\mathscr{H}_{el}(\mathbf{R}) = U_0(\mathbf{R}) + \sum_{j=1}^{N_e} \mathscr{E}_j(\mathbf{R}) \, c_j^\dagger \, c_j. \tag{1}$$

In eqn (1), N_e is the number of active electrons of the system, $E_j(\mathbf{R})$ is the jth energy eigenvalue, and $c_j^\dagger$ and c_j are the fermionic creation and annihilation operators for eigenstate j of the following one-electron Hamiltonian:

$$\mathscr{H}_{el}^1(\mathbf{R}) = [U_1(\mathbf{R}) - U_0(\mathbf{R})] \Big| a \!>\! <\! a \Big| + \sum_{k=1}^{M} \varepsilon_k \Big| k \!>\! <\! k \Big| + \sum_{k=1}^{M} [V_{ak} \, |a\!>\!<\!k| \, V_{ka} \, |k\!>\!<\!a|] \tag{2}$$

$U_0(\mathbf{R})$ is the multidimensional "diabatic" potential energy surface (PES) describing the interaction of the neutral NO molecule with the gold surface, as a function of the coordinates $\mathbf{R}$ of the N, O, and all of the Au atoms. $U_1(\mathbf{R})$ is the corresponding diabatic PES for the negatively charged NO⁻ molecule. $|a>$ is the one-electron wave function (affinity level) of the negative ion, *i.e.*, in the case of NO, the molecular π^* orbital in which the electron resides to form the negative ion. Thus $U_0(\mathbf{R})$ and $U_1(\mathbf{R})$ are the many-electron PESs when the $|a>$ orbital is unoccupied and occupied, respectively. The wave functions $|k>$ are the M (occupied and unoccupied) one-electron states of the metallic conduction band, corresponding to one-electron energies ε_k. The conduction band states $|k>$ should properly form a continuum, with the sums in eqn (2) replaced by integrals, but in practice we must discretize the continuum. Finally, $V_{ak}(\mathbf{R})$ and $V_{ka}(\mathbf{R})$ are the matrix elements of the off-diagonal interaction,

$$V_{ak}(\mathbf{R}) = \omega_k < a|V(\mathbf{R})|k>, \ V_{ka}(\mathbf{R}) = \omega_k < k|V(\mathbf{R})|a>, \tag{3}$$

where ω_k is a weighting factor to account both for the density of electronic states in the conduction band and for the spacing between discrete energy levels at energy ε_k. Diagonalizing the one-electron Hamiltonian, eqn (2), for fixed positions $\mathbf{R}$ of all the atoms, produces $M + 1$ eigenstates, each of which is a linear combination of the initial basis functions.

The many-electron adiabatic states are obtained by designating N_e one-electron eigenstates to be occupied, leaving the remaining $M + 1 - N_e$ eigenstates empty. Thus, the ground state corresponds to filling the N_e lowest energy one-electron eigenstates up to the Fermi level. Excited states are produced by promoting one or more electrons to unoccupied states above the Fermi level. Multiple low-energy excitations allow for the creation of multiple EHPs, a necessity for properly describing electronic friction effects. An enormous number of excited many-electron states can arise from a very modest number of electrons and one-electron orbitals. For example, for the vibrational relaxation simulations presented below, 30 electrons and 61 orbitals were included. This gives rise to more than 10^{17} occupation permutations. Whereas carrying out surface hopping calculations among 10^{17} PESs may seem daunting, as discussed below, it is actually quite practical.

IESH employs the standard "fewest switches" surface hopping,[29] modified to treat multiple independent electrons. The IESH algorithm is presented in detail in ref. 23. Briefly, the initial conditions of each trajectory require selecting initial electronic occupations from a Fermi function at the appropriate temperature, selecting surface atom positions and momenta from a classical canonical ensemble at the desired temperature, and selecting the positions and momenta of the atoms of the molecule to represent the experimental conditions to be modeled. The nuclei are propagated classically on the initial potential surface while, simultaneously, the electronic state is propagated by the time-dependent Schrödinger equation.

IESH is feasible because the nonadiabatic coupling vector is a one-electron operator that can be computed separately for each electron.[23] Surface hops then correspond to single electron transitions from one adiabatic one-electron eigenstate to another. For each electron at each time step, a random number is compared to the standard fewest switches hopping criterion to determine whether a hop occurs between one-electron eigenstates.[29] When a hop occurs, the standard surface hopping adjustment of the component of momentum in the direction of the nonadiabatic coupling vector is applied to conserve total energy.[29] No hop and no velocity adjustment are made if the hop is prevented by energy conservation. In the applications presented here, large numbers of surface hops occurred, often more than 50 for a single scattering event, in contrast to the relatively few hops typically encountered in standard applications of surface hopping. It should be emphasized that both electron transfer (large energy hops) and EHP transitions (small energy hops) are incorporated in this way on an equal footing.

3. Phonon thermostat

Providing a thermostat for the phonon modes is straightforward, using methods employed previously for standard adiabatic molecule–surface molecular dynamics simulations. Probably the simplest method is to include sufficient numbers of classical mechanical substrate atoms to provide a large enough heat bath by themselves, without additional connection to an auxiliary heat bath. In the calculations reported here, 396 moving gold atoms were employed, providing a heat capacity of approximately 1188 k_B, excluding the smaller electronic contribution. Thus input of 1 eV of energy would raise the phonon temperature by about 10C, a perhaps significant but not enormous amount. (This was not an issue for the direct scattering events simulated previously because the molecule left the surface prior to any significant redistribution and/or return of energy back to the molecule.) Since the computational time of the classical mechanical propagation scales linearly with the number of atoms if short-range interatomic forces are employed, and since the quantum mechanical propagation of the electronic states will typically require more computational time than the classical propagation, adding more substrate atoms may be cost effective for most situations. Nevertheless, when simulating high-energy impact phenomena, highly exoergic chemical reactions or laser-induced processes, it may be more efficient and more accurate to employ a thermostat for the substrate atom motions. One method that is well-suited for this purpose is to add Langevin frictions and random forces to the z-direction (surface normal) motion of the lowest layer (farthest from the surface) substrate atoms.[30] The equation of z-direction motion for a last-layer atom is then

$$M \frac{d^2 z}{dt^2} = -\frac{dV}{dz} - \gamma \frac{dz}{dt} + Z(t), \qquad (4)$$

where γ is the friction constant and the autocorrelation function of the white noise fluctuating force $Z(t)$ is determined by the 2nd fluctuation-dissipation theorem:[31]

$$\langle Z(t)Z(0)\rangle = k_{\mathrm{B}}T\gamma. \tag{5}$$

T is temperature and k_{B} is Boltzmann's constant. The following simple prescription for choosing the friction constant in terms of the Debye frequency has proved satisfactory:[32]

$$\gamma = \pi\omega_{\mathrm{D}}/6 \tag{6}$$

An integration algorithm that accurately treats a white noise random force must be used for this procedure.[30]

The Langevin equation, above, results from a Markov (no memory) approximation. Provided that the number of explicitly included substrate atoms is sufficiently large, this will generally be an accurate approximation. The one situation where this must be used with caution is in simulating the dynamics of molecules with much higher vibrational frequencies than the substrate Debye frequency, ω_{D}. In such cases energy transfer between the molecular vibration and phonons should occur *via* multiphonon processes induced by anharmonic interactions. Application of a Langevin friction and white noise random force produces unphysical high frequencies in the substrate atom motion. Single phonon relaxation can occur to these unphysical high-frequency phonons. While the density of states of unphysical frequencies is quite small and decreases with increasing frequency as ω^{-n}, $n \geq 2$, nevertheless this can dominate vibrational relaxation if other mechanisms are slow.[33] Fortunately, in contrast to insulating surfaces, at metal surfaces vibrational energy transfer *via* electronic transitions is likely to be sufficiently fast to dominate any unphysical phonon contribution. In any case, it is straightforward to include memory effects if needed.[34]

4. Electronic thermostat

A phonon thermostat is not sufficient. Although the electronic temperature will eventually equilibrate with that of the phonons, the several picosecond time scale for this energy equilibration is much slower than the actual sub-picosecond electronic equilibration times. Electron–phonon coupling is only one of the mechanisms for electronic energy dissipation, and usually not the most important. Electron–electron collisions can rapidly (<100 fs) produce a thermal Fermi distribution of electronic state occupations, but with a transient temperature that might be quite different from that of the phonons. This is the origin of the two-temperature model often invoked to describe electron-induced dynamics at surfaces.[35,36] The electron–electron collision mechanism does not require nonadiabatic coupling, or nuclear motion at all. In addition, energy deposited in a local region *via* molecule impact or chemical reaction, or in the near-surface region *via* absorption of light, can rapidly diffuse into the bulk, again on the sub-picosecond timescale. To properly describe these processes, a mechanism for direct dissipation of electronic energy must be introduced into the IESH method.

The need for an electronic thermostat is illustrated in Fig. 1a, a plot of the IESH-computed loss of vibrational energy as a function of time for an NO molecule adsorbed on the Au(111) surface and prepared in its $v = 1$ vibrational state. The molecule is initially positioned in its most favorable 3-fold binding site and the surface temperature is 0 K. Vibrational relaxation was computed with the IESH method, without dissipation, using the NO/Au(111) interactions described below. The computed relaxation time is much too slow. By examination of trajectories, it is clear that, through nonadiabatic coupling, NO vibrational energy begins to be transferred to electronic excitations quickly, <1 ps. However, the electronic energy has nowhere to go except back to NO vibration or, more slowly, to phonons. The 10 ps timescale for relaxation shown in Fig. 1a is, incorrectly, determined by the latter. It is thus critical to introduce a realistic procedure for removing and/or equilibrating electronic energy on the proper timescale.

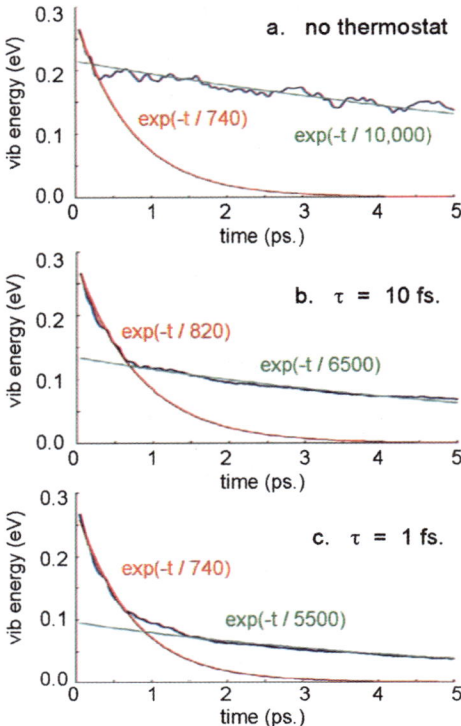

Fig. 1 Vibrational energy as a function of time of the NO molecule adsorbed on the (111) face of gold at its lowest energy binding site in the 3-fold hollow. The surface temperature is 0 K and the NO molecule was excited at time zero with energy corresponding to its $v = 1$ vibrational state. **a.** No electronic thermostat employed. **b.** Electronic thermostat with $\tau = 10$ fs. **c.** Electronic thermostat with $\tau = 1$ fs.

There is an even more pressing need when simulating ultrafast laser-induced chemistry at metal surfaces. In typical experiments the laser radiation is absorbed, not by the adsorbate molecule directly, but by the electrons of the metal, thereby exciting EHPs. The excited electrons then transfer energy to the adsorbate *via* nonadiabatic coupling. This energy transfer competes with dissipation of electronic energy into the bulk which occurs on a timescale comparable to or faster than transfer to the adsorbate.

The most obvious way to partially address this issue is to employ a larger number of discrete states with which to represent the metallic continuum of states. This would not produce dissipation into the bulk; the system is still closed and, in absence of a phonon thermostat, will conserve total energy. However, it would increase the heat capacity of the electronic subsystem, thereby slowing the transfer of energy back to nuclear motion. A serious deficiency of this approach is that it could greatly increase the computational time of the IESH simulations. Surface hopping is properly carried out in the adiabatic representation, requiring diagonalization of the many-state Hamiltonian at every time step, a procedure for which the computational time scales with the number, N, of electronic states as N^3. As this is generally the most time-consuming step in the simulation, employing very large numbers of discrete states can become impractical.

We propose here an alternative procedure for thermalizing the populations of electronic levels and/or removing energy from (or adding energy to) the electronic subsystem. The method employs standard Metropolis Monte Carlo.[37] As noted above, the IESH method invokes an independent-electron description of the many-electron wave function; at any instant in time, each electron is assigned to a single adiabatic orbital j with orbital energy ε_j. The procedure we propose here requires specification of a single adjustable parameter, the mean time τ of Monte Carlo attempts. At every time step a decision is made whether to make a Monte Carlo attempt based on comparing the ratio $\Delta t/\tau$ with a uniform random number, $0 < r_1 < 1$, where Δt is the integration time step. An attempt consists of selecting, at random, one occupied orbital j and one unoccupied orbital k. If $\varepsilon_k < \varepsilon_j$, then the move is accepted, *i.e.*, orbital j becomes unoccupied and orbital k becomes occupied. If $\varepsilon_k > \varepsilon_j$, then the usual Metropolis criterion is used: the occupations are switched if

$$r_2 < \exp(-(\varepsilon_k - \varepsilon_j)/k_B T), \qquad (7)$$

where r_2 is a second uniform random number, $0 < r_2 < 1$. This electronic thermostat algorithm has several useful features, the most important of which are that it is simple to apply, involves negligible computational time, and rigorously approaches a Fermi distribution at the chosen temperature. In addition, it can be generalized to non-thermal distributions, if required, such as an initial nonthermal excitation of electronic energy levels *via* absorption of laser light. The attempt time τ, although an adjustable parameter, can be estimated from measurable properties of the substrate. Moreover, as shown in the examples discussed below, results appear to be quite insensitive to the value of the attempt time, provided that it is chosen in a reasonable range.

Two deficiencies of this approach must be mentioned. First, while enforcing approach to equilibrium as desired, the electronic thermostat does so at the expense of altering the dynamics. This is the objective, of course, and it is true as well of other thermostats used widely in molecular dynamics simulations, notably the Nose thermostat.[38,39] Nevertheless, if we wish to obtain dynamical information, we need to apply the method with caution. A second issue arises from the fact that dissipation into the bulk of the crystal should involve changes in occupation of only the discrete diabatic states that represent the continuum. Localized states on the molecule should not be affected directly. However, as described above, IESH is based on the adiabatic representation. As a result, at every instant in time, each adiabatic orbital has an occupation of either 0 or 1. These adiabatic orbitals contain some contributions from localized molecular states. Thus switching the occupations of two states does have some unwanted direct influence on the localized states of the molecule. In the examples presented below, this effect is unnoticeably small. Furthermore, it can be reduced by increasing the number of discrete levels, albeit at the expense of computational effort.

5. Interactions for NO–Au(111)

Application of the IESH formalism described above requires specification of the neutral and negative ion diabatic PESs and off-diagonal couplings. For the initial application to the NO/Au(111) system, a pragmatic approach that requires only ground state DFT calculations was developed.[24] The Vienna *ab initio* simulation package (VASP) plane wave (periodic boundary) DFT code[40] with the PW91 density functional[41] was employed. The computed binding energies and geometries were in close agreement with previous reports.[42,43] In order to extract the needed diabatic interactions from ground state calculations only, it was assumed that the NO–Au interaction could be adequately represented by two manifolds of diabatic PESs, neutral and ionic. This is an oversimplification, of course, but it encompasses the electron transfer dynamics believed responsible for the large nonadiabatic behavior observed in this system. At every choice of atomic positions sampled, DFT calculations were performed to obtain two quantities, the ground state energy and the effective charge on the NO molecule. In addition, the DFT calculations were repeated with a small external electric field imposed normal to the plane of the gold surface (both positive and negative directions). This allowed calculation of the change of the effective charge on the molecule with respect to the electric field strength. The ground state energy, charge and variation of charge were sufficient to uniquely specify the three unknown diabatic Hamiltonian matrix elements.[24] For NO/Au(111), the resulting 3 elements were quite accurately fit to relatively simple analytic forms, similar to those used in previous simulations of NO–metal dynamics.[44] Diagonalization of the 2×2 matrix accurately reproduced the DFT ground state energy and effective charge at all geometries for which the DFT results were considered to be reliable. Construction of the $(M + 1) \times (M + 1)$ Newns–Anderson one-electron Hamiltonian matrix from the 2×2 diabatic matrix was then a straightforward task under the assumptions of uniform density of states and uniform coupling strength. These NO/Au(111) potential surfaces and couplings published previously were used unchanged for the calculations reported here.[23,24]

6. Application 1. vibrational relaxation

Our first test of the electronic thermostat procedure proposed above was to simulate vibrational relaxation of the NO molecule adsorbed on the Au(111) surface. We employed the IESH method using the interaction potentials and nonadiabatic couplings described in the previous sections. We carried out the simulation for a 0 K surface with all atoms initially placed in their lowest energy positions: the N and O atoms were located 1.247 and 2.454 Å, respectively, above the 3-fold site of the Au(111) surface. The gold atom positions were fully relaxed in the presence of the NO molecules. The molecule was then given a vibrational energy of 0.279 eV, the energy of the $v = 1$ vibrational state of the molecule with inclusion of zero-point energy. (The vibrational frequency of the adsorbed molecule according to our interaction potential is 1503 cm^{-1}.) We approximated the continuum with 60 discrete states, with unequal spacings as determined by the optimization scheme reported previously.[27] With the addition of one NO affinity level, the total number of 1-electron orbitals was 61. The 30 lowest of these orbitals were initially occupied, corresponding to $T = 0$ K. No phonon thermostat was employed; even if all of the vibrational energy were transferred to phonons, with 396 moving Au atoms this would result in a temperature rise of <2 K. An ensemble of 200 trajectories was computed, differing only in the random numbers that control surface hops. Fig. 1a shows the vibrational energy as a function of time for the case where no electronic thermostat was employed. As mentioned earlier, the vibrational lifetime is artificially long. Energy transfers initially on the proper time scale from the vibrational motion of the molecule into electronic excitations. However, the excited electron can be quickly de-excited, returning the energy to molecular vibration. We note that, in addition to increasing the efficiency of the calculations, the optimal choice of electronic energy spacings employed has the desirable effect of minimizing artificial recurrences that might occur if equally spaced levels were used. Nevertheless, the fact that the total energy of the system, nuclear plus electronic, is conserved precludes accurate determination of the vibrational relaxation rate. The slow vibrational decay shown in Fig. 1a results from the more gradual transfer of energy from electronic excitation to phonons and to lower frequency motions of the molecule.

Fig. 1b and 1c show the results of 200 trajectories with all initial conditions and interactions chosen to be the same as for Fig. 1a, but with the electronic thermostat invoked. The attempt times τ were chosen to be 10 fs and 1 fs for Fig. 1b and 1c, respectively. Note that a bi-exponential behavior is exhibited. The initial faster decays of 820 fs (Fig. 1b) and 740 fs (Fig.1c) correspond to the desired vibration-to-EHP relaxation rate. The slowdown at longer times is largely a result of finite spacings between electronic levels. Once the vibrational energy of the molecule becomes comparable to the spacings of levels near the Fermi level, EHP-promoted relaxation becomes hindered. This deficiency can be reduced or largely eliminated by employing more closely spaced discrete levels in the vicinity of the Fermi level. The

This journal is © The Royal Society of Chemistry 2012

results of Fig. 1b and 1c differ only in the choice of the attempt time τ, which differs by a factor of 10. The resulting relaxation times differ by only 11%, demonstrating the insensitivity of vibrational relaxation to choice of attempt frequency.

7. Application 2. ultrafast laser desorption

As a second application to test the proposed electronic thermostat we simulated ultrafast non-resonant laser desorption of NO from the Au(111) surface. As above, we used the NO/Au interactions developed previously.[23,24] The initial surface temperature was taken to be 300 K, and the adsorbed NO molecule equilibrated near its binding site at this temperature. The Monte Carlo attempt time τ was chosen to be 10 fs. As in the prior application, results were found to be relatively insensitive to the choice of this parameter. The non-resonant laser pulse is assumed to rapidly excite electrons in the conduction band of the metal. For this initial application, fast electron–electron collisions are assumed to rapidly equilibrate the electronic populations so that they can be represented by an instantaneous temperature ("hot electron model"). The laser heating is modeled by a time-dependent electronic temperature profile which governs the Metropolis acceptance probability, eqn (7). A number of workers have estimated the electronic temperature profile *via* numerical solution of the macroscopic diffusion equation using the thermal properties of the metal and the known form of the laser pulse.[45,46] We use the following form for the time-dependent electronic temperature:

$$T_{electronic} = At^2/(1 + Bt^3) \qquad (8)$$

with $A = 0.493$ K fs^{-2} and $B = 5.926 \times 10^{-7}$ fs^{-3}. Eqn (8) with these parameters produces a maximum temperature of 4000 K at 150 fs, as shown in Fig. 2, and is typical of the time-varying temperature profiles used previously.[46,47] The rapid laser heating and subsequent slower but still rapid cooling of the substrate determine the timescale and efficiency of laser-induced desorption. The histogram shown in Fig. 2 gives the relative number of desorbing trajectories that achieved a molecule–surface separation of 4 Å from the surface during the time window indicated. The molecule–surface interaction potential has become sufficiently weak at 4 Å that molecules move essentially freely from that point on. Note that a phonon thermostat was

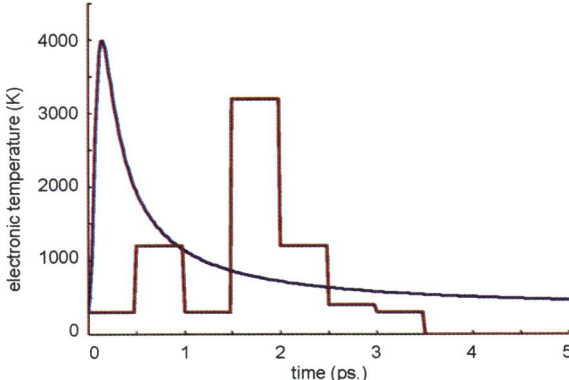

Fig. 2 Time dependence of the electronic temperature imposed in the simulations of ultrafast laser induced desorption of NO from Au. The electronic thermostat attempt time $\tau = 10$ fs. The histogram denotes the fraction of desorbing trajectories (un-normalized) that have reached a distance of 4 Å from the surface plane by that time, a distance at which the molecule–surface forces are almost vanishing. Note that most trajectories require more than 1.5 ps to become sufficiently energized by hot electrons to desorb.

deemed not necessary for this simulation; desorption, if it occurs, is essentially complete prior to significant energy transfer from excited electrons into phonons.

Of the 1000 trajectories run, 69 resulted in desorption. This gives a rough estimate of the desorption yield of 6.9% ± 0.8%, in approximate accord with measurements and electronic friction simulations on related systems.[46,47] The calculated mean energy in vibration of laser desorbed molecules was 0.488 eV, corresponding to a vibrational temperature of about 5800 K; *i.e.*, somewhat higher than the 4000 K maximum electronic temperature assumed to be produced by the laser pulse. This reflects the fact that the small subset of those molecules that desorb experience a larger than average amount of energy transfer from the hot electrons. Since nonadiabatic coupling is dominated by molecular vibrational motion, it is not surprising that a large fraction of energy ends up in vibration. Experimental studies of ultrafast laser desorption of NO from Pd under conditions similar to those assumed in our simulations produced a measured NO vibrational temperature of about 3000 K.[46] By contrast, the calculated rotational energy of the laser-desorbed molecules was very low, about 140 K, again in qualitative accord with experiment.[46]

The calculations reported here are very preliminary. Better statistics and a comprehensive exploration of parameter space will be required to make meaningful comparisons with experiment and to extract trends with confidence. Nevertheless, the present results demonstrate the feasibility of modeling laser-induced chemistry at metal surfaces with the IESH method, and highlight the necessity of employing the electronic thermostat.

8. Summary

The IESH theory is proving to be a viable approach for simulating the dynamics of molecular motion at metal surfaces in cases where the weak coupling based electronic friction model is not adequate. The approach can be applied in a predictive manner with *ab initio* interaction potentials and nonadiabatic couplings. The phonon and electron thermostats proposed here broaden the range of applicability of IESH. In particular, the proposed Monte Carlo electronic thermostat is a natural extension of the stochastic Monte Carlo fewest switches surface hopping algorithm[29] that is the basis of the IESH method. Although the electronic thermostat introduces an empirical parameter, the timescale τ controlling energy transfer to the bulk, this parameter can be estimated from experiments or other calculations. Furthermore, based on the preliminary findings reported here, results do not appear to be very sensitive to the choice of the parameter τ.

The Metropolis Monte Carlo method introduced here to account for contact of the electronic subsystem with the bulk can easily be generalized to describe nonthermal electronic energy distributions. As suggested by Gadzuk,[48] ultrafast laser chemistry at surfaces may well be influenced by the decidedly nonthermal energy distribution of the electrons that is initially prepared by the laser, prior to thermalization produced by electron–electron collisions. This behavior could be modeled in the framework of the present electronic thermostat by employing a non-Metropolis Monte Carlo algorithm.

The results presented here are very preliminary, with the objective of demonstrating the practicality of the method. General application of IESH will require calculation of ground and excited state molecule–metal interaction potentials and casting these results in the form of a independent electron Hamiltonian. While the NO/Au(111) system is a successful example of this strategy, more rigorous and accurate methods are needed.

Acknowledgements

The authors acknowledge helpful discussions with Dr Alexander Kandratsenka. J. T. and initial work by N. S. was supported by award DE-FG02-05ER15677 of the U. S. Department of Energy, Office of Basic Energy Sciences. N. S. was also supported as part of the UNC EFRC: Center for Solar Fuels, an Energy Frontier Research Center funded by the U.S. Department of Energy, Office of Science, Office of Basic Energy Sciences under Award Number DE-SC0001011.94

References

1 M. Born and E. Oppenheimer, *Ann. Phys.*, 1927, **84**, 457.
2 A. M. Wodtke, J. C. Tully and D. J. Auerbach, *Int. Rev. Phys. Chem.*, 2004, **23**, 513.
3 E. Hasselbrink, *Curr. Opin. Solid State Mater. Sci.*, 2006, **10**, 192.
4 G. J. Kroes, *Science*, 2008, **321**, 794.
5 E. G. d'Agliano, P. Kumar, W. Schaich and H. Suhl, *Phys. Rev. B*, 1975, **11**, 2122.
6 A. Nourtier, *J. Physique*, 1977, **38**, 479.
7 K. Schonhammer and O. Gunnarsson, *Phys. Rev. B*, 1980, **22**, 1629.
8 H. Kasai and A. Okiji, *Surf. Sci.*, 1991, **242**, 394.
9 M. Head-Gordon and J. C. Tully, *J. Chem. Phys.*, 1995, **103**, 10137.
10 M. Persson and B. Hellsing, *Phys. Rev. Lett.*, 1982, **49**, 662.
11 J. R. Trail, D. M. Bird, M. Persson and S. Holloway, *J. Chem. Phys.*, 2003, **119**, 4539.
12 V. Krishna and J. C. Tully, *J. Chem. Phys.*, 2006, **125**, 054706.
13 S. Monturet and P. Saalfrank, *Phys. Rev. B*, 2010, **82**, 075404.
14 J. C. Tully, M. Gomez and M. Head-Gordon, *J. Vac. Sci. Technol., A*, 1993, **11**, 1914.
15 S. Roy, N. Shenvi and J. C. Tully, *J. Phys. Chem. C*, 2009, **113**, 16311.
16 M. Head-Gordon and J. C. Tully, *J. Chem. Phys.*, 1992, **96**, 3938.
17 M. Morin, N. J. Levinos and A. L. Harris, *J. Chem. Phys.*, 1992, **96**, 3950.
18 N. H. Nahler, J. D. White, J. Larue, D. J. Auerbach and A. M. Wodtke, *Science*, 2008, **321**, 1191.
19 H. Nienhaus, H. S. Bergh, B. Gergen, A. Majumdar, W. H. Weinberg and E. W. McFarland, *Surf. Sci.*, 2000, **445**, 335.
20 H. Nienhaus, *Surf. Sci. Rep.*, 2002, **45**, 3.
21 B. Mildner, E. Hasselbrink and D. Diesing, *Chem. Phys. Lett.*, 2006, **432**, 133.
22 X. Z. Ji, A. Zuppero, J. M. Gidwani and G. A. Somorjai, *Nano Lett.*, 2005, **5**, 753.
23 N. Shenvi, S. Roy and J. C. Tully, *J. Chem. Phys.*, 2009, **130**, 174107.
24 S. Roy, N. A. Shenvi and J. C. Tully, *J. Chem. Phys.*, 2009, **130**, 174716.
25 N. Shenvi, S. Roy and J. C. Tully, *Science*, 2009, **326**, 829.
26 R. Cooper, C. Bartels, A. Kandratsenka, I. Rahinov, N. Shenvi, K. Golibrzuch, Z. Li, D. J. Auerbach, J. C. Tully and A. M. Wodtke, *Angew. Chem., Int. Ed.*, 2012, **51**, 4954.
27 N. Shenvi, J. R. Schmidt, S. T. Edwards and J. C. Tully, *Phys. Rev. A*, 2008, **78**, 022502.
28 D. M. Newns, *Phys. Rev.*, 1969, **178**, 1123.
29 J. C. Tully, *J. Chem. Phys.*, 1990, **93**, 1061.
30 J. C. Tully, G. H. Gilmer and M. Shugard, *J. Chem. Phys.*, 1979, **71**, 1630.
31 R. Kubo, *Rep. Prog. Theor. Phys.*, 1966, **29**, 255.
32 S. A. Adelman and J. D. Doll, *J. Chem. Phys.*, 1976, **64**, 2375.
33 A. Nitzan, M. Shugard and J. C. Tully, *J. Chem. Phys.*, 1978, **69**, 2525.
34 J. C. Tully, *J. Chem. Phys.*, 1980, **73**, 1975.
35 S. I. Anisimov, B. L. Kapeliovich and T. L. Perel'man, *Sov. Phys. JETP*, 1974, **39**, 375.
36 H. L. Dai and W. Ho, *Laser, Spectroscopy and Photochemistry on Metal Surfaces*, World Scientific, Singapore, 1995.
37 N. Metropolis, A. W. Rosenbluth, A. H. Teller and E. Teller, *J. Chem. Phys.*, 1953, **21**, 1087.
38 S. Nose, *J. Chem. Phys.*, 1984, **81**, 511.
39 W. G. Hoover, *Phys. Rev. A*, 1985, **31**, 1695.
40 G. Kresse and J. Furthmuller, *Phys. Rev. B*, 1996, **54**, 11169.
41 J. P. Perdew, J. A. Chevary, S. H. Vosko, K. A. Jackson, M. R. Pederson, D. J. Singh and C. Fiolhais, *Phys. Rev. B*, 1992, **46**, 6671.
42 D. Torres, S. Gonzalez, K. M. Neyman and F. Illas, *Chem. Phys. Lett.*, 2006, **422**, 412.
43 W. H. Zhang, Z. Y. Li, Y. Luo and J. L. Yang, *J. Chem. Phys.*, 2008, **129**, 134708.
44 C. W. Muhlhausen, L. R. Williams and J. C. Tully, *J. Chem. Phys.*, 1985, **83**, 2594.
45 J. A. Prybyla, T. F. Heinz, J. A. Misewich, M. M. T. Lay and J. H. Glownia, *Phys. Rev. Lett.*, 1990, **64**, 1537.
46 R. R. Cavanagh, D. S. King, J. C. Stephenson and T. F. Heinz, *J. Phys. Chem.*, 1993, **97**, 786.
47 C. Springer, M. Head-Gordon and J C. Tully, *Surf. Sci.*, 1994, **320**, L57.
48 J. W. Gadzuk, *Chem. Phys.*, 2000, **251**, 87.

Reaction dynamics at a metal surface; halogenation of Cu(110)

A. Eisenstein, L. Leung, T. Lim, Z. Ning and J. C. Polanyi*

Received 15th February 2012, Accepted 6th March 2012
DOI: 10.1039/c2fd20023f

Scanning Tunnelling Microscopy (STM) is opening up a new field of reaction dynamics, followed one-molecule-at-a-time, only recently applied to reaction at a metal surface. Here we combine experiment with theory in studying the motions involved in the successive breaking by electron-induced reaction of the two carbon–halogen bonds, C–Cl or C–I, in physisorbed p-dihalobenzene, to form chemisorbed halogen-atoms and organic residue on Cu(110) at 4.6 K. We characterize the geometry of the physisorbed initial state, p-dichlorobenzene (pDCB) and p-diiodobenzene (pDIB), at the copper surface, as well as the successive final states of both chemisorbed reaction products: electron #1 giving rise to the first halogen-atom and a chemisorbed halophenyl and electron #2 giving a second halogen-atom and a chemisorbed phenylene. The major findings reported are (a) the distance and angular distributions of the chemisorbed reaction products relative to the physisorbed reagent molecule, (b) an approximate *ab initio* calculation, coupled with classical molecular dynamics (MD), of the repulsion between the products on the excited potential-energy surfaces, pes*, following excitation by electrons #1 or #2, and subsequently MD on the ground-state pes with inclusion of inelastic surface-interaction as a means to understanding the above, (c) observation of the changing dynamics with the chemistry of the halogen-atom, and (d) characterization of the effects of secondary encounters among the reaction products in the constrained space of the more highly localized reaction of pDIB. Item (d) shows clear evidence of high reactivity in surface-aligned collisions with restricted impact parameter, termed Surface Aligned Reaction, SAR, characterized by STM.

1. Introduction

The present Faraday Discussion covers the full sweep of 'Molecular Reaction Dynamics in Gases, Liquids and Interfaces'. In doing so it makes the point that in all three media the breaking of old bonds and resultant making of new ones depends upon a perturbing collision which should have features in common. With the advent of quantum mechanics it became possible to calculate the changes in potential energy in the course of a collision[1]. A first text on reaction dynamics in the quantum era followed[2]. Metals as reagent played an important part in the early gas-phase studies, because of their large reactive efficiency. It became evident that the metal donated an electron from a distance, a process termed 'harpooning'. The resultant large reactive cross-sections enabled the pioneers of crossed molecular-beam chemistry to detect reaction-products in the numerous single-collision events occurring at the crossing point of their beams.[3,4] Beam scattering[3,4] and infrared chemiluminescence[5] led, for example, to a broad categorization of reactions

Department of Chemistry, University of Toronto, 80 St. George St., Toronto, Canada. E-mail: jpolanyi@chem.utoronto.ca as either 'attractive' or 'repulsive', depending upon whether the energy-release was predominantly as reagents approached or products separated. Repulsive energy-release plays a central role in the single-molecule reaction dynamics characterized by STM, described here.

This study is a contribution to an emerging field of surface reaction dynamics in which, by STM, reactive encounters can be followed a-molecule-at-a-time. A pioneering study was made by Avouris and Wolkow on silicon[6] (for a review see ref. 7) Recently single-molecule studies have been made on the surface of metals.[8,9]

For STM studies on smooth metals liquid helium temperature is required to immobilize the initial reaction products. The reaction is induced, in the referenced work[8,9] and in the study reported here, by energetic electrons coming from the STM tip. The system under study is p-dihalobenzene, pDXB with X = Cl or I, at Cu(110). The objective is to characterize, by STM, the location of the physisorbed reagent molecule in relation to the underlying metal surface and the locations of all the chemisorbed products following electron-induced reaction, thus providing a basis for theory, also presented here, to infer the molecular dynamics.

This type of study depends upon measurement of the *initial* distribution of the products following the reactive event, as done here. Prior work showed the severing of the carbon–halogen bond in halobenzenes on copper by electron-induced reaction[10,11] as well as documenting electron-induced isomerization.[12] The dynamics of halogenation reactions at Cu(110) have been the subject of two recent studies from this laboratory[13,14] which showed for the first time the highly-localized nature of the electron-induced reaction of the first and the second I-atoms in monomers and in linear polymers of p-diiodobenzene (pDIB). This paralleled at a smooth metal surface the phenomenon of Localised Atomic Reaction noted in earlier studies of the thermal, photo-induced and electron-induced reactions of halides at two faces of silicon (Si(111) and Si(100)).[15–21] The dynamics of bond-breaking observed for pDIB on Cu(110)[13,14] resembled those reported for reaction at silicon surfaces in a second respect; the favoured direction of motion for the reacting halogen atom was found to be along a continuation of the linear extension of the C–X bond.[16,22]

The present study goes beyond refs. 13 and 14 which dealt with pDIB, since we report a first study of the dynamics of the electron-induced reaction of p-dichloro-benzene, pDCB, at Cu(110). We compare the reaction dynamics for the reaction releasing Cl-atoms, *i.e.* the new pDCB findings, with the prior findings for the parallel reaction of pDIB releasing I-atoms. We report the dynamics for first C–X bond-breaking by the first electron, and also the dynamics for second C–X bond-breaking by a second electron, with respect to their 'localization' and 'directionality'. A significant finding is the greater distance from its original location travelled by Cl *en route* to final chemisorption, than the distance travelled by I. This is true for Cl released in breaking the first C–Cl bond and also for Cl released in breaking the second C–Cl bond. It is attributed, in our dynamical model, in a large part to the greater repulsion operative in the carbon–chlorine anti-bond of the negatively-charged ionic intermediate (potential energy surface, pes*), than in the corresponding pes* of the carbon–iodine bond. A similar increase in repulsive force in a carbon–chlorine anti-bond as compared with carbon–iodine has been observed in theoretical studies of gaseous methyl halides.[23]

2. Methods

2.A. Experimental

The STM experiments were performed in a commercial low-temperature UHV-STM (Omicron) with a base pressure $<3 \times 10^{-11}$ mbar. All STM images were recorded in the constant-current mode. Bias voltages are quoted for the sample. The Cu surface was cut and polished to within 0.5° of the (110) plane as determined by Laue diffraction. Cleaning was by repeated cycles of Ar$^+$ bombardment (0.6 keV, 7 μA) followed by annealing at 800 K, until no contamination was detected by STM. *para*-Dichlorobenzene and *para*-diiodobenzene (99%) were degassed under vacuum conditions by several pump cycles. The compounds were dosed from a capillary tube directed at the copper crystal. The crystal reached a maximum temperature of 8.4 K during dosing. Dissociation of the C–X bond was electron-induced by placing the tip over the middle of the feature and maintaining a constant bias voltage with the current feedback loop disabled, for several hundreds of milliseconds. Threshold biases were used to ensure that only a single C–X bond broke each time. In the case of *p*DCB, biases of +1.9 V and +2.3 V were used to induced dissociation of the first and second C–Cl bonds respectively. For *p*DIB, the first and second C–I bond-breaking occurred at +1.4 V and +1.6 V, respectively.

2.B. Theory

Density Functional Theory (DFT) based calculations simulated the ground-state of the physisorbed molecule, the products of the two electron-induced reactions, a nudged elastic band (NEB) calculation of the minimum energy path for reaction, and the electron-induced molecular dynamics (MD) trajectories. These used the Vienna *ab initio* Simulation Package (VASP).[24] The electronic structures were performed using DFT within the generalized gradient approximation (GGA) with the Perdew–Burke–Ernzerhof (PBE) functional.[25] The projector augmented waves (PAW) scheme was used to describe the electron–ion interaction as implemented in VASP. Grimme's semi-empirical Van der Waals corrections to DFT were included throughout all the calculations in this work.[26] The Cu(110) surface was modelled using a periodically repeated slab model. The (6 × 6) supercell consisted of 180 copper atoms in 5 atomic layers plus the adsorbed molecule with a vacuum region of at least 15 Å. The supercell was constructed with the theoretically optimized Cu lattice constant of 3.6385 Å. All atoms in the adsorbates and the top three layers of copper were allowed to relax until the force on each atom was less than 0.01 eV Å⁻¹. We used the gamma point only in the surface Brillouin zones of the system, for structural optimizations and molecular dynamics calculations. A denser K-mesh (2 × 3 × 1) was used to evaluate the binding energies given in this work. In order to avoid the over estimation of the binding energy caused by the standard PBE functional, we have used revPBE.[27,28] The STM image-simulations were generated by bSKAN[29] using the calculated wave functions of the optimized initial and final states obtained from VASP.

An ionic PAW pseudopotential[30] was used to simulate the electron attachment that led to the formation of the molecular anion. The ionic pseudopotential was constructed by transferring electrons from an inner core shell to the valence shell. For example, to generate an ionic pseudopotential for Cl⁻, rather than taking the real ionic configuration [He] $2s^2 2p^6 3s^2 3p^6$, we used the configuration [He]$2s^2 2p^5 3s^2 3p^6$ in which an electron from the 2p shell was placed into the valence shell 3p.[31] This approach was originally applied to a VASP study of light-induced core excitations.[31,32] However, pseudopotentials can be more generally applied to valence bonding problems since the valence part of the wave function is not significantly affected by the core and the chemical properties are dominated by the valence electrons. The electron-induced reaction was modelled by adding negative charge totalling an electron into the anti-bonding orbital of the C–X bonds. This was done by placing the negative charge into the valence shell of the X-atoms, through the use of the appropriate X– pseudopotentials in the species being modelled.

The image–charge correction was included in calculating the ionic state. The ionic charge in the anionic atom was treated as a point charge Q above the copper surface at a distance d determined by the position of the anionic atom, calculated in each MD step. The image charge $-Q$ is under the surface at a distance d. The force due to the image charge is: $F = -Q^2/(16\pi\varepsilon_0 d^2)z$, and the correction to the total energy is $E_{\text{image}} = -Q^2/(8\pi\varepsilon_0 d)z$, where z is the surface-normal vector. The forces induced by the image charges were applied to the anionic atoms at each MD step while the system stayed in the ionic potential-energy surface, pes*.

The first-principles MD simulations were performed by solving the equations of motion while preserving the number of atoms (N), the volume of the system (V) and the total energy (E). Conservation of total energy was obtained using a time-step in MD of 0.5 fs, which gave an average drift of total energy of less than 0.01 eV ps^{-1}. The forces on the ions were calculated at each time-step, using the Hellmann-Feynman theorem as implemented in VASP. The electron-induced dynamical process was considered in two sequential steps:

(i) **Forces in the ionic state.** Electron injection by an STM tip causes the molecule to be excited to an ionic potential-energy surface (pes*). The excitation is assumed to persist for a time τ. During τ, the negative ion of DIB is driven away from the initial equilibrium configuration of the neutral ground-state, the molecule being distorted and gaining momentum on the upper surface.

(ii) **Subsequent forces in the ground state.** At τ, we assume a vertical downward transition occurs corresponding to the loss of an electron. The system returns to the neutral ground pes, conserving the distorted coordinates and additional momenta acquired on the ionic surface, pes*. Due to the momentum transfer between the molecule and surface, the molecule eventually comes to a rest on the ground pes, having undergone re-orientation and sometimes chemical reaction.

3. Results and discussion

3.A. Experiment

An example of intact physisorbed pDCB on Cu(110) is shown in the top panel of Fig. 1a. The pDCB is imaged as a slightly asymmetric protrusion, with its long axis oriented at angles between 0° and 7° from the [001] direction of the surface. The adsorption geometry of pDCB on Cu(110) was also computed by DFT. In the most stable configuration the adsorption energy is 0.60 eV with the molecule lying flat on the surface, the two Cl-atoms lying approx. 0.5 Å before short bridge sites (S) at either end, and the Cl$\cdots$Cl axis at 16° away from the [001] direction. If the long axis of pDCB lies along the [001] direction (as observed in the experiment), the computed adsorption energy is 0.55 eV. These calculated adsorption energies are half of that calculated for pDIB (1.10 eV).

The products of the first electron-induced reaction of pDCB are a Cl-atom #1 and a chlorophenyl residue (ClPh), both chemisorbed at the surface, as shown in the experimental STM image of the top panel of Fig. 1b. The Cl-atom #1 can be seen to have recoiled along the C–Cl bond direction (termed the 'forward' direction, a nomenclature used for the atomic reagent in molecular beam scattering experiments). The ClPh residue recoiled backward and rotated in the surface plane, bonding at the dangling bond on the benzene ring to a nearby Cu atom. The forces causing the forward motion of Cl and the backward recoil of ClPh will be shown, according to our model, to be repulsive; the principal force causing the turning of ClPh is likely to be the attraction between the dangling bond on the benzene ring and the copper surface.

The second electron-induced reaction in ClPh gives a chemisorbed Cl-atom #2 and a chemisorbed phenylene residue (Ph'), as shown in Fig. 1c. The Cl-atom #2, coming from ClPh, recoiled along the direction of its rotated C–Cl bond, as will be discussed in detail later. The phenylene residue, Ph', anchored by its first C–Cu bond, can be seen to have little or no displacement from its position in ClPh. The adsorption geometries of the products from the first and second electron-induced reactions given in Fig. 1 have been calculated by DFT; their simulated images are seen to be in good agreement with experiment (Fig. 1b and 1c). We note that the

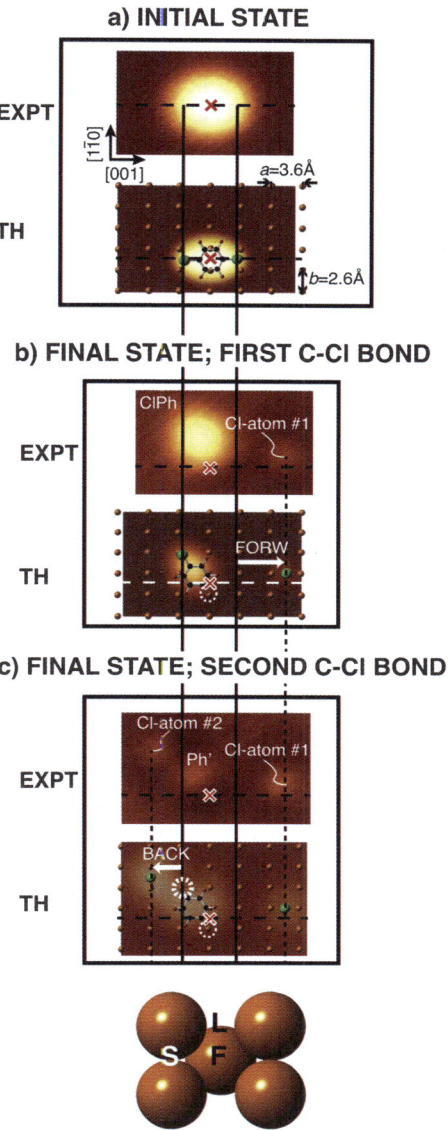

Fig. 1 Dissociation of the first and second C–Cl bond. a) Initial states, experiment (EXPT) and theory (TH), for *p*DCB adsorbed on Cu(110) at 4.6 K. The long axis of the molecule is parallel to the [001] direction with the benzene ring placed over a short bridge site. The dashed line across all images denotes the [001] direction and the red cross the centre of the intact molecule. The vertical lines refer to the position of the Cl-atoms in *p*DCB. b) The products of the first C–Cl bond breaking, Cl-atom #1 and ClPh are shown. The Cl-atom #1 is located to the right of *p*DCB (FORW), whereas the ClPh is to the left. The dashed vertical lines indicate the position of the Cl-atom after reaction. c) The products of the second C–Cl bond breaking, Cl-atom #2 and Ph′, are shown, with the second Cl-atom to the left of *p*DCB (BACK). The STM images were obtained at $I = 0.2$ nA and $V = -1.0$ V. The corresponding simulated images were calculated from geometries (overlaid) computed by DFT. Top view of the Cu(110) unit cell; short bridge (S), long bridge (L), and four-fold hollow (F) sites are marked.

bright feature in ClPh coincides with the location of the C–Cl bond; in the text we shall refer to this as the location of the Cl-atom in ClPh.

The angular and distance distribution for the products of 196 electron-induced *first-bond* C–Cl dissociation events in *p*DCB have been plotted from the centre of the physisorbed *p*DCB molecule as origin in Fig. 2a, with the Cu substrate shown below; the products are Cl-atom #1 in red and the ClPh residue in blue. For comparison, the corresponding distribution for the products from 117 electron-induced first-bond C–I dissociation events of *p*DIB (data taken from our earlier study [14], not, however, previously analyzed in this fashion) have also been plotted in the same way in Fig. 2b. The Cl-atom #1 scatters preferentially along the [001] direction exclusively into the 'forward' hemisphere, moving on average 5.8 Å from its location in the parent *p*DCB molecule. (The concentric circles in Fig. 2 are separated by one lattice constant; 3.6 Å).

The molecular product ClPh is seen to bind to the surface following a short recoil. The blue markings show the location of the bright Cl features seen in the STM images after *p*DCB dissociation. These features are situated approximately one ClPh length from the origin of the plot, a distance indicated in the figure by a black marker-line. The dangling bond at the phenyl end of ClPh would appear to be sigma bound to a Cu atom near the centre of the plot. The majority of the ClPh (76%) have their Cl-end (blue marks; Fig. 2a) along three preferred directions in the backward hemisphere. A smaller fraction (24%) of ClPh has rotated around its C–Cu pivot-point so as to direct its Cl-end into the forward hemisphere.

There is a qualitative similarity between the scatter-plots for the breaking of the first C–Cl bond and the first C–I bond, the latter being shown in Fig. 2b. The qualitative explanation in the latter case is similar; the I-atoms (red) scatter exclusively into the forward hemisphere, whereas the IPh radical, with its I-end shown in blue, points its I-end along a number of radial directions, each displaced by one IPh distance away from the centre of the plot where the phenyl–Cu bond can be presumed to be located. As for ClPh, the IPh fragments direct their halogen-atom

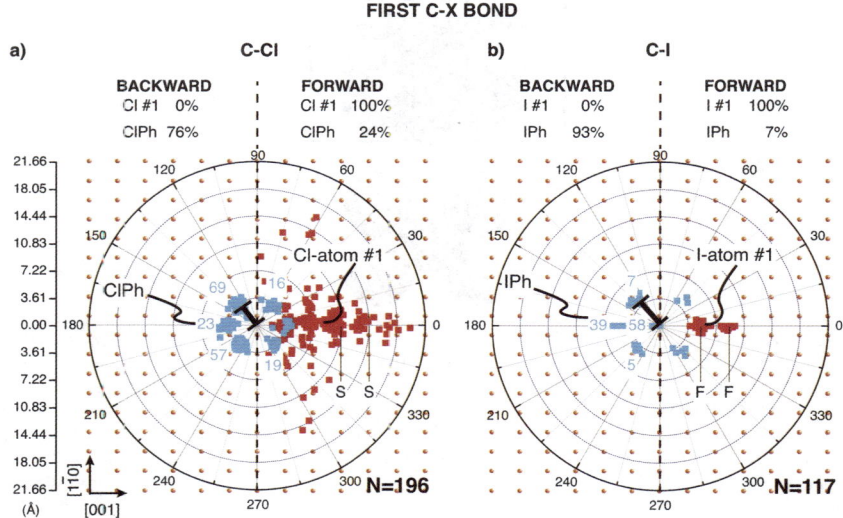

Fig. 2 Distribution of the first X-atom (red) and the X-end of the XPh residue (blue) after electron-induced reaction for first bond-breaking (a) *p*DCB and (b) *p*DIB physisorbed on Cu(110) at 4.6 K. The percentages of the products found in the forward and backward hemispheres are indicated above the plot for both a) the C–Cl and b) the C–I bond. The circles are one lattice constant of 3.61 Å apart. The Cu surface is shown as a guide.

end predominantly (93% for the case of IPh) into the backward hemisphere. The most striking difference between the plots is in the extent of I-atom scattering which is on average only 2.9 Å forward along [001], rather than 5.8 Å in the same direction for atomic chlorine. In the theory section, below, we shall give evidence that this difference is due to an increase in the repulsive energy-release between the fragments XPh and X-atom #1 when X = Cl rather than X = I, as well as to a mass-effect which channels more kinetic energy in the lighter recoiling halogen, Cl.

The greater recoil distance of Cl-atom #1 along the C–Cl ([001]) direction of the *p*DCB parent molecule is seen in Fig. 2a to be accompanied by a broader angular distribution than is the case for I-atom #1 (Fig. 2b). The increased breadth of the angular distribution for Cl-atom #1 as compared with I-atom #1 appears likely to be a consequence of the increased momentum of the Cl-atoms along [001] which, on scattering, leads to increased deflected momenta.

A further, more subtle, difference in dynamics is evident in comparing the preferred location of the atomic-halogen red marks in Fig. 2a and 2b, and is clearly evident in the individual STM images. In Fig. 2a the Cl-atoms #1 are seen to be predominantly on the rings of the polar plots, and to a lesser extent between the rings. This is indicative of a predominance of Cl-atom #1 reaction to yield chemisorbed Cl on short-bridge (S) sites (62%), with a lesser population (38%) in four-fold hollows (F). By contrast, in Fig. 2b the I-atom #1 red marks are largely located at the mid-points between rings, corresponding to the F-site location (87%); only approx 13% of the I-atoms #1 had S as their chemisorbed final state.

Similarly, in Fig. 3 we have plotted the distance and angular distribution of the products of electron-induced reaction of the *second* C–X bond (present in the halophenyl fragment coming from the first reaction) to yield chemisorbed X-atom #2 and the phenylene residue Ph′. The centre of the plot is, as before, located at the centre of the original physisorbed *p*DCB molecule. In Fig. 3a the green dots show the final state chemisorbed locations of 155 cases of Cl-atom #2 product of ClPh

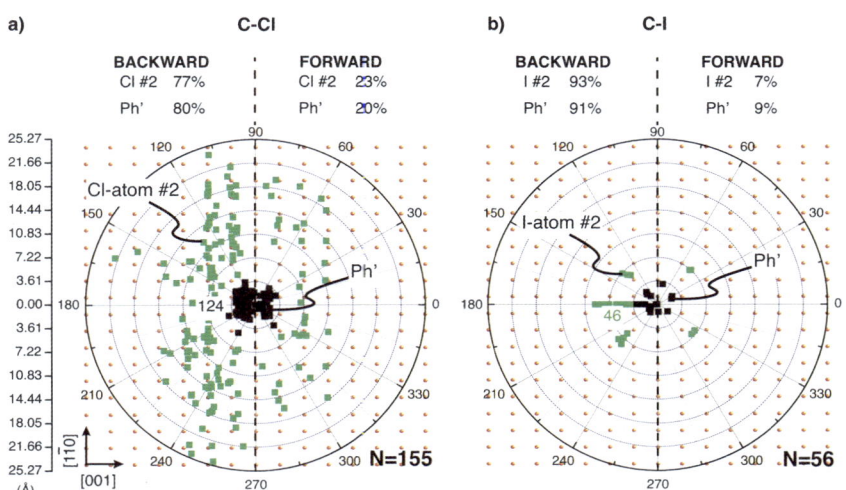

Fig. 3 Distribution of the second X-atom (green) and Ph′ residue (black) after electron-induced reaction for the second bond breaking (a) ClPh and (b) IPh adsorbed on Cu(110) at 4.6 K. The percentages of the products found in the forward and backward hemispheres are indicated above the plot for both a) the C–Cl and b) the C–I bond. The circles are one lattice constant of 3.61 Å apart. The Cu surface is shown as a guide.

dissociation, and the same number of overlapping black dots (not individually visible) give the location of the chemisorbed phenylene.

Comparison of Fig. 3a (electron-induced reaction of ClPh) with 3b (reaction of IPh) shows similarities and differences. What is similar is that the halogen recoils further than the phenylene: What is different is that Cl-atom #2 recoils further than does I-atom #2; an average recoil of 9.6 Å in the former case and 2.0 Å in the latter, from the parent XPh. In Fig. 3 for both ClPh and IPh, the Ph′ ended up close to its starting position, suggesting that the C–Cu bond formed in the first dissociation event does not break during the second carbon–halogen rupture.

Comparison of Fig. 3a with Fig. 2a (Cl-atom #2 compare with Cl-atom #1) shows a marked absence of Cl-atom #2 scattering along [001]; instead Cl-atom #2 (recoiling from ClPh) in Fig. 3a is widely scattered. Since the preference for scattering Cl-atom #1 or I-atom #1 along [001] was ascribed to the observed alignment of its parent molecule, pDCB or pDIB, we examined the possibility that the scattering of Cl-atom #2 in Fig. 3a may be due to preferred directionality of the C–Cl axis in its parent, ClPh. Calculations by DFT shows two stable alignments for ClPh, one along [001], and, significantly, an additional one at 46° to [001]; the latter is computed to be more stable by 60 meV.

How can ClPh re-align at ±46° off the [001] and [00$\bar{1}$] directions? As noted above, the recoiling ClPh during the first bond-break will experience a torque as the dangling bond is attracted to a Cu atom away from the original Cl···Cl axis to form a 3 eV bond (see Theory). We surmise that this torque, while leaving the majority of the terminal Cl-atoms of ClPh still in the backward hemisphere, can re-align ClPh to the stable off-axis angles of ±46°, away from [001]. The location of the Cl-feature in the image of each ClPh permits us, on the basis of experimental observation, to assign an orientation to the ClPh fragment. Since the distance to the centre of the plot corresponds to the length of the XPh molecule, as shown in Fig. 3a and b for ClPh and IPh, we locate the C–Cu end of the molecule in the region of the centre of the polar-plot, thus giving an observed alignment for ClPh.

In Fig. 4 we examine 132 cases of observed off-axis Cl-atoms in ClPh fragments. The ClPh extends from the centre of each plot to the blue circle which marks the Cl-end (the blue circle covers the area of the blue dots in Fig. 2, which grouped themselves into four off-axis directions). The four off-axis alignments of ClPh are shown individually in Fig. 4a–d. Examination of the STM images indicates that the four differing alignments of ClPh shown in Fig. 4 (pre-figured in Fig. 2a) correspond to four stable configurations in which the ClPh is rotated about a C–Cu bond and the terminal Cl is brought into juxtaposition directly over a Cu atom.

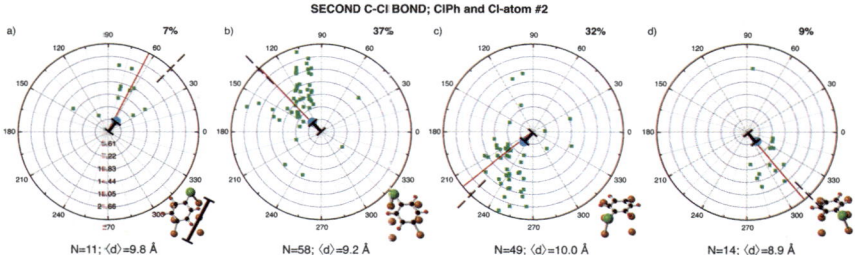

SECOND C-Cl BOND; ClPh and Cl-atom #2

N=11; ⟨d⟩=9.8 Å   N=58; ⟨d⟩=9.2 Å   N=49; ⟨d⟩=10.0 Å   N=14; ⟨d⟩=8.9 Å

Fig. 4 Distributions of the Cl- atom #2 at four angles away from the [001] direction. The black line and blue circle indicate the length and orientation of the ClPh residue in the four observed orientations. The position of the product Cl-atom #2 was obtained by experiment for the designated orientation of ClPh, is shown in a, b, c and d. The dashed line denotes the most stable angle as computed by DFT for the ClPh residue. The average distance travelled by the Cl-atom #2 is quoted from its origin in ClPh. The red line represents the average angle of the Cl-atom #2. The adsorption geometry of the ClPh is shown besides each distribution.

We now enquire whether these four observed off-axis alignments of ClPh correlate with the observed locations of Cl-atoms #2 formed by breaking the ClPh carbon–chlorine bond, in 'second-bond' breaking. From the 132 individual STM images we can identify the terminal positions of the #2 Cl-atoms (green in Fig. 4) that came from the individual ClPh fragments with each of the four alignments. It is evident from inspection of Fig. 4. that the direction of the C–Cl bond in the ClPh parent correlates with the angle to [001] at which the reaction product Cl-atom #2 is found chemisorbed on the copper surface.

As is evident in Fig. 2a, in addition to the off-axis ClPh discussed above, we found 15% on-axis ClPh. Electron-induced dissociation of these ClPh gave largely off-axis scattering of Cl, which we attribute to the fact that this [001] alignment is metastable by ~60 meV with respect to 46°.

The histograms of Fig. 5 enable a comparison of the distances travelled by I-atom reaction-products as compared with Cl-atom products. In these plots all distances are measured from the position of the halogen atom in the parent molecule. For I the parent molecule in Fig. 5a is pDIB (breaking of the first C–I bond), and in Fig. 5b is IPh (breaking of the second C–I bond). For Cl the parent molecule in Fig. 5a is pDCB (for the first C–Cl bond), and in 5b it is ClPh (breaking the second C–Cl). The I-atoms recoil on average less than one lattice constant, whereas the Cl-atoms recoil by approximately one and a half lattice constants (Fig. 5a) away from their location in pDCB, and three lattice constants away from ClPh (Fig. 5b).

3.B. Theory

Following the pattern of a recent study of p-diiodobenzene, pDIB, on Cu(110) in this laboratory[14] we have modelled the electron-induced reaction for both pDIB and its chlorine analogue, pDCB (the latter being reported here for the first time) by a combination of DFT and classical MD. The model involved as initial-state physisorbed dihalobenzene. This was on a five-layer copper slab (see Methods), designated the ground potential-energy surface, pes. To model the breaking of the first C–X bond the system was transferred, without alteration of configuration, to

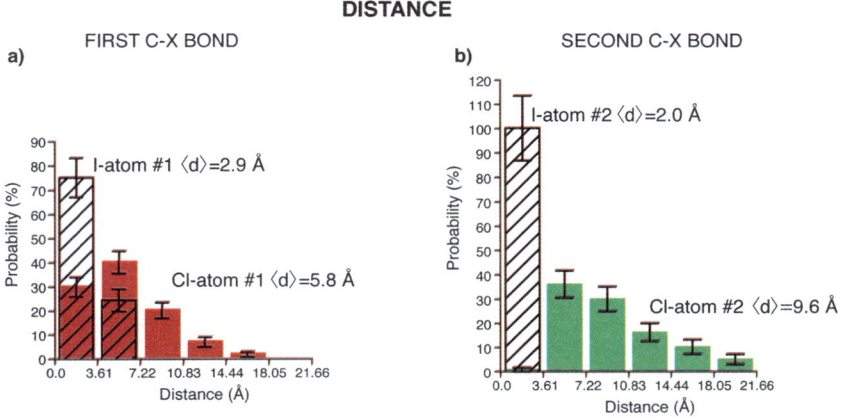

Fig. 5 Distance travelled by the I- and Cl-atoms from their original position in the parent molecules. The distance recoiled by I-atom #1 is given by the two striped histogram bars of panel a), and I-atom #2 is given the single histogram bar in panel b). The distance recoiled by Cl-atom #1 is shown in red in panel a), and in green in panel b). For each halogen atom, the average distance travelled from its origin is indicated in the figure. The bin size is 3.61 Å. The errors represent the square root of the probability.

a negatively-charged ionic state, pes*, by adding a total of one electron to the anti-bonding orbitals of the two C–X bonds. This was done by way of an ionic pseudo-potential applied to the halogen atoms, which distributed a total of one negative charge over their valence shells.

Since the addition of the first electron has been shown experimentally[14] to sever only one C–X bond, the added charge in pes* was placed predominantly at one halogen site. The asymmetric charge distribution was selected as in previous work on pDIB[14], to be 0.8 e$^-$ on one X-atom and 0.2 e$^-$ on the other, resulting in a strong repulsion between C and X in the more highly-charged C–X, which was found to react with the copper surface. A second parameter governing the dynamics, was the time spent in pes*. Following the idea of Seideman and co-workers[33], the mean lifetime of pes* for pDIB was estimated using the Non-Equilibrium Green's Function (NEGF) method to be of the order of 10 fs. To ensure reaction across the energy-barrier on the ground pes, and also to obtain dynamics with recoil of the halophenyl, XPh, as previously observed for pDIB[14], and once more observed here for chlorophenyl (ClPh), the residence time on pes* was set, as previously, at 80 fs.

As discussed when using these parameters earlier[14], this implies a probability of $\sim 10^{-4} \times$ the mean lifetime obtained by the procedure of Gadzuk.[34] However, this is much larger than the observed electron-efficiency for reaction which, for both reactions yielding Cl #1 and #2 studied in the present work, was $\sim 10^{-10}$. By adopting the same charge-asymmetry and pes* ionic-state residence time as those employed in our previous study of pDIB, we are able to test the model's ability to account for major differences in the observed reaction dynamics between the two reagents.

The motion during the 80 fs on the repulsive pes* was obtained by MD. Thereafter the system was transferred to the ground pes, on which MD was continued starting with the coordinates and momenta resulting from the forces operative on pes* and terminating when the reaction products, X-atom #1 and XPh, reached their chemisorbed states.

A similar consistency in input parameters was employed in modelling the electron-induced reaction of the second C–X bond, located in the halophenyl residue coming from the first reaction. In this case the entire additional single-electron charge was (necessarily) placed on the single halogen atom, X-atom #2. The time spent in the pes* ionic state was 70 fs, as in the earlier calculation of the dynamics of second-bond breaking in iodophenyl (IPh).[14] This enables comparison of the dynamics of I-atom #2 formation[14] with that of Cl-atom #2.

The most notable differences in the observed reaction dynamics between the p-dichloro reagent, pDCB, studied here and p-diiodo, pDIB, studied previously in the first C–X bond-breaking were in the greater displacement of the chlorine atom product along the reagent C–Cl axis, [001], and the greater rotation of the chlorophenyl product away from the [001] direction (Fig. 2–5). Since both fragments are subject to recoil as the carbon–halogen bond severs, these findings suggest a stronger recoil when the halogen atom being released is chlorine.

The reaction, however, takes place successively across two pes. The second pes, the ground potential-energy surface, is significantly exothermic, for pDCB releasing 0.80 eV in going from the physisorbed initial state to the chemisorbed pair of products. The corresponding exothermicity for the case of pDIB is slightly smaller; 0.75 eV. Though both energy-releases will contribute to the energy of the products, it appears highly unlikely that this small difference accounts for the marked increase in product momentum for the pDCB reaction.

The striking difference between the halogen-atom reactions, favouring increased recoil energy in the case of Cl-atom #1 and ClPh, is to be found in the magnitude of the repulsion operating in the ionic upper state, pes*. This is pictured in Fig. 6. During the 80 fs that pDCB spends on pes* the energy release as the system relaxes toward a new equilibrium configuration is almost four times larger than the

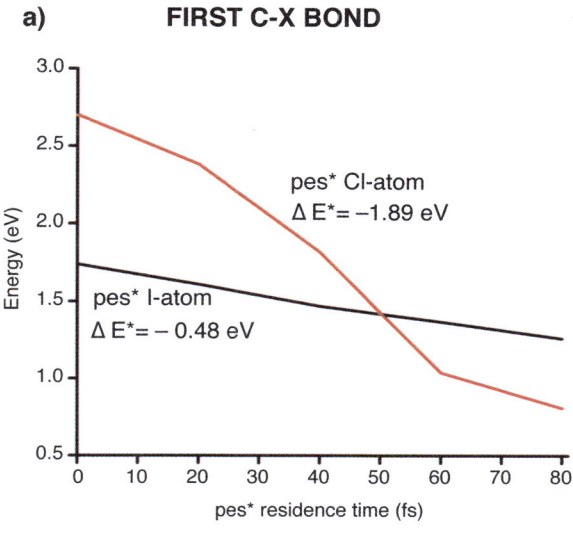

a) **FIRST C-X BOND**

pes* Cl-atom
$\Delta E^* = -1.89$ eV

pes* I-atom
$\Delta E^* = -0.48$ eV

Energy (eV)

pes* residence time (fs)

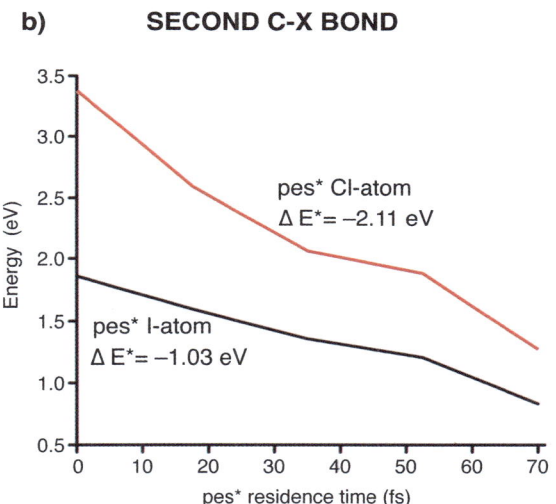

b) **SECOND C-X BOND**

pes* Cl-atom
$\Delta E^* = -2.11$ eV

pes* I-atom
$\Delta E^* = -1.03$ eV

Energy (eV)

pes* residence time (fs)

Fig. 6 Energy release in the ionic state, pes*, for first and second bond-breaking. The energy release in the first bond-breaking is shown in panel a) and that for the second bond-breaking in panel b). Red curves are for the release for Cl-atoms and black curves are for the release of I-atoms. The initial energy for each curve is that of the anti-bonding orbital of the C–X bond. The energy release is indicated for each curve.

corresponding energy release for pDIB. The greater repulsion in the antibonding state of a molecule with a C–Cl bond than in a molecule with a C–I bond, has been noted in previous theoretical studies.[23] For the cases of the Cl-atom #1 and #2 the experimentally measured threshold energies, 1.9 eV and 2.3 eV, do not match the excitation energies at the origins of the red curves in Fig. 6. Instead they match the energy required for excitation of the benzene ring. An intersystem crossing has been postulated for chlorobenzene from the benzene excited state to the antibonding state of the C–Cl bond.[35]

There are two sources of the observed rotation in ClPh. First the strong repulsion on pes* operating between ClPh and Cl-atom #1. Second, the attraction between the dangling bond at the phenyl end of ClPh and a nearby Cu-atom. A factor favouring the channelling of repulsion into Cl-atom translation, as compared with the same repulsion operating on a recoiling I-atom, is a kinematic one; greater velocity must be imparted to Cl (which has $0.28\times$ the mass of I) in order to conserve momentum.

Fig. 6b also shows an enhanced repulsion in pes* for ClPh yielding Cl-atom #2 and phenylene (Ph'), as compared with IPh yielding I-atom #2 and Ph'. The 2.11 eV energy release on pes* for the ClPh case is to be compared with the 1.03 eV energy release on pes* for the IPh case. The magnitude of the energy release, together with the kinematic factor favouring translation in the lighter halogen atom, contribute to the recoil of Cl-atom #2 by several lattice constants, in contrast to I-atom #2 recoiling by only half a lattice constant.

A factor that has not been separately considered is the effect of differing inelasticity of Cl and I moving across Cu(110). With increasing distance along a given trajectory on the ground-state pes, as modelled here, the translational energy decreases until the atom encounters a site with binding energy greater than its residual translation. The main binding sites found experimentally are short-bridge (S) for Cl-atoms and four-fold hollows (F) for I-atoms (pictured in Fig.1). DFT calculations (revPBE, see section 2.B.) gives for Cl a binding energy of 3.47 and 3.77 eV at F and S respectively, suggesting why S may be preferred over F as the final chemisorption site by recoiling Cl-atoms. For I the computed binding energies are 3.0 and 2.9 eV for F and S, slightly favouring F, as observed for recoiling I-atoms. The inelasticity would be expected to be greater for the more strongly-bound atom, which is Cl at both F and S binding sites. Since Cl-atoms are found to recoil further across the surface than I-atoms, it is unlikely that the difference in inelasticity is a major factor governing the different distances of travel.

Fig. 7 compares a pair of trajectories, originating in pDCB at the left and in pDIB at the right, computed with identical input parameters; 0.8 e$^-$ and 0.2 e$^-$ charge on the halogen atoms for the first bond-breaking, and 1.0 e$^-$ on the single halogen for the second bond-breaking. The residence times in pes* were 80 fs for the first bond-breaking and 70 fs for the second bond-breaking. The trajectories are very different, embodying in points (i)–(iv) below the principal observations from experiment (the halogen-atom #1 recoil direction is designated as 'Forward'): (i) the Cl-atoms, #1 and #2, recoil forward and backward respectively, both Cl-atoms #1 and #2 moving roughly $3\times$ as far as the I-atoms, (ii) the first halogen atom, whether Cl or I, moves along the C–X bond direction, (iii) the organic species (halophenyl or phenylene) recoils minimally, (iv) the organic residue XPh, turns in the first bond-break, especially so for X = Cl, (v) the direction of motion of X-atom #2 is along the bond-axis of the (rotated) C–X bond.

The highly-localized nature of the product distribution for pDIB results effectively in one-dimensional reaction, with the exception of a small conversion of the recoil from the first bond-break into IPh rotation. Fig. 8 shows three STM images –initial state, first-bond broken and second-bond broken– with the computed geometries superimposed. These structures have been animated by adding in Fig. 8b (first electron causing the first C–I bond-break) the locus of the first repulsion, and an image of the I-atom #1 that this repulsion acts on. Fig. 8c shows at the left the second repulsion (i.e. the effect of the second electron causing the second bond-break) and an image of the I-atom #2 that this repulsion acts on. The recoiling iodine atoms and their respective recoil distances after each bond-breaking event are labeled in Fig. 8b and 8c.

From the STM image of Fig. 8c it is clear that in the course of the second C–I bond-break which causes the recoil of I-atom #2 to the left, I-atom #1 has been propelled a lattice-spacing, 3.6 Å, further to the right. This is due to a secondary encounter between recoiling phenylene and I-atom #1. It is of interest since it is a reactive event; the Ph' + I-atom #1 collision causes an I–Cu bond at a four-fold hollow, with a computed bond-strength of 3.0 eV

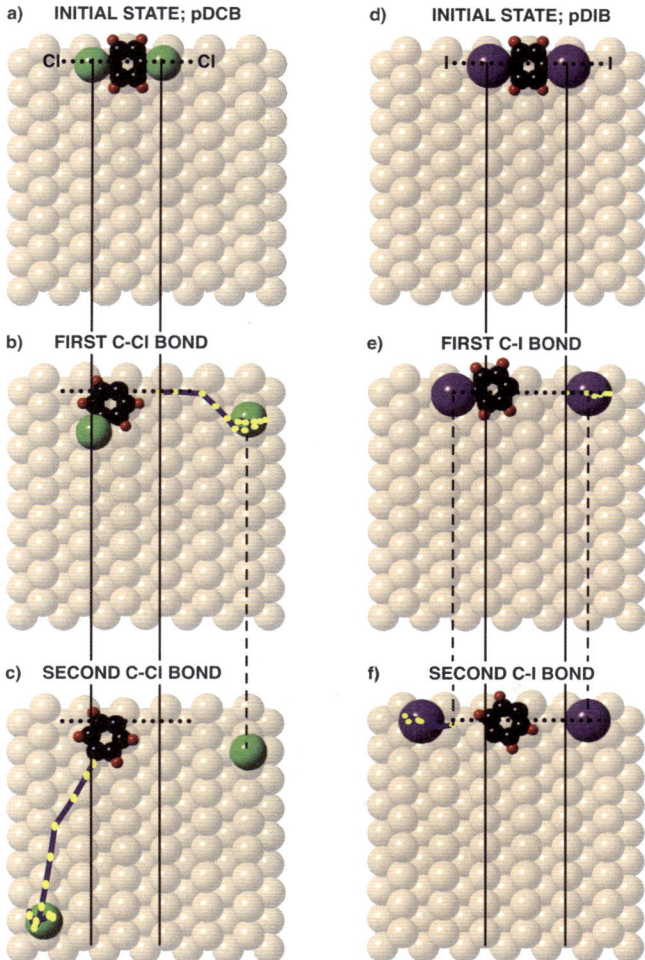

Fig. 7 Computed trajectories of first and second bond breaking of *p*DCB on the left and *p*DIB on the right. The dotted line indicates the initial orientation of the physisorbed molecule, *p*DCB or *p*DIB. Panel a) shows the physisorbed geometry of *p*DCB. Panel b) gives the computed trajectory for the first Cl-atom at every 100 fs until 800 fs and, at every 200 fs thereafter. Panel c) gives the computed trajectory for the second Cl-atom at every 100 fs until 900 fs and, at every 200 thereafter. Panel d) shows the physisorbed geometry of *p*DIB. Panel e) gives the computed trajectory for the first I-atom at every 400 fs. Panel f) gives the computed trajectory for the second I-atom at every 400 fs. The vertical lines indicate the position of the halogen atoms in the physisorbed molecules, and the dashed lines indicate the position of the halogen atoms following reaction.

(obtained by revPBE, see section 2.B.) to break, and a new I–Cu bond of the same strength to form at a further four-fold hollow along [001]. The activation energy for this reaction is 0.1 eV (as calculated by an NEB using the PBE functional). Because of favourable alignment and impact parameter this reaction occurred in every one of the 20 cases examined. Highly-efficient 'Surface Aligned Reactions' have been postulated,[36–41] but not previously observed at the atomic level.

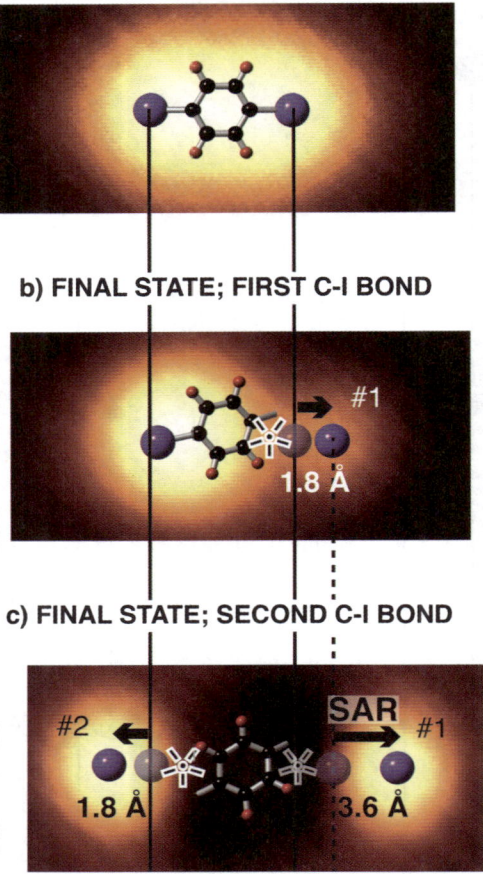

a) INITIAL STATE; pDIB

b) FINAL STATE; FIRST C-I BOND

#1
1.8 Å

c) FINAL STATE; SECOND C-I BOND

#2
1.8 Å

SAR
#1
3.6 Å

Fig. 8 STM images of the pDIB adsorbed on Cu(110) at 4.6 K following the first and second bond-breaking. Images were obtained at $I = 0.5$ nA and V = +1.0 V. The calculated geometries from DFT for each image are overlaid. The vertical lines indicate the position of the I-atom prior to the electron-induced reaction. The motion of the each I-atom is represented by arrows from their initial to their final positions. The greater repulsion initiates the dissociation of the C–I bond, whereas the smaller repulsion initiates the surface aligned reaction, SAR.

4. Conclusion

Here we have studied the electron-induced reaction at Cu(110) of p-dichlorobenzene (pDCB) and have extended earlier work[14] on p-diiodobenzene (pDIB), the results being obtained in each case by STM at 4.6 K. The physisorbed reagent molecules, pDIB and pDCB, were shown to lie with their halogen–halogen axis along the [001] direction, perpendicular to the copper rows. For pDIB the I-atoms in the parent molecule were situated on short bridge (S) sites, $i.e.$ at the mid-point of pairs of adjacent Cu atoms along [1$\bar{1}$0]. For the shorter molecule pDCB the terminal Cl's fell 0.5 Å short of these short bridge sites.

Electron-induced reaction of C–Cl bonds #1 and # 2 of p-dichlorobenzene, pDCB, were found in the present work to occur at threshold energies of 1.9 and 2.3 eV. In a recent study by this laboratory the corresponding thresholds for the reaction of the two successive C–I bonds, #1 and #2, were measured as 1.4 and

1.6 eV.[13,14] Over a hundred (155) STM experiments on pDCB, reported here for the first time, gave the initial distribution at Cu(110) of the four chemisorbed products; from the first bond-breaking, Cl-atom #1 and chlorophenyl, and from the second bond-breaking, Cl-atom #2 and phenylene. Reaction dynamics was inferred from the initial and final states, which were linked by a semi-empirical two-state model described below. These dynamics were compared with the corresponding findings from our prior studies of the electron-induced reaction of p-diiodobenzene, pDIB.

For the iodination reaction, pDIB on Cu(110), as previously noted,[14] all four products (two from the first and two from the second bond-breaking) were found to be chemisorbed at a distance of approximately half a lattice-spacing from their location in the parent molecule, evidencing marked 'localization' of reaction, even at this smooth metal surface. The reaction was also characterized by preferred 'directionality' of the reaction products, along the continuation of the direction of the bond being broken. The dynamics of chlorination, studied here, exhibited differences as well as similarities to those of iodination. A purpose of this paper is to compare chlorination and iodination.

The major novelty found for the new case of chlorination was that for electron-induced reaction of pDCB, on the same surface, Cu(110), at the same temperature, 4.6 K, as used for pDIB, the chemisorbed chlorine-atom products (both chlorine-atom #1 and #2) were less localized to the neighbourhood of the parent molecule, evidencing surface diffusion by an average distance of 5.8 Å for Cl-atom #1 and 9.6 Å for Cl-atom #2 measured from the location of the Cl-atom in the parent molecule. The second attribute of the dynamics noted for the case of iodination, namely directionality of the reaction products, remained pronounced in the case of chlorination.

The diffusion of the chlorine-atom products of electron-induced reaction in contrast to the iodine-atom products is attributed (below) principally to a large increase in the repulsive energy-release for pDCB in its ionic state, pes*. Though this repulsion operates equally on the polyatomic organic products, chlorophenyl and phenylene, it would appear to be channelled into their internal excitation, with the result that their reaction is found to be localized to the neighbourhood of the pDCB parent molecule.

These findings were interpreted in terms of a model using an *a priori* five-layer ground potential-energy surface obtained by DFT (denoted pes). A comparable semi-empirical ionic state (pes*) added charge by way of pseudopotentials to the anti-bonding orbitals of the halogen atoms. The reactive event was assumed to consist of a brief promotion of the system to this repulsive anti-bonding pes*, where classical Molecular Dynamics (MD) were employed to describe the initial distortion of pDIB[14] and pDCB, in which the halogen-atom started to recoil from the organic residue. The system was then transferred, with its coordinates and momenta unchanged, to the DFT ground pes where MD were followed for up to a picosecond, until the products reached their final chemisorbed bonding configurations.

Comparison of pDIB with pDCB showed a much greater energy-release in the ionic state pes* for the case of pDCB. This had the result that the atomic chlorine had a higher translational energy at the time of transfer to the ground pes (for Cl-atom #1, 1.37 eV compared with 0.23 eV for I-atom #1; for Cl-atom #2, 1.11 eV compared with I-atom #2, 0.27 eV). The low momentum of the I-atom tended to leave it in the first four-fold hollow that it encountered, therefore exhibiting 'localized' reaction, whereas the higher momentum of the Cl-atom carried it past some potential-wells, with energy-loss, terminating in one of the numerous S sites, following on average a diffusion of $\langle d \rangle$ = 5.8 Å preferentially along [001] for Cl-atom #1 and $\langle d \rangle$ = 9.6 Å for Cl-atom #2. The peak direction of motion was, in every case, Cl and I, (#1) and (#2), along the continuation of the direction of the carbon–halogen bond being broken. Persistence of directionality remained a valuable guide to the unfolding dynamics.

The high degree of localization in the products of electron-induced reaction from first and second bond-breaking in pDIB, favoured secondary encounters among the products. For Cl recoiling energetically from pDCB or from ClPh, longer distances of travel and wider angular scattering mitigated against secondary encounters. The effect of secondary encounters in the case of iodination was to reveal, at the molecular level, the phenomenon of Surface Aligned Reaction (SAR). In SAR a collision occurs with restricted alignment and impact parameter between a pair of adsorbates. If these collision parameters are optimal, reaction will occur with high probability. Second bond-breaking in pDIB led to secondary encounters that fulfilled these requirements. The reactive collision was between recoiling phenylene from the breaking of the second C–I bond colliding with an adjacent chemisorbed I-atom #1 at a four-fold hollow. This encounter led to the invariable outcome that the chemisorption bond, I–Cu, of dissociation energy 3.0 eV was severed, and a new chemisorption bond, I–Cu of the same energy was formed at an adjacent four-fold hollow site along the collision-axis. Calculation by DFT showed that this involved surmounting an energy-barrier of 0.1 eV. Surface-aligned-reaction, SAR, ensured that the system surmounted this barrier at every collision.

Acknowledgements

This work was funded in part by the Natural Sciences and Engineering Research Council of Canada (NSERC), the Xerox Research Centre Canada (XRCC) and the Canadian Institute for Advanced Research (CIFAR). We are grateful to I. R. McNab for many helpful discussions. Computations were performed on TCS at SciNet HPC Consortium funded by the Canada Foundation for Innovation.

References

1 F. London, *Z. Elektrochem.*, 1929, **35**, 552–555.
2 M. Polanyi, *Atomic Reactions*, Williams and Norgate, London, 1932.
3 D. R. Herschbach, *Angew. Chem., Int. Ed. Engl.*, 1987, **26**, 1221–1243.
4 Y. T. Lee, *Angew. Chem., Int. Ed. Engl.*, 1987, **26**, 939–951.
5 J. C. Polanyi, *Science*, 1987, **236**, 680–690.
6 Ph. Avouris and R. Wolkow, *Phys. Rev. B*, 1989, **39**, 5091–5100.
7 I. R. McNab and J. C. Polanyi, *Frontiers of Nanoscience, Vol. 2: Atomic and Molecular Manipulation*, ed. A. J. Mayne and G Dujardin, Elsevier, 2012, ch. 4, pp.79–120.
8 S. W. Hla, G. Meyer and K. H. Rieder, *ChemPhysChem*, 2001, **2**, 361–366.
9 W. Ho, *J. Chem. Phys.*, 2002, **117**, 11033–11061.
10 G. S. McCarty and P. S. Weiss, *J. Phys. Chem. B*, 2002, **106**, 8005–8008.
11 D. B. Dougherty, J. Lee and J. T. Yates Jr, *J. Phys. Chem. B*, 2006, **110**, 20077–20080.
12 V. Simic-Milosevic, J. Meyer and K. Morgenstern, *Phys. Chem. Chem. Phys.*, 2008, **10**, 1916–1920.
13 L. Leung, T. Lim, J. C. Polanyi and W. A. Hofer, *Nano Lett.*, 2011, **11**, 4113–4117.
14 L. Leung, T. Lim, Z. Ning and J. C. Polanyi, *J. Am. Chem. Soc*, 2012, **134**, 9320–9326.
15 P. H. Lu, J. C. Polanyi and D. Rogers, *J. Chem. Phys.*, 1999, **111**, 9905–9907.
16 P. H. Lu, J. C. Polanyi and D. Rogers, *J. Chem. Phys.*, 2000, **112**, 11005–11010.
17 S. Dobrin, K. R. Harikumar and J. C. Polanyi, *Surf. Sci.*, 2004, **561**, 11–24.
18 K. R. Harikumar, I. D. Petsalakis, J. C. Polanyi and G. Theodorakopoulos, *Surf. Sci.*, 2004, **572**, 162–178.
19 D. Dobrin, X. Lu, F. Y. Naumkin, J. C. Polanyi and S. Y. Yang, *Surf. Sci.*, 2004, **573**, L363–L368.
20 K. R. Harikumar, I. R. McNab, J. C. Polanyi, A. Zabet-Khousousi and W. A. Hofer, *Proc. Natl. Acad. Sci. U. S. A.*, 2011, **108**, 950–955.
21 K. R. Harikumar, L. Leung, I. R. McNab, J. C. Polanyi, H. Lin and W. A. Hofer, *Nat. Chem.*, 2009, **1**, 716–721.
22 S. Dobrin, K. R. Harikumar, C. F. Matta and J. C. Polanyi, *Surf. Sci.*, 2005, **580**, 39–50.
23 D. Ajitha, M. Wierzbowska, R. Lindh and P. A. Malmqvist, *J. Chem. Phys.*, 2004, **121**, 5761–5766.
24 G. Kresse and J. Furthmuller, *Phys. Rev. B: Condens. Matter*, 1996, **54**, 11169–11186.
25 J. P. Perdew, M. Ernzerhof and K. Burke, *J. Chem. Phys.*, 1996, **105**, 9982–9985.
26 S. Grimme, *J. Comput. Chem.*, 2004, **25**, 1463–1475.
27 Y. Zhang and W. Yang, *Phys. Rev. Lett.*, 1998, **80**, 890.

28 M. Dion, H. Rydberg, E. Schroder, D. C. Langreth and B. I. Lundqvist, *Phys. Rev. Lett.*, 2004, **92**, 246401.

29 K. Palotas and W. A. Hofer, *J. Phys.: Condens. Matter*, 2005, **17**, 2705–2713.

30 G. Kresse and D. Joubert, *Phys. Rev. B: Condens. Matter Mater. Phys.*, 1999, **59**, 1758–1775.

31 L. Kohler and G. Kresse, *Phys. Rev. B: Condens. Matter Mater. Phys.*, 2004, **70**, 165405.

32 W. Ji, Z. Y. Lu and H. J. Gao, *Phys. Rev. Lett.*, 2006, **97**, 246101.

33 N. L. Yoder, N. P. Guisinger, M. C. Hersam, R. Jorn, C. C. Kaun and T. Seideman, *Phys. Rev. Lett.*, 2006, **97**, 187601.

34 J. W. Gadzuk, *Phys. Rev. B: Condens. Matter*, 1991, **44**, 13466–13477.

35 Y. Liu, P. Persson and S. Lunell, *J. Phys. Chem. A*, 2004, **108**, 2339–2345.

36 E. B. Bourdon, P. Das, I. Harrison, J. C. Polanyi, J. Segner, C. D. Stanners, R. J. Williams and P. A. Young, *Faraday Discuss. Chem. Soc.*, 1986, **82**, 343–358; J. C. Polanyi and R. J. Williams, *J. Chem. Phys.*, 1988, **88**, 3363–3371.

37 W. D. Mieher and W. Ho, *J. Chem. Phys.*, 1989, **91**, 2755–2756; W. D. Mieher and W. Ho, *J. Chem. Phys.*, 1990, **92**, 5162–5163; D. V. Chakarov and W. Ho, *J. Chem. Phys.*, 1991, **94**, 4075–4077; W. D. Mieher and W. Ho, *J. Chem. Phys.*, 1993, **99**, 9279–9295.

38 T. Yamanaka, Y. Inoue and T. Matsushima, *Chem. Phys. Lett.*, 1997, **264**, 180–185; T. Yamanaka, Y. Inoue and T. Matsushima, *J. Chem. Phys.*, 1999, **110**, 2597–2605.

39 J. V. Setzler, H. Guo and G. C. Schatz, *J. Phys. Chem. B*, 1997, **101**, 5352–5361; J. V. Setzler, J. Bechtel, H. Guo and G. C. Schatz, *J. Chem. Phys.*, 1997, **107**, 9176–9184.

40 C. Tripa and J. T. Yates Jr, *Nature*, 1999, **398**, 591–593; C. Tripa and J. T. Yates Jr, *J. Chem. Phys.*, 2000, **112**, 2463–2469.

41 M. E. Vaida and T. M. Bernhardt, *ChemPhysChem*, 2010, **11**, 804–807.

Kinematics and dynamics of atomic-beam scattering on liquid and self-assembled monolayer surfaces

William A. Alexander,[a] Jianming Zhang,[a] Vanessa J. Murray,[a] Gilbert M. Nathanson[b] and Timothy K. Minton[*a]

Received 23rd February 2012, Accepted 4th April 2012

DOI: 10.1039/c2fd20034a

We have conducted investigations of the energy transfer dynamics of atomic oxygen and argon scattering from hydrocarbon and fluorocarbon surfaces. In light of these results, we appraise the applicability and value of a kinematic scattering model, which views a gas-surface interaction as a gas-phase-like collision between an incident atom or molecule and a localized region of the surface with an effective mass. We have applied this model to interpret the effective surface mass and energy transfer when atoms strike two different surfaces under identical bombardment conditions. To this end, we have collected new data, and we have re-examined existing data sets from both molecular-beam experiments and molecular dynamics simulations. We seek to identify trends that could lead to a robust general understanding of energy transfer processes induced by collisions of gas-phase species with liquid and semi-solid surfaces.

Introduction

Collisional energy transfer is the essential first step in determining the mechanisms of gas-liquid interfacial reactions and solvation. Knowledge of the principles governing the initial redistribution of collision energy is of fundamental importance to a full and predictive understanding of such gas-surface interactions. The community has embraced the use of synthetically-tunable self-assembled monolayer (SAM) surfaces as good models for more structurally complicated liquid surfaces. Over the past few decades, an impressive body of work has involved study of energy transfer processes during atomic and molecular collisions with liquid and self-assembled monolayer surfaces.[1-6] These studies reveal that a complicated interplay of kinematics, supramolecular structures, and potential energy landscapes influence energy transfer. The differences in atom or molecule scattering dynamics on surfaces of different composition suggest the possibility of using beam-surface scattering techniques as an analytical probe of liquid surfaces,[7,8] but the practical implementation of such techniques would benefit from straightforward methods to interpret the data.

The scattering dynamics of atoms or molecules on liquid or SAM surfaces strongly indicate a picture of gas-phase-like collisions with localized regions on a rough surface. In an early study of atomic-beam scattering from a SAM surface, Naaman and co-workers noted that there was significant tangential momentum transfer when He or Ar atoms and O_2 or NO molecules scattered impulsively from hydrocarbon or fluorocarbon SAM surfaces.[1] In addition, they observed

[a]Department of Chemistry and Biochemistry, Montana State University, 103 Chem/Biochem Bldg., Bozeman, MT 59717, USA. E-mail: tminton@montana.edu; Fax: +1 406-994-6011; Tel: +1 406-994-5394

[b]Department of Chemistry, University of Wisconsin, Madison, WI 53706, USA

increased energy transfer on the hydrocarbon surface and attempted to correlate the mass ratio, $\mu = m_g/m_s$, of the impinging gaseous species and an effective surface mass with the extent of energy transfer to the surface. In subsequent experiments, energy transfer was also shown to be higher on hydrocarbon surfaces than fluorocarbon surfaces, reinforcing the idea that effective mass or "stiffness" of a surface has meaning.[2,9,10] One of these studies of rare-gas atom scattering on hydrocarbon and fluorocarbon liquids also provided a compelling view of gas-liquid scattering in terms of sphere–sphere collisions by demonstrating that impulsive energy transfer depends only on the angle of deflection and not on the incidence or final angle alone.[9] Indeed, Rettner *et al.* pointed out in an earlier study of Xe atoms scattering on a Pt(111) surface that such scattering behavior, where the energy lost by an incident atom increases with deflection angle, is indicative of hard-sphere scattering.[11] A later experiment on the dynamics of hyperthermal F-atom scattering on a fluorinated silicon surface also exhibited impulsive energy transfer that was dependent on the deflection angle, and the average fractional energy transfer was fit well by a hard-sphere model, where effective surface masses of ~45 and ~35 amu were obtained for scattering with F-atom incidence energies of 284 and 544 kJ mol^{-1}, respectively.[12] These results implied a "lighter" effective surface mass for higher incidence energies, suggesting the possibility that collisions are faster and become more localized at higher velocities and shorter interaction times, although multiple-bounce scattering was still assumed to occur. The hard-sphere scattering model was shown to break down for scattering of O atoms on hydrocarbon surfaces.[3,13] However, a soft-sphere model, in which the scattered product and recoiling effective surface mass could gain internal energy, was able to provide a good fit to the fractional energy transfer as a function of deflection angle. With the idea of a collision between a gas-phase atom or molecule and a finite mass on the surface, a Newton diagram can be used to deduce the effective surface mass, the collision energy in the center-of-mass (c.m.) frame, the total energy that goes into translation of the scattered product and recoiling surface fragment, and the total internal energy that goes into these two recoiling products (see Appendix).[3,13]

The success of a simple kinematic model in explaining many aspects of the scattering dynamics suggests that a basic picture of atom-molecule scattering, borrowed from gas-phase reaction dynamics, may capture the essence of the complex interactions that undoubtedly occur in gas-surface collisions. This picture appears to be valid when relating fractional energy transfer to deflection angle and in Newton diagrams that show a relationship between laboratory and c.m. reference frames reminiscent of crossed-molecular-beams scattering data. However, the analogy with gas-phase scattering must be reconciled with what is known about gas-phase and gas-surface scattering. For a given collision energy, C_2F_6 molecules absorb significantly more energy than C_2H_6 molecules in collisions with Ar,[14] whereas a fluorocarbon surface absorbs much less energy than a hydrocarbon surface in collisions with Ar and other atoms or molecules.[2,9,10,15–18] Enhanced internal excitation in C_2F_6 *vs.* C_2H_6 is attributed to the more efficient energy transfer to the lower frequency modes in the heavier species.[19] When fluorocarbon and hydrocarbon chains condense and organize to form surfaces, the inter-chain motions on the hydrocarbon surface have lower frequencies than those on the fluorocarbon surface. Therefore, larger energy transfers are expected to occur in atom or molecule collisions with the hydrocarbon surface. The enhanced energy transfer in collisions on organic surfaces with lower frequency inter-chain vibrations has been discussed in the context of several theoretical studies, where it has been observed that more vibrational modes are involved in atom-surface collisions on hydrocarbon SAMs than fluorocarbon SAMs.[15,16,20,21] The excitation of more low-frequency modes in a collision on a hydrocarbon *vs.* fluorocarbon surface is in accord with the experimental observation of higher energy transfer on a hydrocarbon surface. Yet, the dependence of energy absorption on vibrational motions involving large numbers of surface atoms would tend to discount the notion of a localized collision with a finite or "effective" surface mass. Still, the concept of effective surface mass has great appeal, and is part of the mass-ratio parameter that has been used in non-atomistic models of gas-surface scattering, including the "washboard with moment of inertia" (WBMI) model.[22,23] The parameters in this model can be adjusted to give excellent fits to experimental data and simulations. One outcome of this model in the comparison of Ne scattering on surfaces of a hydrocarbon SAM and squalane is that the surface mass of squalane is higher than that of the SAM.[23] This result was explained by the higher "stiffness" of the squalane, as a result of its higher frequency interchain motions compared with those of the SAM. Thus, a correlation between higher frequency surface modes, reduced energy transfer, and increased surface mass is implied by the WBMI results on Ne + hydrocarbon systems. This correlation is consistent with a hard-sphere scattering model, which predicts that a higher effective surface mass must lead to higher recoil energies of the incident atoms and, consequently, less energy transfer to the surface. The idea that lower energy transfer efficiency must be correlated with a higher surface mass is, however, at odds with a modified hard-sphere ("soft-sphere") kinematic model (see Appendix), which has been used earlier with reasonable success.[3] This model allows for the counterintuitive possibility that incident atoms may lose more energy on surfaces with higher effective surface masses.

Here we investigate the merit of a simple kinematic picture of gas-surface scattering and explore the meaning of an effective surface mass within this soft-sphere picture. The kinematic model has been described earlier in various forms, but for completeness and to ensure consistent use of terminology we summarize the model in the Appendix to this paper. We have combined existing and new data on atomic oxygen and argon scattering from hydrocarbon and fluorocarbon liquid surfaces, and we have cast them in the context of the kinematic model. Classical trajectory studies of Ar-atom scattering on hydrocarbon and fluorocarbon SAM surfaces give validity to the simple model while still reinforcing the complex nature of collisions at the interface between gaseous atoms and liquid or SAM surfaces. In particular, the effective surface mass deduced from a soft sphere model is found to be greater for hydrocarbon than for fluorocarbon surfaces, despite the lower mass density and lighter atom structure of the former. This surprising result is supported by the simulations, which show that energy rapidly flows into neighboring hydrocarbon chains during the gas-surface collision and involves more surface atoms in the interaction, thus creating a higher effective surface mass.

Experimental details

The experiments were performed with the use of a crossed-molecular-beams apparatus configured for beam-surface scattering.[13,24] Atomic beams were directed at continually renewed liquid surfaces of squalane and perfluoropolyether (PFPE) at selected angles of incidence, θ_i, and the inelastically scattered atoms were detected by a rotatable mass spectrometer detector as a function of their final angle, θ_f, and time of flight (TOF) from a known reference point to the electron-bombardment ionizer of the detector. Both angles are measured with respect to the surface normal. The scattered atoms are detected in the plane defined by the incident beam and surface normal. The TOF distributions represent number density as a function of time, $N(t)$.

The data presented and analyzed here come from three experiments, involving incident Ar or O atoms, which were carried out at different times. Data for Ar scattering on squalane and PFPE surfaces with an incidence energy of 80 kJ mol^{-1} were collected as part of an early study of rare-gas scattering on liquid surfaces.[9] A later study[13] on the inelastic and reactive scattering of hyperthermal O atoms on a squalane surface also generated data that are used in the current paper. New data were recently collected on the inelastic scattering of O atoms on PFPE, allowing the dynamics of O-atom scattering on squalane and PFPE surfaces to be compared.

The experimental details on the Ar + squalane/PFPE data and the O + squalane data have been described in detail previously and will not be described further here. The new experiment on O + PFPE was conducted under conditions that were nearly identical to those of the O + squalane experiment. As in the earlier experiment on O + squalane scattering, the average O-atom translational energy for the O + PFPE experiment was 504 kJ mol^{-1} and the full width at half maximum in the energy distribution was 125 kJ mol^{-1}. The pulsed, hyperthermal O-atom beam, was directed at a Krytox 1618 [general formula, F–(CF(CF$_3$)–CF$_2$–O)$_n$–CF$_2$CF$_3$, with an average molecular weight of 3130 amu] surface at $\theta_i = 60°$, 45°, and 30°. The surface temperature was held at 293 K. TOF distributions were collected for a total of 800 beam pulses at each final angle. The data were analyzed through an inversion procedure under the assumption of a monoenergetic beam, which is the same procedure used to analyze the data in ref. 13.

Computational details

To provide insight into the atomic-scale scattering dynamics, we examined the results of classical trajectory simulations of Ar collisions with various SAM models. The potential-energy surfaces employed to evolve the trajectories have been described in detail in prior work on Ar + CH$_3$– and CF$_3$-SAM collisions.[25,26] We divide the potential into two terms. The organic monolayer is described by the OPLS (optimized potential for liquid simulations) force field, as this standard force field has shown the ability to describe well the experimental structure of the SAMs.[27,28] The Ar-SAM interaction is described using two-body Buckingham potentials derived from high-quality *ab initio* calculations of rare-gas/hydrocarbon pairs developed in previous work.[25] From comparisons with molecular beam scattering experiments, it has been noted that analytical intermolecular potentials based on high-accuracy *ab initio* data are required to achieve quantitative agreement.[29] Two categories of analysis have been done in the computational portion of this work. We have investigated (1) the behavior of the recoiling Ar atom and (2) the response of the SAM surface as a result of a collision and how these two aspects correlate with a simple kinematic picture of scattering.

An extensive amount of previous theoretical work on Ar scattering from both a CH$_3$-SAM (composed of straight-chain, methyl-terminated alkylthiolates) and a CF$_3$-SAM (composed of straight-chain alkylthiolates whose outer CH$_3$ groups have been substituted with a CF$_3$ group) was completed but remained unpublished. We have re-analyzed much of this work in order to relate the results to the simple kinematic picture described in the Appendix. Results of classical trajectories of Ar-SAM collisions at collision energies over the range of 40–157 kJ mol^{-1} were subjected to our kinematic analysis.[26] We focus herein mainly on detailed analysis of 50 kJ mol^{-1} Ar scattering from the two SAM surfaces, for which there are 9000–10000 total trajectories to examine, thus providing good statistical averaging. Trajectories at other collision energies typically were run in 2000–3000 trajectory batches. Details of these simulations are extensively described in previous work.[26,29] Although the general trends are similar, the poorer statistics for these other collision energies make comparison with the kinetic model more tenuous. Examination of the atomistic details of the individual collisions has allowed us to gain insight into the microscopic phenomena which give rise to the observed behavior of gas-phase-like scattering.

The second category of analysis concerns quantifying the collision response of the SAM surface and linking this response with the concept of an effective surface mass. For these analyses, a limited number of new classical trajectories were propagated to simulate Ar colliding with three different SAM models, described below. Unlike the previous work, in which a hybrid united-atom/all-atom description was used such that the atomic substituents on only the outer two carbon atoms of each chain were explicitly considered, each of these three SAMs were built with a full explicit-atom model of the surface. Comparative studies in the Hase group have shown that better agreement with gas-surface scattering experiments can be achieved with an explicit all-atom description of the SAM chains over united-atom models.[30]

All SAM models were constructed of a slab of 64 twelve-carbon alkylthiol chains arranged in a hexagonal lattice to mimic the experimental SAM structure. This relatively large slab size was chosen in order to minimize the influence of periodic boundary conditions in our surface motion analyses. As has been done previously,[26] the sulfur atoms were held fixed at the optimal lattice spacing locations; no Au layer was employed in this work. The relevant details of each of the three SAM models used in this work follow.

For the H-SAM, we have used the all-atom potentials developed for the hybrid UA/AA model used extensively in previous work. The label, "H-SAM", is often referred to interchangeably as "CH_3-SAM" in previous work, depending on the research group. Here we use "H-SAM" to indicate the all-atom model and "CH_3-SAM" to indicate the hybrid united-atom/all-atom model. Further details of the SAM surface and Ar/SAM potentials can be found in ref. 26.

To simulate a more massive surface, we have built two SAM models, which we will refer to as "F-SAM" and "Heavy-SAM". While the H-SAM model has a direct experimental analogue, SAMs built from *fully* fluorinated alkylthiols are experimentally difficult to produce.[31] As such, the best experimental analogues for these model surfaces are partially fluorinated SAMs which include an ethyl group linker between the sulfur and perfluoroalkane chain.[31-33] While some previous work on scattering from these SAM surfaces has explicitly modeled this ethyl linker,[34] a larger body of work exists for simulations of gas-surface collisions with fully fluorinated SAM chains.[16,35,36] For computational convenience, we have followed this latter work and have chosen not to model the ethyl linker. Thus, the simulated "F-SAM" in the present work refers to fully fluorinated SAM chains. In the F-SAM model, designed to mimic experimentally-accessible perfluorinated SAM surfaces, the thiol lattice spacing is set at the 5.78 Å nearest-neighbor lattice spacing determined experimentally for the SAM with the ethyl linker in the chains.[37,38] The Ar/F-SAM intermolecular potential is described by the pairwise analytical potential developed from *ab initio* calculations of the Ar/CF_4 system.[25] The force field to describe the F-SAM surface is built from the all-atom OPLS parameters of Watkins and Jorgensen,[28] as updated in the TINKER 6 distribution.[39] In the Heavy-SAM, the density, along with the Ar/SAM and SAM potentials are exactly as in the H-SAM (nearest neighbor spacing of 4.98 Å), but the mass of the H atoms is artificially increased to 19 amu to mimic the mass of F atoms. This artificial isotopic substitution has proven useful in the past to study kinematic effects in gas-surface collision simulations.[15,16,26]

We determined initial conditions (atomic positions and momenta) for the gas-surface trajectory calculations by first running a long (1.0 ns, with a 2.0 fs time step) canonical simulation of the SAM at a surface temperature of 300 K. After this initial equilibration run, we continued the thermal simulation and recorded the instantaneous coordinates and momenta of the SAM atoms at 1.0 ps intervals to use as initial surface conditions in the Ar scattering simulations. The Ar atom was directed at the surface at an angle of $\theta_i = 30°$ or $60°$ from the surface normal. The initial azimuthal angle (angle defined by the initial velocity vector and the SAM chain-tilt direction) was randomly sampled from a uniform distribution, and the aiming point of impact was randomly selected from within a region of the surface twice the size of the SAM unit cell. At the beginning of each trajectory, Ar was placed at least 10 Å away from the closest surface atom along its incident velocity vector. The trajectories were stopped post-collision when the Ar atom either recoiled to a distance of 12 Å from the closest atom of the surface, or (always in the case of the control trajectories), after 2.4 ps. This short trajectory cutoff time was chosen because we are concerned in this study only with the initial collision dynamics, and animation of trajectories indicated that this short time was sufficient for up to a few interactions (hops) of the Ar atom on the surface.

For each SAM model, we ran two corresponding batches of trajectories. In the first batch, we ran trajectories in the absence of the Ar atom projectile and recorded the coordinates and momenta of the SAM atoms every 100 time steps (we used a 0.24 fs time step for all trajectories). These trajectories were used as "control" trajectories to determine the equilibrium behavior of the surface. In the second batch of trajectories, we simulated Ar collisions with exactly the same surface conditions used in the control group (including the same random number seeds). Again, the coordinates and momenta of the SAM atoms in the collision trajectories were recorded every 100 time steps. Examination of the coordinates and momenta of the surface atoms during the control and collision trajectories was used to provide a mechanistic understanding of the effective-surface-mass concept.

Experimental results

Representative TOF distributions of O atoms scattered inelastically from PFPE are shown in Fig. 1. As in our earlier studies of gas-surface scattering, we analyzed the TOF distributions in terms of two limiting cases, direct inelastic (or impulsive) scattering (IS) and thermal desorption (TD).[9,12,13] The slower (TD) component was assumed to correspond to O atoms that exited the surface in a Maxwell-Boltzmann (MB) speed distribution given by the surface temperature. This MB distribution, represented as a number density distribution as a function of time, $N(t)$, was matched to the slow tail of the TOF distribution, and the difference between the overall TOF distribution and the MB component was taken to be the TOF distribution of the IS component. The IS component is a manifestation of O atoms that likely suffer one to a few bounces on the surface before being scattered back into the gas phase. The TOF distributions for the IS component were directly inverted to give translational energy distributions. Fig. 2 shows translational energy distributions, $P(E_f)$, corresponding to the TOF distributions in Fig. 1. The IS component

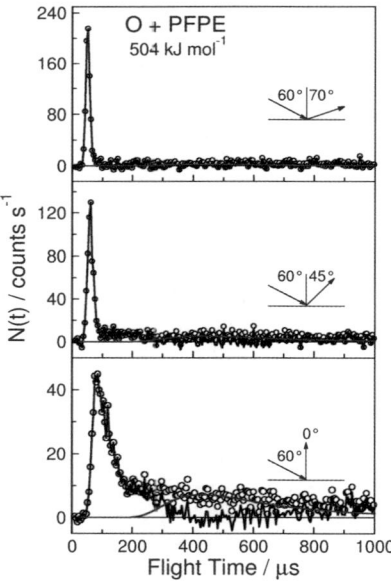

Fig. 1 Representative time-of-flight distributions of O atoms scattered from a PFPE surface following impact with $E_i = 504$ kJ mol^{-1} and $\theta_i = 60°$. Detector (final) angles for each distribution are given in each respective panel. Time zero is the nominal time at which the incident O-atom beam pulse strikes the surface.

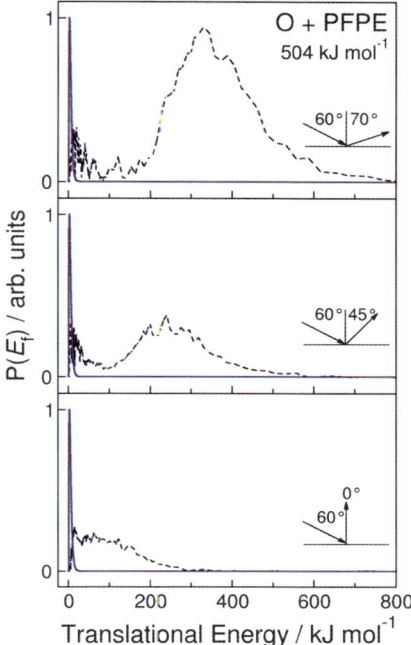

Fig. 2 Translational energy distributions of scattered O atoms derived from the time-of-flight distributions in Fig. 1. Each distribution is divided into a thermal desorption (TD, solid blue curve) and an impulsive scattering (IS, dashed black curve) component. Within each panel, the TD component is normalized to 1.0, and the areas under the IS and TD curves are proportional to the relative fluxes of the scattered O atoms that contribute to these components.

accounts for most of the scattered flux at all final angles. The angular distributions of scattered O-atom flux are shown in Fig. 3. O atoms scatter impulsively from PFPE over a broad range of angles, with a maximum just past the specular angle for the

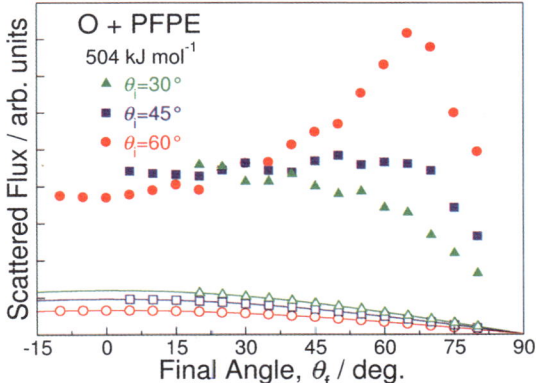

Fig. 3 Flux-weighted angular distributions of O atoms scattered from PFPE as a function of final angle, with $E_i = 504$ kJ mol^{-1} and the three incidence angles used. The solid symbols refer to impulsive scattering (IS) and the open symbols refer to thermal desorption (TD). The curves through the TD points have the functional form, $\cos\theta_f$. All experimental scattering measurements were made in the plane defined by the incident beam and the surface normal. Positive values of θ_f refer to scattering in the "forward" direction, which is on the opposite side of the surface normal from the incident beam.

largest incidence angle used. The small fraction of O atoms that give rise to the TD scattering component follow a $\cos\theta_f$ distribution. These angular distributions are similar to the angular distributions for O-atom scattering on squalane,[13] although the distributions are broader for scattering on PFPE. The broad angular distributions might be a result of surface roughness, which is expected to be higher on PFPE than squalane.[40]

The translational energy distributions corresponding to the IS components were used to calculate average final translational energies for the various combinations of incidence and final angles used in the experiment, and average fractional energy transfers were calculated from these values: $\Delta E/E_i = (E_i - E_f)/E_i$. Average fractional energy transfers for O atoms scattering from squalane and PFPE with $E_i = 504$ kJ mol^{-1} are shown in Fig. 4A and 4B, respectively, while average fractional energy transfers for Ar scattering from squalane and PFPE with $E_i = 80$ kJ mol^{-1} are shown in Fig. 4C and 4D, respectively. In all four gas-surface scattering systems, the average fractional energy transfer in the scattering plane is a function of deflection angle, $\chi = 180° - (\theta_i + \theta_f)$, and not of the incidence or final angles alone. These data

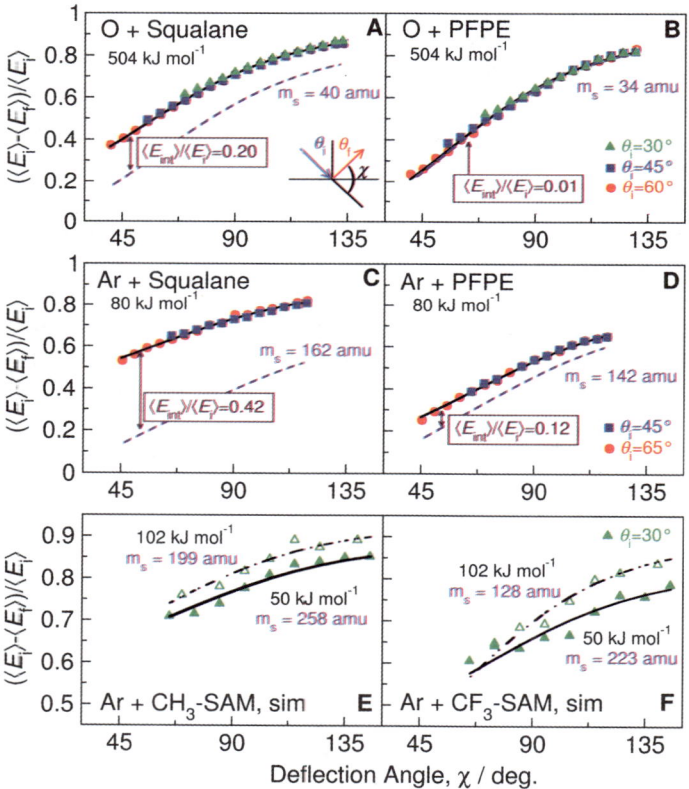

Fig. 4 Average fractional energy transfer as a function of deflection angle for impulsively scattered (IS) O or Ar atoms on the various surfaces indicated. Incidence energies and angles are indicated on the plots. The solid lines in (A) – (D) give the prediction of the "soft-sphere" model, where the incident atom interacts with a region of the surface with effective mass, m_s, and this region may absorb internal energy in the center-of-mass reference frame. The dashed lines in (A) – (D) give the prediction of the hard-sphere model with the same respective m_s but where the interacting surface fragment is assumed to absorb no internal energy in the center-of-mass frame. In (E) and (F), the lines are predictions from the "soft-sphere" model at the indicated m_s values with internal energy as indicated in the text.

are fit well with the kinematic model described in the Appendix (eqn (7)) with the optimized values of the effective surface mass, m_s, and the energy ratio, E_{int}/E_i, shown on each plot. The fit is very sensitive to these parameters, and the estimated uncertainty in m_s is ± 1 amu. The plots in Fig. 4A–D show a comparison between the predictions of the kinematic model with the same m_s and with and without the inclusion of internal excitation, E_{int}, of the interacting surface partner in the c.m. frame (where $E_{int} = 0$ is the hard-sphere limit). Except in the case of O-atom scattering on PFPE, the hard-sphere limit severely underestimates the general magnitude of the energy transfers. Attempts to fit the data with different values of m_s within the hard-sphere limit fail because the curvature does not match the experimental dependence of $\Delta E/E$ vs. χ.

Classical trajectory results

Fig. 4E and 4F contain plots of average fractional energy transfer vs. deflection angle that are derived from trajectory calculations for scattering of 50 and 102 kJ mol^{-1} Ar atoms from CH$_3$- and CF$_3$-SAMs. To connect with the experimental data, which only contain information about atoms that scatter impulsively in the plane defined by the molecular beam and surface normal, we have selected trajectories in which Ar scatters from the SAM surfaces with azimuthal scattering angles of $\pm 15°$ from the scattering plane and experiences one encounter with the surface (sometimes referred to as "one-hop trajectories"). As is common practice, we define an "encounter" as an inner turning point (ITP) in the Ar-atom trajectory (i.e., the slope, $\Delta z/\Delta t$, changes from negative to positive), where the z-axis is parallel to the surface normal. While it has been shown that the IS component may include atoms and molecules that scatter after a few (perhaps up to ~ 5) encounters with organic surfaces (and also that it is possible for an atom to "thermalize" with the surface in a single encounter),[41] restricting our analysis to single inner-turning-point (ITP = 1) trajectories is a simple, convenient, and illustrative way to interrogate the behavior of the IS scattering component. As shown in Fig. 4, the calculated fractional energy transfer for Ar scattering on SAM surfaces increases with increasing deflection angle, similar to what was observed in the experiments. The kinematic model described in the Appendix provides reasonably good fits to the classical trajectory results, and the same trend is seen in the effective surface mass as was seen in the experimental results: larger m_s values are found for Ar scattering on the CH$_3$-SAM than the CF$_3$-SAM. In addition, lower collision energies result in lower average fractional energy transfers and higher m_s values, consistent with earlier experimental results.[3,12,13] These trends are seen in similar analyses of trajectory calculations with other incidence energies in the 40–157 kJ mol^{-1} range (not shown). The calculations yielded $\langle E_{int}\rangle/\langle E_i\rangle$ ratios of 0.58 and 0.57 for the CH$_3$-SAM and 0.42 and 0.29 for the CF$_3$-SAM at 50 and 102 kJ mol^{-1}, respectively. These results indicate that the fraction of the incidence energy that goes into internal excitation of the surface collision partner (in the c.m. frame) decreases with increasing incidence energy and that this decrease is much more pronounced on the fluorocarbon surface than the hydrocarbon surface. This conclusion is consistent with earlier results on hyperthermal O-atom scattering on squalane, which show little difference in the $\langle E_{int}\rangle/\langle E_i\rangle$ ratio for the two incidence energies used. The pronounced reduction in the $\langle E_{int}\rangle/\langle E_i\rangle$ ratio with increasing incidence energy on the fluorinated surface is implied by the experimental results (Fig. 4B and ref. 12), which suggest that atom-surface scattering on fluorinated surfaces reaches the hard-sphere limit at high incidence energies.

The use of an azimuthal angular range within $\pm 15°$ of the scattering plane for trajectories with one ITP was necessary to attain reasonable statistics from the scattering calculations for comparison with the IS experimental data. For the CH$_3$-SAM system with $E_i = 50$ kJ mol^{-1}, out of 9000 Ar trajectories analyzed, $\sim 40\%$ of trajectories scatter after only one ITP, fewer than 20% scatter within $\pm 15°$ of the scattering plane, and only $\sim 10\%$ of the total number of trajectories meet both criteria. For the

CF$_3$-SAM system, the situation is only slightly better, with $\sim$16% meeting both the ITP and in-plane criteria. Narrowing the azimuthal acceptance window further, while maintaining good statistical averaging, is thus prohibitive.

We can approach an understanding of what the truly in-plane scattering behavior might be by examining trends as an increasing number of out-of-plane trajectories are examined. In Table 1, we present the results of the kinematic analysis of ITP = 1 trajectories for 50 kJ mol^{-1} Ar scattering on the CH$_3$-SAM, calculated for various azimuthal ranges. Note that the kinematic model led to good fits (not shown) of the fractional energy transfer *vs.* deflection angle, even when larger ranges of azimuthal angles were included. In addition, the inclusion of more out-of-plane scattering angles led to increasing fractional energy transfers, as has been seen in earlier calculations.[41] From the results, it is clear that out-of-plane Ar scattering correlates with a higher effective surface mass than does in-plane scattering. An extrapolation of the effective surface masses derived for the various ranges of azimuthal angles suggests that the m_s value for trajectories within $\pm 1°$ of the scattering plane would be on the order of $\sim$100 amu. Comparison of this value with the experimental result for Ar + squalane is difficult, because the experimental data include IS collisions with ITP > 1, the squalane surface is less dense than the SAM model surface,[42] and the Ar incidence energy is 50 kJ mol^{-1} in the simulations as opposed to 80 kJ mol^{-1} for the Ar + squalane experiment. Multiple-bounce collisions, lower densities, and higher incidence energies would be expected to lead to increased fractional energy transfers. For a given surface, conditions that lead to a higher fractional energy transfer appear to lead to a lower m_s, so one might conclude that the experimental m_s would be smaller than that from the simulations. However, hydrocarbon SAM surfaces absorb more energy from a collision than squalane despite their higher densities, presumably because of their more flexible intermolecular couplings.[21,42] Thus, it is not obvious how to predict the net effect of the differences between the experimental and simulated conditions on m_s. Nevertheless, the theoretical estimate of $m_s \approx 100$ amu would seem to compare favorably with the experimental result of $m_s = 162$ amu.

The theoretical calculations permit an investigation into the robustness of the kinematic model with respect to collisions with multiple ITPs. Fig. 5 shows the Ar average final energy, E_f, and derived m_s as a function of the number of ITPs, where all azimuthal angles were accepted to enable good statistical averaging. The average final energy decreases with increasing number of ITPs, and the effective surface mass increases proportionally. In terms of average final energy, the Ar atom appears to be able to transfer enough energy to the CH$_3$-SAM surface to reach thermal equilibrium within about 3–5 surface encounters, while thermal equilibrium on the CF$_3$-SAM surface requires at least 5 encounters. It is interesting to note that until the Ar atoms appear to reach thermal equilibrium with the surface, m_s does not increase

Table 1 The influence of the azimuthal acceptance window on the effective surface mass resulting from the kinematic analysis of Ar scattering from the CH$_3$-SAM with $E_i = 50$ kJ mol^{-1}, $\theta_i = 30°$

Azimuthal angle with respect to scattering plane/$\pm$deg.	m_s/amu
15	258
20	327
30	376
40	399
60	415
ALL	510

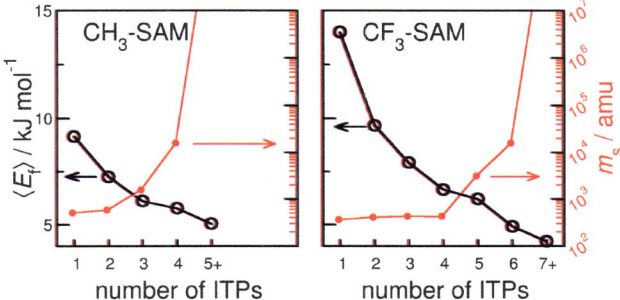

Fig. 5 Average final translational energy (left axes) and the effective surface mass, m_s, (right axes) resulting from kinematic analysis of trajectories as a function of the number of inner turning points (ITPs) during the simulation.

substantially. This result suggests that collisions leading to non-thermal scattering with multiple ITPs correlate with the same effective surface mass as collisions with a single ITP. As with the results for single-ITP trajectories over a narrow azimuthal range, a consideration of all non-thermal trajectories results in derived m_s values that are higher for the CH$_3$-SAM ($\sim$500–600 amu for ITP < 3) than for the CF$_3$-SAM ($\sim$350–425 amu for ITP < 5), although the m_s values are higher in general when all azimuthal angles are considered.

Discussion

The near perfect fit of the fractional energy transfer data by the kinematic model function is a compelling aspect of the model. In addition, the kinematic analysis of the data yields a decreasing effective surface mass with increasing incidence energy, which is a reasonable result because the effective surface mass would be expected to decrease when the atom-surface interaction time becomes shorter. Furthermore, the internal excitation, E_{int}, calculated for a given E_i follows a consistent trend, with higher fractional energy transfers correlating with higher internal energy excitations of the interacting surface region. However, if the simple kinematic model is to provide any real physical insight, then the higher effective surface mass derived for scattering on hydrocarbon surfaces *vs.* fluorocarbon surfaces must be reconciled with the experimental fact, supported by theory, that the average fractional energy transfer is higher on hydrocarbon surfaces. As mentioned in the Introduction, the WBMI results for Ne scattering on squalane and H-SAM surfaces would imply that the stiffer surface, with higher frequency modes, has a higher effective mass,[22,23] in contradiction to our experimental observations. Examination of individual atom-surface scattering trajectories reveals how the complex gas-surface collisions can still resemble gas-phase collisions and explains why a simple kinematic model, born from the ideas of gas-phase scattering dynamics, captures essential features of atom or molecule collisions with liquid and SAM surfaces.

We have used classical trajectory calculations to probe the differences in the atomistic response of model hydrocarbon and fluorocarbon surfaces to collision events. Specifically, we have compared the instantaneous atomic positions and momenta of equilibrium SAM control surfaces to corresponding surfaces which have been subjected to a collision with an Ar atom. This approach is related to the analysis of the surface modes involved in a collision that has been reported earlier,[15,16,21,43] but here the focus is on the number and range of displaced atoms. After every 100 time steps (24 fs) during a trajectory, we calculated the displacement of each atom (2432 atoms total) relative to the atom's position on an identical control surface with no Ar-atom bombardment. Thus we focused on the displacement of surface atoms caused by Ar-atom collisions. All atoms that experienced an instanta-neous displacement of 0.1 Å from their positions on the control surface were logged after each 100 time steps. The total mass of all these displaced atoms at each 100-time-step point was summed. The resulting number and total mass of the displaced atoms could thus be correlated in time with the position and momentum of the projectile atom.

Fig. 6 provides a plot of the time-dependent surface displacement behavior along with the Ar z-axis coordinate value during two representative trajectories, one on the H-SAM and the other on the F-SAM. Correlation between the first ITP and the displacements of the surface atoms is indicated by the vertical dashed line. The first ITP of the Ar collision with the H-SAM corresponds to involvement of 135 atoms in the collision, compared to a total of only 15 atoms being displaced at the time of the first ITP when Ar collides with the F-SAM. These displaced groups of atoms have total masses of 506 amu for the Ar/H-SAM ITP, and 248 amu for the Ar/F-SAM ITP. After the Ar atom impacts the surface and rebounds with its final velocity, energy transfer processes dissipate the energy which is initially localized in the impacted surface volume, so every atom will eventually be displaced from its equi-librium position. At the long-time limit after the collision, the total mass of the dis-placed atoms on the fluorinated surface must be greater than that on the hydrocarbon surface, because of the substituent masses involved. However, only the response of the surface during the collision event dictates the scattering dynamics of the Ar atom. Once the impinging atom reverses direction after collision and rebounds from the surface, the final velocity of the atom will be set by the effective mass of the atoms that collectively recoiled from the atom, as well as the necessary

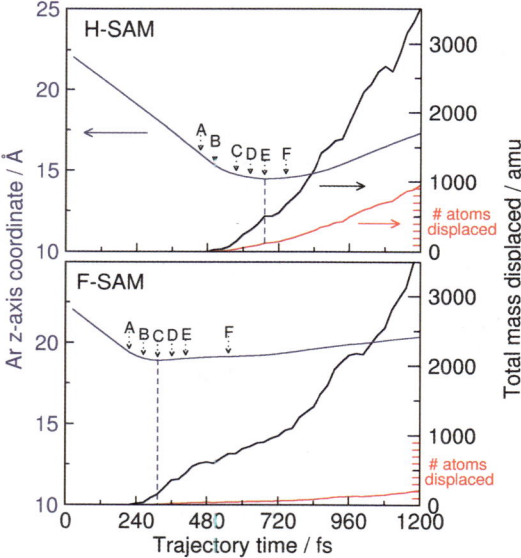

Fig. 6 (Left axis) Ar atom height above the sulfur-atom lattice plane as a function of time for representative trajectories on H-SAM and F-SAM surfaces. The outer-most carbon atoms of the SAM chains are located at 12.9 ± 0.3 and 15.6 ± 0.2 Å above the sulfur-atom plane. The H-SAM trajectory leads to an Ar atom that is scattered sideways (out-of-plane) with respect to the plane of the atomic beam and surface normal. The F-SAM trajectory leads to an Ar atom that is scattered in-plane in the forward direction. (Right axis) Total number and total summed mass of atoms displaced by more than 0.1 Å from their positions on a control surface as a function of time during the simulation. The letters A–F in each panel correspond to the snapshots that are presented in Fig. 7 and 8.

constraints of momentum and energy conservation. Analysis of all of our new Ar/ SAM trajectories in this way yields averages for the total displaced mass, $<m_{disp}>$, at the time of the first ITP. For the H-SAM, Heavy-SAM, and F-SAM at $\theta_i = 30°$, resulting $<m_{disp}>$ values were 515, 147, and 275 amu, respectively. At $\theta_i = 60°$, $<m_{disp}> = 163$ and 132 for collisions with the H-SAM and F-SAM, respectively. We note that $<m_{disp}>$ for the H-SAM is in near perfect agreement with the m_s value of 510 amu obtained from kinematic analysis of all ITP = 1 Ar scattering trajectories on the CH$_3$-SAM. This agreement lends validity to the approach of estimating effective surface masses by tracking surface displacement at the first ITP. Overall, the $<m_{disp}>$ values are consistent with the m_s values obtained from our kinematic analysis of experimental and theoretical scattering studies and demonstrate that m_s is larger on the hydrocarbon surfaces because the hydrocarbon surfaces elicit a faster and larger collective response to the collision event.

The collective response of the surface atoms can be appreciated by snapshots of the trajectories used to generate the plots in Fig. 6. These snapshots are shown in Fig. 7 and 8, where the panel labels correspond to the points along the trajectories that are labelled in Fig. 6. Atoms that have been displaced by more than 0.1 Å from their control surface position are highlighted in pink. Fig. 7 illustrates the collision of an Ar atom with the H-SAM surface, with its initial parallel velocity moving to the right. The Ar atom collides with the surface, making its first contact with an H atom 216 fs before experiencing its first momentum reversal in the surface normal direction (ITP). Within this 216 fs, the Ar atom has caused displacements of many atoms from many different SAM chains, in a roughly hemispherical volume, which expands during the 216 fs impact event. The portion of the surface involved in the collision grows as a function of time and reaches its effective mass near the ITP. At this point, the net motion of all the displaced atoms presumably gives rise to a momentum that is equal and opposite to that of the recoiling Ar atom; thus, the surface atoms close enough to the point of impact to induce an ITP in the Ar-atom trajectory contribute to the effective surface mass and, consequently,

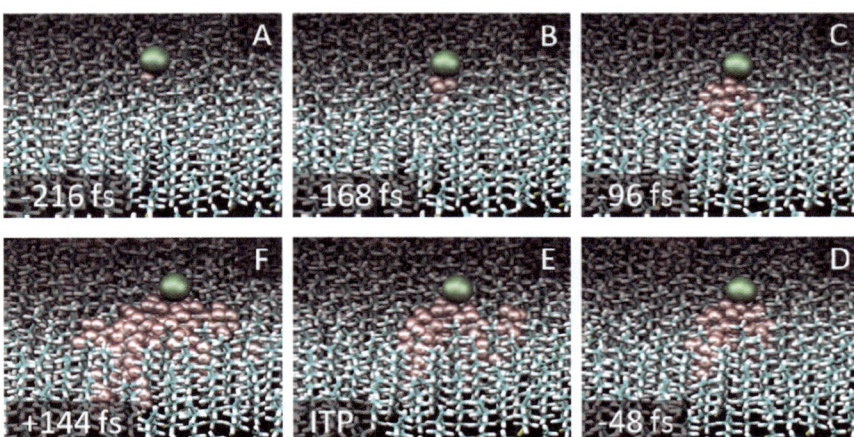

Fig. 7 Snapshots of the Ar + H-SAM collision trajectory represented in Fig. 6. Each panel corresponds to the labeled time in Fig. 6. In each panel, we have highlighted in pink those atoms whose displacement from their positions on a control surface exceeds 0.1 Å. Atom colors correspond to: Ar, green; C, blue; H, white. Time labels are relative to when the Ar atom experiences its first inner turning point (ITP). After Ar makes its first point of surface contact (A) with an H atom at −216 fs, it continues on its trajectory toward the surface (B–D) and the number and total mass of the displaced atoms increases in a quasi-hemispherically-shaped volume. After Ar experiences an ITP (E) it recoils from the surface, and (F) energy transfer processes continue to propagate the atomic displacements throughout the surface.

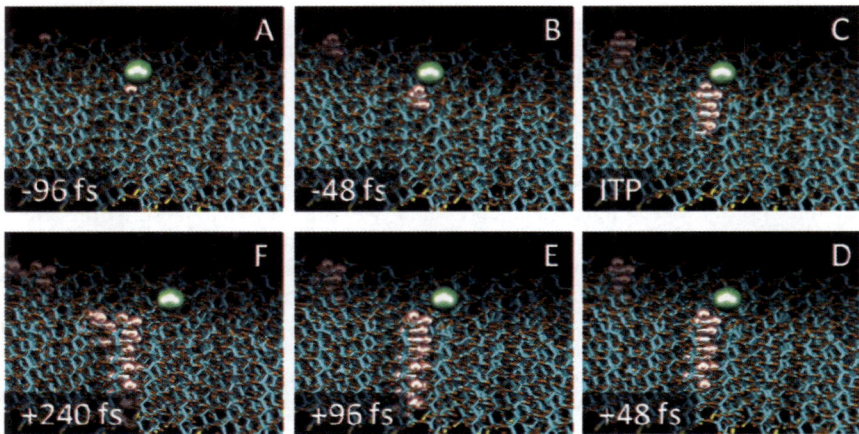

Fig. 8 Snapshots of the Ar + F-SAM collision trajectory represented in Fig. 6. Panel labels and displacement criteria are as in Fig. 7. Atom colors correspond to: Ar, green; C, blue; F, orange. After Ar makes its first point of surface contact (A) with an F atom on a SAM chain at −96 fs, it continues on its trajectory toward the surface (B) and the number and total mass of the displaced atoms increase only within the same SAM chain. After Ar experiences an ITP (C) it recoils from the surface. Intra-chain energy transfer processes continue (D–E) to propagate the atomic displacements down the fluorocarbon chain, before, at a relatively long time after the first ITP, inter-chain energy transfer processes begin to become involved (F). Pink highlighted atoms seen in the background arise from the use of periodic boundary conditions in the simulations and are not considered in our analysis.

determine the final velocity of the Ar atom. In contrast to the collision on the H-SAM, when Ar collides with the F-SAM (Fig. 8), the displacement propagates along a single fluorocarbon chain on the surface and involves far fewer atoms. After the point of first contact with an F atom, the Ar atom quickly experiences its first momentum reversal in 96 fs. The atomic displacement does not extend all the way down the chain at this point, and other SAM chains remain unperturbed. After the Ar is relatively far from the impact site, energy transfer processes eventually begin to induce inter-chain energy transfer, as seen 240 fs after the first ITP. This particular Ar + F-SAM trajectory is a case in which the Ar atom directly hits an F atom nearly along the chain-axis direction. However, we find the flow of energy along the chains to be a general trend in the collisions with F-SAMs. In a case where the Ar atom impinges on the surface in-between three adjacent SAM chains, we see that the displacement proceeds along the three adjacent chains simultaneously. The important feature of the F-SAM collisions is that intra-chain energy transfer appears to proceed throughout the chain before any extensive inter-chain energy redistribution occurs. On the H-SAM surface, these two processes occur on the same time scale. Even in trajectories in which the Ar collides head-on with a terminal methyl group along the chain-axis direction on the H-SAM, we always see multiple chains being displaced immediately. Furthermore, by the time of the ITP, the intra-chain displacement has only travelled about half-way down the fluorocarbon chain, whereas in Ar + H-SAM trajectories with similar impact geometries we generally observe full intra-chain energy transfer as well as extensive inter-chain energy transfer by the first ITP. The motions of the fluorocarbon chains have higher frequencies than those of the hydrocarbon chains, and energy transfer to the fluorocarbon surface is less efficient,[15] Therefore, the collision with the incoming Ar atom is not communicated as efficiently with neighboring atoms, and a minimal number of atoms are involved in forcing an ITP in the Ar atom trajectory. Collisions on the fluorocarbon surface are thus relatively short, fewer surface atoms are involved, and the energy transfer is minimal. On the H-SAM surface, the collision time is relatively long as a result of the significant deformation of the surface. The labile hydrocarbon chains are able to accept and redistribute energy on the time scale of the atom-surface impact; hence, a larger surface volume can participate in the collision. During this longer and more perturbative collision, more energy can be absorbed by the interacting surface volume. The difference in the participating surface volumes on the two surfaces is enough that many more light (H and C) atoms sum together in a cooperative way to yield a larger effective surface mass on the H-SAM than the combined masses of the much heavier but fewer displaced (C and F) atoms on the F-SAM.

By comparing the Heavy-SAM to the H-SAM, we can investigate the surface response in the absence of all influences other than increased mass (recall that the only difference in the Heavy- and H-SAM surfaces is that the hydrogen atoms on the Heavy-SAM have been artificially given a mass of 19 amu). When Ar atoms scatter from the Heavy-SAM instead of the H-SAM, we see a substantial decrease in m_s and a concomitant decrease in energy transfer to the surface: the same qualitative result that was observed when comparing scattering from the H-SAM and the F-SAM. With the proper potential interactions and surface density for the F-SAM, we observe an m_s value intermediate to the m_s values for the H- and Heavy-SAMs. While the potentials are different, it has been shown that they do not vary enough to have a significant influence on the scattering dynamics.[26] As they have the same actual mass, the deviation in m_s between the Heavy- and F-SAMs is attributed to the higher frequency of the surface modes on the Heavy-SAM compared with the F-SAM. As the surface mass density increases, the large-amplitude, low-frequency modes responsible for much of the collisional energy transfer increase in energy,[15,20,34] and energy transfer becomes less efficient, which, as discussed above, results in shorter collisions involving fewer surface atoms.

Conclusion

We have applied a simple kinematic model, often used to describe gas-phase scattering, to the interpretation of atomic beam-surface inelastic scattering dynamics on liquid and SAM surfaces. Three atom-surface scattering pairs were considered: (1) atomic Ar scattering on hydrocarbon and perfluoropolyether liquids, with an incidence energy of 80 kJ mol^{-1}, (2) atomic oxygen scattering on hydrocarbon and perfluoropolyether liquids, with an incidence energy of 504 kJ mol^{-1}, and (3) atomic Ar scattering on hydrocarbon and fluorocarbon SAM surfaces, with an incidence energy of 50 kJ mol^{-1}. The first two pairs were studied with atomic beam-surface scattering experiments, and the third pair was studied with classical trajectory calculations. Energy transfers for impulsively scattered atoms were larger on the hydrocarbon surfaces than the fluorocarbon surfaces, and for a given surface and incident atom, an increase in incidence energy led to an increase in the fractional energy transfer. The dependence of average fractional energy transfer on the deflection angle, $\chi = 180° - (\theta_i + \theta_f)$, of the incident atom was described well by the kinematic model in all cases. With this model, an impulsive collision is viewed as a collision between the incident atom and a localized region of the surface. The effective surface mass for scattering on the fluorinated surface is lower than that for scattering on the hydrocarbon surface. This surprising result contradicts the often-expressed belief that denser or "stiffer" surfaces necessarily have larger effective surface masses. The theoretical calculations reveal the atomic nature of effective surface mass and show that many more atoms become displaced during a collision on a hydrocarbon surface than on a fluorocarbon surface. The perturbed surface atoms initially recoil with a collective momentum that is equal and opposite to the rebounding atom, and after the collision, the energy imparted to the surface atoms dissipates throughout all the surface modes. Although atom-surface collisions are complex and involve a large number of atoms in the surface, the kinematic model appears to provide a robust picture of the average nature of impulsive atom-surface scattering as gas-phase-like, thus giving meaning to an effective surface mass and validating the picture of localized gas-phase-like collisions when atoms scatter on a liquid or SAM surface without coming into thermal equilibrium.

Appendix: kinematic model

The kinematic model is based on a treatment of hard-sphere scattering (often referred to as the "Baule formula"), in which a gas-phase, hard-sphere projectile, with an incidence translational energy of E_i, undergoes an elastic collision with a hard sphere on the surface in the c.m. frame.[44] This model can be extended by assuming that the collision may be inelastic in the c.m. frame, which means that the c.m. collision energy may be partitioned into total translational energy, E_T, and total internal excitation, E_{int}, of the two recoiling spheres. Because the spheres may acquire internal excitation, the extended model describes "soft-sphere" scattering. Fig. 9 provides a diagram of the model, where m_1 and m_2 are the masses of the incident and surface spheres, respectively, and v_0 is the incidence velocity of m_1. The surface mass is assumed to be at rest initially in the laboratory frame. After the collision, m_1 scatters away from the surface with velocity, v_1, and m_2 recoils into the surface with velocity, v_2. χ is the angle through which the incident sphere is deflected from its initial direction, and θ_2 is the angle between the recoil direction of m_2 and the initial direction of m_1. It is assumed that the collision will impart instantaneous translational and internal energy to m_2, which will determine the final translational and internal (if applicable) energy of m_1, but soon after the collision the energy associated with m_2 will be dissipated throughout the surface modes.

The derivation of an expression for fractional energy transfer comes from considerations of conservation of momentum and energy. Conservation of momentum in the initial direction of m_1 requires

$$m_1 v_0 = m_1 v_1 \cos\chi + m_2 v_2 \cos\theta_2 \tag{1}$$

and conservation of momentum perpendicular to the initial direction of m_1 requires

$$m_1 v_1 \sin\chi = m_2 v_2 \sin\theta_2 \tag{2}$$

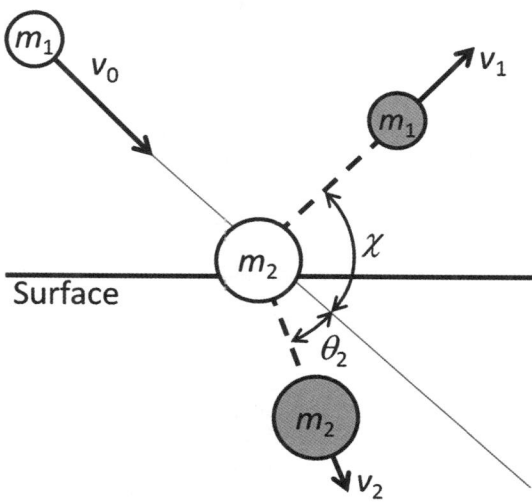

Fig. 9 Sphere-sphere scattering model of gas-surface collisions.

Assuming the initial energy of the surface is negligible compared to the translational energy of m_1, conservation of energy requires

$$\frac{1}{2} m_1 v_0^2 = \frac{1}{2} m_1 v_1^2 + \frac{1}{2} m_2 v_2^2 + E_{\text{int}} . \tag{3}$$

Using eqn (2) and 3, an expression for the ratio of the squares of the final and initial velocities of m_1 may be obtained,

$$\frac{v_1^2}{v_0^2} = \frac{1 - \dfrac{E_{\text{int}}}{E_i}}{1 + \mu \dfrac{\sin^2\chi}{\sin^2\theta_2}} , \tag{4}$$

where μ is the mass ratio, m_1/m_2. Eqn (1) and 2 allow the derivation of an expression for the ratio, $\sin^2\chi/\sin^2\theta_2$,

$$\frac{\sin^2\chi}{\sin^2\theta_2} = 1 + \frac{v_0^2}{v_1^2} - 2\frac{v_0}{v_1}\cos\chi, \tag{5}$$

which may be used in eqn (4) to find the ratio of the squares of the final and initial velocities of m_1. This ratio is related to the fractional energy transfer through

$$\frac{\Delta E}{E_i} = \frac{E_i - E_{f,1}}{E_i} = 1 - \frac{v_1^2}{v_0^2}, \tag{6}$$

where $E_{f,1}$ is the final translational energy of m_1. eqn (4), (5), and (6) may manipulated into the form,

$$\frac{\Delta E}{E_i} = \frac{2\mu}{(\mu+1)^2} \left[1 + \mu\sin^2\chi + \frac{E_{\text{int}}}{E_i}\left(\frac{\mu+1}{2\mu}\right) - \cos\chi\sqrt{1 - \mu^2\sin^2\chi - \frac{E_{\text{int}}}{E_i}(\mu+1)} \right]. \tag{7}$$

The terms involving E_{int} become zero in the limit of hard-sphere scattering. This function has two parameters, μ and E_{int}, which may be found by optimizing a fit of this functional form to a plot of $\Delta E/E_i$ vs. χ.

This kinematic model may be viewed in velocity space with the use of a Newton diagram,[45] which relates the laboratory and c.m. reference frames. Fig. 10 shows an example of a Newton diagram for direct inelastic scattering of O atoms from squalane, with an average incidence energy of 504 kJ mol^{-1} and an incidence angle of 60°.

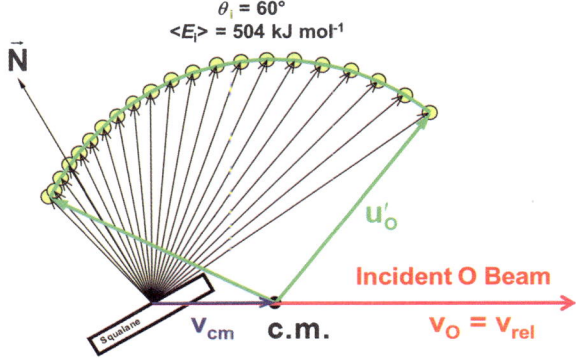

Fig. 10 Newton diagram for impulsive scattering of hyperthermal O atoms on a squalane surface, with E_i = 504 kJ mol^{-1} and θ_i = 60°.

Assuming that the thermal velocity of the tethered surface mass (m_2 in the derivation above) is negligible compared with the velocity of the incident atom (m_1 in the derivation above), the c.m. moves in the laboratory frame in the same direction as the incident atom. The surface mass is fixed in the laboratory frame, so the relative velocity, v_{rel}, between the incident atom and the surface is the velocity of the incident atom, which, in a beam-surface scattering experiment, is the measured beam velocity. There is no *a priori* knowledge of the velocity of the c.m., because the effective surface mass is unknown. The kinematic model predicts, however, that vectors corresponding to the final velocities of the scattered atoms should lie on a point on the relative velocity vector and that this point is the tip of the velocity vector of the c.m. in the laboratory reference frame, v_{cm}.

The effective surface mass ($m_s = m_2$) follows directly from v_{cm} because the mass of the incident atom (m_1) and its velocity (v_0) are known:

$$m_2 = m_1 \left(\frac{v_0}{v_{cm}} - 1 \right). \tag{8}$$

Indeed, the data show that the average final velocities of scattered O atoms at various in-plane scattering angles lie on a circle whose center is on the relative velocity vector. After the effective surface mass has been determined, the collision energy in the c.m. frame can be calculated as follows:

$$E_{coll,cm} = \frac{1}{2} \mu_r v_{rel}^2, \tag{9}$$

where μ_r is the reduced mass of the incident atom and the surface collision partner. The total energy that goes into translation of the scattered product and the recoiling surface collision partner in the c.m. frame is

$$E_T = \frac{1}{2} \mu_r' v_{rel}'^2 = \frac{1}{2} (m_1' + m_2') \frac{m_1}{m_2} u_1'^2. \tag{10}$$

where the primes refer to post-collision quantities, and u_1' is the velocity of the scattered product in the c.m. frame. Note that this model applies to both inelastic and reactive scattering. In the latter case, the product masses would be different from the incident and effective surface masses. Implicit in the kinematic analysis is the assumption that the average final velocity, u_1', and the total translational energy release, E_T, are independent of the scattering angle. This assumption is supported by the near perfect fit of a circular arc to the average final velocities over a c.m. angular range of 97°, as seen in Fig. 10. The difference between the c.m. collision energy and the total energy in translation yields the product internal excitation, $E_{int} = E_{coll,cm} - E_T$. Thus, when cast in the form of a Newton diagram, the kinematic model allows for the straightforward determination of the effective surface mass, the c.m. collision energy, the c.m. translational energy release, and the total internal excitation of the products. The mass ratio, μ, and the internal energy, E_{int}, from an analysis of the final velocities of scattered products should match these same quantities when they are derived from a fit of a plot of $\Delta E/E_i$ *vs.* χ. In our experiments, we do find agreement between the mass ratios and internal energies determined by the two different methods.

Acknowledgements

This work was supported by National Science Foundation Grant No. CHE-0943639 as part of the *Center for Energetic Non-Equilibrium Chemistry at Interfaces*. The authors are grateful to Dr Bohan Wu, who assisted with the collection of the O + PFPE scattering data. W. A. A. acknowledges computational resource support from the Texas Advanced Computing Center (TACC) at The University of Texas

This journal is © The Royal Society of Chemistry 2012

at Austin under the National Science Foundation XSEDE grant number TG-CHE110109. Computational resources provided by NSF Grant No. CHE-0741927 have also contributed to the research results reported within this paper. The authors would like to thank Prof. Diego Troya for supplying the molecular dynamics code and for providing access to previously unreported data and Prof. Bill Hase for insightful discussions.

References

1 S. R. Cohen, R. Naaman and J. Sagiv, *Phys. Rev. Lett.*, 1987, **58**, 1208.
2 M. E. Saecker, S. T. Govoni, D. V. Kowalski, M. E. King and G. M. Nathanson, *Science*, 1991, **252**, 1421.
3 T. K. Minton and D. J. Garton, in *Chemical Dynamics in Extreme Environments: Advanced Series in Physical Chemistry - Vol. 11*, ed. R. A. Dressler, World Scientific, Singapore, 2001, pp. 420–489.
4 G. M. Nathanson, *Annu. Rev. Phys. Chem.*, 2004, **55**, 231.
5 D. Troya and G. C. Schatz, *Int. Rev. Phys. Chem.*, 2004, **23**, 341.
6 L. F. Phillips, *Int. Rev. Phys. Chem.*, 2011, **30**, 301.
7 C. Waring, P. A. J. Bagot, J. M. Slattery, M. L. Costen and K. G. McKendrick, *J. Phys. Chem. Lett.*, 2010, **1**, 429.
8 C. Waring, P. A. J. Bagot, M. L. Costen and K. G. McKendrick, *J. Phys. Chem. Lett.*, 2011, **2**, 12.
9 M. E. King, G. M. Nathanson, M. A. Hanning-Lee and T. K. Minton, *Phys. Rev. Lett.*, 1993, **70**, 1026.
10 P. A. J. Bagot, C. Waring, M. L. Costen and K. G. McKendrick, *J. Phys. Chem. C*, 2008, **112**, 10868.
11 C. T. Rettner, J. A. Barker and D. S. Bethune, *Phys. Rev. Lett.*, 1991, **67**, 2183.
12 T. K. Minton, K. P. Giapis and T. A. Moore, *J. Phys. Chem. A*, 1997, **101**, 6549.
13 J. Zhang, D. J. Garton and T. K. Minton, *J. Chem. Phys.*, 2002, **117**, 6239.
14 U. Tasic, P. Hein and D. Troya, *J. Phys. Chem. A*, 2007, **111**, 3618.
15 S. A. Vazquez, J. R. Morris, A. Rahaman, O. A. Mazyar, G. Vayner, S. V. Addepalli, W. L. Hase and E. Martinez-Nunez, *J. Phys. Chem. A*, 2007, **111**, 12785.
16 J. J. Nogueira, S. A. Vazquez, O. A. Mazyar, W. L. Hase, B. G. Perkins, Jr, D. J. Nesbitt and E. Martinez-Nunez, *J. Phys. Chem. A*, 2009, **113**, 3850.
17 J. W. Lu, W. A. Alexander and J. R. Morris, *Phys. Chem. Chem. Phys.*, 2010, **12**, 12533.
18 J. W. Lu and J. R. Morris, *J. Phys. Chem. A*, 2011, **115**, 6194.
19 T. Lenzer, K. Luther, J. Troe, R. G. Gilbert and K. F. Lim, *J. Chem. Phys.*, 1995, **103**, 626.
20 B. S. Day, J. R. Morris, W. A. Alexander and D. Troya, *J. Phys. Chem. A*, 2006, **110**, 1319.
21 U. Tasic, B. S. Day, T. Yan, J. R. Morris and W. L. Hase, *J. Phys. Chem. C*, 2008, **112**, 476.
22 T. Yan, W. L. Hase and J. C. Tully, *J. Chem. Phys.*, 2004, **120**, 1031.
23 Y. Peng, L. Liu, Z. Cao, S. Li, O. A. Mazyar, W. L. Hase and T. Yan, *J. Phys. Chem. A*, 2008, **112**, 20340.
24 W. A. Alexander, J. P. Wiens, T. K. Minton and G. M. Nathanson, *Science*, 2012, **335**, 1072.
25 W. A. Alexander and D. Troya, *J. Phys. Chem. A*, 2006, **110**, 10834.
26 W. A. Alexander, B. S. Day, H. J. Moore, T. R. Lee, J. R. Morris and D. Troya, *J. Chem. Phys.*, 2008, **128**, 014713.
27 W. L. Jorgensen, D. S. Maxwell and J. Tirado-Rives, *J. Am. Chem. Soc.*, 1996, **118**, 11225.
28 E. K. Watkins and W. L. Jorgensen, *J. Phys. Chem. A*, 2001, **105**, 4118.
29 B. S. Day, J. R. Morris and D. Troya, *J. Chem. Phys.*, 2005, **122**, 214712.
30 T.-Y. Yan and W. L. Hase, *J. Phys. Chem. B*, 2002, **106**, 8029.
31 M. Graupe, T. Koini, V. Y. Wang, G. M. Nassif, R. Colorado, Jr, R. J. Villazana, H. Dong, Y. F. Miura, O. E. Shmakova and T. R. Lee, *J. Fluorine Chem.*, 1999, **93**, 107.
32 M. Graupe, M. Takenaga, T. Koini, R. Colorado Jr and T. R. Lee, *J. Am. Chem. Soc.*, 1999, **121**, 3222.
33 Y. Yuan, C. M. Yam, O. E. Shmakova, R. Colorado, Jr, M. Graupe, H. Fukushima, H. J. Moore and T. R. Lee, *J. Phys. Chem. C*, 2011, **115**, 19749.
34 D. Troya, *Theor. Chem. Acc.*, 2012, **131**, 1072.
35 J. J. Nogueira, S. A. Vazquez, U. Lourderaj, W. L. Hase and E. Martinez-Nunez, *J. Phys. Chem. C*, 2010, **114**, 18455.
36 J. J. Nogueira, Z. Homayoon, S. A. Vazquez and E. Martinez-Nunez, *J. Phys. Chem. A*, 2011, **115**, 23817.

37 N. Camillone, C. E. D. Chidsey, P. Eisenberger, P. Fenter, J. Li, K. S. Liang, G. Y. Liu and G. Scoles, *J. Chem. Phys.*, 1993, **99**, 744.
38 G. Y. Liu, P. Fenter, C. E. D. Chidsey, D. F. Ogletree, P. Eisenberger and M. Salmeron, *J. Chem. Phys.*, 1994, **101**, 4301.
39 J. W. Ponder and F. M. Richards, *J. Comput. Chem.*, 1987, **8**, 1016.
40 M. E. King, M. E. Saecker and G. M. Nathanson, *J. Chem. Phys.*, 1994, **101**, 2539.
41 T.-Y. Yan, W. L. Hase and J. R. Barker, *Chem. Phys. Lett.*, 2000, **329**, 84.
42 D. Kim and G. C. Schatz, *J. Phys. Chem. A*, 2007, **111**, 5019.
43 T. Yan, N. Isa, K. D. Gibson, S. J. Sibener and W. L. Hase, 2003, **107**, p. 10600.
44 J. Harris, in *Dynamics of Gas-Surface Interactions*, ed. C. T. Rettner and M. N. R. Ashfold, Royal Society of Chemistry, Cambridge, UK, 1991, pp. 1–46.
45 J. I. Stenfield, J. S. Francisco and W. L. Hase, *Chemical Kinetics and Dynamics*, 2nd Edition, Prentice Hall, NJ, 1998.

General discussion

Professor Tully opened the discussion of the paper by Professor Beck: The selectivity that you observe is spectacular. Usually, people struggle to find ways to enhance marginal amounts of selective reactivity. In your case of nearly 100% selective bond breaking, it might be informative to find ways to reduce or destroy the selectivity. For example, exploring incident angle dependence and/or initial rotational state dependence might provide valuable insights.

Professor Beck answered: The strong C–H bond selectivity observed is clear evidence that there is little or no vibrational energy redistribution between C–H and C–D stretching modes due to molecule–surface interactions *via* surface induced intramolecular energy redistribution (IVR) during the sub-picosecond time approach of the incident methane molecules towards the Pt(111) surface. The suggestion to increase the interaction time by reducing the incident speed or by changing the incident angle towards grazing incidence and to monitor selectivity changes could help to reveal the time scale of surface induced IVR processes. However, transition state access for methane molecules excited with a single C–H stretch quantum requires a substantial amount of translational energy along the surface normal since the reaction barrier for CH_4 dissociation on Pt(111) is close to twice the 36 kJ mol^{-1} energy of a quantum of C–H stretch. We plan to use C–H stretch overtone excitation of slower incident methane molecules to extend the molecule–surface interaction time in order to probe if the observed bond selectivity is reduced by surface induced IVR on a longer timescale. Still longer timescales are accessible at sufficiently low enough incident speed and surface temperature where the methane can be trapped intact on the surface in a physisorbed precursor state with a trapping lifetime that depends only on the surface temperature. We have excited physisorbed CH_4 on the Pt(111) surface at T_s=75 K by irradiating the surface with 100 mW of IR radiation resonant with either the ν_1 or ν_3 vibration of the physisorbed CH_4 but did not observe any evidence for dissociation consistent with a reaction barrier much higher than the 36 kJ mol^{-1}. In future experiments, we plan to excite C–H/C–D stretch combination vibrations and C–D stretch overtones of the incident molecules to probe the degree of bond selectivity afforded by other vibrational states of the different methane isotopologues.

Dr Mackenzie asked: These are elegant and conclusive experiments, demonstrating exceptional mode specificity. As you indicated in your summary, the key to this bond selective chemistry must lie in the rate of surface enhanced IVR compared with the (sub-picosecond) timescale of the collision. In order to determine the former, I wonder to what extent you could control the latter though either a glancing collision with the metal surface, or by using other carrier gases (*e.g.*, Ne)?

Are you aware of any similar experiments ever having been performed that react vibrationally excited molecules with isolated metal clusters? Would you expect any difference in the degree of mode selectivity or reactions with neutrals or charged clusters (with longer range interactions)?

Professor Beck responded: For the first part of your question, I would refer you to the response earlier to Professor Tully. In answer to the second part, we are not aware of any quantum state-resolved reactivity measurements using isolated metal clusters or metal clusters supported on a substrate. These would be challenging but also very interesting experiments which could probe the catalytic properties of size selected metal clusters as a function of cluster size and structure. For highly reactive clusters or cluster ions the degree of bond selectivity might be reduced compared to what we observed on the (111) terraces of a Pt single crystal.

Professor Truhlar remarked: Early modeling studies[1] showed that the dissociating C–H bond is stretched to about 1.37 Å at the transition state for dissociative chemisorption of methane on a metal surface, which is similar to the bond length at the transition state for the Cl + CH_4 reaction;[2] however, more recent (and probably more accurate) studies[3] for CH_4 specifically on Pt(111) show a significantly larger C–H transition state distance of about 1.48–1.53 Å. Would you comment on whether this larger distortion would make one expect a significant amount of vibrational energy transfer among C–H stretching modes in a reactive event prior to reaching the dynamical bottleneck for reaction?

1 S. E. Wonchoba and D. G. Truhlar, *J. Phys. Chem. B*, 1998, **102**, 6842.
2 J. C. Corchado, D. G. Truhlar, and J. Espinosa-Garcia, *J. Chem. Phys.*, 2000, **112**, 9375.
3 S. Nave, A. K. Tiwari, and B. Jackson, *J. Chem. Phys.*, 2010, **132**, 54705.

Professor Beck replied: Our experiments prepare the vibrationally excited methane reactants in a single ro-vibrational eigenstate by narrowband infrared laser pumping in the molecular beam far from the Pt(111) surface. Following laser excitation, no intramolecular vibrational redistribtion (IVR) occurs in the isolated molecules prepared in a stationary eigenstate. Only when the incident methane comes sufficiently close to the Pt surface, "surface-induced IVR" starts to occur if the interactions between the CH_4 molecule and the metal surface start to change the Hamiltonian so that the initially prepared state is no longer an eigenstate of the combined molecule/surface system but can be described as a time dependent superposition of eigenstates of the isolated molecule. Halonen *et al.*[1] have modeled this surface-induced IVR process as vibrationally adiabatic energy flow between different C–H stretching states of CH_4 during the approach to the surface. Different initial states (ν_1 *vs.* ν_3) were shown to lead to different vibrational energy localization in the proximal *vs.* distal C–H bonds offering an explanation for the mode-specific methane reactivity observed experimentally.[2] While there may well be vibrational energy flow between the C–H stretching states of methane during a CH_4–surface collision, our observation of strong bond selectivity dissociation for the C–H stretch excited, partially deuterated methane isotopologues show that there is no significant vibrational energy transfer between the C–H stretch and C–D stretch modes. This absence of energy flow during the collision time of less than a picosecond is likely to be due to the large energy difference between the C–H and C–D stretch modes.

1 L. Halonen, S. L. Bernasek, and D. J. Nesbitt, *J. Chemical Physics*, 2001, **115**, 5611.
2 P. Marori *et al.*, *Phys. Rev. Lett.*, 2005, **94**, 246104.

Professor Neumark asked: From Fig. 3 in your paper, there appears to be less reaction in lower translational energy collisions with the laser on than in higher energy collisions with the laser off. Is this correct?

Professor Beck answered: The RAIRS data shown in Fig. 3 alone does not correctly reflect the relative reactivity of the laser-off and laser-on experiments since the IR laser pumping did not necessarily excite the full cross section of the molecular beam which can lead to different areas covered by the reaction products on the Pt(111) surface for laser-off and laser-on experiments. With proper calibration of the RAIRS absorption signal intensities we find a vibrational efficacy of 0.6 for $CH_4(\nu_3)$ excitation, indicating that ν_3 vibrational energy is less effective than normal translational energy in promoting the dissociation of CH_4. We have not yet quantified the vibrational efficacies for the partially deuterated methane isotopologues.

Professor Dong remarked: In your work presented here, the activation of the C–H bond in CHD_3 has been investigated. I think it would be interesting to investigate the activation of the C–D bond in CDH_3 and compare the two results. It is, of course, desirable to deposit about the same amount of energy into the C–H and the C-D bonds. Is such an investigation possible experimentally?

Professor Beck said: One quantum of C–D stretch vibration contains between 2100–2200 cm^{-1} of vibrational energy for the deuterated methane isotopologues whereas the C–H stretch quanta are in the range of 2880–2940 cm^{-1}. Therefore we cannot prepare iso-energetic vibrational states with either pure C–D or C–H excitation. However, we can use different amounts of incident translational energy for the C–D and C–H stretch excited methane to compare the bond selectivity of iso-energetic reactants (translation + vibration). Furthermore, we can also excite the first overtone of the C–D stretching modes and obtain reactants with roughly 4300 cm^{-1} of C–D stretch vibrational energy and compare to either a single quantum or two quanta of C–H stretch excitation. These experiments are currently being performed in our laboratory.

Professor Tully enquired: As I understand it, your results suggest that vibrational and translational energy are roughly equally effective in promoting bond breaking. I might have expected that vibrational energy might be more effective. Do you think it is possible that some of the vibrational excitation is quenched by electron-hole excitations prior to the molecule reaching the transition state? Can you suggest experiments to shed light on this possibility?

Professor Beck responded: The question if there is significant vibrational relaxation of the incident state prepared methane molecules *via* electron–hole pair excitation on the way to the transition state is an important question which has not been definitively answered to my knowledge. So far, all theoretical models for hydrogen and methane dissociation have treated the reaction as electronically adiabatic, neglecting the possibility of electron–hole pair excitation and the analysis of our state-resolved methane reactivity has included this assumption. The fact that both H_2 and CH_4 don't include a deep molecular chemisorption well in their potential energy surface and their negative electron affinities seem to favor electronic adiabaticity. A state-to-state surface scattering experiment using IR laser tagging in combination with bolometric detection of scattered methane molecules could be used to measure vibrational energy transfer rates in order to shed light on the question of electron–hole pair excitation. We plan to perform such experiments in the near future.

Professor Sibener asked: Most of the work you have addressed today has involved pre-excitation of the incident gas-phase molecule before it encounters the substrate followed by prompt dissociation upon collision with the surface. I would like to ask what we know at this time about molecules which do not directly react upon collision with the surface, but rather which initially stick without prompt dissociation. Do these vibrationally excited molecules continue to have preferred reactivity, or do relaxation processes win out where either the energy leaks out to the substrate or perhaps where the energy randomises within the molecule, again leading to a loss of state-selected chemistry? Further, what is known at this time if we directly pump adsorbed molecules as opposed to molecules incident from the gas phase? Do they exhibit either preferential or enhanced reactivity, or, again, do energy randomisation processes involving the substrate occur on a competitive timescale?

Professor Beck said: In response to the first part of your question, we have observed no evidence for a precursor mediated dissociation channel for CH_4 chemisorption on Pt(111) at low incident translational energy and low surface temperature (E_t=13 kJ mol^{-1} and T_s=150 K) either with or without vibrational excitation of the incident CH_4. On the more corrugated Pt(110) surface, where there is evidence for a precursor mediated dissociation pathway for $CH_4(\nu=0)$,[1,2] we have measured the state resolved reactivity of $CH_4(2\nu_3)$ but observed no vibrational activation for the precursor mediated reaction pathway. Presumably, because fast vibrational energy transfer from the physisorbed CH_4 to electronic excitations of the metal (electron–hole pairs) outcompetes the slower precursor pathway. This is consistent with the fact that in previous work[3] on SiH_4 chemisorption on Si(100), we did observe vibrational activation for the precursor mediated chemisorption pathway. On this semiconductor surface, the vibrational lifetimes are in the nanosecond range because vibrational energy transfer to electron–hole pairs is absent due to a bandgap that is larger than the vibrational energy of the incident SiH_4. Concerning the second part of your question, I would like to report that we have also performed experiments at lower surface temperature T_s =75 K where the incident methane (E_k=9 kJ mol^{-1}) transiently physisorbs on Pt(111) without dissociating. The RAIR spectrum of $CH_4(ads)$ shows both a strong antisymmetric C–H stretch mode ν_3 at 2997 cm^{-1} as well as a weaker absorption at 2882 cm^{-1} due to the ν_1 symmetric C–H stretch mode which becomes infrared active for $CH_4(ads)$ due to the reduced symmetry compared to the gas phase CH_4. We have irradiated $CH_4(ads)$ on the Pt(111) surface with more than 100 mW single-mode cw IR radiation (45 degree angle of incidence, p-polarization) tuned to the center of either one of these two absorption lines for up to 35 min but could not detect either dissociation into $CH_3(ads)$ + H(ads) nor resonant desorption of $CH_4(ads)$. Dissociation would have been detected with high sensitivity by the appearance of the RAIRS signal for the symmetric CH_3 stretch signal at 2880 cm^{-1} which would persist when the surface temperature is raised above 80 K while the absorptions due to $CH_4(ads)$ disappear due to desorption of $CH_4(ads)$. Desorption by vibrational excitation of $CH_4(ads)$ with the resonant IR radiation would have been detected by a reduction of the $CH_4(ads)$ RAIRS signals which was not detectable. In conclusion, we observe neither enhanced reactivity nor enhanced rate of desorption upon vibrational excitation of CH_4 on Pt(111).

1 A. V. Walker and D. A. King, *Phys. Rev. Lett.*, 1999, **82**, 5156.
2 R. Bisson, M. Sacchi and R. D. Beck, *J. Chem. Phys.*, 2001, **132**, 094702.
3 R. Bisson, T. T. Dang, M. Sacchi and R. D. Beck, *J. Chem. Phys.*, 2008, **129**, 81103.

Professor Wodtke questioned: You showed IR spectra of CO background gas that adsorbed on your surface appearing as a doublet. This reminds us that CO binds to terrace and step edge sites and more to the point that reactivity at specific surface sites can vary dramatically. What is the evidence that your methane dissociation reaction is taking place at terrace sites and not at step edges? Moreover, do you see promising experimental approaches to studying the reactivity of methane at different surface sites?

Professor Beck answered: The fact that we only observe the symmetric C–H stretch mode in the RAIR spectrum of the $CH_3(ads)$ products of the dissociative chemisorption of CH_4 on Pt(111) is consistent with a local C_{3v} adsorption symmetry with a three-fold axis along the surface normal indicating adsorption on the terrace sites.[1] The symmetric $CH_3(ads)$ stretch occurs at 2880 cm^{-1} and a weaker peak observed at 2756 cm^{-1} is due to the overtone of the antisymmetric bending mode. If CH_3 were adsorbed on a lower symmetry step edge or other defect sites we should also observe the antisymmetic C–H stretch mode of CH_3 of E symmetry which is not detected in our RAIR spectra. The time sequence of the RAIR spectra shows that any residual CO which is present in traces in our molecular beam and the UHV background first adsorbs on the step edge sites (RAIRS peak at 2070 cm^{-1}) followed by adsorption on the terrace site (RAIRS peak at 2078 cm^{-1}). From the intensity of the CO RAIRS signals we estimate that less 0.1% of the surface sites on our Pt(111) surface are step edge sites and that these are quickly passivated by residual CO. Studies of the adsorption on step edges can be done by using surface samples that are cut to produce a larger density of steps such as Pt(533).[2] Since our RAIR spectra are sensitive to the adsorbate structure and adsorption site they can be used to distinguish adsorption on different surface sites.

1 M. Trenary, *Langmuir*, 1994, **10**, 3649.
2 A. Kleyn *et al.*, *J. Chem. Phys.*, 2003, **118**, 3334.

Professor Whitaker opened the discussion of the paper by Professor Nesbitt by commenting: Your interpretation of your elegant experiment and the observation that the rotational temperature of scattered CO_2 molecules is sensitive to the nature of the anion rather than the cation of the dissolved salt is that there is a propensity for anions to preferentially accumulate close to the gas-liquid interface. But if the spatial alignment of the quadrupole due to the presence of a charged species very close to the surface was the only factor leading to rotational excitation wouldn't one point charge behave much as another? Are not other factors potentially at play, for example, the "size" of the anions with respect to the cations, in the sense of their relative polarizability, and could this not explain the relative difference between chloride and iodide salts and the insensitivity to the nature of the counter ion? Put another way, were the surface to contain a roughly equal number of positively and negatively charged ions the effect of changing the salt could be more manifest in the nature of the anion than the cation simply because the former are "bigger" and that the rotational excitation happens at short range. Indeed the fact that they are bigger is the basis of your argument for the anions rising to the surface, which in itself seems a reasonable proposition, but I'm questioning that the evidence, or rather non-evidence, presented in the paper proves that there is a surface excess of negative charge. Of course you go on in your paper to present some simulations of alignment effects that would distinguish between an anionic and cationic surface excess and I wonder what hope there is of realizing these predictions experimentally?

Professor Nesbitt responded: The size of the anion is indeed related to the polarizability. Thus your suggestion that the effects observed might be due to the fact that iodide anions are simply "bigger targets" than chloride anions (with both anions large with respect to counter cations) is an interesting one and worth considering. The reason why I don't think this is a likely contribution is that the Coulomb field and field gradient interactions with the anions/cations extend out to appreciably large distances from the surface. From our simple trajectory calculation model, the characteristic distance scale on which the CO_2 projectiles become rotationally aligned with respect to the ion is in the order of 10 Å away from the surface. In essence, this translates into a much larger cross section for such electrostatic interactions than the physical size of the anion or cation in question. As emphasized in the paper, this length scale tends to make the collisions overly rotationally adiabatic, which in turn makes it easy to have strong alignment effects near the surface be greatly minimized by subsequent recoil from the surface. Nevertheless, I think this is clearly a relevant issue that should be resolvable with better molecular dynamics with realistic potentials, which I am hopeful will be feasible in the near future.

Dr Mackenzie remarked: At the risk of making challenging experiments impossible, I wonder whether you think a similar approach might be used to study that most "elementary" of negative charges, the hydrated (or solvated) electron, at liquid interfaces. There is, of course, significant evidence for surface-bound electrons on clusters,[1,2] thin ices[3] and liquid jets.[4]

1 N. I. Hammer, J. W. Shin, J. M. Headrick, E. G. Diken, J. R. Roscoli, G. H. Weddle and M. A. Johnson, *Science*, 2004, **306**, 675.
2 J. R. R. Verlet, A. E. Bragg, A. Kammrath, O. Cheshnovsky and D. M. Neumark, *Science*, 2005, **307**, 93.
3 C. Gahl, U. Bovensiepen, C. Frischkorn, and M. Wolf, *Phys. Rev. Lett.*, 2002, **89**, 107402.
4 K. R. Siefermann, Y. Liu, E. Lugovoy, O. Link, M. Faubel, U. Buck, B. Winter, and B. Abel, *Nat. Chem.*, 2010, **2**, 274.

Professor Nesbitt answered: I believe Robert Browning wrote, "Ah, but a man's reach should exceed his grasp? – or what's a heaven for?" I welcome the friendly suggestion (indeed, even from a former postdoc of mine) that our current fledgling scope of gas–liquid scattering experiments might be lacking in sufficient challenge! The idea of looking at solvated electrons is an excellent one, one of many fascinating directions into which we are hoping to develop. We are, in particular, inspired by the superb molecular beam studies of Nathansan, Minton and coworkers, that used Na atom projectiles to release free electrons into, and monitor subsequent chemical reaction products emerging from, the gas–glycerol interface.[1] Other potential schemes would involve use surfactants such as *t*-butyl ammonium iodide, the anion of which can be photodetached to yield electrons exclusively localized in the interfacial surface layer. But then again, our reach may continue to exceed our grasp until a certain Oxford professor decides to come back for a second postdoctoral experience at JILA!

1 W. A. Alexander, J. P. Wiens, T. K. Minton, and G. M. Nathanson, *Science*, 2012, **335**, 1072.

Professor Neumark said: I have two questions. First, is there any theoretical work concerning the relative abundance of chloride and iodide at the glycerol interface? Also, in halide–CO_2 clusters, the CO_2 is known to be bent owing to a combination of electrostatic and charge-transfer effects. So I wonder if the collision of a CO_2 molecule with a surface halide could produce bend-excited CO_2. Is there any evidence for this in your experiments?

Professor Nesbitt replied: Those are both great questions. Arguably the best theoretical simulations for alkali halide salts in hydrogen bonded liquids have been performed by Jungwirth and Tobias, who showed quite clearly this trend for more polarizable anions (such as iodide *vs.* chloride) to exhibit excess density in the interface for H_2O and ethylene glycol.[1] Unfortunately, although glycerol is probably quite similar to ethylene glycol in terms of the hydrogen bonding environment, it is also considerably more viscous and thus much more challenging to achieve comparable quality calculational efforts. As for the possible role of a transient CO_2-anion complex in the collision dynamics, this is a very interesting observation and indeed the reason why we chose to use CO_2 as the projectile. From previous studies without salts, we had demonstrated that collisional vibrational excitation of the CO_2 bend at 667 cm^{-1} to be readily measurable but basically quite inefficient.[2-4] Our interpretation has been that this was the result of vibrationally adiabatic dynamics, due to a mismatch in the fast time scales for CO_2 bending (50 fs) *vs.* the much slower time scales (1–10 ps) for collisions with the surface. The thought in the present studies was that partial electron donation from the anion into the C atom and subsequent bending of the OCO framework would be a more "sudden" collision event and provide a potentially more efficient mechanism for exciting the degenerate bend vibration. However, despite the clear influence of dissolved alkali halide salt on the rotational distributions, we see essentially no difference in the vibrational bending states populated. This could imply still adiabatic collision behavior with respect to vibrational time scales. However, detailed trajectory simulations with realistic *ab initio* potentials would clearly be quite useful in correctly interpreting the role of such CO_2-ion complex dynamics in the scattering process.

1 P. Jungwirth and D. J. Tobias, *Chem. Rev.*, 2006, **106**, 1259.
2 J. J. Nogueira, S. A. Vazquez, O. A. Mazyar, W. L. Hase, B. G. Perkins, D. J. Nesbitt, and E. Martinez-Nunez, *J. Phys. Chem. A*, 2009, **113**, 3850.
3 B. G. Perkins and D. J. Nesbitt, *J. Phys. Chem. B*, 2006, 110, 17126.
4 B. G. Perkins and D. J. Nesbitt, *J. Phys. Chem. A*, 2009, 113, 4613.

Professor Tully queried: David (Professor Nesbitt), have you studied the dependence on the salt concentration? Specifically, because negative ions repel each other, the concentration of anions at the surface might saturate. Do you see evidence for this saturation?

Professor Nesbitt said: That is a very interesting question. The simple answer is that we have not seen any saturation in terms of the rotational distributions as a function of anion concentration. Our initial expectations were that we might indeed see such a concentration dependence but for a different reason. For a single anion on the surface, the E field experienced by an incoming projectile increases roughly as $1/r^2$. For many ions at the interface, the E field will approach a constant value far from the surface, much like for a uniform plate of charge. (This will be modified by the counter-cations). Our thoughts were that the torque exerted on molecules might therefore decrease to zero in the limit of both low and high surface charge densities. The fact that we see a steady increase with no such saturation could mean that we are still in the low surface charge limit, but clearly more work needs to be done to ensure the correct interpretation.

Dr Glowacki noted: In your conclusion, you state that cations are less inclined to cluster at the surface than anions. I'm wondering whether you think this is specific to glycerol and closely related liquids, or if it can be generalised to a range of common liquids? Can one describe the relative propensity for anions (and not cations) to cluster at the surface using free energy concepts?

Professor Nesbitt responded: This is still a topic of some controversy, which is why a completely independent probe of such properties using quantum state resolved gas–liquid scattering may be of some value. Most certainly this is an equilibrium system, with any such propensities for anion *vs.* cation location presumably quite well described with free energy concepts. The best quality simulations for NaI solutions in water and ethylene glycol do reveal this propensity for peaking of the polarizable I^- in the top interfacial layer, with the Na^+ cation peaking close behind further into the liquid.[1] This preference for anions at the surface is sensitive to the inclusion of polarizability in the model. The qualitative interpretation is that large polarizable anions (like I^-) near the interface can minimize free energy by moving into the Gibbs dividing region. This permits the interfacial solvent molecules to get closer and establish more hydrogen bonds, while still allowing the polarizability of the anion to interact favorably with the surface solvent layer. I therefore would expect this to be generalised to strongly hydrogen bonded systems, though such high level simulations are not yet available for these more viscous liquids, such as glycerol.

1. P. Jungwirth and D. J. Tobias, *Chem. Rev.*, 2006, **106**, 1259.

Dr Jordan asked: The presence of ions in the surface layer must be governed by free energy considerations. Is the key quantity determining the location of the ion the difference between the free energy of the ion–water interaction and the free energy of the water–water interaction?

I think I can rationalize the enthalpy of solvation – surface water molecules have fewer hydrogen bonds so that water molecules in the bulk will have a lower enthalpy. If the ion–water interaction is less favourable than the water–water interaction, then the ion will be enthalpically driven to the surface. Is this interpretation correct? How is the entropy term effected by the ion?

Professor Nesbitt amswered: It is definitely the case that differences in free energies must govern any preference for cation *vs.* anion density at or near the interfacial surface. Furthermore, Professor Jordan's intuition is certainly right that there is a stronger enthalpic interaction between a pair of hydrogen bonded solvent molecules than between a large, polarizable anion and a single solvent molecule. Indeed, I tend to invoke a similar physical picture of these interactions, *i.e.*, for a sufficiently large, polarizable anion (*e.g.* I⁻) to be solvated requires breaking up the equivalent of all the missing hydrogen bonds that have been displaced. This increase in enthalpy is of course partially compensated by anion–dipole attractions with the local solvent dipoles, but as a result of electron cloud overlap, at a larger distance and therefore not so effectively for the larger anions. Conversely, anions sitting in the interfacial layer avoid the enthalpic penalty of disrupting hydrogen bonds, but at the same time profit from the dipole–induced dipole attractions between the dipolar solvent and an easily polarizable anion species. But despite the appeal of a simple enthalpic physical picture, the free energy effects that dictate cation *vs.* anion spatially dependent densities must be substantially more subtle. For example, if purely additive enthalpies were the only consideration, the above argument, in a *reductio ad absurdam* limit, would predict that all anions above some critical polarizability would "float" to the surface rather than solvate into the bulk, which is clearly not experimentally the case. Indeed, it is only in the last dozen or so Å into the solvent interface that any imbalance of ion densities is even postulated. The free energy changes driving this putative sequestration near the interface are clearly small, and thus, as Professor Jordan correctly notes, must involve some delicate balance between enthalpic and entropic contributions, both additive and non-additive. The chemical physics community may perhaps be forgiven for a lack of quantitative intuition, since Onsager's first predictions were that both cation and anions would be excluded from the interfacial region.[1] As further indication of the degree of subtlety involved, molecular dynamics simulations that assume pure electrostatics qualitatively agree with these Onsager predictions. It is only when polarizability is incorporated into the model that these charge dependent sequestration effects in the interfacial region are seen to arise.[2]

1 L. Onsager and N. N. T. Samaras, *J. Chem. Phys.*, 1934, **2**, 538.
2 P. Jungwirth and D. J. Tobias, *Chem. Rev.*, 2006, **106**, 1259.

Professor Wodtke addressed Professor Nesbitt and Professor Neumark :I would like to follow up on the question raised by Professor Neumark concerning the observation of CO_2 bending excitation in collisions with chloride at the interface of RTILS. As Professor Neumark pointed out, the chloride anion may introduce a bending force to the CO_2 molecule in a gas phase collision; hence vibrational excitation might result from Translation–Vibration (T–V) coupling if the CO_2 molecule were to interact with the interfacial chloride at the RTIL interface. I would however like to comment that the present experiment appears to me to be in a rather bad position to observe this and Professor Nesbitt has already alluded to some of the difficulties. Let me reiterate his brief allusion in more detail. Since one needs translational energy for T–V coupling, one needs to heat the nozzle from which the molecular beam is formed, even if this might only mean "heating to room temperature". For CO_2 bending excitation, even room temperature samples have substantial $v = 1$ population that does not fully relax in the expansion. This population undergoes full rotational redistribution in the surface scattering producing a background in any experiment designed to detect population in $v = 1$ that is not due to vibrational excitation in the molecule–surface collision. So, one is left in the unenviable position of wanting to produce beams from low nozzle stagnation temperatures with high kinetic energies. We have also encountered similar problems in many experiments and in general this can make observation of vibrational excitation challenging. I would like to point out that one can reveal very similar physics in a somewhat different experiment using IR overtone excitation. Using suitable laser sources (for example pulse amplified ring dye lasers with follow-on IR non-linear optical conversion) one can pump overtones in many molecules. For example, we now routinely pump NO to $v = 3$ (i.e. $v = 0–3$ excitation) for beam surface scattering studies. In these experiments one may then look at the vibrational relaxation of $v = 3–2$ or 1 in addition to vibrational excitation to $v = 4$. In such an experimental configuration one has much more freedom to heat the nozzle to obtain high kinetic energies necessary for the T–V conversion. More importantly, one may directly observe the V–T relaxation and by adjusting delays between pump and probe, obtain the translational energy before and after the collision.[1] Certainly similar laser systems could be used to excite CO_2 via its overtones to a variety of states. It would be extremely exciting to combine the exquisite IR absorption methods of Professor Nesbitt's laboratory with these double resonance ideas exploiting overtone excitation.

1 For examples, see: I. Rahinov, R. Cooper, C. Yuan, X. Yang, D. J. Auerbach and A. M. Wodtke, *J. Chem. Phys.*, 2008, **129**, 214708; R. Cooper, I. Rahinov, C. Yuan, X. Yang, D. J. Auerbach, and A. M. Wodtke, *J. Vac. Sci. Technol, A*, 2009, **27**, 907.

Professor Nesbitt responded: Professor Wodtke makes several excellent points and suggestions, which may be further clarified with the following additional information. The supersonic jets we use in the gas- liquid scattering studies are indeed axisymmetric, with the usual $1/r^2$ loss in density with distance downstream. This is quite intentional, as this permits the densities to be low enough and mean free paths large enough at the gas–liquid interface to facilitate collision free detection of the final scattered quantum states. This rapid drop off in number of cooling collisions of course inhibits the relaxation of the bending mode at 667 cm^{-1}. The incident vibrational distribution is characterized by a 0.63% fractional population in each of the lambda doubled bending states, or if you like, a vibrational "temperature" of 190 K. It is against this backdrop of incident beam population that T–V vibrational excitation or relaxation efficiency would need to be measured. Despite such backdrop, we nevertheless do see modest albeit clear evidence for a vibrational warming of the CO_2 on scattering at the gas–liquid interface, with population increasing from 0.63% up to 1.5% and translating into a positively "balmy" 230 K.[1,2] It is interesting that this T–V excitation is dramatically less efficient than T–R excitation, which achieves temperatures in the order of 800–1000 K, and does not even correspond to warming up to the temperature of the liquid itself. Thus, although we clearly do have the sufficient sensitivity and dynamic range to see such changes in vibrational populations, it is clear that for CO_2, the bending mode clearly behaves more nearly like a "spectator" in the dynamics, at least for ionic salt conditions and collision energies explored to date. It is worth noting, however, that we do have evidence in another experimental apparatus with higher sensitivity LIF/REMPI detection for substantially more T–V excitation dynamics.[3,4] Specifically, in even moderate energy (8 kcal mol^{-1}) collisions of benzene with Au(111) surfaces, we see a number of the low frequency (400–900 cm^{-1}) vibrational modes "light up" in the scattered flux, corresponding to over an order of magnitude increase in population from that of the incident jet cooled beam. However, for the correspondingly higher vibrational modes (2000–3000 cm^{-1}) in diatomic species such as NO and HCl, we see essentially no vibrational excitation, even for collision energies up to 20 kcal mol^{-1}. The simple adiabatic ideas of collisional interaction time scales needing to be on the order of or faster than the vibrational period seem to be holding up nicely. The proposed experiments by Professor Wodtke are very elegant indeed – and clearly offer yet another direction to be taken with such gas–liquid scattering studies. The best system accessible to the laser capabilities we currently have would probably be $2 \leftarrow 0$ first overtone vibrational excitation of incident jet cooled HCl, monitoring the HCl ($v = 1$, J) after the scattering event with 2+1 REMPI followed by velocity map imaging (VMI).[3] The advantage here is that there are much fewer initial quantum states to pump, fewer final states to probe, and the data yield information on correlated internal state and velocity vector degrees of freedom. As a final comment, it is also interesting to point out that slit supersonic expansions used extensively in other experiments in the group have a much slower 1/r drop off in density.[5] This yields a much greater number of cooling collisions, which quite efficiently relax low frequency vibrations for molecules such as CO_2. Indeed, as a function of probe distance downstream of the slit jet orifice, we find the temperatures associated with each of the rotational, (bending) vibrational and transverse translational degrees of freedom in CO_2 to cool down uniformly and indeed remain in equilibrium.

1 B. G. Perkins, T. Haber, and D. J. Nesbitt, *J. Phys. Chem. B*, 2005, **109**, 16396.
2 B. G. Perkins and D. J. Nesbitt, *J. Phys. Chem. B*, 2006, **110**, 17126.
3 J. R. Roscioli and D. J. Nesbitt, *Faraday Disc.*, 2011, **150**, 471.
4 M. P. Ziemkiewicz, J. R. Roscioli, and D. J. Nesbitt, *J. Chem. Phys.*, 2011, **134**, 234703.
5 C. M. Lovejoy, M. D. Schuder, and D. J. Nesbitt, *J. Chem. Phys.* 1987, **86**, 5337.

Professor Orr-Ewing commented: In Professor Beck's experiments on scattering from metal surfaces, it is clear where his surface is. In your experiments on scattering from liquid surfaces, the interface is more diffuse and the surface is therefore not as sharply defined. What is the length scale at the interface for evolution from vapour above the liquid to bulk liquid? Which parts of this interface are your scattered CO_2 molecules sampling, and might the cations or anions ions modify the local structure of the solvent molecules in this region?

Professor Nesbitt communicated in reply: For liquids, the analogous quantity is the Gibbs dividing surface. Due to capillary wave motion in a liquid, there is time dependence to the interfacial height, as well as structure to the molecular subunits themselves, both of which lead to a slower roll off in mean density with distance away from the interface than for a crystalline solid. Some useful guidance to these length scales in complex liquids like water and ethylene glycol come from large scale molecular trajectory calculations, such as by Jungwirth and Tobias, which suggest a roll off in density from 90% to 10% over a length scale of $\sim$3–4 Å that a molecular projectile such as CO_2 might be predominantly sampling.[1] The dividing surface in glycerol is probably similar in length scale, though such liquids are much more viscous and therefore still challenging to simulate with molecular dynamics. Your question concerning the locations of anions and cations in this roll off region is particularly astute. In the water and ethylene glycol simulations with NaI salts, there is a strong peak in I^- density predicted to be centered directly around this Gibbs dividing surface region with a similar width. There is also a corresponding excess density peak for Na+ another $\sim$5 Å further into the liquid, reminiscent of a local first and second layer structure for ion sites in the liquid interfacial region.

1 P. Jungwirth and D. J. Tobias, *Chem. Rev.*, 2006, **106**, 1259.

Dr Chandler asked: David (Professor Nesbitt), can you discuss the possibility of scattering dipolar molecules that might be aligned as they approach the charged surface? Could scattering of molecules such as HCl from charged liquids offer a way of seeing the surface charge?

Professor Nesbitt answered: This is an excellent suggestion, one toward which we have been working with DCl molecules more accessible to diode laser detection. There are several reasons why we feel dipolar species such as DCl might in principle make better projectiles for sensing the surface charge. The first is, as you suggest, that the charge–dipole interactions are more long range and stronger. The second is that DCl has a center of mass close to the Cl atom, which would suggest much larger rotational excitation efficiency for collisions pointing "D atom down" (into an anion) *vs.* "Cl atom down" (into a cation). However, there are several potential countervailing influences as well. One is that the B constant of DCl is more than 10 fold larger than CO_2, which makes it harder to mix free space rotational wavefunctions into an oriented pendular state. Another is that the length scale for achieving such a molecular orientation from charge–dipole *vs.* charge–quadrupole forces is larger, potentially resulting in an even more rotationally adiabatic behavior. We now have preliminary data for DCl + glycerol scattering, and are quite optimistic that these can be extended to ionic salt solutions of glycerol.

Professor Wodtke said: I would like to comment on the fascinating challenges presented by studies of molecular interactions at liquid surfaces. As we all know, this has become an extremely active field of research and it clearly stimulates the imagination to use molecular beam scattering methods to probe the atomic scale dynamics of gas liquid interfaces. Of course, in our field it is extremely important to carefully consider the proper choice of model system, not only one that is reproducible but – eventually at least – also one that can be compared to atomic scale and hopefully first principles theories. Beam–liquid surface scattering is particularly challenging in this regard for a variety of reasons, most especially due to the difficulty in accounting for the atomic scale structure of the sample, which may not even be static on the time scale of the experiment. We see in this paper, for example, that it is a difficult question to answer as to whether in RTILs the chloride anions are available at the interface for CO_2 to collide with. Certainly there are approaches to answering such questions and I don't want to suggest that RTILs are not an interesting model system for molecular interactions at interfaces. I would, however, like to point out that there is another set of model systems that we have recently been exploring that might be a promising approach to this challenge. It is well known that many metals exhibit surface pre-melting.[1] Here Pb(110) surface pre-melting takes place more than 50 K below the bulk melting temperature. Indeed, we found that even Au(111) exhibits this phenomenon[2] – and in the course of extending some of our work to higher surface temperatures, we observed scattering signals from premelted liquid surfaces of Au. This work is still in its infancy but appears to provide a variety of practical advantages that aid the design of well controlled experiments probing molecular interactions at liquid interfaces that could be compared to theory.

1 See, for example, J. W. M. Frenken and J. F. van der Veen, *Phys. Rev. Lett.*, 1985, **54**, 134.
2 See, A. Hoss, U. Romann, M. Nold, P. v Blanckenhagen and O. Meyer, *Europhys. Lett.*, 1992, **20**, 125.

Professor Nesbitt remarked: First of all, I would like to clarify that these solutions are not room temperature ionic liquids (RTILs), but rather moderate (0–5 M) concentrations of ionic salts dissolved in a hydrogen bonding liquid such as glycerol. RTILs, which are purely ionic salts with low melting points, represent yet another fascinating class of liquids, interesting in their own right, and which we have also studied with CO_2 and NO supersonic beam–liquid interfacial scattering.[1-3] Nevertheless, Professor Wodtke's point is completely valid that these gas–liquid interfaces are complex dynamical objects, with microscopic surface roughening, capillary wave excitations and heterogeneity in surface composition, and dynamics of surface motion occurring even on the time scale of the collisional event. Though our group's work has historically been on much simpler and more pristine (often gas phase) molecular systems, I personally find this complexity of the gas–liquid interface rather exciting, offering some fundamentally new and important challenges to the physical chemistry community. These are early efforts and there is clearly much to be done, in particular combining the tools of both theory and experiment. The suggestion of looking at surface premelting dynamics by gas–surface scattering techniques is a fascinating one, one of the many which we have to date not been able to pursue. We have looked at scattering of NO from molten metals such as liquid Ga, and have wanted very much to compare solid *vs.* liquid scattering dynamics for the same material.[4,5] One problem for this system is that Ga supercools indefinitely, which makes it inconvenient to move reversibly in finite time above and below the melting point. Professor Wodtke's idea of using a surface liquid layer on an ordered metallic surface offers an excellent alternative strategy.

1 M. P. Ziemkiewicz, A. Zutz, and D. J. Nesbitt, *J. Phys. Chem. C*, 2012, DOI: 10.1021/jp212336a.
2 J. R. Roscioli and D. J. Nesbitt, *J. Phys. Chem. Lett.*, 2010, **1**, 674.
3 J. R. Roscioli and D. J. Nesbitt, *J. Phys. Chem. A*, 2011, **115**, 9764.
4 M. P. Ziemkiewicz, J. R. Roscioli, and D. J. Nesbitt, *J. Chem. Phys.*, 2011, 134.
5 L. F. Phillips and D. J. Nesbitt, *Chem. Phys. Lett.*, 2012, **536**, 61.

Professor Gruebele said: This technique is a very promising twist to probe ions near liquid surfaces by quadrupole–charge interaction. The quadrupolar interaction range is shorter than dipolar or Coulomb, so higher resolution can be obtained from the scattering results.

A system that may turn out to be very fruitful would be composite membranes made from 2 or more lipids, including such as POPG, POPC (for charge control) and cholesterol (to produce rafts). These systems can be applied to a glass slide, have low vapour pressure, controllable charge in the head group, and undergo sharp phase transitions from liquid to liquid crystal phases that could have dramatic effects on scattering. There are also many outstanding biophysical question about membrane dynamics as a function of membrane composition. It might even be possible to maintain a thick water film on the membrane that will not evaporate away, but even without, such studies could be very illuminating.

Professor Nesbitt responded: That is a wonderful suggestion, which we will certainly keep in mind! As a potentially simpler first step in this direction, we have been looking into using fast liquid crystal switching technologies locally developed at the University of Colorado Boulder. The idea would be to use low voltage E fields to orient long phospolipid-like molecules at the gas–liquid interface, and thereby probe the surface anisotropy by molecular beam scattering techniques.

Professor Gruebele opened the discussion of the paper by Professor Sibener by asking: One useful way to model the imaging data of polymer dislocation pair annihilation would be an anisotropic Smoluchowski equation with a creation (by thermal fluctuation) and annihilation term. The dislocations have relative diffusion coefficients that differ parallel and perpendicular to the dislocation axis on the polymer surface, with the perpendicular diffusion coefficient being smaller. Likewise, the potential of mean force (or free energy surface) in the Smoluchowski treatment could be anisotropic. You indicated that the potential is net attractive, so adjacent dislocations have negligible probability of passing one another without annihilation. Are there enough observations of isolated pair annihilation events in your SPM time lapse movies that these could be used to fit an anisotropic Smoluchowski model? This could provide further insight into the nature of the short range attractive effective force between dislocations. It may even turn out that low temperature experiments, or those with different charge polymers, would reveal a longer range repulsive barrier.

Professor Sibener answered: We agree, and in many aspects are working towards the goal of observing a sufficient number of events with high structural fidelity to extract further details of defect mobility, pair-annihilation, and the energetics for such processes. Our use of lithographically-generated channels is proving to be an excellent platform for observing the relative motion of an isolated pair of interacting defects in an otherwise nearly perfect structural environment. Energetic parameters are currently being extracted for polymeric motion along, and perpendicular to, the locally aligned microdomains of the diblock copolymer systems we are studying.

Professor Wodtke addressed Professor Sibener: There is a long history of outstanding work using He scattering in the single phonon excitation regime, where surface temperatures are kept low, beam energies are kept low to obtain longer de Broglie Wavelengths *etc*. As you of all people would know, such studies yield exquisite information on the phonon structure of solids. In the experiments you presented, presumably due to the nature of the sample most data was obtained in the multi-phonon excitation regime. Could you briefly discuss the state of the art in exploiting such multi-phonon regime data to potentially obtain similar levels of experimental detail as that to which one has grown accustomed for single phonon regime data?

Professor Sibener responded: As you properly note, single phonon inelastic scattering based on atomic beam methods has proven a powerful tool for examining the dynamics of clean, stepped, alloyed, and molecular surfaces including important extensions to molecular thin films. These advances have occurred due to the co-development of exquisitely sensitive experimental instrumentation along with quantitatively accurate analytic theory and numerical simulations. Similar advances have been made in exploring the systematic evolution from rather gentle single phonon scattering interactions to few and ultimately multiphonon processes. We have in particular exploited analytic theories such as noted in Ref. 13 of our paper, as well as molecular dynamics calculations that employ chemically accurate condensed phase and gas-surface potentials, to extract precise dynamical information on the nature of polymer thin films, polymer thins that undergo structural phase transitions, and, most recently, methyl-decorated semiconductor interfaces. Questions that have been successfully addressed, for example, in the polymer thin films include the fascinating topic of how spatial confinement on the nanoscale modifies the surface dynamics of the true surface (low energy atomic beams do not penetrate into the film, in contrast to complementary probes such as X-ray and neutron scattering). The methyl-decorated Si(111) surface offers a superb testbed for such studies due to its near structural perfection, commensurate structure, low-defect density. Moreover, the vibrational density of states and dispersion relations are fully calculable, giving entree to the entire span of single-, few-, and multiple-phonon scattering events and the information thus revealed about the dynamical properties of both the scattering itself as well as the sought after properties of the interface. In summary, the field is increasingly well-equipped to undertake studies of more complex interfaces of both crystalline and amorphous character. It has clearly developed to the point where extant analytic theory and numerical methods can model quantitatively the gas-surface encounter, and, in so doing, allows such atomic beam methods to fully join its more mature xray and neutron counterparts as a unique probe of interfacial, thin film, and condensed matter systems.

Professor Nesbitt opened the discussion of the paper by Professor Tully by saying: These are clearly landmark efforts. I am wondering whether you still have enough "headroom" in these calculations to also include spin orbit inelastic collision dynamics between the $NO(^2\Pi_{1/2})$ and $NO(^2\Pi_{3/2})$ rovibrational state manifolds. You say you need the equivalent of 10^{11} surface channels involved in order to treat the electron–hole pair continuum adequately – how much harder does adding this extra two-fold increase in spin orbit dimension make these calculations? We would be very interested in such capabilities for comparing with recent results on scattering NO from molten metals such as liquid gallium, where the effective electronic spin orbit "temperature" of the scattered NO have been observed to vary with incident energy between slightly cooler and much hotter than the surface.

Professor Tully replied: The calculations reported here assume a single manifold of NO neutral states and therefore cannot shed light on spin-orbit populations. Adding another manifold of states will complicate the dynamics simulations, but probably not prohibitively. The bigger obstacle is calculating the required NO–surface interaction potentials. We are exploring a method based on "constrained density functional theory" that might be capable of calculating the molecule–surface interaction potentials separately for the $NO(^2\Pi_{1/2})$ and $NO(^2\Pi_{3/2})$ states. If this effort is successful, we can examine spin–orbit effects in the NO/Au(111) system. We should then be able to extend this to NO on molten gold and later to other molten metals. However, we hope that by studying spin-orbit and Λ-doubling transitions in NO/Au(111) we may be able to uncover the underlying factors that control these processes without needing to repeat this type of calculation for many different systems.

Professor Zare questioned: Am I correct in suggesting that the calculation of the excited-state potentials is the most challenging aspect of your procedure? In that regard, might it be necessary to include the effects of Λ-doubling as well as spin-orbit interaction in considering the scattering of NO from a metal surface?

Professor Tully answered: Professor Zare is definitely correct; the most difficult and most critical aspect of simulating chemical dynamics at metal surfaces is obtaining the required ground and excited potential energy surfaces and the nonadiabatic couplings among them. There do not currently exist any practical *ab initio* procedures that properly describe the coupling of localized excited states to the metallic continuum and are of sufficient quantitative accuracy. Progress in this area is a high priority. In the specific case we have examined, NO scattering from an Au(111) surface, we invoked a very severe approximation: we assumed that there were only two continuous manifolds of diabatic potential energy surfaces, one describing the interaction of neutral NO with the surface and the other of the NO^- anion with the surface. These two sets of parallel potential energy surfaces and their off-diagonal couplings were extracted from density functional calculations, carried out with and without application of an external electric field. However, the NO ground state is a doublet Π state that, except when oriented perpendicularly to the surface, splits into two states of different energy. This splitting critically effects both Λ-doubling and spin-orbit transitions. Our calculations assumed only a single neutral state and clearly cannot describe these effects. It will be very instructive both to extend the calculations to include these effects as well as to measure them in the laboratory.

Professor Truhlar asked: It has been argued by several groups, including Thachuk *et al.*,[1] Granucci and Persico,[2] and Jasper and myself,[3] that the trajectory surface hopping method is improved by including decoherence between regions of strong interaction. This can be especially important when there are a large number of hops, as in your problem. Can you comment on whether your method would be improved by adding decoherence or switching to a formalism with decay of mixing,[4] which also corrects for this defect of surface hopping?

1 M. Thachuk, M. Y. Ivanov, and D. M. Wardlaw, *J. Chem. Phys.*, 1998, **109**, 5747.
2 G. Granucci and M. Persico, *J. Chem. Phys.*, 2007, **126**, 134114.
3 A. W. Jasper and D. G. Truhlar, *J. Chem. Phys.*, 2007, **127**, 194306.
4 C. Zhu, S. Nangia, A. W. Jasper, and D. G. Truhlar, *J. Chem. Phys.*, 2004, **121**, 7658.

Professor Tully answered: First, treatment of decoherence is not a defect of the surface hopping method, it is an integral part of it. The original 1990 "fewest switches" paper was formulated in density matrix notation with an explicit decoherence factor included. Having said this, there is no consensus on the optimal way to include decoherence, even for small-molecule nonadiabatic dynamics. It will require further study to evaluate the importance of decoherence in simulations of dynamics at metal surfaces, a different regime where electronic states are infinitesimally spaced and nonadiabatic couplings are delocalized.

Professor Martinez remarked: You mention that many surface hops (often more than 50) occur in the IESH simulations. Previously, *e.g.* in your 1990 J. Chem. Phys. paper[1] introducing fewest switches surface hopping, you have stressed that surface hopping becomes similar to an Ehrenfest approach when the trajectories hop between surfaces frequently. Is it correct to conclude that the IESH method is effectively an Ehrenfest dynamics treatment of the coupled electronic states? Would it not be computationally convenient to solve the Ehrenfest dynamics equations instead of surface hopping? What impact does this have on the dynamics, especially as concerns energy surfaces which are highly dissimilar (where surface hopping ameliorates the failings of Ehrenfest approaches, *e.g.* population of energetically forbidden channels)?

1 J. C. Tully, *J. Chem. Phys.*, 1990, **93**, 1061–1071.

Professor Tully said: The relationship between the self-consistent field Ehrenfest method and surface hopping, in cases involving a continuum of electronic states and multiple surface hops, is interesting and not well understood. In the case of a closed shell molecule approaching a metal surface, where the only excited states correspond to electron-hole pair excitations, it may be that Ehrenfest and surface hopping give similar results. Indeed, the weak coupling limit of Ehrenfest, the Electronic Friction model, appears to describe translational and vibrational inelasticity due to conduction electron excitations with quite reasonable accuracy. The surface hopping approach we have implemented for vibrationally inelastic scattering of NO from Au(111) gives almost identical results as Electronic Friction, using the same Hamiltonian, for the probability of de-excitation of $NO(v = 1)$ to $NO(v = 0)$. In this case the NO^- negative ion state is not accessed significantly. However, for de-excitation of highly vibrationally excited NO scattered from Au(111), where the negative ion resonance plays a dominant role, Electronic Friction is inadequate. It is critical in this case for motion to evolve on potential energy surfaces of a specific character, notwithstanding the large density of electronic states of different character. In fact, with Ehrenfest the scattered NO molecule would be partly neutral and partly ionic, a deficiency that is cured by surface hopping.

Professor Polanyi asked: Would Professor Tully care to comment on the degree to which his IESH theory has been subjected to experimental test? This many-electron many-state model has the great virtue of accounting for the observed large transfer of vibration to and from NO in collisions with gold. How well does it do (for example) in describing the translational-energy dependence of that transfer?

Professor Tully responded: The only experimental tests to date of the IESH theory relate to the experimental findings of Wodtke and coworkers on vibrationally inelastic scattering of NO from Au(111). Vibrational energy loss of vibrationally excited NO molecules as well as excitation of $v = 0$ molecules are in quite good agreement with experiments. Trapping probabilities and rotational energy distributions are also in good accord, as are many qualitative trends. As just mentioned by Professor Wodtke in this Faraday Discussion, however, translational energy distributions are not in good agreement. Further tests on different systems are certainly needed. Moreover, it should be noted that the calculated results depend on the accuracy of the interaction potentials and couplings employed, so agreement or disagreement does not necessarily reflect the accuracy of the surface hopping dynamics.

Professor Wodtke commented: Professor Tully presented his recent work on the IESH theory as applied to the electronically nonadiabatic interactions of NO with Au(111). I would like to comment that this theory has gone a long way to settling the essential mechanism of nonadiabatic interaction in this system, namely that it is indeed induced by an electron transfer event (or electron transfer events). Despite all of the reasonable criticisms that have been pointed out by many of the theorists in the audience, quantitative comparison to experiment shows that, while improvements to the model are still important to pursue, many of its assumptions must have some significant validity. I would particular point out a recent paper of ours that we had the pleasure to publish together with Professor Tully[1] that shows comparison of the IESH predictions to a benchmark set of experimental data concerning vibrational excitation of NO(v=0–1, 2). This comparison is over a wide range of surface temperatures and incidence energies. Agreement is not perfect, but it is remarkably good, and moreover electronic friction theory fails. It is especially important to note that the IESH was developed to account for vibrational relaxation of NO(v=16) in collisions with Au(111). We applied it (perhaps naively) to the vibrational excitation of NO(v=0–1,2) and without adjustment of the theory obtained good quantitative agreement with experiment. Lest one incorrectly conclude that all is settled, I would also like to point out that we have recently been studying the translational inelasticity of collisions of NO on Au(111) and comparisons to the IESH theory are also possible here. The methods used for these measurements are described in the literature.[2,3] Briefly, IR overtone excitation is used to tag a single state of NO and some 2 cm away REMPI is used on the population produced in the upper state of the overtone transition to record TOF data. This allows us to see, for example, to what extent NO translation is a spectator in the electronically nonadiabatic vibrational energy exchange at the Au surface. It is almost a perfect spectator but not completely! We may also compare to the predictions of IESH. In general we find that the potential surface is too soft. That is, less translational inelasticity is seen in experiment that in theory. This is particularly troubling as IESH apparently yields good agreement with measured trapping probabilities as a function of incidence energy of translation and vibration. We are unsure how the model can give accurate trapping probabilities with inaccurate translational inelasticity. Clearly further work is needed to clarify this problem and potentially improve the theory.

1 R. Cooper, C. Bartels, A. Kandratsenka, I. Rahinov, N. Shenvi, K. Golibrzuch, Z. Li, D. J. Auerbach, J. C. Tully, A. M. Wodtke, *Angew. Chemie.*, 2012, **124**, 5038.
2 I. Rahinov, R. Cooper, C. Yuan, X. Yang, D. J. Auerbach, A. M. Wodtke, *J. Chem. Phys.*, 2008, **129**, 214708.
3 R. Cooper, I. Rahinov, C. Yuan, X. Yang, D. J. Auerbach and A. M. Wodtke *J. Vac. Sci. Technol. A*, 2009, **27**, 907.

Professor Tully replied: The fact that our IESH simulations predict too much translational inelasticity and yet satisfactorily reproduce experimental trapping probabilities is indeed troubling, as noted by Professor Wodtke, since trapping and loss of translational energy should go hand in hand. Experience suggests that translational energy transfer and trapping do not depend strongly on electron–hole pair excitations. Rather, they are dominated by energy transfer to phonons which, in turn, is controlled largely by the surface phonon spectrum, the steepness of the molecule–surface interaction potential, the extent of corrugation of the gas–surface potential, and the adsorbate binding energy. In particular, an accurate trapping probability might occur fortuitously if an overestimated translational inelasticity was compensated by an underestimated positionally averaged binding energy.

Dr Glowacki interjected: Professor Tully, I have two questions:
(1) How many different states did you actually have to fit from the DFT calculations before running the surface hopping simulations? This strikes me as potentially

an extremely expensive procedure if it is necessary to fit lots of surfaces. Is it possible to define any sort of convergence criteria - *i.e.*, how many surfaces do you need before your results stabilize?

(2) You say in the paper that your fits are a function of the coordinates of NO and all 396 gold atoms in the system, which is on the order of ~1200 degrees of freedom. I'm curious as to the functional form you actually used to fit over that many degrees of freedom.

Professor Tully responded: (1) Our procedure for NO/Au(111) required fitting only the three independent elements of a symmetric 2 × 2 matrix, as a function of all relevant nuclear positions. This simplification results from assuming that all members of the continuum of potential energy surfaces that describe a neutral NO molecule interacting with the surface, with various electron–hole pair occupations, are parallel, reflecting the assumption that a delocalized electron–hole pair excitation does not significantly alter the local forces on the adsorbate molecule. The same approximation was invoked for the NO^- anion–surface interaction. (2) The interactions between two gold atoms are taken to be pairwise additive, with parameters chosen to reproduce the bulk phonon spectrum. The N–Au and O–Au interactions are different for the neutral, negative ion and coupling terms but include a pairwise sum over gold atoms. The analytical forms used are standard, except that they include a molecular orientation dependence and, for the ionic state, an image potential term.

Professor Martinez asked: You include dissipation appropriate to the nuclear and electronic degrees of freedom in the IESH method using thermostats and a Monte Carlo occupation switching procedure, respectively. Yet the Ehrenfest-like Schrodinger equation that is solved along the trajectories in order to determine the hopping probabilities corresponds to pure state evolution at absolute zero (*i.e.*, it is a Schrodinger equation and not a Liouville–Von Neumann type equation applicable to mixed states). Can you comment on potential inconsistencies and limiting behaviors of such a mixed dissipative/non-dissipative treatment?

Professor Tully said: Tough question. The surface hopping calculations we report here were not carried out at zero temperature. Initial electronic populations were selected from a Fermi distribution at the experimental temperature. Furthermore, an electronic thermostat was introduced to account for electronic thermalization. Sampling many trajectories with different initial quantum states and classical variables introduces additional averaging. Including decoherence factors may also help. Nevertheless, inconsistencies are inevitable in any mixed quantum-classical scheme, and their consequences in complex, multidimensional systems are not understood.

Professor Neumark opened the discussion of the paper by Professor Polanyi by asking: In the gas phase, low energy electron collisions with phenyl halides result in dissociative attachment to form phenyl plus the halide anion. Is it correct to assume that this is what happens initially in your experiments on metal surfaces (*i.e.* on the repulsive pes*), and that the ground state for dissociated products corresponds to charge transfer from the halide to the metal surface? If this is indeed the case, what are the underlying dynamics that determine when charge transfer occurs?

Professor Polanyi responded: Professor Neumark draws a parallel between our surface-studies of electron-induced reaction, and the gas-phase work that paved the way for such experiments. In both phases the first step is to attach a low-energy electron to a molecule. For gaseous phenyl halides this gives rise first to a gaseous anionic molecule, and then, rapidly, to dissociation into a gaseous phenyl radical and halide anion. The corresponding process at a metal surface is the transformation of an intact physisorbed phenyl halide into a chemisorbed phenyl and a chemisorbed halogen atom. Both reagent and product are visualised by low-temperature STM as to their geometry and their position with respect to the underlying surface metal atoms; this represents quite a lot of detailed information. At some point in this electron-induced dissociation the electron is lost from the adsorbed anionic phenyl halide, and joins the sea of electrons in the metal. Regrettably our experiments do not directly tell us the average time, t^*, for which the anion exists; we have to infer this from a simple model. The model is described in our paper in this Discussion. Briefly it involved a repulsive anionic state, pes*, and an uncharged ground-state pes. The molecular dynamics (MD) was followed on the upper state for a time t^*, after which the system was transferred to the ground state for further MD leading to chemisorbed products. The time t^* was customarily chosen as the minimum time required to energise the organic halide sufficiently that when transferred to the ground pes it reacted .This, we found, led to positions for the chemisorbed products that adequately match our major experimental observations, suggesting that our t^* of some tens of femtoseconds was reasonable. Minor reactive pathways could be modeled by minor variations in t^*. What is missing from this picture is an *a priori* calculation of the coupling between the anionic and ground states, governing the lifetime t^*. Prof, Neumark asks what physics determines "when charge transfer occurs" . We don't have the answer. Some theorist will, we hope, calculate the coupling and hence mean lifetime of the adsorbed ionic state with respect to the ground pes. That would not quite be the end of the calculation, since the value of t^* is substantially longer than the mean lifetime in order to allow time for sufficient energization of the adsorbate to give reaction subsequently on the ground pes. Since our measured yield of product is only one reactive event in 10^{10} electrons, an exponential decay in probability of the anionic state up to a time in the region of t^* can readily be accommodated.

Professor Orr-Ewing said: You have gone to extraordinary lengths to initiate dynamics in single molecules that are immobilized in very well defined geometries in the laboratory, and the paper discusses a model for the recoils exhibited by Cl atoms from dissociation of *p*-dichlorobenzene and I atoms from *p*-diiodobenzene. The measurements of Cl atom recoil (for both Cl atom #1 and #2) show a distribution of radial displacements whereas the model suggests a single valued energy release and hence distance of travel before trapping on a favorable surface site. Is your distribution of Cl atom final distances a consequence of a range of kinetic energies for the nascent Cl atoms, balanced by deposition of energy into the vibrational modes of the polyatomic co-product? This would suggest dynamics of individual molecular dissociations that differ because the zero point motions of the parent molecule prior to the electron attachment lead to a range of starting conditions for the dissociative events.

Professor Polanyi responded: Prof. Orr-Ewing is right in thinking that we start our experimental studies by establishing the location and geometry of the physisorbed initial state, as precisely as possible – STM is a marvelous tool for that. He is correct too in stating that our simple model of the reaction computes a single value for the repulsive energy-release in the adsorbate's ionic state, leading , therefore, to a unique trajectory across the ground pes. This, as he notes, falls short of describing the observed dynamics which include a range of outcomes. The question that he raises is the nature of the variable (or variables) likely to be responsible for the observed range of outcomes. We don't know, at present, but experiment gives a suggestive hint. (Details are to be found in the paper in this Discussion, and for *p*-diiodobenzene in a recent paper by Leung *et al.*[1]). Reaction is highly localized in the experiments on *p*-diiodobenzene, but, as always, with some breadth in the distances to which the I atom recoils. The breadth of I atom recoil distances is substantially larger for the first I atom then for the second (resulting from the breaking of the first and second C–I bonds). The explanation may well lie in the greater range of configurations available to the singly-bound iodophenyl radical (following the first bond scission) than to the doubly sigma-bound phenylene which attaches in a restricted fashion to an identifiable pair of copper atoms in the surface below. This is, however, merely a comment on the availability of variable outcomes for the organic moiety in the two cases; it does not address Orr-Ewing's question as to the modification that our model will require if it is to encompass a such a range of outcomes, rather than merely the most probable one. There is an obvious candidate as the mainspring of variability, which is the time that the system spends in the ionic state. We have customarily employed the shortest time needed if the system is to subsequently react successfully on ground pes. In reality longer times will also be represented, albeit with a somewhat reduced probability (since our times already exceed the expected mean ionic-state lifetime at a metal surface, of a few femtoseconds). These longer ionic lifetimes provide a ready explanation of stronger impulses and broadened product distributions. We did not include them in our model, in order not to over-elaborate at this stage. (Devoted readers of this Discussion will note that I hinted at this variation in ionic lifetime in my reply, earlier, to Professor Neumark) In reply to Prof. Orr-Ewing, we did test the effect of including some salient zero-point energies in our ionic state, but found that this resulted in no more than a minor increase in breadth of the computed outcomes.

1. L. Leung *et al.*, *J. Am. Chem. Soc.*, 2012, **134**, 9320.

Professor Wodtke asked: Could electron hole pair excitation be important in the translational cooling of the hot halogen atoms produced in the electron induced dissociation events observed in your work?

Professor Polanyi answered: In both the anionic excited state and the ground state the molecular dynamics were calculated for the recoiling fragments (recoiling halogen atom and organic radical) using DFT to obtain the energies and then solving the equations of motion to obtain the trajectories across a 180 atom slab five copper atoms deep with the top three layers free to move. Translational cooling in the ionic state will be enhanced by our inclusion of image–charge interaction with the substrate. Though the interaction with the substrate will be less strong on the ground potential-energy surface, the time the system spends on that surface is an order-of-magnitude longer. The recoiling fragments are translationally hot (by an eV or so) by the end of their brief stay on the ionic surface, so much of the cooling occurs on the ground pes. As for the role of electron–hole pair excitation, DFT will presumably result in some charge-transfer to the electronegative halogen atom. We don't know what part that plays in promoting translational cooling.

Professor Ashfold commented: How should I picture the electron transfer from the tip to the σ^* orbital of the adsorbate molecule? I envisage the tip as large compared with the size of the adsorbed molecule, current flows (at a constant bias) for hundreds of milliseconds – corresponding to the passage of some 10^{10} electrons – and at some instant one electron populates the σ^* orbital and the carbon-halogen bond starts to break. Is this right, and can you expand on this picture?

Professor Polanyi replied: This is much the way that we picture the event. Though the tip is large there must be an aspersion that allows a degree of sub-molecular resolution – so that we can see where the halogen atoms are located in the physisorbed molecule. The experiment is performed at a selected bias voltage with the feed-back loop disconnected. An abrupt change in resistance due to bond-breaking gives rise to a corresponding change in the measured current. For our nanoamp currents a bond breaks in these halides after about a second, *i.e.* with an efficiency of 10^{-10}. According to our model of the repulsive anionic state, the carbon–halogen bond begins to break during the tens of femtoseconds for which that state persists. We then transfer the system, with its accumulated distortions and momenta, to the ground pes where, because of the impulse in the ionic state, it can now sail over the barrier and form reaction products. The greater part of the dynamics leading to products unfolds during the much longer time that the system spends on the ground pes. Of course, the idea is to compare the distance and angular distributions observed for the products with those given by the model. Experiment gives a broader distribution of products than does theory, but the theory is very revealing of the median behaviour.

Professor Zare noted: I am intrigued by the lovely results you have presented. I wonder if you have considered using in place of your *p*-dichlorobenzene molecules 1,3,5 trichlorobenzene or the like?

Professor Polanyi answered: We too are intrigued by the possibilities, especially for uncovering new patterns of reaction dynamics. The highly symmetric molecule that you mention has been on our list (is there simultaneous release of all three halogen atoms, and is there a symmetric outcome?) but at present we are in pursuit of examples in which the departing atoms have a clearer opportunity to interact with one another, as their bonds break. But who knows; your proposed 1,3,5 halobenzene may exhibit an element of that. Thanks for the suggestion.

Professor Gruebele asked: I would like to propose an alternative model for the excitation that leads to halide dissociation: instead of single-step, a highly stepwise excitation of the C-X bond by superexchange-mediated vibrational ladder climbing.

A few years ago we studied dissociation of Si–H bonds on Si(100) surfaces by voltage–current dependent scanning tunneling microscopy.[1] We found that we could not explain the observed I–V curves by a rare single tunneling event (*e.g.* 1 : 10^{10} electrons in our measurements), or a $\Delta v = 1$ ladder climbing on the ground state C–X stretching surface. Instead, at low tunneling voltages (<2 V), we required a model in which each of the 1010 electron tunnels (superexchanges) through the antibonding Si–H electronic state, visiting the anionic state with only a very short lifetime. This short lifetime is sufficient for the Si–H$^-$ stretching vibrational wavepacket to move slightly towards dissociation before the system returns to the Si–H ground electronic surface as the electron tunnels into (or from) the surface. The process then repeats itself (superexchange *via* antibonding Si–H$^-$ state, brief stretching of bond, return to ever more vibrationally excited ground state) many times until dissociation occurs. To our surprise, we found that Franck–Condon factors "stolen" from silane were able to produce quantitative agreement with our STM dissociation experiments (I–V dependence), and that climbs with $\Delta v > 1$ were not uncommon. We expect that the single-step model is more likely at large tunneling voltages, whereas the vibrational ladder climbing *via* a virtual Si–H$^-$ state is more likely at low voltages. It may be that for C–X reactions, especially at low voltages, the multi-step process is also more appropriate than the single-step process; or a bond breaking process between these extremes may be obtained.

1. V. Wong and M. Gruebele, *Chem. Phys. Lett.*, 2002, **363**, 182-188.

Professor Polanyi replied: We believe that we have the possibility of two regimes for electron-induced dissociation of our halides, though we only spoke of one of these in our contribution to this Discussion. The most important regime is a single-electron process occurring above a specific threshold- energy (specific to the bond being broken). This single-electron process we associate with excitation to an ionic state. In ongoing experiments we have been able to discern a second, roughly three orders-of-magnitude less efficient, process that involves excitation by several successive electrons. This we believe to be the vibration-induced reaction pathway identified in other laboratories (those of Wilson Ho, Maki Kawai and

Karina Morgenstern come to mind). Interestingly it leads to different product distributions from the higher probability direct-excitation pathway.

Professor Gruebele's intriguing multiple electronic-excitation mechanism is new to us. It could represent a third reactive path. It is, however, unlikely to apply to the single-electron events that we discuss in our paper.

Professor Wodtke said: Congratulations on such beautiful experiments. Using the tunneling bias voltage, one can imagine that you might inject the electron into the molecule in different ways; that is, possibly through different molecular orbitals. This might result in different single molecule dynamics. Have you observed such effects?

Professor Polanyi responded: I shall transmit your kind comments to those who did the work. As for the possibility of exciting individual molecular orbitals, this has been achieved recently in other systems. Our dihalobenzenes do not seem to lend themselves to localised excitation – we have no evidence, at present, that the location of the tip governs the choice of C–X bond to be broken.

Professor Martinez asked: Does the polarity of the STM affect the observed reactivity, *i.e.* whether the voltage is positive or negative? Does the dependence of reactivity on STM polarity say anything about the mechanism of the reaction, *i.e.* whether anionic or cationic states of the adsorbed species are involved and whether these are "real" or "virtual" states in the sense often used when discussing Raman spectroscopy?

Dr Tingbin Lim replied: We have not tried to induce the reaction of dihalogenated benzenes on copper with negative surface polarity, but based on what we know from other experiments that we performed on silicon, we expect the molecules also to react under negative surface polarity, though not necessarily at the same bias voltage (and therefore electron energy). As regards the second question, since we have found that our threshold voltages correspond to actual energy levels of adsorbate-plus-surface electronic states, we believe that the electrons are entering actual electronic surface states in the molecule–surface system.

Professor Truhlar opened the discussion of the paper by Timothy K. Minton by commenting: When O is far from the surface, it is in a triplet state. When O is close to the surface, the ground state is expected to be a singlet state so it is governed by a different potential energy surface. Furthermore the spin state of O after the collision can be either a singlet or triplet. A classical model brakes down when multiple potential energy surfaces are involved. What roles do spin states and spin transitions play in these collisions, and is the effective mass model capable of modeling their role?

Professor Minton responded: Certainly, the nature of the atom-surface interaction must be governed by the potential energy surface, and there might be multiple potential energy surfaces involved. Neither our current experiments on O and Ar atom collisions on liquid and SAM surfaces nor our calculations on Ar atom collisions on SAM surfaces are capable of providing information on the role of spin in the atom–surface collisions that we have studied. Our model is a kinematic model, and it has utility in the way that a Newton diagram has utility in helping us to infer gas-phase scattering dynamics in molecular beam experiments. In crossed-beams experiments, we know the masses and velocities of the colliding partners, and we focus on the behavior of the scattered products to learn about the interaction potential of the colliding partners. In our beam-surface scattering experiments, we rely on the behavior of the scattered products to derive the mass of one of the partners (*i.e.*, the effective surface mass). Thus, the effective surface mass inherently contains information about the interaction potential. Nevertheless, we expect the effect of the potential to be small, because our model, which does not even contain a rudimentary description of an interaction potential (although it could), produces curves that almost perfectly match our data on fractional energy transfer as a function of deflection angle. The apparently good description of the energy transfer with a classical model may be the result of the relatively high translational energies used in our experiments, which tend to sample the repulsive part of the potential. In order to deduce information about the atom–surface interaction potential from the observed scattering behavior, the model would have to be extended and go beyond simple kinematics. Perhaps this is possible, in which case the entire differential scattering cross section, and not just the average final velocity as a function of scattering angle, would need to be incorporated into the model.

Professor Nesbitt asked: I wonder if you could comment on whether the data reveal any substantial dependence in the effective mass on incident kinetic energy, and if so, whether this can be predicted with this elegantly simple "Newton circle" kinetic model. Also, might there be any interesting predictions based on this model that could be tested for the dependence of the effective masses on temperature of the self assembled monolayer?

Professor Minton answered: The effective surface mass does depend on incidence energy. For example, the effective surface masses derived for O atom scattering on squalane are 109, 43, and 40 amu with incidence energies of 47, 297, and 504 kJ mol^{-1}, respectively. The decrease in effective surface mass with increasing incidence energy is consistent with the picture afforded by the model, because a collision at higher incidence energy would be expected to occur faster and lead to involvement of fewer surface atoms in the collision, thus decreasing the effective surface mass. The three experimental data points that we have suggest that the energy dependence of the effective surface mass is larger when the incidence energies are lower. But to be able to find a quantitative relationship between effective surface mass and incidence energy would require further systematic investigations of the same scattering system with a variety of incidence energies.

In an earlier study of the temperature dependence of Ar-atom scattering dynamics on PFPE with $E_i = 71$ kJ mol^{-1},[1] it is evident that the energy transfer associated with impulsively scattered atoms remains unchanged when the surface temperature is varied from 280 to 359 K. In the picture provided by our model, we would assume that the effective surface mass would not change for a given scattering system if the energy transfer remains the same. Indeed, in unpublished work on the inelastic scattering of 520 kJ mol^{-1} O atoms on an ionic liquid surface, [emim][Tf$_2$N], we derived the same effective surface masses from impulsively scattered O atoms (32 amu, within experimental error) for surface temperatures of 283 and 323 K. These preliminary data support the prediction that the effective surface mass derived from impulsive atom-surface scattering is, at most, only weakly dependent on surface temperature.

1 M. E. King, K. M. Fiehrer, G. M. Nathanson, and T. K. Minton, *J. Phys. Chem. A*, 1997, **101**, 6556

Professor Sibener commented: Your presentation was very clear in delineating and addressing the issues related to the number of atoms that actually participate in energy accommodation during gas-surface encounters. I note that self-assembled monolayers of differing lengths may offer another complementary avenue to addressing this question for organic systems. I believe that simulations by Bill Hase and coworkers, and perhaps others, have examined gas-surface collisional energy transfer for standing-up alkylthiol SAM phases that have 6, 8, 10 and more carbon atoms. Convergence in the observed scattering occurs in the range of 6 to 8 C atoms, giving one another window on how the initial excitation occurs in the chain that was directly impacted, followed by simulations that examine how inter-chain interactions subsequently complete the energy accommodation process. These time-scales can be compared to the duration of the actual encounter for thermal, super-sonic, and hyperthermal scattering regimes.

Professor Minton communicated in reply: Yes, Bill Hase has published extensively on the surface modes involved in collisions of atoms with SAM films, and these studies are complementary to our work. The effective surface mass must be related to the number of modes involved in the collision, and the time scale of the collision will affect the number of modes that are excited and thus the effective surface mass. As we have seen, higher incidence energies lead to reduced effective surface masses, which would be expected because the duration of the collision would be shorter. It is an interesting idea to consider the time scale of the collision when attempting to gain insight about the surface from the effective surface mass.

Professor Nesbitt communicated: We find in our studies of CO_2 rotational excitation by molecular beam scattering from room temperature ionic liquids (such as al-kylmethylimmidazolium cation with various anions) that there is a strong dependence on the length of the C_n alkyl chain.[1,2] In particular, for chain lengths below $n < 4$, the surface scattering dynamics is still quite sensitive to the identity of the anion. However, for $n > 6$, there is a saturation of this effect, with essentially no further sensitivity to choice of anion, which we have interpreted as complete coverage of the surface by the alkyl chain hydrocarbon layer. In further support of this, we find these CO_2 rotational quantum state distributions to be essentially indistinguishable from scattering at comparable incident energies from pure hydro-carbon liquid squalane interfaces. I wonder if you see any similar saturation effect with respect to alkyl chain length in your translational distributions for the self assembled monolayer surfaces?

1 J. R. Roscioli and D. J. Nesbitt, *J. Phys. Chem. Lett.*, 2010, **1**, 674 -678.
2 J. R. Roscioli and D. J. Nesbitt, *J. Phys. Chem. C.*, 2011, **115**, 9764-9773.

Professor Minton responded: We have not investigated the relationship between effective surface mass and alkyl chain length of SAM surfaces. However, we have con-ducted experiments on hyperthermal O atoms scattering on two ionic liquids of different alkyl chain length, [emim][Tf$_2$N] and [C$_{12}$mim][Tf$_2$N],[1] and we observed that the translational energy distributions of inelastically scattered O and reactively scattered OH are much more similar (reflecting more energy transfer) to those for O atoms scattering from the pure hydrocarbon, squalane, when the alkyl chain on the imidazolium ring is longer. We concluded, like you, that the alkyl chains tend to domi-nate the surface as they become longer. We have only derived an effective surface mass for the [emim][Tf$_2$N] so far (32 amu), but we would expect the effective surface mass to increase as the alkyl chain becomes longer and approach that of squalane (40 amu).

1 B. Wu, J. Zhang, T. K. Minton, K. G. McKendrick, J. M. Slattery, S. Yockel, and G. C. Schatz, *J. Phys. Chem. C*, 2010, **114**, 4015.

Professor Wodtke said: In your kinematic analysis, it would seem that you might expect a higher degree of energy uptake by the liquid when the effective mass is larger. I say this since your beautiful picture of the effective mass being related to the number of interacting atoms suggests it. That is, when the effective mass is large, one has many vibrationally coupled atoms, hence energy flow into the vibrational degrees of freedom of the liquid should also be enhanced. Do you see such a positive correlation of energy uptake with effective mass?

Professor Minton replied: Yes, in all the examples we have studied, increased energy transfer to the surface correlates to a higher effective surface mass. The explanation is as you described: there is more energy transfer when more vibrationally coupled atoms on or near the surface are involved in the collision.

Professor Crim asked: What are the prospects of building more detailed models of energy transfer to the surface than contained in the effective mass picture for the scattering? One can imagine building models analogous to those used in isolated collisions in which the Fourier transform of the driving force and the frequencies of the transitions in the driven system determine the efficiency of energy transfer. Is it possible than those sort of ideas can shed light on the effects your data reveal?

Professor Minton communicated in reply: Your point is well taken. Now that we have a picture of an incident atom interacting with a localized region of the surface with an effective mass, it would be interesting to build on this simple model such that we could relate the effective surface mass to fundamental physical properties of the surface that are independent of incidence energy, surface temperature, or even incident projectile. A good starting point might be to explore how we could extract information about surface modes from the effective surface mass and the energy transfer to the surface that the derived surface mass implies. The answers to your questions will require significant additional work.

Vibrationally resolved transition state spectroscopy of the F + H$_2$ and F + CH$_4$ reactions†

Tara I. Yacovitch,[a] Etienne Garand,‡[a] Jongjin B. Kim,[a] Christian Hock,[a] Thomas Theis[a] and Daniel M. Neumark*[ab]

Received 28th January 2012, Accepted 15th February 2012
DOI: 10.1039/c2fd20011b

The transition state regions of the F + *para*-H$_2$, F + *normal*-H$_2$, F + CH$_4$ and F + CD$_4$ reactions have been studied by slow electron velocity-map imaging (SEVI) spectroscopy of the anionic precursor clusters *para*-FH$_2^-$, *normal*-FH$_2^-$, FCH$_4^-$ and FCD$_4^-$. The F + H$_2$ results improve on previously published photoelectron spectra, resolving a narrow peak that appears in the same position in the *para*-FH$_2^-$ and *normal*-FH$_2^-$ spectra, and suggesting that additional theoretical treatment is necessary to fully describe and assign the experimental results. A small peak in the *para*-FH$_2^-$ results is also identified, matching simulations of a product resonance in the $v' = 3$ vibrational level. SEVI spectra of the $^2P_{3/2}$ bands of FCH$_4^-$ and FCD$_4^-$ show extended structure from transitions to the entrance valley van der Waals region and the reactant side of the F + CH$_4$ transition state region. Much of this structure is attributed to bending or hindered rotation of the methane moiety and may be a spectroscopic signature of reactive resonances.

I. Introduction

The F + H$_2$ → FH + H reaction and its various isotopologs have been extensively studied as the quintessential bimolecular reaction.[1-4] The F + CH$_4$ → FH + CH$_3$ reaction has evolved into a benchmark polyatomic reaction and provides insight into how additional degrees of vibrational freedom affect chemical reaction dynamics.[5-7] Exploration of the potential energy surfaces for these reactions can be accomplished by crossed molecular beam experiments[8,9] where the characteristics of the surface are gleaned from the cross sections, angular distribution and states of the final products. Negative ion photoelectron (PE) spectroscopy[10] can complement these studies by directly accessing the transition state region *via* photodetachment of an anion with appropriate geometry. In the case of the F + H$_2$ and F + CH$_4$ reactions, the corresponding FH$_2^-$ and F$^-$(CH$_4$) anions have good geometric overlap with the neutral transition state region[11,12] and hence directly probe the spectroscopy of this critically important region of the reactive surface. In this work, we employ slow electron velocity-map imaging (SEVI), a high resolution variant of anion PE spectroscopy, to obtain significantly improved results for both systems.

[a]Department of Chemistry, University of California, Berkeley, CA 94720, USA. E-mail: dneumark@berkeley.edu; Fax: +1 510 642 3635; Tel: +1 510 643 3850
[b]Chemical Sciences Division, Lawrence Berkeley National Laboratory, faraBerkeley, CA 94720, USA

† Electronic supplementary information (ESI) available: sample data workup for FH$_2^-$ showing parallel and perpendicular spectra. Original data sets for FCH$_4^-$ and FCD$_4^-$ showing invariance of peak positions. See DOI: 10.1039/c2fd20011b
‡ Current address: Sterling Chemistry Lab, Yale University, New Haven, CT 06520, USA

Negative ion photoelectron (PE) spectra of the FH_2^- anion, have been reported previously.[11, 13-15] These spectra show resolved hindered rotor progressions associated with the $F + H_2$ transition state region. The structure of these progressions depends on whether the F^- is complexed to *para* or *normal* H_2. Given that the anion is linear, the observation of a hindered rotor progression suggests that the neutral transition state (TS) is bent. This interpretation is supported by extensive theoretical work on the $FH_2 \leftarrow FH_2^-$ system,[11,16,17] which reproduced the main experimental features *via* exact quantum scattering calculations on the best potential energy surfaces available at the time. In the original PE spectra, the peak widths were about 19 meV wide, with a good fraction of this arising from the experimental resolution of around 12 meV, raising the question of what additional structure might be seen at higher experimental resolution.

To address this point, Russell and Manolopoulos (RM) performed time-dependent wavepacket simulations of FH_2^- photoelectron spectra[16] on the *ab initio* potential energy surface calculated by Stark and Werner (SW-PES).[18] These simulations predict sharp resonance peaks and broader direct scattering peaks that are clearly separable at a resolution of 1 meV, but not at the lower resolution of the experimental PE spectra. Hartke and Werner[17] (HW) subsequently made adjustments to the potential energy surface (HSW-PES) to include spin–orbit coupling, and new photoelectron spectroscopy simulations using a more realistic anharmonic potential for the anions were performed. Since then, a number of highly sophisticated surfaces have been published.[19-23] These post-SW surfaces have proven accurate in the description of various crossed-molecular beam results, but have not yet been used to simulate the $FH_2 \leftarrow FH_2^-$ negative ion photoelectron spectra.

Liu and co-workers[7] have performed extensive experimental work on the $F + CH_4$ reaction and its isotopologs, and several potential energy surfaces for this reaction have been published recently.[24-28] Recent negative ion photoelectron spectra of FCH_4^- have been published by Cheng *et al.*[12] These spectra show two broad bands which were assigned to the atomic fluorine $^3P_{3/2}$ and $^3P_{1/2}$ spin orbit (SO) states, shifted to a larger splitting by the proximity of the methane. The experimental spin orbit (SO) splitting was compared to SO splittings taken from calculated energies for the ground and excited state neutral surfaces, revealing that the photodetachment accesses a region on the ground state $F + CH_4$ potential between the reactant van der Waals (vdW) well and the TS. Simulated photoelectron spectra for this system have not been calculated, though Bowman and coworkers have recently published an *ab initio* surface for both the FCH_4^- anion[29] and the $F + CH_4 \rightarrow FH + CH_3$ neutral surface.[28,30] Their anion calculations support previous findings[31] of a $F^-\cdots H-CH_3$ C_{3v} equilibrium geometry. Their neutral calculations reveal an early TS with a bent $F\cdots H-CH_3$ geometry, as well as two reactant vdW structures with C_{3v} geometries, with the deeper minimum corresponding to a $F\cdots H_3CH$ structure.

In this paper, we investigate the $F + H_2$ and $F + CH_4$ systems *via* SEVI of FH_2^- and FCH_4^- to see what additional structure can be observed at higher resolution than in previously reported work. Fig. 1 shows potential energy diagrams of both systems along the hydrogen transfer reaction coordinate, illustrating the geometric overlap between the anion and neutral surface. In both cases, the anion has good overlap with the neutral TS region and, possibly, bound vdW states in the reactant valley. We thus expect to observe vibrational progressions in one or more modes perpendicular to the reaction coordinate in the TS region,[32] sharp structure from the vdW states,[33] and, possibly, reactive resonances. The latter are from quasibound quantum states, localized along the reaction coordinate, such as those seen in the $I + HI \rightarrow IH + I$ reaction *via* zero electron kinetic energy (ZEKE) spectroscopy on IHI^-.[34]

The observation and identification of reactive resonances is a key goal in transition state spectroscopy as these relatively long-lived states are very sensitive to the transition state region of the potential energy surface.[32,35] There is considerable

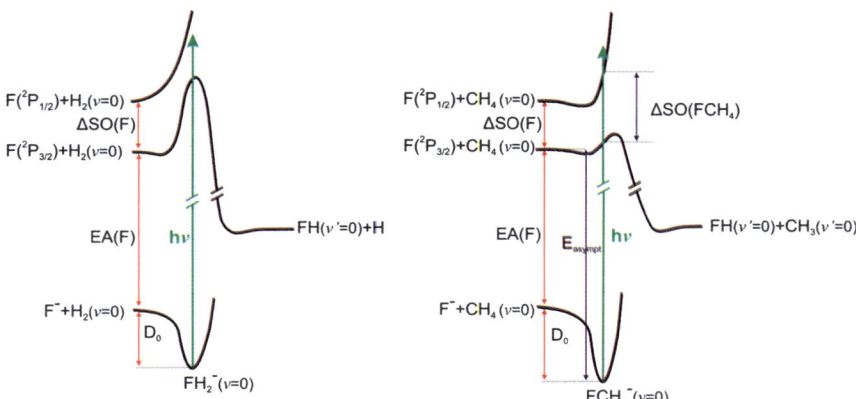

Fig. 1 Potential energy diagram (not to scale) for the $F + H_2$ and $F + CH_4$ systems. Vibrationally adiabatic curves (potential energy + zero point energy) are shown for the anion and neutral along with a green arrow ($h\nu$) indicating the approximate regions accessed by photodetachment. Relevant energies are labeled: the anion well depth, D_0; fluorine atom electron affinity, EA(F) and spin orbit splitting, ΔSO(F); neutral FCH_4 spin orbit splitting at the anion geometry, ΔSO(FCH_4). The electron binding energy to the ground state reactant asymptote is labeled E_{asympt} and is used as an energetic reference.

evidence from scattering experiments and calculations supporting the presence of resonances for the two reactions considered here. It was only in 2000 that Skodje *et al.* definitively identified a reactive resonance in the integral cross section of the $F + HD$ reaction.[36] This resonance corresponds to a state with three vibrational quanta in the HF product and all other quantum numbers zero: $HF(\nu' = 3) + D$. Subsequent studies on differential cross sections and product state distributions validate this assignment and further characterize this resonance.[23,37-39] Around this same time, Shiu *et al.* reported the first experimental signs of a reactive resonance in the $F + CH_4 \rightarrow FH + CH_3$ reaction with similar origins as the one in the $F + HD$ results: $FH(\nu' = 3) + CH_3$.[5] Evidence for this same resonance then emerged for the $F + CHD_3$ and $F + CD_4$ isotopomers.[40,41] The SEVI work presented here offers the opportunity to observe the direct spectroscopic signature of resonances associated with these reactions as sharp spectral features.

II. Experimental methods

SEVI is a high-resolution variant of negative-ion photoelectron spectroscopy and has been described in detail previously.[42,43] Briefly, negative ions are photodetached with a tunable laser, and the slow electrons are selectively detected using a low-voltage extraction velocity-map imaging (VMI) setup.[44] By varying the detachment wavelength, a number of high-resolution scans over limited energy windows are obtained.

Anions were produced by expanding an appropriate gas mixture, at a stagnation pressure of 250–400 psi, into the source vacuum chamber through an Even-Lavie pulsed valve.[45] FH_2^- anions were formed using a grid discharge source,[46] and FCH_4^-/FCD_4^- anions using a circular ionizer and water-cooled jacket for the valve.[47] *Normal*-FH_2^- (n-FH_2^-) anions were produced using a gas mix containing trace NF_3 and 10% to 24% H_2 in a balance of argon. A similar mix using parahydrogen (p-H_2) was used to make the *para*-FH_2^- ions (p-FH_2^-). p-H_2 was produced in the setup of Theis *et al.*[48] to a purity of 95%. In order to prevent back-conversion to *ortho*-H_2, all stainless steel except for the interior surface of the pulsed valve was removed from the gas manifold. FCH_4^- anions were produced with trace NF_3, 10% CH_4 and a balance of argon. FCD_4^- was similarly made with trace NF_3 in 3% CD_4/argon mix.

Ions were mass-selected[49] and directed to the detachment region by a series of electrostatic lenses and pinholes. They were then photodetached between the repeller and the extraction plates of the VMI assembly by the frequency-doubled output of a Nd:YAG-pumped tunable dye laser. The resulting photoelectron cloud was coaxially extracted down a 50 cm flight tube and mapped onto a detector comprising micro-channel plates coupled to a phosphor screen.[50,51] Events on the screen were collected by a 1024×1024 Charge-Coupled Device (CCD) camera and sent to a computer, where they were summed, centered, smoothed, quadrant-symmetrized and transformed using the inverse-Abel[52] or pBasex[53] methods. Centering of images is crucial for high resolution results after transformation[54] so a F^- or Cl^- image was used as a reference.

Photoelectron kinetic energy spectra were obtained by angular integration of the transformed images. SEVI spectra are plotted with respect to electron binding energy (eBE), defined as the difference between the photodetachment photon energy and the measured electron kinetic energy (eKE). In each SEVI image, better energy resolution was obtained for slower electrons. Hence, by varying the photodetachment laser wavelength, overview and detailed scans can be obtained.

SEVI also provides information on the photoelectron angular distribution (PAD).[51] The anisotropy of the photoelectron images is determined by the angular momenta of the photoelectron partial waves, which, in turn, reflect the shape of the orbital from which detachment occurs. s-wave ($\ell = 0$) detachment results in isotropic images, while p-wave ($\ell = 1$) detachment peaks at $0°$ and $180°$ relative to the laser polarization axis.

The apparatus was calibrated by acquiring SEVI images of atomic Cl^- at several different photon energies.[55,56] VMI repeller voltages of -350 and -200 V were used in this study. For an ideal system of atomic transitions, with s-wave character, intense ion signal, and well-spaced easy-to-center peaks, the SEVI instrument can easily surpass 1 meV resolution. For a Cl^- peak using -350 V VMI repeller voltage, fwhm of 3.7 cm^{-1} (0.46 meV) at 22 cm^{-1} from threshold and 9.4 cm^{-1} (1.2 meV) further out at 161 cm^{-1} from threshold are achieved. Using a -200 V VMI repeller voltage, a fwhm of 2.5 cm^{-1} (0.31 meV) is achieved 16.0 cm^{-1} from threshold. This is our instrumental resolution. However, linewidths in the FH_2^- spectra are limited because, as shown below, photodetachment occurs mainly by emission of p-wave electrons ($\ell = 1$). The photodetachment cross section σ drops off severely close to threshold for p-wave detachment according to the Wigner threshold law, $\sigma \propto (eKE)^{\ell + \frac{1}{2}}$.[57] One must therefore collect data at photon energies further from the threshold than for s-wave ($\ell = 0$) detachment, limiting the ultimate resolution of the SEVI spectra.

III. Results and analysis

Smoothed and symmetrized photoelectron images at 311 nm for p-FH_2^- and 332 nm for FCH_4^- are shown in Fig. 2. These images are dominated by p- and s-wave detachment, respectively, with the former leading to a PAD peaked along the laser polarization and the latter to an isotropic PAD. While analysis of the FCH_4^- image is straightforward, extraction of the photoelectron eBE spectrum from the p-FH_2^- image using the inverse Abel algorithm by Hansen and Law[52] is problematic since the centerline noise falls along the parallel axis and obscures the region with the most signal. The pBasex inversion method[53] is thus advantageous since it produces center point noise. An additional complication arises because photodetachment of FH_2^- not only accesses the reactive $F + H_2$ $^2\Sigma_{1/2}$ state, but also the repulsive, spin–orbit excited $^2\Pi$ states,[15] resulting in lower eKE photoelectrons near the center of the image with a more isotropic PAD. This latter contribution overlaps energetically with the high eBE signal from the transition to the ground electronic state. In an attempt to separate out contributions from the excited states to the photoelectron eBE spectrum, a "parallel" spectrum was constructed in order to capture the

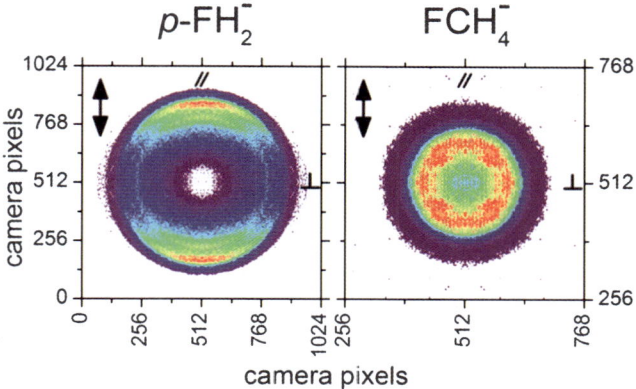

Fig. 2 SEVI images of p-FH$_2^-$ and FCH$_4^-$ showing the p-wave and s-wave nature of the peaks, respectively. The p-FH$_2^-$ image was taken at 311 nm and the FCH$_4^-$ image at 332 nm with laser polarization as indicated by the arrow. Parallel (//) and perpendicular ($\perp$) regions of the image are labeled.

maximum contribution from a p-wave atomic transition (maxima at 0° and 180° with respect to the laser field) by integrating over two 45° angular slices. A corresponding "perpendicular" spectrum containing contributions from primarily s-wave results was constructed with two 45° slices at 90° and 270° with respect to the laser field. The perpendicular results were subtracted from the parallel results to yield a p-wave only FH$_2^-$ spectrum (See Fig. ESI1 in the Electronic Supplementary Information†). The pBasex image transformation and parallel-perpendicular slice subtraction were done using modified code from Kornilov *et al.*[58]

High-resolution SEVI spectra for n-FH$_2^-$ and p-FH$_2^-$ are shown in Fig. 3 underneath the previous PE results.[11] Fig. 4 shows overview spectra for FCH$_4^-$ and FCD$_4^-$, while high-resolution spectra of the more intense band around 29900 cm^{-1} are shown in Fig. 5. Peak labels, positions and splittings are listed in Tables 1 and 2.

The overview SEVI spectrum for p-FH$_2^-$ (top black trace in Fig. 3) is dominated by three major bands: a relatively narrow band (**A**) at 29180 cm^{-1}, an intense broad band at 29780 cm^{-1} (**B**) and a weak broad band at 30550 cm^{-1} (**C**). The n-FH$_2^-$ spectrum (Fig. 3) similarly has 3 major bands: a narrow band (**D**) at 29130 cm^{-1} and two intense broad bands at 29490 cm^{-1} (**E**) and 30060 cm^{-1} (**F**). A final very weak band appears around 31000 cm^{-1} (**G**). The overview spectra are similar to previous PE spectra of these anions[11,15] but the peaks are better-resolved, especially for n-FH$_2^-$. Specifically, band **D** is newly resolved from the neighboring intense band **E**.

As the detachment laser is tuned to the red, and higher resolution, closer-to-threshold images are taken, we observe major relative decreases in peak intensities of the higher eBE features due to the sharp drop-off in photodetachment cross-section for the slowest p-wave electrons owing to the Wigner threshold law, as discussed in Section II. Some fine structure also appears on top of the broad bands: these peaks and shoulders are labeled in lower case and reported in Table 1. We note that the pBasex transformed and subtracted images shown in Fig. 3 show oscillatory noise in low signal-to-noise (S/N) regions like band **G** in the n-FH$_2^-$results. For this reason, the origin of much of the fine structure is ambiguous, with the notable exception of peaks **a**, **b** and **c** in the p-FH$_2^-$ results, and peak **x** in the n-FH$_2^-$ results, which is essentially a narrower version of band **D**. These weak peaks appear in all the data, including spectra that were transformed using the inverse Abel method and where slices were not subtracted as for the plotted results. Peak **a** appears in the higher resolution scans at 29136 cm^{-1}, nearly lining up with peak **x** at 29121 cm^{-1}. Peaks **c** and **d** appear at 29260 and 29430 cm^{-1} respectively. We

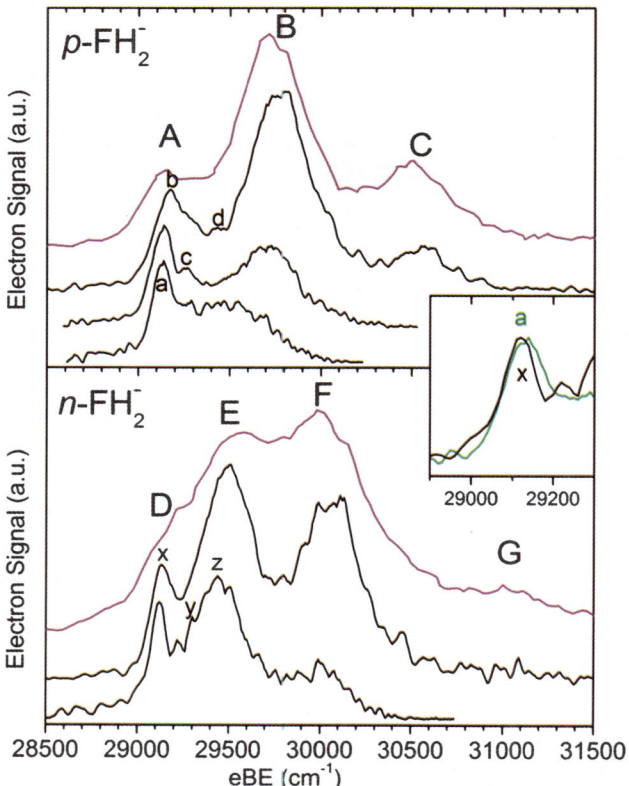

Fig. 3 High resolution SEVI spectra of the p-FH_2^- and n-FH_2^- anions taken at laser energies of 311 nm (top black trace), 325 nm (next trace) and 329 nm (lower trace, *para* only) showing pBasex transformed and subtracted images (see text). Previous PE results[11] are shown in magenta above the SEVI results. An inset compares the highest resolution scans available for peak **a** (green trace) and peak **x** (black trace).

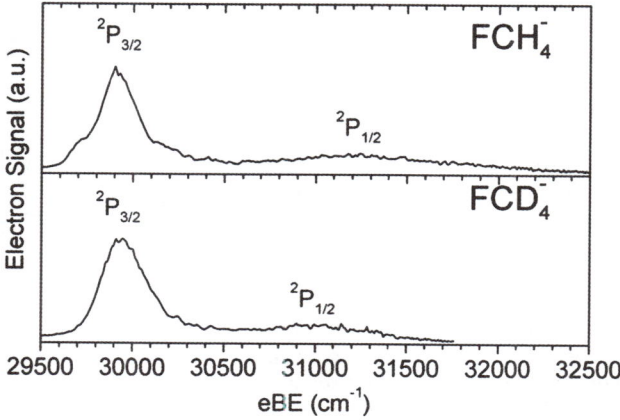

Fig. 4 Overview spectra of the FCH_4^- and FCD_4^- anions taken at 307 nm and 315 nm, respectively, showing transitions to the $^2P_{3/2}$ and $^2P_{1/2}$ spin-orbit states.

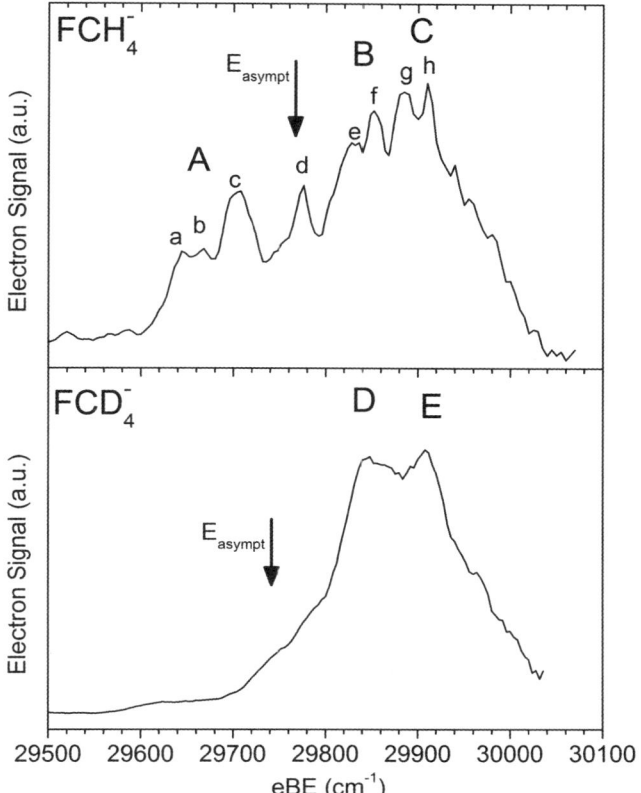

Fig. 5 High resolution SEVI spectra showing transitions to the $^2P_{3/2}$ spin orbit state of FCH_4 and FCD_4. Spectra are spliced together from several high resolution traces taken at 333 nm (see text). Arrow indicates the approximate calculated energy of the reactant asymptote (E_{asympt}).

Table 1 Fine structure of p-FH_2^- and n-FH_2^- showing position of peaks and the splitting between them

p-FH_2^-				n-FH_2^-			
Band	Label	Position (cm^{-1})	Splitting	Band	Label	Position (cm^{-1})	Splitting
A	a	29136		D	x	29121	
	b	29180	44			29225	104
	c	29260	80	E	y	29300	75
	d	29430	170		z	29438	138
B		29768	338	F		30044	606
C		30542	774	G		30163	281

also note that some fine structure of ambiguous origin labeled **y** and **z** appears in n-FH_2^- at close to the same frequencies as peaks **c** and **d**: 29300 cm^{-1} and 29438 cm^{-1}, respectively. Peak **b** appears at 29180 cm^{-1} in the lowest resolution SEVI trace but is never resolved from the neighboring transitions.

SEVI overview spectra for FCH_4^- and FCD_4^- are plotted in Fig. 4. Two broad bands are visible for these low resolution spectra, correlating to the $^2P_{3/2}$ (lower eBE) and $^2P_{1/2}$ (higher eBE) spin–orbit states of the fluorine atom. The maximum of the $^2P_{3/2}$ peak in the FCH_4^- ion appears at 3.709 eV (fwhm = 0.035 eV) while the $^2P_{1/2}$ peak maximum appears 0.160 eV higher at (fwhm = 0.035 eV). The corresponding peaks in the FCD_4^- isotopomer appear at the same position within experimental uncertainty at 3.713 eV (fwhm = 0.032 eV) and 0.128 eV higher at 3.841 eV (fwhm = 0.048 eV). A summary of the peak maxima and spin orbit (SO) splittings appear in Table 2. The splitting of the two bands is similar to that seen previous by Cheng et al.,[12] but the $^2P_{3/2}$ band is considerably narrower in our SEVI overview spectrum. The slight shoulder at 29715 cm^{-1} in the FCH_4^- trace in Fig. 4 is the first indication of the rich rovibrational fine structure investigated at higher resolution.

Fig. 5 shows high resolution SEVI spectra of the $^2P_{3/2}$ band of FCH_4^- and FCD_4^-. There are three major bands in the FCH_4^- spectrum: band **A** spanning the 29610–29740 cm^{-1} range, and bands **B** and **C** centered around 29850 cm^{-1} and 29910 cm^{-1} respectively. Bands **A**, **B** and **C** exhibit progressions of closely-spaced peaks, with individual peaks indicated by lower-case letters. The FCD_4^- spectrum comprises two major broad bands **D** and **E** peaking at around 29841 cm^{-1} and 29910 cm^{-1}. For FCH_4^-, the spacing between the two major bands **B** and **C** is ~58 cm^{-1} (peaks **f** and **h**), while it is ~64 cm^{-1} for bands **D** and **E** in the FCD_4^- spectrum. The fine structure comprising **B** and **C** bands averages a 26 cm^{-1} splitting but varies between 20 and 33 cm^{-1}. Upon deuteration, any fine structure overlaying bands **D** and **E** bands is no longer distinguishable at our resolution. This observation suggests that these features are due to motions involving the substituted deuterium atoms, and not due to low S/N artifacts. The fine features in the FCH_4^- spectrum have a fwhm $\approx$ 30 cm^{-1} (based on a convolution of all the peaks in FCH_4^- bands **A**, **B** and **C**.

The SEVI spectrum for FCH_4^- is spliced together from 4 averaged spectra at −200 V or −350 V VMI, all taken at wavelengths around 332 nm and varying by about 0.2 nm image-to-image to attempt to eliminate systematic noise. Intensities of the spliced spectra were adjusted to match the averaged envelope shape for all spectra taken at −350 V VMI. The appearance of the fine structure present in Fig. 5 varied in intensity and resolution depending on the day and VMI voltage, but never in energy. The FCD_4^- results were similarly constructed from the spliced and intensity scaled portions from 4 averaged spectra. Figs. ESI2 and ESI3 in the Electronic Supplementary Information show the separate traces and the invariance of peak positions.†

IV. Discussion

The FH_2^- and FCH_4^- anions are superficially similar species, though comparison of their SEVI spectra shows quite different characteristics. Relatively little structure is visible for the smaller p-FH_2^- and n-FH_2^- species while the larger FCH_4^- and

Table 2 Fine structure of the $^2P_{3/2}$ band of FCH_4^- and FCD_4^- showing position of peaks and the splitting between them.^{a}

FCH$_4^-$				FCD$_4^-$		
Band	Label	Position (cm^{-1})	Splitting	Band	Position (cm^{-1})	Splitting
A	a	29644		D	29844	
	b	29668	24	E	29908	64
	c	29700	32			
	d	29776	76			
B	e	29832	56			
	f	29852	20			
C	g	29885	33			
	h	29910	25			

^{a} Peak positions are ±2 cm^{-1}.

FCD$_4^-$ ions show broad structure with regular progressions of closely-spaced peaks for FCH$_4^-$. The explanation for this effect lies in the second major difference between the SEVI results: the overall anisotropy of the images differs greatly between systems, with FH$_2^-$ yielding p-wave images and FCH$_4^-$ yielding isotropic s-wave images (see Fig. 2).

The ejected electron anisotropies are characteristic of the electronic state of the resulting neutral: $^2\Sigma_{1/2}$ for F + H$_2$ and $^2A_1$ for F + CH$_4$.[15,28] Owing to the Wigner threshold law[57] (Section II), we can obtain good signal-to-noise on features in the FCH$_4^-$ much closer to threshold, where the resolution is best. Thus, for example, the fwhm of peak **d** in the FCH$_4^-$ spectrum is 17 cm^{-1} (2 meV) when measured at 162 cm^{-1} from threshold. On the other hand, the narrowest peak measured in the FH$_2^-$ system was peak **x** (fwhm = 90 cm^{-1} = 11 meV) which was taken much farther from threshold at 1647 cm^{-1}.

1. *p*-FH$_2^-$ and *n*-FH$_2^-$

Fig. 6 compares SEVI spectra taken at 325 nm for p-FH$_2^-$ and n-FH$_2^-$ with the simulated spectra of Russell and Manolopoulos[16] (RM, left panels) and Hartke and Werner[17] (HW, right panels). The simulated spectra assumed an energy resolution of 1 meV, which, as discussed above, is better than that of the SEVI spectra for these anions. In making Fig. 6, anion dissociation energies D_0 of 0.205 and 0.1938 eV were used for RM and HW, respectively, as per the original publications.[11,17] In the following discussion, simulated quantum states are labeled according to their origin from the product (primed) or reactant side (unprimed), and numbered according to the quanta of excitation in the H–H (ν) or H–F stretch (ν'), the hindered H$_2$ (j) or HF (j') rotation, and the vdW stretch coordinate F–H$_2$ (t) or H–HF (t').

We first compare the positions of the broader experimental bands **A–F** to the simulations. The relevant selection rules for photodetachment from p-FH$_2^-$ and n-FH$_2^-$ have been discussed in detail previously.[15] According to RM, the dominant contributions to peaks **A**, **B**, and **C** in the p-FH$_2^-$ photoelectron spectrum are from transitions to the direct scattering states ($\nu = 0, j = 0, 2, 4$), where ν refers to the H$_2$ vibrational quantum number and j to the FH$_2$ hindered rotor state (very close to a H$_2$ free rotation). The first three such transitions are marked by an * in Fig. 6. The remaining features are due to resonance states localized in the reactant or product valleys of the F + H$_2$ surface, which are generally somewhat narrower

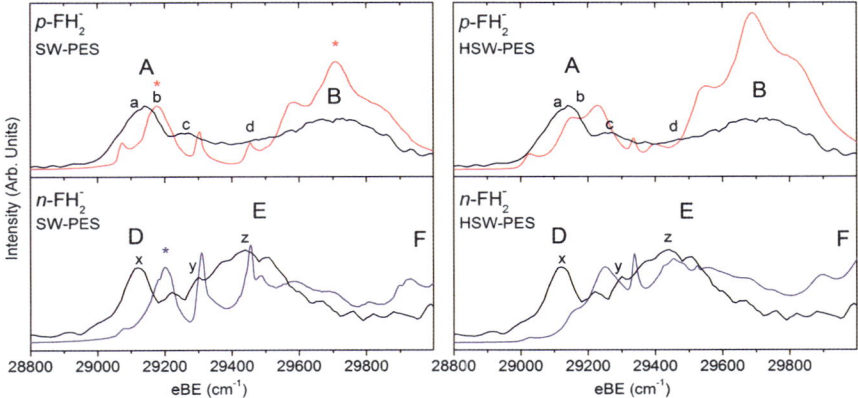

Fig. 6 Comparison of p- and n-FH$_2^-$ SEVI spectra taken at 325 nm (black traces) with simulated results using the SW-PES[1] and the HSW-PES[2]. Red traces correspond to simulated p-FH$_2$ spectra and blue to n-FH$_2^-$ (a 3 : 1 ratio of *ortho* : para). Simulated direct scattering states are indicated with an * (see text).

than the peaks associated with direct scattering states. Upon first glance, the n-FH_2^- bands **E**, **F** and **G** would then be assigned to the ($v = 0, j = 1, 3, 5$) direct scattering states by their strong intensities and by their positions in relation to the even j peaks for p-FH_2^-. In the n-FH_2^- simulations however, the direct scattering transition to the ($v = 0, j = 1$) state appears between peaks **D** and **E**, the ($v = 0, j = 3$) state contributes to peak **F**, and several resonance features are also seen. The ($v = 0, j = 1$) state lies only 3 meV above the ($v = 0, j = 0$) state in the simulations by RM. The main difference in the HW simulations is that the resonance peaks are reduced relative to the direct features, an effect attributed to the slightly larger barrier on the HSW neutral surface used in their simulations (bent barriers for SW and HSW are 535 cm^{-1} and 668 cm^{-1}, respectively[59]). Peaks in the HW simulations are also at slightly higher eBE, in general. HW also note that the near degeneracy of the ($v = 0, j = 0$) and ($v = 0, j = 1$) states reflects a small tunneling splitting due to the double-minimum well at the TS of the SW-PES. The positions of bands **A**, **B**, **C** and **E**, **F** provide a benchmark for the simulation and testing of the hindered rotor potential at the transition state of FH_2.

We next focus on comparing the finer structure seen in the SEVI data to the simulations, particularly the spectral region around **A** and **D**. This is complicated by the noise level in the experimental spectra, but the features labeled with lower-case letters offer some interesting points of comparison. For example, peak **c**, in the p-FH_2^- spectrum and the shoulder **y** in n-FH_2^- line up with a feature in the RM simulations assigned to the ($v' = 3, j' = 0, t' = 0$) product resonance state. In the simulated p–FH_2^- spectrum, this product resonance is also the most intense, and happens to be well-separated from the neighboring direct scattering peaks, making it the best candidate for experimental detection. This is the same resonance that has been definitively identified in crossed beam experiments on F + HD.[23,36–39] The neighboring ($v' = 3, j' = 2, t' = 0$) product reactive resonance then lines up with **d** in the top SEVI p-FH_2^- trace and **z** in the bottom n-FH_2^- trace of Fig. 3.

However, the most interesting comparison involves peaks **a**, **b**, and **x**. Although peaks **a** and **b** do not appear as two distinct features at any photon energy, Fig. 3 shows that as the photon energy is lowered, the band **A** maximum shifts to lower eBE and becomes narrower. If peak **b** were simply due to an unresolved contribution of peak **a** and **c**, we would expect either a shifted broad maximum, or a non-shifted narrower peak with a shoulder. Since we observe both a shift and a shoulder, it seems that band **A** might comprise *three* unresolved transitions, **a**, **b** and **c**, with an increase in relative intensity of peak **a** close to threshold. Such an effect could be due to differences in anisotropy with peak **a** (and possibly **c**) having more s-wave character and peak **b** having more p-wave character, causing peak **a** to persist even close to threshold. Unfortunately, the peaks comprising band **A** are not fully resolved and so we are unable to distinguish between the anisotropies (in the form of beta parameters[51]) of these peaks. Differences in intrinsic peak width could also cause the observed intensity behavior of peaks **a**, **b** and **c**, with narrow transitions retaining significant intensity close to threshold while the higher eBE portion of a broader peak disappears. Differences in peak widths are indeed expected for this system, with direct-scattering states being broader than resonance states.[16]

Comparing the n-FH_2^- and p-FH_2^- SEVI spectra, we find that at the lowest photodetachment energy, the Gaussian-fit center of peak **a** occurs 15 cm^{-1} *higher* than peak **x**. However, peak **a** is significantly broader than peak **x** (118 cm^{-1} *versus* 90 cm^{-1}), and the rising edges of both **x** and **a** match up exactly, as shown in the inset in Fig. 3. This comparison suggests that peak **a** is not yet fully resolved from the neighboring peak **b**. There is little evidence for the analog of peak **b** in the n-FH_2^- spectrum.

Peaks **a** and **x** each fall between two features in the simulated spectra. Peak **a** lies between the ($v = 0, j = 0, t = 0$) reactant resonant state at 29050 cm^{-1} and the ($v = 0, j = 0$) direct scattering peak. Peak **x** lies between the same resonance peak (which is much lower in intensity as it originates from p-FH_2^-) and the direct scattering peak dominated by the ($v = 0, j = 1$) state. These comparisons suggest two possible assignments for peaks **a**, **b**, and **x**. First, peak **b** could be the ($v = 0, j = 0$) direct scattering state, with the ($v = 0, j = 1$) state lying considerably higher in energy, perhaps at the band **E** maximum. A larger splitting between these states would be expected for nearly free rotation of the H_2 moiety in the FH_2 complex. Under these circumstances, peaks **a** and **x** would be assigned to resonance states, with their approximately equal binding energies and strong intensities in both p-FH_2^- and n-FH_2^- spectra suggesting they might be product as opposed to reactant resonances. This assignment agrees with earlier analyses of lower resolution PE results.[11]

An alternate assignment arises if peak **b** is not a true peak, but instead simply the maximum of the convolution of peaks **a** and **c**. This would then require that peak **a** be the ($v = 0, j = 0$) state, which would appear with some intensity in the n-FH_2^- spectrum as peak **x**. The ($v = 0, j = 1$) direct scattering state in the n-FH_2^- spectrum would then be significantly higher in energy, at peak **E**. This assignment is appealing as it does not require the presence of any other reactant or product resonances than **c**, for which there is experimental evidence, and attributes all the most intense features to direct scattering states. However, if peak **x** (band **D**) originates only from the presence of p-FH_2^- in the n-FH_2^- spectrum, we would also expect stronger contributions from the other p-FH_2^- bands, especially in the 311 nm n-FH_2^- SEVI trace where the 311 nm p-FH_2^- results show intense **B** and **C** bands. These contributions are clearly missing, most notably in the 29800 cm^{-1} region of the 311 nm n-FH_2^- SEVI trace (Fig. 3).

Overall, the comparison between the SEVI data and simulated spectra indicates that there are discrepancies between experiment and theory in the low eBE spectra where our signal-to-noise is best. It would be of considerable interest to test the proposed assignments of the peaks in this region by performing simulations based on the more recent potential energy surfaces available for the F + H_2 reaction.[19–23]

2. FCH$_4^-$ and FCD$_4^-$

In this section, we consider the highly structured FCH$_4^-$ and FCD$_4^-$ SEVI spectra and attempt to interpret them in light of past experimental and theoretical work. Infrared spectroscopy and electronic structure calculations[29,31] indicate that FCH$_4^-$ has C_{3v} symmetry with a linear F–H–C bond arrangement. The most recent calculations[29] find $r_{CF} = 2.955$ Å and $D_0(FCH_4^-) = 0.290$ eV. On the recent F + CH$_4$ surface reported by Czakó et al.,[28] there are two vdW structures in the entrance valley lying 40 and 160 cm^{-1} below the reactant asymptote, and a saddle point 240 cm^{-1} above the reactants. The more weakly bound vdW structure, which is actually a saddle point, has the same C_{3v} structure as the anion but with a considerably longer r_{CF} of 3.6 Å. The other structure, representing a true minimum, also has C_{3v} symmetry with the F atom bound directly to the C atom (i.e. back-side bonding) with $r_{CF} = 2.940$ Å. At the TS, calculations at the highest level of theory find a nonlinear F–H–C bond with $\angle$FHC = 152.3° and $r_{CF} = 2.732$ Å, but the potential energy with respect to this angle is very flat; the optimized TS geometry with $\angle$FHC = 180° is higher in energy by less than 10 cm^{-1}. Based on these results, as pointed out by Cheng et al.,[12] photodetachment of FCH$_4^-$ will probe the reactant side of the F + CH$_4$ reaction. The Franck–Condon region accessible via photodetachment should overlap best with the more strongly bound vdW structure and the reactant side of the transition state region, as indicated in Fig. 1.

We can distinguish the contributions from these two regions to the spectrum by comparing the eBE of each feature to the reactant asymptote E_{asympt}, illustrated in Fig. 1 and defined by

$$E_{asympt} = D_0(FCH_4^-) + EA(F).$$

We find $E_{asympt} = 3.691$ eV ($\sim$29770 cm^{-1}) based on the electron affinity of fluorine[60] (EA(F) = 3.401 eV) and $D_0(FCH_4^-)$ given above. Below this energy, only those states residing within the entrance channel vdW well are expected. In the FCH_4^- spectrum, all features associated with band **A** fall below this asymptote and are thus assigned to bound reactant vdW states. Bands **B** and **C** fall above E_{asympt}, and are attributed to states with energies partway between the reaction asymptote and the TS.[12] The sharp peak **d** matches E_{asympt} almost exactly. In the FCD_4^- spectrum, E_{asympt} is expected to be slightly lower (by $\sim$22 cm^{-1}, difference based on optimized MP2/aug-cc-pVDZ structures) due to the isotope effect on D_0, and only the unresolved low eBE tail extending to $\sim$29600 cm^{-1} falls below it.

The vdW band **A** shows some fine structure labeled **a–d** in Fig. 5, although the inclusion of peak **d** with this band rather than band **B** is somewhat arbitrary. As indicated above, r_{CF} for the anion and more strongly bound reactant vdW structure are quite close, but the orientation of the CH_4 relative to the F atom changes from a linear F–H–C bond conformation in the anion ($\angle FCH = 0°$) to a back-side bonded F–C–H bond conformation in the neutral ($\angle FCH = 70.5°$).[29,28] Under these circumstances, it is reasonable to assign peaks **a–c**, and possibly **d**, to intermolecular bending vibrations or hindered rotation of the CH_4 moiety of this vdW complex; the latter would be consistent with the varying peak spacing with eBE. The lowest eBE feature, peak **a**, lies only 124 cm^{-1} below E_{asympt}, consistent with the calculated binding energy of the lowest energy vdW structure (160 cm^{-1}) relative to F + CH_4.

Intensity in the vdW region of the FCD_4^- spectrum (low eBE tail) is less pronounced. Since the methyl orientations in the anion and lowest energy neutral vdW structure are so different, the Franck–Condon intensity will be governed by an imperfect overlap of the wavefunctions, aided by the shallow neutral potential in the $\angle FCH$ coordinate and the spread-out wavefunctions residing within. We expect the lowest energy vdW state to be less intense in the deuterated species than in the hydrogenated species due to a small decrease in zero-point energy (ZPE) causing a large decrease in wavefunction spread in the shallow neutral potential. Fig. 7 presents a highly schematic version of potential energy *versus* $\angle FCH$ in the anion and neutral. We also expect the best overlap with the anion levels to occur at a higher quantum number state in the deuterated species than in the hydrogenated species due to the smaller level spacing. In any case, it is unlikely that the usual Franck–Condon/harmonic oscillator picture for analyzing photoelectron spectra will be sufficient to interpret the vdW features in the SEVI of spectra of the two isotopologs; a full multidimensional quantum treatment along the lines of that used to simulate the ClH_2^- SEVI spectrum will probably be needed.[33]

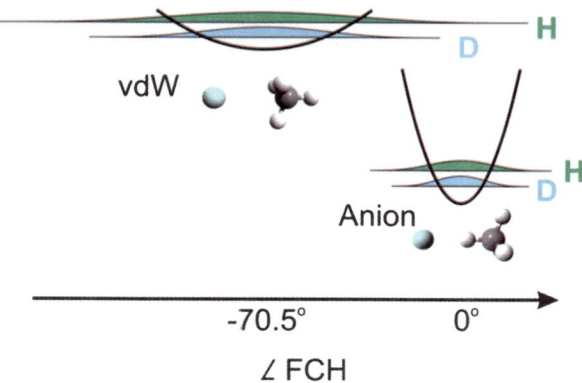

Fig. 7 Schematic potential energy diagram as a function of F–C–H angle ($\angle FCH$) showing comparative wavefunction spread for hydrogenated and deuterated species in the shallow vdW well and in the deep anion well.

The remaining region of the spectrum (bands **B** and **C**; **D** and **E**) shows progressions with an average spacing of 26 cm^{-1} in FCH$_4^-$. These regions exhibit two intensity maxima separated by 58 and 64 cm^{-1} for the two isotopologs. All of this structure lies above the reactant asymptote (E_{asympt}), peaking at a vertical detachment energy (VDE) of E_{asympt} + 126 cm^{-1} (see Fig. 5) and extending to about 300 cm^{-1} above E_{asympt} (see Fig. 4). These values are near to the previous VDE[12] estimate of E_{asympt} + 100 cm^{-1} and the calculated TS energy[28] of 240 cm^{-1} above E_{asympt}, respectively, indicating that any structure underlying **B**, **C**, **D** and **E** corresponds either to direct scattering states or resonances near the transition state but primarily on the reactant side. Taking the calculated saddle point geometry as a reference point, there are significant changes upon photodetachment in $\angle$FCH and r_{CF}, as mentioned above, which would lead to low frequency progressions in intermolecular bending modes and the intermolecular F$\cdots$CH$_4$ stretch. To first order, these modes are perpendicular and parallel, respectively, to the reaction coordinate in the entrance valley. Under this assumption, direct scattering states would be quantized with respect to the intermolecular bend or hindered CH$_4$ rotation (along with all the other higher frequency vibrational modes), whereas resonances states, as in F + H$_2$, would be additionally quantized along the intermolecular stretch frequency.

The fine structure comprising bands **B** and **C** can be compared to the available calculated frequencies for the TS of the neutral. The TS of FCH$_4$ has an imaginary intermolecular stretching frequency at 357i(246i) cm^{-1}, and two intermolecular bending vibrations at 40(277) cm^{-1} and at 117(344) cm^{-1} with a' and a'' symmetry, respectively.[28] These values are all harmonic, calculated from the PES fundamental (or from *ab initio* structures). None of the calculated TS frequencies are in particularly good agreement with peak spacings in the SEVI spectra. However, treating the intermolecular bend as a harmonic vibration is unlikely to be very accurate, as the bend potential at the transition state is very anharmonic and flat, with shallow minima separated by small barriers. Hence, one might expect a considerably lower frequency for the intermolecular bend than the harmonic value, or, alternatively, hindered rotor structure with peaks spaced much more closely than the harmonic bend frequency. Both bending and hindered rotor levels would display a substantial isotope effect in going from CH$_4$ to CD$_4$ that would be much less pronounced for an intermolecular stretching mode. These considerations suggest assigning the 26 cm^{-1} progression to large amplitude intermolecular bend vibrations or hindered rotations.

The question then arises as to whether the fine structure in the spectrum represents direct scattering states or resonance structure. It is difficult to resolve this issue in the absence of high level calculations. However, the experimental peaks are narrower (fwhm $\approx$ 27 cm^{-1}) than the direct scattering features in the FH$_2^-$ simulations (fwhm $\approx$ 76 cm^{-1}), suggesting that they arise from resonance states. If this is the case, then the ~58 cm^{-1} spacing underlying the fine structure might represent a progression along the reaction coordinate, primarily involving the intermolecular F$\cdots$CH$_4$ stretch. The calculated frequency[29] of the corresponding mode in the anion is 200 cm^{-1}, and a substantially lower value for the corresponding quasi-bound mode in the neutral is reasonable. Moreover, this mode should exhibit a minimal isotope effect as it primarily involves heavy atom motion, consistent with the observation of similar spacings in the SEVI spectra of FCH$_4^-$ and FCD$_4^-$.

The interpretation of the results shown here is necessarily simplistic. For example, the assignment of a single internal coordinate as the reaction coordinate is a questionable approximation, given that r_{CH}, r_{HF}, and $\angle$FCH, where H refers to the hydrogen atom being transferred, all evolve along the minimum energy path in the reactant valley.[28] What is unambiguous from our experiment is the observation of narrower features than have ever been seen in any of our transition state spectroscopy experiments with the exception of IHI$^-$, where we clearly were observing resonances.[34]

The SEVI spectroscopy of the FCH$_4$ ← FCH$_4^-$ system can improve the understanding of the bimolecular reaction by providing experimental features with which to test the currently available F + CH$_4$ and FCH$_4^-$ PES.[28-30] Crossed beam experiments on the F + CH$_4$ reaction have shown results such as product selectivity due to reactant vibrational excitation that depend on the exact location of the TS. The identification of resonances on the reactant side of the TS (*versus* the previously identified reactive resonances in the HF($\nu' = 3$) product region)[5,41] could have implications on the dynamics of the bimolecular reaction. Additionally, the properties of the reactant vdW region are especially important in low collision energy cross beam experiments, as suggested by the increased FD product selectivity in the F + CHD$_3$($\nu_1 = 1$) reaction at low collision energies.[61] The vdW region has proven particularly difficult to simulate,[28] and so the observed vdW structure, which shows multiple peaks and a strong isotope dependence, may well provide the experimental motivation for a detailed exploration of this surface including accurate frequencies and barriers to interconversion of the two different vdW structures.

V. Conclusions

Here we present a set of high resolution negative ion photoelectron spectra for p-FH$_2^-$, n-FH$_2^-$, FCH$_4^-$ and FCD$_4^-$ ions, improving on previously published results and showing transitions to states at or near the TS of the F + H$_2$ or F + CH$_4$ bimolecular reactions. The higher resolution spectra for p-FH$_2^-$ resolve a new peak **a** observable only close to threshold, while the spectra for n-FH$_2^-$ completely resolve a previously observed shoulder **x**. These two lowest energy peaks are significantly narrower than the other features and occur at the same position in both spectra, disagreeing with previous assignments. A narrow peak is identified and tentatively assigned to the ($v' = 3, j' = 0, t' = 0$) product resonance state, the same resonance that has been seen in crossed beam experiments on F + HD.[23,36-39] The experimental results suggest that improvements are needed in the SW-PES and HSW-PES, in particular for the bending potential in the TS region. High resolution spectra of the FCH$_4^-$ ^{2}P$_{3/2}$ states reveal a rich fine structure, with peak progressions spaced by 26 cm^{-1}. We attribute this structure to the bending or hindered rotation of the methyl moiety with respect to the fluorine atom. A band attributed to bound vdW states is also observed in the FCH$_4^-$ spectrum, and shows significant changes upon deuteration. We attribute this to the isotope effect, which decreases the neutral vdW cluster ZPE and level spacings resulting in differing Franck–Condon overlap of the anion and neutral states.

Much is left to be unraveled in the SEVI spectra of these TS systems. We hope that, given the availability of increasingly accurate PES and the documented procedures for simulation of photoelectron results including quantum effects like reactive resonances,[16,17] theoretical comparisons might soon be available for the FCH$_4^-$ and FCD$_4^-$ spectra. Treatment of the hindered rotation in such a system is also an interesting problem all on its own. Conversely, we also hope that these SEVI results will allow for further testing and validation of both the F + CH$_4$ and F + H$_2$ PES by providing a direct probe of the TS region and complementing the numerous available crossed beam studies.

Acknowledgements

This work was supported by the Air Force Office of Scientific Research under grant numbers FA9550-09-1-0343 and F49620-03-1-0085. We thank Oleg Kornilov for use of his program to perform the pBasex transform and wedge subtractions. We also thank David Manolopoulos and Gabor Czakó for helpful discussions. T.I.Y. and E.G. thank the National Science and Engineering Research Council of Canada (NSERC) for post graduate scholarships. C.H. is supported by a postdoctoral scholarship from the German Academic Exchange Service (DAAD).

This journal is © The Royal Society of Chemistry 2012

Notes and references

1 D. M. Neumark, A. M. Wodtke, G. N. Robinson, C. C. Hayden and Y. T. Lee, *J. Chem. Phys.*, 1985, **82**, 3045–3066.
2 D. E. Manolopoulos, *J. Chem. Soc., Faraday Trans.*, 1997, **93**, 673–683.
3 K. P. Liu, *Annu. Rev. Phys. Chem.*, 2001, **52**, 139–164.
4 W. F. Hu and G. C. Schatz, *J. Chem. Phys.*, 2006, **125**, 132301.
5 W. Shiu, J. J. Lin and K. P. Liu, *Phys. Rev. Lett.*, 2004, **92**, 103201.
6 W. Q. Zhang, H. Kawamata and K. P. Liu, *Science*, 2009, **325**, 303–306.
7 J. Zhou, J. J. Lin and K. Liu, *Mol. Phys.*, 2010, **108**, 957–968.
8 N. Balucani, G. Capozza, F. Leonori, E. Segoloni and P. Casavecchia, *Int. Rev. Phys. Chem.*, 2006, **25**, 109–163.
9 K. P. Liu, *J. Chem. Phys.*, 2006, **125**, 132307.
10 D. M. Neumark, *Phys. Chem. Chem. Phys.*, 2005, **7**, 433–442.
11 D. E. Manolopoulos, K. Stark, H. J. Werner, D. W. Arnold, S. E. Bradforth and D. M. Neumark, *Science*, 1993, **262**, 1852–1855.
12 M. Cheng, Y. Feng, Y. K. Du, Q. H. Zhu, W. J. Zheng, G. Czakó and J. M. Bowman, *J. Chem. Phys.*, 2011, **134**, 191102.
13 S. E. Bradforth, D. W. Arnold, R. B. Metz, A. Weaver and D. M. Neumark, *J. Phys. Chem.*, 1991, **95**, 8066–8078.
14 A. Weaver and D. M. Neumark, *Faraday Discuss. Chem. Soc.*, 1991, **91**, 5–16.
15 S. E. Bradforth, D. W. Arnold, D. M. Neumark and D. E. Manolopoulos, *J. Chem. Phys.*, 1993, **99**, 6345–6359.
16 C. L. Russell and D. E. Manolopoulos, *Chem. Phys. Lett.*, 1996, **256**, 465–473.
17 B. Hartke and H. J. Werner, *Chem. Phys. Lett.*, 1997, **280**, 430–438.
18 K. Stark and H. J. Werner, *J. Chem. Phys.*, 1996, **104**, 6515–6530.
19 M. Hayes, M. Gustafsson, A. M. Mebel and R. T. Skodje, *Chem. Phys.*, 2005, **308**, 259–266.
20 M. H. Qiu, Z. F. Ren, L. Che, D. X. Dai, S. A. Harich, X. Y. Wang, X. M. Yang, C. X. Xu, D. Q. Xie, M. Gustafsson, R. T. Skodje, Z. G. Sun and D. H. Zhang, *Science*, 2006, **311**, 1440–1443.
21 C. X. Xu, D. Q. Xie and D. H. Zhang, *Chin. J. Chem. Phys.*, 2006, **19**, 96–98.
22 G. L. Li, H. J. Werner, F. Lique and M. H. Alexander, *J. Chem. Phys.*, 2007, **127**, 174302.
23 Z. Ren, L. Che, M. Qiu, X. Wang, W. Dong, D. Dai, X. Wang, X. Yang, Z. Sun, B. Fu, S.-Y. Lee, X. Xu and D. H. Zhang, *Proc. Natl. Acad. Sci. U. S. A.*, 2008, **105**, 12662–12666.
24 J. F. Castillo, F. J. Aoiz, L. Banares, E. Martinez-Nunez, A. Fernandez-Ramos and S. Vazquez, *J. Phys. Chem. A*, 2005, **109**, 8459–8470.
25 D. Troya, *J. Chem. Phys.*, 2005, **123**, 214305.
26 T. S. Chu, X. Zhang, L. P. Ju, L. Yao, K. L. Han, M. L. Wang and J. Z. H. Zhang, *Chem. Phys. Lett.*, 2006, **424**, 243–246.
27 J. Espinosa-Garcia, J. L. Bravo and C. Rangel, *J. Phys. Chem. A*, 2007, **111**, 2761–2771.
28 G. Czakó, B. C. Shepler, B. J. Braams and J. M. Bowman, *J. Chem. Phys.*, 2009, **130**, 084301.
29 G. Czakó, B. J. Braams and J. M. Bowman, *J. Phys. Chem. A*, 2008, **112**, 7466–7472.
30 G. Czakó and J. M. Bowman, *Phys. Chem. Chem. Phys.*, 2011, **13**, 8306–8312.
31 Z. M. Loh, R. L. Wilson, D. A. Wild, E. J. Bieske and M. S. Gordon, *Aust. J. Chem.*, 2004, **57**, 1157–1160.
32 D. M. Neumark, *Acc. Chem. Res.*, 1993, **26**, 33–39.
33 E. Garand, J. Zhou, D. E. Manolopoulos, M. H. Alexander and D. M. Neumark, *Science*, 2008, **319**, 72–75.
34 I. M. Waller, T. N. Kitsopoulos and D. M. Neumark, *J. Phys. Chem.*, 1990, **94**, 2240–2242.
35 J. C. Polanyi and A. H. Zewail, *Acc. Chem. Res.*, 1995, **28**, 119–132.
36 R. T. Skodje, D. Skouteris, D. E. Manolopoulos, S. H. Lee, F. Dong and K. Liu, *J. Chem. Phys.*, 2000, **112**, 4536–4552.
37 F. Dong, S. H. Lee and K. Liu, *J. Chem. Phys.*, 2000, **113**, 3633–3640.
38 R. T. Skodje, D. Skouteris, D. E. Manolopoulos, S. H. Lee, F. Dong and K. P. Liu, *Phys. Rev. Lett.*, 2000, **85**, 1206–1209.
39 W. Dong, C. Xiao, T. Wang, D. Dai, X. Yang and D. H. Zhang, *Science*, 2010, **327**, 1501–1502.
40 J. G. Zhou, J. J. Lin and K. P. Liu, *J. Chem. Phys.*, 2004, **121**, 813–818.
41 G. Czakó, Q. A. Shuai, K. P. Liu and J. M. Bowman, *J. Chem. Phys.*, 2010, **133**, 131101.
42 A. Osterwalder, M. J. Nee, J. Zhou and D. M. Neumark, *J. Chem. Phys.*, 2004, **121**, 6317–6322.
43 D. M. Neumark, *J. Phys. Chem. A*, 2008, **112**, 13287–13301.
44 A. Eppink and D. H. Parker, *Rev. Sci. Instrum.*, 1997, **68**, 3477–3484.

45 U. Even, J. Jortner, D. Noy, N. Lavie and C. Cossart-Magos, *J. Chem. Phys.*, 2000, **112**, 8068–8071.
46 E. Garand, T. I. Yacovitch and D. M. Neumark, *J. Chem. Phys.*, 2009, **130**, 064304–064307.
47 J. Zhou, E. Garand, W. Eisfeld and D. M. Neumark, *J. Chem. Phys.*, 2007, **127**, 034304.
48 T. Theis, P. Ganssle, G. Kervern, S. Knappe, J. Kitching, M. P. Ledbetter, D. Budker and A. Pines, *Nat. Phys.*, 2011, **7**, 571–575.
49 W. C. Wiley and I. H. McLaren, *Rev. Sci. Instrum.*, 1955, **26**, 1150–1157.
50 D. W. Chandler and P. L. Houston, *J. Chem. Phys.*, 1987, **87**, 1445–1447.
51 A. Sanov and R. Mabbs, *Int. Rev. Phys. Chem.*, 2008, **27**, 53–85.
52 E. W. Hansen and P.-L. Law, *J. Opt. Soc. Am. A*, 1985, **2**, 510–519.
53 G. A. Garcia, L. Nahon and I. Powis, *Rev. Sci. Instrum.*, 2004, **75**, 4989–4996.
54 V. Dribinski, A. Ossadtchi, V. A. Mandelshtam and H. Reisler, *Rev. Sci. Instrum.*, 2002, **73**, 2634–2642.
55 U. Berzinsh, M. Gustafsson, D. Hanstorp, A. Klinkmuller, U. Ljungblad and A. M. Martenssonpendrill, *Phys. Rev. A: At., Mol., Opt. Phys.*, 1995, **51**, 231–238.
56 C. Blondel, *Phys. Scr.*, 1995, **T58**, 31–42.
57 E. P. Wigner, *Phys. Rev.*, 1948, **73**, 1002–1009.
58 O. Kornilov, C. C. Wang, O. Bunermann, A. T. Healy, M. Leonard, C. Peng, S. R. Leone, D. M. Neumark and O. Gessner, *J. Phys. Chem. A*, 2010, **114**, 1437–1445.
59 J. F. Castillo, B. Hartke, H. J. Werner, F. J. Aoiz, L. Banares and B. Martinez-Haya, *J. Chem. Phys.*, 1998, **109**, 7224–7237.
60 C. Blondel, C. Delsart and F. Goldfarb, *J. Phys. B: At., Mol. Opt. Phys.*, 2001, **34**, L281–L288.
61 G. Czakó and J. M. Bowman, *J. Am. Chem. Soc.*, 2009, **131**, 17534–17535.

The last mile of molecular reaction dynamics virtual experiments: the case of the OH(N = 1–10) + CO(j = 0–3) reaction†

Antonio Laganà,[*a] Ernesto Garcia,[b] Alessandra Paladini,[‡a] Piergiorgio Casavecchia[a] and Nadia Balucani[a]

Received 6th March 2012, Accepted 17th April 2012

DOI: 10.1039/c2fd20046e

By exploiting the potentialities of a recently implemented grid empowered molecular simulator based on the combination of collaborative interoperable service oriented computing and the usage of high performance – high throughput technologies, the results of crossed molecular beam experiments have been virtually simulated and compared with the real (measured) laboratory data for the reactive system OH + CO. The direct comparison of theoretically predicted laboratory angular distributions with experimental raw data avoids possible uncertainties associated with the analysis of crossed beam experiments, in which trial centre-of-mass functions are tested until the best-fit of the experimental data is achieved. To make such a comparison as accurate as possible, the rotational distributions of the OH radicals employed in previous crossed beam experiments have been characterized by laser-induced-fluorescence. The capability of performing massive calculations using grid-distributed technologies has allowed the running of quasiclassical trajectory calculations for all the initial rotational states of the OH radicals present in the beam (from the ground rotational state $N_{OH} = 1$ up to $N_{OH} = 10$) on three different potential energy surfaces and the comparison of related outcomes.

1 Introduction

Data produced by crossed molecular beam (CMB) experiments are an ideal test for the outcomes of theoretical molecular reaction dynamics studies.[1-4] This is particularly true if measurements are performed using a narrow distribution of reactant collision energy (E_c) around its nominal value and a single or a small set of reactant rovibrational states (v, j) (unprimed quantities refer to reactants, primed quantities to products). The main observable of CMB experiments is the product number density $N_{LAB}(\Theta',t')$ measured in the laboratory (LAB) frame as a function of the scattering angle, Θ', and time of flight, t'.[3] This quantity can be evaluated directly from first principles, if one knows the experimental conditions and machine characteristics.[4] Yet, no single stream procedure has ever been developed to standardize

[a]Dipartimento di Chimica, Università degli Studi di Perugia, E-mail: lag@unipg.it; nadia.balucani@unipg.it; Fax: +39 075 5855606; Tel: + 39 075 5855527; + 39 075 5855507
[b]Departamento de Quimica Fisica, Universidad del Pais Vasco (UPV/EHU), Vitoria, Spain. E-mail: e.garcia@ehu.es

† Figures reporting PADs and PTDs for various combinations of N_{OH} and j_{CO} on the YMS and LTSH PESs. See DOI: 10.1039/c2fd20046e
‡ Present Address: CNR-IMIP, Istituto di Metodologie Inorganiche e dei Plasmi, Sede di Potenza, I-85050, Tito Scalo (PZ) Italy.

such a process and offer it as a service on the web. The reason for this resides in the fact that one needs:

a) a well automated procedure allowing the interconnecting of theoretical (quantum chemistry and molecular dynamics) predictions with the experimental quantities (considering the appropriate experimental conditions);

b) free access to the vast quantity of computing resources needed to carry out related computational procedures.

In most CMB experiments, experimentalists convert $N_{LAB}(\Theta', t')$, that depends on the experimental conditions (such as crossing angle, velocity distributions of the reactants, detector resolution) into the centre-of-mass (CM) differential cross section that corresponds to the CM product flux $I_{CM}(\theta', u')$ and is a function of the CM scattering angle, θ', and velocity, u'. This quantity can, then, be compared with the results of other experimental data measured under different experimental conditions, as well as with the outcome of theoretical calculations. In principle, it is possible to directly convert the measured $N_{LAB}(\Theta', t')$ into the $I_{CM}(\theta', u')$. However, such a direct inversion procedure is in practice never used. Because of the finite resolution of experimental conditions (*i.e.*, finite angular and velocity spread of the reactant beams and angular resolution of the detector), in fact, the LAB → CM transformation is difficult to achieve and, therefore, the analysis of the laboratory data is carried out by a forward convolution trial-and-error procedure, in which tentative CM angular and velocity distributions are assumed, averaged and transformed to the LAB distributions for comparison with the experimental data, until the best-fit is achieved (with this procedure it is trivial to account for the averaging over the experimental conditions). Furthermore, $I_{CM}(\theta', u')$ is normally expressed as a product of angular (PAD) and translational energy (PTD) distributions under the assumption that its angular and velocity (or translational energy) dependences are largely uncoupled. This approximation is not always warranted (see, for instance, Ref.s 5–14) and the coupling between PAD and PTD can be accounted for in the trial-and-error procedure. The best-fit CM PAD and PTD are affected by some uncertainty, that is usually rendered in graphic terms as an error bar comprising the best fit functions and delimiting all the CM functions that provide an acceptable fit of the experimental data. The best-fit CM PAD and PTD are interpreted by associating their shapes with the reaction mechanism under guidance of some accredited model behaviours.[3,4] Yet, only a comparison of the experimental information with the outcomes of a dynamical calculation performed on an accurate potential energy surface (PES) can provide a solid ground for understanding the underlying reaction mechanism and the role played by atomic and molecular interactions. Such a comparison can be performed at two different levels. In most cases, the comparison is done directly between calculated and experimental PADs and PTDs (see, for instance, Refs. 15–17). Nevertheless, because of the uncertainties associated with the derivation of the experimental PADs and PTDs from the measured raw data, a more straightforward comparison can be performed by transforming the theoretical PADs and PTDs into the LAB distributions, so that a direct comparison with the raw laboratory data can be obtained (see, for instance, Refs. 6–9,18,19) and uncertainties associated to the derivation of the best fit PADs and PTDs avoided. Such a transformation has to be carried out by taking into account the experimental conditions (crossing angle, beam velocities) and the averaging over the experimental parameters (beam velocity distributions, angular divergences, detector aperture). This approach represents the most accurate evaluation of the quality of an *ab initio* PES and/or dynamical calculations. To achieve this goal, different research groups with different expertise are normally involved, as *ab initio* electronic structure and dynamical calculations are required, as well as the forward convolution routine with the appropriate averaging.[6-9]

Some of the present authors have developed a single stream procedure to calculate from first principles the $N_{LAB}(\Theta', t')$ distributions, including the "last mile" (from $I_{CM}(\theta', u')$ to $N_{LAB}(\Theta', t')$) step by implementing a suitable computational procedure that converts the calculated CM functions into LAB distributions to be compared directly with the experimental results. This procedure has been automated and offered as a web service. All this has become feasible thanks to the use of the European Grid Infrastructure (EGI),[20] the Grid Empowered Molecular Simulator (GEMS)[21] and the Service Oriented Architecture (SOA)[22] approach for the development of scientific software applications.

In this paper we report on the case of the OH + CO reaction, as an example of the use of this procedure. An accurate comparison of the theoretical predictions on three different PESs available in the literature with the CMB experimental data of Casavecchia and collaborators[23,24] will be presented, as obtained by considering all the rotational levels of OH and CO involved in the experiments. The reaction OH + CO is of paramount importance in combustion and atmospheric chemistry, being the final stage of conversion of hydrocarbons and CO into CO_2. For this reason numerous dynamical calculations have been performed by using quasiclassical trajectory (QCT) or approximate quantum dynamical methods on different PESs.[17,25–48] Amongst them, the most recent and most used are those derived by Yu et al. (YMS),[33] Lakin et al. (LTSH)[17] and Valero et al. (Leiden).[34] The YMS and LTSH PESs are built on ab initio accurate electronic energy values fitted using a Many-Body Expansion (MBE) functional[49] plus various Gaussian terms to enforce the reproduction of some local structures of the ab initio data. The Leiden PES instead is formulated using the Shepard interpolation of a set of 1250 high level (linear combination of DFT and CCSD(T) energies) ab initio potential energy values together with their first and second derivatives generated iteratively using the Collins' GROW program.[50] Accordingly, in the Leiden PES the potential energy associated with an arbitrary geometry is calculated as a weighted sum of Taylor expansions (truncated to the second order) each centered on one of the ab initio data points. This makes the computation associated with the Leiden PES significantly more expensive (about 100 times) than that associated with the YMS and LTSH ones. While writing this manuscript, a new HOCO PES has been published by Bowman and coworkers, but it has not been tested in dynamical calculations yet.[51]

Some results of dynamical calculations on the three PESs have been already compared with the best-fit CM functions derived in CMB experiments at $E_c = 8.6$ and 14.1 kcal mol^{-1}, but before our recent paper[52] they have never been compared directly with the LAB raw data. Using the GEMS simulator, we have compared QCT calculations on the three YMS, LTSH and Leiden PESs with laboratory data by considering the ground rovibrational states of OH and CO.[52] In addition, we have explicitly considered the possible effects of the PAD and PTD coupling to establish whether there was an effect on the simulated LAB data.[52] However, this work has not been conclusive yet, because of the missing information on the internal rotational states of the reactants. To better judge the quality of PESs within the QCT approach, indeed, we should consider the possible involvement of the higher OH rotational levels present in the beams employed in the CMB experiments. The OH beams were produced in a radiofrequency (RF) discharge beam source,[53,54] where the cooling of the internal degrees of freedom is incomplete. As shown by Lakin et al. on the LTSH PES,[17] which treats accurately the entrance channel with the inclusion of two van der Waals wells, the rotational excitation of OH can have a significant impact on the global reactivity of the system. Therefore, it is important to determine the experimental rotational state distributions of OH and characterize the dynamics of rotationally excited OH from scattering calculations, because the state of the comparison between experimental and theoretical results could be affected by their neglect.

In the experimental part of this paper, we present the results of the characterization of the OH rotational population in the beams used for the CMB experiments, obtained by means of a dedicated laser induced fluorescence (LIF) apparatus described in Refs. 55 and 56. As we are going to see, rotational levels up to

$N_{OH} = 10$ have been found to be populated. Once achieved the rotational distribution of OH, an accurate inclusion of the contributions from each rotational level has required additional calculations relative to the reactions involving each populated OH rotational level. In other words, QCT calculations have been performed for each initial rotational state of the OH reactant (from $N = 1$ to $N = 10$). Due to the classical nature of the calculations, it has been not possible to reproduce the quantum number of the rotational states of OH and the setting $j_{OH} = N_{OH} - 1$ has been used. The first four rotational levels of the CO reactant have also been considered in the calculations. These four rotational levels are the most populated at the estimated rotational temperature of the CO beam.[23] Therefore, reactive differential cross sections have been derived by QCT calculations for 36 combinations of N_{OH} and j_{CO}. For the Leiden PES, only the effect of N_{OH} was considered because of the much higher computer time needed for the calculations.

The GEMS strategy is the ideal tool to undertake this kind of massive calculations (2.2×10^9 trajectories were run on each YMS and LTSH PESs and 0.14×10^9 trajectories were run on the Leiden PES). All the contributions to the global reaction from the different internal states of the reactants have been considered to simulate the experimental $N_{LAB}(\Theta', t')$ distributions, so that the comparison between theoretical quantities and experimental observables has been made as accurately as possible to assess the quality of the published PESs and the effects of OH rotational excitation.

The paper is articulated as follows: the features of direct grid based *ab initio* molecular simulation are illustrated in Section 2; the experimental outcomes of the OH + CO reaction use case and OH beam characterization are described in Section 3; the comparison between theoretical and experimental data is presented on Section 4; conclusive remarks are given in Section 5.

2 The direct calculation of $N_{LAB} (\Theta', t')$

As already said, the direct calculation of $N_{LAB} (\Theta', t')$ from first principles and considering all the initial rotational states of the reactants has been made possible by the impressive evolution of networked computing. In recent years, in fact, networked computing has been developed either as (small clusters) of supercomputers (High Performance Computing or HPC) or as a large grid of off-the-shelf computers (High Throughput Computing or HTC) in order to increase by orders of magnitude the computing power offered to research and industry. HPC networks (like PRACE[57] in Europe and TeraGrid[58] in the United States) offer a suitable distributed infrastructure for tightly coupled calculations requiring large memory sizes, MPI libraries,[59] a high speed interconnection network, large and high throughput storage devices. The interconnection of the different supercomputers is mainly intended in this case for facilitating the job management and offering unified resources (like storage). On the other hand HTC grids (like that of EGI[20]) offer a highly cost-effective computational platform exploiting the concurrent elaboration of a huge number of small-to-middle sized computers most often of the rackable CPU cluster type. The typical job exploiting the advantage of this infrastructure consists of a large amount of substantially uncoupled computational tasks which are distributed on independent CPUs. Although HPC and HTC infrastructures have been developed separately (and sometimes even conflictingly so that related strategies are based on different views of concurrent computing, rely on different types of Middleware and also target two different classes of applications, numerical algorithms and computational approaches) we have developed a strategy combining them to the end of enabling the accurate modeling of real-like systems as well as virtual reality simulations based on multi-scale and multi-physics approaches. For this reason we have established within the European Grid Infrastructure EGI (with the support of the Italian Grid Infrastructure IGI[60]) a Virtual Organization (VO)[61] named COMPCHEM[62,63] whose mission is that of fostering collaborative work among molecular and materials science researchers and provide them with combined access to HPC-HTC resources.

The main product of COMPCHEM (that counts at present more than 100 members and accesses a large portion of the machines of the EGI platform) is the already mentioned grid empowered molecular simulator GEMS that connects in a single workflow the building blocks of an *ab initio* virtual experiment to be provided, eventually, as a service to the Grid users (references and more details on workflows and grid services will be given in the next section). GEMS was initiated by developing SIMBEX[64] (a simulator of atom diatom elementary processes occurring on CMB apparatuses and based on classical trajectory techniques). Then, following the strategy outlined in Ref. 65 and 66 GEMS evolved as a more general simulator incorporating *ab initio* packages (to calculate molecular electronic structure), both quantum and classical dynamics programs (to calculate the time evolution of the system) and, finally, the necessary statistical averaging over unobserved variables (to work out quantities of experimental interest). The first complete application of GEMS was carried out to perform a full *ab initio* quantum investigation of the reactive properties of $H + H_2$ for which also the problem of developing standard quantum chemistry data formats was considered.[66] GEMS is articulated into the following highly cooperative computational blocks:

- INTERACTION, that is devoted to the *ab initio* calculations determining at various levels of accuracy the electronic structure of the small molecular systems (within the Born–Oppenheimer approximation) and is meant to provide an extended point-wise representation (including all the molecular geometries appreciably contributing to the evolution of the reactive process) of the potential energy surface when this is not (or incompletely) available from the literature. To this end GEMS provides the option of choosing among different suites of codes (some of which are also commercial packages). The possibility of skipping this block for accessing dynamics programs (when using an on the fly approach) and of assembling empirical force field simulations is also considered.

- FITTING, that is devoted to either globally or locally interpolate *ab initio* data using a suitable functional form (also this block is skipped when an on the fly dynamical approach is considered or a force field is adopted or a suitable PES is already available from the literature). Such a process is often iterated (possibly after adding further input from the previous block or from empirical considerations) after a first calculation providing an overall view of the PES (worked out using a fast package at the best affordable level of accuracy). Next steps make use of higher level calculations, more detailed determination of critical geometries, symmetry enforcing, a graphical analysis of the PES and, possibly, also some preliminary dynamics studies. Higher level *ab initio* calculations are also performed when aiming at optimizing the force field parameters of large systems for small subsets of atoms.

- DYNAMICS, that is devoted to carry out dynamics calculations on the PES worked out in the FITTING block (or on the individual *ab initio* points worked out by *ab initio* packages when using on the fly approaches). For few atom (at present only atom diatom) systems the dynamics problem is dealt with by using full dimensional quantum mechanics techniques which are the elective complement to the *ab initio* calculations of the PES especially when convergence with total angular momentum quantum number J is needed. Approximate quantum approaches and classical mechanics methods can also be used, especially when the number of atoms is large. Possible iterations going back to the INTERACTION block are foreseen as a result of the execution of this block.

- OBSERVABLES, that is the "last mile" block of GEMS devoted to carry out the statistical (and model) treatments of the outcomes of the theoretical calculations (including the incorporation of the physical parameters of the machine) necessary to provide a realistic estimate of the signals detected by the apparatus. In this case, we have used the routine normally employed for the conversion of trial CM functions into the corresponding LAB distributions (see below).

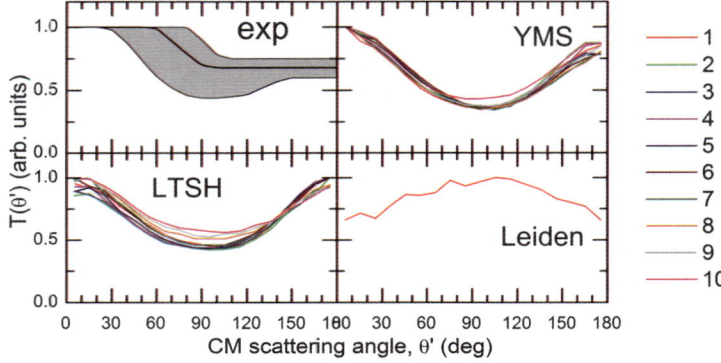

Fig. 1 PADs for different OH rotational levels for the reactions $OH(N) + CO(j = 0) \rightarrow CO_2 +$ H at E_c of 8.6 kcal mol^{-1}. Top left-hand side panel: experimental best-fit; top right-hand side panel: predictions on the Leiden PES; bottom left-hand side panel: predictions on the YMS PES; bottom right-hand side panel: predictions on the LTSH PES.

Obviously iterations are also considered when agreement between theoretical outcomes and experimental data is not satisfactory and checks for possible inaccuracies introduced in the various blocks of GEMS become necessary.

The key innovative feature of a grid approach is, indeed, its intrinsic cooperative nature of which GEMS is just an example pertaining to the Molecular and Materials simulation field. In particular, the GEMS simulation of the OH + CO reactive properties measured in the related CMB experiment (see next section for a detailed discussion) is, indeed, a typical example of how collaboration is necessary in Grid computing. In it, in fact, *ab initio* information and PESs fitted to them produced at different times by various geographically dispersed laboratories[17,33,34] have been incorporated into the procedures of the DYNAMICS block by the Spanish coauthor of the present paper who ran also related dynamics calculations on the grid. To this end the subset of the EGI grid platform accessible by the COMPCHEM VO, managed by the computational coauthors of Perugia, was used and the grid distribution tools developed also by them were modified for a more efficient usage. Results collected on the Grid storage were compared with the experimental data produced by the experimental coauthors of Perugia. Obviously, the high level of cooperation provided by the Grid relies on several layers of interoperability the most basic of which is the one offered by the common middleware installed on all

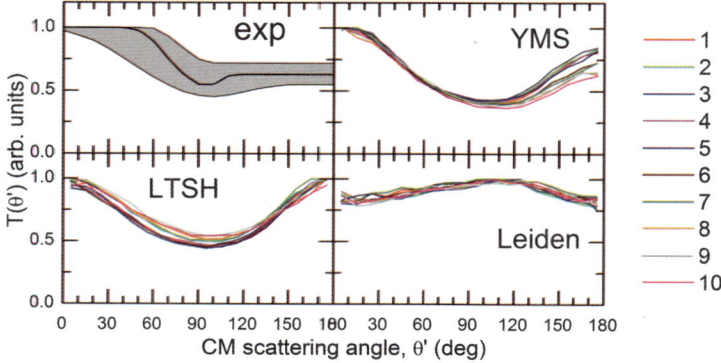

Fig. 2 Same as in Fig.1, but at $E_c = 14.1$ kcal mol^{-1}.

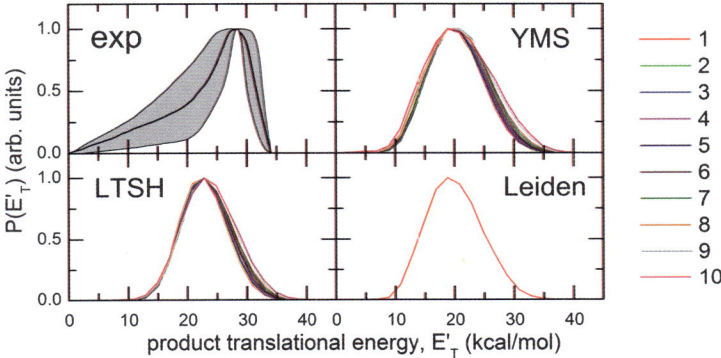

Fig. 3 PTDs for different OH rotational levels for the reactions $OH(N) + CO(j = 0) \rightarrow CO_2 + H$ at E_c of 8.6 kcal mol^{-1}. Top left-hand side panel: experimental best-fit; top right-hand side panel: predictions on the YMS PES; bottom left-hand side panel: predictions on the LTSH PES; bottom right-hand side panel: predictions on the Leiden PES (only for $N_{OH} = 1$).

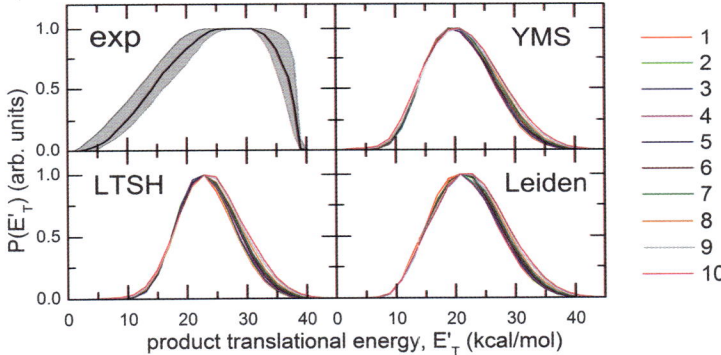

Fig. 4 Fig. 2 Same as in Fig.3, but at $E_c = 14.1$ kcal mol^{-1}.

the interconnected machines. This basic level of interoperability offers the possibility of continuously expanding the platform thanks to the conferring of users' and providers' machines. Most of the European grid adopts the Glite[67] middleware. Despite the popularity of Glite among Grid operators (other important middleware are Arc,[68] Dcache[69] and Unicore[70]), the need to use it might represent a barrier to access the grid for the generic chemistry user. Such a barrier is circumvented by the usage of GriF,[71] a Java-based[72] framework. GriF is made of two Java servers and one Java client. The two Java servers (one acting as an information registry for available services and one acting as a container providing Web Services) describe services and reference self-describing interfaces which are published in platform-independent documents. The Java client, instead, takes care of running the jobs on the associated User Interface and supports the correct interfacing of the offered Web Services. Accordingly the client can manage the basic Grid operations like checking the status of the launched jobs, search for the available services (including available programs, their compilation and execution, retrieving results, *etc.*) and operate the desired changes. As a matter of fact GriF hides the difficulty of handling the grid and puts the users at the centre of the action as a point of generation, integration, control and evaluation of the computational activities. A second step towards bypassing the above mentioned barrier on the side of platforms access

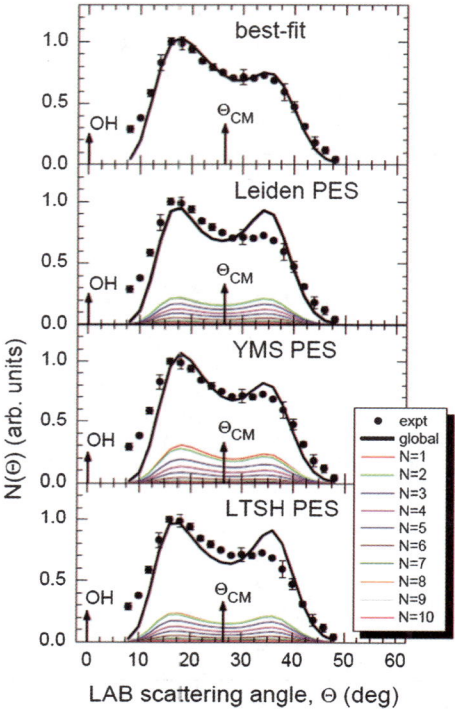

Fig. 5 Laboratory angular distribution recorded at mass-to-charge ratio $m/z = 46$ ($CO^{18}O^+$, water with isotopycally labeled ^{18}O was used to produce $H^{18}O$ with the aim to reduce the background signal at the mass of the CO_2 product) for the reaction $OH(N) + CO(j) \rightarrow CO_2 + H$ at $E_c = 14.1$ kcal mol^{-1}. Top panel: experimental points (solid circles) and best-fit distribution (black solid line) obtained by using the best-fit functions of Fig. 2 and 4. Second panel from the top: experimental points against the prediction of QCT calculations on the Leiden PES (black solid line: global simulation, colored solid lines: partial contributions from each initial N_{OH} considered, see legend). Third panel from the top: experimental points against the prediction of QCT calculations on the YMS PES (black solid line: global simulation, colored solid lines: partial contributions from each initial N_{OH} considered, see legend). Bottom panel: experimental points against the prediction of QCT calculations on the YMS PES (black solid line: global simulation, colored solid lines: partial contributions from each initial N_{OH} considered, see legend).

has been made by equipping GriF with the possibility of bridging GLite systems to machines adopting different middlewares. Of particular importance for the quantum chemistry community is the ability of the new version of GriF to bridge applications running on the Grid to platforms running on Unicore middleware (a first experiment has been already carried out with CINECA).[73] Both steps have been inspired by SOA. SOA approaches are based on a radically innovative strategy that is revolutionizing our way of carrying out research. The SOA approach builds software applications by making use of services (implementations of well-defined functionalities that can be consumed by clients in various applications and automatic processes) available on the public network by promoting loose coupling among software components to enhance their reuse. The basic rules of SOA are: a) services expose well defined independent interfaces (the what) separated from their implementation (the how) implemented by producers and used by consumers; b) services are self contained (perform predetermined tasks) and are dynamically searchable; c) composite services can be assembled out of simpler ones. This is, indeed, the true innovative nature of GriF that has been used to assemble the composite services

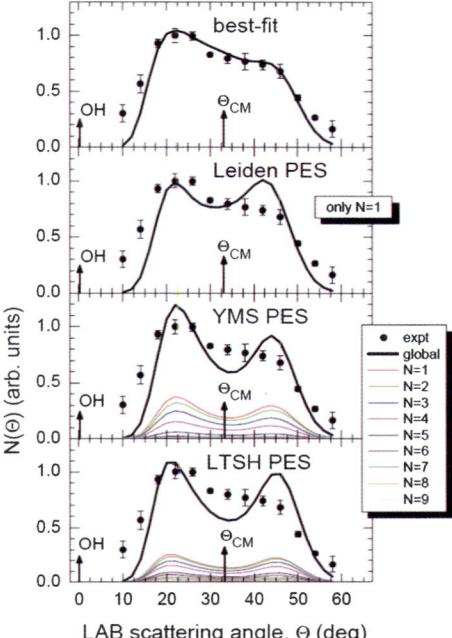

Fig. 6 As in Fig. 5, but at $E_c = 8.6$ kcal mol^{-1}. For the Leiden PES only $N_{OH} = 1$ was derived from QCT calculations.

needed to launch the components of GEMS on the Grid. Moreover, the SOA nature of GriF represents a solid basis for the introduction of a metric to evaluate the contributions of the members of a community on its behalf and for the establishing, as well, of terms of exchange[74] between the work done and the reward deserved because of its efficacy as service. In the adopted SOA approach this is quantified in terms of Quality of User (QoU) and Quality of Service (QoS) which are the pillars of a being developed system of credit awards.[75]

The forward convolution routine requires several inputs related to the experimental conditions, such as the velocity distributions, v_1 and v_2, and masses of the reactants, the crossing angle and its spread, the tentative (or theoretically derived) CM functions and other several pieces of information related to the time-of-flight system (flight length, ion flight time, chopper wheel rotation frequency and diameter). The experimental angular and time-of-flight (TOF) distributions are also needed, as the program scales the simulated data with respect to them, in such a way that both experimental and simulated LAB angular distributions have the same area, whereas the TOF spectra are scaled in such a way that their area corresponds to the intensity (in arbitrary units) of that angle in the LAB (both experimental and simulated) angular distributions. The routine considers and sums up the results of simulations over all possible Newton diagrams compatible with the reactant velocity distributions and crossing angle spread. Each Newton diagram furnishes a contribution weighted by a factor depending on the probability that a collision with that specific kinematics takes place. These weighting factors depend on the probability that the reactants have specific velocities, $f(v_1)$ and $f(v_2)$, and a specific crossing angle, $f(\Gamma)$ (the angular spreads of the beams are such that small deviations from a crossing angle of $\Gamma = 90°$ are possible).

When more than one contribution needs to be considered to reproduce the signal recorded at a specific mass-to-charge ratio (m/z), the routine repeats the simulations and the averaging for all the contributions, which are then summed using weighting factors depending on the abundance of that specific reactant in the beam multiplied by its relative cross section. For instance, in the present case, the contributions associated with each rotational level N_{OH} (in the QCT calculations the $j_{OH} = N_{OH} - 1$ relationship has been adopted) have been considered and weighted by a factor which is obtained by multiplying the integral cross section (ICS) associated to that specific rotational level by its relative abundance in the beam. In other words, the CM flux for the CO_2 product has been obtained according to

$$I\left(\theta', u'\right) = \sum_{N_{OH}=1}^{10} w_{N_{OH}} T\left(\theta'\right)_{N_{OH}} \times P(u')_{N_{OH}}$$

where $w_{N_{OH}}$ are the relative weights for each OH rotational levels (obtained by multiplying its abundance in the beam by its ICS) and the $T(\theta')_{N_{OH}}$ and $P(u')_{N_{OH}}$ are the values assumed by the PAD and PTD for each OH level at the angle θ' and velocity u'. In the convolution routine we have not used the QCT distributions calculated for different values of j_{CO} because the related PADs, PTDs and ICSs are substantially identical from $j_{CO} = 0$ up to $j_{CO} = 3$ (see below).

3 Experimental

The CMB experimental data have been already published.[15,23,24] We remind that the experiments have been performed at two collision energies ($E_c = 8.6$ and 14.1 kcal

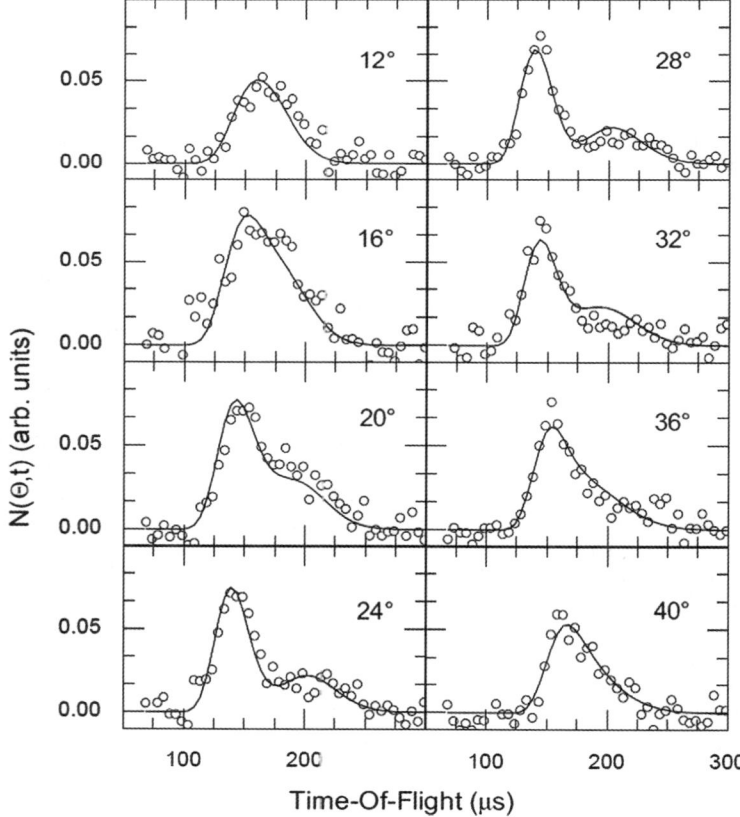

Fig. 7 Laboratory (open circles) and best fit (solid lines) TOF spectra at selected angles at $E_c = 14.1$ kcal mol^{-1}.

mol^{-1}) by employing two different OH beams. The analysis of measured data was performed as usual (see the Introduction) *via* a CM → LAB forward convolution in the assumption that the differential cross section, $I_{CM}(\theta',u')$ which is usually converted to $I_{CM}(\theta',E_T')$ where E_T' is the product translational energy, can be factorized as a product of CM PAD and PTD. In previous work using the same GEMS computational approach,[52] it has been demonstrated that PAD and PTD can be treated as uncoupled and there is no need to consider their coupling in the data simulation. In Fig. 1–4 the hatched area represents the ensemble of PADs and PTDs that provides a fit to an acceptable level at $E_c = 8.6$ and 14.1 kcal mol^{-1}. The best fit CO_2 PADs[15,23,24] exhibit a bimodal forward-backward structure with a preference for scattering in the forward direction (with respect to the incoming OH reactant), which is more pronounced in the case of the higher E_c experiment. This was interpreted in terms of the osculating model of chemical reactions[76] and the lifetime, τ, of the HOCO intermediate estimated accordingly ($\tau = 0.6$ and 1 ps for the higher and lower E_c experiments, respectively). The average fraction of energy released as product translational energy, $<f'_T>$, was found to be quite high (0.64 and 0.70 for the higher and lower E_c experiments, respectively).[23,24] For both experiments the average product translational energy, $<E_T'>$, was ~25 kcal mol^{-1}, very close to the value of the barrier height in the exit channel. The solid lines superimposed on the experimental points of the top-left panels of Fig. 5 and 6 and in Fig 7 and 8 are the curves calculated when using the best-fit CM functions.

As already mentioned in the Introduction, the OH radical beams have been produced by means of a RF discharge source starting from a mixture of water in He or He/Ne. The RF discharge beam source has been designed in such a way that the supersonic expansion takes place soon after the plasma is formed.[54–56] The collisional cooling of the excited states, normally achieved in supersonic expansions, is not expected to be complete here because, to minimize radical recombination, the plasma is formed directly behind the nozzle through which the supersonic expansion takes place. In other words, the number of collisions which could quench the excited internal states of OH is small. We remind that, in previous work using a magnetic analysis of the beams,[54] the thermal estimate of the relative populations of the electronic states produced in the same RF discharge beam source for several atomic species (O, N and Cl) was proved to be incorrect. In addition, the recent LIF characterization of the internal states of C_2 and CN,[55,56] produced in the same radical beam source, has demonstrated that high rotational and vibrational levels survive the supersonic expansion. It has to be noted, however, that both C_2 and CN are formed in the plasma *via* the reaction of atoms or ions produced in the discharge, starting from mixtures of CO(1.5%)/O$_2$(0.8%)/He and CO$_2$(0.8%)/N$_2$(2.5%)/He. These reactions can form highly internally excited C_2 and CN products. In the case of the OH radical beam, instead, OH is simply formed by the

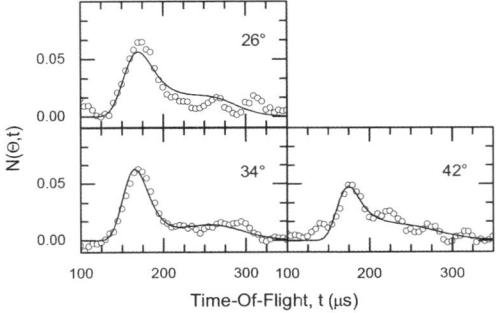

Fig. 8 As in Fig. 7, but at $E_c = 8.6$ kcal mol^{-1}.

dissociation of water, its molecular precursor, induced by collisions with metastable excited He or Ne. It is reasonable, therefore, to expect in this case a much lower internal excitation, especially as far as the vibrational excitation is concerned.

Expecting some internal excitation of the OH radicals in the beams, after the theoretical study of Lakin et al.[17] which demonstrated the effect of OH rotational excitation, we have decided to characterize the population of the internal states. The characterization was performed with the same experimental apparatus used to characterize the internal states of CN and C_2.[55,56] Briefly, it consists of a RF discharge source identical to that employed in the CMB machine and used for reactive scattering experiments. The radical beam is subsequently collimated using a skimmer before passing into a vacuum chamber where the OH beam crosses the light beam of a pulsed tunable laser. The probe laser beam, obtained by doubling the output of a dye laser (Quanta System, D100) pumped by the second harmonic (at 532 nm) of a Nd-YAG laser (Quanta System, HYL 101E) operating at 10 Hz, propagates in the vacuum chamber at right angles to the radical beam after passing through baffles to reduce the detection of scattered laser light and window fluorescence. The optical detection system, which is placed at right angles to the plane containing the radical beam and probe laser beam, comprises two convex lenses to collect and focus fluorescence from the excited radicals onto the photocathode of a photomultiplier tube (PMT) (THORN EMI, 9816 QB). A narrow band interference filter is placed immediately prior to the PMT to minimize the detection of scattered light which mainly originates from the discharge source.

The characterization of the OH beam has been carried out by analyzing the rotational structure of the well characterized (0,0) vibrational band and also the off-resonance (0,1) band in the electronic transition ($A^2\Sigma \leftarrow X^2\Pi_{3/2,1/2}$) for the two spin–orbit states $^2\Pi_{3/2}$ and $^2\Pi_{1/2}$ (we recall that the energy difference between the excited and ground spin–orbit state is 126 cm^{-1}). As expected, the rotational temperature of OH is higher than those normally observed in room temperature expansions of stable diatomic species: a rotational temperature of 300 K well represents the population of low rotational levels, whereas a rotational temperature of 600 K is necessary to reproduce the population of $N_{OH} = 9,10$ (we remind that, at low rotational levels, N_{OH} is not a good quantum number because of Hund's case (a) contamination; nevertheless we have chosen it as a label of the OH rotational states at the ground adiabatic state). The rotational OH distribution is reported in Table 1 for the ground state spin–orbit level. The $^2\Pi_{1/2}/^2\Pi_{3/2}$ relative population is nearly statistical. We have explored the effect of RF and stagnation pressure variation in a limited range of values. The pressure has been varied between 150–180 mbar and the RF power between 200–300 W. When increasing the RF

Table 1 Rotational population of the spin orbit ground state of OH radicals produced in two beams under conditions resembling those of the scattering experiments performed at $E_c = 8.6$ and 14.1 kcal mol^{-1}

N_{OH}	relative population (carrier He)	relative population (carrier He/Ne)
1	1.00	1.00
2	0.93	0.88
3	0.66	0.70
4	0.45	0.55
5	0.32	0.36
6	0.19	0.30
7	0.12	0.23
8	0.06	0.17
9	0.04	0.13
10	0.04	Not determined

power, we have observed a slight increase of the population of the higher rotational states and a decrease of the population of $N_{OH} = 1,2$. This observation is in line with the higher plasma temperature reached when increasing the RF power. By increasing the pressure, instead, there is a slight decrease of the population of large N_{OH}, probably because there are more collisions able to quench the rotational states. Nevertheless, the difference in the populations are never pronounced in the RF power and pressure ranges explored. The OH − CO scattering experiments were performed using a larger stagnation pressure (300 mbar), but with a smaller orifice in the quartz nozzle ($\phi = 0.23$ mm rather than 0.28 mm). The conditions are comparable and we can assume that the results obtained in this characterization are representative of the beam employed in the scattering experiment.

We have also characterized the beam produced by expanding a mixture of 2.5% water vapour in He/Ne (70/30), under conditions similar to those of the experiment at $E_c = 8.6$ kcal mol^{-1}. The results are rather similar to those of the previous case, with some increase in the population of higher N_{OH} states, which is in line with the fact that the plasma temperature is higher in the presence of Ne (see Table 1). In all cases the N_{OH} states most populated are $N_{OH} = 1$–4. Since the integral cross sections increase with the value of N_{OH} in the LTSH PES, there could be an effect in the LAB simulated data with respect to the simulation performed considering only $N_{OH} = 1$. In particular, since the simulations with only $N_{OH} = 1$ were not able to reproduce the wings of the LAB angular distributions, the inclusion of the contribution of all populated N_{OH} levels could help to recover the missing intensity in the wings.

Vibrational excitation of OH is, instead, very small: we have estimated that only $\leq 1\%$ of the OH radicals are in the first vibrational level. This result is very different with respect to those of the C_2 and CN radicals, for which significant vibrational excitation was observed.

4 From centre-of-mass to laboratory frame comparison of theoretical and experimental data

For the first block of the simulator, INTERACTION, we took from our repository the three full dimensional PESs, YMS,[33] LTSH[17] and Leiden,[34] available in the literature. As to the second block of GEMS, DYNAMICS, we employed the program VENUS[77] to perform QCT calculations. Some details of the calculations follow. The value of the maximum impact parameter was set to 2.8 Å for the YMS and LTSH PESs and to 2.4 Å for the Leiden PES. For all trajectories initial and final distances were set at 8.0 Å, a distance large enough to consider as negligible the interaction between the fragments of the related channels. An integration step of 0.24 fs was set for the YMS and LTSH PESs while for the Leiden PES it was chosen to be 0.012 fs. All remaining parameters (vibrational phases and spatial orientation of molecules) were selected randomly. Calculated values of the cross sections were scaled using the statistical factor of 1/2 (to account for the double degeneracy of the $^2\Pi$ electronic state). The calculations do not include any zero point energy (ZPE) correction or any other quantum-like effects because the use of pseudoquantization methods like Gaussian binning[78] and one-dimensional Gaussian binning[79] were found to hardly improve the standard QCT results.[80] To carry out this study bearing high accuracy, a large number of trajectories was integrated in order to obtain a number of reactive events statistically significant (a hundred thousand of reactive trajectories). Moreover, since the trajectories calculated on the YMS and LTSH PESs showed to be affected by some numerical instabilities and led to a poor conservation of total energy, a suitable tolerance limit was introduced for energy deviation and all the trajectories exceeding that limit were discarded. When using a tolerance limit of 0.04 kcal mol^{-1} (the same value chosen for cutting the long range tail of the potential) 13–17% of the reactive trajectories calculated at

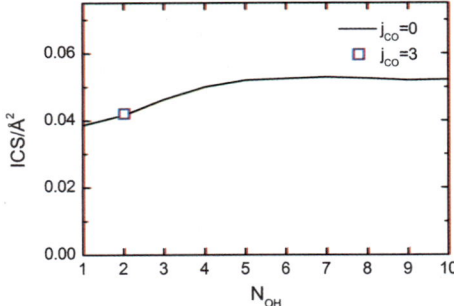

Fig. 9 ICS plotted as a function of the OH rotational level for the reactions $OH(N) + CO(j) \rightarrow CO_2 + H$ at $E_c =$ of 14.1 kcal mol^{-1} on the Leiden PES.

$E_c = 14.1$ kcal mol^{-1} were discarded in the case of the YMS PES whereas in the case of the LTSH PES the fraction of rejected trajectories increased to 61–66%. These values are, respectively, 21–33% and 65–72% at $E_c = 8.6$ kcal mol^{-1}. Obviously, in order to be left with a still statistically significant number of reactive events (about one hundred thousand) a larger batch of trajectories had to be integrated.

In Fig. 9–11 are reported the ICS for each rotational state of CO (up to $j_{CO} = 3$) as a function of the OH rotational levels for the three PESs employed at the two

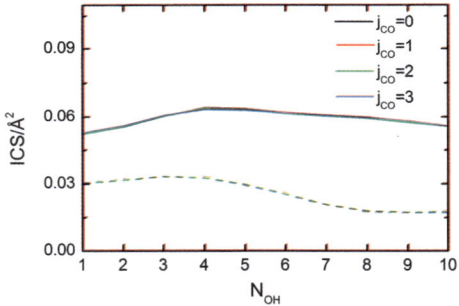

Fig. 10 ICS plotted as a function of the OH rotational level for the reactions $OH(N) + CO(j) \rightarrow CO_2 + H$ at the two E_c of 8.6 (dashed lines) and 14.1 kcal mol^{-1} (continuous lines) on the LTSH PES.

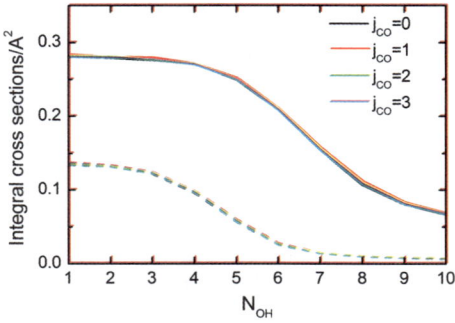

Fig. 11 ICS plotted as a function of the OH rotational level for the reactions $OH(N) + CO(j) \rightarrow CO_2 + H$ at the two E_c of 8.6 (dashed lines) and 14.1 kcal mol^{-1} (continuous lines) on the YMS PES.

collision energies investigated. As is well visible, ICSs increase with E_c and the effect of the rotational excitation of CO is totally negligible in all PESs. On the contrary, the three PESs show a different behavior as far as the effect of the OH rotational excitation is concerned. Clearly, QCT calculations predict a decreasing reactivity with increasing values of N_{OH} for the YMS PES, while in the case of the LTSH PES the ICS slightly increases from N = 1 up to N = 3 or 4 for the higher and lower collision energy, respectively, and then slowly decreases with the increase of N_{OH}. The calculations on the Leiden PES were performed only for $j_{CO} = 0$ because these were very time-consuming calculations. ICS was calculated also in the case of $N_{OH} = 2$ and $j_{CO} = 3$ to explore whether there is an effect of the CO rotational excitation. As visible in Fig. 9, this is not the case and once again the CO rotational excitation appears to have no influence on the value of ICS. The effect of OH rotational excitation is less pronounced than in the YMS and LTSH PES. ICS appears to slightly increase with N_{OH} and reaches a plateau from $N_{OH} = 6$ to $N_{OH} = 10$. For the lower E_c considered, the calculations were performed considering only $N_{OH} = 1$.

Clearly, some details in the entrance channels are responsible for these different behaviours in the three PESs. Lakin et al.[17] have already analyzed this aspect, by comparing the effect of the entrance channel of the LTSH PES with or without including the two van der Waals wells associated with OH$\cdots$CO and OH$\cdots$OC. In particular, at low collision energy (<7 kcal mol^{-1}) the OH rotational excitation clearly reduces the value of ICS (from $N_{OH} = 1$ to $N_{OH} = 5$), while for higher E_c the effect is reversed and ICS increases with N_{OH}. These results were interpreted as originating from a combined effect of the van de Waals wells, deflecting the trajectories from the critical configuration of the transition state, and the increase of the total energy which helps in overcoming the reaction barrier.

The (normalized at the peak) QCT CM product angular distributions calculated on the three PESs for each OH rotational level from 1 to 10 and $j_{CO} = 0$ are shown in Fig. 1 and 2, for the lower and the higher E_c experiments. CM PTD distributions were obtained using a binning method with a box of 2.0 kcal mol^{-1} The QCT functions for each OH rotational level from 1 to 10 and $j_{CO} = 1$–3 are reported in the Fig. S1–S4 of the Electronic Supplementary Information. As clearly visible, the CM PADs are essentially identical for different values of j_{CO} at fixed values of N_{OH}, while there are some differences for different values of N_{OH} at fixed values of j_{CO}. For instance, in the case of the YMS PES, the intensity ratio at the two poles, $I(\theta' = 180°)/I(\theta' = 0°)$, might vary from 0.6 to 0.8 without a clear trend, with only the $N_{OH} = 10$ having always the smallest $I(\theta' = 180°)/I(\theta' = 0°)$ value at both E_cs. For the LTSH PES, instead, the differences with varying N_{OH} are mostly related to the degree of polarization, that is $I(\theta' = 90°)/I(\theta' = 0°)$, which varies from $\sim$0.5 (for $N_{OH} = 1$) up to 0.6 (for $N_{OH} = 9,10$) at both E_cs. The Leiden PES PADs do not vary much with N_{OH}. Differently from the case of the two other PESs, the Leiden PADs are almost isotropic with a slight preference for sideways scattering ($\theta' = 100°$–$120°$).

The (normalized at the peak) QCT CM product translational energy distributions calculated on the three PESs for each OH rotational level from 1 to 10 and $j_{CO} = 0$ are shown in Fig. 3 and 4, for the lower and the higher E_c experiments. CM PADs were obtained using a binning method with a box of 10°. The QCT functions for each OH rotational level from 1 to 10 and $j_{CO} = 1$–3 are reported in the Fig. S5–S8 of the Electronic Supplementary Information. Once again, the CM PTDs are essentially identical for different values of j_{CO} at fixed values of N_{OH}, while there are some differences for different values of N_{OH} at fixed values of j_{CO}. In particular, because of the higher total energy available, the peak and the tail of the PTDs extends to larger energy values with increasing N_{OH}. Quite interestingly, there is no difference in the rise of PTDs with the value of N_{OH}.

The experimental and simulated LAB angular distributions are shown in Fig. 5 and 6, respectively, for the higher and lower E_c experiments in the case of the three PESs. In Fig. 7 and 8 the LAB TOF data with the best-fit reproduction are shown,

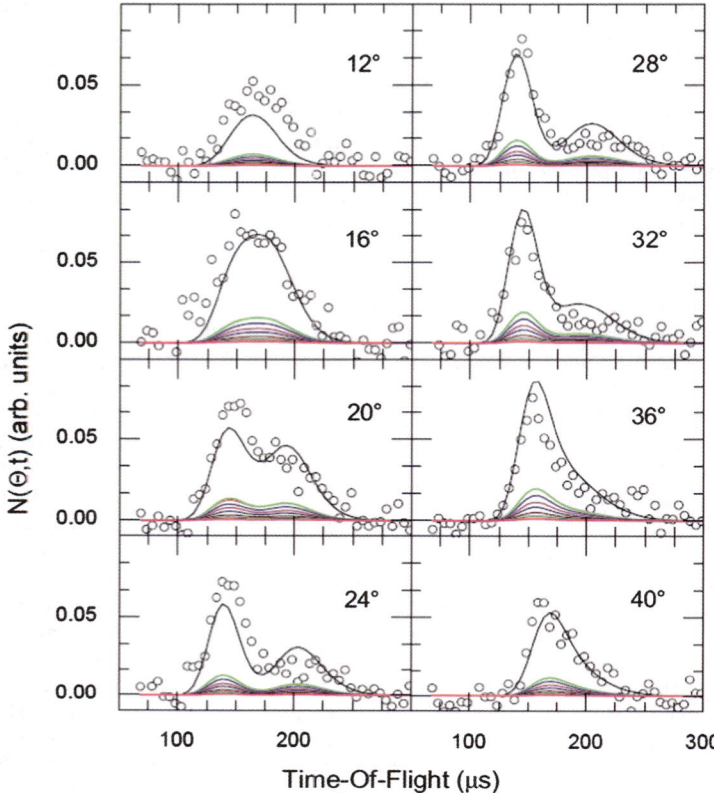

Fig. 12 Laboratory (open circles) and Leiden PES QCT simulated (solid lines) TOF spectra at selected angles at $E_c = 14.1$ kcal mol^{-1}.

while in Fig. 12–17 the TOF spectra simulated by the QCT calculations on the three PESs are shown. From the analysis of those figures the following considerations can be made.

The state of comparison for the experiment at $E_c = 14.1$ kcal mol^{-1} is not very different with respect to the case where only $N_{OH} = 1$ was considered.[52] There is little improvement on the reproduction of the wings in the three cases: this is due to the fact that the PTDs of high N_{OH} states have a tail and a peak displaced towards

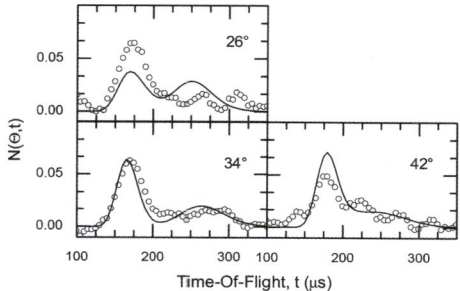

Fig. 13 As in Fig. 12, but at $E_c = 8.6$ kcal mol^{-1}.

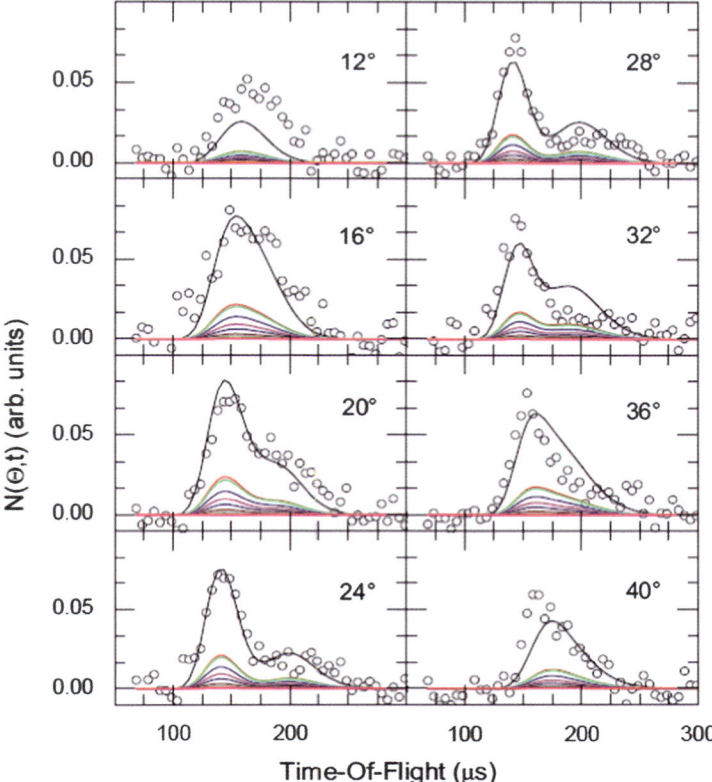

Fig. 14 Laboratory and YMS PES simulated TOF spectra at selected angles at $E_c = 14.1$ kcal mol^{-1}.

higher value of E_c, but the rise is roughly the same for all N_{OH} levels. LAB angular and TOF distributions are mostly affected by the rise of the PTDs and peak position, with little sensitivity to the tails.

At $E_c = 14.1$ kcal mol^{-1} YMS calculations are able to reproduce most of the features of the LAB angular distribution, such as the relative height between the peak around $\Theta = 16°-18°$ and the shoulder at $\Theta = 35°$. However, the discrepancies in the wings of the angular distribution did not vary and the consideration of the

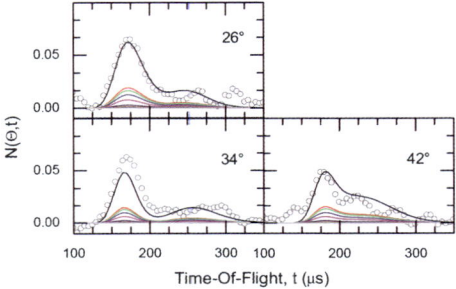

Fig. 15 Laboratory and YMS PES simulated TOF spectra at selected angles at $E_c = 8.6$ kcal mol^{-1}.

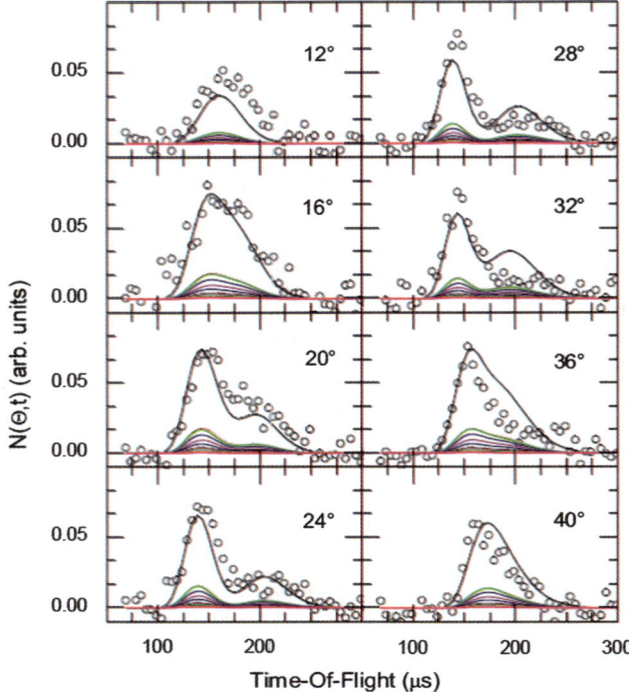

Fig. 16 Laboratory and LTSH PES simulated TOF spectra at selected angles at $E_c = 14.1$ kcal mol^{-1}.

high N_{OH} contributions did not help, especially because for this PES the ICS decreases with the increase of N_{OH}. We remind that the fraction of energy released as product translational energy $<f'_T>$ is only $\sim$0.54 in the YMS PES, while the best-fit value is larger and also the detailed shape of the experimental and YMS PTDs is different. This explains why, although the comparison with the best-fit PAD is very good, the simulated LAB angular distribution is somewhat disappointing. The comparison with the more detailed observables, the TOF spectra, is generally good, with the exception of the TOF recorded at $\Theta = 12°$, $32°$, $36°$ and $40°$.

As to the LTSH PES, the comparison with the experimental LAB angular distribution at 14.1 kcal mol^{-1} is less good, with the shoulder a $\Theta = 35°$ as high as the

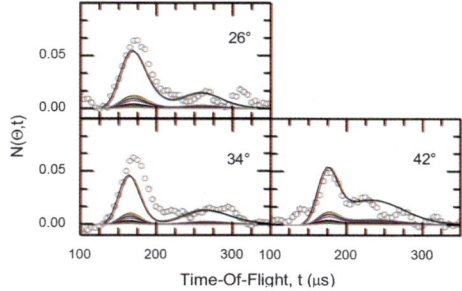

Fig. 17 Laboratory and LTSH PES simulated TOF spectra at selected angles at $E_c = 8.6$ kcal mol^{-1}.

peak. Also in this case, there is not much difference with respect to the case where only $N_{OH} = 1$ was considered.[52] The wings of the angular distribution are better reproduced than in the case of the YMS calculations. This is related to the higher $<f'_T>$ value derived from this PES, that is 0.6, close to the best-fit one. This is reflected also in the simulation of the TOF spectra. In this case, it is the shape of the CM PAD, largely backward-forward symmetric, to cause the observed discrepancies in the LAB angular distribution that, instead, clearly indicates some preference for forward scattering.

The same kind of discrepancies occurs also for the Leiden PES calculations: the shoulder around $35°$ is as high as the peak around $16–18°$, at variance with the measured distribution. It might seem striking that, while by comparing the CM PADs the situation looks quite different for the Leiden and LTSH results, the difference is largely mitigated when moving to the LAB frame. This is due to a combined effect of the characteristics of PTDs and PADs and is an excellent example of how much more significant is the comparison of theoretical predictions directly with CMB LAB data.

The data at 8.6 kcal mol^{-1} have never been compared before with theoretical functions converted to the LAB frame. The general trend is similar to the one observed for the higher E_c experiment, but the disagreement with the LAB angular distributions is more pronounced. In particular, all the PESs predict LAB angular distributions with a pronounced dip around the centre-of-mass angle, Θ_{CM}, which is absent in the experimental results. This seems to be due to the fact that the YMS and LTSH PADs are too polarized with respect to the best-fit one. The almost isotropic Leiden PAD, in fact, produces the smallest dip at Θ_{CM}, but it fails to reproduce the backward-forward asymmetry. Also at the level of TOF spectra, the comparison between experimental and theoretical results is not good with the experimental ratio between the intensity of fast peak and slow shoulder never reproduced. Overall, the LTSH PES produce the best simulations of TOF spectra at both E_c. This might be taken as an evidence that the exit barrier is correctly described along this PES.

5 Conclusive remarks

The exploitation of service oriented approaches and the construction of a recently implemented grid empowered molecular simulator based on the combination of networked and high performance computing have allowed the assemblage of *ab initio* simulations of the virtual crossed molecular beam experiments to be offered as a service to the experimentalists. As a demo use case, the OH + CO reactive system has been considered in the present paper. The direct comparison of the simulated results with experimental raw data has avoided the possible uncertainties built-in into the usual procedures developed to carry out the analysis of crossed beam experiments. To make such a comparison as accurate as possible, the rotational distributions of the OH radicals employed in previous crossed beam experiments have been characterized by laser-induced-fluorescence. The capability of running massive calculations with a grid-distributed approach has also been exploited when running quasiclassical trajectory calculations for all the initial rotational states of the OH radicals present in the beam on three different potential energy surfaces.

By comparing the results of QCT predictions on the three most recently used HCO$_2$ PESs directly with experimental observables of CMB experiments, we can assess the status of theory *versus* experiment for the title system. Unfortunately, the inclusion of the higher N_{OH} levels has not improved the status of the comparison and all the three tested PESs appear to be affected by some shortcomings. It is true that the QCT method is not as accurate as full dimensional quantum scattering calculations that have finally started to appear.[81–84] Nevertheless, for all the simpler systems investigated so far, QCT calculations have proved to be good at describing most of the characteristics of the reaction dynamics.[7,9,18] In the cases where the same PES has been used for both quantum and QCT calculations, significant quantum effects have been noted in a limited number of systems.[85,86] In conclusion, for this system none of the three PESs seem to be accurate enough to reproduce the CMB experimental results. The YMS PES is the one that better reproduces the preference for forward scattering, but the fraction of energy released as product E'_T is too small with respect to the experimental value. The LTSH PES, instead, better reproduces the product energy release, but fails to reproduce the preference for forward scattering. Finally, the Leiden PES, which compares worst with best-fit CM functions, once transformed into LAB distributions furnishes a prediction similar to the LTSH one. It would be interesting to test in the same way the new PES developed by Bowman and coworkers.[51]

A general conclusion of this paper is that to assess the quality of a PES it is much more instructive to compare the theoretical predictions with the raw data in the LAB reference frame. The example of the Leiden PES results illustrates this point very convincingly. In this respect, the effort to provide a single stream calculation method that leads from first principle to the LAB data can be of great help for many theoreticians and also experimentalists that can use the same approach as a Grid service with no need for being an expert in distributed computing and theoretical treatments.

Acknowledgements

The authors acknowledge financial support from the COST action CM901 (*Detailed Chemical Models for Cleaner Combustion*), the EGI-Inspire project (contract 261323), MIUR PRIN 2008 (2008KJX4SN 003), MIUR PRIN 2007 (2007H9S8SW-004) the ESA-ESTEC contract 21790/08/NL/HE, the Phys4entry FP7/2007-2013 project (contract 242311), ARPA and MICINN (CTQ2008-025878/BQU). Computations have been supported by the use of Grid resources and services provided by the European Grid Infrastructure (EGI) and the Italian Grid Infrastructure (IGI) as well as by CESGA, CASPUR, CINECA and the COMPCHEM VO of EGI. Partial financial support from Spanish MICINN (CTQ-2008-02578/BQU) is also acknowledged. This work was also partially supported by PRIN 2009SLKFEX. AP, PC and NB acknowledge some help by G. Restani in the LIF experiments.

References

1 P. Casavecchia, *Rep. Prog. Phys.*, 2000, **63**, 355.
2 N. Balucani, G. Capozza, F. Leonori, E. Segoloni and P. Casavecchia, *Int. Rev. Phys. Chem.*, 2006, **25**, 109.
3 Y. T. Lee, in *Atomic and Molecular Beam Methods*, ed. G. Scoles, Oxford University Press, New York, 1987, vol. 1.
4 R. D. Levine, *Molecular Reaction Dynamics*, Cambridge University Press, New York 2005.
5 A. Bergeat, L. Cartechini, N. Balucani, G. Capozza, L. F. Phillips, P. Casavecchia, G. G. Volpi, L. Bonnet and J.-C. Rayez, *Chem. Phys. Lett.*, 2000, **327**, 197.
6 N. Balucani, P. Casavecchia, L. Banares, F. J. Aoiz, T. Gonzalez-Lezana, P. Honvault and J. M. Launay, *J. Phys. Chem. A*, 2006, **110**, 817.
7 N. Balucani, G. Capozza, E. Segoloni, A. Russo, R. Bobbenkamp, P. Casavecchia, T. Gonzalez-Lezana, E. J. Rackham, L. Banares and J. F. Aoiz, *J. Chem. Phys.*, 2005, **122**, 234309.
8 N. Balucani, G. Capozza, L. Cartechini, A. Bergeat, R. Bobbenkamp, P. Casavecchia, F. J. Aoiz, L. Banares, P. Honvault, B. Bussery-Honvault and J. M. Launay, *Phys. Chem. Chem. Phys.*, 2004, **6**, 4957.
9 N. Balucani, P. Casavecchia, J. F. Aoiz, L. Banares, J. M. Launay, B. Bussery-Honvault and P. Honvault, *Mol. Phys.*, 2010, **108**, 373.
10 F. Leonori, R. Petrucci, N. Balucani, P. Casavecchia, M. Rosi, D. Skouteris, C. Berteloite, S. D. Le Picard, A. Canosa and I. R. Sims, *J. Phys. Chem. A*, 2009, **113**, 15328.
11 C. Berteloite, S. D. Le Picard, A. Canosa, I. R. Sims, M. Rosi, F. Leonori, R.P., N. Balucani, X. Wang and P. Casavecchia, *Phys. Chem. Chem. Phys.*, 2011, **13**, 8485.
12 J. J. Lin, J. Shu, Y. T. Lee and X. Yang, *J. Chem. Phys.*, 2000, **113**, 5287.

13 J. J. Lin, Y. T. Lee and X. Yang, *J. Chem. Phys.*, 2000, **113**, 1831.
14 S.-H. Lee, W.-K. Chen and W.-J. Huang, *J. Chem. Phys.*, 2009, **130**, 054301.
15 M. Alagia, N. Balucani, P. Casavecchia, D. Stranges and G. G. Volpi, *J. Chem. Soc., Faraday Trans.*, 1995, **91**, 575.
16 B. Fu, Y.-C. Han, J. M. Bowman, L. Angelucci, N. Balucani, F. Leonori and P. Casavecchia, submitted.
17 M. J. Lakin, D. Troya, G. C. Schatz and L. B. Harding, *J. Chem. Phys.*, 2003, **119**, 5848.
18 D. Skouteris, H.-J. Werner, F. J. Aoiz, L. Banares, J. F. Castillo, M. Menendez, N. Balucani, L. Cartechini and P. Casavecchia, *J. Chem. Phys.*, 2001, **114**, 10662.
19 N. Balucani, D. Skouteris, L. Cartechini, G. Capozza, E. Segoloni, P. Casavecchia, M. H. Alexander, G. Capecchi and H.-J. Werner, *Phys. Rev. Lett.*, 2003, **91**, 013201.
20 http://www.egi.eu/.
21 A. Costantini, O. Gervasi, C. Manuali, N. Faginas Lago, S. Rampino and A. Laganà, *J. Grid Comput.*, 2010, **8**, 571.
22 T. Erl, Service-Oriented Architecture: Concepts, Technology, and Design, Prentice Hall PTR Upper Saddle River, NJ, USA ©2005 ISBN:0131858580.
23 M. Alagia, N. Balucani, P. Casavecchia, D. Stranges and G. G. Volpi, *J. Chem. Phys.*, 1993, **98**, 8341.
24 P. Casavecchia, N. Balucani and G. G. Volpi, in *Advanced Series in Physical Chemistry: Vol. 6, The Chemical Dynamics and Kinetics of Small Radicals*, ed. A. F. Wagner and K. Liu, World Scientific, Singapore, 1995, p. 365.
25 E. Garcia, A. Saracibar and A. Laganà, *Theor. Chem. Acc.*, 2011, **128**, 727.
26 E. Garcia, A. Saracibar, L. Zuazo and A. Laganà, *Chem. Phys.*, 2007, **332**, 162.
27 H. Y. Sun and C. K. J. Law, *Mol. Struc. – THEOCHEM*, 2008, **862**, 138.
28 J. M. Bowman and G. C. Schatz, *Annu. Rev. Phys. Chem.*, 1995, **46**, 169.
29 K. Kudla and G. C. Schatz, in *Advanced Series in Physical Chemistry: Vol. 6, The Chemical Dynamics and Kinetics of Small Radicals*, ed. A. F. Wagner and K. Liu, World Scientific, Singapore, 1995, p. 438.
30 D. C. Clary and G. C. Schatz, *J. Chem. Phys.*, 1993, **99**, 4578.
31 M. I. Hernandez and D. C. Clary, *J. Chem. Phys.*, 1994, **101**, 2779.
32 E. M. Goldfield, S. K. Gray and G. C. Schatz, *J. Chem. Phys.*, 1995, **102**, 8807.
33 H. G. Yu, J. T. Muckerman and T. J. Sears, *Chem. Phys. Lett.*, 2001, **349**, 547.
34 R. Valero, M. C. van Hemert and C. J. Kroes, *Chem. Phys. Lett.*, 2004, **393**, 236.
35 D. H. Zhang and J. Z. H. Zhang, *J. Chem. Phys.*, 1995, **103**, 6512.
36 N. Balakrishnan and G. D. Billing, *J. Chem. Phys.*, 1996, **104**, 4005.
37 F. N. Dzegilenko and J. M. Bowman, *J. Chem. Phys.*, 1998, **108**, 511.
38 G. D. Billing, J. T. Muckerman and H. G. Yu, *J. Chem. Phys.*, 2002, **117**, 4755.
39 R. Valero and G. J. Kroes, *J. Chem. Phys.*, 2002, **117**, 8736.
40 D. A. McCormack and G. J. Kroes, *Chem. Phys. Lett.*, 2002, **352**, 281; Erratum, ibid, 2003, 373, 648.
41 D. M. Medvedev, S. K. Gray, E. M. Goldfield, M. J. Lakin, D. Troya and G. C. Schatz, *J. Chem. Phys.*, 2004, **120**, 1231.
42 Y. He, E. M. Goldfield and S. K. Gray, *J. Chem. Phys.*, 2004, **121**, 823.
43 R. Valero, D. A. McCormack and G. J. Kroes, *J. Chem. Phys.*, 2004, **120**, 4263.
44 R. Valero and G. J. Kroes, *Phys. Rev. A*, 2004, **70**, 040701.
45 R. Valero and G. J. Kroes, *J. Phys. Chem. A*, 2004, **108**, 8672.
46 R. Valero and G. J. Kroes, *Chem. Phys. Lett.*, 2006, **417**, 43.
47 S. Zhang, D. M. Medvedev, R. M. Goldfield and S. K. Gray, *J. Chem. Phys.*, 2006, **125**, 164.
48 X. Song, J. Li, H. Hou and B. Wang, *J. Chem. Phys.*, 2006, **125**, 094301.
49 J. N. Murrell, S. Carter, S. C. Farantos, P. Huxley, A. J. C. Varandas, *Molecular Potential Energy Functions*. Wiley, Chichester, 1984.
50 M. Yang, D. H. Zhang, M. A. Collins and S. Y. Lee, *J. Chem. Phys.*, 2001, **115**, 174.
51 J. Li, Y. Wang, B. Jiang, J. Ma, R. Dawes, D. Xie, J. M. Bowman and Hua Guo, *J. Chem. Phys.*, 2012, **136**, 041103.
52 A. Laganà, N. Balucani, S. Crocchianti, P. Casavecchia, E. Garcia and A. Saracibar, *Lect. Notes Comput. Sci.*, 2011, **6784**, 453.
53 S. J. Sibener, R. J. Buss, C. Y. Ng and Y. T. Lee, *Rev. Sci. Instrum.*, 1980, **51**, 167.
54 M. Alagia, V. Aquilanti, D. Ascenzi, N. Balucani, D. Cappelletti, L. Cartechini, P. Casavecchia, F. Pirani, G. Sanchini and G. G. Volpi, *Isr. J. Chem.*, 1997, **37**, 329.
55 F. Leonori, R. Petrucci, K. M. Hickson, E. Segoloni, N. Balucani, S. D. Le Picard, P. Foggi and P. Casavecchia, *Planet. Space Sci.*, 2008, **56**, 1658.
56 F. Leonori, K. H. Hickson, S. Le Picard, X. Wang, R. Petrucci, P. Foggi, N. Balucani and Casavecchia, *Mol. Phys.*, 2010, **108**, 1097.
57 http://www.prace-project.eu/.
58 http://www.teragrid.org/.

59 http://www.mcs.anl.gov/research/projects/mpi/.
60 IGI: Italian Grid Infrastructure http://grid.infn.it/about.
61 http://www.metacentrum.cz/en/VO/.
62 Computational Chemistry (COMPCHEM) Virtual Organization, http://www3.compchem.unipg.it.
63 A. Laganà, O. Riganelli and O. Gervasi, *Lect. Notes Comput. Sci.*, 2006, **3980**, 665.
64 O. Gervasi and A. Laganà, *Future Gener. Comput. Syst.*, 2004, **20**, 703.
65 A. Laganà, Towards a Grid Based Universal Molecular Simulator, in A. Laganà, G. Lendvay, ed. in *Theory of the Dynamics of Elementary Molecular Reactions*, 363–380 (Kluwer, 2004).
66 S. Rampino, A. Monari, S. Evangelisti, E. Rossi, A. Laganà, Chem. Phys., in press; DOI DOI: 10.1016/j.chemphys.2011.04.0287.
67 http://glite.web.cern.ch/glite/.
68 http://www.nordugrid.org/.
69 http://www.dcache.org/.
70 http://www.unicore.eu/.
71 C. Manuali and A. Laganà, *Future Gener. Comput. Syst.*, 2011, **27**, 315.
72 http://www.java.com/en/.
73 C. Manuali, A. Costantini, A. Laganà, M. Cecchi, A. Ghiselli, M. Carpenè and E. Rossi, Lecture Notes in Computer Science, in press.
74 A. Laganà, S. Crocchianti, N. Faginas Lago, A. Riganelli, C. Manuali, S. Schanze, From Computer Assisted to Grid Empowered Teaching and Learning Activities inHigher Chemistry Education, in: *Innovative Methods in Teaching and Learning Chemistry in Higher Education*, I. Eilks and B. Byers Eds, RSC Publishing (2009) p. 153–190.
75 C. Manuali and A. Laganà, *Lect. Notes Comput. Sci.*, 2011, **6784**, 397.
76 G. A. Fisk, J. D. Mc Donald and D. R. Herschbach, *Discuss. Faraday Soc.*, 1967, **44**, 228.
77 W. L. Hase, R. J. Duchovic, X. Hu, A. Komornicki, K. F. Lim, D. Lu, G. H. Peslherbe, K. N. Swamy, S. R. Van de Linde, A. J. C. Varandas, H. Wang and R. J. Wolf, *QCPE Bull.*, 1996, **16**, 43.
78 L. Bonnet and J. Espinosa-Garcia, *J. Chem. Phys.*, 2010, **133**, 164108.
79 G. Czako and J. M. Bowman, *J. Chem. Phys.*, 2009, **131**, 244302.
80 E. Garcia, J. C. Corchado, J. Espinosa-Garcia, Comput. Theor. Chem., in press, DOI: 10.1016/j.comptc.2011.09.039.
81 C. Xiao. X. Xu, S. Liu, T. Wang, W. Dong, T. Yang, Z. Sun, D. Dai, X. Xu, D. H. Zhang and X. Yang, *Science*, 2011, **333**, 440.
82 S. Liu, X. Xiu and D. H. Zhang, *J. Chem. Phys.*, 2011, **135**, 141108.
83 S. Liu, X. Xiu and D. H. Zhang, *Theor. Chem. Acc.*, 2012, **131**, 1068.
84 D. H. Zhang , *et al.*, Faraday Disc. **157**, in press.
85 N. Balucani, L. Cartechini, G. Capozza, E. Segoloni, P. Casavecchia, G. G. Volpi, F. J. Aoiz, L. Banares, P. Honvault and J. M. Launay, *Phys. Rev. Lett.*, 2002, **89**, 013201.
86 R. T. Skodje, D. Skouteris, D. E. Manolopoulos, S.-H. Lee, F. Dong and K. Liu, *Phys. Rev. Lett.*, 2000, **85**, 1206.

Surface-aligned femtochemistry: Photoinduced reaction dynamics of CH_3I and CH_3Br on MgO(100)

Mihai E. Vaida and Thorsten M. Bernhardt*

Received 9th May 2012, Accepted 16th May 2012
DOI: 10.1039/c2fd20104f

Methyl iodide and methyl bromide molecules were adsorbed at submonolayer coverages on an ultrathin MgO(100) film on Mo(100) and photoexcited by 266 nm femtosecond-laser irradiation. The subsequent photodissociation and desorption dynamics were probed by time delayed multi photon ionization mass spectrometric detection of the emerging reaction products. The pronounced difference in the appearance times of the methyl radical fragments from methyl iodide and bromide is discussed on the basis of the different molecular adsorption geometries on magnesia. The surface adsorption structure also defines the alignment of the encounter complex for the observed bimolecular formation of the halogen molecules I_2 and Br_2 within about 1 and 2 ps, respectively. Finally, photoexcitation of co-adsorption layers of CH_3I and CH_3Br resulted in a heteronuclear bi-molecular reaction yielding IBr molecules.

1 Introduction

Pump–probe mass spectrometry in combination with molecular beams is a widely applied experimental technique of fundamental importance to gas phase femtochemistry investigations.[1] In surface femtochemistry, different experimental approaches are employed to study molecular reaction dynamics, such as two photon photoemission, time-resolved sum frequency generation, or two pulse correlation spectroscopy.[2] In the present contribution, we report results of a different experimental approach for direct mass resolved monitoring of surface transition states and products adapting the gas phase pump–probe methodology.

These investigations were motivated by a seminal contribution from John C. Polanyi and Ahmed H. Zewail in 1995 in which they proposed the investigation of surface transition state dynamics *via* direct detection of reaction intermediates and products in a femtosecond (fs) pump–probe schema similar to gas-phase femtochemistry investigations.[3] They termed this new approach that relies on a well defined reactant adsorption geometry 'surface aligned femtosecond photoreaction'. A great deal of both surface aligned reaction studies (not time-resolved)[4] as well as real-time surface femtochemistry investigations[2,5] have provided insight into surface reaction dynamics. Nevertheless, the direct time- and mass-resolved detection of the transition state and the product formation dynamics of a bimolecular surface reaction as proposed by Polanyi and Zewail has been realized only very recently employing the new approach of surface pump–probe femtosecond-laser mass spectrometry.[6,7]

This technique relies on the combination of time-of-flight mass spectrometry with non-resonant or resonance enhanced multi-photon ionization (REMPI) detection

Institute of Surface Chemistry and Catalysis, University of Ulm, Albert-Einstein-Allee 47, 89069 Ulm, Germany. E-mail: thorsten.bernhardt@uni-ulm.de; Fax: +49 731 5025452; Tel: +49 731 5025455

and fast surface preparation by a pulsed molecular beam.[7] Ultrathin oxide films serve as versatile substrates that interact only weakly with adsorbed molecules. Yet, the molecular adsorption structure on the substrate provides the geometrical alignment that determines the reaction dynamics subsequent to photoexcitation. Escaping fragments might, *e.g.*, directly desorb into the gas phase or inelastically exchange energy *via* collisional interaction with the substrate surface depending on the adsorption geometry. Furthermore, collisional encounters with neighboring adsorbate molecules might lead to the formation of new reaction products.

In the present contribution we extended our previous investigations of methyl iodide adsorbed at submonolayer coverages on magnesia[6,8] by studying the dynamics with different probe wavelengths and by comparing it to the photoinduced reactions of methyl bromide. In addition, for the first time, the photoreactions of co-adsorbed layers of two different molecules were investigated and, most interestingly, the formation of IBr molecules was detected after photoexcitation of co-adsorbed CH_3I and CH_3Br on MgO.

2 Experimental setup

A detailed account of the experimental setup for surface pump–probe femtosecond-laser mass spectrometry has been published recently elsewhere.[7,8] This technique builds on earlier approaches to study the dynamics of laser desorption from surfaces *via* mass spectrometry[9] and is strongly motivated by the capabilities of gas phase fs pump–probe mass spectrometry.[10,11] In brief, the experiments were carried out in an ultra high vacuum (UHV) surface science chamber equipped with standard tools for surface preparation and investigation (base pressure $<1 \times 10^{-10}$ mbar) including temperature programmed desorption (TPD) mass spectrometry.[7] A schematic view of the experimental arrangement for the time resolved laser investigations is shown in Fig. 1. The Mo(100) single crystal was attached to a liquid nitrogen cryostat manipulator and could be resistively heated up to 1200 K. A further temperature increase up to 2000 K could be achieved by direct electron bombardment. The cleaning procedure for the crystal consisted of an initial heating to 1200 K in 2×10^{-7} mbar of oxygen and a subsequent flash to 2000 K. The cleanliness and the quality of the Mo(100) surface as well as the thin film composition were routinely checked by Auger electron spectroscopy (AES) and low electron energy diffraction (LEED) employing a four-grid LEED-optics.

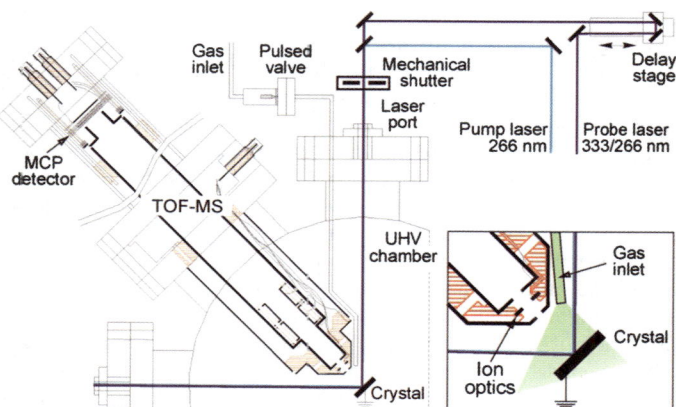

Fig. 1 Schematic layout of the experimental arrangement employed for the investigation of ultrafast reactions at oxide surfaces (see text for more details). The inset shows a magnified view of the TOF-MS head and of the surface position. Note that the relative dimensions in the illustrations are not to scale.

The MgO ultrathin films were prepared, according to the procedure reported in the literature,[12] by evaporation of magnesium metal in an atmosphere of 2×10^{-7} mbar of oxygen. During the deposition the Mo(100) crystal was held at 600 K.[13] The Mg evaporation rate was 0.15 monolayers (ML) min^{-1}. A magnesia layer thickness dependent fs-laser photoemission study of this system was reported by our laboratory recently.[14] In the present investigations a comparably thick magnesia film of 8 ML was employed to ensure the prevalence of the insulating bulk properties of MgO.[15]

Methyl iodide and methyl bromide (both obtained from Sigma-Aldrich, >99.5%, and additionally purified by several freeze–pump–thaw cycles) were dosed onto the magnesia surface at 90 K *via* an UHV compatible pulsed valve (General Valve Ser. 99) that was connected to a stainless steel tube ending close to the surface. For the preparation of the co-adsorption layers two such pulsed valves were employed. In all presented time resolved experiments an adsorbate coverage of 0.25 ML was employed as determined by TPD.[7]

The fs-laser light was produced by a Ti:Sapphire oscillator continuously pumped by a 6 W Spectra Physics Millennia Nd:YVO$_4$ laser. Pulse amplification was carried out by a Nd:YLF laser pumped Ti:Sa amplifier to yield 35 fs pulses at a repetition rate of 1 kHz. The amplified near-infrared laser pulses (central wavelength near 800 nm) were converted to 266 nm by frequency tripling in a homebuilt third harmonic generator. Other wavelengths that were employed for multi photon ionization detection were produced using a commercial optical parametric amplifier (OPA-800; Spectra-Physics). In the present experiment the laser pulses were p-polarized and the temporal pulse width was about 80 fs.

The mass of the ionized reaction products after photoexcitation was analyzed with a homebuilt time-of-flight mass spectrometer (TOF-MS, see Fig. 1).[7,8] The grounded Mo crystal served as the repeller electrode of a Wiley–McLaren-type acceleration lens arrangement.[16] The pump and probe laser beams collinearly irradiated the crystal surface at an angle of 45°. Reaction products and intermediates that were ionized by the probe laser pulse were immediately removed from the surface and directed into the time-of-flight mass spectrometer by the static electric field between the substrate surface and the first acceleration electrode of the mass spectrometer. The ions subsequently passed a field free drift tube with different velocities according to their mass to charge ratio and were finally detected by a microchannel plate (MCP) amplifier arrangement as a function of their flight time.

To obtain the transient evolution of the product ion mass signals the mass peaks were averaged over 2500 laser pulses for a fixed pump–probe delay time. The initial coverage was subsequently restored by admitting an identical amount of molecules to the surface with the pulsed valve. Subsequently, the procedure was repeated for a new pump–probe delay time.[7] Several of the thus obtained transients were averaged to yield the data shown.

3   Results and discussion

The results of TPD measurements as a function of the adsorbate coverage will be presented first. Subsequently, the laser ionization mass spectra and the transient evolution of the different product mass signals will be described for submonolayer coverages of each molecule, methyl iodide and methyl bromide, separately. The interpretation of the observed dynamics will be based on the gas phase electronic structure of the weakly adsorbed molecules. Finally, the product mass spectrum resulting from the photoexcitation and ionization detection of a co-adsorption layer of the two molecules will be presented.

3.1   Temperature programmed desorption

TPD spectroscopy was employed to characterize the interaction of the adsorbate molecules with the surface as a function of the molecular coverage. Both investigated molecules exhibit very similar desorption properties as can be seen from the graphs in Fig. 2.

Fig. 2(a) displays a series of TPD spectra that were recorded at different methyl iodide dosages between 0.3 and 1 L (Langmuir; $1\,L = 1 \times 10^{-6}$ torr s). As reported previously,[8,17] methyl iodide is adsorbed molecularly on the MgO(100) surface and desorbs without decomposition. The shift of the maximum of the desorption intensity in the TPD spectra to lower temperatures with increasing the methyl iodide coverage is attributed to a repulsive adsorbate–adsorbate interaction due to an adsorbate structure with the methyl iodide symmetry axis and hence the permanent dipole moment of the molecules aligned parallel to each other and parallel to or slightly tilted from the surface normal. This is in accordance with earlier results obtained on the (100) surface of an MgO single crystal.[17,18] On magnesia ultrathin films on Mo(100) a similar orientation was reported previously with the methyl group heading toward the substrate surface.[3] The adsorption geometry is illustrated schematically in the inset of Fig. 2(a). However, it has to be noted that neither the adsorption site, nor the exact tilt angle of the molecules with respect to the surface normal are known.[6]

In Fig. 2(b) a series of TPD spectra are shown that have been recorded after dosing different amounts of CH_3Br onto an 8 ML MgO/Mo(100) surface at 90 K. No evidence for methyl bromide dissociation subsequent to adsorption on the MgO/Mo(100) substrate was found. No reaction products such as ethane or molecular bromine, which could be formed due to methyl bromide dissociation on surface, were observed. Hence, methyl bromide also adsorbs molecularly on the MgO/Mo(100) surface and desorbs without decomposition. Also in the case of methyl bromide the maximum of the desorption signal shifts to lower temperatures with increasing coverage indicating a repulsive adsorbate–adsorbate interaction between the molecules. The same behavior had been observed previously for CH_3Br adsorbed at low coverages on bulk MgO(100)[19] and on LiF(100).[20] However, angular distributions of methyl fragments emerging from CH_3Br photodissociation on MgO(100) lead to the assignment of an adsorbate structure in which the C–Br axes of the adsorbed CH_3Br molecules lay close to the plane of the substrate surface as indicated by the inset in Fig. 2(b).[19]

3.2 Photoinduced surface reaction dynamics of methyl iodide

A product mass spectrum that has been obtained by photoexcitation with 266 nm (1 mW cm^{-2} average power) and 2 ps delayed multi photon ionization detection with a 333 nm laser beam (600 mW cm^{-2} average power) is shown in Fig. 3(a).[6] 266 nm radiation (pump) excites methyl iodide molecules to the dissociative A-band states as will be detailed below.[6,7] The time delayed detection (probe) laser

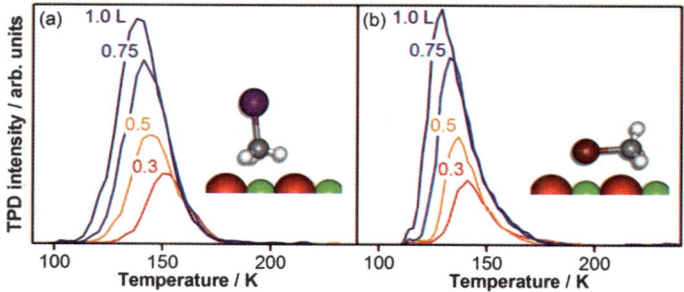

Fig. 2 TPD spectra of the molecular desorption of (a) CH_3I and (b) CH_3Br from submonolayer coverages on 8 ML MgO(100)/Mo(100) (heating rate: 2 K s^{-1}). The numbers indicated on the spectra represent the exposure in units of Langmuir (1 L = 0.25 ML for both molecules[17,19])

This journal is © The Royal Society of Chemistry 2012

was tuned to sensitively detect the methyl fragments employing (2 + 1) resonance enhanced multi photon ionization (REMPI) *via* the $3p^2A_2''$ Rydberg state.[21] Yet, other product signals are also apparent in the mass spectrum. Especially pronounced is the peak of molecular iodine, I_2^+, apart from minor signals of I^+ and CD_3I^+. In this particular experiment deuterated methyl iodide had been employed. However, no differences between the results for CD_3I and CH_3I have been observed in these investigations.

The temporal evolution of the methyl ion signal as a function of the pump–probe delay time is depicted in Fig. 3(b). This signal had been reported before.[8] The open circles in Fig. 3(b) represent the experimental data. The transient signal consists of a peak structure starting at 0 fs with a maximum reached at about 130 fs followed by a delayed exponential rise. The solid line represents the best fit of a "rise and decay" – followed by a "delayed exponential rise" – model (convoluted with the pump–probe autocorrelation function) to the experimental data.[7,8,10] Fitting of this kinetic model to the data results in time constants for the rise and the decay of the peak structure of $\tau_1(CD_3) = \tau_2(CD_3) = 90 \pm 10$ fs. The exponential rise starting with a delay of 170 ± 40 fs exhibits a time constant of $\tau_3(CD_3) = 680 \pm 50$ fs. The peak and the delayed exponential rise structures can be attributed to methyl ions that originate from two different elementary processes.[6]

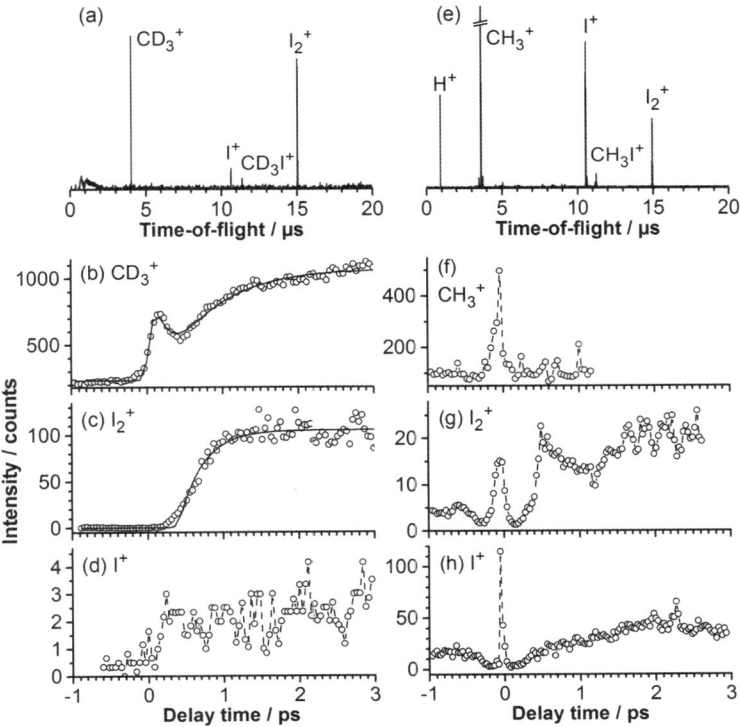

Fig. 3 The graphs in the top row display time-of-flight fs-laser excitation/ionization mass spectra obtained from 0.25 ML of methyl iodide at a pump–probe delay time of 2 ps with different probe wavelengths. The graphs below show the temporal evolution of the different mass signals as indicated. Left column (a–d): Pump beam: 266 nm, 1 mW cm^{-2}; probe beam: 333 nm, 600 mW cm^{-2}. Right column (e–h): Pump wavelength 266 nm, 2 mW cm^{-2}, probe beam: 266 nm, 25 mW cm^{-2}. The solid lines in (b) and (c) represent the best fit of kinetic models to the data as explained in the text.

The assignment of these two dynamic processes is illustrated in Fig. 4 in which the relevant potential energy curves of the methyl iodide ground state and A-band are depicted.[22,23] As the exact influence of the surface on the methyl iodide potentials is not known the gas phase potentials shown are assumed to be almost unperturbed by the surface to a first approximation. The 266 nm pump photon excites the CD_3I molecule almost exclusively to the $^3Q_{0+}$ state of the A-band and prepares the initial wave packet at time zero. The inset of Fig. 4 shows that close to the point of the initial excitation, the $^3Q_{0+}$ state leading to CD_3 and spin orbit excited I* crosses the $^1Q_1$ potential that correlates diabatically with CD_3 and ground state I. A bifurcation of the wave packet is possible at this conical intersection.

In contrast to the gas-phase where the dissociation proceeds directly leading to a decay of the transition state within less than 40 fs,[10,24] in the presence of the surface the adsorption geometry depicted in Fig. 2(a) causes the emerging methyl radical to head toward the MgO substrate (illustrated by the dashed potential S in Fig. 4). This results in an inelastic collision which prevents the direct dissociation. Consequently, the dissociative transition state of the excited CD_3I is trapped by the surface leading to a considerably prolonged dissociation time. The peak structure of the transient methyl signal at 130 fs in Fig. 3(b) is therefore assigned to the ionization of the trapped $CD_3I^{*\ddagger}$ transition state. The non-resonant ionization by two 333 nm photons was confirmed by power dependence measurements.[7,8]

The subsequent delayed exponential rise of the methyl ion signal (Fig. 3(b)) is attributed to the (2 + 1)-REMPI detection (as confirmed by power and wavelength dependent measurements[7]) of CD_3 fragments emerging from the molecular dissociation after the inelastic interaction with the magnesia surface. The 170 fs delay and the 680 fs rise time constant of this part of the methyl ion signal (compared to 80 fs delay and an instantaneous rise for the gas-phase photodissociation[26]) strongly support the theory that there is an inelastic interaction with the surface.

Whereas the initial delay of 170 fs is interpreted to reflect the minimal time needed for the liberation of the methyl fragments from the molecular force field and from the force field of the magnesia surface, the subsequent growth of the methyl signal

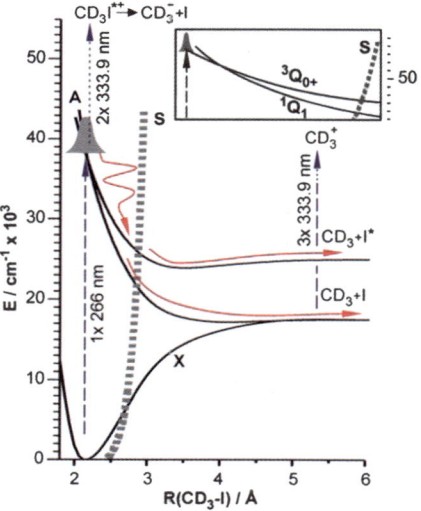

Fig. 4 Potential energy curves of free methyl iodide.[23] The vertical arrows represent the excitation and the detection laser pulses. The curved red arrows illustrate the propagation of the wave packet. The dashed potential S reflects the Lennard-Jones interaction between the methyl group and the magnesia.[25] The inset shows a magnification of the CD_3I-A-band excitation region.

(680 fs) is due to the average lifetime of all trajectories leading from the transition state to the release of the methyl fragment through the inelastic interaction with the surface and the slowly moving, heavy iodine atom.[8]

The temporal evolution of the I_2^+ mass signal is displayed in Fig. 3(c).[7] The open symbols again represent the experimental data. Iodine molecules are detected after a pump–probe delay time of about 200 fs. Subsequently, the signal rises exponentially and reaches its maximum value at 1.2 ps. The fit of an exponential growth model (convoluted with the pump–probe autocorrelation function, solid line in Fig. 3(c)) to the experimental data yields a rise time constant of $\tau_1(I_2) = 310 \pm 30$ fs. Important to the interpretation of this signal is the probe laser power dependence, which shows that the I_2^+ signal is due to a two-photon ionization process, because X- and A-states of I_2 are both not accessible for two-photon ionization with 333 nm. It is therefore assumed that the I_2^+ signal originates from the ionization of iodine molecules in the electronically excited B-state which can be formed in a bimolecular reaction of emerging spin–orbit excited I* with ground state I atoms (cf. Fig. 4) via a favorable surface aligned geometry that enables the required transition state.[6]

This interpretation is further supported by the fact that the I_2^+ signal only stays constant up to about 3 ps delay time and subsequently decays again with a decay time constant of $\tau_2(I_2) = 5.0 \pm 0.3$ ps as reported earlier by our laboratory.[6] This decay is attributed to a coupling between the B-state and a $^1\Pi_u$ dissociative state of molecular iodine which is induced by the presence of the surface in the vicinity of I_2 and which leads to predissociation on the ps timescale.[27]

The measured transient signal of atomic iodine shown in Fig. 3(d) is very weak with a small signal to noise ratio. It apparently starts to increase close to zero pump–probe delay time and seems to remain at constant height after a few hundred fs. Such a transient evolution would be in accordance with the proposed adsorption geometry (Fig. 2(a)) in which the iodine atoms are immediately heading away from the surface upon photodissociation.[7]

The photodissociation of adsorbed methyl iodide was also studied in a one color experiment with a probe wavelength of 266 nm. The corresponding product mass spectrum obtained at 2 ps pump–probe delay time is shown in Fig. 3(e). In general, the same product signals are observed as with a probe wavelength of 333 nm, yet with different relative intensities. The strongest signal originates from the methyl radical. However, the temporal evolution of the CH_3^+ peak (Fig. 3(f)) basically only reflects the autocorrelation function of pump and probe laser pulses (fragmentation resulting from direct multi photon ionization) with a considerable background originating from pump- and probe-only ionization. This leads to the conclusion that the methyl radical formation dynamics are not detectable at 266 nm probe wavelength.

A multi photon ionization signal is also observed in the case of the product I_2^+ at zero delay time with 266 nm detection (Fig. 3(g)). In addition, the I_2^+ signal intensity increases again after a delay time of a few hundred fs. Due to the relatively weak transient signal intensity and a small signal to noise ratio it is however hard to assign a defined kinetic model to this transient signal. Also, no probe power dependent measurements were possible. Nevertheless, the I_2^+ transient is in general rather similar to the one shown in Fig. 3(c). Yet, in striking difference to the I_2^+ signal detected with 333 nm probe wavelength, the transient in Fig. 3(g) does not decrease again within 100 ps. This might be an indication that a major part of this signal results from the detection of ground state molecular iodine formed as a consequence of methyl iodide photodissociation.

The I^+ transient signal detected at 266 nm probe wavelength (Fig. 3(h)) is peculiar in that, apart from the coherent ionization signal at zero time delay, it exhibits an additional signal increase rather close to zero time delay which is completed after about 1.5 ps. The interpretation of this signal is also not straightforward because no probe power dependent measurements have been possible so far.

3.3 Photoinduced surface reaction dynamics of methyl bromide

Fig. 5(a) shows a time-of-flight mass spectrum obtained at 2 ps pump–probe delay time from methyl bromide adsorbed at submonolayer coverage (0.25 ML) on 8 ML MgO/Mo(100). The pump laser wavelength was again 266 nm and the probe laser was tuned to the center wavelength of 333 nm to sensitively detect the methyl fragments.[21] Interestingly, the only reaction product observed in the mass spectrum in this case was the methyl fragment. No other reaction products were detected regardless of laser intensity or pump–probe delay time.

In order to monitor in real-time the photodissociation dynamics of methyl bromide molecules adsorbed at submonolayer coverage on the magnesia surface, the methyl cation signal intensity was also recorded as a function of the pump–probe delay time and the result is shown in Fig. 5(b) (open circles). No transient signal was observed up to 150 fs. Subsequently, the CH_3^+ signal presents an exponential rise with the maximum reached around 1.2 ps. Through fitting of the experimental data by a 'delayed exponential rise'-model (solid line in Fig. 5(b)) a time constant of $\tau = 320 \pm 60$ fs was obtained for the rise of the methyl cation signal.

The known electronic structure of the free CH_3Br molecule will now be considered here as well to discuss the molecular surface photodissociation because to a first

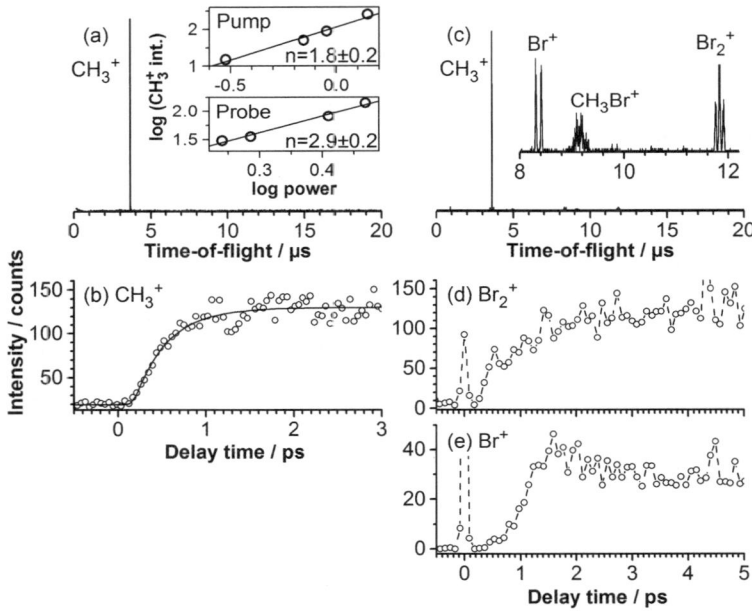

Fig. 5 (a) Time-of-flight mass spectrum obtained after photoexcitation of methyl bromide adsorbed at submonolayer coverage (0.25 ML) on 8 ML MgO/Mo(100) at 266 nm (6 mW cm^{-2}) and probing at 333 nm (600 mW cm^{-2}, 2 ps pump–probe delay time) through (2 + 1)-REMPI *via* the $3p^2A_2''$ Rydberg state.[21] Upper inset: Methyl cation signal intensity as a function of the 266 nm pump power. Lower inset: Methyl cation signal intensity as a function of the 333 nm probe laser power (both also measured at 2 ps pump–probe delay time, both in a double logarithmic representation). The slopes n of the linear fits to the data are indicated in the plots. (b) Temporal evolution of the methyl cation signal as a function of the pump–probe delay time (open circles). The solid line represents the best fit of a single exponential rise model to the data (see text for more details). (c) Mass spectrum obtained in a 266 nm one color experiment (also 2 ps pump–probe delay time, pump 2 mW cm^{-2}, probe 25 mW cm^{-2}). The inset shows a magnification of the 8–12 µs range. The graphs below show the transient evolution of the corresponding mass signals of (d) Br_2^+ and (e) Br^+ obtained in the 266 nm one color experiment.

approximation the methyl bromide molecules are only weakly disturbed by the interaction with the magnesia surface as evidenced by TPD (Fig. 2(b)).

In contrast to methyl iodide, which was extensively studied in the past, few reports are available on methyl bromide photodissociation. Previous investigations[28] assumed that the potentials of methyl bromide and methyl iodide molecules are similar, but due to the shorter C–Br bond length of methyl bromide (1.939 Å) compared to the C–I bond length of methyl iodide (2.1396 Å), the A-band maximum of CH_3Br is shifted to higher energy. The A-band of CH_3Br, i.e. the first absorption continuum, can be accessed in the 170–270 nm spectral range with an absorption maximum around 200 nm.[29]

The excitation of methyl bromide to the A-band also leads to two dissociative channels: (i) the Br-channel giving rise to the formation of a methyl radical and a bromine atom in the ground state ($Br(^2P_{3/2})$), and (ii) the Br*-channel giving rise to methyl and a bromine atom in the spin–orbit excited state ($Br(^2P_{1/2})$). Five dissociative electronic states are correlated with these two channels. Using the Mulliken notation,[30] these states are labeled in ascending energy order $^3Q_2$, $^3Q_1$, $^3Q_{0+}$, $^3Q_{0-}$, and $^1Q_1$. The $^3Q_1$ and $^1Q_1$ states can be accessed from the ground state via single photon excitation in a perpendicular electric dipole transition and they adiabatically correlate with the Br-channel. The $^3Q_{0+}$ state can be accessed in a parallel transition by single photon excitation and it is correlated with the Br*-channel. The single photon excitation of the $^3Q_2$ and $^3Q_{0-}$ states is electric dipole forbidden. A curve crossing occurs between the $^1Q_1$ and $^3Q_{0+}$ states[31,32] which is possible due to the different symmetries of these states. However, in contrast to the methyl iodide molecule, where the A-band absorption occurs mainly via a parallel transition to the $^3Q_{0+}$ state which correlates with I* (cf. Fig. 4), in the case of methyl bromide, magnetic-circular-dichroism investigations show that the $^1Q_1$ state is more strongly populated than the $^3Q_1$ and $^3Q_{0+}$ states.[33]

In order to clarify the initial excitation mechanism that eventually leads to the dissociation of methyl bromide adsorbed on the magnesia surface, the CH_3^+ yield at 2 ps pump–probe delay time was monitored as a function of the pump laser power. The result is displayed in the upper inset of Fig. 5(a) and clearly shows that the photodissociation process requires more than one single photon. The slope value of 1.8 ± 0.2 of a linear fit to the measured power dependence data in the double logarithmic plot indicates that the photodissociation is most likely initiated by two pump photons.

The prospective dissociative pathways due to an excitation with two photons at a central wavelength of 266 nm are depicted in Fig. 6, which displays the diabatic potential energy curves corresponding to the $^1A_1$ ground state and to the A-band of a free methyl bromide molecule (Fig. 6(a)) as well as the 5s–5p Rydberg states with consideration of spin–orbit interactions (Fig. 6(b)). When spin–orbit interactions are included, an ^XE state in Fig. 6(a) becomes $^XQ_\Omega$, where $\Omega = 0, 1, 2$ is the projection of the total angular momentum (orbital + spin), i.e. ^3E becomes $^3Q_2$, $^3Q_1$, $^3Q_{0-}$ (A_2-symmetry), and $^3Q_{0+}$ (A_1-symmetry), and ^1E becomes $^1Q_1$. The $^3A_1$ state give rise to two states corresponding to $\Omega = 0$ and $\Omega = 1$.

The most probable two-photon excitation channel for the CH_3Br molecule which leads to dissociation is indicated by the vertical arrows in Fig. 6(a). According to ref. 31, the $^3A_1$ dissociative state can be accessed near the C–Br equilibrium distance by two-photon excitation at 266 nm. However, the 5p Rydberg states are also located at 9.3 eV (2×266 nm) above the ground state (cf. Fig. 6(b)). Yet, the dissociation directly from these Rydberg states is excluded because these states do not have a repulsive character. However, theoretical investigations[34] indicate that a crossing between one particular state of the 5p Rydberg series and a repulsive valence-character state (corresponding to $^3A_1$) is avoided and thus a potential barrier is generated through which tunneling and consequently transition from a Rydberg to a valence state can occur. This photodissociation channel is also excluded because of the predicted long lifetime of the Rydberg states of about 10 ps[34] which would be reflected in the appearance time of the CH_3^+ reaction products. Our data presented in Fig. 5(b) show, however, a methyl appearance time of only about 320 fs (see below).

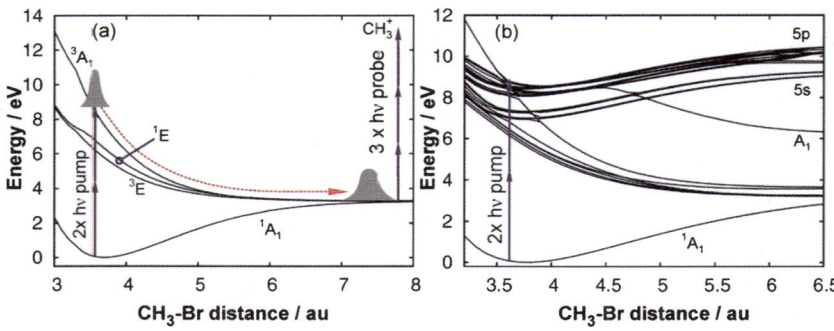

Fig. 6 Potential energy curves of the free CH_3Br molecule. (a) Ground state ($^1A_1$) and A-band (^3E, ^1E, $^3A_1$) (adapted from ref. 31); (b) Ground state ($^1A_1$), A-band and the 5s–5p Rydberg states with consideration of spin–orbit interaction (adapted from ref. 34). The dashed arrows illustrate the propagation of the wave packet. The vertical arrows represent the excitation and the detection laser pulses. The distance is given in atomic units (au).

Consequently, we conclude that the excitation of CH_3Br molecules at 266 nm by means of a two-photon transition most likely leads to A-band dissociation *via* the $^3A_1$ state, always under the assumption that the gas-phase potentials of CH_3Br are only weakly perturbed on the MgO surface.

The emerging methyl radicals are subsequently detected *via* (2 + 1) REMPI at a wavelength of 333 nm. This is confirmed by the probe power dependence measurements shown in the lower inset of Fig. 5(a). Similar to CH_3I/MgO, the initial coherent delay of $\Delta t_0 = 150 \pm 50$ fs of the methyl ion signal emerging from CH_3Br (Fig. 5(b)) is proposed to reflect the liberation of the methyl fragments from the molecular force field and from the force field of the magnesia surface. The subsequent growth of the methyl signal with a time constant of $\tau = 320 \pm 60$ fs is consequently interpreted as the average lifetime of all trajectories leading from the initial excitation to the release of the methyl fragment.

Interestingly, the time constant of $\tau = 320 \pm 60$ fs for the rise of the CH_3^+ signal in Fig. 5(b) is considerably shorter compared to the one obtained for the exponential rise of the methyl cation signal in the case of CH_3I photodissociation on MgO (680 ± 50 fs, see Section 3.2). As previously mentioned, no direct measurement of the CH_3Br photodissociation time has been performed until now. However, based on the method provided by Sander and Wilson[35] to calculate the lifetime of a dissociative state using the anisotropy parameter β which is deduced from the angular distribution of the photofragments, Gougousi *et al.*,[36] estimated that the lifetime of the CH_3Br A-band is 120 ± 40 fs. If this method is applied to CH_3I, an upper limit for the lifetime of the A-band of 70 fs can be estimated.[37] This means that in the gas-phase the methyl bromide A-band photodissociation takes longer than the methyl iodide A-band photodissociation. However, in the present investigation methyl bromide appears to photodissociate faster than methyl iodide, if adsorbed on MgO.

By means of angular resolved time-of-flight mass spectrometry, Polanyi and co-workers have investigated the 193 nm photodissociation of CH_3Br adsorbed on a bulk MgO(100) surface.[19] Depending on the coverage, this experiment indicates that the emerging methyl photofragments can be either directly liberated into the gas-phase or they can lose a significant part of their total kinetic energy released from the A-band photodissociation in a collision process with neighboring molecules. This latter mechanism exclusively occurs at submonolayer coverage, where all escaping methyl fragments have to experience a collision with adjacent molecules because of an adsorption geometry in which the C–Br axis is nearly parallel to the surface (*cf.* inset in Fig. 2(b)). These results by Polanyi and coworkers[19] provide the basis for a reasonable explanation why the methyl photofragment appearance time on MgO, measured in our experiment, is much faster for methyl bromide

(320 ± 60 fs) compared to methyl iodide (680 ± 50 fs) despite the theoretical prediction of a longer dissociation time for free CH_3Br.

The different appearance time can be connected to the different adsorption geometries of the methyl halide molecules on the magnesia surface. As discussed in Section 3.1 it is proposed that methyl iodide molecules are adsorbed with the methyl-end heading toward the surface. This leads to a trapping of the methyl group between the MgO surface and the I atom subsequent to the photodissociation and the liberation of the methyl fragment from the surface takes a considerable time in the case of methyl iodide. In contrast, according to Polanyi and coworkers,[19] the methyl bromide molecules are adsorbed with their C–Br axis almost parallel to the MgO substrate. Therefore, the methyl photofragment can escape much easier and faster, suffering in most cases only one collision with the adjacent molecule, which would explain the faster appearance time of 320 ± 60 fs for CH_3Br.

The results of a 266 nm one color pump–probe experiment for methyl bromide on magnesia are displayed in the right hand column of Fig. 5. The mass spectrum shown in Fig. 5(c) demonstrates that Br^+, CH_3Br^+, and Br_2^+ can also be detected at this wavelength. The temporal evolution of the methyl cation signal (not shown) only consists of the coherent multi photon ionization autocorrelation signal at zero time delay similar to Fig. 3(f). In contrast, the molecular bromine signal additionally exhibits an exponential rise which starts at about 200 fs, reaches the maximum intensity around 2.5 ps delay time, and stays constant afterwards (Fig. 5(d)). Thus, also in the case of the methyl bromide layer a photoinduced reaction of the halogen fragments occurs that leads to the formation of the observed Br_2 molecules.

Interestingly, the signal of bromine atoms (Fig. 5(e)) starts to rise a few hundred fs after the signal of molecular bromine (Fig. 5(d)), but then rapidly reaches a maximum value around 1.5 ps. This dynamical behavior would be in accordance with the CH_3Br adsorption structure shown in Fig. 2(b) and with the dynamics detected using 333 nm probe wavelength as discussed above. Due to the adsorption with the C–Br axes almost parallel to the surface (Fig. 2(b)) the Br atoms are not expelled directly away from the surface upon photodissociation but have to first collide with neighboring molecules which leads to the delayed rise of the Br^+ signal. In contrast, Br_2 molecules are apparently detected as soon as they are formed. Yet, the elucidation of the detailed molecular reaction mechanism will require additional power and wavelength dependent measurements, which are in progress.

3.4 Photoinduced surface reactions of co-adsorbed methyl iodide and methyl bromide

To investigate the possibility of observing a reaction of fragments emerging from different types of molecular adsorbates on the substrate, co-adsorbed layers of methyl iodide and methyl bromide were prepared on MgO/Mo(100). These layers were excited with 266 nm laser light and the emerging reaction products were also detected with a 266 nm laser beam. A mass spectrum that was obtained at a pump–probe delay time of 5 ps is shown in Fig. 7. The most prominent signal is again the methyl cation. Furthermore, all product peaks that were observed under similar laser conditions for layers of the pure molecules are also present when the co-adsorbed layer is irradiated. Most interestingly, however, an additional mass signal of IBr^+ appears as can be seen in the magnified part of the spectrum in the inset of Fig. 7. This signal clearly demonstrates that a reaction of halogen atoms originating from the different adsorbates occurs.

Currently, experiments are in progress to optimize the coverage conditions for the measurement of the transient evolution of the IBr signal with different detection laser wavelengths. It is expected that a detailed investigation of the dynamics of all of the observed product signals in this case will provide a comprehensive picture of the photoinduced reaction behavior exhibited by different adsorbate molecules on the oxide substrate.

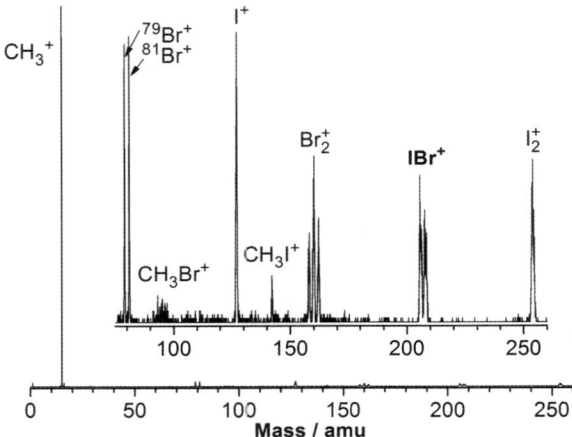

Fig. 7 Mass spectrum obtained by irradiation of a co-adsorbed layer of CH_3I and CH_3Br on 8 ML MgO/Mo(100). Pump laser: 266 nm, 0.7 mW average power; probe laser: 266 nm, 4 mW average power; delay time: 5 ps. The total coverage amounts to 0.5 ML with a CH_3Br/CH_3I ratio of approximately 5 to account for the different photodissociation power dependencies of CH_3I and CH_3Br. Please note that the time-of-flight scale has already been converted to the corresponding mass scale. The inset displays a magnification of the mass region between 75 and 260 amu.

4 Conclusion

In this contribution it was demonstrated that fs-laser induced excitation of molecular adsorbate layers on oxide surfaces in conjunction with time delayed direct ionization and mass spectrometric detection of reaction products on the surface can provide insight into the molecular surface reaction dynamics. As examples the photoreactions of methyl iodide and methyl bromide adsorbed on ultrathin MgO(100) films on Mo(100) were studied. A pronounced difference in the appearance times of the methyl radical fragments was observed for the two different methyl halides. This observation was related to the different molecular adsorption structures on magnesia. Furthermore, the adsorption geometry on the surface also defined the alignment of the encounter complex for the observed bimolecular formation of the halogen molecules I_2 and Br_2. Most interestingly, the irradiation of co-adsorption layers of CH_3I and CH_3Br lead to the detection of IBr molecules, which confirmed the occurrence of a reaction of fragments from different adsorbates. Further investigations, also including different co-adsorbate molecules like O_2, N_2O and CO_2 in addition to the methyl halides, are in progress.

Acknowledgements

The authors acknowledge financial support from the Deutsche Forschungsgemeinschaft (DFG), the Fonds der Chemischen Industrie (FCI), and by the Deutscher Akademischer Austauschdienst (DAAD).

References

1 A. H. Zewail, *J. Phys. Chem. A*, 2000, **104**, 5660.
2 C. Frischkorn and M. Wolf, *Chem. Rev.*, 2006, **106**, 4207.
3 J. C. Polanyi and A. H. Zewail, *Acc. Chem. Res.*, 1995, **28**, 119.
4 J. C. Polanyi and Y. Zeiri, in *Laser Spectroscopy and Photochemistry on Metal Surfaces*, ed. H.-L. Dai and W. Ho, World Scientific Publishing, Singapore, 1995, p. 1241.
5 H.-L. Dai and W. Ho, *Laser spectroscopy and photochemistry on metal surfaces*, World Scientific, Singapore, 1995; H. Petek and S. Ogawa, *Annu. Rev. Phys. Chem.*, 2002, **53**,

507; W. Ho, *Surf. Sci.*, 1996, **363**, ⁞66; J. A. Misewich, A. Kalamarides, T. F. Heinz, U. Hoefer and M. M. T. Loy, *J. Chem. Phys.*, 1994, **100**, 736; P. Szymanski, A. L. Harris and N. Camillone III, *J. Phys. Chem. C*, 2008, **112**, 15802; M. Bonn, S. Funk, C. Hess, D. N. Denzler, C. Stampfl, M. Scheffler, M. Wolf and G. Ertl, *Science*, 1999, **285**, 1042; J. A. Prybyla, H. W. K. Tom and G. D. Aumiller, *Phys. Rev. Lett.*, 1992, **68**, 503; I. M. Lane, D. A. King, Z.-P. Liu and H. Arnolds, *Phys. Rev. Lett.*, 2006, **97**, 186105; P. Szymanski, S. Garrett-Roe and C. B. Harris, *Prog. Surf. Sci.*, 2005, **78**, 1; C. Reuss, I. L. Shumay, U. Thomann, M. Kutschera, M. Weinelt, T. Fauster and U. Hofer, *Phys. Rev. Lett.*, 1999, **82**, 153; M. Bauer, C. Lei, K. Reak, R. Tobey, J. Gland, M. M. Murnane and H. C. Kapteyn, *Phys. Rev. Lett.*, 2001, **87**, 025501; C. D. Lindstrom and X.-Y. Zhu, *Chem. Rev.*, 2006, **106**, 4281; A. Foehlisch, P. Feulner, F. Hennies, A. Fink, D. Menzel, D. Sanchez-Portal, P. M. Echenique and W. Wurth, *Nature*, 2005, **436**, 373.

6 M. E. Vaida and T. M. Bernhardt, *ChemPhysChem*, 2010, **11**, 804.

7 M. E. Vaida and T. M. Bernhardt, *Rev. Sci. Instrum.*, 2010, **81**, 104103.

8 M. E. Vaida, P. E. Hindelang and T. M. Bernhardt, *J. Chem. Phys.*, 2008, **129**, 011105.

9 N. Eleftheriadis, H.-P. Ludescher, M. Sässbauer and D. von der Linde, *Appl. Surf. Sci.*, 1990, **46**, 284; P. Nuernberger, D. Wolpert, H. Weiss and G. Gerber, *Proc. Natl. Acad. Sci. U. S. A.*, 2010, **107**, 10366.

10 D. P. Zhong and A. H. Zewail, *J. Phys. Chem. A*, 1998, **102**, 4031.

11 D. Zhong, T. M. Bernhardt and A. H. Zewail, *J. Phys. Chem. A*, 1999, **103**, 10093.

12 M.-C. Wu, J. S. Corneille, C. A. Estrada, J.-W. He and D. W. Goodman, *Chem. Phys. Lett.*, 1991, **182**, 472; M.-C. Wu, J. S. Corneille, J.-W. He, C. A. Estrada and D. W. Goodman, *J. Vac. Sci. Technol., A*, 1992, **10**, 1467; U. Heiz, F. Vanolli, L. Trento and W.-D. Schneider, *Rev. Sci. Instrum.*, 1997, **68**, 1986.

13 M. Sterrer, E. Fischbach, M. Heyde, N. Nilius, H.-P. Rust, T. Risse and H. J. Freund, *J. Phys. Chem. B*, 2006, **110**, 8665.

14 M. E. Vaida, T. Gleitsmann, R. Tchitnga and T. M. Bernhardt, *J. Phys. Chem. C*, 2009, **113**, 10264.

15 S. Schintke, S. Messerli, M. Pivetta, F. Patthey, L. Libioulle, M. Stengel, A. De Vita and W.-D. Schneider, *Phys. Rev. Lett.*, 2001, **87**, 276801.

16 W. C. Wiley and I. H. McLaren, *Rev. Sci. Instrum.*, 1955, **26**, 1150.

17 V. P. Holbert, S. J. Garrett, J. C. Bruns, P. C. Stair and E. Weitz, *Surf. Sci.*, 1994, **314**, 107.

18 D. H. Fairbrother, K. A. Trentelman, P. G. Strupp, P. C. Stair and E. Weitz, *J. Vac. Sci. Technol., A*, 1992, **10**, 2243; K. A. Trentelman, D. H. Fairbrother, P. G. Strupp, P. C. Stair and E. Weitz, *J. Chem. Phys.*, 1992, **96**, 9221.

19 S. J. Garrett, D. V. Heyd and J. C. Polanyi, *J. Chem. Phys.*, 1997, **106**, 7847.

20 S. J. Garrett, D. V. Heyd and J. C. Polanyi, *J. Chem. Phys.*, 1997, **106**, 7834.

21 J. W. Hudgens, T. G. DiGiuseppe and M. C. Lin, *J. Chem. Phys.*, 1983, **79**, 571.

22 M. Tadjeddine, J. P. Flament and C. Teichtel, *Chem. Phys.*, 1987, **118**, 45.

23 A. B. Alekseyev, H.-P. Liebermann, R. J. Buenker and S. N. Yurchenko, *J. Chem. Phys.*, 2007, **126**, 234102.

24 J. Durá, R. De Nalda, J. Álvarez, J. G. Izquierdo, G. A. Amaral and L. Banares, *ChemPhysChem*, 2008, **9**, 1245.

25 J.-Y. Fang and H. Guo, *Chem. Phys. Lett.*, 1995, **235**, 341.

26 R. de Nalda, J. Durá, A. Garcia-Vela, J. G. Izquierdo, J. González-Vázquez and L. Banares, *J. Chem. Phys.*, 2008, **128**, 244309.

27 E. D. Potter, Q. Liu and A. H. Zewail, *Chem. Phys. Lett.*, 1992, **200**, 605.

28 G. N. A. Van Veen, T. Baller and A. E. De Vries, *Chem. Phys.*, 1985, **92**, 59; W. P. Hess, D. W. Chandler and J. W. Thoman Jr, *Chem. Phys.*, 1992, **163**, 277.

29 W. S. Felps, K. Rupnik and S. P. McGlynn, *J. Phys. Chem.*, 1991, **95**, 639; L. T. Molina, M. J. Molina and F. S. Rowland, *J. Phys. Chem.*, 1982, **86**, 2672.

30 R. S. Mulliken, *J. Chem. Phys.*, 1940, **8**, 382; R. S. Mulliken, *Phys. Rev.*, 1942, **61**, 277; R. S. Mulliken and E. Teller, *Phys. Rev.*, 1942, **61**, 283.

31 C. Escure, T. Leininger and B. Lepetit, *J. Chem. Phys.*, 2009, **130**, 244305.

32 D. Ajitha, M. Wierzbowska, R. Lindh and P. A. Malmqvist, *J. Chem. Phys.*, 2004, **121**, 5761.

33 A. Gedanken and M. D. Rowe, *Chem. Phys. Lett.*, 1975, **34**, 39.

34 C. Escure, T. Leininger and B. Lepetit, *J. Chem. Phys.*, 2009, **130**, 244306.

35 R. K. Sander and K. R. Wilson, *J. Chem. Phys.*, 1975, **63**, 4242.

36 T. Gougousi, P. C. Samartzis and T. N. Kitsopoulos, *J. Chem. Phys.*, 1998, **108**, 5742.

37 M. Dzvonik, S. Yang and R. Bersohn, *J. Chem. Phys.*, 1974, **61**, 4408.

Crowding effects on the small, fast-folding protein λ_{6-85}

Sharlene Denos,[a] Apratim Dhar[b] and Martin Gruebele[*abc]

Received 22nd January 2012, Accepted 15th February 2012
DOI: 10.1039/c2fd20009k

The microsecond folder λ_{6-85} is a small (9.2 kDa = 9200 amu) five helix bundle protein. We investigated the stability of λ_{6-85} in two different low-fluorescence crowding matrices: the large 70 kDa carbohydrate Ficoll 70, and the small 14 kDa thermophilic protein SubL. The same thermal stability of secondary structure was measured by circular dichroism in aqueous buffer, and at a crowding fraction $\varphi = 15 \pm 1\%$ of Ficoll 70. Tryptophan fluorescence detection (probing a tertiary contact) yielded the same thermal stability in Ficoll, but 4 °C lower in aqueous buffer. Temperature-jump kinetics revealed that the relaxation rate, corrected for bulk viscosity, was very similar in Ficoll and in aqueous buffer. Thus viscosity, hydrodynamics and crowding seem to compensate one another. However, a new fast phase was observed in Ficoll, attributed to crowding-induced downhill folding. We also measured the stability of λ_{6-85} in $\varphi = 14 \pm 1\%$ SubL, which acts as a smaller more rigid crowder. Significantly greater stabilization (7 to 13 °C depending on probe) was observed than in the Ficoll matrix. The results highlight the importance of crowding agent choice for studies of small, fast-folding proteins amenable to comparison with molecular dynamics simulations.

1. Introduction

Unimolecular reactions of small molecules in the aqueous phase usually have a fairly straightforward dependence on solvent condition, given by Kramers' model.[1] What about large unimolecular reactants, such as a protein folding from its denatured state to its native state? When large proteins are crowded in aqueous solution by other large molecules (*e.g.*, high molecular weight carbohydrates), interesting things can happen. For example, the reaction can speed up upon crowding, even though the viscosity increases[2] (and the reaction is not in the Kramers anomalous regime). But when is a protein really a 'large' molecule? In recent years, the fast folding kinetics of many 'small' (20–100 amino acid) proteins have been studied, and the results can now be compared directly with full atom molecular dynamics simulations.[3] Until now, the folding kinetics of such 'small' proteins have not been examined in crowding environments.

When proteins fold, they traverse an energy landscape that can be characterized overall by two parameters: its energetic bias towards the native state, and its energy roughness.[4] If we perform the Legendre transform from energy to free energy (relevant in the laboratory where temperature, not entropy, is the adjustable parameter)

[a]Center for Biophysics and Computational Biology, University of Illinois, 600 South Mathews Avenue, Urbana-Champaign, IL 61801. E-mail: mgruebel@illinois.edu
[b]Department of Chemistry, University of Illinois, 600 South Mathews Avenue, Urbana-Champaign, IL 61801
[c]Department of Physics, University of Illinois, 600 South Mathews Avenue, Urbana-Champaign, IL 61801

a continuous range of scenarios emerges. At one extreme, if the bias is weak and the roughness is large, folding intermediates accumulate. This is the likely scenario for large proteins with many opportunities for non-native contacts. Work on smaller proteins has demonstrated that folding intermediates are not obligatory for folding: small proteins usually fold over a single barrier,[5] analogous to most chemical reactions.[6] Going even further to the other extreme, the native bias can become so large, and the roughness so small that some small proteins fold without any significant barrier at all, analogous to many ion–molecule, excited electronic state, or radical reactions.[7] Extensive experimental and computational evidence for downhill folding in solution has been accumulated in recent years.[8–24]

Downhill folding can be detected thermodynamically:[9] When there is a substantial barrier to heat-induced unfolding, a single melting (unfolding) temperature T_m is obtained by all spectroscopic probes. Without such a barrier, tuning the temperature yields different T_m for different probes.[9,15,19] Downhill folding can also be detected dynamically (Fig. 1A): When there is a substantial barrier, a single activated rate coefficient k_a for folding is measured. Upon lowering the barrier, k_a at first simply increases; but when the population diffusing through the transition state reaches a detectable size, the rate coefficient becomes time-dependent at short times.[11] In this limit, reaction kinetics becomes reaction dynamics. The time it takes for the rate coefficient to settle into a constant plateau value k_a is termed 'molecular rate' or $k_m = \tau_m^{-1} \gg k_a$.[6] It can be measured to extract absolute folding barriers $\Delta G^{\dagger}$ directly from experiment:[11]

$$\Delta G^{\dagger} = k_B T \ln(k_m/k_a) \tag{1}$$

Because of large protein size and solvent viscosity, k_m is much greater for proteins (typically > 100 ns) than for small molecule reactions (typically < 1 ps).

Very recently, single trajectory molecular dynamics simulations on the Anton supercomputer have confirmed downhill folding of small proteins in solution computationally *via* observation of repeated folding and unfolding events.[3] Sampling was sufficient to compute one dimensional potentials of mean force (PMFs, coordinate-dependent free energies) using Hummer's coordinate reduction

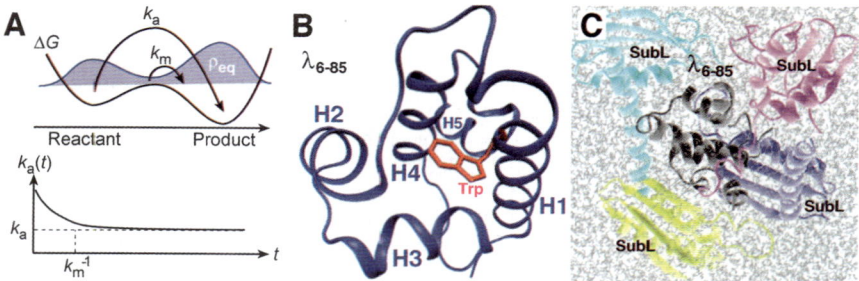

Fig. 1 (A) Connection between kinetics and dynamics. If the unimolecular activation barrier is very low (middle of ΔG curve), a measurable activated population exists (middle of the shaded equilibrium population ρ_{eq}). This population reacts by direct diffusion with a rate $k_m \gg k_a$. Thus the rate coefficient $k_a(t)$ does not immediately settle into a constant value k_a: the usual unimolecular concentration change $\sim \exp[-k_a t]$ is preceded by a short 'burst' of very fast reaction, allowing a direct independent measurement of the prefactor k_m in the equation $k_a = k_m \exp[-\Delta G^{\dagger}/RT]$ for the activated rate coefficient k_a. The size of the short burst is proportional to the activated population, and becomes unobservable when the activation barrier is very high. (B) Monomer of lambda repressor fragment 6-85 showing the tryptophan fluorescence probe as a stick structure, and helices 1 through 5 as ribbon structures. (C) Model of λ_{6-85} in 200 mg ml^{-1} SubL, to indicate the nature of the crowding effect. Water molecules are shown as small gray sticks. Plots were created using the free visualization software VMD.[62]

method.[25] For example, Anton simulations and thermodynamic experiments agree that a certain protein BBL mutant has a negligible activation barrier even at T_m.[3,13] For the largest protein folded by a single trajectory to date (Fig. 1B), a mutant of the 5 helix bundle λ_{6-85}, experiment yielded an absolute barrier of ≈ 4 kJ mole^{-1} at 65 °C,[11] and a melting point range of $T_m = 68$ °C (fluorescence detection) to 73.5 °C (circular dichroism detection). The Anton simulation yielded an identical barrier height of ≈ 4 kJ mole^{-1} at 77 °C. There is good reason to believe that globular protein domains (usually <200 amino acids) will be foldable in solution routinely by single trajectory,[26] multi-trajectory[27] or replica[21] molecular dynamics techniques to Ångstrom accuracy within a decade, as calibration against experiments further improves the accuracy of empirical force fields, yielding better stabilities, kinetic mechanisms, and structures.[28]

As a result, experimental work to study protein folding in more realistic environments has accelerated during the past several years. Proteins are now studied on the ribosome,[29,30] in densely packed crowding agents,[31–37] and even in living cells directly.[38,39] Significant changes of protein stability and folding kinetics are expected in such complex environments, based on the aqueous solution results outlined above: free energy barriers and stabilities of proteins are small (10s of kJ mole^{-1}), and thus easily modulated by interactions of the protein with its solvation environment. For technical reasons, crowding studies have focused on large proteins so far; no experimental results are available for small sub-millisecond folders suitable for all atom simulation, the closest being the small heme protein cytochrome c that folds in under a second.[40]

Here we study λ_{6-85}, already discussed above, in two very different crowding environments: Ficoll 70, a 70 kDa semi-rigid cross-linked carbohydrate that acts as a roughly spherical crowder in aqueous solution; and SubL, a 14 kDa thermophilic protein that serves as a smaller and more rigid crowder than the commonly used large molecular weight carbohydrates (Fig. 1C). We find that the stability of λ_{6-85} in Ficoll deviates only little from that in aqueous solution. A small probe dependence of the melting temperature in aqueous buffer (circular dichroism *vs.* fluorescence) disappears in Ficoll. T-jump kinetics of λ_{6-85} in Ficoll likewise reveals similar rates in Ficoll as in aqueous buffer, although a molecular phase 'burst' precedes the activated unimolecular kinetics in Ficoll 70. To enhance crowding effects, we propose the new protein crowder SubL.[41] It is smaller and more rigid than Ficoll, thermophilic ($T_m > 90$ °C), and has very low fluorescence, making it an ideal match for crowding small, fast-folding proteins such as λ_{6-85} by fluorescence techniques. We find that SubL has a larger effect on λ_{6-85} stability at volume fractions comparable to those for Ficoll. We conclude that SubL matrices provide a better size match and more rigid crowding environment for small, fast folding proteins such as lambda repressor fragments.

2. Methods

2.1 Sample preparation

The fluorescent lambda repressor fragment λ_{6-85}^* ($T_m = 61$ °C by circular dichroism = CD[42]) and its Glu33Tyr/Ala37,49Gly mutant λ_{6-85}^*YG ($T_m = 54.5$ °C by CD[42]) were expressed and purified as described previously.[42] A tyrosine 22 to tryptophan mutation makes both proteins strongly fluorescent when excited at 290 nm.[43] λ_{6-85}^*YG tryptophan fluorescence monitors a specific Trp22-Tyr33 tertiary interaction[23] whose thermodynamics and kinetics we studied in Ficoll, while λ_{6-85}^* has a strongly decreasing fluorescence upon unfolding, ideally suited for studies in the slightly fluorescent SubL crowder.

A pET-19b plasmid containing the gene for protein SubL from the thermophile *Methanococcus jannashii* (obtained from Prof. Gary Olsen) was transformed and expressed in BL21(DE3) Rosetta cells. Protein was harvested by French Press and purified by heating to 62 °C to precipitate non-thermophilic cell lysate, followed by column chromatography. The chromatography protocol consisted of Ni-NTA affinity purification followed by size exclusion purification and a second Ni-NTA affinity purification. The His purification tag was not removed and the resulting protein had a mass of 14028 Da. Purified protein was dialyzed against double deionized water and checked for purity by MALDI and ESI mass spectrometry prior to lyophilization.

λ_{6-85}^*YG crowded by Ficoll 70 was prepared by mixing concentrated protein buffers with carbohydrate buffers (50 mM NaH$_2$PO$_4$, pH 7.0). Volume fractions in the range 15 ± 1% Ficoll (240 mg ml^{-144}) were used. λ_{6-85}^* samples crowded by SubL were prepared by adding lyophilized protein to 2 mL of buffer (300 mM NaCl, 50 mM NaH$_2$PO$_4$, pH 7.0), and then concentrating 4 times in <200 mTorr vacuum overnight. The buffer had no effect on SubL stability below 90 °C. A crowding fraction of $\varphi = 14 \pm 1\%$ SubL (220 mg ml^{-1}) was used for SubL. These crowding fractions avoided precipitation of the protein in the matrices. Although we did not centrifuge samples, it has been shown by dynamic light scattering that such saturated protein solutions can experience a 'liquid–liquid' phase separation into a highly crowded and a less crowded phase.[45] This can be remedied simply by centrifugation and collection of the denser phase.

Final concentration of lambda repressor in the matrices were determined using UV/Visible absorption at 280 nm and an extinction coefficient of 1750 M^{-1}cm^{-1} for SubL, 6970 M^{-1}cm^{-1} for λ_{6-85} mutants and ≈ 0 for Ficoll. The lambda repressor concentrations in the matrix were kept at least 10 times lower than the matrix concentration to avoid strong interaction of lambda repressor with itself.

2.2 Circular dichroism and fluorescence measurements

Samples were overlaid with mineral oil to prevent evaporation. Larger temperature steps and minimal collection time for acceptable signal-to-noise ratio were used to reduce sample denaturation at higher temperatures.

Fluorescence spectra of lambda repressor, in aqueous buffer, crowded by Ficoll, or crowded by SubL, were measured as a function of temperature using a Varian (Palo Alto, CA) Cary Eclipse spectrofluorimeter with a 4-position Peltier multicell changer and a PCB-150 circulating water bath. Emission spectra were obtained from 300 to 400 nm upon 290 nm excitation of the tryptophan 22 residue, using slit widths of 5 nm for both emission and excitation monochromators to maximize signal. Temperature titrations were performed using a nitrogen purge of the sample chamber and the temperature probe placed directly in the buffer blank cuvette.

For Ficoll crowding only, it was also possible to measure circular dichroism spectra and melts of lambda repressor on a JASCO J-715 spectropolarimeter equipped with a Peltuer temperature controller. We simultaneously collected total fluorescence intensity (320–450 nm) during hose scans. No CD melts could be obtained for lambda repressor in the SubL matrix because SubL itself has a large CD spectrum and is at much higher concentration than the probe protein.

CD thermal melts dilute SubL solutions, to measure its stability, were obtained on a JASCO J-715 spectropolarimeter. The samples were 5 μM SubL in 2 M GuHCl denaturant at pH 4.5, and 264 mg ml^{-1} SubL in 2 M NaCl, 200 mM NaH$_2$PO$_4$ buffer at pH 7.

Integrated fluorescence and CD data were normalized and plotted as a function of temperature. We used a simple cooperative thermodynamic model to obtain T_m by fitting fluorescence and CD intensities. Table 1 summarizes the results of the model fits. The cooperative model is

$$[N] = \frac{K_{eq}}{1 + K_{eq}} = 1 - [D], \quad K_{eq} = e^{-\Delta G(T)/RT}, \quad \Delta G(T) = C^{(1)}(T - T_m) \quad (2)$$

In this expression, [N] and [D] are the native and denatured concentrations as detected by either intensity or wavelength shift, K_{eq} is the folding equilibrium constant, $\Delta G(T)$ is the folding free energy as a function of temperature. We approximate it here by a just the linear Taylor series expansion, which was sufficient to fit all the data: $C^{(1)}$ measures how quickly the population switches from native to denatured state at T_m. (A heat capacity-based model from ref. 46 provides equally satisfactory fits.)

2.3 Temperature-jump kinetics in Ficoll

We measured T-jump kinetics of λ^*_{6-85}YG crowded by Ficoll 70 under the same buffer conditions used to measure stability. The experiment has been described in detail before.[47] Briefly, a train of 280 nm UV pulses from a frequency-tripled, mode-locked Ti:S laser is used to excite the sample tryptophan fluorescence in a 0.025 diameter by 0.4 mm long sample region every 12.5 ns. A 10 °C T-jump is then induced in the sample by a Raman-shifted (1.9 μm) Nd:YAG laser collimated at the sample to a ~1 mm diameter spot. The change in tryptophan fluorescence lifetime is recorded as folding kinetics progress, and plotted as "χ" normalized from 1 (just after the jump) to 0 (500 μs after the jump). λ^*_{6-85}YG relaxes in $ca.$ 50 μs, significantly slower than the <5 μs relaxation rates observed for the fastest downhill-folding mutants of lambda repressor.[15,23] Thus we use the two-state approximation $k_a \approx k_{obs}K_{eq}/(1 + K_{eq})$ to determine the activated folding rate coefficient from the observed rate coefficient k_{obs} and folding equilibrium constant K_{eq} (determined from fluorescence measurements in 2.2). The molecular phase k_m (downhill folding) and activated phase k_a have been discussed extensively elsewhere,[11,15,23,48] and here we plot only k_a to see if the crowder has an effect on the small folding barrier.

To correct folding rates for bulk solvent viscosity, we measured the viscosity of Ficoll solutions with a Bohlin Instruments High Resolution C-VOR Torque Rebalance in viscometry mode, using a double gap couette cell, steel solvent trap and a Polyscience recirculating water bath. Measurements were performed at 20, 45, 48, 50, 53, 55, 58, 60 and 63 °C, scaled to the known viscosities at 20 °C, and globally fitted with water ($X = 0$)[49] viscosities to yield the following model equations:

$$\eta_{H_2O}(T) = 0.226 + 1.0723e^{-(T-10)/33} \qquad (3a)$$

$$\eta_{Ficoll}(T,X) = (1 + 0.3843X - 0.00508X^2)[1 + (1.966 + 0.234X)e^{-0.0235T} \\ -(1.275 - 0.0283X)e^{-0.00367}] \qquad (3b)$$

where T is in °C, X is in mg ml^{-1} (mass of crowder/total solution volume), and η is the viscosity in centiPoise units.

3. Results

3.1 Crowding effect of Ficoll 70 on λ_{6-85} stability

The 70 kDa carbohydrate polymer Ficoll 70 acts as a ≈ 5.1 nm Stokes radius semi-rigid crowding agent in aqueous solution. We studied the stability of λ^*_{6-85}YG (9.2 kDa, hydrated radius of gyration $ca.$ 1.2 nm[50]) in a Ficoll crowding matrix of volume fraction $\varphi \approx 15\text{–}16\%$ in phosphate buffer at pH 7. A Trp22-Tyr33 interaction acts as a probe of non-local tertiary contact formation in λ^*_{6-85}YG.[23]

The melting point obtained by circular dichroism (CD) is $T_m = 54 \pm 1$ °C in Ficoll, identical within measurement uncertainty with $T_m = 55 \pm 1$ °C in buffer, which agrees with the previous literature value of 54.5 °C.[42] Furthermore, the cooperativity parameter $C^{(1)}$ from eqn (2) was 502 ± 28 J mol^{-1} K^{-1} in Ficoll and 521 ± 39 J mol^{-1} K^{-1} in buffer. Thus concentrations of over 200 mg ml^{-1} of Ficoll 70 have no appreciable effect on the thermal stability of lambda repressor secondary structure compared to that in buffer. Ficoll does affect the shape of the CD spectrum, with more random coil signal (deeper peak at 208 nm relative to 222 nm[51]) than in aqueous buffer (Fig. 2A inset).

The fluorescence intensity probe reveals a larger difference between Ficoll crowding and aqueous buffer (Fig. 2B): T_m remains 54 ± 1 °C in $\varphi = 15\%$ Ficoll, but it decreases to 49 °C in aqueous buffer. This difference between CD and fluorescence probes has been observed before for other lambda repressor fragment mutants,[15,52] and is a signature for deviations from two-state folding over a single barrier. Thus thermal stabilization of λ^*_{6-85}YG by 225–240 mg ml^{-1} Ficoll is in the range of -1 to 5 °C, depending on the probe.

We also carried out measurements in sucrose for reference, and obtained melting points of 54 and 51 °C by CD and fluorescence. Sucrose occupies an intermediate ground between aqueous buffer and Ficoll, also not strongly stabilizing λ^*_{6-85}YG relative to aqueous buffer.

3.2 Folding relaxation kinetics of λ_{6-85} in Ficoll 70

We investigated the folding kinetics of λ^*_{6-85}YG to see if a larger kinetic effect could be obtained. Temperature jump relaxation rates k_{obs} were measured around T_m, between 47 and 60 °C final temperatures. Representative jumps to a final temperature of 54 °C are shown in Fig. 3A in aqueous buffer, $\varphi = 15\%$ sucrose and $\varphi = 15\%$

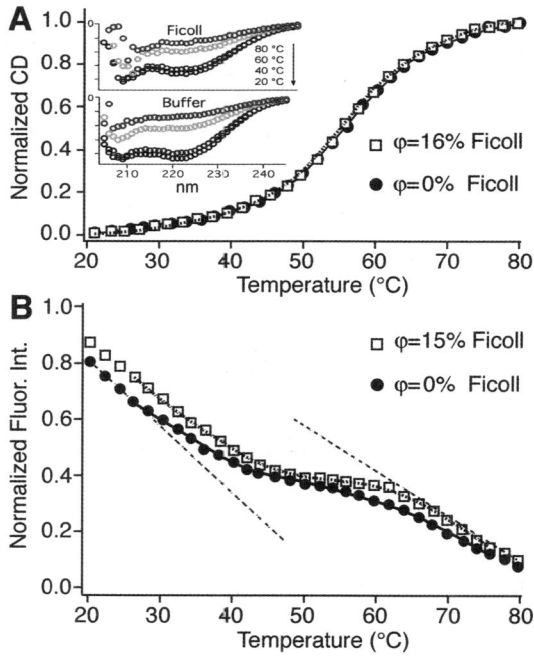

Fig. 2 Stability of λ^*_{6-85}YG in aqueous buffer and Ficoll. Protein concentration was 7.1–7.2 μM for the protein in Ficoll, and 7.2 μM for protein in buffer (50 mM phosphate buffer at pH 7). Markers are measured, solid and dotted curves are fits to eqn (2) (A) Circular dichroism-detected thermal melts with 240 mg ml^{-1} Ficoll and without Ficoll. The insets show the CD spectra at 4 temperatures, indicating more random coil structure in Ficoll (scale 0 top to −30,000 bottom MRE in ° cm^2 dmol^{-1} residue^{-1}). (B) Fluorescence intensity-detected thermal melt of λ^*_{6-85}YG with 225 mg ml^{-1} Ficoll and in buffer (50 mM phosphate buffer at pH 7). The dashed lines are two of the four fitted baselines. Melting temperatures given in the text were extracted by fitting to eqn (2).

Ficoll. An Arrhenius plot of the observed relaxation rate is shown in Fig. 3B. The buffer, Ficoll, and sucrose reference data differ by less than 0.2 natural log units, or 20% of the rate.

To compare folding rates further, we made the two-state approximation $k_{obs} \approx k_a + k_a/K_{eq} = k_a(K_{eq} + 1)/K_{eq}$ and scaled the folding rate coefficient k_a by $\eta(T)^{-1}$. (We used the equilibrium constant K_{eq} from the fluorescence data in 3.2 because fluorescence was used to detect the kinetics.) The result is shown in Fig. 3B for an average over several data sets taken with $\varphi = 10\text{–}15\%$ Ficoll vs. 0% (buffer). The estimated activated folding rates in buffer and Ficoll are virtually identical over the temperature range we covered. Ficoll neither greatly stabilizes $\lambda^*_{6\text{-}85}$YG, nor does it increase the folding rate significantly.

As seen in Fig. 3A, sucrose and Ficoll have a very fast phase with rate coefficient $k_m \approx (2\ \mu s)^{-1}$. A small fast phase is observed in sucrose, and a significantly larger one in Ficoll, even though the main folding rate coefficient k_a does not speed up appreciably in sucrose and Ficoll. Such phases have been observed for downhill-folding lambda repressor mutants with experimental and calculated barriers <7 kJ mole^{-1}.[11,15] The 'burst' with rate coefficient k_m is not observed for $\lambda^*_{6\text{-}85}$YG in aqueous buffer, in agreement with previous studies.[48] Therefore crowding induces downhill folding of $\lambda^*_{6\text{-}85}$YG, with a transit time across the transition state of approximately 2 microseconds.

3.3 High stability and low background fluorescence of SubL for crowding studies

With Ficoll having a relatively modest effect, we wanted to see if the crowding matrix/$\lambda_{6\text{-}85}$ combination could be optimized to produce large signals and large T_m changes in a small, fast folder like lambda repressor fragment. A small protein could act as a more compact crowding agent, predicted to have a larger effect by excluded volume models.[53] We chose as our matrix candidate the 14 kDa

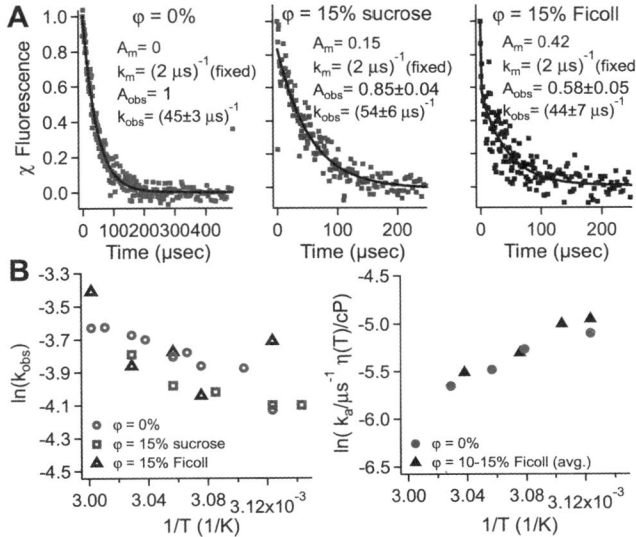

Fig. 3 Temperature jump relaxation kinetics of $\lambda^*_{6\text{-}85}$YG in aqueous buffer, sucrose and Ficoll. (A) $6 \pm 1\ ^\circ$C T-jumps of 83 μM protein to a final temperature of 54 $^\circ$C. Data are shown as dots, fits to a double exponential (rate coefficients k_m and $k_{obs} = k_a + k_a/K_{eq}$) as black lines. The molecular phase k_m (rapid 'burst' of diffusing transition state population) tends to be larger in sucrose and Ficoll jumps. (B) Arrhenius plots of the fitted rate coefficients k_{obs}, and of the viscosity-scaled activated folding rate k_a as described in the text. No significant deviations between Ficoll and aqueous buffer (50 mM phosphate buffer at pH 7) are seen.

thermophilic protein SubL, the L subunit of an archaeal DNA repair protein.[41] As the probe protein, we investigated λ^*_{6-85}, which has a fluorescence-detected thermal stability similar to λ^*_{6-85}YG, but a strongly decreasing fluorescence signal as it unfolds, easy to detect in a slightly fluorescent matrix.

To be suitable as a crowding agent for fluorescence-detected thermal melts, a protein must be easy to purify in large quantities, must be highly thermostable even at high concentration, and must fluoresce as little as possible (Fig. 4). A thermal scan of SubL circular dichroism shows no evidence of unfolding at temperatures below 70 °C. The magnitude of the CD spectrum (mean residue ellipticity at 222 nm) increases with temperature (Fig. 4B), indicating more secondary structure at high temperature. Thus SubL has a much higher melting point than λ^*_{6-85}. Only a combination of 2 molar guanidine and pH 4.5 shows evidence of unfolding beginning below 70 °C. Heating above 70 °C is required before SubL or SubL-λ_{6-85} mixtures show strong scattering (milky solution) due to precipitation.

SubL fluorescence excited at 290 nm is approximately 60 times lower than tryptophan fluorescence from λ^*_{6-85} (Fig. 4A). The small amount of residual fluorescence comes from two tyrosines, and decreases slowly with increasing temperature. Thus SubL contributes less than 14% background fluorescence in experiments with a 1 : 10 ratio of λ^*_{6-85} to crowder. SubL expresses in large quantities, and is easily purified by heating, whereupon only the thermophilic SubL remains in solution (see Methods).

3.4 Stabilization of λ_{6-85} by the SubL matrix

In buffer at pH 7, λ^*_{6-85} had a melting temperature of 50 ± 1 °C detected by total fluorescence intensity (Fig. 5). Fluorescence wavelength shift (Fig. 5A) yielded a slightly lower temperature, 48.5 ± 1.5 °C. Both of these are lower than the previously measured T_m detected by circular dichroism,[42] indicating a deviation from a simple two-state equilibrium.

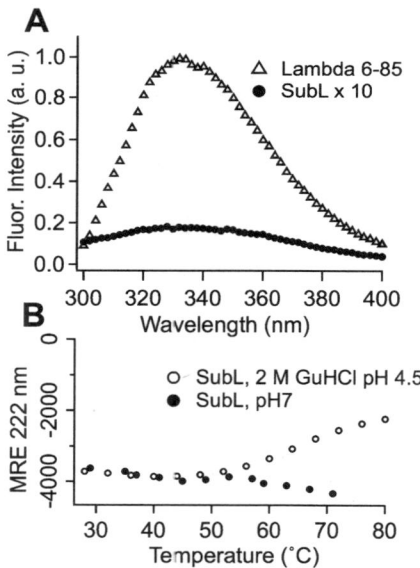

Fig. 4 Fluorescence and stability of SubL. (A) Peak SubL fluorescence is *ca.* 60 times smaller than peak λ^*_{6-85} fluorescence at the same concentration (normalized to 100 μM). (B) Thermophilic SubL at pH 7 shows no evidence of melting below 70 °C, as determined by circular dichroism Mean Residue Ellipticity (MRE in ° cm^2 dmol^{-1}residue^{-1}). Only in acidic guanidinium solution was there any evidence of unfolding below 70 °C (at *ca.* 55 °C).

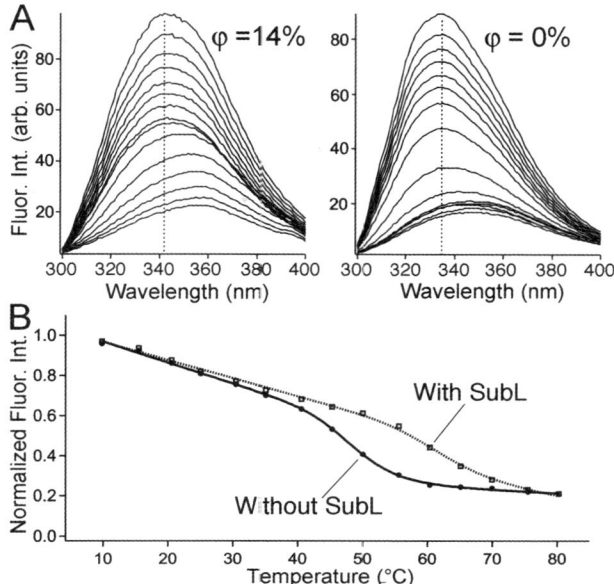

Fig. 5 (A) 12 mg ml^{-1} λ^{*}_{6-85} in 220 mg ml^{-1} SubL (left), and 12 mg ml^{-1} λ^{*}_{6-85} in buffer only (right) fluorescence spectra. Temperatures are (in order of decreasing intensity, $\pm 1\,°C$ absolute calibration): 10 to 80 °C, in 5 °C steps. The decrease of total fluorescence intensity, as well as the red-shift of the peak wavelength from the folded reference wavelength (vertical dashed line) can be seen clearly. (B) Integrated fluorescence intensity as a function of temperature, showing the melting transitions with and without the SubL crowding matrix.

We investigated the effect of crowding on the thermal stability of λ^{*}_{6-85} at matrix concentrations $\varphi = 14\%$ (≈ 220 mg ml^{-1} SubL), similar to the Ficoll concentrations in sections 3.1–3.2 (Fig. 5B). Fluorescence spectra were obtained at 1 nm wavelength intervals and at 5 °C temperature intervals to minimize heating time and aggregation during thermal denaturation. The thermodynamic fit yielded $T_m = 63 \pm 2$ °C, which is 13 °C higher than the $T_m = 50$ °C measured in buffer. The cooperativity in SubL was not appreciably different from buffer, and was globally fitted to $C^{(1)} = 124 \pm 30$ J mol^{-1} K^{-1}. A similar analysis by wavelength shift (Fig. 5A) yielded the same $C^{(1)}$ and a smaller change of T_m between buffer and matrix (7 °C), but still larger than obtained in Ficoll. We could not compare fluorescence and CD probes in SubL because of the large CD signal from the SubL crowding matrix.

4. Discussion

The stability of several proteins has been investigated in Ficoll or similar synthetic crowding matrices, notably apoflavodoxin,[33,37] VlsE,[33] PGK[44] and on the small size end, cytochrome c.[40] The consensus is that synthetic crowders enhance the stability of relatively large proteins. Some proteinaceous crowding matrices, such as lysozyme, have been shown to destabilize probe proteins.[36] Thus there is room for both excluded volume effects (stabilizing), and other effects (e.g. destabilizing electrostatics) of the crowding matrix.

Here we find that SubL acts as a strongly stabilizing matrix for the small protein λ^{*}_{6-85}, whereas the 5 times larger (by mass) Ficoll 70 acts only as a modest stabilizing agent at the same crowding fraction ($\varphi = 14$–16%). It is thus possible for protein crowding matrices to stabilize probe proteins even more so than a synthetic crowding matrix. Indeed, experiments in live cells, where proteins are expected to be important crowding agents, have also shown that probe proteins are stabilized.[39,54]

Simple models that emphasize excluded volume have been developed.[53,55,56] In these models, rigid globular crowders act by restricting the configurational entropy of the probe protein's denatured ensemble. Thereby they destabilize the denatured state and speed up refolding except at the very highest crowder concentrations. Cheung and Thirumalai predict how T_m increases with crowder volume fraction φ_c, assuming a random coil denatured state:

$$\Delta T_m \approx 0.84 T_m(\text{Kelvin})\varphi_c^{1.8} \qquad (4)$$

The general trend is consistent with lysozyme in dextran matrices[57] and PGK in Ficoll matrix,[44] for example.

Of the two cases studied here, eqn (4) adequately describes the temperature shift observed for λ_{6-85}^* in the SubL matrix: we measure an average of 10 °C by two different fluorescence methods (intensity, wavelength), compared to eqn (4) predicting 8 °C for λ_{6-85}^* in 14% crowder. Of course, this agreement indicates only that excluded volume effects can account for the observed order of magnitude of the T_m change: λ_{6-85}^* in SubL has probe-dependent melting temperatures with a 6 °C range (indicating breakdown of the two-state assumption).

The predictions of simple crowding models generally break down for λ_{6-85}^*YG in Ficoll 70: Protein stability is enhanced by at most half the prediction of eqn (4), and the folding rate does not speed up at all. The most likely reason for the breakdown is the size difference between protein and crowding agent. Minton's excluded volume models predict that crowding effects increase for smaller matrix constituents, as observed here for Ficoll vs. SubL.[53] (Of course, this can hold only up to a point: if the carbohydrate crowder is made very small compared to the probe protein, e.g. sucrose compared to λ_{6-85}^*YG, it must act like a cosmotrope rather than a crowding agent.) Ficoll 70 may simply be too large to efficiently crowd λ_{6-85}^*YG. In addition, the highly hydrated, low density Ficoll molecule is likely to be softer and more penetrable than the dense (ca. 1.5 g ml^{-1}) SubL, further reducing its ability to conformationally restrict the dynamics of the small lambda repressor fragment unfolded state.

A model for Ficoll 70 crowding of λ_{6-85}^*YG must explain the following four observations: 1) secondary structure is not stabilized by crowding (CD data); 2) non-local tertiary contacts are stabilized slightly by crowding (Y33-W22 probe[23]); 3) the activated folding rate is not altered by crowding; 4) a new fast phase appears, so crowding favors downhill folding. The following hypothesis accounts for all of these: Ficoll 70 interstitial spaces have higher local viscosity than bulk water, and they are smaller than the length scale of tertiary interactions, but larger than the length scale of the local secondary structure of lambda repressor fragment. If so, Ficoll 70 would stabilize tertiary structure more than secondary structure, and the unchanged activated rate results from a cancellation of higher local viscosity and a lower activation barrier; the lower barrier in turn causes the appearance of the fast molecular phase k_m and downhill folding. There is experimental and computational precedent showing that increased local viscosity reduces folding rates. Mukherjee et al. previously showed that crowding of small model peptides by Ficoll 70 can cause a slow-down of secondary structure formation.[58] In live cells, the local viscosity for folding has been shown to increase by a factor of 2 from aqueous solution,[59] so compensation of a lower barrier by higher viscosity is plausible. Ando and Skolnick calculated that local viscosity variations can result from hydrodynamics, as the drag of aqueous layers flowing past multiple macromolecules changes the local viscosity from the bulk value.[60]

In view of our results, it will be very interesting to develop improved theoretical and computational models for crowders of small fast-folding proteins. Such models should include crowder flexibility, local viscosity, hydrodynamic and electrostatic effects on the probe protein, in addition to the excluded volume effect. Small proteins like λ_{6-85}^* are ideal probes for crowding studies because they can be modeled at atomistic detail with currently available molecular dynamics technology. Computational efforts for small model proteins with rigid crowers are already underway.[40,61] On the experimental side, it will be useful to provide systematic folding kinetics data for fast-folding small probe proteins in different crowding environments. For example, a series of λ^*_{6-85} mutants in Ficoll or SubL could pin down how protein stability, denatured state compactness, and other properties that can be varied by mutation, affect the ability of the matrix to crowd the probe protein.

Acknowledgements

This work was supported by US National Science Foundation grant MCB 1019958. S. D. wishes to thank the NSF for support by a GK-12 Fellowship while this research was carried out.

5. References

1 H. A. Kramers, *Physica*, 1940, **7**, 284.
2 D. Homouz, M. Perham, A. Samiotakis, M. S. Cheung and P. Wittung-Stafshede, *Proc. Natl. Acad. Sci. U. S. A.*, 2008, **105**, 11754–11759.
3 K. Lindorff-Larsen, S. Piana, R. O. Dror and D. E. Shaw, *Science*, 2011, **334**, 517–520.
4 J. N. Onuchic, P. G. Wolynes, Z. Luthey-Schulten and N. D. Socci, *Proc. Natl. Acad. Sci. U. S. A.*, 1995, **92**, 3626–3630.
5 S. E. Jackson and A. R. Fersht, *Biochemistry*, 1991, **30**, 10428–10435.
6 D. Chandler, *J. Chem. Phys.*, 1978, **68**, 2959–2970.
7 I. R. Sims and I. W. M. Smith, *Annu. Rev. Phys. Chem.*, 1995, **46**, 109–137.
8 J. Sabelko, J. Ervin and M. Gruebele, *Proc. Natl. Acad. Sci. U. S. A.*, 1999, **96**, 6031–6036.
9 M. Garcia-Mira, M. Sadqi, N. Fischer, J. M. Sanchez-Ruiz and V. Muñoz, *Science*, 2002, **298**, 2191–2195.
10 A. Cavalli, U. Haberthur, E. Paci and A. Caflisch, *Protein Sci.*, 2003, **12**, 1801–1803.
11 W. Y. Yang and M. Gruebele, *Nature*, 2003, **423**, 193–197.
12 H. Ma and M. Gruebele, *Proc. Natl. Acad. Sci. U. S. A.*, 2005, **102**, 2283–2287.
13 M. Sadqi, D. Fushman and V. Munoz, *Nature*, 2006, **442**, 317–321.
14 R. B. Dyer, *Curr. Opin. Struct. Biol.*, 2007, **17**, 38–47.
15 F. Liu and M. Gruebele, *J. Mol. Biol.*, 2007, **370**, 574–584.
16 A. Fung, P. Li, R. Godoy-Ruiz, J. M. Sanchez-Ruiz and V. Munoz, *J. Am. Chem. Soc.*, 2008, **130**, 7489–7495.
17 A. N. Naganathan and V. Munoz, *Biochemistry*, 2008, **47**, 6752–6761.
18 S. S. Cho, P. Weinkam and P. G. Wolynes, *Proc. Natl. Acad. Sci. U. S. A.*, 2008, **105**, 118–123.
19 R. Godoy-Ruiz, E. R. Henry, J. Kubelka, J. Hofrichter, V. Munoz, J. M. Sanchez-Ruiz and W. A. Eaton, *J. Phys. Chem. B*, 2008, **112**, 5938–5949.
20 F. Liu, D. Du, A. A. Fuller, J. Davoren, P. Wipf, J. Kelly and M. Gruebele, *Proc. Natl. Acad. Sci. U. S. A.*, 2008, **105**, 2369–2374.
21 J. W. Pitera, W. C. Swope and F. F. Abraham, *Biophys. J.*, 2008, **94**, 4837–4846.
22 S. J. DeCamp, A. N. Naganathan, S. A. Waldauer, O. Bakajin and L. J. Lapidus, *Biophys. J.*, 2009, **97**, 1772–1777.
23 F. Liu, Y.-G. Gao and M. Gruebele, *J. Mol. Biol.*, 2009, **397**, 789–798.
24 C. Dumont, T. Emilsson and M. Gruebele, *Nat. Methods*, 2009, **6**, 515–519.
25 R. B. Best and G. Hummer, *Proc. Natl. Acad. Sci. U. S. A.*, 2005, **102**, 6732–6737.
26 D. E. Shaw, P. Maragakis, K. Lindorff-Larsen, S. Piana, R. O. Dror, M. P. Eastwood, J. A. Bank, J. M. Jumper, J. K. Salmon, Y. B. Shan and W. Wriggers, *Science*, 2010, **330**, 341–346.
27 D. L. Ensign, P. M. Kasson and V. S. Pande, *J. Mol. Biol.*, 2007, **374**, 806–816.
28 P. L. Freddolino, C. B. Harrison, Y. X. Liu and K. Schulten, *Nat. Phys.*, 2010, **6**, 751–758.
29 J. P. Ellis, P. H. Culviner and S. Cavagnero, *Protein Sci.*, 2009, **18**, 2003–2015.
30 K. G. Ugrinov and P. L. Clark, *Biophys. J.*, 2010, **98**, 1312–1320.
31 A. P. Minton and J. Wilf, *Biochemistry*, 1981, **20**, 4821–4826.
32 B. van den Berg, R. Wain, C. M. Dobson and R. J. Ellis, *EMBO J.*, 2000, **19**, 3870–3875.
33 M. Perham, L. Stagg and P. Wittung-Stafshede, *FEBS Lett.*, 2007, **581**, 5065–5069.
34 A. Samiotakis, P. Wittung-Stafshede and M. S. Cheung, *Int. J. Mol. Sci.*, 2009, **10**, 572–588.
35 Y. Wang, C. Li and G. J. Pielak, *J. Am. Chem. Soc.*, 2010, **132**, 9392–9397.

36 A. C. Miklos, M. Sarkar, Y. Wang and G. J. Pielak, *J. Am. Chem. Soc.*, 2011, **133**, 7116–7120.
37 L. Stagg, A. Christiansen and P. Wittung-Stafshede, *J. Am. Chem. Soc.*, 2011, **133**, 646–648.
38 Z. Ignatova and L. M. Gierasch, *Proc. Natl. Acad. Sci. U. S. A.*, 2004, **101**, 523–528.
39 S. Ebbinghaus, A. Dhar, J. D. McDonald and M. Gruebele, *Nat. Methods*, 2010, **7**, 319–323.
40 A. Christiansen, Q. Wang, A. Samiotakis, M. S. Cheung and P. Wittung-Stafshede, *Biochemistry*, 2010, **49**, 6519–6530.
41 C. I. Reich, L. K. McNeil, J. L. Brace, J. K. Brucker and G. J. Olsen, *Extremophiles*, 2001, **5**, 265–275.
42 W. Y. Yang and M. Gruebele, *Biochemistry*, 2004, **43**, 13018–13025.
43 S. Ghaemmaghami, J. M. Word, R. E. Burton, J. S. Richardson and T. G. Oas, *Biochemistry*, 1998, **37**, 9179–9185.
44 A. Dhar, A. Samiotakis, S. Ebbinghaus, L. Nienhaus, D. Homouz, M. Gruebele and M. S. Cheung, *Proc. Natl. Acad. Sci. U. S. A.*, 2010, **107**, 17586–17591.
45 M. Muschol and F. Rosenberger, *J. Chem. Phys.*, 1995, **103**, 10424–10432.
46 I. Nishii, M. Kataoka, F. Tokunaga and Y. Goto, *Biochemistry*, 1994, **33**, 4903–4909.
47 R. M. Ballew, J. Sabelko, C. Reiner and M. Gruebele, *Rev. Sci. Instrum.*, 1996, **67**, 3694–3699.
48 W. Yang and M. Gruebele, *Biophys. J.*, 2004, **87**, 596–608.
49 R. H. Perry and C. H. Chilton, *Perry's Chemical Engineer's Handbook*, Mc Graw-Hill, New York, 1973.
50 S. J. Kim, C. Dumont and M. Gruebele, *Biophys. J.*, 2008, **94**, 4924–4931.
51 N. Berova, K. Nakanishi and R. W. Woody, *Circular dichroism: principles and applications*, Wiley, New York, 2000.
52 M. B. Prigozhin and M. Gruebele, *J. Am. Chem. Soc.*, 2011, **133**, 19338–19341.
53 M. Minton, *Biophys. J.*, 2005, **88**, 971–985.
54 A. Dhar, K. Girdhar, D. Singh, H. Gelman, S. Ebbinghaus and M. Gruebele, *Biophys. J.*, 2011, **101**, 421–430.
55 A. P. Minton, *Biophys. J.*, 2000, **78**, 101–109.
56 M. S. Cheung, Klimov, Dmitri and D. Thirumalai, *Proc. Natl. Acad. Sci. U. S. A.*, 2005, **102**, 4753–4758.
57 K. Sasahara, P. McPhie and A. P. Minton, *J. Mol. Biol.*, 2003, **326**, 1227–1237.
58 S. Mukherjee, M. M. Waegele, P. Chowdhury, L. Guo and F. Gai, *J. Mol. Biol.*, 2009, **393**, 227–236.
59 A. Dhar, S. Ebbinghaus, Z. Shen, T. Mishra and M. Gruebele, *Biophys. J.*, 2010, **99**, L69–L71.
60 T. Ando and J. Skolnick, *Proc. Natl. Acad. Sci. U. S. A.*, 2010, **107**, 18457–18462.
61 A. Samiotakis and M. S. Cheung, *J. Chem. Phys.*, 2011, **135**.
62 W. F. Humphrey, A. Dalke and K. Schulten, *J. Mol. Graphics*, 1996, **14**, 33–38.

Ligand-field symmetry effects in Fe(II) polypyridyl compounds probed by transient X-ray absorption spectroscopy†

Hana Cho,[ab] Matthew L. Strader,[a] Kiryong Hong,[b] Lindsey Jamula,[c] Eric M. Gullikson,[d] Tae Kyu Kim,[*b] Frank M. F. de Groot,[e] James K. McCusker,[c] Robert W. Schoenlein[a] and Nils Huse[*af]

Received 28th February 2012, Accepted 11th May 2012
DOI: 10.1039/c2fd20040f

Ultrafast excited-state evolution in polypyridyl Fe(II) complexes is of fundamental interest for understanding the origins of the sub-ps spin-state changes that occur upon photoexcitation of this class of compounds as well as for the potential impact such ultrafast dynamics have on incorporation of these compounds in solar energy conversion schemes or switchable optical storage technologies. We have demonstrated that ground-state and, more importantly, ultrafast time-resolved X-ray absorption methods can offer unique insights into the interplay between electronic and geometric structure that underpins the photo-induced dynamics of this class of compounds. The present contribution examines in greater detail how the symmetry of the ligand field surrounding the metal ion can be probed using these X-ray techniques. In particular, we show that steady-state K-edge spectroscopy of the nearest-neighbour nitrogen atoms reveals the characteristic chemical environment of the respective ligands and suggests an interesting target for future charge-transfer femtosecond and attosecond spectroscopy in the X-ray water window.

Introduction

Transition metal-based polypyridyl complexes represent a large and important class of inorganic compounds. Historically, compounds of Ru(II), Os(II), and Re(I) have garnered most of the attention from researchers. However, potential limitations of such systems in more applied contexts due to the intrinsic scarcity of these elements has refocused attention on complexes based on the more earth-abundant members of the first transition series; of these, Fe(II)-based chromophores are among the most widely studied in terms of their photophysical properties. In particular, poly-pyridyl Fe[II] complexes are considered as potential candidates for dye-sensitized solar cells.[1,2] Initially, photo-perturbation and laser temperature jump measurements on

[a]Ultrafast X-ray Science Laboratory, Chemical Sciences Division, Lawrence Berkeley National Laboratory, Berkeley, California, USA
[b]Department of Chemistry and Chemistry Institute of Functional Materials, Pusan National University, Busan, Republic of Korea. E-mail: tkkim@pusan.ac.kr
[c]Department of Chemistry, Michigan State University, East Lansing, Michigan, USA
[d]Center for X-Ray Optics, Lawrence Berkeley National Laboratory, Berkeley, California, USA
[e]Department of Chemistry, Utrecht University, Utrecht, Netherlands
[f]Max Planck Research Department for Structural Dynamics at the University of Hamburg & Center for Free Electron Laser Science, Hamburg, Germany. E-mail: nils.huse@mpsd.cfel.de

† Electronic supplementary information (ESI) available. See DOI: 10.1039/c2fd20040f

polypyridyl Fe[II] spin-crossover (SCO) complexes with a low-spin (LS) $^1A_1$ ground state were used to characterize the transient high-spin (HS) $^5T_2$ state and to determine the intramolecular kinetics of ground-state recovery, i.e., the $^5T_2 \rightarrow {}^1A_1$ relaxation process.[3,4] This HS $\rightarrow$ LS relaxation can generally be described in terms of a thermally activated non-adiabatic multi-phonon process in the strong coupling limit.[5] At sufficiently low temperatures, deactivation of these promoting modes results in light-induced excited spin-state trapping (LIESST)[6–8] with tunnelling rates for the high spin-to-low spin conversion process as low as 10^{-6} s^{-1}, a phenomenon that has generated considerable interest in potential applications for these molecules in storage devices.[9]

Ultrafast electronic absorption measurements revealed the sub-picosecond dynamics associated with formation of the $^5T_2$ state following photo-excitation of low-spin Fe(II) complexes[10,11] as well as a sub-100 fs time scale of charge transfer-to-ligand field-state conversion.[2] A definitive combined electronic/resonance Raman study detailed a ca. 200 fs time scale for establishment of the $^5T_2$ transient electronic state.[12] A transient infrared spectroscopic study has reported formations times of the $^5T_2$ state and subsequent intramolecular energy redistribution consistent with these findings.[13] Ultrafast optical studies by Chergui and co-workers provided important additional mechanistic information on the ultrafast spin-state conversion with evidence for ultrafast ^{1}MLCT $\rightarrow$ ^{3}MLCT before subsequent population of the $^5T_2$ high-spin state.[14] Interestingly, no signatures of vibrational coherence of the symmetric breathing mode have been observed, although this mode is likely coupled to the LS $\rightarrow$ HS conversion due to the ca. 0.2 Å increase in bond length that characterizes the transition. Fast population of the $^5T_2$ high-spin state appears to be impulsive on the time-scale of some vibrational modes of the $^5T_2$ state,[15] having been assigned as the origin of vibrational coherences observed upon formation of the transient high-spin state. A detailed theoretical description of these dynamics has yet to be reported, however, recent studies by de Graaf and Sousa suggest that ultrafast MLCT $\rightarrow$ $^5T_2$ relaxation may occur due to favourable potential energy surface crossings[16,17] and Veenendaal's group showed that for certain energy level configurations ultrafast spin-cross over between two states could be mediated by a third state in the presence of vibrational manifolds.[18,19]

Complementary to optical techniques, core-level spectroscopy techniques have provided information on the structural dynamics and the accompanying changes in valence charge distribution underlying the SCO transition.[20–26] Compared to time-resolved optical spectroscopy, ultrafast X-ray spectroscopy is chemically very specific due to the highly localized initial state of the core-level transition. Well-separated transition energies allow for probing specific atomic species with information on molecular structures and valence electronic configurations in the excited state.[27–31] Transient extended X-ray absorption fine structure (EXAFS) spectroscopy has been employed to study the structural changes around the metal centre via the Fe K-edge of [Fe(tren(py)$_3$)]$^{2+}$ and [Fe(bpy)$_3$]$^{2+}$, providing information on the geometric changes (e.g., ~0.2 Å dilation of the Fe–N bonds) subsequent to optical excitation.[20,21] Transient X-ray diffraction from solid-state SCO complexes has revealed a thermal SCO transition following the initial photo-induced SCO and thermalization of the optical excitation.[32] Transient X-ray absorption and emission spectroscopy of first-row transition metals can report on spin-state changes on ultrafast time-scales and can reveal electronic changes associated with valence orbitals of specific symmetry.[24–26] For instance, we have recently demonstrated picosecond and femtosecond XAS in the soft X-ray range of solvated [Fe(tren(py)$_3$)]$^{2+}$, which provided a detailed picture of the changes in valence electron distributions in the 3d manifold of the transient high-spin state in an Fe[II] complex subsequent to charge-transfer excitation.[24,26] These results underscore the potential of ultrafast L-edge spectroscopy for the study of transition metal chemistry in solution. Furthermore, recent advances in ab initio modelling of X-ray absorption near-edge structures (XANES)[33] show very good agreement with experimental Fe K-edge

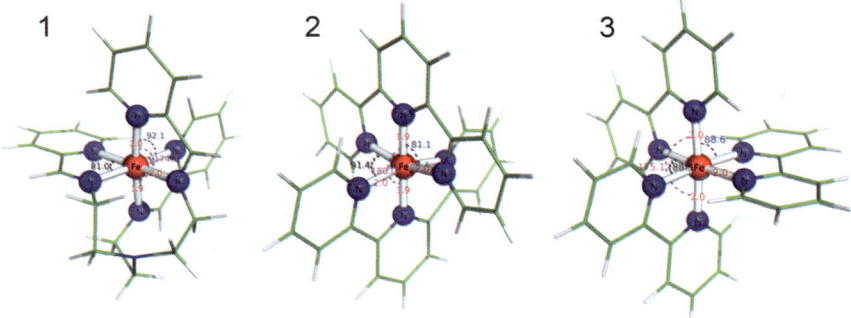

Fig. 1 DFT-derived molecular structures of the investigated polypyridyl Fe^{II} compounds. **1**: $[Fe(tren(py)_3)]^{2+}$. **2**: $[Fe(terpy)_2]^{2+}$. **3**: $[Fe(bpy)_3]^{2+}$. All three compounds feature a pseudo-octahedral arrangement of six nitrogen atoms around the central Fe^{II} ion. The deviations from octahedral symmetry vary but are substantial in all three compounds.

spectra, allowing for a more detailed and confident interpretation of XANES spectra in general.

In this work, we aim to understand changes of valence charge density in polypyridyl Fe^{II} complexes upon spin-crossover for different, but related, coordination environments of the metal centre. Specifically, we studied the three compounds displayed in Fig. 1 to examine the extent to which steady-state and time-resolved X-ray absorption spectroscopy techniques could provide information concerning small variations in the local ligand field of the Fe^{II} ion. The compounds in Fig. 1 are often approximated as possessing O_h symmetry; in reality, the three pyridine and three imine nitrogen donors of compound **1** effectively reduce this to C_3 symmetry. And while compounds **2** and **3** all present pyridine nitrogen donors to the metal centre, distortions due to the geometric constraints of the ligands result in symmetry reductions to C_2 and D_3 symmetry, respectively. In principle, even a slight reduction of molecular symmetry formally lifts some of the orbital degeneracies, which may in turn noticeably affect electronic interactions between the metal centre and the ligands and give rise to detectable perturbations in the X-ray absorption properties of both the ground- and photo-induced excited states of these compounds.

Methods

Synthesis

All three of these Fe(II) complexes have been previously reported in the literature. $[Fe(tren(py)_3)](PF_6)_2$ was prepared from $FeCl_2 \cdot 4H_2O$ reacting with the condensate of three equivalents of 2-pyridine-carboxaldehyde and tris(2-aminoethyl)amine in MeOH solution under a nitrogen atmosphere as described elsewhere.[2] Both $[Fe(terpy)_2](PF_6)_2$ and $[Fe(bpy)_3](PF_6)_2$ were prepared in a similar fashion using $FeCl_2 \cdot 4H_2O$ and appropriate stoichiometric equivalents of $2,2':6',2''$-terpyridine and $2,2'$-bipyridine, respectively. Sample identities and purities were confirmed using elemental analysis, electrospray mass spectroscopy, and comparison of their optical and electrochemical properties with known samples.

Spectral measurements

Solution-phase X-ray absorption spectra of the ground state and transient excited state of the samples were measured at the ultrafast soft X-ray beamline of the Advanced Light Source (ALS) while the ground-state spectra of the crystalline compounds were recorded at the EUV calibration beamline of the ALS. Fig. 2

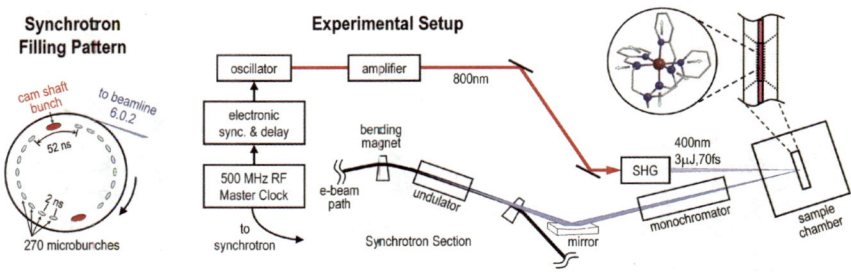

Fig. 2 Schematic of the laser-pump/X-ray-probe experiments at the ultrafast X-ray facility of the Advanced Light Source. The tuneable X-ray pulses produced by the isolated cam shaft electron bunch are focused onto the sample with an imaging monochromator to probe sample transmission changes after laser excitation. The sample is held in a 2 μm thin liquid cell inside a soft X-ray compatible experimental chamber.

depicts the general layout of the time-resolved experiment: The ALS synchrotron has a filling pattern with one electron bunch separated from all other bunches by 50 ns gaps. The 70 ps X-ray pulses generated in the beamline undulator by this so-called cam-shaft bunch are recorded by gated detection. A pulsed Ti:Sapphire laser oscillator with 62.5 MHz repetition rate is synchronized to the 500 MHz radio frequency (RF) with which the electron bunches are driven inside the synchrotron. Appropriately selected and amplified 800 nm pulses are converted to 400 nm pulses *via* second harmonic generation (SHG) and overlapped with the X-rays in the plane of the sample.

Experimental setup

The details of the sample cell and the experimental chamber have been described elsewhere.[24] Briefly, a 2 μm-thick liquid film of a 100 mM acetonitrile solution of polypyridyl Fe^{II} complexes was held between two 100 nm-thick silicon nitride membranes and the solid samples were deposited on a thin silicon nitride substrate with a sample thickness of a few hundred nanometres. The liquid samples were excited by 400 nm laser pulses[34] with 3 μJ pulse energy at a repetition rate of 1 kHz and probed with tuneable X-ray pulses at the iron $L_{2,3}$-edges at around 700 eV by recording changes in the transmitted X-ray intensity at 2 kHz (Fig. 2). Transient differential X-ray absorption spectra at fixed time delay and transients at fixed X-ray photon energy were obtained by taking the ratio of X-ray transmitted intensities of the unexcited and laser-excited samples.[30]

Density functional calculations

The low-spin ground state of $[Fe(bpy)_3]^{2+}$, $[Fe(terpy)_2]^{2+}$, and $[Fe(tren(py)_3)]^{2+}$ in the gas phase were calculated using density functional theory (DFT).[35,36] All geometries of these complexes were fully optimized using the hybrid functional of Perdew, Burke, and Ernzerhof (PBE0).[37] We used the effective core potential (ECP) to treat the scalar relativistic effect for Fe. The Los Alamos effective core potential with corresponding valence basis set and polarization f functions, LANL08(f)[38,39] and 6-31G(d,p) basis sets were used for Fe and the other atoms (C, H, and N), respectively. To obtain the zero point energy (ZPE), we carried out vibrational frequency calculations by using the same level of theory. We used the natural population analysis (NPA)[40] for characterization of atomic charges and electronic structures. We also performed the same level of calculations (PBE0/LANL08(f) + 6-31G(d,p)) for the three Fe^{II} complexes in acetonitrile solution. The solvent calculations used the integral equation formalism variant of the polarizable continuum model (IEFPCM).[41,42] The Gaussian 09 program[43] was used for all DFT calculations.

Results and discussion

We have studied the ground and lowest-energy electronic excited states of the three compounds shown in Fig. 1 by time-resolved X-ray absorption spectroscopy at the iron $L_{2,3}$-edges. The ground-state electronic absorption spectra of the compounds are plotted in Fig. 3A, along with the probing scheme where electrons are excited from spin-orbit-split Fe-2p core-levels to the unoccupied Fe-3d valence orbitals. This type of core-level spectroscopy can be exploited to extract information on changes of energy levels, valence charge delocalization and bonding, and spin-state changes during the photo-induced intersystem crossing. In particular, the absorption features at the Fe $L_{2,3}$-edges, arising from dipole-allowed resonant Fe-2p transitions, are sensitive to the local electronic structure and dynamic changes in ligand field split Fe-3d levels. In the following, we present metal (Fe L-edges) and ligand (N K-edge) X-ray absorption spectra, and deduce consequences of structural variation of the ligand cage.

The photo-excited *vs.* the chemically stabilized high-spin state

Unlike compound **1** which has a low-spin $^1A_1$ ground state at all temperatures, an analogue in which the protons at the 6-positions of the pyridine rings are replaced with methyl groups exhibits a thermal low-spin to high-spin transition at ~210 K.[44,45] For this reason, $[Fe^{II}(6\text{-Me-py})_3\text{tren}](PF_6)_2$ has served as a room-temperature high-spin analogue (HSA) in previous studies, allowing for a facile comparison of the HSA ground-state structure and properties with those of the transient high-spin state of the photo-excited compound **1**. Previous results from time-resolved visible and EXAFS spectroscopy at the Fe K-edge,[2,20] clearly illustrated that the structure of the photoexcited high-spin state is very similar to the ground state of $[Fe(tren(6\text{-Me-py})_3)]^{2+}$, thereby validating its use as a proxy for the transiently formed $^5T_2$ state of the photoexcited low-spin species. The recent *ab initio* study by van Kuiken and Khalil[33] also reported substantial structural and electronic resemblance between the transient and chemically stabilized high-spin states. In the following we present experimental results from steady-state and transient X-ray absorption spectroscopy measurements at the Fe $L_{2,3}$-edges which complement and extend the information from previous experimental and *ab initio* results.

Steady-state X-ray absorption spectra of the two solvated compounds at the Fe $L_{2,3}$-edges are displayed in Fig. 4A. Shifts to lower energy and distinct spectral reshaping can be observed for the high-spin analogue as compared to the low-spin $L_{2,3}$-edges of compound **1**. The HSA L_3-peak is 1.8 eV lower in energy than the

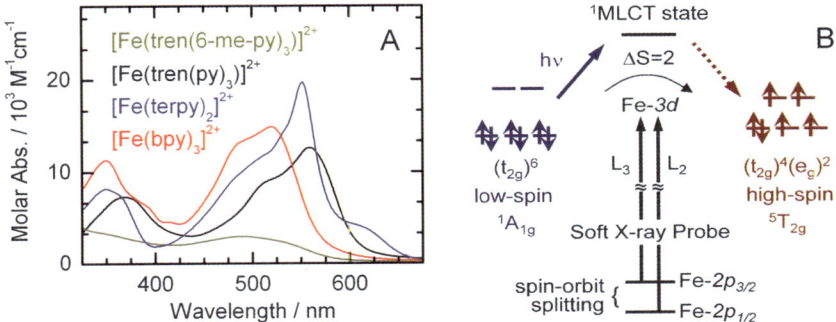

Fig. 3 **A** Electronic absorption spectra of polypyridyl Fe^{II} compounds in acetonitrile solution. **B** Simplified orbital diagram for an Fe^{II} compound in O_h symmetry, illustrating the probing of the low-spin ground state and the transient high-spin excited state *via* Fe 2p → 3d dipole-allowed core-level transition.

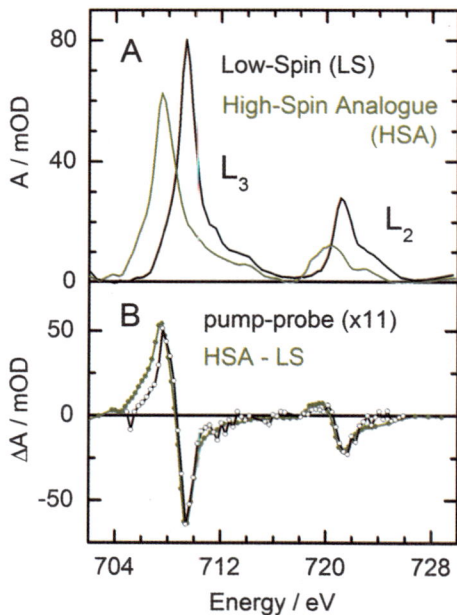

Fig. 4 **A** Steady-state L-edge spectra of low-spin (LS) [Fe(tren(py)$_3$)]$^{2+}$ and its high-spin analogue (HSA), both in acetonitrile solution. **B** A comparison of the calculated difference spectrum of the steady-state L-edge spectra from panel A (HSA-LS, mustard-coloured line) and the experimental transient difference spectrum (black line) of the photo-excited high-spin and the ground-state low-spin spectrum recorded 150 ps after excitation of the sample at 400 nm.

LS L$_3$-peak and significantly broader while the centre of the HSA L$_2$-peak is shifted to lower energy by <1 eV with substantially less intensity than the LS L$_2$-peak. Great care was taken in the energy calibration of the beamline monochromator to ensure a precision of 0.1 eV during consecutive measurements of the LS and HSA spectra.

Next, we compare the steady-state difference spectrum derived from the data in Fig. 4A with the pump–probe spectrum in Fig. 4B. The rescaling factor of the pump–probe spectrum reflects the fraction of excited molecules $x_{exc} = 1/11$ which matches well with our estimates based on 400 nm excitation fluence and the molar extinction coefficient. The similarity of the two difference spectra is quite striking. All features are reproduced within the signal-to-noise ratio. The most pronounced difference exists at the L$_3$-edge. A larger absorption increase of the dispersive HSA-LS spectrum below 707.6 eV indicates that the red-shift of the HSA L$_3$-edge from that of the LS spectrum is slightly larger than the red-shift of the transient HS L$_3$-edge. We note that this observation is independent of the uncertainties in the relative absorption of LS and HSA spectra (due to variations in sample concentration and sample thickness) as varying the magnitude of the HSA absorption within experimentally reasonable boundaries shows. Only the maximum absorption loss and gain are affected, not their energetic position.

To understand these spectra we will briefly review our previous work:[24] We concluded from comparison of the L$_2$ and L$_3$ absorption line shapes with model systems and multiplet calculations that in the low-spin $^1A_1$ ground state, strong π-back-bonding results in delocalization of the Fe 3d(t$_{2g}$) electron into the π*-orbitals of the tren(py)$_3$ ligands. The high-spin $^5T_2$ excited state is characterized by suppressed π-back-bonding and attenuated σ-donation from the tren(py)$_3$ ligand relative to the low-spin ground state. Both effects lead to more localized N-2p and Fe-3d orbitals while keeping the integrated occupancy of the metal-based e$_g$ and

This journal is © The Royal Society of Chemistry 2012

t_{2g} orbitals roughly constant, that is, the overall Fe-3d charge density is unchanged while the character of the metal–ligand bonding becomes more ionic in nature. These findings show some agreement with a recent *ab initio* study of the Khalil group.[33] From the similarity of the difference spectra in Fig. 4B it is apparent that the HSA ground state is very similar to the transient HS state of compound **1**.

The slightly larger red shift of the high-spin analogue's L_3 peak by $\sim$0.1 eV is a manifestation of the reported small differences in structure[20] and calculated valence charge density[33] between the transient HS state and the steady-state HSA ground state; the latter exhibits slightly larger bond distances and increased structural distortion compared to octahedral symmetry. Origins of the red shift of the transient HS spectrum are the reduction in energy splitting of the Fe t_{2g} and e_g orbitals with increasing bond lengths and increased charge localization.[24] Hence, the spectral difference in Fig. 4B is experimental evidence of the predicted electronic structure of the HSA.

The L_2-edges of transient HS and HSA are essentially identical. The larger absorption increase at the HSA L_3-peak signals a redistribution of absorption from the L_2 to L_3-edge which is commonly a sign of a higher spin-state or spin admixture but the spin-state has been reported as pure. We think that this behaviour is a result of the larger deviation from octahedral symmetry of the HSA ground state structure compared to the transient HS state as predicted by van Kuiken and Khalil.[33] We will discuss changes in the relative L_2 and L_3 intensity in the context of ligand variations in the next section.

Effects of octahedral distortion

The ligand field in which a transition metal ion is placed has a profound influence on virtually all of the physical properties of the molecule that relate to electronic structure including geometry, optical and magnetic properties, and chemical reactivity. In this context, the question arises what effects ligand variation has on the transient HS state in polypyridyl Fe^{II} complexes, and if metal L-edge spectroscopy can detect meaningful differences between the different compounds. The three Fe^{II} polypyridyl complexes we are examining provide a convenient platform on which to begin examining these issues given their similar, but nevertheless distinct, symmetry and compositional characteristics.

Ground state spectra of the three Fe^{II} complexes are shown in Fig. 5A. Absolute absorption differences are not identical due to variation in solubility and sample thickness on the sub-micron scale; after correcting for sample concentration the peak absorption exhibited variances of only $\sim$10%. Accordingly, the ground-state absorption spectra in Fig. 5 have been normalized to the L_3-peak of [Fe(tren-(py)$_3$)]$^{2+}$ at 709.4 eV for comparative purposes (black line). The overall spectral shapes are very similar. Compound **1** displays a more distinct shoulder on the low-energy side of the L_3-lineshape (Fig. 5B) which is weaker in compounds **2** and **3**. Given that compound **1** is distinct in possessing two different types of nitrogen donors (pyridine- and imine-based), we believe that this difference is a direct reflection of this compositional variation. Conversely, compound **3** displays the narrowest absorption features which we attribute to the fact that this molecule possesses the highest point-group symmetry of the three compounds studied. Compounds **1** and **2** show identical peak positions within the spectral precision of $\sim$0.1 eV. In contrast, compound **3** is significantly shifted to lower energy by $\sim$0.2 eV. This is somewhat surprising as one would expect that the ligand systems of compounds **2** and **3** with only pyridine groups have very similar spectral positions.

The branching ratio of the integral $L_{2,3}$-intensities due to Fe 2p $\rightarrow$ 3d transitions as defined by Thole and co-workers,[46] $r_3 = I(L_3)/[I(L_2) + I(L_3)]$, reports on the spin-state, electrostatic interactions between the core–hole and the valence charge density, and the spin–orbit (SO) interactions between the core–hole and the metal-3d manifold as well as within the latter.[46] To properly account for the bound–bound

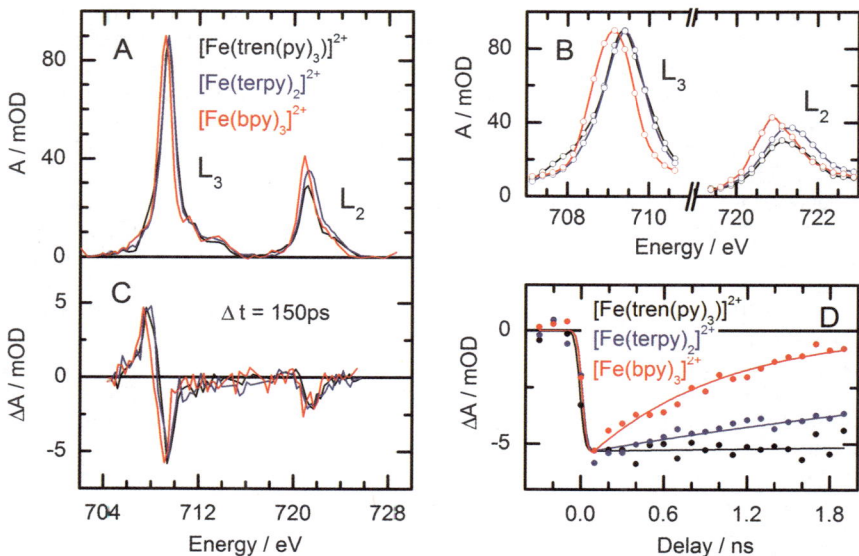

Fig. 5 Comparison of [Fe(tren(py)₃)]²⁺, [Fe(terpy)₂]²⁺, and [Fe(bpy)₃]²⁺ in 100 mM acetonitrile solution **A** Steady-state X-ray absorption spectra at the Fe $L_{2,3}$-edges. **B** Expanded view of the absorption peaks at the Fe $L_{2,3}$-edges (colour coding as in panel A) **C** Corresponding transient differential X-ray absorption spectra at 150 ps pump–probe delay after excitation at 400 nm. **D** Pump–probe delay scans at 709.4 eV (dots) and fits of the data to mono-exponential decay models using previously published excited-state lifetimes. See text for further details.

core-level transitions, we have subtracted the continuum edges which are modelled by two arctangents at the $L_{2,3}$-edges with a lifetime broadening of 0.4 eV and 0.2 eV, respectively. Some residual absorption at around 713 eV stems from a Gaussian-like feature in the EXAFS spectrum of the PF_6^- fluorine K-edge. It accounts for 8% of the L_3-edge intensity and we have corrected the branching ratio accordingly. The branching ratio r_3 equals 0.67 for both the pyridine compounds **2** and **3** which contain terpy and bpy ligands, respectively, while it amounts to 0.71 in compound **1** containing the tren(py)₃ ligand. For compounds with singlet ground states, the branching ratio is typically ~0.6 while in the absence of electrostatic and SO inter-actions, the branching ratio is statistical, amounting to $^2/_3$. We conclude from empir-ical charge-transfer multiplet calculations (see supplementary material for details†) that the chemical differences in the ligand cage between compound **1** on one hand and compounds **2** and **3** on the other lead to variations in electrostatic and SO inter-actions that manifest to a measurable extent in the branching ratio of the ground-state Fe-2p spectra as observed in Fig. 5A,B. *Ab initio* multiplet calculations would be very beneficial to gain more detailed insight into the two interactions discussed. While currently no such program code exists, recent developments may soon provide *ab initio* descriptions of third-row transition metal 2p spectra.[47]

Fig. 5C shows the normalized changes upon 400 nm excitation at a time delay of 150 ps. The spectral features are identical within the signal-to-noise ratio (~0.5 mOD r.m.s.) of the experiment. This empirical observation implies that the differences in geometry noted in the ground states of these compounds are reflected in their respective high-spin excited states, as well.

Fig. 5D shows pump–probe delay scans at a fixed probe energy of 709.4 eV, the region of maximal ground state bleaching. These transients are probing ground-state recovery dynamics, *i.e.*, the $^5T_2 \rightarrow {}^1A_1$ relaxation process subsequent to photo-induced formation of the transient high-spin state. The differences in lifetimes inferred from these data are readily understandable within the context of non-radiative decay theory and known variations in high-spin/low-spin zero-point energies across this series; this has been discussed at length in the literature (see Hauser *et al.*[48] and references therein) and will not be explored further here. We briefly list the published room-temperature $^5T_2$ lifetimes of the three compounds in acetonitrile solution. **1**: $\tau_{HS} = 60 \pm 5$ ns, **2**: $\tau_{HS} = 5.4 \pm 0.1$ ns, and **3**: $\tau_{HS} = 980 \pm 50$ ps. It can be seen that the data in Fig. 5D can be considered wholly consistent with the corresponding optical data over this temporal range, indicating that these X-ray measurements are reliable probes of ground-state recovery, as well.

Chemical signatures of nearest-neighbour nitrogen atoms

An additional approach to more detailed information concerning metal–ligand interactions in coordination compounds is to focus on the atoms bound to the metal centre(s). These lighter elements such as carbon, nitrogen, and oxygen can reveal the 'ligand perspective' by probing their 1s $\rightarrow$ 2p (and higher energy) core-level transitions. The former transition probes the LUMO of the metal centre's nearest neighbours while core-level excitation at higher energy reveal spectrally broader continuum resonances which can also be sensitive to the chemical environment and changes thereof.[49] In the following we present data on crystalline films of PF_6-salts of the three compounds in Fig. 1 that we recorded at the Advanced Light Source' EUV calibration beamline 6.3.2 to explore the potential of nitrogen K-edge spectroscopy to elucidate the local chemical environment.

Fig. 6 shows the nitrogen K-edge absorption spectra of compounds **1**, **2**, and **3** around 400 eV. The left panel A features the 1s $\rightarrow$ 2p. The bipyridine and terpyridine ligands have identical 1s $\rightarrow$ 2p transitions at 399.5 eV with additional absorption structure characteristic of each ligand. While these two ligand types feature only pyridine ligands, the tren(py)$_3$ ligand cage contains two types of nearest-neighbour nitrogen atoms (those from the pyridine and the imine groups) which lead to two unresolved transitions centred at 399.2 eV, resulting in a seemingly broader lineshape that reflects the two chemically distinct nitrogen species. These observations are commensurate with the observations of differing branching ratios of compound **1** *versus* the purely pyridine-based compounds **2** and **3**. However, the latter two compounds show distinct differences in the N K-edge spectra above 400 eV which makes this type of XANES spectroscopy an interesting and chemically specific target.

The same spectra are plotted in Fig. 6B over a wider energy range. In all three spectra a broader absorption feature that starts to rise at 403 eV at energies above the core level ionization threshold as a continuum resonance can be found. The latter may be interpreted as shape resonances, *i.e.* they can be thought of as unoccupied

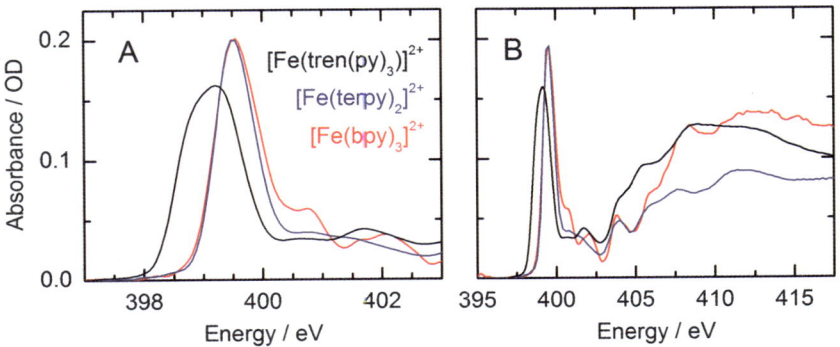

Fig. 6 Ground-state nitrogen K-edge spectra of thin solid films of [Fe(tren(py)$_3$)](PF$_6$)$_2$, [Fe(terpy)$_2$](PF$_6$)$_2$, and [Fe(bpy)$_3$](PF$_6$)$_2$. **A** Nitrogen 1s $\rightarrow$ 2p transitions. **B** Broader spectral energy range illustrating both Nitrogen 1s $\rightarrow$ 2p and σ^* shape resonances.

molecular orbitals embedded in the continuum to which the core-excited electron is promoted. However, there is still some dispute over the nature of these absorption features.[50] Generally, transient nitrogen K-edge edge spectroscopy could prove useful in clarifying whether these resonances provide a measure of bond length changes during chemical reactions, *e.g.* shifting to higher energy with decreasing chemical bonds. More specifically, this type of spectroscopy on solvated transition-metal complexes could prove very useful in understanding additional details of the metal–ligand interactions. For instance, to address the question of why these complexes, when excited to the MLCT manifold relax to the high-spin $^5T_2$ state with time-constants of $\sim$100 fs without a clear signature of involving metal-centred ligand-field states that are energetically intermediate between the MLCT manifold and the high-spin $^5T_2$ state. One advantage of K-edge spectroscopy in general has been that the core-excited electron can be treated in a one-electron picture which allows for *ab initio* methods to simulate core-level spectra (in contrast to L-edge spectra of first-row transition metals in which multiplets due to strong spin–orbit interactions have complicated the development of *ab initio* methods for core-level spectroscopy. For transition metals heavier than those of the first row, multiplet effects are relatively weak allowing for reliable *ab initio*-based simulations of transient L-edge spectra in analogous molecular systems[51]). *Ab initio* studies of 1s transitions of light elements would provide direct comparison with experiment and make details of the evolving valence charge density of the nearest neighbour atoms in transition-metal complexes accessible. Last we note that rapid progress in laser-based high-harmonic sources has led to laser technology with photon energies in the soft X-ray regime.[52] It can be anticipated that in the next years laser-based ultrashort X-ray sources (<30 fs) with sufficient flux at absorption edges in the lower soft X-ray range will allow for femtosecond and possibly attosecond spectroscopy in solution to study charge migration and atomic rearrangement on their natural time scales.

Conclusions

We have studied polypyridyl Fe^{II} complexes in solution with emphasis on the effects of ligand variation on the valence charge density using time-resolved core-level spectroscopy at the Fe $L_{2,3}$-edges. In particular, the transiently excited high-spin state of $[Fe(tren(py)_3)]^{2+}$ and its high-spin analogue have very similar Fe-2p spectra which points to the great similarity of their respective valence charge densities, complementing previous experimental findings[2,20] and supporting detailed predictions from recent *ab initio* calculations.[33] We further conclude from comparing ground-state and transient iron L-edge spectra of three similar polypyridyl Fe^{II} complexes that the metal valence charge densities in these compounds are similar. However, varying chemical composition and ligand-field symmetry influence the valence charge densities of ground and excited states as probed by transient L-edge spectroscopy. Symmetry effects could be an important factor in influencing the dynamics that lead to high-spin state formation. Recent theoretical work by van Veenendaal's group[18,19] proposes fast high-spin formation upon MLCT excitation due to energetic proximity of a third state which mediates ultrafast spin-cross over *via* spin-orbit interactions. De Graaf and Sousa relate to this work, suggesting that deviations from octahedral symmetry might by relevant in this context.[17] Hence, systematic ligand variation in conjunction with ultrafast L-edge spectroscopy could shed more light on ligand-field symmetry. Additionally, ultrafast spectroscopy at the K-edges of the nearest-neighbours of metal centres in solvated transition-metal compounds is proposed to probe the 'ligand view' of the metal–ligand interactions. Such measurements will provide complementary information on the valence charge density of short-lived intermediates in combination with newly developed *ab initio* methods for core-level spectroscopy. We anticipate that advances in laser technology will allow for 'table-top' experiments in the foreseeable future with time-resolution approaching the fundamental time-scales of charge transfer. Such techniques could shed further light on elementary chemical processes beyond the prototypical iron complexes studied in this work.

Acknowledgements

This work was supported by the Director, Office of Science, Office of Basic Energy Sciences, the Chemical Sciences, Geosciences, and Biosciences Division under the Department of Energy, Contract No. DE-AC02-05CH11231 (N.H., H.C., and R.W.S.) and Grant No. DE-FG02-01ER15282 (J.K.M.), as well as the National Research Foundation of Korea (NRF) funded by the Korea government (Grant No. 2009-0068446, 2010-0006570, and 2007-0056330), MEST, PAL, and the Korean XFEL project (H. C., K. H., and T.K.K.).

References

1 S. Ferrere and B. A. Gregg, *J. Am. Chem. Soc.*, 1998, **120**, 843.
2 J. E. Monat and J. K. McCusker, *J. Am. Chem. Soc.*, 2000, **122**, 4092.
3 C. Creutz, M. Chou, T. L. Netzel, M. Okumura and N. Sutin, *J. Am. Chem. Soc.*, 1980, **102**, 1309.
4 J. J. McGravey and I. Lawthers, *J. Chem. Soc., Chem. Commun.*, 1982, 902.
5 E. Buhks, G. Navon, M. Bixon and J. Jortner, *J. Am. Chem. Soc.*, 1980, **102**, 2918.
6 S. Decurtins, P. Gütlich, C. Köhler, H. Spiering and A. Hauser, *Chem. Phys. Lett.*, 1984, **105**, 1.
7 C. L. Xie and D. N. Hendrickson, *J. Am. Chem. Soc.*, 1987, **109**, 6981.
8 P. Gütlich and A. Hauser, *Coord. Chem. Rev.*, 1990, **97**, 1.
9 A. Hauser, A. Vef and P. Adler, *J. Chem. Phys.*, 1991, **95**, 8710.
10 J. K. McCusker, K. N. Walda, R. C. Dunn, J. D. Simon, D. Magde and D. N. Hendrickson, *J. Am. Chem. Soc.*, 1992, **114**, 6919.
11 J. K. McCusker, K. N. Walda, R. C. Dunn, J. D. Simon, D. Magde and D. N. Hendrickson, *J. Am. Chem. Soc.*, 1993, **115**, 298.
12 A. L. Smeigh, M. Creelman, R. A. Mathies and J. K. McCusker, *J. Am. Chem. Soc.*, 2008, **130**, 14105.
13 M. M. N. Wolf, R. Groß, C. Schumann, J. A. Wolny, V. Schünemann, A. D, H. Paulsen, J. J. McGarvey and R. Diller, *Phys. Chem. Chem. Phys.*, 2008, **10**, 4264.
14 W. Gawelda, A. Cannizzo, V.-T. Pham, F. van Mourik, C. Bressler and M. Chergui, *J. Am. Chem. Soc.*, 2007, **129**, 8199.
15 C. Consani, M. Prémont-Schwarz, A. ElNahhas, C. Bressler, F. van Mourik, A. Cannizzo and M. Chergui, *Angew. Chem., Int. Ed.*, 2009, **48**, 7184.
16 C. de Graaf and C. Sousa, *Chem.–Eur. J.*, 2010, **16**, 4550.
17 C. de Graaf and C. Sousa, *Int. J. Quantum Chem.*, 2011, **111**, 3385.
18 M. van Veenendaal, J. Chang and A. J. Fedro, *Phys. Rev. Lett.*, 2010, **104**, 067401.
19 J. Chang, A. J. Fedro and M. van Veenendaal, *Phys. Rev. B*, 2010, **82**, 075124.
20 M. Khalil, M. A. Marcus, A. L. Smeigh, J. K. McCusker, H. H. W. Chong and R. W. Schoenlein, *J. Phys. Chem. A*, 2006, **110**, 38.
21 W. Gawelda, V.-T. Pham, M. Benfatto, Y. Zaushitsyn, M. Kaiser, D. Grolimund, S. L. Johnson, R. Abela, A. Hauser, C. Bressler and M. Chergui, *Phys. Rev. Lett.*, 2007, **98**, 057401.
22 C. Bressler, C. Milne, V.-T. Pham, A. ElNahhas, R. M. van der Veen, W. Gawelda, S. Johnson, P. Beaud, D. Grolimund, M. Kaiser, C. N. Borca, G. Ingold, R. Abela and M. Chergui, *Science*, 2009, **323**, 489.
23 S. Nozawa, T. Sato, M. Chollet, K. Ichiyanagi, A. Tomita, H. Fujii, S.-i. Adachi and S.-y. Koshihara, *J. Am. Chem. Soc.*, 2010, **132**, 61.
24 N. Huse, T. K. Kim, L. Jamula, J. K. McCusker, F. M. de Groot and R. W. Schoenlein, *J. Am. Chem. Soc.*, 2010, **132**, 6809.
25 G. Vankó, P. Glatzel, V.-T. Pham, R. Abela, D. Grolimund, C. N. Borca, S. L. Johnson, C. J. Milne and C. Bressler, *Angew. Chem., Int. Ed.*, 2010, **49**, 1.
26 N. Huse, H. Cho, K. Hong, L. Jamula, F. M. de Groot, T. K. Kim, J. K. McCusker and R. W. Schoenlein, *J. Phys. Chem. Lett.*, 2011, **2**, 880.
27 L. X. Chen, W. J. H. Jager, G. Jennings, D. J. Gosztola, A. Munkholm and J. P. Hessler, *Science*, 2001, **292**, 262.
28 C. Bressler and M. Chergui, *Chem. Rev.*, 2004, **104**, 1781.

29 P. Wernet, G. Gavrila, K. Godehusen, C. Weniger, E. Nibbering, T. Elsaesser and W. Eberhardt, *Appl. Phys. A: Mater. Sci. Process.*, 2008, **92**, 511.
30 N. Huse, H. Wen, D. Nordlund, E. Szilagyi, D. Daranciang, T. A. Miller, A. Nilsson, R. W. Schoenlein and A. M. Lindenberg, *Phys. Chem. Chem. Phys.*, 2009, **11**, 3951.
31 N. Huse, H. Wen, R. W. Schoenlein and A. M. Lindenberg, *J. Chem. Phys.*, 2009, **131**, 234505.
32 M. Lorenc, J. Hébert, N. Moisan, E. Trzop, M. Servol, M. Buron-Le Cointe, H. Cailleau, M. Boillot, E. Pontecorvo, M. Wulff, S. Koshihara and E. Collet, *Phys. Rev. Lett.*, 2009, **103**, 028301.
33 B. E. Van Kuiken and M. Khalil, *J. Phys. Chem. A*, 2011, **115**, 10749.
34 Although this wavelength corresponds to excitation into a higher energy absorption feature than we have previously exploited, the sub-picosecond conversion to the high-spin $^5T_2$ state characteristic of these compounds implies that the same electronic excited state is being probed on the 10s of picosecond time scale relevant for the present study.
35 P. Hohenburg and W. Kohn, *Phys. Rev.*, 1964, **136**, B864.
36 W. Kohn and L. J. Sham, *Phys. Rev.*, 140, A113.
37 C. Adamo and V. Barone, *J. Chem. Phys.*, 1999, **110**, 6158.
38 P. J. Hay and W. R. Wadt, *J. Chem. Phys.*, 1985, **82**, 299.
39 A. Ehlers, M. Böhme, S. Dapprich, A. Gobbi, A. Höllwarth, V. Jonas, K. Köhler, R. Stegmann, A. Veldkamp and G. Frenking, *Chem. Phys. Lett.*, 1993, **208**, 111.
40 E. Reed, R. B. Weinstock and F. Weinhold, *J. Chem. Phys.*, 1985, **83**, 735.
41 B. Mennucci, E. Cancès and J. Tomasi, *J. Phys. Chem. B*, 1997, **101**, 10506.
42 E. Cancès, B. Mennucci and J. Tomasi, *J. Chem. Phys.*, 1997, **107**, 3032.
43 M. J. Frisch, G. W. Trucks, H. B. Schlegel, G. E. Scuseria, M. A. Robb, J. R. Cheeseman, G. Scalmani, V. Barone, B. Mennucci, G. A. Petersson, H. Nakatsuji, M. Caricato, X. Li, H. P. Hratchian, A. F. Izmaylov, J. Bloino, G. Zheng, J. L. Sonnenberg, M. Hada, M. Ehara, K. Toyota, R. Fukuda, J. Hasegawa, M. Ishida, T. Nakajima, Y. Honda, O. Kitao, H. Nakai, T. Vreven, J. A. Montgomery, Jr., J. E. Peralta, F. Ogliaro, M. Bearpark, J. J. Heyd, E. Brothers, K. N. Kudin, V. N. Staroverov, R. Kobayashi, J. Normand, K. Raghavachari, A. Rendell, J. C. Burant, S. S. Iyengar, J. Tomasi, M. Cossi, N. Rega, J. M. Millam, M. Klene, J. E. Knox, J. B. Cross, V. Bakken, C. Adamo, J. Jaramillo, R. Gomperts, R. E. Stratmann, O. Yazyev, A. J. Austin, R. Cammi, C. Pomelli, J. Ochterski, R. L. Martin, K. Morokuma, V. G. Zakrzewski, G. A. Voth, P. Salvador, J. J. Dannenberg, S. Dapprich, A. D. Daniels, O. Farkas, J. B. Foresman, J. V. Ortiz, J. Cioslowski and D. J. Fox, *GAUSSIAN 09 (Revision B.01)*, Gaussian, Inc., Wallingford, CT, 2009.
44 K. M. Kadish, C. H. Su, D. Schaeper, C. L. Merrill and L. J. Wilson, *Inorg. Chem.*, 1982, **21**, 3433.
45 A. J. Conti, C. L. Xie and D. N. Hendrickson, *J. Am. Chem. Soc.*, 1989, **111**, 1171.
46 B. T. Thole and G. van der Laan, *Phys. Rev. B*, 1988, **38**, 3158.
47 P. Wernet, private communication.
48 A. Hauser, C. Enachescu, M. L. Daku, A. Vargas and N. Amstutz, *Coord. Chem. Rev.*, 2006, **250**, 1642.
49 J. M. Garcia-Lastra, P. L. Cook, F. J. Himpsel and A. Rubio, *J. Chem. Phys.*, 2010, **133**, 151103.
50 M. Piancastelli, *J. Electron Spectrosc. Relat. Phenom.*, 1999, **100**, 167.
51 B. E. van Kuiken, N. Huse, H. Cho, M. L. Strader, M. S. Lynch, R. W. Schoenlein and M. Khalil, *J. Phys. Chem. Lett.*, 2012, 1695.
52 M.-C. Chen, P. Arpin, T. Popmintchev, M. Gerrity, B. Zhang, M. Seaberg, D. Popmintchev, M. M. Murnane and H. C. Kapteyn, *Phys. Rev. Lett.*, 2010, **105**, 173901.

General discussion

Professor Bowman opened the discussion of the paper by Professor Neumark by asking: I wonder if it would be possible to extend your experiments in two ways to make contact with some of the molecular beam experiments. One way is to vibrationally excite the anion precursor and the second is to partially deuterate it?

Professor Neumark replied: With our new ion trapping and cooling capability, it is possible to vibrationally excite ions inside the trap, extract them, and measure their SEVI spectrum. This would be a non-trivial undertaking, however. In 1991, we reported the photoelectron spectrum of FDH^-, which we think is the lowest energy structure from the addition of HD to F^-.[1] We have not tried analogous experiments with partially deuterated CH_4 but these would clearly be of interest.

1 A. Weaver and D. M. Neumark, *Faraday Discuss. Chem. Soc.*, 1991, **91**, 5.

Mr Welsch addressed Professor Neumark and Professor Bowman :In your paper you assign the second peak to come from the spin–orbit excited state. Recently, our group in collaboration with Juliana Palma did some full-dimensional quantum dynamics calculations of the low-resolution photodetatchment spectra of FCH_4^-. Although the calculations only considered the ground state potential energy surface of FCH_4, the second peak appears in these calculations and was assigned to an excitation of the bending modes of methane. However, the width and height of the second peak are smaller than in the experiment. This could be due to the neglect of the electronic excited state, but also due to limitations in the employed potential energy surface (PES).

Professor Bowman responded: There is an interesting coincidence in the energies of the vertical spin–orbit excited state and the bend fundamental of the anion. It would of course require a calculation including that spin-orbit excited PES to sort out the relative contributions from both. At this point, the fact that the absence of that PES in your calculations leads to a smaller peak than seen experimentally is at least in the "right direction".

Certainly the excited PES should be considered in order to make a quantitative comparison with experiment, notwithstanding any possible limitations in the accuracy of the ground state PES you employed.

Professor Neumark added: I would not be surprised if the second peak comprises electronic and vibrational contributions. Transitions to the spin-orbit excited state of the neutral FCH_4 complex should be present and, according to Professor Bowman's work, in about the right place. Also, this feature is very broad, even in our SEVI spectrum, which might suggest that it corresponds to a transition to a repulsive electronic state.

Professor Wester questioned: CH_3^- is a weakly-bound, but stable negative ion. What are the prospects for clustering this anion to HF inside your radiofrequency ion trap using three-body collisions at low temperature? How much new information could be gained on the $F + CH_4$ reaction by accessing the product side of the transition state with photoelectron spectroscopy in this way?

Professor Neumark said: If there were a stable $HF \cdot CH_3^-$ structure then such an experiment would be of interest. This would clearly be a higher energy structure than FCH_4^-, and the question is whether there is at least a local minimum that could be populated in our ion trap. This higher energy structure has not been observed experimentally and, as far as I know, has not been considered theoretically, either.

Professor Nesbitt questioned: Just as H_2 has unquenchable *ortho* (J = odd) and *para* (J = even) species, CH_4 has three different nuclear spin states (A,F,E), which will rotationally cool down in your jet into associated lowest quantum numbers of J = 0,1,2, respectively. What is the barrier to interconversion between various conformers of $CH_4–F^-$ and might there be any residual influence of these unquenchable nuclear spin states in the transition state dynamics?

Professor Neumark answered: The FCH_4^- anion is best thought of as a symmetric top with C_{3v} symmetry. The barrier to interconversion is likely to be around 0.2 eV, *i.e.* on the order of the dissociation energy of the anion to F^- + CH_4. So the nuclear spin statistics appropriate to a three-fold symmetric top are most relevant here. As a result, even at T = 0 K, some excited K levels of the anion will be populated, and these may have a non-negligible effect on the SEVI spectrum.

Professor Bradforth said: I am wondering whether you have tried the argon clustering approach to cooling off the parent anions, which was an effective strategy for narrowing up the photoelectron spectrum of $BrHI^-$ and IHI^- in earlier work[1] from your lab? Now you resolve under the PE band, could it be that some of this structure is due to hot bands, perhaps in the hindered rotor or $F\cdots HX$ stretch of the anion?

1 Z. Liu, H. Gomez, and D. M. Neumark, *Chem. Phys. Lett.*, 2000, **332**, 65.

Professor Neumark replied: At the resolution of our SEVI experiment, addition of an Ar atom would probably complicate the observed spectrum because we would see low frequency progressions in van der Waals vibrations involving the Ar atom, similar to those seen in the zero electron kinetic energy (ZEKE) spectra of $ArCl^-$ and related systems.[1] The ion trapping and cooling capability recently implemented on our instrument is a more desirable way to cool the anions in a SEVI experiment.

1 T. Lenzer, I. Yourshaw, M. R. Furlanetto, G. Reiser, and D. M. Neumark, *J. Chem. Phys.*, 1999, 110, 9578.

Professor Bradforth asked: For FH_2^-, have SEVI experiments been performed at photodetachment energies relevant to the F* spin orbit excited channel, and do these reveal any similar structure?

Professor Neumark said: In our paper, we mention that there is a broad, isotropic contribution to the FH_2^- photoelectron images at low electron kinetic energy that comes from photodetachment to the neutral spin-orbit excited state. This contribution is subtracted from the image shown in Fig. 2 in the paper. Neither the work reported here nor our previous photoelectron spectra[1] show vibrational structure from the spin-orbit excited state, implying that this excited state is quite repulsive in the Franck–Condon region accessed by photodetachment.

1 S. E. Bradforth, D. W. Arnold, D. M. Neumark, and D. E. Manolopoulos, *J. Chem. Phys.*, 1993, **99**, 6345.

Professor Ashfold commented: One photon detachment from FH_2^- yields a *p*-wave electron, which limits the ultimate resolution and signal to noise ratio of the SEVI spectrum of this system. Two photon induced detachment should presumably yield *s*- and or *d*-wave electrons, with the former dominating at threshold. What are the prospects for measuring the two photon induced SEVI spectrum of FH_2^-?

Professor Neumark responded: This is an interesting idea. However, I suspect that non-resonant two-photon detachment under our operating conditions is unlikely to be a favorable process. If we focus the laser tightly enough for this process to compete with one-photon detachment, it is likely that the number of ions in the interaction volume will be too small to be of much use, since typical ion densities are on the order of 10^5 per cc.

Professor Kable asked: At the end of the discussion in your paper you speculate as to the origin of the narrow resonances in the slow electron velocity map imaging spectrum. The suggestion is that they "might represent a progression along the reaction coordinate, primarily involving the $F \cdots CH_4$ stretch". I understand how resonances in vibrational coordinates orthogonal to the reaction coordinate can give rise to resonances. Would you explain how motion in the reaction coordinate could give rise to a resonance as it has an imaginary frequency?

Professor Neumark answered: There are some semantics at play here, and some physics. In most of our photodetachment experiments on transition states, we observe vibrational structure associated only with motion orthogonal to the reaction coordinate. These features can be quite broad. However, in IHI^-, and possibly in FH_2^- and FCH_4^-, there is additional finer structure associated with trapped vibrational motion along the reaction coordinate. This structure arises, to first order, from minima in vibrationally adiabatic curves that can be present even if the potential energy surface itself does not have a well at the transition state.[1] It is only this finer structure that we refer to as resonances in our papers; we can assign quantum numbers to these states corresponding to vibrational motion perpendicular and parallel to the reaction coordinate, in contrast to "direct scattering" features for which one can only specify the perpendicular vibrational modes. While this terminology is not universally accepted, we find it useful in classifying features in our photodetachment experiments.

1 See M. S. Child, *Molecular Collision Theory*, Academic Press (London and New York), 1974, p 224.

Dr Glowacki opened the discussion of the paper by Antonio Laganà by saying: I have a general comment which I think Professor Laganà's paper hints at. I don't necessarily have the answer to this question, but I am interested whether any of the experts have insight to offer.

Compared to other fields of chemistry, gas phase dynamics and kinetics has extremely rich sets of experimental data. And alongside this, it also has a rich and well-established hierarchy of theoretical approaches from which to draw. Within other well-established fields of chemistry, like thermodynamics[1] and synthetic organic chemistry,[2] people are beginning to accumulate rich data sets into databases and then use network theories to analyze the self-consistency of the data, determine which components are subject to the largest uncertainties, and identify statistical outliers.

I'm wondering whether gas phase dynamics and kinetics are at a stage where one can begin to build such databases and link them together in a similar fashion. In principle it might then be possible to start performing self-consistency tests – for example, to determine whether a series of PESs, calculated rate coefficients, measured product state distributions, kinetic energy releases, *etc.* are all consistent with one another, and whether they suggest errors or new insight. If one was able to establish such databases and subsequently analyze them, I wonder whether it would be possible to perform systematic sensitivity analyses on potential energy surfaces, and decipher (based on how well they fit the experimental data) which regions of the potential are most error prone, in much the same way that sensitivity analyses are often performed in large scale kinetic modelling. This may help to pin down those parameters to which a model fit is most sensitive. Often, one chases the holy grail of a perfect global potential energy surface; however, as Professor Zhang's paper showed, it is often not always necessary to represent every single degree of freedom. Sensitivity analyses of the sorts outlined above might help to pin down the most error prone regions of a potential, and thereby focus efforts on improving those parts of the models and performing decisive experiments.

1 Branko Ruscic et al., J. Phys.: Conf. Ser. 2005, 16, 561, DOI:10.1088/1742-6596/16/1/078.
2 K. J. M. Bishop, R. Klajn, B. A. Grzybowski, Angew. Chem. Int. Ed., 2006, **45**, 5348.

Professor Bowman communicated in reply: As you note databases do exist for thermochemical properties and perhaps to a lessor extent on rate coefficients. Both of these, perhaps the former especially, are useful in developing PESs, e.g., by incorporating the correct reaction exoergicity/enthalpy. Rate coefficients unfortunately don't supply the same quality of information on say a barrier height, since tunneling, adiabatic barriers, re-crossing, etc. affect rate coefficients. Finally, the point on what aspects of the PES are important to experiment obviously depends on the experiment. While rate coefficients for simple barrier-dominated reactions focus attention on the barrier, experiments that probe the final states of products may also be sensitive to post-transition state features of the PES. There are other examples as well.

Professor Truhlar added: A generic difficulty with sensitivity analysis is that it is almost necessarily based on small perturbations from an existing best estimate or working hypothesis. For example, a sensitivity analysis of the potential energy surface may be based on making localized changes to the given potential energy surface. Often though, when one considers larger perturbations, the entire mechanism changes, and the new mechanism may have entirely different sensitivities. Consider, for example, a set of reaction rates used to simulate a complex process, like combustion. If a reaction rate leading to product B, for example, is underestimated, the mechanism will not be sensitive to reaction rates of products that arise only downstream from B, and it may not even be sensitive to small perturbations in the incorrect rate constant for producing B. But when that rate constant is corrected, the whole set of sensitivities may be changed qualitatively. Thus sensitivity analysis has its valid uses, but it is often called for inappropriately.

Professor Balucani communicated:†Because of the importance of the HO + CO → H + CO_2 reaction in combustion[1] and atmospheric chemistry,[2] numerous kinetic and dynamical studies have been reported in both theoretical[3-4] and experimental fronts.[5-6] Nevertheless, there remain important unresolved issues. Among these is the poor agreement between experimental and theoretical results that is to a great extent attributable to the insufficient accuracy of the adopted PES. Indeed, this is the case of the most popular PESs[7-9] which are based on a small number of ab initio points, and are scarcely accurate in representing the full six-dimensional configuration space and providing support for dynamical calculations.

To improve the situation, some of us have recently developed a new global PES for the HO + CO → H + CO_2 reaction based on a large number of ab initio points at the CCSD(T)-F12b/AVTZ level of theory.[10,11] The analytical fit of the ab initio points used the permutation invariant form of Bowman and coworkers.[12] The PES has been shown to faithfully reproduce stationary points along the reaction path as well as the entire configuration space within chemical accuracy. The calculated vibrational frequencies of the HOCO isomers and the photodetachment spectrum are in good agreement with experimental results.[13]

† On behalf of J. Li, H. Guo, J. M. Bowman, N. Balucani, P. Casavecchia and A. Laganà.

An important test of the reliability of the PES is the comparison of the theoretical and experimental data for the scattering process at the state-to-state level. Using this PES, we have recently reported quasi-classical trajectory (QCT) calculations for the title reaction.[11] (A QCT study of the reverse reaction has also been carried out on a slightly modified PES and most results are in good agreement with the existing experimental data.[14]) The calculated total integral cross section and rate constant for the title reaction are in reasonably good agreement with experimental data. However, there are significant discrepancies between theory and crossed molecular beam (CMB) experiments at the state-to-state level in the centre-of-mass product angular and translational distributions.[15,16] As some of us have pointed out recently,[17] there might be some ambiguity in converting raw experimental angular distributions in the laboratory frame to the centre-of-mass frame, which is reflected in the associated error bars. To avoid this ambiguity, we report a direct simulation of the laboratory angular and time-of-flight distributions using the Observables block of the GEMS approach presented in the paper by Laganà *et al.*[18] To simulate CMB laboratory distributions, the Observables block of GEMS makes use of the forward convolution routine usually employed to analyze the results of CMB experiments, with the difference that, rather than using a trial-and-error approach, the results of dynamical calculations (namely, centre-of-mass angular and product

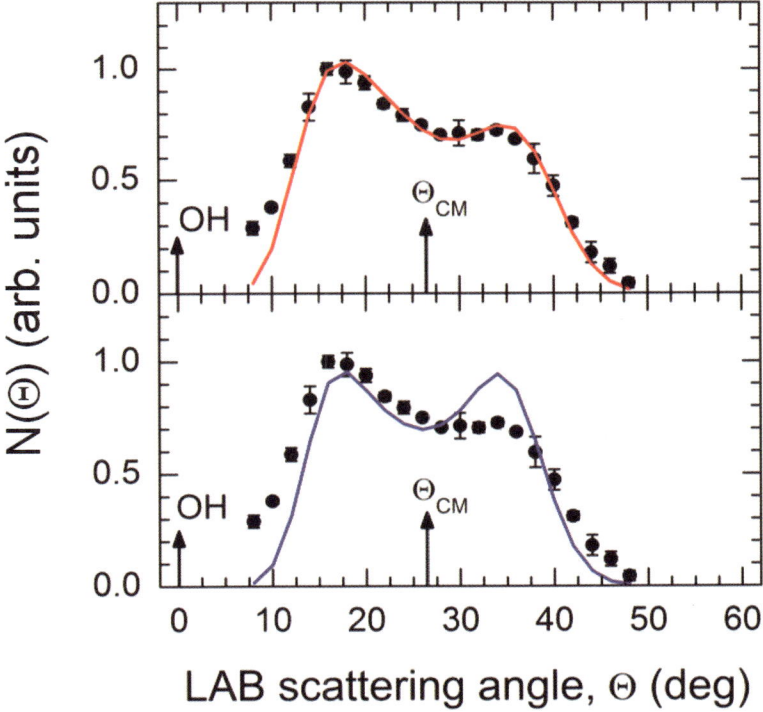

Fig. 1 Laboratory angular distribution recorded at $m/z=46$ ($CO^{18}O^+$, water with isotopically labeled ^{18}O was used to produce ^{18}OH with the aim to reduce the background signal at the mass of the CO_2 product) for the reaction $OH(N) + CO(j) \rightarrow CO_2 + H$ at $E_c=14.1$ kcal mol^{-1}. Top panel: experimental points (solid circles) and best-fit distribution (red line) obtained by using the experimental best-fit functions.[15,16,18] Bottom panel: experimental points against the prediction of QCT calculations (blue line) on the new *ab initio* PES (only $N_{OH} = 1$ and $j_{CO} = 0$ are considered in the QCT calculations).

translational distributions) are directly introduced in the forward convolution program, converted to the laboratory frame and averaged over the experimental distributions. As illustrated in other cases,[19-23] this is the best way to judge the quality of an *ab initio* PES and/or approximate dynamical calculations. Differently from the QCT calculations for the three previous PESs (YMS,[7] LTSH[8] and Leiden[9]) reported by Laganà *et al.*,[18] in the present case QCT calculations have been carried out only for both OH and CO reactants in their ground rotational levels. As illustrated by Laganà *et al.*,[18] however, the OH radicals produced in the radiofrequency discharge beam source are characterized by some rotational excitation.

The simulated laboratory angular distributions at the collision energies of 14.1 and 8.6 kcal mol^{-1} are reported in Fig. 1 and 2 below (bottom panels), together with the experimental points and the best-fit simulation (top panels) obtained in the trial-and-error procedure during the analysis of the data.[15,16]

By comparing the simulated laboratory distributions with the experimental ones, we observed that at $E_c = 14.1$ kcal mol^{-1} the shoulder around 35° is as high as the peak around 16–18°, at variance with the measured distribution, while at $E_c = 8.6$ kcal mol^{-1} the QCT simulated laboratory angular distribution fails to reproduce the asymmetry around Θ_{CM}. Overall, the state of comparison is quite similar to the case of the QCT calculations on the Leiden PES,[17,18] which, indeed, also produced similar centre-of-mass functions. As in the case of the Leiden calculations, however, the discrepancies noted while comparing the centre-of-mass functions are largely mitigated when moving to the laboratory frame. As already commented in the below refs. 17 and 18 this is due to a combined effect of the characteristics of centre-of-mass angular and product translational energy distributions and is an excellent example of how much more significant the comparison of theoretical predictions directly with crossed molecular beam data is.

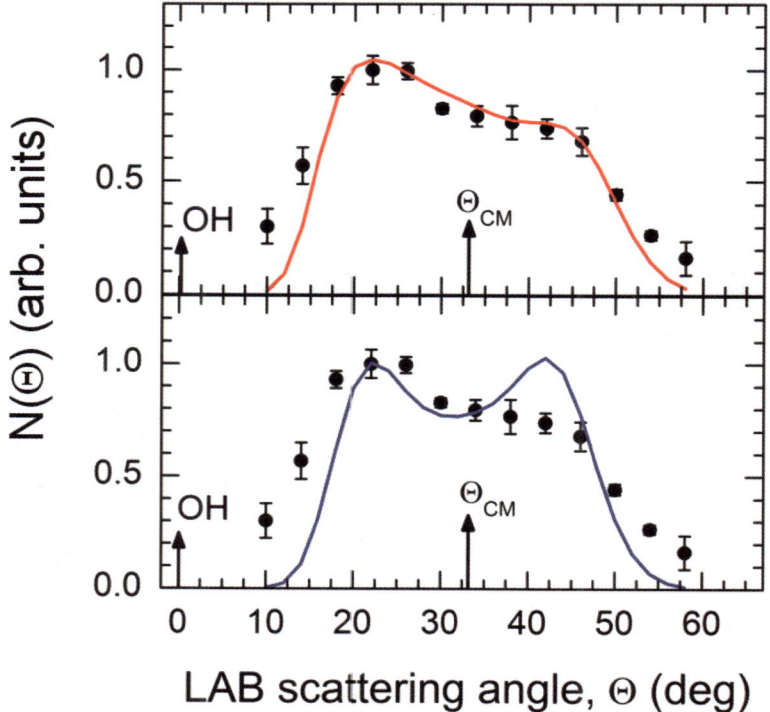

Fig. 2 As in Fig. 1, but at $E_c = 8.6$ kcal mol^{-1}.

Nevertheless, to make such a comparison as accurate as possible, the effect of OH rotational excitation remains to be explored for the new HOCO PES.

1 J. A. Miller, R. J. Kee and C. K. Westbrook, *Annu. Rev. Phys. Chem.*, 1990, **41**, 345.
2 R. P. Wayne, *Chemistry of Atmospheres*, Oxford University Press, Oxford, 2000.
3 J. M. Bowman and G. C. Schatz, *Annu. Rev. Phys. Chem.*, 1995, **46**, 169.
4 H. Guo, *Int. Rev. Phys. Chem.*, 2012, **31**, 1.
5 P. Casavecchia, *Rep. Prog. Phys.*, 2000, **63**, 355.
6 P. Casavecchia, N. Balucani and G. G. Volpi, *Annu. Rev. Phys. Chem.*, 1999, **50**, 347.
7 H. G. Yu, J. T. Muckerman and T. J. Sears, *Chem. Phys. Lett.*, 2001, **349**, 547.
8 M. J. Lakin, D. Troya, G. C. Schatz and L. B. Harding, *J. Chem. Phys.*, 2003, **119**, 5848.
9 R. Valero, M. C. van Hemert and C. J. Kroes, *Chem. Phys. Lett.*, 2004, **393**, 236.
10 J. Li, Y. Wang, B. Jiang, J. Ma, R. Dawes, D. Xie, J. M. Bowman and H. Guo, *J. Chem. Phys.*, 2012, **136**, 041103.
11 J. Li, C. Xie, J. Ma, Y. Wang, R. Dawes, D. Xie, J. M. Bowman and H. Guo, *J. Phys. Chem. A*, 2012, **116**, 5057.
12 J. M. Bowman, G. Czakó and B. Fu, *Phys. Chem. Chem. Phys.*, 2011, **13**, 8094.
13 J. Ma, J. Li and H. Guo, *Phys. Rev. Lett.*, 2012, **109**, 063202.
14 C. Xie, J. Li, D. Xie and H. Guo, *J. Chem. Phys.*, 2012, **137**, 024308.
15 M. Alagia, N. Balucani, P. Casavecchia, D. Stranges and G. G. Volpi, *J. Chem. Phys.*, 1993, **98**, 8341.
16 P. Casavecchia, N. Balucani and G. G. Volpi, in *Chemical Dynamics and Kinetics of Small Free Radicals*, ed., A. F. Wagner and K. Liu, World Scientific, Singapore, 1995, Vol. I, pp. 365.
17 A. Laganà, N. Balucani, S. Crocchianti, P. Casavecchia, E. Garcia and A. Saracibar, *Lect. Notes Comput. Sci.*, 2011, **6784**, 453.
18 A. Laganà, E. Garcia, A. Paladini, P. Casavecchia and N. Balucani, *Faraday Disc.*, 2012, DOI: 10.1039/c2fd20046e.
19 D. Skouteris, H.-J. Werner, F. J. Aoiz, L. Banares, J. F. Castillo, M. Menendez, N. Balucani, L. Cartechini and P. Casavecchia, *J. Chem. Phys.*, 2001, **114**, 10662.
20 N. Balucani, P. Casavecchia, L. Banares, F. J. Aoiz, T. Gonzalez-Lezana, P. Honvault and J. M. Launay, *J. Phys. Chem. A*, 2006, **110**, 817.
21 N Balucani, G. Capozza, E. Segoloni, A. Russo, R. Bobbenkamp, P. Casavecchia, T. Gonzalez-Lezana, E. J. Rackham, L. Banares and J. F. Aoiz, *J. Chem. Phys.*, 2005, **122**, 234309.
22 N. Balucani, G. Capozza, L. Cartechini, A. Bergeat, R. Bobbenkamp, P. Casavecchia, F. J. Aoiz, L. Banares, P. Honvault, B. Bussery-Honvault and J. M. Launay, *Phys. Chem. Chem. Phys.*, 2004, **6**, 4957.
23 N. Balucani, P. Casavecchia, F. J. Aoiz, L. Bañares, J. F. Castillo and V. J. Herrero, *Mol. Phys.*, 2005, **103**, 1703.

Professor Bowman commented: The OH + CO reaction is amenable to a full-dimensional quantum treatment, and I believe that is being contemplated by Hua Guo, using the new potential energy surface that Professor Balucani mentioned in her comment. Given the various entrance and exit channel barriers tunneling and resonances could play a major role in this reaction and in particular influence the angular distributions.

Professor Laganà responded: The above comment points out possible sources of inaccuracy for QCT treatments of the OH + CO reaction. Obviously, an extended analysis of the differences between quantum and quasiclassical results (especially for product angular distributions) would require the calculation of detailed S matrix elements converged with the value of the total angular momentum quantum number J. Unfortunately such results are not available and we had to limit the analysis to J = 0 results.[1]

The analysis of the J = 0 QCT total reaction probabilities calculated for reactants in the ground and the first excited vibrational states, shows that they reproduce reasonably well the exact QM values for collision energy values raging up to 0.4 eV. The key differences between the two sets of results are the fine oscillatory structure of the quantum probability (that is not reproduced by the QCT one) and its high threshold energy (that is higher than the QCT one). We also found that the agreement between quantum and QCT values largely depends on how strictly the conservation of the total energy is enforced on classical trajectories and how non energy-conserving trajectory outcomes are treated. As a matter of fact, when discarding trajectories whose energy conservation is worse than a few percents of the total energy results are similar to those obtained without any filtering and agree well with the QM results. Whereas, when enforcing of a more severe energy conservation (trajectories with energy deviation larger than 0.002 eV are discarded) the reactivity is excessively penalized. In other words, the adoption of increasingly more drastic energy conservation criteria builds into the dynamical outcome a non reactive bias. At the same time, however, if no filtering criterion is adopted, it is impossible to obtain an acceptable accuracy when evaluating reactive probabilities.

A similar effect was found when considering QCT results for which the continuous classical internal energy has been discretized. The adoption of a filter discarding trajectories deviating a few percents of the total energy has led to an acceptable reproduction of the detailed QM probabilities in the whole range of the investigated collision energy values although most of the QM fine structure goes lost. More detailed QM product properties, such as the product rotational (PRD) and vibrational (PVD) distributions, are also acceptably reproduced by QCT calculations. QCT-PRDs, in fact, exhibit a similar Boltzmann-like shape although, as expected, the QM oscillatory structure gets smoothed down. Similarly, QCT PVDs predict a preference to populate vibrational states with excited bending as the QM results do, although they fail to reproduce the QM population of the vibrational ground state of CO_2.

1 E. Garcia, F. J. Aoiz, A. Laganá, *Theor. Chem. Acc.*, in press.

Professor Martinez asked: With the availability of greatly enhanced computing power through a "computational grid" approach and the ease of using different potential energy surfaces provided by the "distributed workflow" you have described, is it possible and/or appropriate to start thinking about some of these problems in qualitatively different ways? For example, you described a system which allows us to specify experimental observables and then to compute predicted values of these using many available analytic potential energy surfaces. Could one instead determine the combination of all the available potential energy surfaces (possibly weighted according to the region of configuration space) which best fits the observables? In this way, one might be able to effectively invert the experimental data or at least to identify regions of the potential energy surface where more accurate electronic structure calculations are merited. Are there other ways to leverage a "high-throughput simulation engine," or is this just a faster and easier way to do the same simulations that were done (more laboriously and slowly) before?

Professor Laganà answered: The possibility of using the enhanced power of networked computing and the collaborative propensity of distributed workflows to "effectively invert the experimental data or at least to identify regions of the potential energy surface where more accurate electronic structure calculations are merited" is an interesting suggestion that is likely to find its way as an alternative approach to the traditional use of computing resources for determining interactions in Molecular and Materials Sciences and Technologies (MMST).

This approach, in fact, finds support on the extra computing power made available by the Grid (we refer here mainly to the organization of the European Grid Infrastructure (EGI) http://www.egi.eu/ and the Italian Grid Infrastructure (IGI) http://www.italiangrid.it/) that has an on-demand and distributed nature as opposed to the centralized and grant-based system adopted by most of the computer centers administering large scale facilities. Therefore, apart from the fraction of local computing resources self-administered by scientific users, computational researchers have to rely on calls for proposals (followed by a scientific evaluation of the presented applications) issued by the computer centres and subject to ranking criteria based on their policies (maximize the overall turn around, import the most popular packages, experiment new architectures, etc.). In our experience this policy does not foster the development of new theoretical approaches and, more than that, is difficult to adapt to the work habits of MMST users.

In the Grid, instead, resource users and providers concur at various levels (operations, middleware, user support, services, dissemination, management) to make the platform sustainable. As a result, cooperation is an angular stone of Grid communities and activities are characterized by a Service Oriented Approach (SOA).[1] This makes research activities truly community-driven with the user being classed either as simple users (active if they implement their own program on the Grid and passive if they use just what is already implemented on it) or software providers (active or passive depending on whether the implemented and maintained software works interfaced with other programs or not). A similar distinction can be made for hardware providers.[2]

The cooperative approach implies, in addition to the SOA structure of the applications, a strong characterization of the MMST applications in terms of interoperability and codes composition. Such characterization is obtained through a continuous iteration of a stream of designing, implementing, running and validating data and software components (going from simpler to more complex and realistic descriptions of chemical structures and processes) accompanied by the aggregation of codes and data of different origin. This fosters the establishing of Virtual Organizations (VO) and Virtual Research Communities (VRC) based on the building of a Grid economy enabling the objective recognition of the contributions (and the consequent award of related credits) of its members within collaborative/competitive endeavours. To this end quality evaluators (QoU, quality of users and QoS, Quality of Services) have been developed.[3]

As a matter of fact, EGI supports the formation of VOs and VRCs (currently, the main EGI VRCs are WeNMR (structural biology), LSGC (life sciences), HMRC (hydro-meteorology) and WLCG (high energy physics), AA (astrophysics) and COMPCHEM (computational chemistry); see http://www.egi.eu/community/vrcs/) not only through a general coordination effort but also through the provision of a variety of services like those already implemented in EGI (Dashboards, User interfaces and Frameworks, Workflows, Gateways, Data management).

The Grid, however, that is a distributed computing (or high throughput computing, HTC) resource, is at present fast developing also in the direction of storage capabilities and computer performances (usually provided by high performance computing (HPC) on which most of the MMST computational applications are implemented and run). As to the first aspect, due to the fact that an increasing amount of research is being carried out using large amounts of stored data, distributed repositories and databases have been developed. As to the second aspect, due to the increasing size of the systems considered an answer to the request of running large applications is being sought by bridging the HTC platform to the Grid.[4]

On these grounds, among the activities of the COMPCHEM VO[5] we have built the Grid Empowered Molecular Simulator (GEMS)[6] that has been used for generating the virtual signal of the crossed beam experiments illustrated in the paper. In this way, in the Born Oppenheimer Approximation scheme, the procedure is solidly anchored to the highest level accessible by the adopted *ab initio* quantum method for all the potential energy surfaces intervening in the process. My understanding is that when using the "blind"-like inversion procedure proposed in the comment while the power of the "high-throughput simulation engine" is fully exploited the unicity of the solution and most of its stability may get lost.

1 C. Gresty, G. Thanos, T. Dimitrakos, P. Warren, *Service oriented infrastructure: Meeting customer needs*, *BT Technology Journal*, Jan 2008, **26**, issue 1.
2 A. Laganà, A. Riganelli, O. Gervasi, *Lecture Notes Computer Science*, 2006, **3980**, 665-674.
3 C. Manuali, A. Laganà, *Lecture Notes Computer Science*, 2011, **6784**, 397–411.
4 C. Manuali, A. Costantini, A. Laganà, E. Rossi, M. Carpenè, A. Ghiselli, M. Cecchi, *Lecture Notes Computer Science*, 2012, **7333**, 345–357.

5 A. Costantini, O. Gervasi, C. Manuali, N. Faginas Lago, S. Rampino, A. Laganà, *J. Grid Comp.*, 2012, **8**, 571–586.

6 C. Manuali, A. Laganà, *Future Generation Computer Systems*, 2011, **27**, 315–318.

Professor Bowman commented :The IR spectroscopy of *cis* and *trans* HOCO are still active research topics. We are in the process of developing *ab initio* potential and dipole moment surfaces that can accurately describe these isomers and the barrier separating them.

Professor Wodtke addressed Professor Laganà and Dr Glowacki :I'd like to return to Dr Glowacki's important question on importance sampling. His question as I understood it drove at the quite central question of applying dynamics methods to more complex problems, certainly a direction the entire community encourages and is striving for. We all realize that the traditional approach to theoretical dynamics involving electronic structure calculations and fitting becomes quite challenging as the number of degrees of freedom increases. It is perhaps unrealistic to think that the traditional approach can be applied to systems with significantly more than 7 dimensions, although various people are attempting to push this envelope. Nowhere is this question more important than in the study of molecular interactions at surfaces. Even modeling the dissociation of a diatomic molecule at a surface requires a 6D potential energy surface (PES) (unlike in the gas phase where it is a 1D problem). And even here, the "simplicity" of obtaining a 6D PES and fitting it is only possible when one completely neglects the motion of the atoms in the solid, a startling approximation (by the way often made) that removes the dynamic coupling of the molecule to the solid. For many of us, this is exactly the phenomena we would like to better understand. I would like to point out that there is an approach that is being employed that Dr Glowacki might be encouraged by; namely, "on the fly dynamics" sometimes called *ab initio* molecular dynamics. Here electronic structure calculations, usually employing DFT, are done as part of the classical trajectory calculation. Only points in geometry space relevant to the actual trajectory are calculated. Its greatest strength relates to the fact that one can carry out full dimensional calculations. For molecular collisions at surfaces this means one may include surface atom motion in an unbiased way. It remains to be seen if these methods provide realistic representation of the coupling of molecular motion to phonons, but it is clearly a promising approach. I am certain that many of the theoreticians here will raise criticisms about on the fly dynamics and there are many potential problems. One very practical one is that these are still very "heavy" calculations. Obtaining a statistically significant number of trajectories requires quite large computer power. But it seems to me that the clever minds with vision – like those of Dr Glowacki – might find ways to improve the practical implementation of these methods with potentially great future impact.

Dr Glowacki responded: Professor Wodtke is certainly correct that *ab initio* dynamics are a promising route forward, and I very much agree with much of what he says. However, for large systems – *e.g.*, occurring in biochemistry, at interfaces, or in solution – then *ab initio* dynamics will remain a significant challenge for a long time to come. However, what encourages me is that, for a range of chemical systems, it is not usually necessary to treat every part of the system uniformly, *i.e.*, at the same level of theory. This idea that the system can be treated using a hierarchy of approaches is very much what drives QM/MM type approaches, or the sort of work we have recently done to understand condensed phase reaction dynamics.[1-3] The point at which a simulation becomes "on-the-fly" is when its gradients and energies are calculated using *ab initio* approaches; however, for large systems, this can be overkill and simulations using classical force fields (often fitted to *ab initio* theory) can be very good.

Of course, fully automated dynamics at very high levels of theory would be great, but it seems to me that there is much insight to be gained from understanding and deriving general frameworks for identifying which parts of a system require very accurate treatments. This also saves on computational expense which ultimately cuts down on computer time, energy usage and CO_2 emissions.

So far as the present paper is concerned, it strikes me that it may be possible to perform large scale data analysis techniques on experimental dynamics data, and actually determine which portions of a potential energy surface are most critical to the reaction outcome, thereby helping us to understand how to optimize hierarchical approaches in theoretical treatments of chemical dynamics.

1 S. J. Greaves et al., Science, 2011, **331**, 1423–1426.
2 D. R. Glowacki et al., Nat. Chem., 2011, **3**, 850–855.
3 D. R. Glowacki et al., J. Chem. Phys., 2011, **134**, 214508.

Professor Laganà added: The central problem brought back to discussion in the comment of Professor Wodtke (from the question of Dr Glowacki) is concerned with his suggestion of dismissing the traditional approach to molecular dynamics involving first an extended electronic structure ab initio calculation and second a fitting of the calculated values using a multidimensional functional form to generate a suitable PES for the system considered in favor, especially for large systems (like systems including surfaces), of "on the fly" techniques.

A point to accord to this view is that while usual ab initio calculations of electronic energies are prevalently considered either as a means of interpreting structural molecular properties (and are therefore focused on the determination of molecular geometries close to equilibrium with limited distortion of the involved bonds) or as a means for rationalizing molecular processes in a transition state like fashion by focusing on the geometries associated with stationary points sitting on the minimum energy path(s), "on the fly" techniques are more democratically oriented towards all molecular geometries relevant to the process considered even if they are far away from stationary geometries. As a result, they can be considered as better tailored to suite the aims of dynamics studies and to provide a more comprehensive picture of the regions of the phase space relevant to the characterization of the evolution in time of the system under the conditions considered.

However, a negative score of the "on the fly" methods, as also admitted in Professor Wodtke's comment, is the huge waste of computer time. This is apparent for dynamical calculations of large systems which are carried out by integrating a statistically significant number of classical trajectories by repeating the evaluation of the potential energy (and its derivatives) at each time step for almost coincident (or even identical) molecular geometries without exploiting reusability or closeness to the already produced information. Such waste of time occurs also when carrying out quantum calculations which by occurring usually on a fixed predetermined grid of points can be evaluated at the beginning once for ever.

The separation of the evaluation of the interaction governing dynamical processes in a first step concerned with the calculation of an extended set electronic energies followed by a second step concerned with their fitting using a multidimensional functional form to generate a PES (together with the possible iteration of such sequence to better qualify the sample of geometries considered, raise the level of the ab initio treatment and compare with other available information in a multi-property analysis) is instead the true asset of the traditional approach. First of all, this allows the proposed PES to be checked for accuracy and continuity (both for the energy values and their derivatives) and eventually improved and integrated either by the authors themselves or by other researchers. This feature of the traditional approach, that bears a higher level of validation of the PES, is scientifically robust and is opposed to the complete transparency of the "on the fly" methods which may not spot occasional numerical inaccuracies and inconsistencies of the ab initio calculations.

Moreover, the generation of a PES as a separate step is fully in line with modern service oriented policies[1,2] which are a solid pillar of the virtual experiment reported in our paper and are an important component of the computational machinery of the grid empowered molecular simulator (GEMS)[3] used in our study. The GEMS computational machinery is, in fact, articulated in four distinct blocks (INTERACTION, FITTING, DYNAMICS and OBSERVABLES) with each of them having, as typical of services, clearly defined input and output suited for a quality (quality of service or QoS) control through a well a defined metric. The service oriented nature of GEMS does not only guarantee the delivery of the desired QoS but also fosters the formation of specific know how for all the blocks of the simulator and facilitates the design of a credit system through which a grid economy[4] is being built in the COMCPHEM virtual organization[5] of the European Grid Infrastructure (EGI).[6]

Such organization has allowed in the past the development of a variety of theoretical treatments for dealing with elementary chemical processes. We would like to point out here the enrichment of the FITTING block of GEMS using the so called Bond Order (BO) coordinates (more suited than internuclear distances for describing molecular interactions) and related BO polynomial formulations of the PES.[7-9]Using similar coordinates, a more sophisticated polynomial functional representation has been proposed in the literature.[10] Our plans are to apply, in collaboration with Professor Bowman and coworkers, the same functional form to other 4 atom systems like the $N_2 + N_2$ one for which a preliminary fit of the PES obtained using the computational machinery of ref 11 below has been already published.[12]

1 T. Erl, *Service-Oriented Architecture: Concepts, Technology, and A Design*, 2005, Prentice Hall PTR Upper Saddle River, NJ, USA, ISBN:0131858580.
2 C. Manuali, A. Laganà, *Future Generation Computer Systems*, 2011, **27**, 315–318.
3 A. Costantini, O. Gervasi, C. Manuali, N. Faginas Lago, S. Rampino, A. Laganà, *Journal of Grid Computing*, 2010, **8**, 571–576.
4 C. Manuali, A. Laganà, *Lecture Notes Computer Science*, 2011, **6784**, 397–411.
5 A. Laganà, A. Riganelli, O. Gervasi, *Lecture Notes Computer Science*, 2006, **3980**, 665–674.
6 European Grid Infrastructure, http://www.egi.eu/
7 E. Garcia, A. Laganà, *Mol. Phys.*, 1985, **56**, 621–627.
8 E. Garcia, A. Laganà, *Mol. Phys.*, 1985, **56**, 629–639.
9 E. Garcia, A. Laganà, *J. Chem. Phys.* 1995, **103**, 5410–5416.
10 J. M. Bowman, B. J. Braams, S. Carter, C. Chen, G. Czakó, B. Fu, X. Huang, E. Kamarchik, A. R. Sharma, B. C. Shepler, Y. Wang, and Z. Xie, *J. Phys. Chem. Lett.*, 2010, **1**, 1866–1874.
11 A. Aguado, C. Tablero, M. Paniagua, *Comput. Phys. Comm.*, 2001, **134**, 97.
12 M. Verdicchio, L. Pacifici, A. Laganà, *Lecture Notes Computer Science*, 2012, **7333**, 371–386.

Professor Bradforth opened the discussion of the paper by Professor Bernhardt by saying: These are very elegant experiments. To achieve a pump–probe measurement, you rely on the reproducibility of re-dosing the surface with 0.25 ML between optical delays. How reliably can you achieve the same dose?

Professor Bernhardt replied: Thank you very much for your encouraging comment. The re-dosing of the surface with exactly the same amount of adsorbate molecules for each delay time is indeed crucial to the pump–probe measurement, *i.e.* to the signal to noise ratio of the transient data. Therefore, the magnetic valve that is employed to admit a pulse of the molecules to the surface has been carefully calibrated by means of temperature programmed desorption spectroscopy.[1] After the first dosage the surface is irradiated with sufficient laser pulses to essentially dissociate and desorb all adsorbate molecules in the irradiated surface area. Subsequently, the laser beams are blocked again and the surface is covered with an identical amount of molecules as before. This procedure is of course only applicable if the substrate surface is not changed during transient recording procedure due to molecular fragments remaining on the surface, *i.e.* it works very reliably for weakly interacting substrates like the MgO surface employed in the presented experiments. A change of the substrate during the measurement has been observed, *e.g.*, for the methyl iodide photodissociation on a gold surface.

1 See also M. E. Vaida and T. M. Bernhardt, *Rev. Sci. Instrum.*, 2010, **81**, 104103.
2 M. E. Vaida, R. Tchitnga, and T. M. Bernhardt, *Beilstein J. Nanotechnol.*, 2011, **2**, 618.

Professor Orr-Ewing asked: You report rise times for formation of Br atoms and Br_2 molecules from photolysis of methyl bromide that I would expect to be dependent on the surface coverage. Have you looked for such a dependence? If the experimental data for Br and Br_2 formation shown in Fig. 5 in your paper were taken under the same conditions of CH_3Br coverage on your MgO surface, why does the Br_2 signal rise more quickly than that of the Br atoms? If collision induced surface reactions are leading to Br_2 formation, I would expect the rise of the Br_2 signal to be controlled by, and therefore slower than or the same as, the rate of production of Br atoms.

Professor Bernhardt answered: The investigation of the methyl bromide coverage dependent dynamics is a very important suggestion. In particular as the molecular adsorption geometry depends on the coverage and thus a distinct change in the dissociation and desorption dynamics with increasing coverage is to be expected. These experiments are currently in progress.

With respect to the transient evolution of Br and Br_2 (Fig. 5d and e in the paper), it has to be emphasized that the detection in these cases was non-resonant. Therefore, the unambiguous assignment of the observed temporal evolution is considerably more difficult than in the case of a resonant detection. Nevertheless, the fact that the signal of the bromine atoms starts to rise only after a few hundred femtoseconds and takes 1.5 ps to reach its maximum value would be in accordance with the CH_3Br adsorption structure with the C–Br axes almost parallel to the surface because then the Br atoms are not expelled directly away from the surface upon photodissociation but have to first collide with neighboring molecules. This could explain the delayed rise of the Br^+ signal.

The observation that the Br_2 molecules are detected more quickly that the Br atoms is indeed curious, but might be related to the non-resonant detection schema in these experiments which apparently causes the Br_2 molecules to be detected as soon as they are formed (*i.e.* potentially before desorption from the surface). However, as stated in the paper, this interpretation is only speculative and the elucidation of the detailed molecular reaction mechanism will require additional power and wavelength dependent measurements, which are in progress.

Professor Wodtke addressed Professor Bernhardt: I am concerned that the assignment of the time profiles measured in your work might be difficult without additional data. Professor Orr-Ewing pointed out for example the curiosity that Br_2 seems to appear before Br in your study of CH_3Br on MgO. I wonder if you could obtain ion imaging data as a function of delay time, if this might not be a much more tractable problem. Could you comment on whether such experiments are possible and what problems one must overcome to realize this suggestion?

Professor Bernhardt responded: We completely agree that the assignment of the time profiles poses some difficulties and that additional information is essential to their interpretation. Therefore, whenever possible, we tried to resonantly ionize the molecular surface reaction products. In this case it is assured that the products are detected as soon as they are liberated from the force field of the surface and the neighboring molecules. In this respect, the laser power and wavelength dependence measurements are crucial to the assignment of the photoexcitation and detection mechanisms. Yet, additional experimental information would be highly desirable and the suggestion of ion imaging detection is highly appreciated. We do not have the capability so far to obtain such data, but this technique should be readily applicable to our surface aligned adsorbate systems. As a first step in this direction, we were able to resolve initial kinetic energy differences of the different methyl cations released during methyl iodide photodissociation on MgO (peak at 130 fs and exponential rise at later times in Fig. 3b in the paper) by recording the detailed time profile of the time-of-flight mass peak of the methyl signal.[1]

1 See M. E. Vaida, T. M. Bernhardt, *ChemPhysChem*, 2010, **11**, 804.

Professor Wester asked: Is it possible to perform vibrational state-selected REMPI of the methyl radical as it leaves the surface? What do you know about the arrangement and the spacing of the molecules on the surface in the combined CH_3I/CH_3Br study? Is it possible to differentiate direct IBr formation and association following stochastic migration of the I and Br atoms on the surface?

Professor Bernhardt replied: As apparent from Fig. 3 below, we are currently simultaneously exciting all vibrational states of the emerging methyl radicals with our probe laser pulses. However, by slightly shifting the probe laser wavelength it should be possible to selectively detect these states. The challenge, of course, is the remaining signal intensity, but we are definitely planning to perform these experiments. Concerning the arrangement and the spacing of the molecules on the surface in the combined CH_3I/CH_3Br study we measured the temperature programmed desorption (TPD) spectra which appear to exhibit a sum signal of the separated molecular signals. No further information is available so far. We are currently investigating the TPD spectra and transient photoexcitation data of the coadsorption layers as a function of the molecular coverages to obtain further insight.

With respect to stochastic migration of photodissociation products on the surface we tend to assume that this process will rather occur on the ns time scale which is (while in principle possible) difficult to observe in the present fs-resolution setup. Therefore we think that, in the transient data of the bimolecular surface reactions, we are observing the direct formation and association of the halogen molecules *via* a favorable surface aligned geometry.

1 J. W. Hudgens, T. G. DiGiuseppe and M. C. Lin, *J. Chem. Phys.*, 1983, **79**, 571.

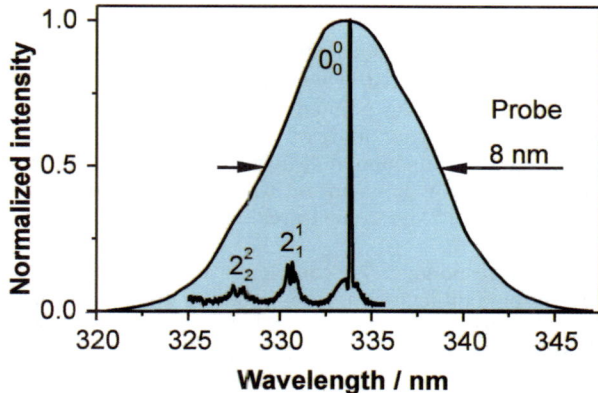

Fig. 3 Two photon resonant ionization spectrum of CD_3 radicals[1] displaying the three vibronic transitions in the probe wavelength region and measured spectrum of the probe laser beam (highlighted in blue).

Professor Bradforth questioned: You report results for metal oxide surfaces and in the photodissociation step infer that the methyl group immediately bounces off the surface. What are the prospects for looking at similar photodissociation induced reactions for softer surfaces? For example could you measure exchange of energy and transient trapping for the methyl radical fragment on the surface?

Professor Bernhardt responded: This is a very interesting suggestion and the investigation of comparable molecular photodissociation reactions on different substrate surfaces is very important. So far we employed MgO thin films[,1] metal substrates,[2] as well as oxide supported metal nano-particles[3] . To specifically address the exchange of energy and transient trapping for the methyl radical fragment on the surface we performed experiments with magnesia thin films on which intentionally a significant amount of surface defect centers (steps, corners, and vacancies) was created. The stronger interaction of the adsorbed molecules with these defect centers had a clear effect on the possible photoexcitation pathways and the resulting photodissociation dynamics.[4]

1 M. E. Vaida and T. M. Bernhardt, *ChemPhysChem*, 2010, **11**, 804.
2 M. E. Vaida, R. Tchitnga and T. M. Bernhardt, *Beilstein J. Nanotechnol.*, 2011, **2**, 618.
3 M. E. Vaida, T. M. Bernhardt, C. Barth, F. Esch, U. Heiz, and U. Landman *Phys. Status Solidi B*, 2010, **247**, 1139.
4 M. E. Vaida, T. Gleitsmann, R. Tchitnga and T. M. Bernhardt, *Phys. Status Solidi B*, 2010, **247**, 1001.

Professor Martinez commented: Professor Wodtke has pointed out the utility of *ab initio* molecular dynamics (AIMD, also known as direct dynamics) approaches where the PESs are solved simultaneously with the integration of the molecular dynamics equations (quantum or classical). In fact, our contribution discussed earlier uses such AIMD approaches while including quantum mechanical effects such as nonadiabatic transitions and a basis set expansion of the nuclear wavefunction.

An often under appreciated but nevertheless important point is that AIMD can be computationally cheaper than the more conventional approach of fitting PESs to *ab initio* data. First, we distinguish two types of error – error in the PESs (interpolation/fitting error) and the statistical error associated with using a finite number of trajectories or nuclear basis functions to describe the dynamics (sampling error). In many potential fitting schemes, e.g. Shepard interpolation,[1,2,3] the fitting error at a given molecular configuration can be related to the distance from the molecular configuration to the nearest data point (where *ab initio* data such as energies and gradients have been calculated explicitly). Thus, the error in the PES at a given molecular configuration will decrease polynomially with the number of data points, N_{data}, as $N_{data}^{-m/d}$, where *m* is a constant related to the order of local polynomial expansions around the data points (in Shepard interpolation based on second-order expansions, *m* is 3) and *d* is the number of degrees of freedom. In contrast, the statistical sampling error is expected to decrease with the number of trajectories or basis functions, N_{traj}, as $N_{traj}^{-1/2}$. When the PES has been fitted, the error can be expected to be dominated by the fitting error.[4] In AIMD, the fitting error vanishes identically and the error in the dynamical results is entirely sampling error. The important point is that the sampling error decreases at the same rate regardless of molecular size, but the fitting error depends on the number of degrees of freedom. Thus, as the molecules under consideration get larger, it is inevitable that AIMD will be the more efficient approach. To make this more concrete, the computational effort will quadruple in order to halve the sampling error in AIMD, but it will grow as $2^{d/m}$ in order to halve the fitting error in an interpolation–fitting approach. For a molecule with 10 atoms and assuming $m=3$, this translates to an astonishing 512-fold increase in the computational effort! And the situation of course becomes much worse very quickly – for a molecule with 50 atoms, the cost to decrease the fitting error by a factor of two is a factor of 1014! In practice, the number of degrees of freedom d may be replaced with a smaller effective number of degrees of freedom if the PES is nearly separable in the coordinate system used to represent the molecular configurations (this is one of the reasons why the choice of coordinate system is so important in fitting/interpolation). This does not change the essence or veracity of the argument – rather, it only changes the associated prefactors (which we do not attempt to estimate here). We have recently demonstrated empirically that this scaling analysis holds as expected for the ethylene molecule.[3]

Of course, apart from the above considerations, there is a further advantage for AIMD in that there is no "startup cost" for each new molecular system, *i.e.* one does not need to fit a new set of PESs for each new molecule of interest. The arguments sketched here are scaling arguments, and the specific merits of a particular approach for a particular system will depend significantly on the prefactors. Thus, one should not be surprised that potential fitting approaches may sometimes be found preferable for small molecular systems (e.g., less than 10 atoms). However, in the long run (as computational power increases and as AIMD methods become more advanced) and/or for larger molecular systems, it should be expected that the fitting/interpolation approach will become less favorable than AIMD approaches.

1 J. Ischtwan and M. A. Collins, *J. Chem. Phys.*, 1994, **100** 8080.
2 M. J. T. Jordan, K. C. Thompson, and M. A. Collins, *J. Chem. Phys.*, 1995, **102** 5647.
3 C. Evenhuis and T. J. Martínez, *J. Chem. Phys.*, 2011, **135**, 224110.
4 Here, we presume that it is very inexpensive to calculate dynamics on a fitted PES, although of course there can be exceptions such as grid-based quantum wavepacket propagation for large molecules.

Professor Bowman added: As a major developer of potential energy surfaces, I use direct dynamics at a variety of total energies and initial configurations to develop the database for the potential fitting. Of course, it is helpful if some stationary points, minima and saddle points, especially, are known in advance so trajectories can be initiated from them.

Before making some general comments about doing direct dynamics *versus* dynamics with a fitted PES, it might be useful to see a few numbers for a moderately complex unimolecular photochemical dissociation of CH_3CHO, where the goal is to determine the product branching ratio. From the dynamics point of view, this is a relatively simple process, compared to the analogous bimolecular scattering event, $O + C_2H_4$, because a single value of the total angular momentum is often sufficient for the unimolecular case. So, the number of trajectories needed to get well-converged branching ratios is about a factor of ten smaller than needed for the bimolecular case. Using a fairly high level of electronic structure theory, and for example, just for the ground electronic state, the following are timings for a single energy calculation on a node of a modern cluster – CCSD(T)/aug-cc-pVTZ: 25 mins; MRCI(Q)/VTZ: 10 mins. To get product branching ratios, we need to run trajectories for roughly 100 ps, which corresponds to roughly 1×10^6 steps. So in a direct dynamics calculation this typical trajectory would require 1×10^6 electronic energies and gradients, requiring roughly 10^7–10^8 minutes. We typically run at least 10^4 trajectories to get reasonable sampling of this initial phase space. So, multiplying these two numbers we get the absurdly large range 10^{11}–10^{12} minutes (ca. 2 million years) for the direct dynamics calculations. Even if we imagine using 1000 10-core nodes instead of the single node-single core the time for this calculation would be 20 years. A PES for this electronic state was developed using roughly 150 000 CCSD(T)/aug-cc-pVTZ and (some) MRCI(Q)/VTZ and trajectories were run as described above. The point of the above exercise clearly reminds of the desirability of have a fitted PES.

No matter what the application, I think we all would gladly use a PES instead of running direct dynamics. The comment by Professor Martinez concerns the scaling of effort in getting fitted PESs, as the number of atoms increases. We currently develop fitting global potentials for up to ten atoms and that is just where our fitting library stops. There are no inherent problems in increasing that number by some modest factor in the next 5–10 years. The approach we use to fit high-dimensional PES absolutely does not rely on direct-product grids, as this requires exponentially increasing amounts of electronic energies. Instead we use "scattered" data (which come naturally from direct dynamics in fact) and simple linear least squares fitting, but using a fitting basis that is invariant with respect to the permutations of all like atoms.[1-4] For larger systems, and of course depending on the problem of interest, there are several fitting strategies that effectively deal with the "curse of dimensionality". One is the n-mode representation of the potential, which partitions the full-dimensional PES into a number of lower dimensional ones. This approach is widely used in vibrational dynamics[5] and has been successfully applied to systems with tens of atoms. There are generalizations of this method that are also quite promising.

My prediction about the future of direct dynamics and fitted PESs is that they will both be done and in fact there will be a close alignment of the two. For example, here is one way I can imagine the synergy between the methods, where "machine learning" will play a major role. Direct-dynamics will be done along with "on-the-fly" fitting. The local fitting (still permutationally invariant) will be used until the first failure of the fit is encountered, in which case direct-dynamics is resumed, and a new fit (either more global or local) is done, etc. In this way, with enough direct-dynamics a PES (or more likely a sequence of PESs) will naturally evolve.

Finally, I would like to propose that the permutational invariance of the PES be incorporated into direct-dynamics codes. This could effectively address the inefficiency of direct-dynamics calculations that "re-visit" the same or permutationally equivalent configuration space.

1 B. J. Braams and J. M. Bowman, *Int. Rev. Phys. Chem.*, 2009, **28**, 577–606.
2 Z. Xie and J. M. Bowman, *J. Chem. Theory Comput.*, 2010, **6**, 26–34.
3 J. M. Bowman, G. Czakó, and B. Fu, *Phys. Chem. Chem. Phys.*, 2011, **13**, 8094–8111.
4 J. M. Bowman, B. J. Braams, S. Carter, C. Chen, G. Czakó, B. Fu, X. Huang, E. Kamarchik, A. R. Sharma, B. C. Shepler, Y. Wang, and Z. Xie, *J. Phys. Chem. Lett.*, 2010, **1**, 1866–1874.
5 S. Carter, S. J. Culik and J. M. Bowman, *J. Chem. Phys.*, 1997, **107**, 10458.

Professor Truhlar communicated: In considering the question of fitting analytic surfaces *vs.* direct dynamics, I think it is important to keep in mind the use of generic parameters, combined approaches, and the rapid advances in density functional theory.

Molecular mechanics may be considered to be an example of fitting surfaces with generic rather than system-specific parameters. When one treats complex systems, these generic parameters are often useful for many of the degrees of freedom, and it makes practical sense to take advantage of them by employing combined quantum mechanical molecular mechanical (QM/MM) methods. This is a particularly efficient form of direct dynamics.

Density functional theory (DFT) has made great strides in the past twenty years, and we can anticipate further progress in the near future. Thus, more and more attention will be turning to direct dynamics algorithms that are specialized for using DFT. An example of where DFT has largely displaced wave function theory is the treatment of complex systems involving transition metals.[1]

Another way to combine theories is to use high-level wave function methods (when affordable) to validate DFT and then use DFT for the direct dynamics. A refinement on this approach is to make small changes in the DFT parameters to achieve even better agreement with one's best estimates; this is an example of using specific reaction parameters (SRPs). We began employing this technique with DFT many years ago[2] and with other electronic structure theories even earlier.[3] It provides an example of "surface fitting" that does not involve analytic potential energy functions. I think that progress in such combined methods will be an increasingly important strategy in the future.

1 C. J. Cramer and D. G. Truhlar, *Phys. Chem. Chem. Phys.*, 2009, **11**, 10757–10816.
2 J. Pu and D. G. Truhlar, *J. Chem. Phys.*, 2002, **117**, 10675–10687.
3 A. González-Lafont, T. N. Truong, and D. G. Truhlar, *J. Phys. Chem.*, 1991, **95**, 4618–4627.

Professor Orr-Ewing returned to discussion of the paper by Professor Neumark by asking: The reaction of Cl atoms with methane to form HCl and CH_3 has been the subject of considerable prior study, as highlighted by Professor Crim in his introductory lecture. Professor Bowman has recently reported a newly developed potential energy surface (PES) and his quasi-classical trajectory calculations demonstrate quantitative agreement with experimental measurements of the asymptotic reactive scattering.[1] Professor Liu and others have reported exquisitely detailed studies of quantum state resolved reactive scattering and signatures of resonances. However, there remain some questions that are not fully resolved, such as the nature of any non-adiabatic dynamics on the PESs that correlate with $Cl(^2P_{3/2})$ and $Cl(^2P_{1/2})$ spin–orbit states[2-5]. You have previously published a SEVI study of the electron detachment from $ClCH_4^-$ anions[6] analogous to the study of FCH_4^- reported in the current paper. Do you plan to investigate this process further and at higher resolution, perhaps using the pre-cooling of trapped ions, to try to resolve some of the remaining questions about the Cl + methane reaction dynamics?

1 G. Czako and J. M. Bowman, *Science* 2011, **334**, 343.
2 B. Retail, J. K. Pearce, C. Murray and A. J. Orr-Ewing, *J. Chem. Phys.*, 2005, **122**, 101101.
3 B. Retail, S. J. Greaves, J. K. Pearce, R. A. Rose and A. J. Orr-Ewing, *Phys. Chem. Chem. Phys.*, 2007, **9**, 3261.
4 B. Retail, J. K. Pearce, S. J. Greaves, R. A. Rose and A. J. Orr-Ewing, *J. Chem. Phys.*, 2008, **128**, 184303.
5 S. M. Remmert, S. T. Banks, J. N. Harvey, A. J. Orr-Ewing and D. C. Clary, *J. Chem. Phys.*, 2011, **134**, 204311.
6 D. M. Neumark, *J. Phys. Chem. A*, 2008, **112**, 13287.

Professor Neumark replied: We certainly plan to look at this system again. The published $ClCH_4^-$ SEVI spectrum that you refer to could certainly benefit from comparison to quantum simulations, and we are hoping that Professor Bowman or another theorist will be able to carry out these calculations in the near future.

Dr Skouteris addressed Professor Neumark: It can be shown from symmetry arguments on the T_d group that, denoting by I the total nuclear spin quatum number of the CH_4 molecule, the $I = 2$ state corresponds to states of A symmetry (including $j = 0$), the $I = 1$ state corresponds to T symmetry (including $j = 1$) and the $I = 0$ state corresponds to states of E symmetry (including one component of $j = 2$), where j is the CH_4 rotational quantum number. Introducing some I-specificity in your experiment (in an analogous way to ortho/para H_2) would presumably go some way towards decongesting the final spectrum. Could you comment on the feasibility of such an approach?

Professor Neumark communicated in reply: I am not aware of successful attempts to produce methane in a single hyperfine spin state. I suspect it is much harder to do this for methane than for H_2, given that the rotational constant for methane is much smaller (5 cm^{-1} *vs.* 60 cm^{-1}).

Dr Chandler opened the discussion of the paper by Professor Gruebele by asking: Martin (Professor Gruebele), can you discuss what you mean by dynamics *versus* kinetics when discussing protein folding. Also are there other types of measurements such as multi-coloured FRET experiments that could offer information on the speed at which proteins move and fold?

Professor Gruebele replied: In protein folding (as in small molecule unimolecular reactions), the distinction I make between kinetics and dynamics measurements is whether only populations of long-lived species are resolved, or whether information at the state level is resolved (*e.g.* time-resolving highly transient species like transition states, or angularly resolving products to infer how internal energy reorganized during the reaction, or FRET measurements you suggest measure site-specific contact rates). Of course, in both cases, there is gray territory: in an ion-molecule reaction, measuring the product formation rate may be tantamount to resolving the actual reaction step, if the energy barrier is negligible. Likewise in proteins that are very stable, measuring the reaction rate may be tantamount to resolving the actual barrier crossing if the barrier is very small.

As chemical physicists, we are ever hungry for dynamical information in addition to kinetics. We want to know what the contacts are that are formed during reaction, and whether they form in a specific order or their order is statistically distributed; we want to time-resolve the actual reaction itself, whether it is breaking of a bond, or formation of the critical contact that allows a protein to cross over its free energy barrier.

In one sense, dynamics measurements for proteins are harder than for small molecules: there are so many more degrees of freedom to characterize. In another sense, they can be easier to do: the FRET labeling you suggest is much less disruptive for a 100,000 atomic mass unit protein in a high-friction solvent than for a small molecule. So there is hope that we can understand protein folding reactions in much more detail than just a rate coefficient connecting a native structure with an initial unfolded structural ensemble. Measuring the time scale of activated populations crossing the barrier, separately from the slower time scale with which the native state is populated, as we were able to do by crowding the protein lambda repressor, is a baby step in that direction.

Professor Nesbitt commented: Professor Polanyi has told us of the gospel of dynamics and the corresponding "good news" that chemical physics concepts have historically brought to the broader scientific community. Your paper brings to mind another excellent example of this – how transition state theory, which was developed for rather simple molecular systems, can be equally powerful in describing multidimensional conformational changes in complex biomolecules. Indeed, biochemists even borrow the language of "late" and "early" barriers in nucleic acid and polypeptide conformational processes when the describing transition states that are more product-like (folded) or reactant-like (unfolded), respectively. If there is any major difference between TST in these two fields, it seems to be in our ability (or perhaps inability) to characterize what a transition state actually looks like in such highly dimensional systems. Specifically, I am wondering if your trajectory calculations give us insight into what general characteristics of the transition state region might look like for these fast folding proteins?

Professor Gruebele responded: We and other groups have done molecular dynamics simulations, to at least see what classical trajectories of proteins look like as they go through the transition state. What was noted by Professor Truhlar for enzymes is also true for folding: the transition states show structural and dynamical heterogeneity. A good example would be the folding of the small beta sheet protein WW domain, so named because its natural form contains two tryptophan (or "W") residues. This protein is a key player in cell signalling, including cell death through apoptosis. Experimental as well as computational analysis (*e.g.* ref. 1) showed that while folding of this protein is mainly rate-limited by formation of "loop 1" connecting beta strands 1 and 2, occasionally it can also fold by forming "loop 2" that connects beta strands 2 and 3. In fact, the two paths communicate with one another: loop 2 can interact with the region that will form loop 1, slowing down the folding reaction. Based on a simulation of the transition state ensemble, we made a mutation in loop 2 to reduce this interaction – and the experimental folding rate of WW domain we measured sped up by a factor of three, very close to the result predicted by simulation.

It seems that the reason we can get away with simple 1D pictures at all, is that different pathways on the free energy landscape differ enough in free energy for Boltzmann factors to hide higher-lying pathways. For example, if the loop 2 path is just 1kT higher in energy, it contributes only ~40% to the data. At 3kT, its contribution is down to 5%. The signal-to-noise ratio of biomolecular kinetics experiments is often no better than 20 : 1, so these minority pathways remain hidden, even though they are not very high up in free energy.

The hidden pathways could be important from the standpoint of protein evolution. For example, a higher energy pathway could be associated with a new function. If a mutation blocks the previously lowest pathway, new function could emerge. Higher pathways could also make the molecule more robust to mutations: a mutation that closes on pathway towards the folded state could leave open another one; the protein may fold less efficiently (more slowly, lower stability), but at least it still folds.

In my mind, these hidden higher energy states of biomolecules are the primary motivation for studying dynamics. I believe that a lot of the interesting biology arises from the rarer excitation of higher energy pathways and states, and a lot of robustness is conferred to biological function and evolution by redundant pathways. Of course, you do not want to have too many of them, either: then their entropic weight would overwhelm the native state. It appears that biomolecules navigate in that subtle free energy range where the native state is just barely stable enough, whereas excited states confer structural flexibility indispensable for function.

1 S. Piano et al., J. Mol. Biol. 2011, **405**, 43–49.

Professor Truhlar communicated: You raised two very interesting issues in the discussion, namely (1) the question of whether enzyme kinetics is controlled by a single saddle point or a range of saddle points and (2) the advantage of having data on fast transient phases as opposed to "only" kinetics data. I would like to add to the discussion of both issues.

The question of multiple transition states for enzyme kinetics should, I believe, be considered to be settled in favor of there being a range of structures. Even for the very "simple" reaction considered in my paper, there are multiple conformers of the transition state. In enzyme kinetics, some key degrees of freedom may be tied down by strong interactions of the substrate with the enzyme (docking), but there are many degrees of freedom available. In ensemble-averaged variational transition state theory,[1,2] we first find an ensemble corresponding to the maximum generalized free energy of activation as a function of a chemical reaction coordinate. Then we sample this ensemble and optimize ensemble members to the nearest transition state; we find a large number of transition states (so large that we do not by any means find them all), many with quite different properties.[3] We conclude that ensemble averaging must be included in any quantitative modeling of enzyme kinetics.

Regarding the second question, I simply want to point out an advantage of conventional kinetics data, namely that is very well defined and is, by definition, essentially independent of initial state (since it refers to the quasisteady time period after the induction interval[4]). Data referring to fast transient phases may, in contrast, depend sensitively on the preparation of the system.

1 C. Alhambra, J. Corchado, M. L. Sánchez, M. Garcia-Viloca, J. Gao, and D. G. Truhlar, J. Phys. Chem. B, 2001, **105**, 11326–11340.

2 D. G. Truhlar, J. Gao, M. Garcia-Viloca, C. Alhambra, J. Corchado, M. L. Sanchez, and T. D. Poulsen, *Int. J. Quantum Chem.*, 2004, **100**, 1136–1152.
3 J. Pu, J. Gao, and D. G. Truhlar, *Chem. Rev.*, 2006, **106**, 3140–3169.
4 C. Lim and D. G. Truhlar, *J. Chem. Phys.*, 1983, **79**, 3296–3306.

Professor Gruebele communicated in reply: You raise two excellent points. Regarding transition state heterogeneity, I would say that experimental tests are woefully behind the computational models. I think it is very clear from your work that enzymatic transition states can be quite heterogeneous, in the sense that many different conformations of the active site with different binding abilities and rates exist. Some hints of how protein environment effects enzyme action have been seen by single molecule experiments from Sunney Xie's group,[1] but what is missing from experiments is structural or 'state' level resolution. In addition to the need for careful ensemble averaging that you mention, necessary to get accurate average rates, I want to highlight also the deviations about the average: in many biological signalling and reaction events, it is the 'deviant' very slow or very fast events that can play a role, for example when allostery (parts of one protein interacting with another to affect function) occurs on a time scale between the slowest and fastest events. Experiments will have to be developed to correlate rate with structure, and to look at the rate outliers, which may have biological relevance. Alternatively, kinetics experiments and atomistic simulations (*e.g.* coupled quantum-classical as you discuss) will have to be closely coupled, so kinetics can calibrate simulation, and simulation can provide the rich dynamical detail that is difficult to obtain from experiments.

Regarding the second point, kinetic and dynamical measurements have their advantages and disadvantages. As you say, kinetics are reliable once the temperature (and other thermodynamic parameters) of the system has been defined, and otherwise do not depend on preparation. Dynamical experiments of course depend on the preparation of the initial state, since they provide state-resolved information at some level. This is definitely true of our data, *i.e.* the fast barrier-crossing phase we have observed in some proteins. That sensitivity can be used to obtain more information about the system, by tuning the initial state. In small molecule experiments, we might adjust the angle or center of mass collision energy of a beam, or the vibrational state prepared in a molecule that is about to collide with a surface. Even for proteins, we can do such tuning, although our tools are currently much cruder than for small molecules. For example, we showed for a different protein than the one I discussed (WW domain, ref. 2) that the barrier crossing time depends on the initial condition of the reaction, when the reaction is carried out at the same final temperature. This additional complexity can provide additional information on how different initial state distributions reach the transition state. So I would say: always do your kinetics first, and make sure the overall features of the reaction are understood; then poke 'under the hood' of the system, and see if the elementary reaction events can be understood in more detail, by tuning initial conditions and watch how the dynamics of the system responds. With proteins, we are just getting to the point where we dare poke around.

1 B. P. English *et al.*, *Nature Chem. Biol.*, 2006, **2**, 87–94.
2 F. Liu, M. Nakaema and M. Gruebele, *J. Chem. Phys.*, 2009, **131**, 195101.

Professor Nesbitt commented: Given the enormous dynamic range of fields that are being considered and discussed here at this meeting, it is perhaps worth noting that there is also an enormous intellectual distance between our (potential) energy surfaces and the 1D pictures of free energy surfaces you show for protein folding. Even to get from an energy landscape to an enthalpy landscape requires calculation of PdV "work" terms, and to get further to a free energy landscape involves calculation of even more subtle TdS entropic contributions. Thus, the source of any free energy barrier is always a fight between enthalpy and entropy contributions. For RNA and DNA folding, these transition state barriers are often entropic in origin, with relatively little enthalpic change in the entrance channel. I was wondering if you could comment on the relative enthalpic *vs.* entropic contributions for your fast folding proteins to i) the transition state barrier and ii) overall folding thermodynamics. Is there a consistent enthalpic *vs.* entropic trend in these contributions for protein folding in general?

Professor Gruebele communicated in reply: The fight is on at all levels! Even for small molecule reactions, we have to worry about such things as vibrational frequencies of modes orthogonal to the reaction coordinate changing along the reaction coordinate, when the full potential energy surface cannot be included in the calculation – although it now can be, for systems with a few atoms. In proteins and RNA, this problem of reaction coordinate *versus* other degrees of freedom is exacerbated many times over. One solution involves looking at multiple probes, for example site-specific FRET pairs as suggested by Professor Chandler. Indeed, experiments have shown that when the reaction rate between two states is slow, different probes yield the same rate; but when the rate is fast, different probes can yield very different results, see ref 1 for example. Presumably on the faster time scales, many types of paths are available to lead from reactant to product, and really a >1D free energy surface should be used to describe the system. An example where this has been attempted is for the small helix bundle protein alpha-3D (ironically with "3D" in its name): the infrared and fluorescence spectroscopy data simply cannot be accounted for by a 1D free energy surface. It will be very important for biomolecule experiments to monitor more reaction coordinates by monitoring more spectroscopic probes: the prevalence of 1D pictures in the literature is largely a consequence of "good enough to rationalize the data we have." With richer dynamics results, higher dimensional surfaces will be needed, and at least at the classical simulation level, we do have mechanisms (such as hidden Markov modeling) that can make the connection between the full-dimensional dynamics, and the low-dimensional kinetics of such systems.

In most of these cases, what you suggest is exactly what happens: there is relatively little change in contact enthalpy when proteins are modified to become very efficient (fast) folders, so the entropic penalty of conformations along the reaction coordinate must have decreased to speed up the reaction. That is why it is so important to consider a free energy surface when reducing the dimensionality of the system. To make matters worse, the entropic contributions can come from flexibility of side-chains in the protein, but also largely from the dynamics of solvent (water) molecules that interact with the protein. I would estimate that about half the folding free energy of a protein could be attributed to water molecules, if the energy were partitioned approximately among solvent and reactant degrees of freedom. After all, the typical folding free energy of a protein is quite comparable to that of just 2 hydrogen bonds among water molecules. For a protein of about 100 residues, the enthalpic and entropic contributions to native state stability are on the order of 40 kJ/mole at room temperature, and cancel at the protein's melting point (usually ~40–100 °C) because the entropic contributions win out at higher temperature.

RNA has a much smaller alphabet than proteins (4 *vs.* 20 residues), and the cancellation of enthalpic and entropic effects is not as good along the reaction coordinate. This can be advantageous from the point of view of studying dynamics because the contributions can be separated better. The nucleic acid systems can also be tuned more independently, for example by varying magnesium ion concentration. For this reason, it is somewhat surprising that so much less work has been done on nucleic acids than on proteins. Part of the reason must be a biological bias: until the 1980s, nucleic acids were considered relatively 'boring' molecules, with simple and few functions and structures. Only since the 1980s has it become clear that nucleic acids may lie at the origin of life, and have as rich an inventory of functions as proteins. I suspect this recognition of the rightful place of nucleic acids will also lead to more kinetics and dynamics measurements on these molecules.

This journal is © The Royal Society of Chemistry 2012

1 F. Liu *et al.*, *J. Chem. Phys.*, 2009, **130**, 061101.

Professor Gruebele opened the discussion of the paper by Professor Huse by asking: Given the newest experimental results, as well as DFT calculations, how well does the classic concept of pi-back bonding hold up as a qualitative explanation of trends in the ultrafast time-resolved X-ray absorption results for these iron complexes? If it fails, is there another simple idea to replace it?

Professor Huse communicated in reply: In collaboration with Munira Khalil's group we are trying to understand ultrafast spin cross-over and the nature of the valence charge distribution by simulating nitrogen K-edge spectra of aqueous $[Fe(bpy)_3]^{2+}$ in the ground state and the quintet ligand-field state which is the excited state populated in light-induced excited spin-state trapping. Preliminary calculations suggest that the populated iron $3d$-t_{2g} orbitals are largely localized on the metal center in the ground state while in the quintet state, the formally anti-bonding half-filled iron $3d$-e_g orbitals display mixed metal–ligand character with orbital populations equally stemming from ligand nitrogen $2p$ and iron $3d$ charge density. These *ab initio* results deviate substantially from our previous findings[1,2] and suggest that back-bonding and donation are not necessarily adequate concepts for describing the valence charge distribution in these systems. Whether this has any general validity would be audacious to say at this point.

1 N. Huse, T. K. Kim, L. Jamula, J. K. McCusker, F. M. de Groot and R. W. Schoenlein, *J. Am. Chem. Soc.*, 2010, **132**, 6809.
2 N. Huse, H. Cho, K. Hong, L. Jamula, F. M. de Groot, T. K. Kim, J. K. McCusker and R. W. Schoenlein, *J. Phys. Chem. Lett.*, 2011, **2**, 880.

Professor Bradforth questioned: The reaction systems that have been studied by time-resolved X-ray probing are mostly transition metal systems where excited state electronic structure calculations, particularly *ab initio* molecular dynamics that can address the curve crossing dynamics, are particularly difficult. Have any excited state theoretical calculations on the iron coordination compounds systems studied in your work been carried out? Your work also suggests that rearrangement of the ligands, for example by nitrogen K-edge, might be followed. What are the prospects for X-ray probing of photo-initiated reactions of compounds made up entirely of light elements, particular organic systems where this community has much greater experience?

Professor Huse communicated in reply: The calculations that relate directly to the polypyridyl iron(II) complexes studied were published by Coen de Graaf and Carmen Sousa.[1,2] The work employed the CASSCF/CASPT2 method as implemented in the MOLCAS 7.2 code to determine the variation of the energy of the electronic states relevant to ultrafast spin crossover as a function of the Fe-ligand distance in $[Fe(bpy)_3]^{2+}$. The work suggests that the potential energy of singlet and triplet metal-to-ligand charge-transfer (MLCT) states is minimal at a similar Fe-ligand distance as the one of the ground state. With reference to theoretical work on a different Fe(II) compound,[3-4] the possibility of spin crossover *via* a quintet MLCT state is discussed to explain the very fast time-scale of ligand-field quintet state formation. Although this scenario as calculated by de Graaf and Sousa seems unlikely due to energetics (the ^{5}MLCT state of $[Fe(bpy)_3]^{2+}$ is calculated to be 0.7eV higher than the initially excited ^{1}MLCT state), coupling to additional vibrational modes may lead to near degenerate MLCT states which could provide an explanation involving the ^{5}MLCT state after all.

Concerning the second part of the question, we have experimentally succeeded in recording transient nitrogen K-edge spectra of $[Fe(bpy)_3]^{2+}$ in aqueous solution as mentioned in the reply to Professor Gruebele, and we are in the process of understanding these spectra. Such an experiment could also be performed on solvated molecules entirely composed of light elements (such as H, C, N, and O). The severe restriction is the strong absorption of solvents containing carbon and nitrogen. Hence, transient nitrogen K-edge spectroscopy in aqueous solutions has been successfully performed by us (though not yet published) but nitrogen K-edge spectroscopy of solutions with other solvents than water would be difficult yet not impossible for high-concentration samples. The complementarity of such measurements would certainly be very beneficial in understanding the photochemistry of many organic systems.

1 Coen de Graaf and Carmen Sousa, *Chem. Eur. J.* 2010, **16**, 4550.
2 Coen de Graaf and Carmen Sousa, *Int. J. Quant. Chem.* **111**, 3385.
3 Michel van Veenendaal, Jun Chang, and A. J. Fedro, *Phys. Rev. Lett.,* **104**, 067401.
4 Jun Chang, A. J. Fedro, and Michel van Veenendaal, *Phys. Rev. B*, **82**, 075124.

Miss Harris asked: Professor Huse, you mention at the end of your paper that you believe that within the foreseeable future the technology required to make the experiments you have outlined here accessible within a laboratory setting should become available. Could you comment of the "state of the art" of this technology as it stands currently and outline the further advances still required as well as the time frame in which you expect them to occur? Could you also comment on the types of system/molecular species that you feel this technology could most usefully be employed to study?

Professor Huse answered: High-harmonic generation typically produces photons up to ~100eV using 800nm Ti : Sapphire technology. The cutoff at high photon energies relates to the driving laser wavelength *via* the ponderomotive potential and scales with the square of the wavelength of the driving laser. Therefore, several groups are currently devising powerful laser systems that use longer wavelengths to extend the laser-based X-ray generation to above 1keV. Published results have so far reached the 1keV limit[1] and I expect laser-based systems producing in excess of 10^8 photons s^{-1} in 1% bandwidth to be developed within the next three years. This would allow for many synchrotron experiments in the soft X-ray range to be performed in a transient manner with at least 30fs time-resolution on essentially a table-top size experiment. Alternatively, compact laser wake-field accelerators have been demonstrated to generate 1GeV electrons which when injected into an undulator could generate short X-rays as well.[2] Since all first-row transition-metals have L-edges in the region and the carbon and nitrogen K-edges are at 285eV and 400eV, respectively, numerous molecular systems, whether in gas-phase or in solution could be targeted.

1 T. Popmintchev *et al.*, *Science*, 2012, **336**, 12878.
2 A. J. Gonsalves *et al.*, *Nat. Phys.*, 2011, **7**, 862.

Professor Neumark said: The paper by Nils Huse raises the point that new light sources based on lasers and accelerators are coming on line in several laboratories and research facilities worldwide. One can now produce femtosecond and even attosecond light pulses in the extreme ultraviolet and X-ray regions of the electromagnetic spectrum. These light sources bring new capabilities that should be of considerable interest to the chemical dynamics community. In time-resolved experiments, these high energy pulses can either initiate or probe dynamics that are inaccessible to more conventional femtosecond laser systems. For example, as exemplified in some of the experimental and theoretical talks given here, time-resolved photoelectron spectroscopy experiments on neutral molecules are often limited by the photon energy of the ionizing (probe) laser pulse, which is typically 5 eV or less. As a result, once an excited molecule undergoes sufficient relaxation, it can no longer be ionized by a single probe photon, complicating the interpretation of the time-resolved photoelectron spectrum. Using a much higher energy photon for the probe photon alleviates this problem, as was first demonstrated in the Leone laboratory for the photodissociation of Br_2.[1] More generally, experiments utililizing very short pulses are starting to be carried out with much higher energy photons (1 keV and higher) at LCLS and similar facilities. These experiments probe new aspects of the interaction of light with atoms and molecules. Thus far, most of this work has been carried out by the atomic/molecular physics community, but I think that we would be very well served by thinking about chemical physics/reaction dynamics experiments that could be performed at these new facilities.

1 L. Nugent-Glandorf, M. Scheer, D. A. Samuels, A. M. Mulhisen, E. R. Grant, X. M. Yang, V. M. Bierbaum, and S. R. Leone, *Phys. Rev. Lett.*, 2001, **87**, 193002.

Professor Huse added: I fully agree with the opinion that the community is well served in thinking about experiments using higher photon energies. This would allow for core-level electrons to be ionized extending element specificity and, in the case of solutions, introducing probing depth by varying the kinetic energy of the ionized photoelectron. Furthermore, the sensitivity of core-levels to valence charge density provides additional information, especially when two core holes on different atoms within a molecule are generated. I hope that with the increasing number of new light sources becoming available in Europe, Asia, and North America, opportunities for chemical reaction dynamics will become more evident, resulting in more experiments and new insight.

Dr Glowacki commented: First, I would like to reiterate what Professors Gruebele, Truhlar, and Nesbitt have said – that, for biological systems, one usually finds a distribution of transition states. For example, we recently showed that it was possible to rationalize a host of controversial enzyme kinetic isotope effect (KIEs) data using some simple multi-state kinetic models,[1,2] which were inspired by coarse-grained master equation approximations. So there's exciting opportunities for this community to apply our collective chemical dynamics expertise to a host of interesting biological systems.

Second, I have a rather general question for Professor Gruebele. Much of your previous work, which I have read closely and have found very useful, has explored relaxation pathways for non-equilibrium perturbations in molecular energy distributions. Presently, you're investigating relaxation pathways for non-equilibrium perturbations in peptide configuration space. I'm curious to hear you comment on whether you think there are any useful connections to highlight between these two apparently different areas.

1 D. R. Glowacki, J. N. Harvey and A. J. Mulholland, *Nat. Chem.*, 2012, **4**, 169–176.
2 D. R. Glowacki, J. N. Harvey and A. J. Mulholland, *Biochem. Soc. Trans.*, 2012, **40**, 515–521.

Professor Gruebele responded: I believe there are useful connections. As you have found in your work, configuration and energy redistribution, even in the unimolecular case, often fail to obey single exponential kinetics. The reason is that there is a whole hierarchy of bottlenecks. In the case of vibrational energy redistribution, there are local and global bottlenecks: local ones arise from the presence or absence of resonances that facilitate energy transport in certain parts of the state space and hinder it in others. (State space is the quantum analog of phase space.) Global ones arise because energy is 'diluted' as it spreads over many localized modes, so effectively the average excitation energy felt by different parts of a molecule is much smaller than the total energy put into the molecule. Both of these can be overcome by off-resonant energy transport ('superexchange'), which however is much less efficient than resonant energy transport. An interesting paper in that connection is by Heller.[1] The hierarchical nature of couplings that results in stretched dynamics is also discussed by Schofield and Wolynes.[2] A simple factorized force field that can be calculated easily for any molecule and yields the hierarchical nature of couplings and power law dynamics, in agreement with theory, was developed by us.[3]

Local and global bottlenecks also exist in the dynamics of solvated biomolecules. When crossing over the barrier, proteins encounter substantial solvent and internal friction, which can be described by motions orthogonal to the reaction coordinate in a multidimensional free energy picture, or as a decreased diffusion coefficient/re-crossing in one dimension. The interactions cover a broad range of strengths. Molecular dynamics simulations have shown that there is a hierarchy of local (e.g. secondary structure) to global (contacts among amino acids far removed in sequence) bottlenecks that produce multiple timescales and stretched kinetics. A good example would be work from Noé's group.[4] The peptide data from Hochtrasser's group,[5] which you have considered in your calculations, are an excellent experimental example.

In both cases, the peptide chain structural rearrangement, and the small molecule vibrational energy flow, the state space of the system is highly structured because the strength of interactions fluctuates randomly. In some regions of state space a few strong interactions cluster and fast dynamics ensues, while in other regions, dynamics remains slow. In fact, even the time evolution is not as different as it may seem at a first glance: both classical and quantum dynamics are symplectic, or area conserving, and can be written in the same form.[6] Stretched dynamics has also been observed in 'hybrid' systems, molecules between the scale of small organics and large peptides. An example from our work is where a transition from simple exponential kinetics to highly stretched kinetics occurs when the reaction temperature is lowered and the system becomes trapped.[7] There is a deep connection between energy flowing through vibrations, and reconfiguration of large molecules, as reconfiguration is just a form of large amplitude vibration. Leitner and coworkers have considered these connections and have found universal scaling exponents for different but related types of dynamics.[8]

1 E. J. Heller, J. Phys. Chem., 1995, 99, 2625–2634.
2 S. Schofield and P. G. Wolynes, J. Phys. Chem., 1995, 99, 2753–2763.
3 D. Madsen et al., J. Chem. Phys., 1997, 106, 5874–5893.
4 F. Noé et al., Proc. Nat. Acad. Sci. USA, 2009, 106, 19011–19016.
5 M. Volk et al., J. Phys. Chem. B, 1997, 101, 8607–8616.
6 S. K. Gray and J. M. Verosky, J. Chem. Phys., 1994, 100, 5011–5022.
7 W. Y. Yang et al., J. Am. Chem. Soc., 2000, 122, 3248–3249.
8 X. Yu and D. M. Leitner, J. Chem. Phys., 2005, 122, 054902.

Professor Truhlar added: I would like to comment further on the remark of Dr Glowacki about a distribution of transition states in biological systems. Our paper at this discussion concerns multiple transition states in gas-phase reactions as treated by multi-path variational transition state theory. For enzyme-catalyzed reactions, we have used an alternative formalism called ensemble-averaged variational transition state theory (EA-VTST),[1,2] and our results have been reviewed[3,4] and discussed.[5] In EA-VTST, we sample an ensemble of transition states rather than sum over all transition states (there are too many to sum). We have successfully calculated many kinetic isotope effects.[1-6] In our experience, ensemble averaging[7] is absolutely essential to getting reasonable results for such systems, as there is a wide variation[4] in results for individual reaction paths.

1 C. Alhambra, J. Corchado, M. L. Sánchez, M. Garcia-Viloca, J. Gao, and D. G. Truhlar, J. Phys. Chem. B, 2001, 105, 11326–11340.
2 D. G. Truhlar, J. Gao, C. Alhambra, M. Garcia-Viloca, J. Corchado, M. L. Sánchez, and J. Villà, Acc. Chem. Res., 2002, 35, 341–349.
3 D. G. Truhlar, J. Gao, M. Garcia-Viloca, C. Alhambra, J. Corchado, M. L. Sanchez, and T. D. Poulsen, Int. J. Quantum Chem., 2004, 100, 1136–1152.
4 J. Pu, J. Gao, and D. G. Truhlar, Chem. Rev., 2006, 106, 3140–3169.
5 D. G. Truhlar, J. Phys. Org. Chem., 2006, 23, 660–676.
6 A. Dybala-Defratyka, P. Paneth, R. Banerjee, and D. G. Truhlar, Proc. Nat. Acad. Sci. U.S.A., 2007, 104, 10774–10779.
7 J. Gao and D. G. Truhlar, Annu. Rev. Phys. Chem., 2002, 53, 467–505.

Reaction dynamics: concluding remarks

Richard N. Zare

Received 10th July 2012, Accepted 10th July 2012
DOI: 10.1039/c2fd20120h

The 157th Faraday Discussion represented a historic turning point in the development of the field of reaction dynamics because it concerned itself with how reactions occur in gases, in liquid, and at interfaces. Never before has the attempt been made to unify the various approaches to reaction dynamics in one Faraday Discussion meeting and to discover what language was common and what was special to these previously distinct subdisciplines. This Discussion also marked a maturation of the field of reaction dynamics in that so much emphasis was placed on what the combination of theory and experiment could tell us about the detailed course of chemical transformations.

I Introduction

This is the first time that a Faraday Discussion had devoted itself to the topic of reaction dynamics in and between all different phases of matter – a brave decision by the organizers of this meeting, which rewarded the participants with many insights into previously unrealized interconnections. Professor F. Fleming Crim, University of Wisconsin, set the tone for this meeting with his superb Introductory Lecture[1] on chemical transformations across phases. This theme was continued by the many presentations and discussions found in this volume. It is not my purpose here to provide "a Readers Digest account" of what transpired. Instead, I offer some of my own perspectives on this field.

I begin by sharing with you a photograph (Fig. 1) I took in sun-drenched Italy in June, which reminds me of the superb location of our meeting in Assisi. In many ways this snapshot captures the field of reaction dynamics – obvious beauty to behold, but many features hidden from view, even behind barred passages. With advances in both experiment and theory the window is being lifted for us to view inside and the field of reaction dynamics is providing us with rich pictures of how chemical reactions do take place. We have come very far from the early times when chemists first started questing after the knowledge of chemical change, collision by collision, and yet so many questions remain unanswered. At the heart of chemistry are chemical transformations of all types and the understanding of how they happen continues to represent a central problem in all chemistry.

Often it is best to decide what comes next by reflecting on the past. The birth of reaction dynamics begins with the study of isolated gas-phase collisions, whether by infrared spectroscopy as pioneered by John Polanyi and co-workers or crossed molecular beams, pioneered by Dudley Herschbach and co-workers. Of course, there were many others who contributed to early ground-breaking experiments but the names of Polanyi and Herschbach stand out as icons in this field. We are so appreciative that John Polanyi could join us for this Discussion and share with us his new work on surface-controlled reactions.[2] We were also so fortunate that the President of the Faraday Society, Prof. Michael Ashfold, was also able to join us and add so much to our conference.

Department of Chemistry, Stanford University, 333 Campus Drive, Stanford, California 94305-5080, USA. E-mail: zare@stanford.edu

Fig. 1 Flowers in a window, taken in Siena, Italy.

II Past, present and future

In the beginning, detection schemes were truly primitive. Isolated gas-phase colli-sions imply necessarily low concentrations of collision partners, which places a chal-lenge on how the reaction products can be detected. We think back on what was called "the Alkali Age" in which only alkali-atom containing molecules could be probed, which was accomplished by means of hot-wire ionization. This was followed by the use of mass spectrometric detection, so powerfully demonstrated by Yuan T. Lee and co-workers among others. This technique provided universal detection of reaction products but was almost completely blind to the quantum state distribution of the products. This situation would dramatically change with the introduction of the laser, which not only provided quantum state resolution but could interrogate reactions on increasingly shorter time scales, today reaching to femtoseconds and at-toseconds. Another remarkable development was ion imaging, pioneered by David Chandler and Paul Houston, and further developed by others such as André Eppink, David Parker, Arthur Suits, Hanna Reisler, and others. At the same time we should not forget what pulsed nozzle beams have made possible for studies of reaction dynamics, pioneered by John Fenn, Ronald Gentry, and others. As time-resolved spectroscopies have been developed and refined, the ability to study reactions in and across various phases of matter by "pump–probe techniques" has dramatically improved, beginning with gas–surface scattering, reactions in liquids, *etc.*

At the same time, theoretical approaches have deepened and reached the stage that every experimentalist must look to theory in helping to interpret reaction dynamics results. In the beginning, reaction dynamics was most successfully pictured by means of hard-sphere collisions, like some game of billiards. As we have seen during this Discussion, these ideas are still alive and well, and progress is also made by considering soft-sphere collisions, which allow energy transfer into the internal degrees of freedom of the reaction products. The most dramatic advances can be attributed to the generation of chemically accurate potential energy surfaces through advances in quantum chemistry, which owe so much to the ingenuity of theorists and the advances in computational power. We learn that in many chemical reactions we are not simply playing a straightforward game of billiards but the billiard table becomes warped as the billiard balls move on the table, which can easily confound some simple intuitive ideas. Most ground-state reactions are considered to occur on a single potential energy surface, but increasingly we realize that nonadiabatic effects cannot be ignored and can in certain cases, particularly in excited states of polyatomic systems through conical intersections, become dominant in controlling the outcome of reactive collisions. In that sense we are not simply playing billiards on a warped billiard table whose contours change in time but rather on interconnected billiard tables whose shapes change in time! In this regard the surface hopping approach pioneered by John Tully[3] and its further elaborations as well as the *ab initio* multiple spawning method pioneered by Todd Martinez[4] are of great interest and hold much promise. We have also been treated by Donald Truhlar to an exposition on a coupled-mode theory called multi-structural variational transition state theory, which includes a treatment of the differences in the multi-dimensional tunneling paths and how they contribute to the reaction rate.[5]

I cannot emphasize enough the power of the question in defining new directions for a field. At this Discussion speakers were kept on their toes by many probing questions, especially from Joel Bowman, David Glowacki, David Nesbitt, Daniel Neumark, and Donald Truhlar. We seek to know how atom-plus-diatom reactions differ from complex reactions involving polyatomic reagents having many degrees of freedom. This statement is particularly true when reagents are internally excited in different ways. We must not let ourselves become locked into imagining all reactions take place close to the minimum energy path or that reactions pass through some one point called simply the transition state. Past and present work on migratory insertion and what is called roaming has amply demonstrated the dangers of becoming intellectually phase-locked into too simple pictures of reaction dynamics that do not allow the possibility of gross rearrangements of the nuclei in the course of a reaction. We need to understand better the role of clusters and of solvents in mediating reaction dynamics. Environmental effects can be of major importance in determining reaction outcomes. We are learning so much from the careful study of vibrationally resolved small radical–molecule reactions and how they compare to the same reactions in the liquid phase, for example. The same is true for ion–molecule reactions.

We should not imagine that the same questions asked of a gas-phase reaction are necessarily what are important in describing reactions occurring in the condensed phase. In my estimation, the one word that most belongs to chemistry is the word 'catalysis'. If I have a criticism, it would be that we did not hear enough about the reaction dynamics of catalytic systems, particularly biological catalysts called enzymes – but hopefully in the next such Discussion we will. Nevertheless, the topics covered in this meeting were amazing in their breadth, spanning elementary reactions to how crowding affects protein folding!

I do suggest we must keep in mind what we are seeking to uncover about nature. To some it is simply how well theory and experiment can be put into agreement. To me, this goal is certainly laudable but I seek more. I seek the ability to learn what features can be applied with some confidence in predicting and understanding the behavior of related but more complex reactions. I leave this meeting energized with the conviction that reaction dynamics has a bright and healthy future. We owe a great debt of gratitude to the organizers, Piergiorgio Casavecchia, David Clary, Peter Hamm, Andrew Orr-Ewing, George Schatz, and Alec Wodtke, who put together a most memorable conference.

References

1 F. Crim, *Faraday Discuss.*, 2012, **157**, DOI: 10.1039/c2fd20123b.
2 A. Eisenstein, L. Leung, T. Lim, Z. Ning and J. C. Polanyi, *Faraday Discuss.*, 2012, **157**, DOI: 10.1039/c2fd20023f.
3 N. Shenvi and J. C. Tully, *Faraday Discuss.*, 2012, **157**, DOI: 10.1039/c2fd20032e.
4 T. S. Kuhlman, W. J. Glover, T. Mori, K. B. Møller and T. J. Martínez, *Faraday Discuss.*, 2012, **157**, DOI: 10.1039/c2fd20055d.
5 J. Zheng and D. G. Truhlar, *Faraday Discuss.*, 2012, **157**, DOI: 10.1039/c2fd20012k.

Poster titles

Molecular dynamics calculations of surface processes involving O atoms on silica surfaces, **M. Rutigliano, C. Zazza, S. Orlandini, N. Sanna, V. Barone and M. Cacciatore**, *CNR-Institute of Inorganic Methodologies and Plasmas, Italy*

Roaming in the dark: Deciphering the mystery of the $NO_3 \rightarrow NO + O_2$ photolysis, **M. P. Grubb, M. L. Warter, H. Xiao, S. Maeda, K. Morokuma and S. W. North**, *Texas A&M University, U.S.A.*

Reaction mechanisms of the growth of polycyclic aromatic hydrocarbons and nitrogen-containing polycyclic aromatic compounds at low temperatures: A view from theoretical calculations of potential energy surfaces, **A. Landera, V. V. Kislov, R. I. Kaiser and A. M. Mebel**, *Florida International University, U.S.A.*

Vibrationally resolved chemical reaction dynamics in solution, **S. J. Greaves, R. A. Rose, F. Abou-Chahine, G. T. Dunning, D. R. Glowacki, J. N. Harvey and A. J. Orr-Ewing**, *University of Bristol, United Kingdom*

X-H/X-Me Photodissociation dynamics of aromatic molecules: Linking gas phase processes with the solution phase, **S. J. Harris, D. Murdock, T. N. V. Karsili, M. J. Grubb, I. P. Clark, G. Greetham, M. Towrie and M. N. Ashfold**, *University of Bristol, United Kingdom*

Dynamics of the $OH^-(H_2O)_n + CH_3I$, $n = 0$–2, reaction, **J. Xie, M. R. Siebert and W. L. Hase**, *Texas Tech University, U.S.A.*

Dynamics of proton transfer in H-bonds: characteristic IR and NMR spectra, **K. Sagarik, S. Chaiwongwattana, C. Lao-ngam and M. Phonyiem**, *Suranaree University of Technology, Thailand*

Transition state based calculation of state-to-state reaction probabilities, **R. Welsch and U. Manthe**, *Universität Bielefeld, Germany*

Imaging energy transfer and hydrogen-bond breaking in the water dimer, **L. C. Ch'ng, A. K. Samanta and H. Reisler**, *University of Southern California, U.S.A.*

A shortcut to introduce the ergodic principle in teaching collision theory, **G. Monaco**, *Università di Salerno, Italy*

Importance of CH_4 rotation for its dissociation on Ni(111), **X. J. Shen, W. Dong and X. H. Yan**, *Ecole Normale Supérieure de Lyon, France*

Competition between the electron- and phonon-excitation channels in the dynamics of nitrogen atoms at metal surfaces, **L. Martin-Gondre, G. A. Bocan, M. Blanco-Rey, J. I. Juaristi, M. Alducin and R. Díez Muiño**, *EHU, Spain*

Probing dynamics of endothermic S_N2-reaction $I^- + CH_3Cl$ using velocity map imaging, **A. H. Kelkar, E. Carrascosa, M. Stei, F. Hochheimer, J. Wildauer, T. Best and R. Wester**, *Universität Innsbruck, Austria*

Comparing the ultraviolet photostability of azole chromophores, **G. M. Roberts and V. G. Stavros**, *University of Warwick, United Kingdom*

Exploring H-atom tunneling dynamics in photoexcited phenols and catechol, **G. M. Roberts and V. G. Stavros**, *University of Warwick, United Kingdom*

Steady states and dynamics of the π-stacked anisole dimer in the gas phase, **F. Mazzoni, M. Pasquini, G. Pietraperzia, M. Becucci**, *University of Florence and LENS, Italy*

Chemical reactions studied at ultralow temperature in liquid helium clusters, **S. Krasnokutski and F. Huisken**, *Friedrich Schiller University of Jena, Germany*

Photodissociation dynamics of mass-selected cluster ions studied by reflectron imaging, **Y. Yamakita, H. Hoshino, Y. Suzuki, M. Saito, K. Koyasu and F. Misaizu**, *Tohohu University, Japan*

Intramolecular charge transfer in some push−pull distyryl furans, thiophenes and pyridines, **B. Carlotti, I. Kikaš, I. Škorić, A. Spalletti and F. Elisei**, *University of Perugia, Italy*

Ionization dynamics of water molecules by metastable neon atoms, **B. Brunetti, P. Candori, D. Cappelletti, S. Falcinelli, F. Pirani, D. Stranges and F. Vecchiocattivi**, *Universita di Perugia, Italy*

Interfacial Collisions Dynamics of OH with organic liquid surfaces **G. Paterson, K. L. King, M. Iljina, G. Rossi, R. Westacott, M. L. Costen and K. G. McKendrick**, *Heriot-Watt University, UK*

Photoinduced isomerization of a small protonated Schiff base: Bond selectivity and quantum yield, **L. Vuković, C. Burmeister, P. Král and G. Groenhof**, *Max-Planck-Institut für Biophysikalische Chemie, Germany*

Study of the dynamics of the OH radical inside a carbon nanotube, **D. Skouteris and A. Lagana**, *Universita degli Studi di Perugia, Italy*

Crossed molecular beam experiments with a Stark-decelerated molecular beam, **X. Wang, M. Kirste, H. C. Schewe, K. Liu, A. van der Avoird, G. C. Groenenboom, G. Meijer and S. Y. T. van de Meerakker**, *Fritz-Haber-Institut der Max-Planck-Gesellschaft, Germany*

Low energy scattering of Stark-decelerated OH radicals with He atoms with high energy resolution, **H. C. Schewe, X. Wang, M. Kirste, L. Scharfenberg, H. Haak, S. Y. T. van de Meerakker, N. Vanhaecke and G. Meijer**, *Fritz-Haber-Institut der Max-Planck-Gesellschaft, Germany*

Electron stimulated processes of astrophysically relevant molecules at solid surfaces, **D. Marchione, J. D. Thrower, A. G. M. Abdulgalil, M. P. Collings and M. R. S. McCoustra**, *Heriot-Watt University, UK*

Molecular stereodynamics of chirality: experiments and theory, **F. Palazzetti, G. S. Maciel, A. Lombardi, G. Grossi and V. Aquilanti**, *Universita di Perugia, Italy*

UV photodissociation of N,N-dimethylformamide studied by velocity-map imaging, **S. H. Gardiner, L. Lipciuc and C. Vallance**, *University of Oxford, UK*

Rotational angular momentum polarisation effects in the inelastic scattering of fully quantum state selected NO(X) with the rare gases, **M. Brouard, H. Chadwick, C. J. Eyles, B. Hornung, B. Nichols, F. J. Aoiz and S. Stolte**, *University of Oxford, UK*

Dissociative double photoionization dynamics of acetylene molecules by linearly polarized light, **M. Alagia, C. Callegari, P. Candori, S. Falcinelli, F. Pirani, R. Richter, S. Stranges and F. Vecchiocattivi**, *University of Perugia, Italy*

Photodissociation of 2,3-dimethyl-2-butene at 193 nm, **D. Stranges and E. Ripani**, *Universita di Roma, Italy*

Coupled-monomers in molecular assemblies: theory and application to the water tetramer, pentamer and ring hexamer, **Y. Wang and J. L. Bowman**, *Emory University, U.S.A.*

Crossed molecular beam study of O(3P) + H3CCCH reaction, **P. Casavecchia, N. Balucani, F. Leonori, V. Nevrly, S. Falcinelli and D. Stranges**, *Universita di Perugia, Italy*

Experimental and theoretical study of O(3P) + C2H4 multichannel reaction dynamics, **F. Leonori, L. Angelucci, N. Balucani, A. Occhiogrosso, R. Petrucci, P. Casavecchia, B. Fu, Y.-C. Han and J. M. Bowman**, *Universita di Perugia, Italy*

Photodissociation dynamics of 2-propyl Radical at 248 nm, **D. Stranges, E. Ripani and G. Scotti**, *Universita di Roma, Italy*

NO(A) + Ne collisions: alignment and differential scattering cross sections, **D. W. Chandler, J. D. Steill, J. Kay, and K. E. Strecker**, *Sandia National Laboratory, U.S.A.*

Molecule-surface scattering with velocity-controlled molecular beams, **D. Engelhart, F. Grätz, H. Haak, D. J. Auerbach, G. Meijer and A. M. Wodtke**, *Georg-August-Universität, Germany*

Diamonds aren't forever: surface chemistry of local hotspots on carbon nanostructures, **D. R. Glowacki, W. J. Rodgers and J. N. Harvey**, *University of Bristol, UK*

Reaction dynamics at a metal surface. Electron-induced reaction of p-dichlorobenzene and p-diiodobenzene on Cu(110), **A. Eisenstein, L. Leung, T. Lim, Z. Ning and J. C. Polyani**, *University of Toronto, Canada*

Crystal surface, solid state reactions, and kinetic isotope effects, **I. Biljan, I. Halasz, I. Huskic, A. Maganjić, I. Šolić, S. Milovac, H. Vancik**, *University of Zagreb, Croatia*

Electronic relaxation and ground state vibrational coupling analysis of the push-pull indoaniline dye phenol-blue by means of transient infrared (TIR) and bidimensional infrared (2DIR) spectroscopies, **E. Ragnoni, M. Lima, M. Di Donato, A. Lapini, F. Terenziani and P. Foggi**, *Universitá di Firenze, Italy*

The Skinner Prize for the best poster was jointly awarded to Miss Stephanie Harris of the University of Bristol, United Kingdom, for her poster on X-H/X-Me Photodissociation dynamics of aromatic molecules: Linking gas phase processes with the solution phase and Miss Lee C. Ch'ng of the University of Southern California, U.S.A, for her poster on imaging energy transfer and hydrogen-bond breaking in the water dimer.

List of participants

Professor V. Aquilanti, *Università degli Studi di Perugia, Italy*
Professor M. Ashfold, *University of Bristol, United Kingdom*
Prof Dr N. Balucani, *Università degli Studi di Perugia, Italy*
Professor R. Beck, *École Polytechnique Fédérale De Lausanne, Switzerland*
Prof Dr T. Bernhardt, *University of Ulm, Germany*
Dr T. Best, *University of Innsbruck, Austria*
Professor J. Bowman, *Emory University, U.S.A.*
Prof S. Bradforth, *University of Southern California, U.S.A.*
Mr R. Brown, *University of Chicago, U.S.A.*
Dr P. Candori, *Università degli Studi di Perugia, Italy*
Dr B. Carlotti, *Università degli Studi di Perugia, Italy*
Professor P. Casavecchia, *Università degli Studi di Perugia, Italy*
Miss H. Chadwick, *University of Oxford, United Kingdom*
Dr D. Chandler, *Sandia National Laboratories, U.S.A.*
Miss L C. Ch'ng, *University of Southern California, U.S.A.*
Dr M. Costes, *Universite de Bordeaux, France*
Professor F. Crim, *University of Wisconsin, U.S.A.*
Dr S. Dixon, *Royal Society of Chemistry, United Kingdom*
Professor W. Dong, *ENS-Lyon, France*
Mr D. Engelhart, *Georg August Universität Gottingen, Germany*
Dr S. Falcinelli, *Università degli Studi di Perugia, Italy*
Professor H. Fielding, *University College London, United Kingdom*
Miss S. Gardiner, *University of Oxford, United Kingdom*
Dr D. Glowacki, *University of Bristol, United Kingdom*
Mr M. Grubb, *University of Bristol, United Kingdom*
Professor M. Gruebele, *University of Illinois at Urbana-Champaign, U.S.A.*
Professor P. Hamm, *University of Zurich, Switzerland*
Miss S. Harris, *University of Bristol, United Kingdom*
Professor N. Huse, *University of Hamburg, Germany*
Dr M. Jordan, *University of Sydney, Australia*
Professor S. Kable, *University of Sydney, Australia*
Dr A. Kelkar, *Universität Innsbruck, Austria*
Dr S. Krasnokutskiy, *Friedrich Schiller University Jena, Germany*
Mr T. Kuhlman, *Technical University of Denmark, Denmark*
Professor A. Lagana, *Università degli Studi di Perugia, Italy*
Dr F. Leonori, *Università delgi Studi di Perugia, Italy*
Dr L. Leung, *University of Toronto, Canada*
Dr T. Lim, *University of Toronto, Canada*
Dr M. Lipciuc, *University of Oxford, United Kingdom*
Dr K. Liu, *Academia Sinica, Taiwan*
Dr S. Mackenzie, *University of Oxford, United Kingdom*
Mr D. Marchione, *Heriot-Watt University, United Kingdom*
Professor T. Martinez, *Stanford University, U.S.A.*
Dr L. Martin-Gondre, *Donostia International Physics Center, Spain*
Mr F. Mazzoni, *University of Florence, Italy*
Professor A. Mebel, *Florida International University, U.S.A.*
Professor T. Minton, *Montana State University, U.S.A.*
Dr G. Monaco, *Universita di Salerno, Italy*
Dr C. Mowatt, *Royal Society of Chemistry, United Kingdom*
Professor D. Nesbitt, *JILA/University of Colorado, U.S.A.*
Professor D. Neumark, *University of California, Berkeley, U.S.A.*

Index of contributors*

* The page numbers in **bold** type indicate papers submitted for discussions.